THE EVOLUTION OF PRIMARY SEXUAL CHARACTERS IN ANIMALS

THE EVOLUTION OF PRIMARY
SEXUAL CHARACTERS IN ANIMALS

Edited by

Janet L. Leonard
Alex Córdoba-Aguilar

OXFORD
UNIVERSITY PRESS
2010

OXFORD
UNIVERSITY PRESS

Oxford University Press, Inc., publishes works that further
Oxford University's objective of excellence
in research, scholarship, and education.

Oxford   New York
Auckland   Cape Town   Dar es Salaam   Hong Kong   Karachi
Kuala Lumpur   Madrid   Melbourne   Mexico City   Nairobi
New Delhi   Shanghai   Taipei   Toronto

With offices in
Argentina   Austria   Brazil   Chile   Czech Republic   France   Greece
Guatemala   Hungary   Italy   Japan   Poland   Portugal   Singapore
South Korea   Switzerland   Thailand   Turkey   Ukraine   Vietnam

Published by Oxford University Press, Inc.
198 Madison Avenue, New York, New York 10016

www.oup.com

Oxford is a registered trademark of Oxford University Press.

Library of Congress Cataloging-in-Publication Data
The evolution of primary sexual characters in animals /
edited by Janet L. Leonard and Alex Córdoba-Aguilar.
    p. cm.
Includes bibliographical references and index.

ISBN 978-0-19-532555-3

1. Generative organs—Evolution. 2. Sexual selection in animals.
I. Leonard, Janet L. (Janet Louise), 1953–II. Córdoba-Aguilar, Alex.
QL876.E96 2010
591.56'2–dc22   2009038212

9 8 7 6 5 4 3 2 1
Printed in the United States of America
on acid-free paper

Contents

Contributors

Baur, Bruno. Conservation Biology Group, Department of Environmental Sciences, University of Basel, St. Johanns-Vorstadt 10, CH-4056 Basel, Switzerland

Cordero, Rivera Adolfo. Grupo de Ecoloxía Evolutiva e da Conservación, Departamento de Ecoloxía e Bioloxía Animal, Universidade de Vigo, E.U.E.T. Forestal, Campus Universitario, 36005 Pontevedra, Spain

Córdoba-Aguilar, Alex. Departamento de Ecología Evolutiva, Instituto de Ecología, Universidad Nacional Autónoma de México, Apdo. Postal 70–275, Mexico, D. F., 04510, Mexico

David, Patrice. Centre d'Ecologie Fonctionnelle et Evolutive, UMR 5175, Campus CNRS, 1919 route de Mende, 34295 Montpellier cedex, France

Eady, Paul. Department of Biological Sciences, University of Lincoln, Lincoln LN2 2LG, UK

Eberhard, William G. Smithsonian Tropical Research Institute, and Escuela de Biología,
Universidad de Costa Rica, Ciudad Universitaria, Costa Rica

Ghiselin, Michael T. Department of Invertebrate Zoology, California Academy of Sciences, 875 Howard Street, San Francisco, CA 94103, USA

Gosliner, Terrence M. Department of Invertebrate Zoology and Geology, California Academy of Sciences, 55 Music Concourse Drive, San Francisco, CA 94118, USA

Gowaty, Patricia Adair. Department of Ecology and Evolutionary Biology, 621 Charles E. Young Drive, Los Angeles, CA 90095, USA and Smithsonian Tropical Research Institute, APO AA 34002

Hodgson, Alan N. Department of Zoology & Entomology, Rhodes University, Grahamstown 6140, South Africa

Hosken, David J. Centre for Ecology and Conservation, School of Bioscience, University of Exeter, Cornwall Campus, Tremough, Penryn TR10 9EZ, UK

Houck, Lynne D. Department of Zoology, Oregon State University, Corvallis, OR 97331, USA

Hubbell, Stephen P. Department of Ecology and Evolutionary Biology, 621 Charles E. Young Drive, Los Angeles, CA 90095, USA, and, Smithsonian Tropical Research Institute, APO AA 34002

Huber, Bernhard A. Alexander Koenig Research Museum of Zoology, Adenauerallee 160, 53113 Bonn, Germany

Jarne, Philippe. Centre d'Ecologie Fonctionnelle et Evolutive, UMR 5175, Campus CNRS, 1919 route de Mende, 34295 Montpellier cedex, France

Koene, Joris M. Department of Animal Ecology, Faculty of Earth & Life Sciences, VU University, De Boelelaan 1085, 1081 HV Amsterdam, The Netherlands

Leonard, Janet L. Joseph M. Long Marine Laboratory, University of California-Santa Cruz, Santa Cruz, CA 95060, USA

Levitan, Don R. Department of Biological Science, Florida State University, Tallahassee, FL 32306, USA

Lotterhos, Katie E. Department of Biological Science, Florida State University, Tallahassee, FL 32306, USA

Machado, Glauco. Departamento de Ecologia, Instituto de Biociências, Universidade de São Paulo, Brazil

Macías-Ordóñez, Rogelio. Departamento de Biología Evolutiva, Instituto de Ecología, A.C., Apartado Postal 63, Xalapa, Veracruz 91000, Mexico

Miller, Edward H. Biology Department, Memorial University, St. John's, NL A1B 3X9, Canada

Montgomerie, Robert. Department of Biology, Queen's University, Kingston, ON K7L 3N6, Canada

Neat, Francis C. Population Biology Group, Marine Scotland, Marine Laboratory, P.O. Box 101, 375 Victoria Road, Aberdeen AB11 9DB, UK

Olsson, Mats. School of Biological Sciences, University of Wollongong, NSW 2522, Australia

Peretti, Alfredo V. Senior Researcher CONICET-Argentina, Associate Professor Universidad Nacional de Cordoba, Reproductive Biology & Evolution Lab–Catedra de Diversidad Animal I, Fac Cien Ex Fis y Nat, UNC, Argentina

Pérez-González, Abel. Grupo de Sistemática e Biologia Evolutiva, Núcleo em Ecologia e Desenvolvimento Sócio-Ambiental de Macaé, Universidade Federal do Rio de Janeiro, Brazil

Pointier, Jean-Pierre. Biologie et Ecologie Tropicale et Méditerranéenne, UMR 5244 CNRS-EPHE-UPVD, 52 avenue Paul Alduy, 66860 Perpignan cedex, France

Shultz, Jeffrey W. Department of Entomology, University of Maryland, College Park, MD 20742, USA

Stuart-Fox, Devi Department of Zoology, University of Melbourne, Victoria 3010, Australia

Taborsky, Michael. Division Behavioural Ecology, Institute of Ecology and Evolution, University of Bern, Wohlenstrasse 50a, CH-3032 Hinterkappelen, Switzerland

Uller, Tobias. Edward Grey Institute, Department of Zoology, University of Oxford, Oxford OX1 3PS, UK

Valdés, Ángel. Department of Biological
Sciences, California State Polytechnic
University, 3801 West Temple Avenue,
Pomona, CA 91768, USA

Verrell, Paul A. School of Biological Sciences,
Washington State University, Pullman,
WA 99164, USA

Wedell, Nina. Centre for Ecology and
Conservation, School of Bioscience,
University of Exeter, Cornwall Campus,
Tremough, Penryn TR10 9EZ, UK

THE EVOLUTION OF PRIMARY SEXUAL CHARACTERS IN ANIMALS

1

Introduction

Celebrating and Understanding Reproductive Diversity

JANET L. LEONARD

"...endless forms most beautiful and most wonderful have been, and are being, evolved."

(Darwin 1859)

One of the great joys of biology is the realization that there is no end of wonders to discover, to describe, and to attempt to understand. Not only is there a multitude of bizarre and fascinating organisms on this planet but every new insight into evolution changes our view of all organisms, deepening our enjoyment of, and our excitement about, this diversity. Very happily, it is clear that this process will go on indefinitely. In this volume, a variety of authors explore new perspectives on one set of wondrous phenomena, the reproductive traits of animals. Phenomena related to sexual reproduction in animals have long been a source of amazement; from Aristotle's description of the love-dart of the garden snail *Helix* to the discovery of egg-laying Australian mammals, sexual cannibalism in spiders and mantids, the identical quadruplets of the Nine-banded Armadillo, complemental males in barnacles, and social control of sex change in fishes. As evolutionary and behavioral ecologists have begun to analyze reproduction in terms of selective forces acting on individuals, simple astonishment at these oddities has turned to wonder and admiration at the solutions that evolution has found to the problems of reproducing in environments filled with complex physical forces, myriad predators and diseases, conspecific competitors, and mates that may be manipulative and/or selfish. The stimulus for this book was the feeling that it was time to get a better overview of the diversity of reproductive

characters in a wide range of animals in light of modern biology.

The first step in understanding diversity is to describe it. Sexual reproduction involves some of the most startling and strange adaptations known. Sex in the Metazoa ranges from simple broadcast spawning, where both sperm and eggs are shed into the water, through spermcast (Bishop & Pemberton 2006) systems where eggs are brooded and sperm are released into the water (see Lotterhos and Levitan, this volume), to internal fertilization involving copulation with fantastically baroque genitalia [see the spiked vaginas of nudibranchs in Valdes et al. (this volume) and the amazing penes in beetles (Eady, this volume) and odonates (Cordero Rivera and Córdoba Aguilar, this volume)], simple hypodermic impregnations where the sperm are injected directly into the body cavity of the female, or indirect sperm transfer where the sperm are transferred from the male gonoduct to another structure for transfer to the female (e.g., arachnids, urodeles, see chapters by Eberhard and Huber, Peretti, Machado et al., and Houck and Verrell, this volume). Moreover, all of these forms of reproduction are found in both species with hermaphroditism and those with separate sexes. Gametes vary from tiny broadcast eggs produced by the zillion, to the huge ostrich egg weighing 1.4 kg, or the kiwi egg, weighing one quarter of the mother's weight, to an amazing variety of sperm including tiny sperm, huge sperm, amoeboid sperm, infertile sperm that seem to go along for the ride (Hodgson, this volume), huge ejaculates, sperm transferred in spermatophores (nutritious or otherwise), sperm transferred singly, etc. Internal fertilization, which has evolved many times in the Metazoa, can be fairly straightforward or it can involve structures, in both males and females, that look like the imaginings of Rube Goldberg. Sexual behavior can also be either complex or simple; matings may be rigorously monogamous (even involving fusion of two individuals) or they may be impersonal and/or promiscuous.

Our knowledge of the amazing range of reproductive phenomena in animals has come gradually over the centuries, mostly from studies of individual species or taxa. There are still many, many groups of animals, particularly among the invertebrates, known only to taxonomists and many bizarre and fascinating reproductive adaptations remain buried in that literature. One of the goals of the current volume is to make some of that information accessible to a wider range of biologists. For practical reasons, the book is divided in two sections: a small group of general chapters, which address issues common to many taxonomic groups (see below) and a larger group of chapters which deal in detail with particular taxonomic groups. There are chapters on the reproductive characters of broadcast spawners (Lotterhos and Levitan); internally-fertilizing "prosobranch" gastropods (Hodgson), and three groups of simultaneously hermaphroditic, internally-fertilizing gastropods, the opisthobranchs (Valdes et al.), the basommatophorans (Jarne et al.) and the stylommatophorans (Baur), along with three chapters on arachnids (spiders by Eberhard and Huber; scorpions by Peretti, and Opiliones by Macias-Ordoñez et al.), a general chapter on insects by Wedell and Hosken and chapters on beetles (Eady) and odonates (Cordero Rivera and Córdoba-Aguilar), and chapters on each of the major groups of vertebrates (fishes, Taborsky and Neat; amphibians, Houck and Verrell; reptiles, Uller et al.; birds, Montgomerie; and mammals, Miller).

The second step in understanding diversity is simply documenting its pattern of evolution; that is, to what extent does the presence or absence of a particular trait reflect phylogeny. Where reproductive traits are characteristic of higher taxa; classes, orders, etc. (e.g., placental mammals, indirect sperm transfer in arachnids, hermaphroditism in euthyneuran gastropods) it is difficult to argue

that the trait reflects active selection forces. Its absence or modification would be more interesting; for example, the loss of indirect sperm transfer and evolution of copulation in some opiliones (Macias-Ordónez et al., this volume). One common theme of many of these chapters is that, in many groups, phylogeny is in a state of flux and uncertainty as workers try to reconcile traditionally important morphological characters with the newer molecular data. In several chapters, attempts are made to map reproductive characters onto phylogenies with interesting results: Hodgson (this volume) discusses the evolution of internal fertilization in "prosobranch" gastropods, now known to be a polyphyletic group, through several independent evolutionary events. Valdes et al. (this volume) describe the parallel evolution of several reproductive traits in an apparently monophyletic group, the opisthobranch gastropods. In reptiles, Uller et al. (this volume) used a phylogenetic analysis to test the hypothesis that sperm competition (see below) is correlated with testis size and conclude that environmental factors are more important. Montgomerie (this volume) controls for phylogeny by focusing on one clade, anseriform birds, to test hypotheses about the evolution of intromittent organs. In most taxa, as Jarne et al. (this volume) conclude, in their discussion of basommatophorans, mapping traits to phylogenies will require data on a wider range of species than have been currently studied in the group. Eady (this volume) suggest that in beetles genital evolution is tightly linked to speciation. In mammals, Miller (this volume) discusses the prevalence of both inter- and intraspecific variation in genital characters.

The emphasis in this volume is on "primary" sexual characters, those traits that are directly associated with sexual reproduction. Since Darwin's time biologists have understood that all traits are basically reproductive traits. Survival and growth are only important in so far as they enhance an organism's lifetime reproductive success, so that the adaptive value of traits that serve in foraging, predator evasion, temperature adaptation, resistance to desiccation, etc., comes from their effect on reproduction. However, some traits are more directly and intimately related to reproduction than others. In this volume, we address the question of the evolution of traits that are very intimately tied to sexual reproduction in multicellular animals, the Metazoa.

In *The Descent of Man* (Darwin 1871) made a distinction between "secondary sexual characters",

characters which were sexually dimorphic, associated with obtaining mates or access to mates, and under sexual selection, such as weapons (antlers, horns, etc.) used to repel rivals, or ornaments (bright plumage, song, etc.) used to attract the opposite sex, and "primary sexual characters", such as genitalia, gametes, etc., that were considered to be under natural selection for efficient reproduction. To provide a framework for the discussion, Mike Ghiselin (this volume) discusses Darwin's ideas and points out, very importantly, that whether a character is "primary" or "secondary" depends to a large extent on how the character is defined. For example, a testis is certainly a primary sexual character in animals but the size of the testis in a species may reflect sexual selection pressures through sperm competition (Smith 1984, but see Uller et al., this volume). Therefore the size of the testis relative to other species may be a secondary sexual character. Ghiselin (this volume) uses the term tertiary sexual character for characters that are under the indirect influence of both natural and sexual selection, and as we see from the chapters in this volume, a majority of sexual characters once considered "primary" may actually have evolved, to some extent, under the influence of sexual selection. In his chapter on stylommatophoran gastropods, Baur (this volume) concludes that most of the reproductive anatomy of land snails may represent such sexual characters. In contrast, Houck and Verrell (this volume) in their chapter on amphibians conclude that the array of traits of gametes, gonads, oviducts and associated structures that they review are true primary sexual characters that serve to increase survival of the young. Even such fundamental properties of animal sexuality as anisogamy or sexual system (dioecy vs. hermaphroditism) may be better thought of as tertiary sexual characters (Leonard, this volume).

As is seen in the array of taxonomic chapters, the study of diversity in "primary" sexual characters has been approached from very different frameworks: a primary emphasis in this volume is to attempt to determine to what extent particular reproductive traits reflect the action of natural versus sexual selection. For almost 100 years, the dogma was that traits such as gametes, genitalia, etc., which are intimately involved in the actual process of sexual reproduction, were, almost by definition, products of natural selection. The first major cracks in this paradigm appeared in the 1970s when Geoff Parker (1970) addressed the issue of

sperm competition in insects and Jonathan Waage (1979) observed the use of a penial appendage for removing rival sperm in dragonflies. The issue of how reproductive traits evolve was broken wide open by Bill Eberhard (1985, this volume), who, expanding on a suggestion by Lloyd (1979), pointed out that since genitalia often differ substantially among congeners in many taxa, particularly those with internal fertilization, they seem to evolve very rapidly, which is more easily explained by sexual than natural selection. That is, it is not clear why natural selection would act on the physiology and anatomy of closely related taxa in such diverse ways. This was a striking departure from the conventional wisdom of that time. The prevailing explanation for the cases of elaborate, species-specific, genitalia, found in so many taxa with internal fertilization, was the "lock-and-key" or mechanical isolation hypothesis (see discussion by Eberhard 1985, this volume, etc.). That is, that elaborate genitalia had evolved, by natural selection, to make it impossible for closely-related species to mate, thereby preventing the formation of sterile hybrids. The "lock-and-key" hypothesis has long been known to be incorrect in many cases and was already rejected as a general mechanism of species isolation by Stebbins in his 1971 textbook (Stebbins 1971). Eberhard and Huber (this volume) argue that it does not explain genital evolution in spiders; and Peretti rules it out as an explanation for spermatophore diversity in scorpions. Also, Eberhard (this volume) argues that it seems inadequate as a general explanation of the evolution of genital diversity, although Cordero-Rivera and Córdoba-Aguilar suggest that the lock-and-key hypothesis is consistent with some patterns in odonates and the authors of some other chapters conclude that it can't be entirely ruled out with the data available for their group.

An alternative hypothesis is that sexual selection explains much of the diversity of "primary" sexual characters. The pattern of evolution of genitalia, and some other reproductive traits, seems often to be similar to that of sexually selected traits. Sexual selection as described by Darwin involves two processes: male–male competition and female choice. Modern behavioral ecology has augmented this picture by describing mating systems as being shaped by male–female conflict; that is a conflict of interests between males acting to enhance their own fitness and females acting to enhance theirs (Andersson 1994). Since the 1970s (see above) sexual selection has been seen as a process that continues after copulation. Lloyd (1979) suggested that not only male–male competition, but also female choice, continues within the insect reproductive tract and that genital morphology may represent the outcome of an evolutionary arms race based on sexual conflict. This field of research took off with the work of Eberhard (1985, 1996). Male–male competition within the reproductive tract was an idea that was widely accepted by the 1980s (e.g., Smith 1984), but Eberhard's emphasis on "cryptic female choice" (Lloyd 1979; reviews in Eberhard 1990, this volume) was, and to a large extent, remains, a novel idea. A third approach to the evolution of reproductive interactions after copulation or insemination has been to consider the process one of sexual conflict. That is, the reproductive processes of a species, reflect, as does the mating system (Andersson 1994), evolution acting on males, and on females, but in different directions. Gowaty and Hubbell (this volume) also discuss the role of evolution on females in shaping reproductive interactions and how male behavior and male clasping structures may influence females to change their decision-making about sperm use, by imposing time constraints.

The possibility of sperm competition and its role in shaping genital morphology and physiology is perhaps the most popular line of investigation into postmating sexual selection. Sperm competition is possible whenever sperm from two or more males have potential access to the eggs of one female. It may occur with internal or external fertilization and is probably extremely common. Jarne et al. (this volume) conclude that sperm competition is important in basommatophoran snails and Cordero-Rivera and Córdoba-Aguilar (this volume) suggest that sperm competition has been a major factor in the evolution of odonate genitalia. Taborsky and Neat (this volume) argue that the intensity of sperm competition is correlated with testis morphology and that these parameters can vary among populations within single species in fishes. Cryptic female choice (female selection of one male's sperm and/or ejaculate over another's) has been presented as an alternative explanation for bias in fertilization success among males and has been more controversial. Distinguishing between the two processes is obviously going to be difficult and there is no reason to think that they are always mutually exclusive (e.g., see arguments in Kokko et al. 2003). There is increasing evidence for cryptic female choice in a variety of taxa. Eberhard and Huber (this volume)

suggest that it is an important selective force in spiders and Taborsky and Neat (this volume) discuss ovarian fluid as a mechanism of cryptic female choice in fishes, even those with external fertilization.

Another hypothesis often cited in studies of postcopulatory sexual selection is that of sexually antagonistic coevolution, in which current reproductive characters reflect an "arms race" between males and females (or male and female function in hermaphrodites), in which each partner seeks to enhance its own reproductive success in a given encounter, perhaps at the cost of reducing the lifetime fitness of its partner (e.g., Holland & Rice 1999). In several chapters, authors conclude that sexual conflict or sexually antagonistic coevolution is important in taxa ranging from stylommatophoran snails (Baur, this volume), to Opiliones (harvestmen and their relatives) (Macias-Ordónez et al., this volume) and anseriform birds (Montgomerie, this volume). Jarne et al. (this volume) discuss the role of partner manipulation in basommatophoran snails and Gowaty and Hubbell (this volume) present a detailed model of how males may be able to manipulate female choice by increasing the duration of sexual encounters or the time between encounters. In other chapters, authors indicate that it is clear, or probable, that postcopulatory sexual selection is operating but unclear as to how, exactly, it operates (Wedell and Hosken, this volume; Eady, this volume). Peretti (this volume) suggests that no single hypothesis can account for the patterns observed in scorpions.

Taken as a whole the chapters here present more questions than answers. The extent to which primary sexual characters have been studied and the types of questions that have been addressed vary widely among groups. For example, Uller et al. (this volume) argue that reptiles have been underrepresented in studies of postcopulatory sexual selection compared to some other vertebrates. On the other hand, insects have been the focus of such studies for more than 30 years now and there is substantial information from several taxa (see chapters by Wedell and Hosken; Eady; and Cordero Rivera and Córdero-Aguilar, this volume). The results from insects and other arthropods (Eberhard, this volume) have contributed to a changing view of the elaborate genitalia of such animals as the stylommatophoran gastropods such that the coevolution of male and female genitalia in these hermaphrodites is seen more in terms of postcopulatory sexual selection than lock-and-key species isolating mechanisms

(Baur, this volume). The spiny vaginas illustrated in the chapter on opisthobranchs by Valdes et al. seem to beg for similar studies. The field of evolution of reproductive characters is clearly at a very exciting stage where new questions are being asked and there is a growing tendency for studies to focus on general evolutionary issues such as natural versus sexual selection, rather than being limited to descriptions of individual taxa. As the focus of studies changes, the need for new types of observations and data emerges. While many of the available data focus on morphology, the chapters by Jarne et al., and Taborsky and Neat, demonstrate the importance of physiological and biochemical characters as has been shown to be the case in *Drosphila* (Holland & Rice 1999), and the artificiality of a distinction between behavioral and other sexual characters is made clear in the chapters by Miller (mammals) and Lotterhos and Levitan (broadcast spawners). It is clear that the wealth of information contained in this volume is only the tip of the iceberg. We are just starting on a very exciting era of evolutionary biology in which we look at not just the patterns of diversity of sexual characters but also their function and the forces that have shaped them. The array of chapters in this volume demonstrates both the tremendous challenge ahead and the exciting new discoveries that await us.

REFERENCES

Andersson, M. 1994. Sexual Selection. Princeton, N.J.: Princeton University Press.

Bishop, J.D.D. & Pemberton, A.J. 2006. The spermcast mating mechanism in sessile marine invertebrates. Integrative and Comparative Biology 46, pp. 398–406.

Darwin, C. 1859. On The Origin of Species 1967 reprint edn. New York: Atheneum.

Darwin, C. 1871. The Descent of Man, and Selection in Relation to Sex. 1981 reprint edn. Princeton, N.J.: Princeton University Press.

Eberhard, W.G. 1985. Sexual Selection and Animal Genitalia. Cambridge, MA: Harvard University Press.

Eberhard, W.G. 1990. Animal genitalia and female choice. American Scientist, 78, 134–141.

Eberhard, W.G. 1996. Female Control: Sexual Selection by Cryptic Female Choice. Princeton, NJ: Princeton University Press.

Holland, B. & Rice, W.R. 1999. Experimental removal of sexual selection reverses intersexual antagonistic coevolution and removes a reproductive load. Proceedings of the National Academy of Sciences USA 96, 5083–5086.

Kokko, H., Brooks, R., Jennions, M.D. & Morley, J. 2003. The evolution of mate choice and mating biases. Proceeding of the Royal Society of London B 270, 653–664.

Lloyd, J.E. 1979. Mating behavior and natural selection. The Florida Entomologist 62, 17–34.

Parker, G.A. 1970. Sperm competition and its evolutionary consequences in the insects. Biological Reviews 45, 525–567.

Smith, R.L. 1984. Human sperm competition. In: Sperm Competition and the Evolution of Animal Mating Systems (Ed. by R.L. Smith), pp. 601–659. Orlando: Academic Press.

Stebbins, G.L. 1971. Processes of Organic Evolution. 2nd edn. Englewood Cliffs, NJ: Prentice-Hall, Inc.

Waage, J.K. 1979. Dual function of the damselfly penis: Sperm removal and transfer. Science, 203, 916–918.

One

GENERAL CONSIDERATIONS

2

The Distinction between Primary and Secondary Sexual Characters

MICHAEL T. GHISELIN

INTRODUCTION

The main goal of this chapter is to explain the difference between primary and secondary sex characters. In doing that however, much else needs to be clarified, explained, and defined. The term 'sex' has several different meanings, and so, for that matter does 'character'. (Here we will follow a common convention among philosophers and use single quotation marks when referring to words, and double ones for other purposes such as "scare quotes" and the direct quotation of what an author says.) It often makes a big difference when one is talking about the word and the thing. Beyond pointing out that terms are ambiguous and that we are using them in one sense rather than another, it is necessary to explain the conceptual background. The language of science consists largely of theoretical terms, and one has to understand the theoretical context if one is to make sense of the words. Darwin (1871) was able to explain the differences between males and females through a combination of natural selection and sexual selection. The former gives rise to the primary sex characters, the latter to the secondary ones. This distinction then becomes the defining property that distinguishes between the two. These theoretical terms only make sense if one understands the theory from which the distinction derives. But knowing what the terms mean does not automatically put us in a position to apply them to a given case. We may not know to what extent a

given phenomenon is the outcome of these two modes of selection, or perhaps to some other evolutionary mechanism such as pleiotropy.

CHARACTERS

The ambiguity of the term 'character' is in part responsible for disputes about such matters as whether there are any "non-adaptive" characters (Ghiselin 1984). Sometimes the word means a part, such as a penis or a spine. Sometimes it means an attribute of a part, as when we say that a spine is sharp. However, we might say, for example, that a penis is spiny. In that case the spine is being used attributively; in other words its presence is an attribute (property). Attributes can be predicated of whole organisms (as when we say that one of them is female), and also of populations (as when we say what the sex ratio is). For good reasons, biologists are more apt to treat an attribute than a part as being non-adaptive, or functionless. Organs are expensive to build and to support, and one would expect them to be lost if not maintained by selection. Vestigial organs, which are of small size and perhaps have a residual or secondary function, are the exception that proves the rule. On the other hand, whether an oviduct passes to the left or the right of some nerve in making its way to the ovipositor probably does not make much difference functionally, so long as it gets there. We do need to be careful in passing judgment about so-called adaptive and non-adaptive characters. There was a period in the 1930s when the diagnostic characters of species were widely claimed to be non-functional. But a closer look at the organisms has all too often shown such claims to be mistaken.

SEX CHARACTERS

In this chapter it should be obvious which of the three usual senses of the word 'sex' is under consideration. We are not concerned with genital union or genetical recombination, except incidentally. Here the topic of interest is the differences between males and females. Bringing hermaphrodites into this picture is usually not difficult. Whether sequentially or simultaneously hermaphroditic, these organisms have the reproductive parts and attributes of both males and females. There is a subtle problem here, however, because once the male and female gametes have been united the zygotes and later ontogenetic stages may receive parental care from the mother, the father, or both. We might better consider sexual parts and attributes as having only to do with prezygotic affairs and fertilization, and all else as reproduction.

The most basic distinction between males and females is of course the production of dimorphic gametes, respectively sperm and eggs. Both males and females have the parts called gonads, but differ with respect to having the attributes "sperm-forming" and "egg-forming". There has been a division of labor at the cellular level, and along with that may go a division of labor among organs within organisms and among organisms within populations. The size differences between sperm and egg exemplify the general phenomenon of one sex playing an active role in getting the parents together, while the other concentrates upon provisioning the young. Possession of ovaries on the one hand, or testes on the other, would definitely be "primary" sex characters. For some authors anything less directly involved in sexual reproductions would be "secondary" sex characters. Darwin (1871) pointed out that this directness is a matter of degree, and, carried to its logical conclusion, only the gonads would be primary sexual characters. That would eliminate such accessory parts as intromittent organs, seminal receptacles and ovipositors, as well as anything used in finding a mate or caring for the offspring. He made the distinction between primary and secondary sex characters on the basis of his theory of sexual selection. Some of the differences between males and females have to do with competition for mates between the members of the same sex, and it was these that he called secondary sex characters. He also recognized a third group of differences between the sexes that have nothing to do with reproduction *per se*. Although he did not do so, he might have called these tertiary sex characters. When the males and females have different habits of life, as when they have different food organisms or live in different habitats, they may evolve differences that are related to reproduction either only indirectly or not at all. Some of the differences between male and female mosquitoes, for example, relate to the former feeding upon nectar and the latter upon blood. Again, the dwarf males that occur on the hermaphrodites or the females in certain cirripedes are degenerate and have lost the ability to feed. How direct the connection is supposed to be is not clear. The dwarf males of cirripedes are small and occupy an epizoic position on their mates as a consequence of a kind of reproductive competition that has been considered a form of sexual selection. Their size reduction might be viewed as a more direct result of sexual selection, whereas their functional and anatomical simplification would be an indirect consequence.

SELECTION: ARTIFICIAL, NATURAL, AND SEXUAL

Darwin made a strong conceptual distinction between three modes of selection: artificial, natural, and sexual. His evolutionary theory was a general one, and each of these modes had a different causal basis and different consequences. All three work as a result of the differential reproduction of the components of populations. Usually that means organisms, though they could be families. It can also mean differential reproduction of gametes. The basic difference lies in the "selective agent" that determines whether one organism or another within a given population will have the more reproductive success.

In artificial selection, the selective agent is the breeder, who "selects" (not necessarily consciously) organisms of one kind or another within a population to be the ones that reproduce. Sexual selection is much like artificial selection, in that the selective agent is an organism, but in this case it is an organism of the same species. The agent acts so as to affect which conspecifics will succeed reproductively. In the case of male combat, the males literally act upon each other by fighting, the winners in such contests thereby achieving greater reproductive success than their rivals. In the case of female choice, the selective agent does not act, but rather selects, or chooses, a mate to whom she is more attracted than the alternatives. The males enjoy reproductive success if they are more attractive than the competing males. For this discussion we will ignore the complications of reversed sex roles and also the possibility of additional modes such as male sequestering and male dispersal (Ghiselin 1974).

In natural selection, there is no selective agent that picks one organism rather than another as breeding stock. Instead, reproductive success depends upon the conditions of existence being such that one organism makes more effective use of environmental resources than do other organisms within the same population. This has to be qualified, however, because in natural selection organisms may indeed play a role in selecting among phenotypic variants. Camouflage may evolve because birds feed upon more conspicuous animals within a population of moths, and in that case it is reasonable to call those birds selective agents. On the other hand the birds' visual acuity is also being maintained (perhaps improved) by selection, as the consequence of the birds being more or less effective in finding their prey and therefore having more or fewer offspring. In both artificial selection and sexual selection the agents that make the choices are acting so as to affect reproductive success as such. The breeder of dairy cattle gains an economic advantage through producing an improved breed of cattle, whereas a lion that feeds upon a slower antelope gains nothing but a good meal. Sexual combat is an effort to monopolize the gene pool.

Darwin referred to sexual selection being "solely with respect to reproduction" between members of the same sex, but that seems an obscure way to put it. What he meant was that in the case of sexual selection the conditions of existence that have to do with obtaining food and other resources and turning them into offspring have nothing to do with what determines reproductive success. All that matters is out-reproducing one's conspecifics of the same sex. To test whether we have a case of natural or sexual selection it helps to ask whether the outcome would be an increase in the size of the population if it is under carrying capacity. If the answer is yes, we have to do with natural selection. If the answer is no, and, especially if the consequence is the opposite and the size of the population decreases, we have sexual selection. But that test is not definitive because it has to be purely a matter of reproduction. Other phenomena might have the consequence of decreasing the size of the population. Cannibalism is a good example.

These conceptual distinctions, which are hard to explain, are often misunderstood. In addition there are serious difficulties when we try to decide which mode of selection is really responsible for what we find in nature. Darwin underscored the point that natural selection and sexual selection accompany one another, so that it is often difficult to determine the relative importance of the two. In some cases it seemed clear, to Darwin at least, that differences between males and females result purely from natural selection, and not from sexual selection. For example, where the males and the females differ in their habitats and ways of life, as in the mosquitoes mentioned above, natural selection should favor different adaptations in the two sexes. There are also some cases in which sexual selection occurs in at least relatively pure form, so that the phenomena of interest are clearly not the result of natural selection. We will consider some of the possibilities in the first part of the next section.

SECONDARY SEX CHARACTERS AS THE RESULT OF SEXUAL SELECTION

We may begin with sexual selection by male combat, in which the males engage in a physical struggle with one another in efforts to monopolize opportunities to mate with the females. Darwin pointed out that in those species in which such contests occur the males, but not the females, have characteristic adaptations that are associated with fighting, such as greater body size, well developed tusks, and antlers. Of course, strength and weaponry can be used in defense from predators and in other kinds of interaction within species. In the case of predator defense, one would expect the males to be larger if they are more actively engaged in that activity than the females are. In animals that live in family groups males might be more active in predator defense than females for the simple reason that as a result of sexual selection they are larger and better armed, as is the case in baboons. However, there are larger males in species wherein the males are not involved in defending their mates or offspring and in which sexual size dimorphism is considerable, indeed extreme. Certain pinnipeds, such as elephant seals, provide particularly good examples. An interesting possibility is that the males might use their weapons only in fighting with conspecific males whereas the females would use them only in defending their offspring from predators. For example, in cattle both sexes have horns, but only the males use them in sexual combat. In that case the males' weapons would be secondary sexual characters (*sensu* parts), whereas those of the females would be primary sexual characters. The trouble is that where there is no sexual dimorphism, it makes no sense to call the presence of such parts "sex characters" at all.

In sexual selection by female choice, the males compete by virtue of making themselves relatively more effective in attracting the females as prospective mates. The females chose among males, picking the ones that they find most attractive. It is crucial, however, that when females chose a mate a necessary condition for sexual selection by female choice has been met, but it is not sufficient. Again, as Darwin put it, it has to be solely with respect to reproduction. Anything that would cause the species to increase in absolute numbers, rather than merely to switch between one alternative and another, would be natural, not sexual, selection. The females might, for example, choose mates that are not close relatives, thereby avoiding the adverse consequences of inbreeding. Further possibilities are listed below. Darwin's basic idea for sexual selection by female choice was a version of what is sometimes called the antecedent stimulus or pre-existing bias theory. He believed that human beings are not the only animals that have aesthetic tastes. It is well known in fact that certain stimuli are more attractive to organisms than others are and they often prefer a hypertelic stimulus. This makes more sense than some commentators have maintained. Darwin provided evidence that many of the features of avian plumage are tightly adapted so as to maximize the visual effect. Darwin's position seems all the more plausible now that it has been shown that the preferences for at least some secondary sexual characters were already present before the characters themselves evolved (Basolo 1990). We human beings seem to be particularly attracted to the color red,

and those who want us to buy their goods in grocery stores take advantage of that preference by using a lot of red on boxes and cans. By means of cladistic techniques the pre-existing bias hypothesis has been shown to give a plausible explanation for the prevalence of sexually-selected red coloration among primates in general (Fernandez & Morris 2007).

PRIMARY SEX CHARACTERS AS THE RESULT OF NATURAL SELECTION

The forgoing considerations suggest that many sex characters that might be thought of as secondary are really primary. The material is here organized so as to begin with characters that have to do with finding a mate and end with those that are involved in parental care.

Finding a Mate   Getting the males together with the females is a necessary condition for mating to occur, and doing so can be difficult. The males tend to play a more active role in this process. Therefore the locomotory organs and the sensory apparatus of males may be more highly developed than that of the females. If so, then sexual selection is involved only to the extent that these properties are advantageous to the males in monopolizing the females. The females may emit signals such as flashes of light or pheromones that allow the males to locate them. They probably are not competing with one another for access to the males. Stridulation and calling for females by males, on the other hand, is more likely to involve sexual selection, for it sets up a situation that favors purely reproductive competition among males.

Evaluating Potential Mates   When the prospective mates have located each other both partners may attempt to establish the suitability of the other before proceeding further. The usual assumption is that the females will be the more fastidious because, *caeteris paribus*, they have more to lose from an incorrect decision than the males do. However, the males, even if their investment is small, do have something to lose. A particularly good example is the male fireflies who fall prey to females of another species that mimic the signals sent out by females of the males' species. This is an extreme case of what can go wrong when an animal makes an incorrect species determination. The effects can of course be less dramatic, but nonetheless quite serious. The outcome of such union may be a failure to fertilize the eggs, a failure of the eggs to develop, or inviable offspring. Or there may be offspring that are viable, but sterile, with the result that providing them with parental care is a waste of resources. Any characteristics that further the ability to make such taxonomic determinations should be favored by natural selection. Therefore "species recognition marks" and the like would be primary sexual characters, not secondary ones.

For similar reasons natural selection should favor making an appropriate taxonomic decision within species. Whether outbreeding or inbreeding is favored depends upon the circumstances. It may or may not be advantageous to mate with an animal of another subspecies. However, there seems to be an optimal level of inbreeding versus outbreeding. Animals may avoid mating with distant relatives, much as they avoid mating with siblings and other close relatives. Bateson (1982) provided evidence that Japanese quail mate preferentially with first cousins rather than with siblings or more distant relatives. Again, such choice, whether it be exercised by the females, the males, or both, is natural selection, not sexual selection, and any sexual characters that result from it are primary, not secondary.

By the same token, a female might chose a male as a mate on the basis of his having the qualities that would tend to make him a good provider, such as the ability, whether realized or not, to defend a feeding territory. Likewise males might prefer females with analogous qualities. If that is the reason why a given sexual character has evolved and is being maintained by selection, that selection is natural, not sexual. The same may be said of any other selection in which either partner (or both) chooses more nearly optimal resources. This includes what are called "good genes." Natural selection might favor females making a choice between males for any of a number of reasons. However, it is not sexual selection by female choice unless there is a competition among the males that is solely with respect to reproduction, as Darwin put it.

Furthering Effective Mating and Fertilization The finding of mates was discussed first in order to maintain a temporal sequence. However it might be better to consider that topic from the point of view of bringing the sperm and eggs together effectively. The point here is that anything that gives greater efficiency to the physiological processes of reproduction should

be favored by natural selection. Marine animals that broadcast both sperm and eggs are apt to achieve considerably less than 100% fertilization and they have all sorts of behavioral and other adaptations furthering gametic union. Even where there is an excess of sperm within the bodies of internally fertilizing organism, there is an advantage to making sure that all of the eggs are fertilized. Sperm transfer organs, spermatophores, oviducts, seminal receptacles, and much else besides may make it easier and less costly to engender offspring and thereby increase reproductive output. Such parts are primary sexual characters, produced by natural selection.

Providing Care for the Offspring A wide variety of adaptations may provide the progeny with nutriment and protection after the eggs have been fertilized. Various secretions of the female reproductive tract can be taken up by the developing offspring and used as food, whereas others may give rise to cocoons. Postpartum there are secretions (such as milk) and behavior (such as predator defense). Some of the adaptations may occur before fertilization, or even before mating. The formation of nests is one example. Any of these adaptations, being the product of natural selection rather than sexual selection, is a primary sex character, not a secondary one.

The notion of a sex character is perhaps most straight-forward where the sexes are separate, dimorphic, and agree with the "idealized" case in which the males are involved only in getting the eggs fertilized while the females specialize in care of the young. Things become more difficult when both the males and the females play the two roles in question, and *a fortiori* where there has been a reversal of the "idealized" sexual roles. Organisms like ourselves, in which the males care for the young, have some analogy with hermaphrodites. In which sex it is that a character occurs does not affect its being primary or secondary. Thus the brood pouches in which male sea horses and their allies care for their offspring are primary, not secondary, sex characters.

THE JOINT EFFECTS OF NATURAL AND SEXUAL SELECTION

So long as a "character" such as a brood pouch or a mammary gland is produced and maintained solely by natural selection, it is primary, not secondary. If, however, such an organ becomes hypertelic as a result of sexual selection, it becomes to that extent a secondary one. However, there is a semantic problem here, because, as mentioned above, 'character' can mean either a part or a property. In the case of mammary glands being larger than necessary for providing milk simply because males find them attractive, it is really the hypertelic condition of the glands, and not the glands themselves, that we are treating as a secondary sex character. Again, there are bird species in which the males compete with one another by making nests that attract the females. Nest-making would be a primary sex character that evolved into, in part, a secondary one. Situations in which a part might take on an additional function, or get co-opted so as to have a new function exemplify the point that natural selection and sexual selection may both be operative in the evolution of sexual characters. The complex reproductive systems of many organisms have features that have sometimes been treated as non-adaptive, or sometimes as isolating mechanisms and therefore the product of natural selection. There is evidence that many of them have been produced by sexual selection (Eberhard 1996). Deciding to what extent each of the two modes might be involved is not easy.

Darwin (1871) considered how natural and sexual selection work together in the case of monogamous birds. According to his model the males arrive first on the breeding grounds and compete with one another for the opportunity to mate with the females as they arrive. The males gain an advantage by virtue of obtaining the most vigorous and healthy females. He noted that the females seemed to prefer, among other things, the more lively males. Darwin was aware that earlier authors, including his grandfather, had thought that male combat was an adaptation that benefited the species, providing for a kind of providential eugenics. Given that Darwin's mechanism of natural selection is based upon reproductive competition between individual organisms, the species-level advantage had to be rejected. Eugenical advantages at an individual level were not thereby ruled out. They did not play an important role in Darwin's own theory, but as "good genes" thinking they now are quite popular, especially in the interpretation of

female choice. Carrying such thinking to its logical conclusion, there may be no such thing as sexual selection, and all sex characters are primary. That seems most unlikely, but at any rate the extent to which we should consider a sex character primary or secondary may be a difficult question to answer.

CONCLUSIONS

Given what has been said, we might adopt the following terminology. *Primary sexual characters* are those that owe their existence to the direct effect of natural selection. *Secondary sexual characters* are those that owe their existence to the direct effect of sexual selection. *Tertiary sexual characters* are those that owe their existence to indirect effects of natural and sexual selection.

Acknowledgments   I am pleased to thank Janet Leonard and two anonymous referees for their advice on the manuscript.

REFERENCES

Basolo, A.L., 1990. Female preference predates the evolution of the sword in sword-tailed fish. Science, 250: 808–810.

Bateson, P., 1982. Preferences for cousins in Japanese quail. Nature, 295: 236–237.

Darwin, C., 1871. The Descent of Man, and Selection in Relation to Sex. London: John Murray.

Eberhard, W.G., 1996. Female Control: Sexual Selection by Cryptic Female Choice. Princeton: Princeton University Press.

Fernandez, A. & Morris, M.R., 2007. Sexual selection and trichromatic color vision in primates: statistical support for the preexisting-bias hypothesis. American Naturalist, 170: 10–20.

Ghiselin, M.T., 1974. The Economy of Nature and the Evolution of Sex. Berkeley: University of California Press.

Ghiselin, M.T., 1984. "Definition," "character," and other equivocal terms. Systematic Zoology, 33: 104–110.

3

The Evolution of Sexes, Anisogamy, and Sexual Systems

Natural versus Sexual Selection

JANET L. LEONARD

INTRODUCTION

The distinction between primary and secondary sexual characters, according to Darwin (1871; Ghiselin, this volume) is whether the characters have evolved through natural selection for improved fertilization efficiency and/or survival of offspring, or are the product of sexual selection, that is to say, competition for mates. For animals, having two sexes with anisogamy seems to be a plesiomorphic character. The most fundamental and primary sexual character in animals is that of sex (= gender). In animals, the two sexes are defined by differences in the morphology and behavior of the two types of gametes and this anisogamy has been seen as the foundation and prerequisite of all sexual selection (e.g., Kodric-Brown & Brown 1987). In this chapter, I explore current theories as to the evolution of sexes, anisogamy, and the sexual systems in which reproduction through these two types of gametes has been divided into two individuals or united in one individual.

WHAT IS SEX?

Sex in eukaryotes is a complex phenomenon that shows astonishing variation among taxa. At its most basic, sex can be thought of as involving three steps: fusion of two haploid nuclei, recombination to produce novel genotypes, and division of diploid cells to form haploid nuclei (Hoekstra 1990). That is, eukaryotic sex requires a life cycle, in which, an, at least briefly, diploid cell undergoes meiosis. Two prominent theories as to the starting point for the evolution of eukaryotic sex involve (a) cell fusion as a mechanism for the horizontal transmission of a selfish genetic element or pathogen (see Hickey & Rose 1988; Cavalier-Smith 1987; Hurst 1995) and (b) meiotic recombination evolving as a mechanism of repairing damaged DNA in the chromosomes (Bernstein et al. 1985, 1988; Holliday 1988; Michod 1990). According to the repair hypothesis, natural selection was the original force leading to the evolution of at least recombination in meiosis. The evolution of the complex machinery of segregation and recombination required for meiosis is beginning to be explored (Solari 2002; Marcon & Moens 2005). Differences between somatic DNA-repair and meiosis suggest, however, that there has been an evolutionary specialization for sex in meiosis (Marcon & Moens 2005).

IN THE BEGINNING WAS MEIOSIS ...

Sex in eukaryotes involves meiosis. There are major eukaryotic clades in which meiosis has not yet been detected, including the choanoflagellates, the sister taxon of the metazoans (Lecointre & Le Guyader 2006; King 2005; but see Maldonado 2004); but in many, if not most eukaryotes, sex is facultative (Dacks & Roger 1999), that is, sex occurs only under certain environmental conditions and most reproduction is asexual (see discussion in Pearse et al. 1989). This means that sex may have been overlooked in many of the unicellular eukaryotes. Modern molecular techniques have found core meiotic genes in at least one representative of the diplomonads, a group of primitive eukaryotes that have been thought to lack sex (Ramesh et al. 2005; Birky 2005), suggesting that meiosis and sex may have evolved very early in eukaryotic evolution and be more widespread than has been thought. The fact that meiosis and sex occur within a wide variety of clades of the eukaryotes also suggests that they evolved early in the eukaryotic line. The taxonomy and phylogenetic relationships of the higher clades of eukaryotes are still in flux (Simpson & Patterson 2006; Lecointre & Le Guyader 2006), but it seems clear that the ability to undergo meiosis and sex, facultatively, were ancestral characters of the metazoa (Dini & Corliss 2001).

A fair meiosis, however, it may have evolved, has the effect of shuffling genes to provide a full complement of genes but in novel combinations. Even with the possibility of repair during synapsis and crossing-over, recombination and the segregation of homologous chromosomes during meiosis acts to produce novel but complete haploid genotypes. Thus, meiosis, even if it is followed by automixis, where the products of a single meiosis fuse to form a zygote, involves a 'cost of recombination' (review in Crow 1988), in that a successful genotype is not replicated but rather one or more new and untested genotypes are formed. There is some evidence that where automixis is common it involves mechanisms to preserve heterozygosity (Hood & Antonovics 2004). It is noteworthy that the two-step meiosis found in metazoans creates less genetic diversity among gametes than would be possible in a one-step meiosis but is less easily invaded by asexual strategies (Archetti 2004; but see Maynard Smith & Szathmáry 1995).

SEX, GAMETES, AND SEXES

The most basic primary sexual characters are haploid nuclei that are capable of fusing with other haploid nuclei to form a diploid zygote. Gametes are haploid cells produced by meiosis that fuse with another gamete (syngamy) to form a zygote, which represents a new individual with a new diploid genotype. In metazoans, plants and some fungi individuals produce haploid gametes by meiosis and then the gametes from different individuals fuse (syngamy) to form zygotes containing genetic information from two individuals. The question of why animals, higher plants, some protists, and some fungi have two and only two sexes, whereas certain fungi and other protists may have large numbers of mating types, is an important one. Famously, R.A. Fisher (1958) pointed out that to understand why so many organisms have two sexes, a biologist must consider the case of having three sexes. Hoekstra (1987) took the question further and asked why one sex isn't enough. That is, why have sexes at all? Why couldn't any haploid cell fuse with any other? Hoekstra (1982, 1987) suggested that a label and a receptor are required to enable cells to fuse with cells from conspecifics only, cells from haploid conspecifics only, or only differentiated gametes of conspecifics. A promiscuous system where any gamete can fuse with any other, may be more derived than a simple system of two mating types. The simplest system will be one with unipolar complementarity in which fusion occurs between a labeled cell and a receptor-bearing cell. Such a unipolar complementarity system would produce two mating types. Bipolar complementarity, in which fusion requires both a label and a receptor on each cell, would produce either a promiscuous mating system in which all cells could fuse with any other or, if different alleles existed for the receptor and the label, a system with multiple mating types. Population genetics models of evolution of two mating types based on a unipolar recognition system or through pheromonal attraction were developed by Hoekstra (1982, 1987).

Theoretical studies have shown that, for isogamous organisms or organisms that do not show morphological sexual differentiation, multiple mating types are advantageous, if and only if, time is of the essence (Iwasa & Sasaki 1987); that is, if it important to encounter a compatible partner quickly. The more mating types exist the more likely it is that a given random encounter will produce successful fusion and zygote production. Therefore, rare mating types will have an advantage in encountering a mate. This may be important in organisms such as ciliates, which are likely to be surrounded by clone members and in those fungi which can only encounter mates by the growth of hyphae. The advantage of having only two mating types came from the likelihood of some mating type alleles being associated with a competitive advantage to the carrier, leading to the loss of some less adaptive alleles, and/or alleles being lost from the population by sampling error (Iwasa & Sasaki 1987). That is, if a need for rapid zygote formation does not give a big advantage to multiple mating types, natural selection and genetic drift will tend to drive the system to the absolute minimum number of mating types, just two, assuming that outcrossing is beneficial (see below). Iwasa and Sasaki did not consider the possibility of a single, promiscuous mating type.

Having two sexes, therefore, is older and more widespread than anisogamy, and can be explained, at the simplest level, by natural selection for the ability of gametes (or cells containing gametic nuclei) to fuse with appropriate partners (Hoekstra 1982, 1987). Ghiselin (1974) suggested that mating types may have begun as a mechanism to prevent inbreeding. This idea was further developed by Power (1976) who considered that a recognition system could promote outcrossing if it were the case that gametes of opposite recognition type were otherwise less similar genetically than gametes of the same recognition type. If outcrossing is advantageous the tendency of mating types to self-aggregate gives an advantage to a binary mating system (Czárán & Hoekstra 2006). A novel mechanism for the stability of two and only two sexes was proposed by Haag (2007), who proposed that the existence of two (and only two) mating types provides a reliable cue for the presence of diploidy.

Another phenomenon that has come to be seen as fundamental to the prevalence of two sexes is the pattern of transmission of organelles to offspring (Hoekstra 1990; Hurst & Hamilton 1992; Hutson & Law 1993, see review in Hurst, 1995). In animals mitochondria are normally [but not invariably, e.g. biparental inheritance of mitochondria in mussels, etc. (Zouros et al. 1992; see reviews in Birky 2001; Barr et al. 2005)] maternally inherited and there is evidence that a pattern of uniparental inheritance of mitochondria (and chloroplasts) is widespread among eukaryotes (Birky 1995, 2001; but see Barr et al. 2005). Consideration of the potential for both genomic conflict between organelle and nuclear genomes and the transmission of selfish genomic elements in the cytoplasm has lead to the hypothesis that uniparental inheritance of organelles should evolve in all organisms in which sex involves the fusion of gametes

(Law & Hutson 1992; see review in Hurst 1995). There are two fundamental, but not mutually exclusive, scenarios. In one, the blending of cytoplasm from two gametes in a zygote provides the opportunity for the spread of a selfish cytoplasmic genetic element or pathogen which would include itself preferentially in offspring, creating a reduction in fitness of the offspring, which could reach 90% (Hastings 1999). In the second, organelles, particularly mitochondria, from two different parents may reduce the fitness of the offspring because of incompatible genetic make-up and mutual interference (Hurst 1990a, b; Hurst & Hamilton 1992). Such an asymmetry would provide a basis for the development of two sexes; one that transmits mitochondria and one that doesn't. The hypothesis that an organism might evolve to exclude its own organelles from gametes as a means of increasing the fitness of the zygote has been found to be implausible on theoretical grounds (Randerson & Hurst 1999, 2001b), while the alternative hypothesis that organisms will evolve mechanisms whereby their gametes kill the organelles of the partner in the zygote has been found to be robust in these same modeling studies. It has been suggested that multiple mating types can only exist in organisms such as ciliates and basiomycete fungi in which sex involves only the transfer of haploid nuclei between "adults" and not the fusion of gametes to form a zygote (Hurst & Hamilton 1992; e.g., Aanen et al. 2004).

If, as Hurst (1995) argued, in most isogamous organisms there is linkage between mating type genes and alleles that coordinate the uniparental, or biased, transmission of organelles, then this hypothesis may provide a basis for understanding the prevalence of two sexes in eukaryotes and the ubiquity of two sexes in animals. This is not to say that the problem of transmission of selfish cytoplasmic elements during sexual fusion has been solved. There are a variety of such elements known (see review in Werren 2005) and the danger of such parasitism represents one of the costs of sex. Also the diversity of mechanisms used to accomplish uniparental inheritance suggests that there are many selective pressures involved (Barr et al. 2005). The likeliest scenario is that some ancestor of the metazoa had isogamy with two sexes (mating types) and unilateral inheritance of mitochondria (see also review in Maynard Smith & Szathmáry 1995). However, sex has not been described in the sister taxon of the animals, the choanoflagellates. While the evolution of multiple sexes can best be explained as the result of sexual selection, the prevalence of two sexes may be the result of natural selection for both efficient uniparental transmission of organelles and efficient cell fusion.

THE ORIGINS OF ANISOGAMY

The next question is why, given that two sexes exist, they should produce morphologically disparate gametes. Animals produce two morphologically distinct types of gametes; eggs and sperm. In general, sexual reproduction in animals involves the union of a large, immobile, egg that contributes cytoplasm, stored energy, and mitochondria to the zygote with a much smaller, flagellated, swimming sperm that contributes only a haploid set of nuclear genes. There are exceptions (Morrow 2004); nematodes are known for their amoeboid sperm, the opiliones have immobile sperm (see Macias-Ordoñez et al. this volume), as do some polychaetes (Berruti et al. 1978); bivalves are known to have biparental inheritance of mitochondria (Zouros et al. 1992; Birky 2001), and so on. The relative size of the gametes also varies greatly, [e.g., normal vs. giant sperm in *Drosophila* spp. (Bjork & Pitnick 2006)], but in animals oogamy, a very large size difference between an immobile egg and a motile sperm, seems to be a plesiomorphic character, although anisogamy is the term usually used and used here. Anisogamy has been seen as a prerequisite for sexual selection (see below; Bateman 1948; Parker et al. 1972). The selective pressures that explain the evolution and prevalence of anisogamy in animals, higher plants and some other eukaryotes, have been the subject of considerable debate. The prevailing theory has been that anisogamy has evolved as a trade-off between selection to produce the largest possible number of gametes from a fixed amount of resource and selection to maximize survival of the zygote (Kalmus, 1932; see reviews in Ghiselin 1974; Bell 1982; Hoekstra 1987). That is, a classic trade-off between quality and quantity creates disruptive selection that drives evolution of two distinct size classes of gametes (Kalmus, 1932; see also Kalmus & Smith 1960; Scudo 1967). The formal model of this process provided by Parker et al. (1972) (the PBS model) has become the textbook explanation and is still widely accepted (see also Charlesworth 1978; Maynard Smith 1978). This model starts with panmictic isogametes

and is based on the assumption that there is a strong advantage to increased zygote size. Models starting from isogametes of two mating types have also been developed (Hoekstra 1980, 1987; Maynard Smith 1982; Bulmer & Parker 2002, etc.; see also Bonsall 2006).

A major alternative approach to that of disruptive selection on size is based on encounter probability (see review in Iyer & Roughgarden 2008b). Shuster and Sigmund (1982) starting from Charlesworth's (1978) formulation of the PBS model, argued that Brownian motion of gametes would favor the encounter of a large and a small gamete more often that that of two large or two small gametes, facilitating the evolution of anisogamy (see also Parker 1971). In a modeling study based on marine green algae Togashi et al. (2007) found that isogamy was favored at high gametic density and anisogamy at lower gametic densities and suggested that this explained the tendency for deep water algae to be anisogamous whereas shallow water taxa are more often isogamous.

Other models have explored the effect of chemotaxis on the evolution of anisogamy (Hoekstra 1984; Cox & Sethian 1985; Dusenberry 2000; review in Dusenberry 2006). If it is the case that gametes use a pheromone to find one another, it seems probable that segregation into a pheromone-producing and a pheromone-detecting mating type will occur. The argument is that: (a) a gamete cannot efficiently both produce pheromone and detect it since receptors on the gamete would become swamped with its own pheromone; (b) the larger a gamete (egg) is the more pheromone it can produce and the greater the sphere of attraction or target size that it will represent; and (c) the ability to produce more gametes is sufficient to explain the advantage of small sperm (see Dusenberry 2002). However, in this model there is an obstacle to the evolution of oogamy in that a relatively large size increase is needed to offer an advantage in pheromone production and get the system moving toward oogamy (see discussion in Dusenberry 2006). Dusenberry (2006) has suggested that if the fertile period is dependent on gamete size, then this threshold can be overcome. Data demonstrating the increase in target size provided by eggs, in a system with chemotaxis of sperm, are provided from abalone, along with a recent review of sperm chemotaxis, by Riffell et al. (2004). Hoekstra et al. (1984) stated that pheromone-mediated chemotaxis is not known in organisms with isogamy (with one exception) but is common in organisms with oogamy. A perceived advantage to the models based on encounter probability with or without pheromones is that the selective advantage of anisogamy is based on gamete size rather than zygote size as in the PBS model (but see Iyer and Roughgarden 2008b). Other recent studies have argued that where a size difference is linked to mating type, isogamy becomes difficult to maintain (Matsuda & Abrams 1999) or that isogamy is stabilized by phototaxis, which creates a situation where gametes encounter each other on a two-dimensional surface (Togashi & Cox 2004). In *Chlamydomonas* there are two modes of gamete formation; in one of which the whole cell becomes a gamete whereas in the other, a single cell divides during meiosis to produce four small gametes (Wiese et al. 1979). Association of these facultative processes with mating type could produce an anisogamy with two sexes without selection on a range of gamete sizes (Wiese 1981).

The plausibility of the assumptions of the various models has been hotly debated (e.g., Randerson & Hurst 2001b; Bulmer et al. 2001; Bonsall 2006; Iyer & Roughgarden 2008b). The PBS model and its derivatives predict that anisogamy will be favored when the fitness of zygotes increase steeply with their size and explain the advantage of large zygote size in the multicellularity of the adult. Searches for support for this hypothesis in other taxa have met with some success in the volvocine algae (Knowlton 1974; Madsen & Waller 1983; Bell 1985; Randerson & Hurst 2001a) but attempts to control for phylogeny show that this trend is sensitive to phylogeny and method of analysis. Also, the Volvocales violate some of the assumptions of the PBS model. There are also a number of exceptions in which unicellular taxa show oogamy (e.g., Sporozoa, Knowlton 1974; some algae, Madsen & Waller 1983). Madsen and Waller made the important suggestion that large zygote size might be advantageous, even in unicellular organisms, in providing the resources necessary to withstand a prolonged and uncertain period of dormancy. Comparative data suggest that anisogamy is more common in environments that are ephemeral (Madsen & Waller 1983). This hypothesis might also explain the prevalence of oogamy in the Sporozoa where the zygote forms an oocyst (Knowlton 1974). A third suggestion comes from Kalmus (1932) who suggested that anisogamy was associated with heterotrophy of the zygote (see also

Iyer & Roughgarden 2008b). The ultimate advantage of the large zygote may be the ability to store resources for either multicellular development or a long hiatus before development to a fully-functional organism. Our spotty knowledge of reproduction and ecology in so many groups of eukaryotes makes it seem unlikely that we will be able to definitively identify an ecological or life history correlate of anisogamy at present but there is certainly reason to believe that the relationship between large zygote size and anisogamy assumed by the PBS theory and its derivatives is plausible (see also Randerson & Hurst 2001a).

Alternative explanations of the evolution of anisogamy from an isogamous population exist (see reviews by Randerson & Hurst 2001b; Bonsall 2006). One hypothesis has been that selective pressure to avoid transmission of intracellular parasites or selfish cytoplasmic elements could favor the evolution of small gamete size, creating a situation where an individual's nuclear genes are selected to exclude cytoplasmic components from the gamete (see review in Hurst 1990b). This hypothesis has been investigated in modeling studies and has been found to be untenable if any cost is associated with it (Randerson & Hurst 1999; see also Randerson & Hurst 2001b). However, once there is no chance [or a very slim chance, since "leakage" of paternal mitochondria can occur in animals (Barr et al. 2005)] of organelles of the "male" gamete being transmitted to the zygote, genomic conflict between nuclear and cytoplasmic genes in the male parent will be reduced and there will be fewer barriers to the evolution of a gamete that contributes only nuclear genes to the zygote (Randerson & Hurst 1999). Additionally Allen (1996) suggested that by restricting gamete motility to sperm the mitochondria of the egg are protected from the genetic damage associated with the production of free radicals by oxidative phosphorylation during locomotion. Under these scenarios, anisogamy would largely be the result of natural selection.

ANISOGAMY AND SEXUAL SELECTION

Anisogamy has been seen as the prerequisite for sexual selection (Bateman 1948; Parker et al. 1972), and it may represent the fundamental source of sexual conflict with regard to mating. The PBS-model involves disruptive selection on a trade-off between gamete number and gamete quality. That is, if eggs represent a strategy whereby an individual insures the survival of its zygote, natural selection is at work. Parker (1982) clearly proposed that sperm competition was a critical factor in maintaining the small size of sperm in the face of an advantage to large zygotes. Under the PBS model (and its derivatives) anisogamy can be seen as the result of a trade-off between the pressures of natural and sexual selection and therefore, is the product of both natural and sexual selection. Under the models involving encounter probability (see above), sexual selection may play an even stronger role in that the advantage of small gamete size is the same as in the PBS model but there is no postulate that increased zygote size is advantageous. If the dichotomy in size between sperm and eggs is a product of improved encounter probability in a three-dimensional environment (Dusenberry 2006) then anisogamy might be thought of as wholly a product of sexual selection, although since even natural selection requires encounter between gametes, this is a debatable point. Chemotaxis of sperm toward a pheromone-producing egg is also likely to evolve through sexual selection or once evolved to inevitably produce sexual selection for better search and stronger pheromone production. Therefore, although anisogamy may be the ultimate determinant of sexual conflict between males and females, its evolution has probably been the product of sexual selection to some extent. Only the scenarios that explain anisogamy as the result of evolution to limit the transmission of genetically damaged mitochondria or selfish cytoplasmic elements rely totally on natural selection.

SEXUAL SYSTEMS: DIOECY OR HERMAPHRODITISM?

There are two common sexual systems in the Metazoa; dioecy (= gonochorism), in which individuals are either males or females during their reproductive lives but not both, and simultaneous hermaphroditism in which individuals can produce and use both eggs and sperm during a breeding season. Jarne and Auld (2006) estimated that about one third of animal species, exclusive of insects, are hermaphroditic (about 5% if insects are included). In animals both dioecy and simultaneous hermaphroditism are common and stable in large taxonomic groups, whereas sequential hermaphroditism and mixed sexual systems, for example gynodioecy and androdioecy, are relatively rare and not fixed in large, old clades.

The question of which selective pressures are responsible for the observed patterns of gender expression is almost as old as the theory of natural selection (Darwin 1888; review in Ghiselin 1974). Whether dioecy or hermaphroditism is the plesiomorphic state for animals has been long debated (see Ghiselin 1969; Jarne & Charlesworth 1993; Iyer & Roughgarden 2008a; etc.). Since the phylogeny of the metazoans is currently in dispute (e.g., Lecointre & Le Guyader 2006; Iyer & Roughgarden 2008a; Dunn et al. 2008) the point is unlikely to be resolved in the near future. Whatever the basal state may have been, it is clear that the present distribution of hermaphroditism and dioecy (see table 3.1), represents many transitions from one type of sexual system to the other and as George Williams (1975) pointed out, in animals, major taxonomic groups, whole phyla and classes (table 3.1; Eppley & Jesson 2008), are often consistently either hermaphroditic and dioecious suggesting that the sexual system has been stable across tens if not hundreds of millions of years, thousands of species, and a variety of ecological conditions (William's Paradox, see review in Leonard 1990; discussion in Jarne & Auld 2006, below).

The most obvious difference between dioecy and hermaphroditism is that hermaphroditism offers the possibility of self-fertilization while dioecy does not. Self-fertilization occurs in most, if not all, major hermaphroditic taxa of metazoans, although it is not always very prevalent (see Jarne & Auld 2006 for a review). Within such simultaneously hermaphroditic groups as the pulmonate gastropods, the capacity to self-fertilize may vary within genus or even among populations within a species (see discussion in Clark 1978; references in Leonard et al. 2007; Baur this volume, etc.; Jarne & Auld 2006; Jarne et al. this volume). Many hermaphrodites have evolved self-incompatibility mechanisms (e.g., Bishop & Pemberton 2006) or other mechanisms to prevent self-fertilization (Jarne & Charlesworth 1993; Hadfield & Schwitzer-Dunlap 1984). Sequential hermaphroditism is also an effective mechanism to prevent self-fertilization, although the size advantage model is the prevalent model to explain sequential hermaphroditism in animals (see below; Muñoz & Warner 2003; Taborsky & Neat, this volume). Therefore selfing cannot be the only advantage of hermaphroditism.

TABLE 3.1 Distribution of dioecy and hermaphroditism in the Metazoa (Taxonomy from Lecointre and Le Guyader 2006; reproductive information from Meglitsch & Schram 1991; Brusca & Brusca 1990; Pechenik 2005; Jarne & Auld 2006)

Phylum Class	Mode(s) of sexuality	Selfing?*	Fixed costs	Comments
Porifera	Usually sequential hermaphrodites	Not reported	Low	Paraphyletic group of phyla
Placozoa	Yes	?	Low	Very poorly known
Cnidaria	Either dioecious or hermaphroditic	Some (data from 26/9000 species)	Low	
Ctenophora	Largely hermaphrodites	?	Low	
Platyhelminthes	Almost exclusively hermaphroditic	Some (data from 16/13,780 species	High except in cestodes	
Nemertea	Dioecious	N/A	Low	
Rotifera	Dioecious	N/A	Low	*Clade Syndermata*
Acanthocephala	dioecious	N/A	Low	*Clade Syndermata; parasitic*
Cycliophora	Dioecious; sessile female and dwarf male#	N/A	Low	*Clade Syndermata;*
Entoprocta	Protandric or simultaneous hermaphrodites	?	Low	
Sipuncula	Dioecious except for one species	N/A; ?	Low	
Mollusca	Primitively dioecious	Some in hermaphroditic taxa; **0.544**		80/117,495 species studied with regard to selfing
Solenogastres	Dioecious			
Caudofoveata	Dioecious			
Polyplacophora	Mostly dioecious, some hermaphrodites		Low	
Monoplacophora	Dioecious		Low	
Gastropoda			Low to very high; see text	See Hodgson, this volume; Jarne et al.; this volume; Baur this volume; Valdes et al., this volume
"Prosobranchia"	Varied sexuality	?		
Heterobranchia	Almost exclusively simultaneous hermaphrodites	Selfing common in Basommatophora; Stylommatophora, one report in opisthobranchs		
Cephalopoda	Dioecious		High	
Bivalvia	Laregly dioecious; some hermaphrodites; various independent events	?	Low	
Scaphopoda	Dioecious		Low	
Annelida		**0.882**±0.229		5/14,360 studied with regard to selfing
Polychaeta	Mostly dioecious		Low	
Oligochaeta	Hermaphroditic		Low–moderate	
Hirudinea	Hermaphroditic		Low–moderate	
Echiura	Dioecious		Low	Included as annelids by Lecointre and Le Guyader
Pogonophora	Dioecious		Low	Included as annelids by Lecointre and Le Guyader
Ectoprocta	Hermaphroditic		Low	Colonial; 3/4500 species studied with regard to selfing
Phoronida	Either dieocious or hermaphroditic		Low	
Brachiopoda	Dioecious, some hermaphrodites		Low	
Chaetognatha	Sinultaneous hermaphrodites ?		Low	
Gastrotricha	Largely hermaphroditic		Low	
Priapulida	Dioecious		Low	
Loricifera	Dioecious		Low to moderate	

Continued

TABLE 3.1 (*Cont*)

Kinorhyncha	Dioecious		Low–moderate	
Nematomorpha	Dioecious		Low	
Nematoda	Dioecious or (rarely) androdioecious	Yes in androdioecious forms	Moderate	Selfing rate from 1 species: 0.162± 0.259
Onychophora	Dioecious		Low to moderate	
Tardigrada	Dioecious		Moderate	
Euarthropoda		**0.237**±0.096		2//956,414 species studied with regard to selfing
Chelicerformes	Dioecious		Low to high	Pycnogonida, Merostomata and Arachnida
Remipedia	Hermaphroditic		Low to moderate	
Cephalocarida	Hermaphroditic		Low to moderate	
Maxillopoda	Dioecious and Hermaphroditic according to subclade		Low to high	Copepods, ostracods, etc., dioecious; Cirripedia (barnacles, largely hermaphroditic)
Branchiopoda	Largely dioecious, some hermaphroditic (notostracans) and androdioecious (chonchostracan) taxa		Moderate	
Malacostraca	Mostly dioecious; some sequential and simultaneous hermaphrodites		Low to high	
Hexapoda	Dioecious		High	Insects, 830,075 species
Myriapoda	Dioecious		Moderate to high	
Mesozoa	Hermaphroditic and dioecious	Yes	Low to moderate	Rhombozoa are hermaphrodites which may self or cross-fertilize; orthonectids dioecious
Echinodermata	Largely dioecious	One hermaphroditic species studied and found to self	Low	Some hermaphrodites among the asteroids, holothuroids and especially ophiuroids
Hemichordata	Dioecious		Low	
Chordata				
Urochordata	Hermaphroditic	Self-incompatibility known in some species; selfing in others 0.690 ±0.259	Low	8/1300 species studied for selfing
Cephalochordata	Dioecious		Low	
Myxinoidea	Sequential hermaphrodites?		Low	Hagfish, etc.†
Petromyzontiformes	Dioecious		Low	Lampreys (Sower 1990)
Chondrichthyes	Dioecious		High	
Actinopterygii	Largely dioecious; some sequential and simultaneous hermaphrodites among teleosts	Reported	Low to high	Includes teleosts, sturgeons, gars, bowfins (see Taborsky & Neat this volume)
Actiniata	Dioecious		Moderate	Coelacanth; internal fertilization‡
Dipnoi	Dioecious		Low–moderate	Lungfishes
Tetrapoda	Dioecious	Selfing?	Low to high	Includes, amphbians, reptiles, birds, mammals

---

* *Rates from Jarne and Auld (2006).*
† http://www.networksplus.net/maxmush/myxinidae.html
‡ http://sacoast.uwc.ac.za/education/resources/fishyfacts/coelacanth.htm

Advantages of Dioecy

Dioecy precludes self-fertilization and, in the botanical literature, dioecy is seen primarily as a mechanism to avoid inbreeding (or promote outcrossing) (e.g., Charlesworth 2001, 2006; Delph & Ashman 2006; Ashman 2006a; Barrett 2002; discussion of alternative theories in Thomson & Barrett 1981; Ashman 2002, 2006b; Vamosi et al. 2007). If the advantage of dioecy lies in avoidance of inbreeding, that would be explained by natural selection, although reliance on outcrossing sets up conditions for sexual selection in that mate choice and/or mate number become important determinants of reproductive success.

In the animal literature the emphasis has been different. Zoologists have tended to accept dioecy as the norm. It has generally been accepted that separation of the sexes offers advantages in terms of increased efficiency in reproduction by division of labor (Muller 1932; reviewed in Ghiselin 1974). Sex allocation theory (Charnov et al. 1976; Charnov 1982) states that dioecy will be favored if fitness curves are concave (figure 3.1), that is, if reproductive success in one sexual role comes at the expense of reproductive success in the other sexual role. The theory is straightforward and appealing. If fitness through each type of gamete is a linear function of investment (allocation) into reproduction through that type of gamete, that is the more investment, the more fitness, then dioecy will be favored because investment in female function will necessarily come at the expense of investment in male function and vice versa. That is, there will be a trade-off between male and female function, creating a concave fitness function (Charnov 1979, 1982; figure 3.1). The factors that would create a concave fitness function have not been explored in detail, probably because theory suggests that, all else being equal, investment in one sexual role will always come at the expense of investment in the other sexual role. One prediction of the model, that "fixed costs" in the form of sex-specific structures would be incompatible with simultaneous hermaphroditism (Heath 1977; Charnov 1979), is clearly refuted by the data (table 3.1; see also discussion in Leonard 2005, 2006). Exclusively hermaphroditic taxa, such as the trematodes and euthyneuran gastropods, possess some of the most complex genitalia known, with some complex structures dedicated to male and others to female function. Simultaneously hermaphroditic species of *Lysmata* shrimp have the

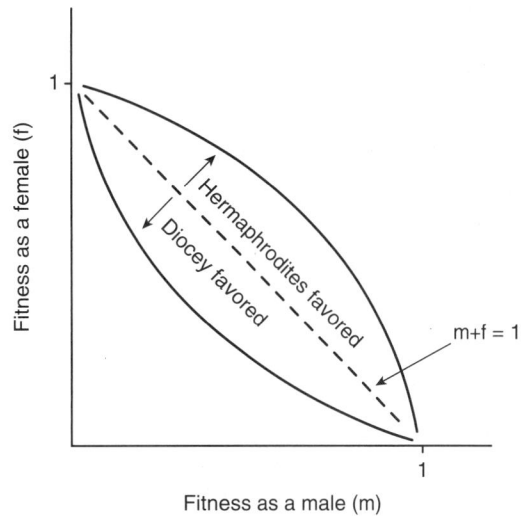

FIGURE 3.1 Charnov's (1982) graph showing possible fitness sets for the trade-off between male and female function for hermaphroditism (convex) and dioecy (concave). Reproduced from Leonard 1990.

full complement of both male and female reproductive structures (Bauer 2006). High fixed costs, per se, then, are not inconsistent with simultaneous hermaphroditism.

In the explanations of early authors (Darwin 1871; Muller 1932, etc.; see also Maynard Smith & Szathmáry 1995) the driving force for dioecy is proposed to be sexual selection, acting on males. That is dioecy is seen as an extension of anisogamy with males specialized for finding mates and females specialized for producing and provisioning young. Puurtinen and Kaitala (2002) suggested that when mate search is efficient and matings in the male role are hard to obtain, a trade-off between the resources available for reproduction and the number of partners obtainable leads to disruptive selection and the evolution of separate sexes (see also Eppley & Jesson 2008; discussion below). In another recent modeling study, Iyer and Roughgarden (2008a), argued that dioecy can evolve from hermaphroditism when there is an advantage to increasing sperm concentrations and a trade-off between the resources available for gamete production and sperm-concentrating capability. An empirical test of the hypothesis that efficient mate search (and/or gamete search) is associated with dioecy in multicellular organisms provides limited support for the hypothesis (Eppley and Jesson 2008; see discusson below).

Advantages of Hermaphroditism

The evolutionary advantage of hermaphroditism in metazoans has been the focus of a great deal of theoretical attention (e.g., Ghiselin 1969, 1974; Williams 1975; Charnov et al. 1976; Charnov 1982; Crowley et al. 1998; reviews in Leonard 1999, 2005; table 3.2). As a result of this work, some general advantages of hermaphroditism, aside from the ability to self, have been identified.

Low Density and Reproductive Assurance Models

The most prevalent model for the initial evolution of hermaphroditism, the Low Density model (Tomlinson 1966) demonstrates that, at low population sizes, either selfing or non-selfing hermaphrodites have a substantially greater likelihood of encountering a compatible mate than do dioecious organisms. Later theoretical work demonstrated that the advantage of hermaphrodites extends to larger populations since, where the sex ratio is unequal, hermaphrodites can supply the limiting gamete (Borgia & Blick 1981). Arguments for the Low-Density model have cited an association between hermaphroditism in animals and (a) sessility; (b) low mobility; (c) low population size; and/or (d) frequent founding effects, where an individual may frequently find itself with no or few conspecifics, either due to dispersal to a new location or change in an ephemeral habitat, which would include parasitism (see discussion in Ghiselin 1969, 1974, 1987; Clark 1978). The Low Density model is consistent with the view that the advantage of hermaphroditism lies in reproductive assurance, which would be driven by natural selection. A problem with the Low Density model is that these factors do not adequately explain the current distribution of hermaphroditism among animals at high taxonomic levels (Williams 1975; Leonard 1990; see below). For example, the Low Density model does not explain why most parasitic platyhelminths are hermaphroditic, whereas the parasitic phylum Acanthocephala is dioecious as are the many parasitic nematodes. Similarly, the sessile urochordates are hermaphroditic but the sessile sea lilies are dioecious, as are the vast majority of taxa

TABLE 3.2 Advantages of hermaphroditism versus dioecy

A. Advantages of Dioecy

1. Reduction of inbreeding depression by prevention of self-fertilization; advantages variable; possible to measure experimentally; driven by **natural selection**.
2. Increased fitness of offspring through outcrossing; demonstrated advantages but variable; **driven by natural selection** but creates **opportunity for sexual selection** because mate choice and/or mate number become important determinants of reproductive success.
3. Increased efficiency through "division of labor"; resolution of trade-off between allocation to male versus female function; strong theoretical support, data from attempts to measure sex allocation mixed; difficult to identify currency; empirical support poor; example from *Schistosoma* (Despres & Maurice 1995; Tchuem Tchuenté et al. 1996); dioecy associated with reduced resource availability (Ashman 2002; Iyer & Roughgarden 2008), **both natural and sexual selection** may play a role.

B. Advantages of Hermaphroditism

1. Reproductive assurance through self-fertilization; phenomenon well-documented; magnitude of advantage variable; not applicable to sequential hermaphroditism; **natural selection**.
2. Higher encounter probability; any individual potentially a mate, effect of doubling effective population size; not applicable to sequential hermaphroditism; some support from comparative data; **natural selection.**
3. Increased efficiency through shared function or division of resources (convex fitness set, Charnov 1979, 1982; figure 3.1); serial egg production (Crowley et al. 1998). **Both sexual and natural selection may play a role.**
4. Reduced covariance of offspring fitness (Lloyd 1982; Leonard 1999); **natural selection.**
5. Reduced variance in fitness (Wilson & Harder 2003); **Natural selection.**
6. Quantitative gender; flexibility in sex allocation with current circumstances (Lloyd 1982; etc.), may occur without significant costs (e.g., Lorenzi et al. 2006); **either natural or sexual selection may be important**.

of echinoderms, a phylum thought to have been primitively sessile. In barnacles, it has been suggested that hermaphroditism is associated with high densities and dioecy with low densities (Yamaguchi et al. 2008).

Variance Reduction

Theoretically, hermaphrodites have a fitness advantage because they can spread the risk associated with offspring or gamete production between two sexual roles, potentially reducing the risk of reproductive failure and thereby increasing fitness. This advantage was first discussed by Lloyd (1982) for plants, based on principles of economic investment theory (Markowitz 1991; first applied to biology by Real 1980). The principle is that all else being equal, a reduction in variance in reproductive success will increase fitness in a finite population because it will be associated with a decreased risk of reproductive failure (Gillespie's Principle; Gillespie 1974, 1977; see discussion in Leonard 1999). Hermaphrodites may have lower variance in reproductive success because eggs and sperm, or offspring produced through eggs versus sperm, have a lower covariance (are less likely to suffer the same risks, because they are distant in space or time) than offspring produced solely through eggs or sperm (Lloyd 1982; Real & Ellner 1992; Leonard 1999). For example, if a hermaphroditic snail mates with a partner and then lays a clutch of eggs which is lost to predation, its offspring through sperm are unlikely to be lost to the same predator, because the partner will have laid its eggs in a different location and perhaps at a different time. A female that put an equivalent total investment solely into eggs might experience total reproductive failure if the eggs experienced predation. Similarly, a male that mated with a partner might lose the entire investment if the partner were eaten before laying a clutch of eggs, or if the clutch were destroyed. Using similar, but independent, logic, Wilson and Harder (2003) in a modeling study, found that dioecious individuals would experience greater variance in reproductive success and hence reduced fitness, relative to hermaphrodites. These arguments predict that there is a natural selection advantage to hermaphroditism, all else being equal (review in Leonard 1999, 2005).

SEX ALLOCATION
AND SEXUAL SYSTEM

In 1976, Eric Charnov, John Maynard Smith, and Jim Bull famously asked the question, "Why be an hermaphrodite?" (Charnov et al. 1976). They answered the question by pointing out that where reproduction through eggs involved a concave fitness curve (figure 3.1), separate sexes should be favored, but if and when there was a positive correlation between success in reproduction through the two types of gametes, simultaneous

hermaphroditism would be favored (Charnov 1979; figure 3.1). Charnov (1979) suggested several scenarios that would produce a convex fitness curve, including flower morphology attracting pollinators that would both pick up and drop off pollen; use of different resources for each sexual function; and temporal segregation of allocation to each sexual function. Other factors that would tend to favor hermaphroditism would stem from non-linear gain curves for one or both sexual roles. That is, if at some point, increased investment in one sexual role no longer results in increased fitness through that role, the gain curve is said to saturate, showing a plateau in fitness at some level. Where this is the case, any additional resources available for reproduction could best be invested in the other sexual role. For example, if female fitness is constrained by the space available for brooding fertilized eggs, an individual cannot gain fitness by producing more eggs, and would benefit by devoting any further resources available for reproduction to male function (Charnov 1979). An example of such a system comes from oysters of the genus *Ostrea*. Individuals produce eggs, which are fertilized and retained in a brood chamber, and then while the embryos are developing, the gonad produces and releases sperm. After the young are released the gonad returns to egg production (Coe 1932; see discussion in Ghiselin 1974; Chaparro & Thompson 1998). This model then provides an explanation for the association of hermaphroditism and brooding (Ghiselin 1969, 1974) that differs from, but is not mutually exclusive with, Clark's (1978) suggestion that brooding is associated with poor dispersal and consequently low encounter probability (Low Density model). In *Ostrea* spp. it would be interesting to determine if the period of brooding is inversely correlated with the degree of outcrossing as Clark's model would predict or whether, as Chaparro et al. (1999) suggest for *Ostrea chilensis*, a short pelagic phase is an adaptation to particularly strong currents in the habitat. Other factors that would tend to saturate a fitness gain curve are Local Mate Competition (Charnov 1980), a variation on the low encounter probability theme, in which male reproductive success is limited, not by the ability to encounter a mate but by the number of eggs that can be produced by the mates within reach; and sibling competition where producing more offspring may not increase fitness because it will reduce the chances of offspring survival. Charnov (1979, 1982) also made a very important contribution to the understanding of hermaphroditism when he explored the role of sexual selection in sex allocation and the stability of hermaphroditism, suggesting that such aspects of sexual selection as sperm competition and sexual conflict over mating decisions may play an important role in the evolution and/or stability of hermaphroditism (reviewed in Leonard 2005, 2006).

Sex allocation theory has offered a useful way of conceptualizing questions of gender and reproductive strategy. However, empirical studies on sex allocation over the last 30 years have muddied rather than clarified the picture. A review of the successes and failures of sex allocation theory is beyond the scope of the current review but the record is mixed. While some studies have found support for the predictions of the theory, others have found contradictions. Major problems stem from the fact that the important currencies in sex allocation are difficult to identify (for discussion in plants see Thomson 2006; Delph & Ashman 2006). A variety of factors go to determine the reproductive success of an individual. Early studies in animals (e.g., Fischer 1981 in a serranine fish) equated the gonad volume devoted to each type of gamete to allocation to that sexual role, although acknowledging the weakness of the assumption. Many of the studies that support the predictions of sex allocation theory have used the strategy of relative measurement of sex allocation, that is, they have identified a parameter that appears to be involved in one of the sexual roles, and determined whether that parameter increases or decreases in magnitude under changed conditions, in a manner consistent with sex allocation theory. For example, Raimondi and Martin (1991), while acknowledging the difficulty in measuring allocation to male function, demonstrated in barnacles that allocation to female function, measured as the ratio of egg mass to body mass, was higher in barnacles held in small groups, and that the ratio of egg mass to that of "male" mass was also higher in small groups as predicted by Local Mate Competition theory (Charnov 1980). More recent studies have demonstrated that in animals, behavior may represent a significant portion of reproductive effort. In a study in the basommatophoran gastropod, *Lymnaea stagnalis*, individuals who had been surgically treated (removal of a portion of the vas deferens) to eliminate male copulatory behavior, showed greater egg production as compared to controls, indicating

both the importance of behavior and a trade-off between male and female function (De Visser et al. 1994). Another experimental study, in the polychaete *Ophryotrocha diadema*, also demonstrated a trade off between egg production and male behaviors, when individuals responded to increased reproductive competition by decreasing egg output and increasing expensive male reproductive behaviors (Lorenzi et al. 2006).

A second problem is that identifying life-history trade-offs can be a complex process (Roff 2002). The theory of sex allocation assumes a pool of resources available for reproduction that can be allocated to male or to female effort. While this is straightforward from a theoretical perspective, in the real world, it is possible that resources for male and female function may stem from different sources (see Harshman & Zera 2006). For example, where there is a separation in time between mating behavior and egg-laying or parturition, it may be the case that male activity is fueled by resources acquired at one season and female investment depends on resources available later in the season (see also Lorenzi et al. 2006). This is true in many hermaphroditic gastropods that mate early in the life cycle, store sperm, and lay eggs later in the season or in their life history, consistent with Charnov's (1979) prediction that temporal separation of function should be associated with hermaphroditism, but it is also true for many dioecious animal species, which may copulate in the fall and bear young in the spring.

It may also be the case that more complex life history trade-offs are involved. In an experimental study, Prevedelli et al. (2006) compared population growth and life history parameters in three species of the polychaete *Ophrotrocha*, in laboratory culture: a simultaneously hermaphroditic species (*O. diadema*), a sequentially hermaphroditic species (*O. puerilis*), and a dioecious species (*O. labronica*), and found that the dioecious species had a significantly higher intrinsic growth rate (λ) than either type of hermaphrodite, due to earlier sexual maturity in the dioecious species. In contrast one of the hermaphroditic species had greater longevity. These data illustrate the complexity of life history parameters that may be involved in a trade-off and suggest an advantage to dioecy through natural selection. Studies in other taxa with similar variation in sexual system among closely related species are needed to determine whether this is a general phenomenon.

SEX ALLOCATION AND QUANTITATIVE GENDER

An important development from tests of sex allocation theory is the realization that sex allocation patterns in hermaphrodites vary within an individual's life span, not only in the form of sex change (see below) but within a simultaneous hermaphrodite (for review see Cadet et al. 2004). Local Mate Competition theory predicts that sex allocation will be skewed to male function at higher population densities which prediction has been borne out in some experimental studies [e.g., the polychaete *O. diadema*, Lorenzi et al. 2005; the flatworm *Macrostomum lignano*, Brauer et al. 2007; observational data from some serranines (recent review in Petersen 2006), but not in others (e.g., the tapeworm *Schistocephalus solidus*, Schärer & Wedekind 2001)]. Sex allocation may vary with size and age in simultaneous hermaphrodites. In some species young animals are males, and become simultaneous hermaphrodites when larger [e.g., *Lysmata* shrimps (see review in Bauer 2007), *Ophryotrocha diadema* (review in Premoli & Sella 1995), or juveniles are females and larger/older individuals are simultaneous hermaphrodites (e.g., some species of *Epiactis* sea anemones (Dunn 1975; review in Edmands & Potts 1997)].

In other hermaphroditic animals sex allocation may change gradually over the life time of the individual as has been described for plants (see review in Klinkhamer & de Jong 1997; Cadet et al. 2004). This may involve a shift of sex allocation in accordance with a shift in functional gender (e.g., *S. solidus*, Schärer et al. 2001) according to the Size-advantage model (figure 3.2; see discussion below, Ghiselin 1969; Taborsky and Neat this volume; recent review in Munday et al. 2006) or it may involve temporal separation of resource allocation to male and female function in a functional simultaneous hermaphrodite (see Policansky 1982). For example, in opisthobranchs and stylommatophorans sperm typically develop in the gonad before eggs. This phenomenon has been described as protandry; however, most cases do not involve functional protandry. That is, in many cases sperm are produced, then exchanged with another "adult male" and the sperm received in that reciprocal mating are used to fertilize eggs at a later date (see discussion in Ghiselin 1965; Hadfield & Schwitzer-Dunlap 1984; Leonard 1991). That is, individuals mate as simultaneous hermaphrodites although the two types of gonad develop sequentially. Tomiyama (2002) found that in the long-lived stylommatophoran *Achatina fulica* older and larger hermaphrodites laid substantially more eggs than smaller, younger animals and younger/smaller individuals competed for opportunities to copulate as males with the larger individuals. In such long-lived species individuals may mate predominantly in the male role early in life and predominantly in the female role later in life, approaching a situation of functional protandry (but see Angeloni et al. 2003). The distinction made in theory between simultaneous and sequential hermaphrodites is clearly an artificial one and a more sophisticated body of theory is being developed (see St. Mary 1997; Angeloni et al. 2002; Cadet et al. 2004).

Another complication is that there may not be a direct correlation between gonadal gender and functional or behavioral gender in simultaneous hermaphrodites. St. Mary (1994, 2000) found that in *Lythrypnus* gobies, individuals that had both ovarian and testicular tissue might act in only one sexual role (St. Mary 1997). This has also been observed in the polychaete *O. diadema* (Lorenzi & Sella 2008). The factors that select for such complex patterns of sex allocation and functional gender in simultaneous hermaphrodites are still poorly understood. However, the role of size and age in fitness through male relative to female function is well-understood through the Size-Advantage model (Ghiselin 1969; Warner 1975; recent review in Munday et al. 2006; Taborsky and Neat this volume; see below and figure 3.2) originally developed to explain patterns of sequential hermaphroditism. The ability to vary sex allocation according to local conditions should offer a fitness advantage to hermaphroditism over dioecy.

Sequential Hermaphroditism

Sequential hermaphroditism, whereby an individual may reproduce as one sex early in its life and the other later on, is a sexual system that is widely distributed in the Metazoa but is not characteristic of any major clade. It occurs mostly commonly in the sponges, crustacea, "prosobranch" molluscs and teleost fishes (see Ghiselin 1974; Warner et al. 1975; Policansky 1982; Allsop & West 2004; Collin 2006; Bauer 2007; Munday et al. 2006; Taborsky & Neat this volume, for reviews). One of the earliest successes in understanding reproductive and life history strategies in terms of selection acting on individuals came from the study of sequential hermaphroditism. Ghiselin's (1969) Size-advantage model (figure 3.2) has been the guiding principle for understanding what pattern of sex change should occur in a species and when individuals should change sex. The fundamental prediction is that individuals should change sex when they can increase their reproductive value by doing so. For example, if gamete production is correlated with body size, but the slopes of the curves are different for eggs and sperm, it may be the case that an individual can produce enough sperm for all practical purposes at a relatively small size, whereas there is a steeper linear relationship between body size and egg production (figure 3.2b). In this case, the model predicts that individuals should start life as males and switch sex to female when the gain curves cross; that is to say, a protandrous life history. This sort of pattern could be explained by natural selection acting on individuals maximizing their fitness. Protandry is characteristic of gastropods of the genus *Crepidula* with solitary species changing from female to male at a fixed size as predicted by the Size-Advantage model, whereas in species that form stacks of individuals, for example C. *fornicata* (Collin 1995), the composition of the social group plays a role in determining size at sex change (Collin 2006), suggesting that sexual selection may also play a role.

Figure 3.2a shows a situation in which the reproductive success of females remains correlated with size, as in figure 3.2b, but male reproductive success starts very low and rises very steeply at a certain size threshold. Such protogynous life histories have been well-studied in fishes over the last 30 years (e.g., blue-headed wrasse, Warner et al. 1975; *Anthias squamipennis*, Shapiro 1979), where it is associated with harem polygyny. In these mating systems large males are able to defend harems of females, and individuals start life as females, live in social groups with a dominant male and typically, the largest female changes sex when the male is removed. This is a striking example of sexual selection acting to determine the sexual system. Social control of sex change in fishes has been a very active field and a wide variety of mating systems have been described (see references in Munday et al. 2006). The precise pattern of sex change is responsive to a variety of factors, for example population density, local sex ratio, and local size distribution (Warner 1988). For example, in the bucktooth parrotfish, *Sparisoma radians*,

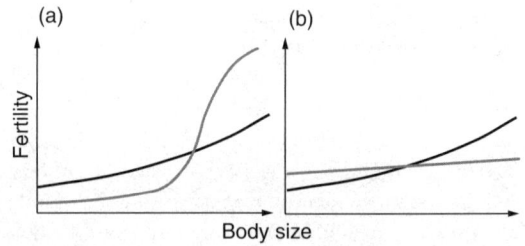

FIGURE 3.2 The size-advantage model (reproduced with permission from Munday et al. 2006). The offspring production expected for females (black line) increases with body size if large females lay more eggs than small females. Expected male offspring production (grey line) may or may not increase with body size, depending on whether large males have an advantage in securing mates. Sex change is favored when the size/age fertility curves of the two sexes cross. Protogyny (a) is predicted when the expected fertility of a male increases more rapidly with size/age than that of a female. Protandry (b) is predicted when the expected fertility of a female increases more rapidly with size than that of a male.

a protogynous, polygynous species, male removal experiments showed that it was not always the largest female that changed sex, contrary to predictions of the Size-advantage model (Muñoz & Warner 2004), a fact that can be explained by the reduction of fitness in dominant males in this species by the high frequency of sperm competition, so that very large females may have higher reproductive value than even harem holding males. Although the original Size-Advantage model focused on difference between the sexes in the correlation between fertility and size, sex change may also be predicted if the two sexes differ in either mortality rate or growth rate over the reproductive period (Iwasa 1991). Sequential hermaphroditism may, therefore, be favored by either natural or sexual selection, or a combination of both.

Most cases of sequential hermaphroditism involve a single age-dependent sex change whereby young/small individuals reproduce as one sex and then change to the other sex as they grow larger/ older. However, there are also cases of successively sequential hermaphroditism, such as in *Ostrea* oysters, where individuals produce sperm while brooding a clutch of fertilized eggs and then change the gonad back to egg production after the larvae have hatched (see Coe 1932; Ghiselin 1974) and *Ophyrotrocha puerilis* (Sella & Ramella 1999) in which the members of a pair trade sexual roles with simultaneous sex changes, sometimes eventually becoming simultaneous hermaphrodites (Berglund 1986). The pattern fits the predictions of the Size-advantage model because male-phase individuals grow more rapidly than individuals producing eggs, so that when the male becomes larger than the female the pair changes sex (Berglund 1990). More recently, facultatively bi-directional sex change has been found in fish, whereby fish that lose a partner or change social status may respond by changing sex in either direction, in the same species (e.g., Nakashima et al. 1996; Kuwamura & Nakashima 1998). Here, the major driving force seems to be simple reproductive assurance. The fish are monogamous and movement to find a mate involves high risk so natural selection will favor the ability to change sex to conform to whatever partner becomes available. In conclusion, either sexual or natural selection may favor sequential hermaphroditism.

Mixed Sexual Systems

Mixed sexual systems in which populations contain a mixture of hermaphrodites and single sex individuals, have evolved many times in animals, but remain rare. Gynodioecy is extremely rare in animals as opposed to angiosperms (review in Charlesworth 2006) but has been reported in corals (Chornesky & Peters 1987), as has pseudogynodioecy in which the "eggs" of hermaphrodites are sterile so that the population is functionally dioecious (Harrison 1988). Another form of "pseudo-gynodioecy" is found in a goby in which individuals mature first as females and some, but not all, change to functional hermaphrodites (Cole & Hoese 2001). The functional significance and adaptive value of this system is not clear, although it may represent a return to dioecy from hermaphroditism. Pseudo-gynodioecy has also been described in the sea anemone *Epiactis prolifera* which begin life as females and become simultaneous hermaphrodites later in life (Dunn 1975).

Androdioecy has evolved many times in animals (see reviews in Pannell 2002; Weeks et al. 2006) and, with the exception of barnacles, usually involves populations of males and hermaphrodites that self but do not mate as males with other hermaphrodites [e.g., *Eulimnadia* clam shrimps (review in Weeks et al. 2006); two species of the nematode

genus *Caenorhabditis* (Braendle & Félix 2006)]. Therefore, androdieocy seems to evolve as a mechanism of reproductive assurance in fundamentally dioecious taxa, suggesting that natural selection is the driving force behind this mode of sexuality. In barnacles, the existence of "complemental" males along with hermaphrodites capable of cross-fertilizing each other represents a sexual system more truly analogous to the androdioecy of plants (see Charnov 1987). One explanation for complemental males in hermaphroditic barnacles is that they represent a transition toward dioecy (Ghiselin 1974); gonochoristic barnacles typically have dwarf males. In barnacles, hermaphroditism seems to be associated with densely packed conditions and dioecy with parasitism and/or low density (Yamaguchi et al. 2008), in contrast to the Low-Density hypothesis of the evolution of hermaphroditism (above). Yamaguchi et al. (2008) have suggested that low resource availability favors the development of dwarf (complemental) males in barnacles (see also Yamaguchi et al. 2007), suggesting that natural selection has been the important selective force. Self-fertilization has been reported in barnacles, although it is not the rule (see discussion in Charnov 1987), so it is not clear why barnacles have evolved dioecy with dwarf males rather than selfing as a means of reproductive assurance. The existence of multiple dwarf males within a hermaphrodite suggests that sperm competition, and hence sexual selection, may be a factor in the mating system. Androdioecy implies a genetic distinction between hermaphrodites and males (Weeks et al. 2006), which seems to be the case for some but not all complemental males in barnacles (Yamaguchi et al. 2007).

"Pseudo-androdioecy", in which populations consist of a mixture of functional males and hermaphrodites, is widely distributed in animals. In some cases, the system involves protandrous simultaneous hermaphroditism, such as *Lysmata* shrimps (Bauer 2007) and the polychaete O. *diadema* (Lorenzi et al. 2006), in which juveniles mature first as males and then become functional simultaneous hermaphrodites. Another form of "pseudo-androdioecy" occurs in some simultaneously hermaphroditic serranine fishes, where, under certain circumstances, large individuals lose ovarian tissue and become males, defending harems of smaller hermaphrodites (Hastings & Petersen 1986; Petersen 2006). In the latter case, sexual selection clearly seems to be the driving force; whereas in the former case, natural selection for early reproduction may play a role.

PHYLOGENY, ECOLOGY, AND SEXUAL SYSTEM

Variation in sexual system in the Metazoa occurs at two levels. At one level sexual systems in higher taxa tend to show little variation. That is, most phyla and classes can be easily characterized as either hermaphroditic or dioecious (table 3.1; see discussion in Williams 1975; Leonard 1990; Iyer & Roughgarden 2008a; Jarne & Auld 2006; Eppley & Jesson 2008). Almost the entire phylum Platyhelminthes is composed of simultaneous hermaphrodites; whereas the phylum Nematoda is dioecious with few exceptions. The entire clade of heterobranch gastropods (pulmonates and opisthobranchs) is simultaneously hermaphroditic (with a couple of possible exceptions), whereas the Hexapoda (insects), the most speciose clade of animals, is entirely dioecious. On the other hand, some animal taxa show substantial variability in sexual system. In the "prosobranch" gastropods (a paraphyletic group, see Hodgson, this volume) there may have been more than 40 evolutionary transitions from dioecy to hermaphroditism (Jarne and Auld 2006). There have also been multiple transitions between dioecy and hermaphroditism in such groups as the polychaetes, the bivalves, the malacostrocan crustacea and the teleost fishes (Ghiselin 1974). In some cases, sexual system may vary among species within a genus or even among populations within a species [e.g., the polychaete genus, *Ophryotrocha* (review in Premoli & Sella 1995); the serranid fish, *Paralabrax maculatofasciatus* (Hovey & Allen 2000), the scyphozoan jellyfish *Cassiopea* sp. in Hawaii (Hofmann & Hadfield 2002); the sea anemone *Epiactis* (Edmands & Potts 1997)].

Most of the work on the relative advantages of dioecy versus hermaphroditism has been theoretical and there have been relatively few attempts, in animals, to test the various hypotheses (table 3.2) as to the relative advantages of dioecy and hermaphroditism empirically. Eppley and Jesson (2008) undertook a very ambitious statistical analysis of the relationship between sexual system and mate-finding ability in multicellular organisms in general (including plants and fungi) at the class level, and found some support for an association

between increased mate search ability and/or low costs of locomotion, and dioecy as predicted by recent theoretical models (Puurtinen & Kaitala 2002; Iyer & Roughgarden 2008a) and suggested on a qualitative level by early authors (Ghiselin 1969; etc.). However, the analysis did not show support for the idea that changes in adult mate search lead to changes in sexual system. The authors conclude that other factors must be involved. One of the problems illustrated by the study, is that by comparing taxa at the level of phyla and class, there are such great differences in morphology and life history that the analysis is of necessity very coarse.

As George Williams pointed out (1975) it is unlikely that the selective pressures that explain the initial evolution, perhaps back in the PreCambrian, of a sexual system characteristic of a phylum or class, can still be at work throughout the adaptive radiation of the group. Attempts to test hypotheses as to the advantage of one sexual system over another are unlikely to be successful at such high taxonomic levels. A similar analysis within a smaller taxon with significant variation in sexual system, for example teleost fishes or gastropods, might be more helpful in understanding the relative importance of hypotheses such as low density, mate search ability, local mate competition, saturating gain cruves, and others (table 3.2), in determining sexual system. A set of such analyses on a variety of taxa with significant variation in sexual system would allow us to identify the relative importance of particular ecological factors or suggest new hypotheses. It might also demonstrate that the determining factors of sexual system vary among taxonomic groups. For example, in *Ophryotrocha* a dioecious species was found to have earlier maturity than hermaphroditic congeners in laboratory culture (Prevedelli et al. 2006). This species (*O. labronica)* seems to live at higher density in the field than the simultaneously hermaphroditic species *O. diadema* (Premoli & Sella 1995; M.C. Lorenzi, personal commication). In barnacles it has been suggested that dioecy is associated with low density ((Yamaguchi et al. 2008), in contrast to the Low-Density model of the advantage of hermaphroditism. The Platyhelminthes are simultaneously hermaphroditic with the famous exception of the family Schistosomidae, which is dioecious. Dioecy in schistosomes has been explained by a division of labor advantage such that the male can be specialized for large size and muscularity to maintain position in the large blood vessels of mammals whereas the female can be small and slender to lay eggs in the capillaries (Despres & Maurice 1995; Tchuem Tchuenté et al. 1996).

Despite the acknowledged stability of sexual systems once evolved, there has been little attention in either theoretical or empirical studies to factors that might oppose a transition from hermaphroditism to dioecy or the reverse. Williams (1975; see discussion in Leonard 1990; Jarne & Auld 2006) suggested that "phyletic inertia" or lack of genetic variation for sexual system must explain this stability. This seems like an explanation of last resort. The nematodes are an overwhelmingly dioecious phylum but it has been demonstrated that two androdioecious species of *Caenorhabiditis* differ from their dioecious congeners by one allele and that it is a different allele in both cases (Braendle & Félix 2006), suggesting that this should be an easy genetic change. It might be possible to test the phyletic inertia hypothesis by looking at cases in which there have been transitions to another sexual system in a generally stable taxon. Are shifts in sexual system associated with increased adaptive radiation as might be expected if phyletic inertia were constraining evolution? A comparative study might help resolve the issue.

Another possibility is that sexual systems tend to be self-perpetuating. One possible factor that would tend to stabilize sexual systems is sexual selection. Certainly, prevailing models of sexual selection in dioecious animals (review in Andersson 1994) would predict that a male that devoted some of its available resources to female function rather than mating as a male, would experience a severe loss of fitness, perhaps explaining the rarity of gynodioecy in animals. A female that devoted part of her available resources to male function would also expect to experience reduced fitness since fitness in females is usually considered to be strongly resource limited. An exception might occur in the case of a female that produced a few sperm for self-fertilization. Her fitness might be increased due to reproductive assurance without substantially reducing egg production. This might explain why (a) androdioecy, while rare, has evolved many times in animals; and (b) most, although not all, androdioecious species in animals consist of males and self-fertile hermaphrodites that mate as females with males but do not mate as males with other hermaphrodites (see review in Weeks et al. 2006).

In hermaphrodites it has been suggested that mating systems based on conditional reciprocity would tend to stabilize hermaphroditism (Axelrod & Hamilton 1981; see discussion Leonard 1990, 2005, 2006; Eppley & Jesson 2008) and that sexual conflict in hermaphrodites will tend to produce reciprocal mating systems (see discussion in Axelrod & Hamilton 1981; Leonard 1990, 2006). The argument is that, if the two sexual roles differ in variance in reproductive success (see discussion in Leonard 2005, 2006), one sexual role may offer potential fitness advantages to a hermaphrodite and sexual conflict between hermaphrodites in a mating encounter, created by a preference for the sexual role with higher fitness, will be expected (Charnov 1979; Leonard & Lukowiak 1984). Axelrod and Hamilton (1981) suggested that where sexual conflict created a preferred sexual role in hermaphrodites, the conflict would be resolved by a reciprocal mating system, which in turn would stabilize hermaphroditism (see reviews in Leonard 1990, 1999, 2005, 2006). Such reciprocal systems have been identified in pair-mating simultaneous hermaphrodites from several taxa (e.g., Fischer 1980; reviews in Leonard 1993, 2006; Petersen 2006, Lorenzi et al. 2006; Webster & Gower 2006) and are predicted to occur *de facto* in sessile and broadcast spawning hermaphrodites (Leonard 1990). Once reciprocal matings sytems have evolved in a hermaphroditic population, an individual that could only mate in one sexual role would be at a strong disadvantage. The ability of hermaphrodites to fine tune sex allocation to current conditions (quantitative gender, see above) and the predicted reduction of variance in reproductive success in hermaphrodites might also contribute to the stability of hermaphroditism as a sexual system. Further study of the factors that might tend to stability of the two sexual systems and/or the costs associated with evolutionary transitions between sexual systems are needed before we accept a "phyletic inertia" explanation.

SEX, ANISOGAMY, AND GENDER

This chapter reviews some of the most fundamental reproductive traits; sex, anisogamy, and sexual system, with a view to understanding the relative importance of natural versus sexual selection on the evolution of these characters in animals. The available evidence and current theories seem to indicate that the ubiquity of two sexes in animals can be explained by natural selection. The hypothesis (review in Hurst 1995) that the fundamental difference between the two sexes in animals may be that one contributes mitochondria to the zygote and the other does not (although there are some clear exceptions, see above), seems appealing. The selective forces involved are a little less clear; but the current hypotheses are based on improved metabolic function of the zygote and so invoke natural selection. Given two sexes or mating types, the available hypotheses; whether based on disruptive selection on size (Parker et al. 1972; Bulmer & Parker 2002), or encounter probability (review in Dusenberry 2006) seem to involve elements of sexual selection; that is efficient search for mates and/or attraction of mates. Anisogamy therefore, seems not so much the prerequisite for sexual selection, as has been classically assumed (e.g., Bateman 1948), as at least partially a product of sexual selection, suggesting that eggs and sperm are at least in part, secondary sexual characters (see Ghiselin, this volume). Similarly, the evidence suggests that either ecological factors, natural selection for reproductive assurance or inbreeding avoidance, or sexual selection may be important in determining the sexual system of animals. However, the basic distribution of hermaphroditism versus dioecy across the Metazoa seems to depend very strongly on phylogeny, suggesting that it is much more stable than explained by current theories. The reasons for this stability are not yet clear. However once either hermaphroditism or dioecy have evolved sexual selection may act to stabilize both sexual systems.

REFERENCES

Aanen, D.K., Kuyper, T.W., Debets, A.J.M. & Hoekstra, R.F. 2004. The evolution of non-reciprocal nuclear exchange in mushrooms as a consequence of genomic conflict. Proceedings of the Royal Society of London Series B 271, 1235–1241.

Allen, J.F. 1996. Separate sexes and the mitochondrial theory of ageing. Journal of Theoretical Biology 180, 135–140.

Allsop, D.J. & West, S.A. 2004. Sex-ratio evolution in sex-changing animals. Evolution 58, 1019–1027.

Andersson, M. 1994. Sexual Selection. Princeton, N.J.: Princeton University Press.

Angeloni, L., Bradbury, J.W. & Charnov, E.L. 2002. Body size and sex allocation in simultaneously hermaphroditic animals. Behavioral Ecology 13, 419–426.

Angeloni, L., Bradbury, J.W. & Burton, R.S. 2003. Multiple mating, paternity, and body size in a simultaneous hermaphrodite, Aplysia californica. Behavioral Ecology 14, 554–560.

Archetti, M. 2004. Loss of complementation and the logic of two-step meiosis. Journal of Evolutionary Biology 17, 1098–1105.

Ashman, T.-L. 2002. The role of herbivores in the evolution of separate sexes from hermaphroditism. Ecology 83, 1175–1184.

Ashman, T.-L. 2006a. The evolution of separate sexes: a focus on the ecological context. In: Ecology and Evolution of Flowers (Ed. by L.D. Harder & S.C.H. Barrett), pp. 204–222. Oxford, UK: Oxford University Press.

Ashman, T.-L. 2006b. The evolution of separate sexes: a focus on the ecological context. In: Ecology and Evolution of Flowers (Ed. by L.D. Harder & S.C.H. Barrett), pp. 204–222. Oxford, UK: Oxford University Press.

Axelrod, R. & Hamilton, W.D. 1981. The evolution of cooperation. Science 211, 1390–1396.

Barr, C.M., Neiman, M. & Taylor, D.R. 2005. Inheritance and recombination of mitochondrial genomes in plants, fungi and animals. New Phytologist 168, 39–50.

Barrett, S.C.H. 2002. The evolution of plant sexual diversity. Nature Reviews Genetics 3, 274–284.

Bateman, A.J. 1948. Intra-sexual selection in Drosophila. Heredity 2, 349–368.

Bauer, R.T. 2006. Same sexual system but variable sociobiology; evolution of protandric simultaneous hermaphroditism in Lysmata shrimps. Integrative and Comparative Biology 46, 430–438.

Bauer, R.T. 2007. Hermaphroditism in caridean shrimps: mating systems, sociobiology, and evolution, with special reference to Lysmata. In: Evolutionary Ecology of Social and Sexual Systems: Crustaceans as Model Organisms (Ed. by J.E. Duffy & M. Thiel), pp. 232–248. New York: Oxford University Press.

Bell, G. 1982. The Masterpiece of Nature. Berkeley, CA: University of California Press.

Bell, G. 1985. The origin and early evolution of germ cells as illustrated by the Volvocales. In: The Origin and Evolution of Sex (Ed. by H.O. Halvorson & A. Monroy), pp. 221–256. New York: Alan R. Liss, Inc.

Berglund,A. 1986. Sex change by a polychaete: effects of social and reproductive costs. Ecology 67, 836–845.

Berglund, A. 1990. Sequential hermaphroditism and the size-advantage hypothesis: an experimental test. Animal Behaviour 39, 426–433.

Bernstein, H., Byerly, H.C., Hopf, F.A. & Mickocl, R.E. 1985. Genetic damage, mutation, and the evolution of sex. Science 229, 1277–1281.

Bernstein, H., Hopf, F.A. & Michod, R.E. 1988. Is meiotic recombination an adaptation for repairing DNA, producing genetic variation, or both? In: The Evolution of Sex (Ed. by R.E. Michod & B.R. Levin), pp. 139–160. Sunderland, MA: Sinauer Associates.

Berruti, G., Ferraguti, M. & Lora Lamia Donin, C. 1978. The aflagellate spermatozoon of Ophryotrocha. Gamete Research 1, 287–292.

Birky, C.W. Jr. 1995. Uniparental inheritance of mitochondrial and chloroplast genes: mechanisms and evolution. Proceedings of the National Academy of Sciences USA 92, 11331–11338.

Birky, C.W. Jr. 2001. The inheritance of genes in mitochrondria and chloroplasts: laws, mechanisms, and models. Annual Review of Genetics 35, 125–148.

Birky, C.W. Jr. 2005. Sex: is Giardia doing it in the dark? Current Biology 15, R56–R58.

Bishop, J.D.D. & Pemberton, A.J. 2006. The spermcast mating mechanism in sessile marine invertebrates. Integrative and Comparative Biology 46, 398–406.

Bjork, A. & Pitnick, S. 2006. Intensity of sexual selection along the anisogamy–isogamy continuum. Nature 441, 742–745.

Bonsall, M.B. 2006. The evolution of anisogamy: the adaptive significance of damage, repair, and mortality. Journal of Theoretical Biology 238, 198–210.

Borgia, G. & Blick, J. 1981. Sexual competition and the evolution of hermaphroditism. Journal of Theoretical Biology 89, 523–532.

Braendle, C. & Félix, M.-A. 2006. Sex determination: ways to evolve a hermaphrodite. Current Biology 16, R468–R471.

Brauer, V.S., Schärer, L. & Michiels, N.K. 2007. Phenotypically flexible sex allocation in a simultaneous hermaphrodite. Evolution 61, 216–222.

Brusca, R.C. & Brusca, G.J. 1990. Invertebrates. Sunderland, MA: Sinauer Associates, Inc.

Bulmer, M.G. & Parker, G.A. 2002. The evolution of anisogamy: a game-theoretic approach. Proceedings of the Royal Society of London Series B 269, 2381–2388.

Bulmer, M.G., Luttikhuizen, P.C. & Parker, G.A. 2001. Survival and anisogamy. Trends in Ecology and Evolution 17, 357–358.

Cadet, C., Metz, J.A.J. & Klinkhamer, P.G.I. 2004. Size and the not-so-single sex: disentangling the effects of size and budget on sex allocation in hermaphrodites. American Naturalist 164, 779–792.

Cavalier-Smith, T. 1987. Evolution of the eukaryotic genome. In: The Eukaryotic Genome: Organisation and Regulation (Ed. by P. Broda, S.G. Olliver & P.F.G. Sims), pp. 333–385. Cambridge, UK: Cambridge University Press.

Chaparro, O.R. & Thompson, R.J. 1998. Physiological energetics of brooding in Chilean oyster Ostrea chilensis. Marine Ecology Progress Series 171, 151–163.

Chaparro, O.R., Thompson, R.J. & Emerson, C.J. 1999. The velar ciliature in the brooded larva of the Chilean oyster Ostrea chilensis (Philippi, 1845). Biological Bullletin 197, 104–111.

Charlesworth, B. 1978. The population genetics of anisogamy. Journal of Theoretical Biology 73, 347–357.

Charlesworth, D. 2001. Evolution: An exception that proves the rule. Current Biology 11, R13–R15.

Charlesworth, D. 2006. Evolution of plant breeding systems. Current Biology 16, R726–R735.

Charnov, E.L. 1979. Simultaneous hermaphroditism and sexual selection. PNAS 76, 2480–2484.

Charnov, E.L. 1980. Sex allocation in barnacles. Marine Biology Letters 1, 269–272.

Charnov, E.L. 1982. The Theory of Sex Allocation. Princeton, NJ: Princeton University Press.

Charnov, E.L. 1987. Sexuality and hermaphroditism in barnacles: A natural selection approach. In: Biology of Barnacles (Ed. by A.J. Southward), pp. 89–103. Rotterdam: Balkema.

Charnov, E.L., Maynard Smith, J. & Bull, J.J. 1976. Why be an hermaphrodite? Nature 263, 125–126.

Chornesky, E.A. & Peters, E.C. 1987. Sexual reproduction and colony growth in the

scleractinian coral *Porites astreoides.*
Biological Bullletin 172, 161–177.

Clark, W.C. 1978. Hermaphroditism as a
reproductive strategy for metazoans; some
correlated benefits. New Zealand Journal of
Zoology 5, 769–780.

Coe, W.R. 1932. Development of the gonads and
the sequence of the sexual phases in the
California oyster (*Ostrea lurida*). Bulletin
of the Scripps Institute of Oceanography 3,
119–140.

Cole, K.S. & Hoese, D.F. 2001. Gonad morphology, colony demography and evidence for
hermaphroditism in *Gobiodon okinawae*
(Telostei, Gobiidae). Environmental Biology
of Fishes 61, 161–173.

Collin, R. 1995. Sex, size and postion: a test of
models predicting size at sex change in the
protandrous gastropod *Crepidula fornicata.*
American Naturalist 146, 815–831.

Collin, R. 2006. Sex ratio, life-history invariants,
and patterns of sex change is a family of
protandrous gastropods. Evolution 60,
735–745.

Cox, P.A. & Sethian, J.A. 1985. Gamete motion,
search, and the evolution of anisogamy,
oogamy, and chemotaxis. The American
Naturalist 125, 74–101.

Crow, J.F. 1988. The importance of recombina-
tion. In: The Evlution of Sex (Ed. by
R.E. Michod & B.R. Levin), pp. 56–73.
Sunderland, MA: Sinauer Associates, Inc.

Crowley, P.H., Cottrell, T., Garcia, T., Hatch, M.,
Sargent, C.S., Stokes, B.J. & White, J.M.
1998. Solving the complementarity dilemma:
evolving strategies for simultaneous hermaphroditism. Journal of theoretical Biology 195,
13–26.

Czárán, T.L. & Hoekstra, R.F. 2006. Evolution of
sexual asymmetry. BMC Evolutionary Biology
4, 34–46.

Dacks, J. & Roger, A.J. 1999. The first sexual
lineage and the relevance of facultative sex.
Journal of Molecular Evolution 48, 779–783.

Darwin, C. 1871. The Descent of Man, and
Selection in Relation to Sex. 1981 reprint edn.
Princeton, N.J.: Princeton University Press.

Darwin, C. 1888. The Different Forms of Flowers
on Plants of the Same Species. 1986 facsimile
reprint of the 1888 edition. Chicago:
University of Chicago Press.

De Visser, J.A.G.M., ter Maat, A. & Zonneveld,
C. 1994. Energy budgets and reproductive
allocation in the simultaneous hermaphrodite

pond snail, *Lymnaea stagnalis* (L.): a trade-off
between male and female function. American
Naturalist 144, 861–867.

Delph, L.F. & Ashman, T.-L. 2006. Trait selection
in flowering plants: how does sexual selection
contribute? Integrative and Comparative
Biology 46, 465–472.

Despres, L. & Maurice, S. 1995. The evolution of
dimorphism and separate sexes in schistosomes. Proceedings of the Royal Society of
London B 262, 175–180.

Dini, F. & Corliss, J.O. 2001. Protozoan
sexuality. In: Encyclopedia of Life
Sciences, pp. 1–11. Hoboken, N.J.:
John Wiley & Sons.

Dunn, C.W., Hejnol, A., Matus, D.Q., Pang, K.,
Browne, W.E., Smith, S.A., Seaver, E.,
Rouse, G.W., Obst, M., Edgecombe, G.D.,
Sørensen, M.V., Haddock, S.H.D.,
Schmidt-Rhaesa, A., Okusu, A., Kristensen,
R.M., Wheller, W.C., Martindale, M.Q. &
Giribet, G. 2008. Broad phylogenomic
sampling improves resolution of the
animal tree of life. Nature 452,
745–750.

Dunn, D.F. 1975. Gynodioecy in an animal.
Nature 253, 528–529.

Dusenberry, D.B. 2000. Selection for high gamete
encounter rates explains the success of male
and female mating types. Journal of
Theoretical Biology 202, 1–10.

Dusenberry, D.B. 2002. Ecological models explaining the success of distinctive sperm and eggs
(oogamy). Journal of Theoretical Biology 219,
1–7.

Dusenberry, D.B. 2006. Selection for high gamete
encounter rates explains the evolution of anisogamy using plausible assumptions about
size relationships of swimming speed and
duration. Journal of Theoretical Biology 241,
33–38.

Edmands, S. & Potts, D.C. 1997. Population
genetic structure in brooding sea anemones
(*Epiactis* spp.) with contrasting reproductive
modes. Marine Biology 127, 485–498.

Eppley, S.M. & Jesson, L.K. 2008. Moving to
mate: the evolution of separate and combined
sexes in multicellular organisms. Journal of
Evolutionary Biology 21, 727–736.

Fischer, E.A. 1980. The relationship between
mating system and simultaneous hermaphroditism in the coral reef fish, *Hypoplectrus
nigricans* (Serranidae). Animal Behaviour 28,
620–633.

Fischer, E.A. 1981. Sexual allocation in a simulta-neously hermaphroditic coral reef fish. American Naturalist 117, 64–82.

Fisher, R.A. 1958. The Genetical Theory of Natural Selection. New York: Dover Publications, Inc.

Ghiselin, M.T. 1965. Reproductive function and the phylogeny of opisthobranch gastropods. Malacologia 3, 327–378.

Ghiselin, M.T. 1969. The evolution of hermaphroditism among animals. Quarterly Review Biology 44, 189–208.

Ghiselin, M.T. 1974. The Economy of Nature and the Evolution of Sex. Berkeley: University of California Press.

Ghiselin, M.T. 1987. Evolutionary aspects of marine invertebrate reproduction. In: Reproduction of Marine Invertebrates: Vol 9, General Aspects: Seeking Unity in Diversity (Ed. by A.C. Giese, J.S. Pearse & V.B. Pearse), pp. 609–665. New York: Academic Press.

Gillespie, J.H. 1974. Natural selection for within-generation offspring numbers. Genetics 76, 601–606.

Gillespie, J.H. 1977. Natural selection for within-generation variances in offspring numbers: a new evolutionary principle. American Naturalist 111, 1010–1014.

Haag, E.S. 2007. Why two sexes? Sex determina-tion in multicellular organisms and protistan mating types. Seminars in Cell & Developmental Biology 18, 348–349.

Hadfield, M.G. & Schwitzer-Dunlap, M. 1984. Opisthobranchs. In: Volume 7, Reproduction (Ed. by A.S. Tompa, N.H. Verdonk & J.A.M. van den Biggelaar), pp. 209–350. Orlando: Academic Press.

Harrison, P.L. 1988. Pseudo-gynodioecy: an unusual bredding system in the scleractinian coral Galaxea fascicularis. Proceedings of the 6th International Coral REef Symposium 2, 699–705.

Harshman, L.G. & Zera, A.J. 2006. The cost of reproductin: the devil in the details. TREE 22, 80–86.

Hastings, I.M. 1999. The costs of sex due to deleterious intracellular parasites. Journal of Evolutionary Biology 12, 177–183.

Hastings, P.A. & Petersen, C.W. 1986. A novel sexual pattern in serranid fishes: simultaneous hermaphrodites and secondary males in Serranus fasciatus. Environmental Biology of Fishes 15, 59–68.

Heath, D.J. 1977. Simultaneous hermaphro-ditism—cost and benefit. Journal of Theoretical Biology 64, 363–373.

Hickey, D.A. & Rose, M.R. 1988. The role of gene transfer in the evolution of sex. In: The Evolution of Sex, pp. 161–175. Sunderland, MA: Sinauer Associates, Inc.

Hoekstra, R.F. 1980. Why do organisms produce gametes of only two different sizes? Some the-oretical aspects of the evolution of anisogamy. Journal of Theoretical Biology 87, 785–793.

Hoekstra, R.F. 1982. On the asymmetry of sex: evolution of mating types in isogamous populations. Journal of Theoretical Biology 98, 427–451.

Hoekstra, R.F. 1984. Evolution of gamete motility differences II. Interaction with the evolution of anisogamy. Journal of Theoretical Biology 107, 71–83.

Hoekstra, R.F. 1987. The evolution of sexes. In: The Evolution of Sex and Its Consequences (Ed. by S.C. Stearns), pp. 59–91. Basel: Birkhäuser Verlag.

Hoekstra, R.F. 1990. The evolution of male-female dimorphism: older than sex? Journal of Genetics 69, 11–15.

Hoekstra, R.F., Janz, R.F. & Schilstra, A.J. 1984. Evolution of gamete motility: I. Relation between swimming speed and pheromonal attraction. Journal of Theoretical Biology 107, 57–70.

Hofmann, D.K. & Hadfield, M.G. 2002. Hermaphroditism, gonochorism, and asexual reproduction in Cassiopea sp. — an immigrant to the islands of Hawai'i. Invertebrate Reproduction and Development 41, 215–221.

Holliday, R. 1988. A possible role for meiotic recombination in germ line reprogramming and maintenance. In: The Evolution of Sex (Ed. by R.E. Michod & B.R. Levin), pp. 45–55. Sunderland, MA, Sinauer Associates, Inc.

Hood, M.E. & Antonovics, J. 2004. Mating within the meiotic tetrad and the maintenance of genomic heterozygosity. Genetics 166, 1751–1759.

Hovey, T.E. & Allen, L.G. 2000. Reproductive patterns of six populations of the Spotted Sand Bass, Paralabrax maculatofasciatus, from Southern and Baja California. Copeia 2000, 459–468.

Hurst, L.D. 1990a. Parasite diversity and the evolution of diploidy, multicellularity and anisogamy. Journal of Theoretical Biology 144, 429–443.

Hurst, L.D. 1990b. Parasite diversity and the evolution of diploidy, multicellularity, and anisogamy. *Journal of Theoretical Biology* 144, 429–443.

Hurst, L.D. 1995. Selfish genetic elements and their role in evolution: the evolution of sex and some of what it entails. Philosophical Transactionf of the Royal Society of London B 349, 321–332.

Hurst, L.D. & Hamilton, W.D. 1992. Cytoplasmic fusion and the nature of sexes. Proceedings of the Royal Society of London [Biology] 247, 189–194.

Hutson, V. & Law, R. 1993. Four steps to two sexes. Proceedings of the Royal Society of London Series B 253, 43–51.

Iwasa, Y. 1991. Sex change evolution and cost of reproduction. Behavioral Ecology 2, 56–68.

Iwasa, Y. & Sasaki, A. 1987. Evolution of the number of sexes. Evolution 41, 49–65.

Iyer, P. & Roughgarden, J. 2008a. Dioecy as a specialization promoting sperm delivery. *Evolutionary Ecology Research*.Vol. 10, pp. 867–892.

Iyer, P. & Roughgarden, J. 2008b. Gametic conflict versus contact in the evolution of anisogamy. Theoretical Population Biology 73, 461–472.

Jarne, P. & Auld, J.R. 2006. Animals mix it up too: the distribution of self-fertilization among hermaphroditic animals. Evolution 60, 1816–1824.

Jarne, P. & Charlesworth, D. 1993. The evolution of the selfing rate in functionally hermaphroditic plant and animals. Annual Reveiw of Ecology and Systematics 24, 441–466.

Kalmus, H. 1932. Über den Erhaltungswert der phänotypischen (morphologischen) Anisogamie und die Entstehung der ersten Geschlechtsunterschiede. Biologisches Zentralblatt 52, 716–726.

Kalmus, H. & Smith, C.A.B. 1960. Evolutionary origin of sexual differentiation and the sex ratio. Nature 186, 1004–1006.

King, N. 2005. Choanoflagellates. Current Biology 15, R113–R115.

Klinkhamer, P.G.L. & de Jong, T.J. 1997. Size dependent allocation to male and female reproduction. In: Plant Resource Allocation (Ed. by F.A. Bazzazz & J. Grace), pp. 211–229. San Diego: Academic Press.

Knowlton, N. 1974. A note on the evolution of gamete dimorphism. Journal of Theoretical Biology 46, 283–285.

Kodric-Brown, A. & Brown, J.L. 1987. Anisogamy, sexual selection, and the evolution and maintenance of sex. Evolutionary Ecology 1, 95–105.

Kuwamura, T. & Nakashima, Y. 1998. New aspects of sex change among reef fishes: recent studies in Japan. Environmental Biology of Fishes 52, 125–135.

Law, R. & Hutson, V. 1992. Intracellular symbionts and the evolution of uniparental cytoplasmic inheritance. Proceedings of the Royal Society of London Series B 248, 69–77.

Lecointre, G. & Le Guyader, H. 2006. The Tree of Life: A Phylogenetic Classification. Cambridge, MA: Harvard University Press.

Leonard, J.L. 1990. The hermaphrodite's dilemma. Journal of Theoretical Biology 147, 361–372.

Leonard, J.L. 1991. Sexual conflict and the mating systems of simultaneously hermaphroditic gastropods. American Malacological Bulletin 9(1), 45–58.

Leonard, J.L. 1993. Sexual conflict in simultaneous hermaphrodites: evidence from serranid fishes. Environmental Biology of Fishes 36, 135–148.

Leonard, J.L. 1999. Modern Portfolio Theory and the prudent hermaphrodite. Invertebrate Reproduction and Development 36, 129–135.

Leonard, J.L. 2005. Bateman's principle and simultaneous hermaphrodites: a paradox. Integrative and Comparative Biology 45, 856–873.

Leonard, J.L. 2006. Sexual selection: Lessons from hermaphrodite mating systems. Integrative and Comparative Biology 46, 349–367.

Leonard, J.L. & Lukowiak, K. 1984. Male–female conflict in a simultaneous hermaphrodite resolved by sperm-trading. American Naturalist 124, 282–286.

Leonard, J.L., Westfall, J.A. & Pearse, J.S. 2007. Phally polymorphism and reproductive biology in *Ariolimax buttoni* (Pilsbry and Vanatta, 1896) (Stylommatophora: Arionidae). American Malacological Bulletin, 23: 121–135.

Lloyd, D.G. 1982. Selection of combined versus separate sexes in seed plants. The American Naturalist 120, 571–585.

Lorenzi, M.C. & Sella, G. 2008. A measure of sexual selection in hermaphroditic animals: parentage skew and the opportunity for selection. Journal of Evolutionary Biology 21, 827–833.

Lorenzi, M.C., Sella, G., Schleicherová, D. & Ramella, L. 2005. Outcrossing hermaphroditic polychaete worms adjust their sex allocation to social conditions. Journal of Evolutionary Biology 18, 1341–1347.

Lorenzi, M.C., Schleicherová, D. & Sella, G. 2006. Life history and sex allocation in the simultaneously hermaphroditic polychaete worm *Ophryotrocha diadema*: the role of sperm competition. Integrative and Comparative Biology 46, 381–389.

Madsen, J.D. & Waller, D.M. 1983. A note on the evolution of gamete dimorphism in algae. The American Naturalist 121, 443–447.

Maldonado, M. 2004. Choanoflagellates, choanocytes, and animal multicellularity. Invertebrate Biology 123, 1–22.

Marcon, E. & Moens, P. 2005. The evolution of meiosis: Recruitment and modification of somatic DNA-repair proteins. BioEssays 27, 795–808.

Markowitz, H.M. 1991. Portfolio Selection, 2nd edn. Cambridge, MA: Blackwell Publishers Ltd.

Matsuda, H. & Abrams, P.A. 1999. Why are equally sized gametes so rare? The instability of isogamy and the cost of anisogamy. Evolutionary Ecology Research 1, 769–784.

Maynard Smith, J. 1978. The Evolution of Sex. Cambridge: Cambridge University Press.

Maynard Smith, J. 1982. In: Evolution and the Theory of Games Cambridge, England: Cambridge University Press.

Maynard Smith, J. & Szathmáry, E. 1995. The Major Transitions in Evolution. Oxford: Oxford University Press.

Meglitsch, P.A. & Schram, F.R. 1991. Invertebrate Zoology, 3rd edn. Oxford, UK: Oxford University Press.

Michod, R.E. 1990. Sex as a conservative force in evolution. In: Organizational Constraints in the Dynamics of Evolution (Ed. by J. Maynard Smith & G. Vida), pp. 235–251. Manchester, UK: University of Manchester Press.

Morrow, E.H. 2004. How the sperm lost its tail: the evolution of aflagellate sperm. Biological Reviews 79, 795–814.

Muller, H.J. 1932. Some genetic aspects of sex. American Naturalist 66, 118–138.

Munday, P.L., Buston, P.M. & Warner, R.R. 2006. Diversity and felxibility of sex-change strategies in animals. Trends in Ecology and Evolution 21, 89–95.

Muñoz, R.C. & Warner, R.R. 2003. A new version of the size-advantage hypothesis for sex change: incorporating sperm competition and size-fecundity skew. American Naturalist 161, 749–761.

Muñoz, R.C. & Warner, R.R. 2004. Testing a new version of the size-advantage hypothesis for sex change: sperm competition and size-skew effects in the bucktooth parrotfish, *Sparisoma radians*. Behavioral Ecology 15, 129–136.

Nakashima, Y., Kuwamura, T. & Yogo, Y. 1996. Both-ways sex change in monogamous coral gobies, *Gobiodon* spp. Environmental Biology of Fishes 46, 281–288.

Pannell, J.R. 2002. The evolution and maintenance of androdioecy. Annual Review of Ecology and Systematics 33, 397–425.

Parker, B.C. 1971. On the evolution of isogamy to oogamy. In: Contributions to Phycology (Ed. by B.C. Parker & R.M. Jr. Brown), pp. 47–51. Lawrence, KS: Allen Press, Inc.

Parker, G.A. 1982. Why are there so many tiny sperm? Sperm competition and the maintenance of two sexes. Journal of Theoretical Biology 96, 281–294.

Parker, G.A., Baker, R.R. & Smith, V.G.F. 1972. The origin and evolution of gamete dimorphism and the male-female phenomenon. Journal of Theoretical Biology 36, 529–553.

Pearse, J.S., Pearse, V.B. & Newberry, A.T. 1989. Telling sex from growth; disolving Maynard Smith's paradox. Bulletin of Marine Science 45, 433–446.

Pechenik, J. 2005. Biology of the Invertebrates, 5th edn. New York: McGraw-Hill HIgher Education.

Petersen, C.W. 2006. Sexual selection and reproductive success in hermaphroditic seabasses. Integrative and Comparative Biology 46, 439–448.

Policansky, D. 1982. Sex change in plants and animals. Annual Review of Ecology and Systematics 13, 471–495.

Power, H.W. 1976. On forces of selection in the evolution of mating types. The American Naturalist 110, 937–944.

Premoli, M.C. & Sella, G. 1995. Sex economy in benthic polychaetes. Ethology, Ecology and Evolution 7, 27–48.

Prevedelli, D., Massamba N'Siala, G. & Simonini, R. 2006. Gonochorism vs. hermaphroditism: relationship between life history and fitness in three species of *Ophryotrocha* (Polychaeta: Dorvilleidae). Journal of Animal Ecology 75, 203–212.

Puurtinen, M. & Kaitala, V. 2002. Mate-search efficiency can determine the evolution of separate sexes and the stability of hermaphroditism in animals. American Naturalist 160, 643–660.

Raimondi, P.T. & Martin, J.E. 1991. Evidence that mating group size affects allocation of reproductive resources in a simultaneous hermaphrodite. American Naturalist 138, 1206–1217.

Ramesh, M.A., Malik, S.-B. & Logadon, J.M. 2005. A phylogenetic inventory of meiotic genes: evidence for sex in Giardia and an early eukaryotic origin of meiosis. Current Biology 15, 185–191.

Randerson, J.R. & Hurst, L.D. 1999. Small sperm, uniparental inheritance and selfish cytoplasmic elements: a comparison of two models. Journal of Evolutionary Biology 12, 1110–1124.

Randerson, J.R. & Hurst, L.D. 2001a. A comparative test of a theory for the evolution of anisogamy. Proceedings of the Royal Society of London Series B 268, 879–884.

Randerson, J.R. & Hurst, L.D. 2001b. The uncertain evolution of the sexes. Trends in Ecology and Evolution 16, 571–579.

Real, L.A. 1980. Fitness, uncertainty, and the role of diversification in evolution and behavior. The American Naturalist 115, 623–638.

Real, L.A. & Ellner, S. 1992. Life history evolution in stochastic environments: a graphical mean-variance approach. Ecology 73, 1227–1236.

Riffell, J.A., Krug, P.J. & Zimmer, R.L. 2004. The ecologicl and evolutionary consequences of sperm chemoattraction. Proceedings of the National Academy of Sciences USA 101, 4501–4506.

Roff, D.A. 2002. Life History Evolution. Sunderland, MA: Sinauer Associates, Inc.

Schärer, L. & Wedekind, C. 2001. Social situation, sperm competition and sex allocation in a simultaneous hermaphrodite parasite, the cestode *Schistocephalus solidus*. Journal of Evolutionary Biology 14, 942–953.

Schärer, L., Karlsson, L.M., Christen, M. & Wedekind, C. 2001. Size-dependent sex allocation in a simultaneously hermaphroditic parasite. Journal of Evolutionary Biology 14, 55–67.

Scudo, F.M. 1967. The adaptive value of sexual dimorphism: I. Anisogamy. Evolution 21, 285–291.

Sella, G. & Ramella, L. 1999. Sexual conflict and mating systems in the dorvilleid genus *Ophryotrocha* and the dinophilid genus *Dinophilus*. Hydrobiologia 402, 203–213.

Shapiro, D.Y. 1979. Social behavior, group structure, and sex reversal in hermaphroditic fish. Advances in the Study of Behavior 10, 43–102.

Shuster, P., and Sigmund, K. 1982. A note on the evolution of sexual dimorphism. Journal of Theoretical Biology, 94: 107–110.

Simpson, A.G.B. & Patterson, D.J. 2006. Current perspectives on high-level groupings of protists. In: Genomics and Evolution of Microbial Eukaryotes (Ed. by L.A. Katz & D. Bhattacharya), pp. 7–30. Oxford, UK: Oxford University Press.

Solari, A.J. 2002. Primitive forms of meiosis: the possible evolution of meiosis. Biocell 26, 1–13.

Sower, S.A. 1990. Neuroendocrine control of reproduction in lampreys. Fish Physiology and Biochemistry 8, 365–374.

St.Mary, C. 1994. Sex allocation in a simultaneous hermaphrodite, the blue-banded goby (*Lythrypnus dalli*): the effect of body size and behavioral gender and the consequences for reproduction. Behavioral Ecology 5, 304–313.

St.Mary, C. 1997. Sequential patterns of sex allocation in simultaneous hermaphrodites: do we need models that specifically incorporate this complexity? American Naturalist 150, 73–97.

St.Mary, C.M. 2000. Sex allocation in *Lythrypnus* (Gobiidae): variations on a hermaphroditic theme. Environmental Biology of Fishes 58, 321–333.

Tchuem Tchuenté, L.-A., Southgate, V.R., Combes, C. & Jourdane, J. 1996. Mating behaviour in schistosomes: are paired worms always faithful? Parasitology Today 12, 231–236.

Thomson, J.D. 2006. Tactics for male reproductive success in plants: contrasting insights of sex allocation theory and pollen presentation theory. Integrative and Comparative Biology 46, 390–397.

Thomson, J.D. & Barrett, S.C.H. 1981. Selection for outcrossing, sexual selection, and the evolution of dioecy in plants. The American Naturalist 118, 443–449.

Togashi, T. & Cox, P.A. 2004. Phototaxis and the evolution of isogamy and 'slight anisogamy' in marine green algae: insights from laboratory observations and numerical experiments. Botanical Journal of the Linnean Society 144, 321–327.

Togashi, T., Cox, P.A. & Bartelt, J.L. 2007. Underwater fertilization dynamics of marine green algae. Mathematical Biosciences 209, 205–221.

Tomiyama, K. 2002. Age dependency of sexual role and reproductive ecology in a simultaneously hermaphroditic land snail, *Achatina fulica* (Stylommatophora: Achatinidae). Venus (Japanese Journal of Malacology) 60, 273–283.

Tomlinson, J. 1966. The advantages of hermaphroditism and parthenogenesis. Journal of Theoretical Biology 11, 54–58.

Vamosi, J.C., Zhang, Y. & Wilson, W.G. 2007. Animal dispersal dynamics promoting dioecy over hermaphroditism. American Naturalist 170, 485–491.

Warner, R.R. 1975. The adaptive significance of sequential hermaphroditism in animals. American Naturalist 109, 61–82.

Warner, R.R. 1988. Sex change in fishes: hypotheses, evidence and objections. Environmental Biology of Fishes 22, 81–90.

Warner, R.R., Robertson, D.R. & Leigh, E.G. Jr. 1975. Sex change and sexual selection. Science 190, 633–638.

Webster, J.P. & Gower, C.M. 2006. Mate choice, frequency dependence and the maintenance of resistance to parasitism in a simultaneous hermaphrodite. Integrative and Comparative Biology 46, 407–418.

Weeks, S.C., Benvenuto, C. & Reed, S.K. 2006. When males and hermaphrodites coexist: a review of androdioecy in animals. Integrative and Comparative Biology 46, 449–464.

Werren, J.H. 2005. Heritable microorganisms and reproductive parasitism. In: Microbial Phylogeny and Evolution (Ed. by J. Sapp), pp. 290–315. Oxford, UK: Oxford University Press.

Wiese, L. 1981. On the evolution of anisogamy from isogamous monoecy and on the origin of sex. Journal of Theoretical Biology 89, 573–580.

Wiese, L., Wiese, W. & Edwards, D.A. 1979. Inducible anisogamy and the evolution of oogamy from isogamy. Annals of Botany 44, 131–139.

Williams, G.C. 1975. Sex and Evolution. Princeton: Princeton University Press.

Wilson, W.G. & Harder, L.D. 2003. Reproductive uncertainty and the relative competitiveness of simultaneous hermaphroditism versus dioecy. American Naturalist 162, 220–241.

Yamaguchi, S., Ozaki, Y., Yusa, Y. & Takahashi, S. 2007. Do tiny males grow up? Sperm competition and optimal resource allocation schedule of dwarf males of barnacles. Journal of theoretical Biology 245, 319–328.

Yamaguchi, S., Yusa, Y., Yamato, S., Urano, S. & Takahashi, S. 2008. Mating group size and evolutionarily stable pattern of sexuality in barnacles. Journal of Theoretical Biology 253, 61–73.

Zouros, E., Freeman, K.R., Ball, A.O. & Pogson, G.H. 1992. Direct evidence for extensive paternal mitochondrial DNA inheritance in the marine mussel *Mytilus*. Nature 359, 412–414.

4

Rapid Divergent Evolution of Genitalia

Theory and Data Updated

WILLIAM G. EBERHARD

INTRODUCTION: WHY THE INTEREST?

The evolutionary forces responsible for the evolution of animal genitalia have a long history of controversy. Why the special interest in genitalia? In addition to the intrinsic interest of organs that are so intimately related to reproduction and fitness, it is because of a classic property of genital evolution: the morphological forms of genitalia are often species-specific, and these forms are often more divergent among closely related species than other traits such as legs, antennae, eyes, etc. In addition, male genitalia often show exuberantly complex forms that seem inexplicable in terms of their sperm transfer function (figure 4.1). This trend toward greater diversity in genitalia than in other structures occurs in at least some subgroups of all major taxonomic taxa with internal fertilization (reviewed in Eberhard 1985).

This widespread, relatively consistent usefulness of genital morphology in distinguishing species can be translated into a statement about evolutionary processes (unless the data are severely biased—see below): genitalia tend to show an evolutionary pattern of sustained, relatively rapid and divergent morphological change (Eberhard 1985). "Rapid" in this sense is in relative terms, with respect to changes in other traits. Genitalia are often much more elaborate than seems necessary for the simple

function of gamete transfer to the female. What could be responsible for such an evolutionary pattern? The objective of this chapter is to review new data and ideas that have appeared since my 1985 book that can help answer this question.

As a result of the sustained exploitation by taxonomists of genital morphology to discriminate closely related species, we surely know more about the evolution of species-level divergence in the morphology of genitalia than any other set of structures in the animal kingdom. For more than 100 years this huge mass of data on genitalia accumulated in nearly complete isolation from the study of sexual selection. The isolation was explicit in the original description of sexual selection by Darwin (1871), in which he specifically excluded genitalia from his discussion of sexual selection: "There are, however, other sexual differences quite unconnected with the primary reproductive organs, and it is with these that we are especially concerned" (p. 567). It ended abruptly, with Waage's path breaking paper (1979) demonstrating that male genitalia are used in sperm competition in damselflies. During this long period of isolation the study of genitalia was the nearly exclusive province of taxonomists, and was largely descriptive. For their part, students of sexual selection did not even begin to recognize the possibility of post-copulatory competition among males until another crucial paper, that of Parker (1970) on

Bombus

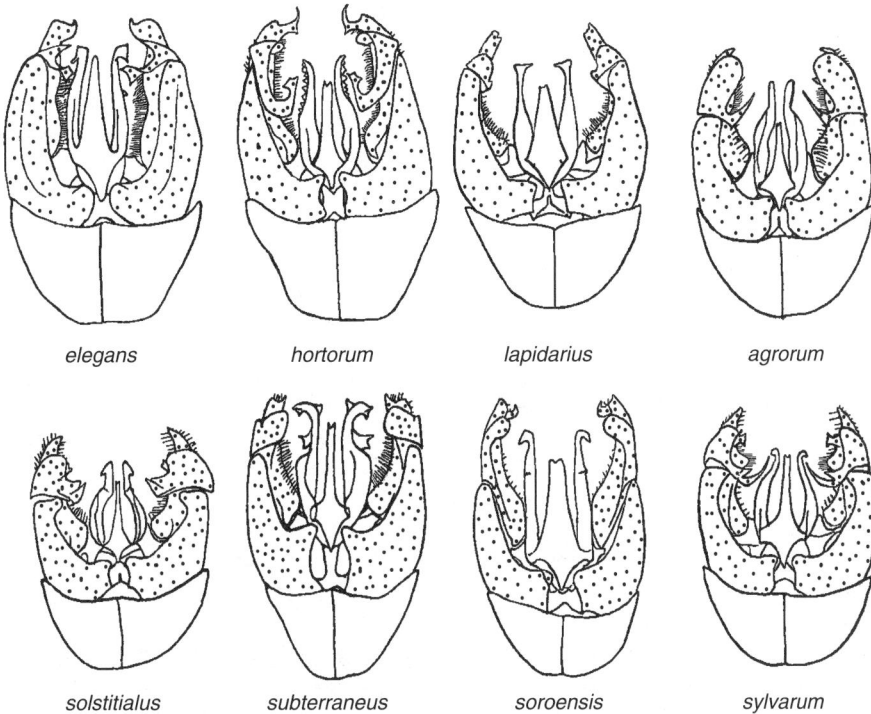

| elegans | hortorum | lapidarius | agrorum |

| solstitialus | subterraneus | soroensis | sylvarum |

FIGURE 4.1 Complex morphology of the male genitalia of different species of bumble bees in the genus *Bombus*, illustrating the pattern of diverse forms among closely related species that is very common in male genitalia. Much of this chapter is dedicated to evaluating hypotheses that attempt to explain why such relatively rapid divergent evolution should typify male structures that are specialized to contact females in sexual contexts. The stippled portions of these male genitalia are thought to contact only the external surface of the female's abdomen, and not to enter her reproductive tract during copulation. The area of the female that they contact is relatively featureless and differs little if at all between species, illustrating a common pattern in genitalia of more rapid morphological divergence in males than females. *Bombus* is especially interesting because it appears not to fit any of the currently popular hypotheses (see "Frontiers" below; drawings after Richards 1927).

sperm competition. Lagging even farther behind was recognition of the possible importance of active female roles in this competition. As interest in female choice surged in the early 1980s, the possibility that it might act on genitalia through cryptic female choice (CFC) (Thornhill 1982) was proposed (Eberhard 1985). More recently genitalia have been mentioned as targets of another type of sexual selection, sexually antagonistic coevolution (SAC) between males and females (Arnqvist & Rowe 2005; Gilligan & Wenzel 2008) (below).

Recent developments in several fields facilitated the linking of genital evolution and sexual selection (Birkhead 1996). The most important advances were: (1) the discovery that the doubts about whether females could gain payoffs from choosing among males, which were based on the theoretical "proof" that no genetic variance could exist among males for traits under selection by female choice, were unfounded; empirical data showed that variance is quite common (summary Andersson 1994); (2) the discovery that multiple mating by females (a prerequisite for post-copulatory sexual selection to occur) is much more common in nature than previously thought; (3) the rediscovery of the importance of sexual selection by female choice; and (4) a gradual disillusionment (for several reasons) with previous, "species isolation" arguments to explain species-specific traits in general (e.g., Paterson 1982), and genital traits in particular (Scudder 1971; Eberhard 1985; Shapiro & Porter 1989). The recent increased emphasis on male–female conflicts during copulation (Parker 1984, 2005; Arnqvist & Rowe 2005) has led to further hypotheses regarding genital evolution on the basis of male–female coevolutionary conflicts.

In this chapter, I will update the search for a unitary explanation for sustained rapid divergent evolution of genitalia. Because of the great generality of the trend, which extends even to non-genital structures that are specialized to contact females in sexual contexts (below), there is probably a very general explanation. The reader should keep in mind, however, that because literally millions of species are involved, it is likely that there may be exceptions to most if not all generalizations. The male genitalia in different groups perform a wide variety of functions, ranging from fighting other individuals ("penis fencing" — Michiels 1998), visual displays (Wickler 1966; Bohme 1983), hooking and holding onto struggling females prior to copulation (Bertin & Fairbairn 2005), plugging the female's reproductive tract (Koeniger 1983; Abalos & Baez 1966; Nessler et al. 2007), prying or squeezing open female ducts and valves (Fennah 1945; Eberhard 1993a; Schulmeister 2001; Sirot 2003; Moreno-Garcia & Cordero 2008), holding on with powerful suction cups (Schulmeister 2001), removing copulatory plugs (Aisenberg & Eberhard 2009), cleaning off detritus from previous copulations (Kumashiro et al. 2006), forming a reserve intromittent structure in case the other is damaged (Kamimura & Matsuo 2001), injecting prostate gland secretion through one aperture and sperm through another in bifid or trifid structures (Merrett 1989; Anthes & Michiels 2007), and rubbing or tapping the female before or after copulation (Otronen 1990; Eberhard 1990, 1994). Whether the structures that perform these different functions all show the same trend toward rapid divergent evolution is not known (the answer might be interesting). Perhaps no single explanation for diversity in form will be correct for all cases.

The line between general and local explanations, and in particular the number and scope of

refutations that are needed to reject a hypothesis as a general explanation, is difficult to determine (Coddington 1987; Shapiro & Porter 1989). I have no magic answers, but believe it is useful to explore the limits of generality of different hypotheses that attempt to explain a widespread phenomenon like this. In keeping with the general focus of this book (and also with the much larger accumulation of data), I will concentrate on the evolution of the morphology of genitalia, rather than that of sperm and other seminal products, even though these also show signs of being under sexual selection (Miller & Pitnick 2002; Holman & Snook 2006; Markow & O'Grady 2005). They are probably crucial for understanding some aspects of the reproductive morphology and physiology, especially of females, as illustrated by the coevolution between the length and the form of sperm cells and female storage organs in *Drosophila* (Miller & Pitnick 2002), diopsid flies (Kotrba 1995, 2006), scathophagid flies (Minder et al. 2005), and featherwing beetles (Dybas & Dybas 1981). Before I begin, I need to make two preliminary points: one concerns non-genital "genitalia"; the other the possibility that the pattern of accentuated diversity in genitalia is an illusion that has arisen from biases in how taxonomists work.

NON-GENITAL CONTACT STRUCTURES

I will discuss in this chapter not only primary genitalia (structures associated with the gonopore), but also secondary genitalia (which receive sperm from the male's gonopore and introduce them into the female), and also non-genital male structures that are specialized for contact with the female (usually in a non-genital part of her body) prior to or during copulation. All three clearly show the same evolutionary pattern of common species-specificity and frequent overly -elaborate form for their relatively simple functions, and thus probably require a similar explanation (Robson & Richards 1936; Eberhard 1985). This pattern in secondary genitalia and non-genital contact structures was discovered long ago by taxonomists in many groups. In fact, entomologists have repeatedly included as "genitalia" some structures which are not associated with the segment on which the genital opening occurs), such as cerci and sternites near the "true" genitalia, in groups in which these structures also show a pattern of rapid divergent evolution and elaborate forms that is typical of more strictly genital structures (e.g., Tuxen 1970; Wood 1991).

Other structures that are even farther from the genitalia and that are specialized to contact non-genitalic parts of the female in sexual contexts also show the same pattern (Eberhard 1985, 2004b; also Darwin 1871; Robson & Richards 1936). Almost any part of the male can be modified in this way, from the sucker-like " bursa" of male nematodes to the cephalothorx, the chelicerae and anterior legs of spiders, the antennae and telson of crustaceans, and the head, mandibles, antennae, pronotum, cerci, legs, and wings of insects (figure 4.2). As pointed

out by Robson and Richards (1936), the mechanical function of many (though not all) of these structures is to grasp the female during copulation; this is the same function that is performed by a large fraction of the male genital structures that are species-specific in form (summaries in Scudder 1971; Eberhard 1985, 2004a). In the end, the line between "true" genital claspers and non-genital claspers is arbitrary (Darwin 1871; Chapman 1969; Eberhard 1985; see also chapters by Leonard & Cordoba-Aguilar and Ghiselin in this book).

Inclusion of non-genital contact organs is especially useful for understanding this evolutionary pattern of rapid divergence because they have two

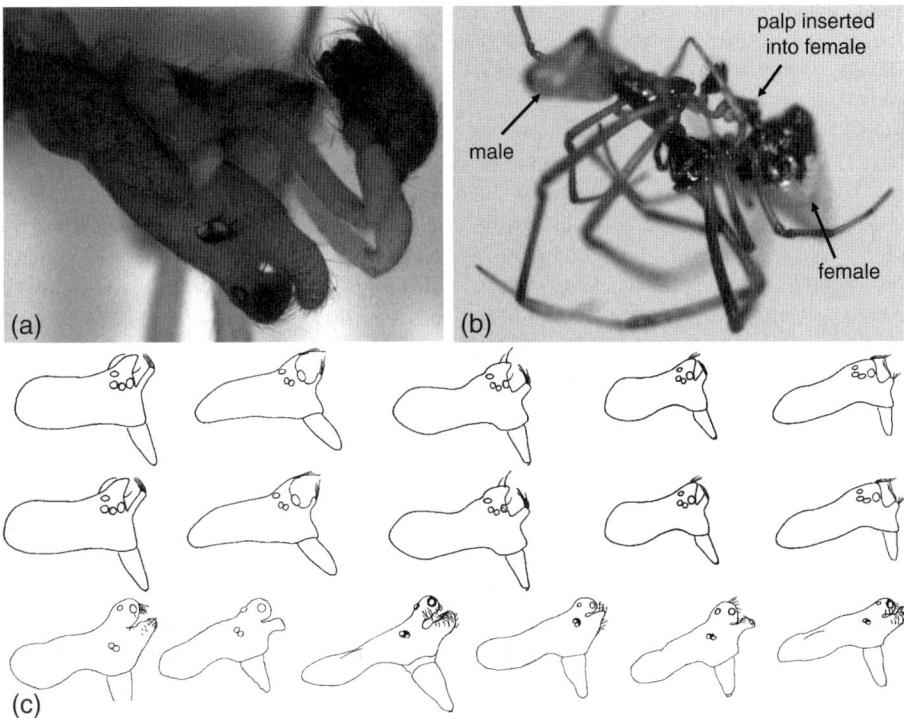

FIGURE 4.2 The elaborate anterior portion of the male cephalothorax of *Argyrodes elevatus* (a) is specialized to contact the female during copulation (b). As with many other non-genitalic contact structures, the forms are elaborate and species-specific (each drawing in (c) is of the male of a different species of *Argyrodes*). SAC explanations for male cephalothorax form based on species-specific female defensive behavior or morphology to avoid damage are unconvincing. The female's mouth area, which contacts the modified area of the male's cephalothorax during copulation (b), does not show any modifications. The female is not physically coerced, as she is free to pull her mouth away from the male at any time during copulation, and thus avoid possible tactile and chemical stimulation; in fact the female pulls away when the gland openings on the male's modified cephalothorax are covered (G. Uhl, personal communication). (a) and (b) courtesy of Gabriele Uhl; (c) from Exline & Levi 1962).

advantages over "true" genitalia: the details of their physical interactions with the female and their possible functions (often grasping the female) are generally better understood; and the female structures that they contact are often more easily studied, because the male organs contact the female's outer surface rather than internal genital structures. I will use the phrase "non-genital contact structures" below to indicate species-specific male structures that are not near his genitalia and that are specialized to contact females in sexual contexts.

IS THE PATTERN OF RAPID DIVERGENT EVOLUTION AN ARTIFACT?

The historical isolation of taxonomic research on genitalia has both advantages and disadvantages in studies of sexual selection. It makes the data more trustworthy in some respects, because they are independent of observer bias with respect to hypotheses about sexual selection. The data are also, however, subject to other possible biases that could result in over-estimating the relative rapidity of genital evolution and divergence (Coddington 1987; Tanabe et al. 2001; Huber 2003, 2004; Mutanen 2005; Mutanen & Kaitala 2006; Song 2006). The trend toward rapid divergent evolution discussed above might be an artifact if taxonomists rely too heavily on genital differences in deciding which groups of individuals should be recognized as species: they might fail to recognize species which differ with respect to other traits but not their genitalia; and they might over-split species if they find genital differences among different populations of the same species, especially in well-studied faunas where discovery of truly new species is rare (Mutanen 2005). Such over-reliance on genitalia could lead to overestimates of the relative rapidity of divergence of genitalia (Shapiro & Porter 1989; Huber 2003; Song 2006). The splitting problem could be particularly important if the amount of intra-specific variation in genitalia is underestimated, and some previous over-splitting mistakes have been documented (Mutanen 2005). Genitalia do vary intraspecifically among geographically distinct populations (Ware & Opell 1999; Sirot 2003; Polihranakis 2006; Song 2006; Gilligan & Wenzel 2008), during the ontogeny of a given individual (Song 2006), between different seasons (Kunze 1959; Vitalievna 1995), and even within a single

population (Mutanen 2005). In addition, male genitalia are polymorphic in some species (Johnson 1995; Huber & Pérez Gonzalez 2001; Mutanen & Kaitala 2006), and there is reason to believe that polimorphism has been underestimated (Huber 2003). Over-reliance on genitalia could be especially damaging when sample sizes are small, an uncomfortably common circumstance in many taxonomic studies (Huber 2003). These problems could lead to overestimates of the rapidity with which genitalia diverge.

Are there reasons to doubt the importance of these taxonomists' self doubt? I think the answer is yes. In the first place, there are data on genitalia that are independent of possible over-reliance on genitalia to distinguish species and that also indicate rapid divergent evolution. The variation in genital morphology at higher taxonomic levels, where uncomfortable questions about what is and what is not a species are not a problem, strongly imply especially rapid, sustained divergence in genitalia. Despite major long-term efforts, homologies have been much more difficult to establish among male genital structures than for other structures (Tuxen 1970; Coddington 1990; Wood 1991; Schulmeister 2001). For instance, Coddington (1990: p. 1) summarizes the situation for araneoid spiders: "On the whole, the century-long effort to homologize the palpal sclerites of male spiders across families and superfamilies seems to have been a rather dismal failure." Similar pessimism characterizes attempts to find homologies within insect orders (Tuxen 1970), and even within a single spider family (Agnarsson et al. 2007). These difficulties testify to divergence that is so rapid and substantial that even highly trained eyes and minds are unable to find and agree upon commonalities.

A second striking aspect of genital evolution is the extraordinarily wide range of groups in which genitalia are thought to constitute especially useful traits for distinguishing species (Eberhard 1985). Perhaps no major group of animals with internal insemination is an exception. Taxonomists working on many different groups have apparently convergently realized that genitalia are especially useful traits in distinguishing otherwise difficult to distinguish taxa. This convergence was not simultaneous; even within the insects, genitalia were used very early in some groups of flies (Dufour 1844 in Shapiro & Porter 1989), and only began to be used later in others, such as papilionid butterflies and sphingid moths (Jordan 1896, 1905), tortricid moths (Dampf 1908 in Gilligan & Wenzel 2008), and certain Hymenoptera (Perez 1894 in Shapiro & Porter 1989), and even later in others such as *Culicoides* flies (Carter et al. 1920 in Jamnbeck 1965).

Could it be that use of genitalia in studies of fly taxonomy induced beetle, snake, rodent, nematode and earthworm taxonomists to concentrate excessively on aedeagi, hemipenes, bacula, spicula and penile spines to distinguish species? Such cross-group imitation is imaginable, but I expect it is relatively unimportant, because I have confidence in the hard-headed independence of taxonomists. Take for example, the likely result of communication among workers on different groups of animals. A worker on group X might begin to examine genital traits after learning that workers in group Y found genitalia to be useful in distinguishing species. But only if the genitalia in group X worked as well or better than the other traits that were previously used to distinguish species in this group, and if the groupings were in at least general agreement with those indicated by other traits, would the taxonomist working on X be likely to adopt them.

There are also other reasons to think that taxonomists in different groups have not been slavishly dependent on others in choosing the traits on which they concentrate. In many subgroups of insects and arachnids, for instance, taxonomists have never used genitalia or have secondarily abandoned their use in particular groups, including most ichneumonid wasps (I. Gauld personal. communication.), lampyrid beetles (Lloyd 1997), field crickets (Alexander et al. 1997), Jerusalem crickets (Tinkham & Renz 1969), tephritid fruit flies (Eberhard 1996), polyctenid bugs (Ferris & Usinger 1939), satyrid butterflies (Cardé et al. 1970), aleyrodid whiteflies (Ossiannilssen et al. 1970), and scorpions (Jacob et al. 2004a) (see Robson & Richards 1936 for others). In some taxa, species that were originally recognized on the basis of non-genital traits were subsequently found to also differ in genitalia (Shapirio & Porter 1989). These data indicate that taxonomists have not been so strongly tradition-bound in choosing characters as the arguments above suggest, and that genitalia do often tend to diverge relatively rapidly.

One further concern (Song 2006, Song & Bucheli 2009) is that the fact that genitalia often evolve slowly enough that their pattern of differences reflect higher-level groupings of different species implies a limited rapidity of genitalic divergence.

Song (2006) found that 94.7% of 89 papers presenting phylogenetic analyses in 19 different arthropod orders concluded that genital characters were phylogenetically informative, and was thus led to the unsurprising conclusion that "genitalia do not evolve chaotically." This pattern does not weaken, however, the possibility that genitalia tend to diverge more rapidly than do other body traits. Rather they probably often evolve rapidly enough to be especially useful compared with other traits in distinguishing closely related species, but nevertheless slowly enough in at least some aspects to also retain a phylogenetic signal.

This is not to say that both improved methods of quantifying genital divergence (e.g., Tanabe et al. 2001; Mutanen & Pretorius 2007) and use of other, independent characters such as molecular differences have not corrected some errors that have resulted from previous over-reliance on genital morphology (e.g., Hedin 1997; Stoks et al. 2005) (such checks have also confirmed distinctions on the basis of genitalia in other taxa—Pizzo et al. 2006a, b). But the general message is that the trend for genitalia to diverge relatively rapidly does exist, although the evidence may not be as conclusive as some have thought. Calls to check species for possible genital polymorphism, and to test for correlations between molecular and genital differentiation (Huber 2003; Jacob et al. 2004a) represent healthy skepticism that promises to help determine the scope of the general trend in particular groups.

WHY RELATIVELY RAPID DIVERGENCE? HYPOTHESES IN DISFAVOR

Many explanations have been proposed to explain the tendency for genitalia to diverge rapidly. One major hypothesis that is generally judged to have failed is Mayr's pleiotropism hypothesis (Mayr 1963). He proposed that genes that are involved in adaptations to other factors such as different ecological conditions also have pleiotropic effects on genital morphology, and that divergent ecological adaptations incidentally resulted in genital divergence. This hypothesis does not explain, however, why such pleiotropic effects should be concentrated in genitalia rather than other structures, or why in groups with other male sperm-transferring structures besides the primary genitalia (e.g., the secondary genitalia of spiders, solfugids, pseudoscorpions, and odonates) it is always the secondary genital structures that show the typical rapid divergent evolution while the primary genitalia do not. Nor does it explain why the genitalia of species with external fertilization show a complete lack of such pleiotropic effects (Eberhard 1985).

A second major explanation, the oldest of all, is the "lock and key" hypothesis. This holds that selection on females to avoid insemination by males of other species has resulted in the evolution of female genital structures that prevent entry or coupling by the male genitialia of other species. Males may also profit from not transferring sperm to heterospecific females, but probably to a lesser degree, given their less costly gametes. The lock and key hypothesis provides a clear explanation for rapid divergence and male species specificity, but it is nevertheless probably in the process of slow death under an accumulation of contrary evidence (Eberhard 1985; Shapiro & Porter 1989). Most notably, the females of many species simply do not have any structures that could act as a "lock" to exclude heterospecific males (summary in Shapiro and Porter 1989; subsequent data in Eberhard & Pereira 1996, Eberhard 2001a–d, 2003, 2004b, c, 2005; Peretti 2003; Ohno et al. 2003; Vanacker et al. 2003; Eberhard and Ramirez 2004; Jagadeeshan and Singh 2006; Briceño et al 2007; Ingram et al. 2008). The existence of mirror image genital dimorphism in one sex of a mantid (Howell & Herberstein in prep.) and a spider (Huber & PerezGonzalez 2001) also argues against the importance of specific fits. And an intra-specific analysis of a water strider showed no effects of the relationship between male and female morphology on male mating success (Arnqvist et al. 1997). There are exceptions (Callahan & Chapin 1960), but the lack of female "locks" is clearly widespread.

In addition, there is often no sign of the character displacement in males that is predicted in zones of sympatry of closely related species (McAlpine 1988, Ware & Opell 1979; Shapiro & Porter 1989). In addition, there is clear evidence of genitalic species-specificity in species that have evolved in complete or nearly complete physical isolation from any close relatives and that thus need no locks and keys, such as those endemic to oceanic islands and caves, and parasites isolated from all close relatives in their different hosts (Eberhard 1985, 1996; Shapiro & Porter 1989; Hedin 1997).

The lock and key hypothesis is still sometimes cited, and a few recent studies present data in favor.

In some noctuid moths male and female genital structures coevolve, as predicted (Mikkola 1992, 2008), but this support is weak because several other hypotheses are also compatible with such coevolution; Mikkola's reason for dismissing cryptic female choice as an alternative explanation is unconvincing, nor is the evidence convincing that female genitalia are designed appropriately to exclude heterospecific males (Eberhard 1996). Lock and key arguments were also given to explain why in cross-specific pairings in *Carabus* beetles, the species-specific male copulatory piece does not fit easily in the a soft sac where it is lodged in the female's reproductive tract, and sometimes causes fatal damage (Sota & Kubota 1998; Usami et al. 2006). These observations show the importance of a mechanical fit between male and female, but do not support the lock and key hypothesis. The hypothesis supposes that females evolve species-specific "locks" in order to gain (from avoiding cross-specific fertilization of their eggs), while in these carabids the female morphology causes them to lose (because of internal damage) when they mate with cross-specific males. Data for another proposed case in millipedes are limited to the genital consequences of differences in the size rather than shape, and are asymmetric (males of the larger species cannot fit into the smaller), and do not explain the diversity of genital forms in this genus (Tanabe & Sato 2008). There are also a few cases of geographic patterns of apparent character displacement in male genitalia that is predicted by lock and key, as in aedeagus length in populations of two closely related species of *Odontolabis* stag beetles that are sympatric at two sites (Kawano 2003). Such patterns are uncommon, however, and species which lack displacement have also been observed (Ware & Opell 1979; Tanabe et al. 2001; Taylor & Knouft 2006). Occasional displacement-like patterns might occur by chance, especially when ranges are not known in great detail or have changed historically (Shapiro & Porter 1989).

Further recent evidence also argues against lock and key. In several different groups female remating frequency is positively correlated with the amount of genital divergence (Eberhard 1985; Dixson 1987, 1998; Roig-Alsina 1993; Arnqvist 1998; Paraq et al. 2006). This correlation is predicted by sexual selection hypotheses (below), but not by lock and key. One recent energetic defense of lock and key in Lepidoptera (especially Noctuidae) involves a major retreat, admitting that the substantial divergence of the male genital structures that remain on the external surface of the female ("external" male genitalia) do not involve lock and key selection, because there are no female "lock" structures; only intromittent male structures are claimed to function as keys in internal female locks. In addition, the frequent divergence between the "internal" male genitalia of allopatric sister species (seen in 34 of 39 pairs of Holarctic noctuid species) is said to be due to drift rather than selection for species isolation. No explanation was given for why genitalia should drift more than all other traits (in only 17 of the 39 pairs did non-genital "habitus" traits differ). Such ad hoc retreats in the face of contradictory evidence are always possible in science, but reduce the credibility of the hypothesis.

A related idea, which is mentioned less often but is less strongly contradicted, is a stimulation version of lock and key: the female uses stimuli from the male's genitalia to determine his species identity, and thus avoids cross-specific insemination (Patterson & Thaeler 1982; Eberhard 1985), or the male uses stimuli from the female to avoid cross-specific sperm transfer (Tanabe & Sota 2008). This hypothesis can explain both species-specificity in males and the frequent lack of coevolution in female morphology in groups with rapid male divergence. It is contradicted, however, by the lack of character displacement in male genital morphology in zones of overlap (Eberhard 1985; Shapiro & Porter 1989), and the clear male divergence in many groups in which cross-specific pairing is impossible because of geographic or ecological isolation (Eberhard 1985, 1996). A stronger test, however, would be to search for female character displacement, as occurs in some other signals such as frog calls (Höbel and Gerhardt 2003). Sensory lock and key also does not necessarily predict the correlation between female polyandry and male genital divergence (Eberhard 1985; Roig-Alsina 1993; Dixson 1987, 1998; Arnqvist 1998; Paraq et al. 2006).

Another hypothesis, motivated by the discovery that male genitalia can remove the sperm of other males from female storage organs in damselflies (Waage 1979) and the inspired speculations of Lloyd (1979), is that direct male–male sexual selection on the male's ability to remove sperm from the female might be responsible for the diversity of male genitalia. Sperm competition (strictly speaking, "the competition within a single female between the sperm from two or more males for the fertilization of the ova" (Parker 1970); more commonly extended to cover direct male effects on the sperm of other males within a female) has subsequently been documented in a variety of species (Simmons 2001). But without selection of some sort that causes rapid evolutionary changes in females that make different male designs better at removing sperm in different, closely related species (as expected under both cryptic female choice and sexually antagonistic coevolution—see below), sperm competition involving male genitalia seems unlikely to result in rapid divergent evolution of males by itself. In addition, in contrast with damselflies, the male genitalia in many (most) of the groups with diverse male genitalia do not reach sperm storage sites inside the female, and thus cannot physically remove sperm there (Eberhard 1985). Nevertheless, in some groups not all sperm from previous males are stored in the spermathecae, and they occur at sites such as the bursa or the vagina that are more accessible for the male. In addition, it can also be imagined (though it has never been convincingly demonstrated) that a male whose genitalia do not reach the sperm stored in the female can nevertheless flush them out with a douche-like spray (Eberhard 1985; Simmons et al. 1996; see Hosken et al. 1999 and Whitney et al. 2004 for refutations of this mechanism in a fly and a shark); physical sperm displacement with the male's spermatophore does occur in one beetle (Förster et al. 1998). In some katydids (von Helversen & von Helversen 1991) and damselflies (Cordoba-Aguilar 1999) the male can induce the female to move sperm from her inaccessible spermathecae to other sites in her reproductive tract that he can reach with his genitalia.

In general, sperm competition could be linked with rapid divergence in two ways. If the species-specific aspects of a male's genitalia allow him to overcome female-imposed barriers to gain access to otherwise inaccessible sperm and displace them, and if the female gains from having barriers by avoiding a cost of the male's actions that reduces her production of offspring, then the male adaptations to overcome female barriers (and the female barriers) could represent adaptations favored by sexually antagonistic coevolution (below). If, on the other hand, these male genital traits serve to increase the male's ability to induce female responses (such as those documented in the katydids and damselflies) that allow him to overcome female barriers and remove other males' sperm, and if the female gains from having the barriers because they enable her to bias paternity so as to obtain sons better able to overcome female barriers in following generations, this could represent a type of cryptic female choice (below).

There is, however, strong evidence against the generality of the sperm removal hypothesis that comes from the many species in which species-specific male structures clearly contact only sites in or on the female where sperm are never present, such as the many non-genital contact structures (Robson & Richards 1936; Eberhard 1985, 2004b). There are also numerous examples of species-specific genital structures that surely never come close to sperm in the female, including male surstyli in sepsid and tephritid flies (Eberhard & Pereira 1995, 1996; Eberhard 2001b), clasping gonocoxae and gonostyli in many dipteran families (reviewed in Eberhard 2004a), the stipes, volsella and squama in male bumble bees (figure 4.1) (Richards 1927), elongated genital setae in *Aelurus* wasps (Eberhard 2004c), and male cerci and associated setae in *Glossina* tsetse flies (Briceño et al. 2007), all of which remain outside the female's body during copulation.

Several newer, more neglected hypotheses have also been proposed. Møller (1998) proposed that the female uses the male's genitalia to judge his ability to resist infection by parasites. Perhaps due to the absence of any obvious reason to suppose that the form of a male's genitalia should be consistently responsive to such infections, this hypothesis has not to my knowledge received further attention. Simmons (2001) proposed a different sexual selection hypothesis involving direct male–male battles: complex male genital morphology diverged under selection to function as holdfast devices that defend copulating males against takeovers by other males. A possible reason for rapid divergent evolution of such holdfast devices, though none were given, would be to counteract the effects of rapid divergence in the behavior that other males use to displace copulating males. To my knowledge, however, no such divergent behavior has ever been documented. This hypothesis also has other serious problems. Many species-specific male genital structures surely do not function as holdfast devices. For example, in a list of functions attributed to 105 male genital structures in 43 species in 22 families of Diptera (Eberhard 2004), nearly half (46.7%) of the attributed functions were for penetrating the female and sperm transfer rather than for clasping. In many other groups with divergent intromittent

male genitalia, such as for instance nematodes, primates, rodents, and bats, the male clasps the female with structures other than his genitalia. In still other groups the male has a very powerful, species-specific clasping device which makes it essentially impossible to displace him from the female, but also has additional species-specific genital structures that enter the female that are not appropriately designed as hold-fast devices (see, e.g., Whitman & Loher 1984 on a grasshopper, Wood 1991 on several groups of flies, Briceño et al 2007 on tsetse flies). Still another problem is that in many groups with divergent male genital structures, displacement battles involving copulating males have never been observed; for instance, in some (probably many) spiders, male fights occur only when they are both out of contact with the female (Rovner 1968; Robinson & Robinson 1980; Eberhard & Briceño 1983; Mendez 2002).

Still another recent proposal is the "mate check" hypothesis of Jocqué (1998). As with Mayr's hypothesis, it supposes that pleiotropic effects on genital morphology are important. Key adaptations to environmental variables are thought to have pleiotropic effects on male genital morphology, and females are thought to use such genital traits as "guarantors" of male fitness. By responding preferentially to males with such morphological traits, the female would be able to increase the chances that her offspring would benefit from these adaptations. This idea suffers from the same serious problems mentioned above in connection with the Mayr's pleiotropism hypothesis, in particular, the unanswered question of why there should be a consistent association between fitness traits and the form of genitalia rather than other body parts (Eberhard 1985). It also fails to explain why "cheater" males lacking the key adaptations but possessing the preferred genital traits would not become common.

Finally, Jagadeeshan and Singh (2006) proposed a "male sex-drive" hypothesis in conjunction with their finding that in four closely related species of the *melanogaster* clade of *Drosophila*, in which male genital morphology is species-specific, a larger size of an evolutionarily derived male genital structure (the posterior process) may facilitate grasping the female oviscape during the first 5–10 min of copulation. A mechanical advantage of this sort may well sometimes be important in the early stages of the evolution of new genital traits; but it is not obvious why it would generally lead to great genitalic diversity. More specifically, why would males of different species of *Drosophila* find that such different posterior process designs are best able to hold the essentially invariant portion of the female's anatomy (Eberhard & Ramirez 2004) that they grasp? More generally, many genital structures have no obvious mechanical grasping function.

WHY RELATIVELY RAPID DIVERGENCE? THE TWO MOST POPULAR HYPOTHESES

The two most popular hypotheses at the moment both invoke sexual selection: cryptic female choice, and sexually antagonistic coevolution. The basic arguments are the following.

Cryptic Female Choice (CFC)

Male genitalia are thought to be courtship devices. Sexual selection by female choice occurs after copulation has begun, with females favoring some male genital designs over others, via biases in post-copulatory processes such as sperm transport, oviposition, remating, etc. (Eberhard 1985, 1996). Male designs can be favored because they result in more effective stimulation of the female, or because they fit better with her genital morphology. The expected sequence of evolution can be outlined as follows.

- Females are inevitably stimulated by male genitalia during copulation in species with internal insemination (and also by non-genital male structures that contact them during sexual interactions). Natural selection on females favors female use of such stimuli to trigger certain reproductive processes, such as sperm transport, ovulation, oviposition, resistance to further copulation, secretion of products to help maintain sperm alive in storage sites, etc., that are otherwise kept inactivated until mating occurs. Triggering these same female processes is, incidentally, favorable to the reproduction of the current male.
- If, as is probably usually the case, females do not give 100% complete responses in all of these post-copulatory processes to every copulation (e.g., they do not ovulate or oviposit all available eggs, do not always dump all of the sperm of previous males, etc.), and if they are not strictly monogamous, then sexual

selection on males will favor the ability to increase the effectiveness of their stimulation of the female during copulation (including stimuli from their genitalia or non-genitalic contact devices) in eliciting more complete female responses.

- Selection on females will favor discrimination that allows them to bias paternity in favor of the males best able to deliver these stimuli, in order to obtain the benefit of sons whose genitalia and non-genital contact structures that are especially effective stimulators. This can result in a runaway process, which will tend to produce sustained, rapid divergent evolution of the corresponding male structures. Females could conceivably benefit from superior sons with respect to both good survivorship genes or good signaling genes, but theoretical expectations suggest a stronger correlation with signaling genes (Eberhard 1985, 1996). Direct empirical tests for a correlation between indicators of male "condition" with measures of genital size have been negative (Schulte-Hostedde & Alarie 2006; House & Simmons 2007). Because there are so many different ways a female may be stimulated, and because many types of stimuli are likely to have effects on triggering a variety of reproductive processes through the highly inter-connected nervous system of the female, divergence in male designs in different populations is likely.

Sexually Antagonistic Coevolution (SAC)

Male genitalia are thought to be devices to manipulate the female in ways that favor the male's reproduction but reduce the female's reproduction; females coevolve to counteract these negative male effects, resulting in an arms race between the sexes (Alexander et al. 1997; Holland & Rice 1998; Chapman et al. 2003; Arnqvist & Rowe 2005; Gilligan & Wenzel 2008). In this view, the sexual selection on males that results from female rejections is a side effect of natural selection on females (Rowe 1994; Arnqvist & Rowe 2005). The expected sequence of evolution can be outlined as follows.

- The male does something to the female with his genitalia or non-genital contact structures that increases his chances of paternity, but at the same time reduces the number of offspring produced by the female. For instance, the male might use spines or a rough surface on his genitalia to scrape a hole in the lining of the female's reproductive tract, thus increasing the ability of his seminal products that induce the female to oviposit by giving them increased access to her body cavity and to her nervous system (figure 4.3). Selection on males could favor this mechanism of inducing rapid oviposition before the female mates with another male, even if it results in a decrease in overall female reproduction because of the physical damage to her reproductive tract, or because such rapid oviposition reduces the survival of her eggs because she was less selective in choosing oviposition sites.

• The female evolves defenses against the damaging effects of male genitalic manipulation. For instance, she might evolve a thicker lining of her reproductive tract in the area that is abraded by the male, reducing the strength of his negative effects on her reproduction.

• The male evolves a way to overcome the new female defense. For instance, he might scrape at a different, unprotected site, or evolve longer or sharper scraping structures or stronger scraping movements at the old site. Sexual selection on the male will favor the development of such male traits, as long as the number of offspring he loses due to damage he inflicts on the female is less than the number of offspring he gains by manipulating her reproductive processes (such as oviposition). This coevolutionary arms race can result in relatively sustained rapid divergent evolution of male genitalia as long as neither sex evolves

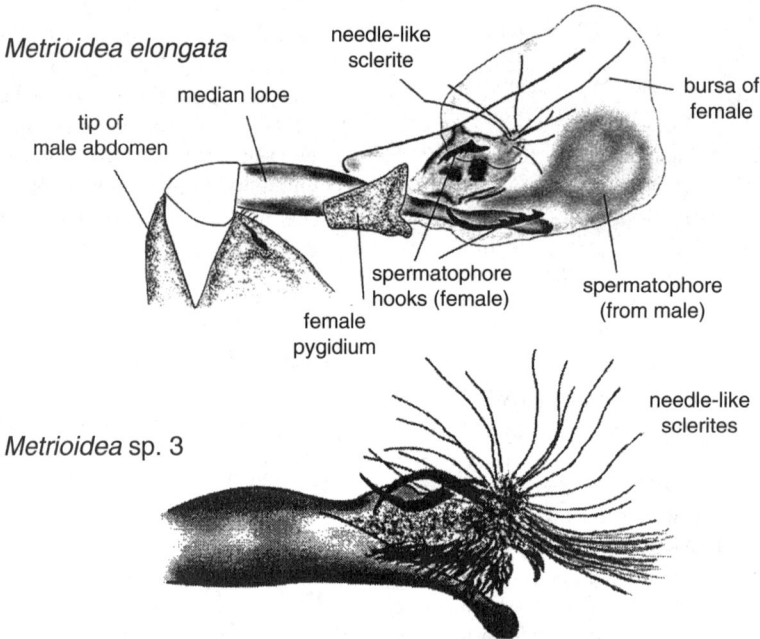

FIGURE 4.3 The needle-like sclerites on the male genitalia of *Metrioidea elongata* beetles flash-frozen in copula perforate the walls of the female bursa (above) (female abdomen dissected away). This damage to the female, perhaps a result of selection on the male to introduce seminal products into the female's body cavity where they will have more effect on her reproductive behavior or physiology, or perhaps to anchor himself more firmly or stimulate her more effectively, typifies the type of damaging male manipulation of the female that could give rise to SAC. The even longer sclerites in *M.* sp. 3 (below) are also thought to perforate the female, because the bursae of field collected females had apparent scars resembling the scars in *M. elongata* (from Flowers & Eberhard 2006).

an unbeatable control mechanism. An outright "win" by one sex, however, would break the coevolutionary spiral, and remove selection on the other sex favoring antagonistic traits. Unassailable female defenses, such as a reproductive tract with a lining too strong for the male spines to perforate (figure 4.3), do not seem difficult to imagine. Thus coevolution might not be consistent and sustained over long periods of time.

SAC could involve different types of genital trait. A "physical coercion" version of SAC involves physical struggles between males and females (Alexander et al. 1997; Arnqvist & Rowe 2002a, b). For instance, males could evolve to seize females with genital claspers, females could then evolve defensive structures that impede such seizures, and males could respond by evolving modified claspers that overcome the female defenses. A second, "stimulation" version of SAC involves sensory traps: the male uses stimuli to which the female has already evolved under natural selection in other contexts both sensitivity and responsiveness (responses which, incidentally, favor the male) (Arnqvist 2006). Such traps are thought to be common in genital evolution, with males exploiting stimuli and the female responses to them that females evolved to control reproductive processes they need to trigger after copulation begins or has occurred (e.g., sperm transport, ovulation, oviposition, etc.) (Eberhard 1996). Sensory traps could be especially important during early stages of male–female evolutionary interactions. Under the stimulation version of SAC, male ability to induce a female response would reduce the female's reproductive output, and thus select for changes in female sensitivity or responses to these stimuli. Female "escape" from these traps, by evolving changed sensitivities or responses, would be constrained by the original advantage of sensing and responding to these stimuli (Arnqvist 2006). The physical coercion version of SAC predicts common coevolution of male and female morphology; in contrast, the stimulation version of SAC does not predict that such easily observed coevolution should be common, because female coevolutionary adjustments could involve her sense organs and properties of her nervous system.

It should be noted that applying the stimulation version of the SAC hypothesis to genital evolution has complications that have not been previously noted.

The constraints on female responses to male manipulations that are posited by Arnqvist's model (2006) are likely to be relaxed in genital evolution. This is because the female response that the male is attempting to alter (e.g., ovulation, oviposition, inhibition of remating, etc.) is the same response under which her sensitivity originally evolved. Or, from the female's perspective, the message the female is under natural selection to obtain from the male's stimuli remains the same: "copulation has occurred." Thus only a small, presumably easy to evolve adjustment in the female's sensitivity would be needed to counteract the disadvantageous effect ("overly emphatic" responses to the male's signal) of a new male stimulus. Female adjustments to new male stimulatory adaptations could presumably be as simple as adding or subtracting a few synapses in her CNS, giving her the ability to retain the original function and also counteract the male-induced damage. This does not eliminate possible male–female SAC, but it implies that the durations of the periods when females are suffering costs from male sensory trap manipulations will tend to be brief. A similar consideration applies to at least some male manipulations by use of hormonal substances (signals) in his semen, unless they are also used in other contexts or have other side-effects in her body (Cordero & Eberhard 2005).

TESTING THE CFC AND SAC HYPOTHESES

The controversy between CFC and SAC explanations of genital evolution is part of a more general controversy currently swirling in discussions of sexual selection that concern many phenomena that were previously attributed to female choice (e.g., Pizarri & Snook 2003; Chapman et al. 2003; Kokko et al. 2003; Cordero & Eberhard 2003, 2005; Arnqvist 2004; Arnqvist & Rowe 2005). The major contrast between the two hypotheses revolves around the payoffs that a female obtains from resisting the sexual attentions of some of the males (Arnqvist & Rowe 2005). CFC presumes that she benefits from obtaining increased offspring quality. Such gains are thought to outweigh possible losses in direct reproduction (numbers of offspring) from male effects and the process of rejection itself; it can result in female behavior and morphology that is designed to give selective *cooperation* with males. SAC, in contrast, presumes that the female gains in the number rather than quality of her offspring from resisting males, and that these gains outweigh potential losses from the process of rejection, and from her inability to screen males and thus increase the quality of her offspring (Arnqvist & Rowe 2005; Cordero & Eberhard 2005); SAC should result in female behavior and morphology that is appropriate for non-selective resistance to males, rather than selective cooperation. On the male side, trait exaggeration under CFC is impelled by female response criteria that evolve to increase offspring quality, while under SAC male trait exaggeration is impelled by the evolutionary responses of the female to these costs (Arnqvist & Rowe 2005). Resolution of the controversy for genitalia may point the way toward more general conclusions regarding sexual selection.

Discriminating between CFC and SAC explanations of genital evolution is difficult. The two hypotheses are not mutually exclusive (Cordero & Eberhard 2003, 2005; Hosken and Stockley 2004; Eberhard 2004b; Arnqvist & Rowe 2005). In addition, some predictions are the same for both, and direct measurements of some crucial variables involved in balancing potential costs and benefits is technically very difficult, if not impossible (Cordero & Eberhard 2003) (see final section of this chapter). The two types of selection can reinforce each other, or act against each other with respect to a given female trait, and they could act at the same time or in sequence on a particular trait (Cordero & Eberhard 2005; Eberhard 2004b; Arnqvist & Rowe 2005). For example, in the SAC example above, the original female payoff from evolving a defense against male genital scraping such as a thicker lining in her reproductive tract could be that it enabled her to avoid damage to her reproductive interests inflicted by his genitalia (a SAC type payoff); but she could also benefit, via superior sons, if the thicker lining also resulted in a bias that favored the males that were more potent manipulators (CFC-type payoffs) (Cordero & Eberhard 2005). Either type of payoff (or both) could be involved, for example, in the tendency for relatively high penile spinosity in male primates to be associated with relatively short durations of female receptivity within the ovarian cycle (Stockley 2002). Even this complex example of male damage to the female reproductive tract may be oversimplified compared with the real world; the females of a bruchid beetle that are damaged in this way also benefit, at least in terms of fecundity, from longer copulations (which may result in more male damage) (Edvardsson & Canal 2006; Eady et al. 2007).

SUPPORT FOR CFC AND SAC

One prediction made by CFC and SAC is that the frequency of female remating in different groups should tend to correlate positively with the rate of genital divergence in that group (Eberhard 1985; Arnqvist 1998). If females consistently mate with only a single male (strict monandry), then CFC among males is not possible. Conflict between male and female that could lead to SAC may also be reduced or eliminated by female monandry, especially if the male is also monogamous (in which case conflict should not occur, and male and female genitalia should not coevolve). The SAC prediction is somewhat less sweeping, however. If females can benefit from polyandry but the polygamous males "impose" monandry via use of their genitalia, then conflict could occur even in a species in which nearly all females are strictly monandrous. In addition, conflict is possible even if monandry is not imposed by the male. For instance, if the male provides the female with some resource that is in short supply (e.g., a large nutritious ejaculate), a polygamous male might provide the monandrous female with less than she wants. Whether this sort of conflict could ever play out in genital morphology (e.g., the female attempts to induce greater male contribution) is not clear, but it has been inferred in non-genital female copulatory courtship in a fly (Ortiz 2002).

Thus CFC clearly predicts that groups with strictly monandrous females should have genitalia that are not species-specific in form, while the SAC expectation is also for a bias toward lack of species-specificity. Possible correlation between female monogamy and genital divergence has been tested in 22 different groups, including termites (males also monogamous) and *Heliconius* butterflies (Eberhard 1985; Arnqvist 1998), bees (Roig-Alsina 1993) primates (males polygynous in some, monogynous in others) (Dixson 1987, 1998; Verrell 1992), *Ischnura* dragonflies (Robinson & Novak 1997; Simmons 2001), mole rats (Paraq et al. 2006), and in 16 other groups of insects (Arnqvist 1998). The predictions of reduced genital divergence were fulfilled in all cases, despite several complications. The predictions concern *rates* of genital divergence, while the data in most cases involved *amounts* of divergence, and in some groups the behavioral data were not complete (e.g., Arnqvist 1998). There were generally no controls for the amount of time since divergence, although Arnqvist's (1998) finding that genitalia but not other structures correlated with the frequency of female remating suggests this was not a problem in his study. Another possible problem is that the particular morphological aspects of genitalia that were studied were chosen at least in part because they were easier to study; there was no guarantee that they are the aspects that most strongly influence female responses (CFC) or do the most damage to her (SAC). These weaknesses make the consistent confirmations even more impressive.

It should also be noted, however, that data from one group, the bumblebee genus *Bombus*, contradict the predictions. The male genitalia are quite complex and strongly divergent among 18 species of *Bombus* (Richards 1927) (figure 4.1), but contrary to the prediction of CFC, females are thought to be strictly monandrous in at least seven of eight species of *Bombus* on the basis of both molecular and behavioral evidence (Schmid-Hempel & Schmid-Hempel 2000). Expectations of the physical coercion version of SAC are also contradicted in *Bombus* because some of the species-specific portions of the male's genitalia contact a relatively featureless portion on the external surface of the female's abdomen (figure 4.1) (Richards 1927; comment by O.W. Richards in Alexander 1964). There is thus no sign of the expected female defensive coevolution that could have selected for the divergence in the males.

Partial confirmation of CFC comes from the correlation between differences in male genital morphology and paternity when a female mates with more than a single male in six species: two in the water strider genus *Gerris* (Arnqvist & Danielsson 1999; Danielsson & Askenmo 1999); two distantly related scarab beetles, *Onthophagus taurus* and *Anomala orientalis* (House & Simmons 2003; Wenninger & Averill 2006) (figure 4.4); the damselfly *Calopteryx haemorrhoidalis* (Cordoba-Aguilar 1999, 2002, 2005); and the chrysomelid beetle *Chelymorpha alternans* (Rodriguez et al. 2004). In addition, experimental modifications in the sepsid fly *Archisepsis diversiformis* of both the morphology of a non-genitalic contact courtship organ on the male's front leg, and of the female's ability to sense this organ reduced the likelihood of female acceptance of copulation (Eberhard 2002a), and experimental modifications of male genital structures and female receptors that they contact during copulation in the tsetse fly *Glossina pallidipes* affected female cryptic choice mechanisms such as ovulation, sperm transfer, and tendency to remate as predicted by CFC theory (Briceño & Eberhard 2009).

These cases support CFC, but possible SAC cannot be ruled out in five of the seven. Very little is known about how the male genital structures that correlate with paternity are used in *Gerris*, and a study of genital function in *O. taurus* failed to even consider the possible role of female stimulation by male structures that pinch her at several sites and are thrust up her rectum (Werner & Simmons 2008). In the oriental beetle, *A. orientalis*, the male sclerite that affected paternity hooks the female just inside her vagina, where it is likely to stimulate her and may also provide purchase for deeper thrusting by other, inflatable portions of his genitalia (Wenninger & Averill 2006). Possible damage to the female was not checked.

In *C. alternans*, the increased paternity associated with greater length of one male genital structure, the effects of experimental shortening this structure, morphological studies of how male genitalia engage the female during copulation, and the dramatic variation in the ducts of females of different species (Rodriguez 1994; Rodriguez et al. 2004), suggest that mechanical fit in the female's rigid, tortuous spermathecal duct, rather than stimulation, may be an important determinant of sperm precedence. Sperm is also deposited, however, outside the duct in the female's bursa, and its significance remains unclear. No male-inflicted damage to female reproduction (as predicted under SAC) is known, and the highly scleritized spermatheca duct seems unlikely to be damaged by the male; nevertheless damage has never been searched for, and might occur deeper in the female (e.g., the spermathecal valve) (D. Windsor personal. communication).

SAC is very unlikely, however, in the two other cases. In the fly *A. diversiformis* the male's clamp fits very precisely with the female's wing (Eberhard 2001a), but experimental modification of the form of the male's clamp did not impair his ability to hold on to the female with his front legs for extended periods, despite shaking behavior by the female (Eberhard 2002a), arguing against a SAC interpretation. Female stress receptors occur in the area

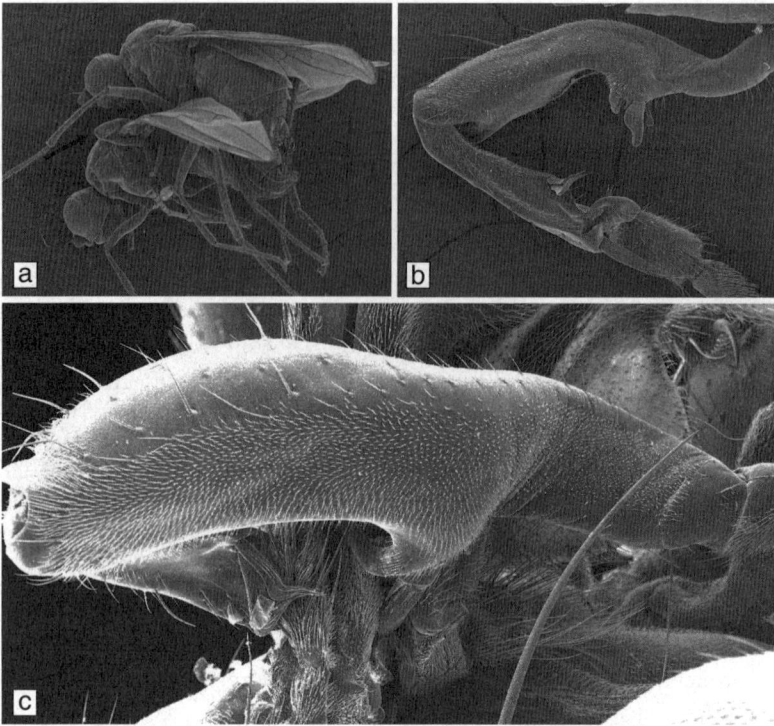

FIGURE 4.4 A detailed understanding of how the spectacularly elaborate, species-specific non-genital male foreleg clasper function permits confident rejection of a SAC explanation for male foreleg morphology in the appropriately named sepsid fly *Themira superba*. The males of this and other sepsid flies clamp the base of the female wing prior to copulation with their modified front legs (arrow in (a)). The form of the male foreleg is elaborate and species-specific (b), and the tibial and femoral modifications fit against the stem and costa veins in the base of the female's wing (c). Despite the striking diversity of male forms, female wing designs are quite uniform in this genus, and indeed throughout the entire family, and they show few signs of resistance structures that might explain the male diversity. Experimental alteration in one species of male foreleg morphology (or of female wing touch receptors) did not reduce the male's ability to hold on; instead, females rejected male copulation attempts (Eberhard 2002). (from Ingram et al., 2008; courtesy of R. Meier).

contacted by the male's front leg in this species (Eberhard 2001a) (as well as in other sepsid species with species-specific male front legs—Ingraham et al. 2008) (figure 4.4), and could thus enable her to sense his grip, supporting a CFC interpretation. The female's wing base is quite sturdy, and there were no signs of damage in *A. diversiformis* (a possible prediction of SAC) (damage inflicted by male claspers to female wings was claimed by Mühlhäuser and Blankenhorn (2002) in another sepsid with similar male grasping devices; but the wing damage that they observed was in other parts of the wing, and likely occurred when female flies beat their wings against the walls of their small glass containers—see Baena and Eberhard 2007). In addition, in only one of the >10 sepsid species that have been checked (in *Archisepsis*, *Microsepsis*, *Palaeosepsis*, *Sepsis*, and *Themira*) is there any even potentially defensive modification of the female's wing in the area where the species-specific modifications of the male's front legs grasp her (Eberhard 2001a, 2005, unpublished; Ingram et al. 2008).

In the damselfly, the male manipulation of the female (he replaces the sperm of a previous male with his own in the female's reproductive tract) is not likely to inflict the types of naturally selected costs (reduction in numbers of offspring) to the female that are specified by SAC theory (Chapman et al. 2003; Arnqvist & Rowe 2005). Some aspects of this case are still puzzling under both SAC and CFC, however. The male genital trait (penis width) shows no sign of the extravagant elaboration that is often associated with genital evolution. In addition, penis width varies both geographically and seasonally in this and in another species in which it also affects the male's ability to remove sperm (Cordoba-Aguilar 2009).

These studies have some possibly important limitations. Except for the chrysomelid, sepsid and damselfly studies, only correlations were demonstrated, rather than cause and effect relationships. The possibility thus remains that paternity was actually affected directly by other, correlated variables rather than by genital form itself. In the chrysomelid study no control was devised for the effects of the operation itself (it was not feasible to cut the male's genitalia and then glue them back together). On the other hand, the tests in all species were conservative in that they did not take into account possible effects of male genitalia on many additional female reproductive processes, such as decreased remating, increased oviposition, etc.

One type of evidence that clearly supports CFC but is incompatible with physical coercion versions of SAC comes from a growing number of observations of genitalia used in ways that are appropriate to stimulate the female, but not to physically coerce her. Probably the genital behavior that is least controversial is stridulation, which has been observed directly in the tipulid fly *Bellardina* sp. (Eberhard & Gelhaus 2009) (figure 4.5), and inferred from male morphology in moths (Forbes 1941; Gwynne & Edwards 1986) and wasps (Richards 1978). Copulation in various mammals also involves genital behavior that is apparently designed to stimulate the female (summary Eberhard 1996; see also Dixson 1998), and some aspects of copulation behavior in rodents correlate with indicators of increased probability of competition with sperm from other males (Stockley & Preston 2004). In addition, the male genitalia of several insects and spiders perform long, highly rhythmic series of taps,

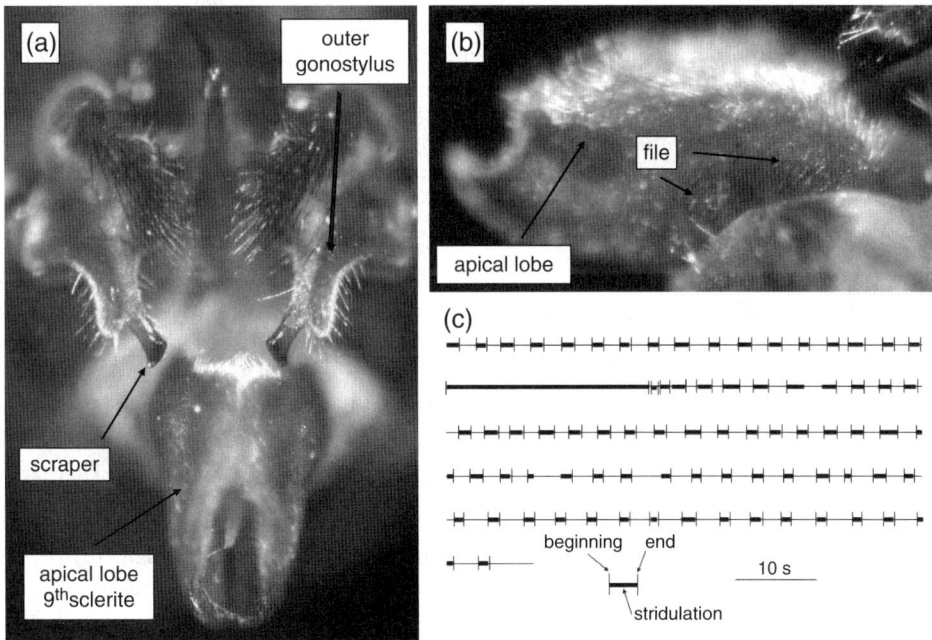

FIGURE 4.5 Male genital structures whose function to stimulate the female seems incontrovertible—the scraper (a) and file (b) of the male genitalia of the tipulid fly *Tipula* (*Bellardina*) sp. Direct behavioral observations show that the scraper is rubbed against the file to produce a highly rhythmic "song" (c) during copulation (from Eberhard & Gelhaus 2010).

or squeezes on membranous portions of the female, that also suggest that stimulation of the female is important; these include a dryomyzid fly (Otronen 1990), a buprestid beetle (Eberhard 1990), a sciarid fly (Eberhard 2001c), sepsid flies in several genera (Eberhard & Pereira 1996; Eberhard 2001b, 2003, 2005), a pholcid spider (Huber & Eberhard 1997; Peretti et al.2006), some scathophagid flies (Hosken et al. 2005), several species of tsetse flies (Briceño et al. 2007; Briceño & Eberhard 2009), and the hesperiid butterfly *Urbanus dorantes* and the katydid *Idiathron* sp. (W. Eberhard unpublished). In those groups in which details of the genital behavior of more than one species are known (the spider, tsetse flies, and the sepsid flies *Microsepsis* and *Archisepsis*), the temporal patterning of squeezes differs among congeneric species, as is likely if this behavior is under sexual selection by CFC (A. Peretti, personal communication; Briceño & Eberhard in 2009; Eberhard 2001b; Eberhard & Pereira 1996). Alternative SAC interpretations based on physical coercion can be discarded in some of these cases. Morphological considerations rule out direct male effects on female internal genital structures with squeezing behavior in the sepsids and the crane fly. Possible external physical damage to the female resulting from male movements may occur in some tsetse flies and the pholcid spider, but not in the sepsids, tsetse flies, or the katydid (data are not sufficient to judge in the others).

The stimulation version of the SAC hypothesis cannot be ruled out, however, because it is possible that male stimulation of the female sometimes leads to reproductive losses to the female, especially when males are using sensory traps (Arnqvist 2006). Female counter-measures to male stimuli could occur in her sense organs or her nervous system, and thus be invisible externally. If such a coevolutionary struggle between males did not "spill over" into battles involving physical coercion, it could not be observable in studies of external morphology.

The strongest support for SAC in genitalia comes from water striders in the genus *Gerris*. Dorsally projecting spines near the female's genitalia are elongated to different degrees in different species, and have independently become especially elongate in *Gerris incognitus* and *G. odontogaster*. Longer female spines impede male attempts to clamp the tip of the female's abdomen with his genitalia (Arnqvist & Rowe 2002a,b; Rowe & Arnqvist 2002) (clamping the female's abdomen helps the male hold on during her energetic struggles to escape after he mounts, and is a necessary prelude to intromission. There is a cross-specific correlation between the relative development of several different male structures, including elongate grasping male genitalia, and the relative development of female defensive structures. An independent contrasts analysis based on a robust phylogeny showed that changes in male and female traits (both genitalic and non-genitalic) probably coevolved. Even in *Gerris* CFC cannot be ruled out, however. The possibility that male genitalia have additional, stimulatory effects on females has never been checked (e.g., by inactivating sense organs at the tip of her abdomen). In addition, the expectation that such a clear case of SAC would lead to morphological diversity in males and females is less clearly fulfilled. The morphological designs of both sexes of *Gerris* differ somewhat among species, but both male and female structures are relatively simple and practical. A morphologically similar abdominal spine that can fend off males also occurs in female *Aquarius paludum*, but its functional interpretation is not clear, because spines also occur in males (where they are proportionally longer), and female fertility is increased rather than decreased in captivity by additional matings (Ronkainen et al. 2005). Finally, it may be that male–female interactions in water striders are not typical of those in other groups, because their essentially 2-dimensional world may make male harassment of females unusually feasible (Eberhard 2006).

There are several other possible cases of less complete support for SAC. In *Lucilia* blowflies, complex, species-specific male genital asperities (Aubertin 1933) rub holes in apparently defensive thickenings in the lining of the female's reproductive tract (Lewis & Pollock 1975; Merrett 1989). Species-specificity in female morphology and the question of whether female reproduction is actually reduced by copulatory damage both remain to be checked, however. In addition, the possibilities remain that stimulation (which seems likely) induces female responses favoring the male, and that females gain by producing superior sons as a result of the thickened lining, so CFC cannot be ruled out.

Summarizing, few species give evidence that compellingly discriminates between the CFC and SAC hypotheses for genital evolution. I think the clearest data favoring CFC over SAC come from the front leg grasping organs of sepsid flies, the cercal claspers of tsetse flies, and from some species with male genitalia that are obviously designed to stimulate the female. The strongest support for SAC comes from *Gerris* water striders, but CFC has not been ruled out in these animals.

DISCRIMINATING BETWEEN SAC AND CFC

There are several other contexts in which SAC and CFC predictions differ. The massive data bank on genital evolution available in the taxonomic literature permits one to utilize huge sample sizes, and tests using these data are to my mind the most powerful evidence available regarding the likely generality of SAC and CFC explanations for genital evolution. Four different tests (all involving >100 species) have been made.

1. Comparing Groups in Which Males Can and Cannot Coerce Females to Mate

The most extensive test of SAC predictions regarding genitalia, in terms of the numbers of species included (up to several hundred thousand, depending on how one adds them up), is based on a prediction formulated by Alexander et al. (1997). They distinguished between coercive and non-coercive circumstances in which males attempt to obtain copulations. Grasshopper males were cited as mating coercively, because they often jump onto females which are engaged in other activities without any preliminaries, and attempt to grasp the female's genitalia with their own. Females often struggle forcefully to dislodge males and to prevent genital coupling. The cricket genus *Gryllus* was cited as not mating coercively, because males produce a calling song and the receptive female, with no overt coercion by the male, approaches the male and positions herself to allow him to couple with her. The female cannot be physically coerced, because she only encounters the male if she seeks him out. She is thus protected from unwanted male attentions. Alexander et al. reasoned that SAC in male and female genitalia would be more likely to occur in a group like grasshoppers in which male and female interests are more clearly in conflict — those in which females are less protected and in which male coercion occurs. Grasshoppers and *Gryllus* fit their prediction: male genitalia are often

species-specific in grasshoppers, while they are not divergent and not useful to distinguish species of *Gryllus* (Alexander et al. 1997).

A sample of two, of course, is not very convincing, and I undertook a larger survey (Eberhard 2004a), using information from the behavioral ecology and taxonomic literatures. Discriminating between SAC and CFC is possible, because CFC suggests that no trend should occur: female use of male genitalia to bias paternity could occur equally well in species with protected or unprotected females (unless unprotected females are more likely to be monandrous due to male manipulations, in which case the prediction would be the opposite— greater genital divergence in non-coercive mating systems).

First, publications on the behavior and ecology of insects and spiders were consulted to determine whether or not females of different groups were likely to be coerced into mating by males. Protection of females from coercion was assumed in species in which males attract females by chemical signals or singing, females attract males with attractant pheromones, females emit light signals at night in response to light signals from the males that allow the male to find them, males form leks or swarms that are not associated with resources needed by females such as oviposition or feeding sites, and in spiders in which males are dwarfs in comparison with females (and the female can thus easily kill a harassing male). In contrast, species in which females are not protected from harassment included those in which males station themselves near oviposition or feeding sites and attempt to mate with arriving females, and those in which males station themselves at sites where females are emerging from pupae and mate with them while they are still relatively defenseless. Second, for each genus in which behavioral evidence suggests that females are consistently either protected or unprotected, the taxonomic literature was then consulted to determine whether male genitalia are or are not useful in distinguishing closely related species.

The data clearly did not conform to the SAC prediction that the male genitalia in groups with unprotected females should diverge more rapidly, and thus that these groups would tend to more often have species-specific male genitalia (figure 4.6a). Analyzed in terms of genera, 75.4% of 223 genera with protected females have species-specific male genitalia, while 68.8% of 105 genera with unprotected females have species-specific male genitalia (data from 113 families in 10 orders). The difference is not significant, $X^2 = 1.82$, d.f. = 1, $p = 0.17$), and is in any case in the opposite direction from that predicted by SAC. Several modified analyses that attempted to correct for possible biases in the data (over-use of genitalia by taxonomists due to custom, under-use of genitalia due to the difficulty of studying them, inadvertent bias in groups that were included in the study, and phylogenetic inertia) also failed to result in the predicted trend (figure 4.6a). More taxonomically restricted analyses of groups, such as the large fly family Chironomidae in which additional behavioral details increase the confidence of the lack of probable male–female conflict, also failed to fit the SAC prediction.

The data in figure 4.6a strongly underestimate the strength of the evidence against SAC, because data from the large order Lepidoptera (which includes something like 250,000 species) were omitted because they are so uniform. Female lepidopterans are nearly all protected from pre-copulatory male coercion, because females throughout the order attract males with long distance attractant pheromones (Phelan 1997). And, contrary to SAC predictions, the genitalia are elaborate and species-

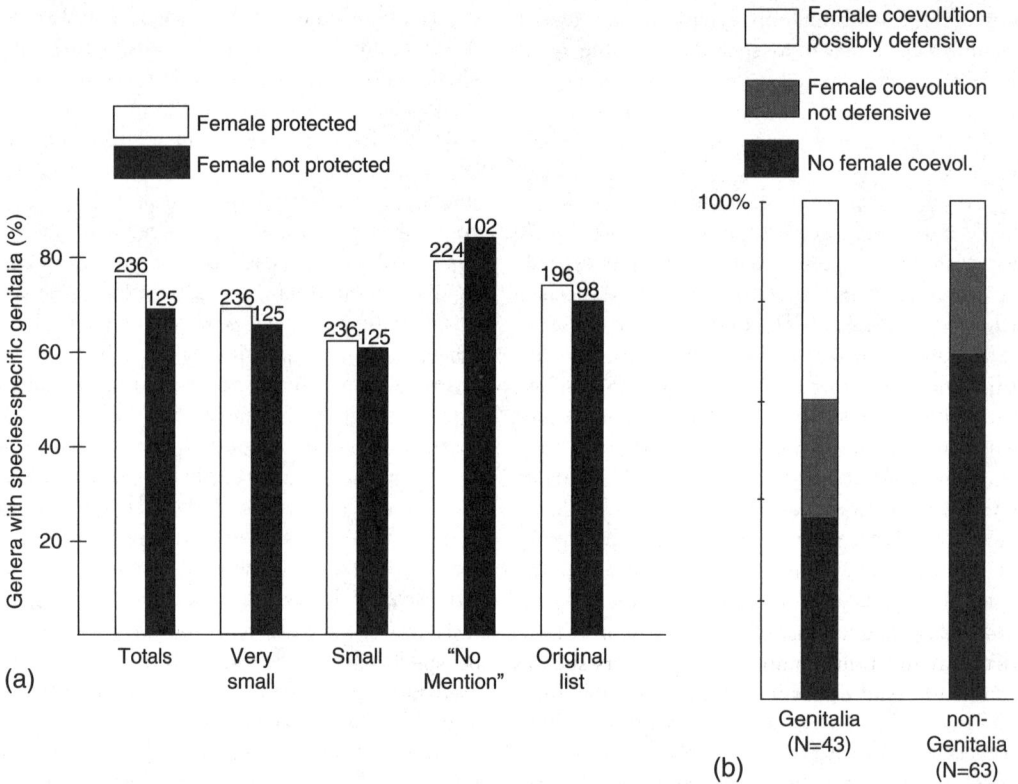

FIGURE 4.6 Summaries of two large survey studies that documented failures to confirm predictions of the SAC hypothesis. (a) Percentages of genera in which male genitalia are and are not species-specific in groups in which non-receptive females are and are not protected from sexual harassment by males. The totals (left pair of bars) include all groups examined; the other pairs of bars represent data that were modified in different ways to attempt to take into account different possible biases in the data against SAC predictions (see text) (numbers at tops of bars are area sample sizes). The SAC prediction that the dark bars would be higher was not confirmed. (b) Conservative estimates of fractions of the 84 taxonomic groups with species-specific male genitalia (left) and non-genital contact structures (right) that did (white) and did not (grey and black) conform to SAC predictions of species-specific defensive female coevolution (a) from Eberhard 2004a; (b) based on data from Eberhard 2004b.

specific in form throughout the order, as shown in taxonomic compendia that review thousands of species in the North American and Palaearctic fauna (Dominick et al., 1971–1998; Amsel et al. 1965–2000; Forster & Wohlfahrt 1952–1981; Huemer et al. 1996).

A possible problem with these results is that the SAC prediction of Alexander et al. (1997) may be overly simple. The reduction in male–female conflict in species with protected females may not be complete, even in species in which no male–female contact occurs unless the female is receptive. This is because once a pair has formed, the male could attempt to manipulate post-copulatory female behavior such as remating or oviposition, and thus reduce female reproduction. Even though a female was receptive to copulation, her reproductive interests might be damaged by such manipulations, and she might evolve to reduce this damage from the genitalia of manipulative males. To estimate how frequently different species-specific male genital structures function in these possibly conflictive ways, I made a separate literature survey of studies of the functional morphology of male genitalia in the order Diptera. The results indicated that the SAC prediction of Alexander et al. is likely to be a strong trend rather than absolute. Of 105 cases in which a function was attributed to species-specific male genital structure (in 43 species in 22 families), the majority (85.7%) were functions in which male–female conflict should be reduced or absent in species with protected females (39.0% apparently function to clasp the female, and 46.7% to facilitate penetration and sperm transfer) (Eberhard 2004a). The precise percentages are probably not especially meaningful, because of several probable biases (including the ease of documentation of these particular functions compared with others such as stimulation; and a bias in the set of possible functions considered by the authors). But the percentages clearly show that the SAC predictions should be met in an appreciable number of genital traits. Even if Diptera are somehow unrepresentative of other insects in this respect (there is no obvious reason to suspect this), the SAC prediction failed when Diptera were analyzed apart from others (Eberhard 2004a). The survey was thus a valid test of SAC predictions.

In sum, data from literally hundreds of thousands of species failed to show the trend predicted by SAC; if anything, the trend was in the opposite direction. The immense number of species in this sample, made possible of course by the huge taxonomic literature on genitalia, is rare in evolutionary studies. A sample of this size should have been sufficient to reveal even a weak trend in the predicted direction, so the lack of this trend constitutes strong evidence against SAC as a general explanation.

2. Female Defensive Coevolution with Males

A second broad survey (Eberhard 2004b) examined a different set of predictions in 61 families, mostly of insects and spiders, in which the functional morphology of species-specific male structures has been studied. Species were only included if morphological studies have determined both the site on the female that is contacted by the species-specific portions of the male structure and the mechanical details of the fit between them. The sample included 43 male genital structures in 34 taxonomic groups, and 63 male non-genital contact structures in 53 taxonomic groups. SAC on the basis of physical coercion (Alexander et al. 1997; Arnqvist & Rowe 2002a, b) makes several clear predictions for these structures: the female morphology should often coevolve with the species-specific aspects of the male; the species-specific female structures of related species should interact mechanically with the species-specific portion of the male; and the designs of the species-specific aspects of the female structures should often be appropriate to defend her against the male, especially against the action of his species-specific structures. Female structures that can hold the male away or impede his access are predicted to be common. Finally, because females under SAC need to mate at least once but resist other males, an especially likely design would be species-specific female structures that can be used facultatively against males. Moveable structures such as erectable spines, inflatable sacs, or sliding barriers that could be moved out of the way to facilitate one (necessary) copulation, but interposed to reject others are expected.

CFC, in contrast, predicts that external female morphology will often (but not always) not vary when females are screening males on the basis of the stimuli they produce. Rather, females are expected to coevolve with respect to their sense organs (sometimes visible externally, as in some damselflies—Roberterson & Paterson 1982; Battin 1993—but often not), and with respect to how their CNS processes information from these sense organs (completely invisible externally) (see below). Females can also screen males on the basis of their morphological fit with the female, and in these cases male–female morphological coevolution is expected. In addition, the designs of females are expected to be often "selectively cooperative" (figure 4.7, below), rather than defensive as is expected with SAC.

The assembled groups were then checked for female traits. Once again, the SAC predictions clearly failed. Of 106 structures in 84 taxonomic groups, in more than half (53.8%) (figure 4.6b) female morphology was inter-specifically uniform while male morphology was species-specific (the respective percentages for genitalic and non-genitalic structures 34.9% of 43, and 68.3% of 63).

In addition, the designs of over half of those female structures that did coevolve with species-specific structures of males did not have the predicted defensive designs: among 49 coevolving female structures in 39 taxonomic groups, they were not even feasible as defensive devices in 55.1% of the structures (57.1% of 28 genital structures and 52.4% of 21 non-genitalic structures). The female designs seem to be selectively cooperative in many species (grooves and furrows used by a male with the appropriate design as sites to support or strengthen their grip on the female) (figure 4.7) rather than defensive. In total, females failed to confirm to SAC predictions in 79.2% of 106 structures (figure 4.6b). This finding that female morphology frequently fails to coevolve with that

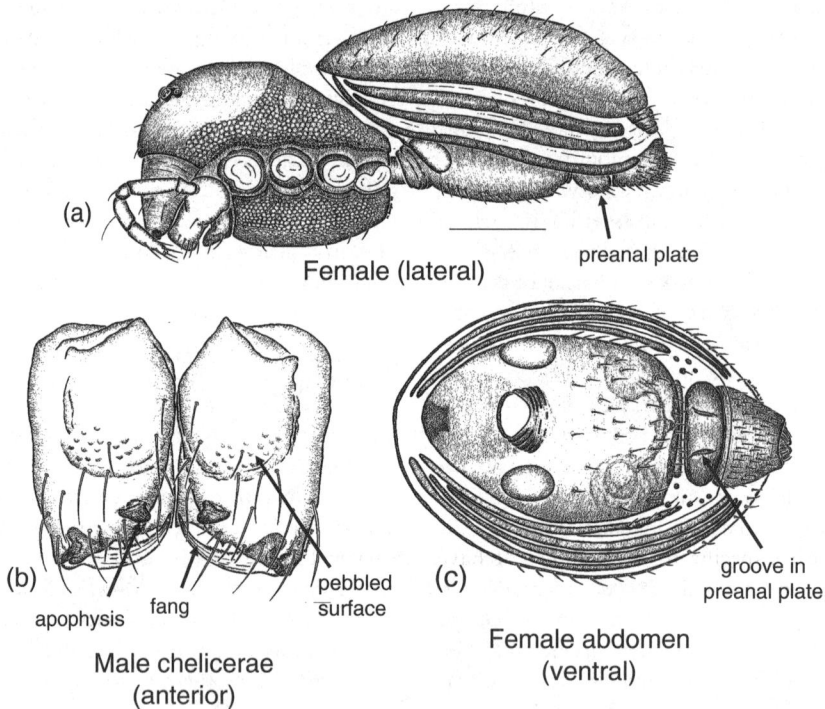

FIGURE 4.7 A recently discovered illustration in the tiny armored spider *Indicoblemma lannaianum* of the general trend for females to have selectively cooperative rather than defensive structures. Males of this genus are distinguished by bearing apophyses and other modifications on the anterior surfaces of their chelicerae (b) that are species-specific in form. Males use these projections to grasp the preanal plate of the female (a) during copulation. The female's preanal plate has "selectively cooperative" grooves (c), which facilitate rather than impede the male's grasp with his fangs. (scale line in (a) = 0.2 mm; drawings and behavioral observations after Burger 2005).

of males echoes the findings of previous surveys performed to test possible lock and key functions in the genitalia of other arthropod groups (Robson and Richards 1936; Kraus 1966; Eberhard 1985; Shapiro & Porter 1989; for further examples, see Djernaes et al. in preparation, and the discussion of lock-and-key above).

In addition, the female design that constituted arguably the strongest prediction by SAC, facultatively defensive structures, was completely absent (0% of 106). A search for defensive designs in an additional, large set of spider species (in which drawings of the female genitalia are routinely included in taxonomic descriptions) failed to reveal a single example of such a defensive device that could be facultatively deployed among the descriptions of thousands of species in general faunal studies and recent reviews (Eberhard 2004b) (see also Eberhard & Huber chapter in this book).

Data on these topics are more limited for other taxonomic groups. The recent discovery of coevolution between male and female genitalia in waterfowl (Brennan et al. 2007) fits SAC predictions better than most of the arthropod data. In some mammals female genital morphology has also coevolved with male penis morphology (Coe 1969; Patterson & Thaeler 1982). Nevertheless, lack of female coevolution with male morphology is common in some other groups with species-specific male genitalia. The bursae of male nematodes and the spermatophores of scorpions are often species-specific in form, but the areas of the female's body that they contact seem not to differ between species (Chitwood & Chitwood 1974; Peretti 2003). Antagonistic female coevolution of female genital morphology with male morphology is also apparently absent in primates, a group with numerous elaborate, species-specific male genitalia: "I have been unable to identify a single case among the primates where the mechanical conflict of interest hypothesis might be applicable" (Dixson 1998: p. 247). Clearly, the predicted defensive female coevolution with males is not a general rule.

It might be possible to rescue the physical coercion version of the SAC hypothesis from these apparently contradictory data in at least some species if it turned out that in the many species in which females that lack species-specific defensive morphology, the females instead use species-specific defensive behavior that selects for diversity in male contact structures (Eberhard 2004b). To my knowledge, however, not a single case of such female

behavior has ever been documented (though female behavior may seldom be studied with sufficient detail). Such a rescue is ruled out by the details of male–female interactions in several of the 84 taxonomic groups (Eberhard 2004b). In 21 genera, species-specific female resistance that could select for the species-specific designs of males is either mechanically impossible or female behavior has been observed with sufficient detail to rule it out (Eberhard 2004b). In nine other genera, it is the female that approaches the male and actively maintains contact with him, rather than vice versa; she is thus free to break away at any time, so female "resistance" behavior is simply not biologically realistic (Eberhard 2004b) (see figure 4.2). One further reason to doubt that as yet unstudied the female behavior will rescue SAC is that it is not clear why females should so often fail to use potential morphological counter-adaptations to males, and rely instead on behavior. Simple spines like those found in some *Gerris* females, for instance, would seem to offer relatively cheap, simple, and effective defenses to females. The stimulation version of the SAC hypothesis is less clearly contradicted, because it predicts only occasional rather than consistent coevolution of the female's morphology, and is thus compatible with the many cases in which such coevolution has not occurred (figure 4.6b).

3. Evolutionary Patterns When Males Inflict Damage on Females

I examined the CFC–SAC controversy over genital evolution from still another angle, that of groups in which current knowledge indicates that male genitalia are especially likely to inflict damage on females. I found 16 groups of insects in which male genital damage to females has evolved independently (Eberhard 2006, plus the recent discoveries of traumatic insemination in mirid bugs—Tatarnic et al. 2005, and *Drosophila* flies—Kamimura 2007). Damage included traumatic insemination (the male punctures the female's exoskeleton and introduces his sperm and seminal fluid into her body cavity), producing perforations of her exoskeleton or internal organs grasping her, or clasping with the genitalia or (in one case) other specialized male structures that increases her susceptibility to predation or decreases her ability to feed. I then consulted the taxonomic literature on these groups to determine whether the male traits that are used to do the damage have undergone sustained divergent evolution, and whether females have evolved defensive morphology against these male traits, as expected under SAC (I made the usual assumption that males can impose at least some copulations on females; the predictions of SAC are weaker to the extent that such coercion is not possible). CFC, in contrast, predicts at least some selectively cooperative female designs in these groups.

The data gave one weak confirmation and two rejections of the SAC predictions. The prediction that male genitalia or grasping organs would evolve relatively rapidly and divergently in these groups was confirmed. Taxonomists of these groups have generally used the morphology of these damage-inflicting structures to distinguish congeneric species (there were two clear exceptions). If one counts (conservatively) any family which has at least a few genera in which genitalia are species-specific as being families that are typified by species-specificity, then 16 of 18 families show rapid divergent genital evolution. (Eberhard 2006). This fraction is higher than that of 71% of 328 genera in the general survey described above (Eberhard 2004a), although the difference is not statistically significant (p = 0.12 with X^2 Test).

Two other predictions, however, were not confirmed. With two clear exceptions (*Lucilia* and *Drosophila* flies), the male structures showed only modest complexity, and relatively small differences between congeneric species, compared with similar structures in groups in which male damage to females has not been documented. The trends to simplicity and small differences were especially clear in two relatively large groups with traumatic insemination, cimicoid bugs and Strepsiptera. The male genitalia of both these groups are secondarily reduced and highly simplified, and have entirely lost structures that were present ancestrally (Eberhard 2006). Male designs are typically utilitarian: for instance, cimicid bugs have simple, sword-like genitalia that are obviously well designed for penetrating the exoskeleton of females. Interestingly, this male evolutionary conservatism contrasts with the evolution of male structures known to function as weapons in male–male battles. Both species-specificity and diversity of design is typical of beetle horns, ungulate horns and antlers, and earwig cerci (Arrow 1951; Geist 1966; Otte & Stayman 1979; Enrodi 1985; Brindle 1976). This contrast is especially striking given the fact that both sets of male traits often function to solve similar mechanical problems, such as grasping and stabbing another animal.

Finally, the SAC prediction that females would possess species-specific defensive structures at sites contacted by males, was clearly not fulfilled. In most groups (with four and possibly five exceptions —*Gerris* water striders, dytiscine water beetles, *Coridromius* plant bugs, *Drosophila* flies, and perhaps *Lucilia* blowflies) female morphology in the area contacted by the male's piercing genitalia or grasping structure was not species-specific (Eberhard 2006; Tatarnic et al. 2005; Kamimura 2007). Female morphology was also generally not defensive in design, in the sense that it lacked design features that could potentially prevent the undoubted physical damage inflicted by traumatic insemination.

Females of the six *Drosophila* species known to have wound-producing male genitalia have small "pockets" into which the penetrating portions of the male genitalia fit, but the clear photos of Kamimura (2007) show no sign of any thickening or sclerotization that would make penetration more difficult, and that would thus select for changes in the male genitalia as predicted under SAC. In fact, the body wall is "especially thin" at the bottom of the pockets in the species complex in which four species of males perform traumatic insemination using divergent genital structures (Y. Kamimura, personal communication.). In some cimicoid bugs and orthoptera, females instead have structures such as grooves or pits that guide the male and give him greater purchase on the female, and thus appear to be "selectively cooperative" instead of defensive as expected under SAC. Female "mimicry" of certain male designs in one cimicid strongly suggest SAC, however (Reinhardt et al. 2007).

In some cimicoid bugs, and perhaps also in strepsipterans, females have diverse internal structures at sites where males penetrate, suggesting that instead, females have evolved internal mechanisms to control sperm (as expected under CFC) or seminal products or invasive pathogens, rather than to avoid the physical damage and infections that result from copulation itself. Lack of external defenses is not be predicted by SAC. The damage to the female comes from the act of insemination (physical injury to the female's tissues, and the increased risk of infection) (Stutt & Siva-Jothy 2001; Morrow & Arnqvist 2003), and to defend against physical damage, females would be expected to evolve defenses against penetration per se. Females could evolve internal defenses against infection at the site of insemination, but such a defense might not set off a coevolved race between males and females, because males would gain nothing (and probably lose) from improving their ability to infect their mates with venereally transmitted pathogens. Coevolution with such internal female defenses could occur, however, if they also killed the male's sperm. This "classic" example of male–female conflict may have a cooperative aspect. Selection on males to cooperate with internal female defenses against infection could explain an otherwise puzzling behavior of males (Siva-Jothy 2006), which insert their hypodermic genitalia just at the site where the female's internal paragenital structures can digest his sperm (Carayon 1966). The possibility that internal female traits like paragenitalia also exercise cryptic female choice by manipulating the sperm and or seminal products within her body has not to my knowledge been tested.

Limitations of this study include the fact that the taxonomic data may biased by a trend for taxonomists to over-utilize genitalia to distinguish species (see above); this bias would favor confirmation of the SAC hypothesis. The sample size was substantially smaller than those in the first surveys (only 114 genera with perhaps 500–1000 species in total), and the traits of many species are undoubtedly not entirely independent among closely related species. Nevertheless, the classic trend for genitalia to diverge relatively rapidly suggests that phylogenetic inertia is not especially strong in genital traits. Finally, the lack of SAC-predicted female defensive morphology could be explained using the same argument regarding yet-to-be-discovered species-specific female defensive behavior.

4. Genital Allometry

If male genitalia are under selection to overcome physical resistance from females, one likely way for males to overcome female resistance is physical force (Lloyd 1979). This expectation, that at least some fraction of male genital structures function as physical weapons in battles with females, yields a strong prediction regarding the allometry of these structures: those male genital structures that are used as weapons should tend to be relatively large in larger individuals when conspecific males of different sizes are compared (they should show relatively high allometric slopes and "positive static allometry"). This prediction is derived from

the well established empirical observation that male structures which are used as weapons in battles with other males usually show positive allometry; the larger males usually have disproportionately large weapons compared with smaller conspecific males in deer antlers, crab claws, spider chelicerae, beetle horns, earwig forceps, and the armed legs of bugs and beetles (summaries in Huxley 1932 [1972] and Kodric-Brown et al. 2006; for exceptions see Bonduriansky 2007). This prediction is especially clear for male structures such as claspers that remain outside the female and are not constrained to act within possibly restrictive female ducts. Thus the expectation of SAC for species in which genital force is used to overcome females is that larger males of a given species should generally have disproportionately large genitalia.

This expectation of positive static allometry in male genitalia is clearly not met. In fact, there is a very strong trend in just the opposite direction, toward lower allometric slopes in the genitalia of insects and spiders: larger individuals almost always have disproportionately smaller genitalia. In 195 of 208 genital structures in 101 species, the allometric slopes was lower than the median allometric slope for other, non-sexually selected, non-genital traits of the same individuals (Eberhard 2009). Counting by species, the median slope for genitalia was lower than the median slope for non-genital structures in 96 of 101 species. "One size fits all" hypotheses that emphasize the importance of physical fits between male and female structures may explain this negative allometry (which also includes female genitalia) (Eberhard et al. 1998; Eberhard 2009). Perhaps some of the genital structures that were measured in these studies do not function to exercise force on the female or are constrained because they must perform in restricted spaces within the female's reproductive tract, and thus may not be expected to follow this SAC weapon prediction. Nevertheless, some structures such as the claspers of 13 species of scathophagid flies (Hosken et al. 2005), two species of sepsid flies (Eberhard et al. 1998; Eberhard 2001b), five species of moth (Ohno et al. 2003; Mutanen & Kaitala 2006; Mutanen et al. 2006) and the parameres of *Onthophagus* and *Macrodactylus* beetles (Palestrini et al. 2000; Eberhard et al. 1998; Eberhard 1993a), perform potentially physically coercive grasping functions; but in 21 of the 22 species they nevertheless showed the typical tendency to negative allometry.

There may also be other possible functions of male genitalia in sperm competition that reduce the numbers of expected offspring for females, as proposed by SAC and that would also show low slopes. Schmitz et al. (2000) mentioned that sperm removal structures might be expected to need to fit precisely with the female; but given the emphasis in SAC theory on male effects on female losses in quantity rather than quality of offspring, SAC seems unlikely to act on a male's sperm removal abilities (see above).

CONCLUSIONS REGARDING CFC AND SAC

In a recent summary, Hosken and Stockley (2004) concluded that current evidence strongly favors sexual selection as the primary force driving rapid divergent evolution of genitalia, but that it is not clear whether SAC or CFC sexual selection mechanisms are responsible. I believe the current balance is more strongly tilted against SAC than they thought. In the first place, further evidence not in accord with expectations of the physical coercion versions of the SAC hypothesis has appeared subsequent to their paper, showing a general lack of female defensive coevolution in groups with species-specific male genitalia and non-genitalic contact devices, and only weak genital diversification in groups with likely intense male–female conflicts (Eberhard 2004b, 2006). Additional extensive data on genital scaling show a strong trend that is opposite to that predicted by the physical coercion version of SAC (Eberhard 2009). Stimulation versions of SAC are also contradicted, though less thoroughly. Females protected from males should also be less subject to damaging male stimulation, yet the especially large sample sizes (Eberhard 2004a) failed to show a trace of the trend predicted by SAC. In addition, one likely female defense against male use of sensory traps with their genitalia (though not the only one—other possibilities include modifications of the female's CNS) would be defensive morphology associated with their genitalia; but arguably the most likely morphological design (facultatively deployable defensive structures) was completely absent. In sum, there is strong evidence against the physical coercion version of SAC, and less conclusive evidence against the stimulation version for both genitalia and non-genital contact structures.

In the second place, Hosken and Stockley argued that the conclusions from the large-scale study of genital evolution in species with females that are and are not protected from male harassment (Eberhard 2004a; see above) were inconclusive, because male–female conflict over fertilization (rather than mating per se) could influence genital evolution even in species with protected females. This possibility is surely reasonable (as noted above, also Eberhard 2004a). But the combination of the immense sample sizes (hundreds of thousands of species, when one includes Lepidoptera in the 2004a study), and the lack of even a trace of the trend in the direction predicted by SAC, means that the SAC effect due to conflict over fertilization, if it exists, must be tiny. The flip side, that there is only relatively modest genitalic diversity in species in which male–female conflict is especially clear, also argues against the importance of SAC. This constitutes evidence against both stimulation and physical coercion versions of SAC. If SAC has acted, it has apparently been brief, weak, or inconsistent; most of the modern diversity of genitalia is apparently due to some other factor.

This is not to argue that SAC, even of the less favored physical coercion type, never occurs on genitalia. Even in cases in which SAC seems especially unlikely to have shaped current morphology, it may nevertheless have played an important role at certain moments in evolution. Take, for instance, the sepsid flies (figure 4.4), a group in which SAC seems especially unlikely to explain the present-day morphology of the modified male front legs that clamp the female's wing (above). Nevertheless, SAC may have played a role in the early stages of the evolution of the clamping structures of male sepsids. Energetic female shaking behavior to dislodge males is widespread in other related flies in which the male's legs are not modified (Crean & Gilburn 1998; Eberhard 2000), as well as in sepsids (Parker 1972; Ward 1983; Eberhard 2005; Ingram et al. 2008). Shaking may thus have originally occurred in sepsids due to male-inflicted losses to females when males began to ride them for long periods at oviposition sites; there is a likely female cost, because a riding male appears to make her less able to avoid predators (personal observation). Early modifications of the male's femur that allowed him to couple his leg more tightly to the female's wing may have represented an antagonistic coevolutionary male response to female shaking behavior, as in some other flies (Dodson 2000). Subsequently, however, the further modifications of the legs of male sepsids that resulted in the great diversity of forms in modern species more likely evolved under sexual selection by female choice.

FRONTIERS

Speculating on where scientific research will go in the future is difficult. I can, however, point to some types of missing data that would help solve presently perceived problems, and also explain why I believe that one currently popular type of research is unlikely to be helpful.

1. Paradoxical Species

Further study of species that seem anomalous under presently popular hypotheses is likely to be especially rewarding. As noted above, the apparently strict monogamy of female bumblebees appears to falsify predictions of the CFC hypothesis, because male genitalia are elaborate and species-specific (figure 4.1). The implications of these data are not, however, entirely conclusive. The evidence for female monogamy is molecular (a single male sires all of a female's offspring), but a female might sometimes make genital contact with other males that are rejected before sperm transfer, or sperm might sometimes be discarded. Claims of strict female monandry based on behavioral data have had a poor track record in other animals (summary in Eberhard 1996). Further observations of events involving genitalia (e.g., attempted couplings that fail) (contact seems to usually lead, however, to copulation—P. Schmid-Hempel, personal communication), and of possible sperm dumping (despite apparent mating plugs—Sauter et al. 2001) might save the CFC hypothesis from these apparently contradicting data. I do not see how to save the physical coercion version of the SAC hypothesis from the lack of female coevolution in the area contacted by the elaborate male genital structures that remain on the outer surface of her body (Richards 1927) (stippled portions in figure 4.1). Or perhaps further understanding of *Bombus* will lead to a new, alternative theory?

Another paradoxical group is the carabid beetle genus *Platynus*, in which changes in female genital traits (development of a dorsal pouch of the bursa, and its subsequent sclerotization and narrowing) apparently preceded rather than occurring in step or following the evolution of associated traits of the male genitalia (various modifications of the tip of his median lobe) (Liebherr 1992). Development of female structures adapted to male structures that have not yet evolved is paradoxical under any of the hypotheses, and this group merits further study.

2. Female Sense Organs

The CFC hypothesis predicts that in all groups lacking rigid species-specific female structures that might be filtering males on the basis of mechanical fit (see for instance the chapter by Eberhard & Huber on spiders in this book), females should have sense organs in the area that is contacted by species-specific portions of male genital structures. Sense organs are also possible, though not necessarily predicted, on rigid female structures that are contacted by species-specific male structures (e.g., the wing bases of sepsid flies—figure 4.4). SAC, on the other hand, is compatible with both the presence and absence of such sense organs regardless of the possible importance of mechanical fit. The CFC sense organ prediction has almost never been tested. Two techniques are available: morphological or histological studies to reveal sense organs; and experimental behavioral studies in which potential female receptors are covered or otherwise inactivated (e.g., Eberhard 2002; Briceño & Eberhard 2009), and then possible changes in female responses to the male are checked. The behavioral technique is especially useful for more difficult to find possible female receptors in membranous areas. Females may utilize generalized receptors that were already present in the area that is contacted by the male, or evolve special sensors that coevolve with the form of the male. Both distributions of sensors are compatible with CFC, because even if there are no receptors located at species-specific sites, differences in female preferences could result from differences in processing of stimuli deeper in their nervous systems.

The only animals I know with data on possible changes in the locations of female sense organs are the damselflies *Enellagma* (Robertson & Paterson 1982), *Coenagrion* (Battin 1993), and *Calopteryx haemorrhoidalis* (Córdoba-Aguilar 2005), sepsid flies in the genera *Archisepsis*, *Themira*, and *Sepsis* (Eberhard 2001a, 2005; Ingram et al. 2008), and four species of cockroaches (Djernaes et al. in preparation). In all cases, female sense organs exist as predicted by CFC, but their placement patterns vary. Female *Enallagma* & *Coenagrion* damselflies have arrays of sense organs whose distribution varies between species in ways that reflect the sites contacted by the species-specific clasping organs of males (Robertson & Paterson 1982; Battin 1993). Female sensory coevolution with male genitalia has also occurred in *C. haemorrhoidalis*, but in the opposite direction. The male trait (increased aedeagus diameter) induces the female to expel larger numbers of sperm stored from previous matings (Córdoba-Aguilar 2005), and the female adjustment to the male has been to *reduce* the number of sensilla that are stimulated by the male's aedeagus (Córdoba-Aguilar 2005). This change could result from either CFC or SAC, as a female mechanism to discriminate in favor of males with an especially thick aedeagus (a CFC explanation), or as a female defense against male manipulation (a SAC explanation).

The wings of the female sepsids and some genital sclerites of the female roaches, in contrast, have stress sensors (campaniform sensilla) that are near but not exactly at the sites where the male's front legs (sepsids) and genitalia (roach) make contact, but they do not reflect the species-specific differences in male form. There are only slight differences between male and female sepsids in the distribution of the sense organs on their wings. Nevertheless, as mentioned above, experimental modifications of either the male's legs or of the female sense organs in one species resulted in sharp increases in female rejections of males (Eberhard 2002), demonstrating that the females can indeed sense the form of the male's front legs (or at least the gross differences in form involved in this experiment) even without a species-specific array of female sensors. Females of this species also appeared to reject mounts by heterospecific males especially vigorously, even though their clamping structures differ only subtly from those of conspecific males (Eberhard 2002). These observations show that the lack of species-specific female sense organs in other groups, such as the roaches, does not rule out the possibility of CFC (Djernaes et al. in preparation).

3. Experimental Manipulations of Male and Female Structures

As noted above, one weakness of most "demonstrations" of the CFC-type effects of male genitalia on female reproductive processes is that they have documented correlations, rather than cause and effect relations. Direct experimentation, such as alterations of species-specific aspects of the male or blockage or alteration of corresponding female sensory traits, is needed to establish cause and effect. Lasers offer a promising technique for altering very small structures (M. Polak personal communication on *Drosophila*). Blocking female sense organs is especially important to control for the possibility in male alteration experiments that changes in female responses are due to changes in the male's behavior that result from changes in his morphology. Experimental ablations of species-specific male genital structures have suggested several possible functions, involving possible natural selection, female choice, and sperm competition (Rodriguez 1993; Rodriguez et al. 2004; Moreno-Garcia & Cordero 2008; H. Brailovsky, personal communication., Takami 2003; Méndez 2002; Méndez & Eberhard in preparation., Nessler et al. 2007) (in none of these species were female sensory structures modified as controls). As with any experimental study, the conclusions that can be drawn are limited by the possible response variables that are measured. This can constitute an especially serious weakness for studies of genital function, because so many different female reproductive responses could be important (more than 20 female mechanisms are known for CFC—Eberhard 1996).

4. Direct (and Indirect) Observations of Genital Behavior

Some simple but nevertheless infrequently used techniques can give insights into how male genitalia are used. Simply observing a copulating pair under a dissecting microscope sometimes leads to surprising discoveries, such as genital stridulation by a crane fly (figure 4.5). In some insects removal of a male's head sometimes results in spontaneous behavior of the male genitalia which can reveal unsuspected functions (e.g., the pushing action of an inflatable sac that gradually inches the male genitalia through the long narrow vagina of the female medfly—Eberhard & Pereira 1995), and rapid, energetic "swimming" motions of inflatable spiny sacs in a tsetse fly (Briceño et al. in preparation). Combining direct observations with studies of musculature can also reveal probable movements of some structures that are hidden from direct view (Briceño et al. 2007). Techniques involving more sophisticated technology include real time phase contrast X-ray imaging of a fly (Briceño et al. in preparation.), and magnetic resonance imaging of human copulation (Schultz et al. 1999) (observing flies rather than humans has the advantage that the subjects are less inhibited by being observed!).

5. Limited Usefulness of Experimental Measurements of Fitness

I do not share the optimism of some of the most outstanding workers on CFC and SAC (e.g., Moore et al. 2003; Pizzari & Snook 2003; Hosken and Stockley 2004; Rice & Chippendale 2001; Orteiza et al. 2005) that laboratory studies of the overall reproductive costs and benefits to females are likely to resolve questions concerning the relative importance of SAC and CFC in the evolution of genitalia (or other traits such as seminal products). Even though the most direct means of resolving the CFC–SAC controversy would be to measure these costs and benefits to females of mating, any direct comparison requires precision in the technically difficult measurements of the magnitudes of both types of fitness; measurements must be accurate enough to correctly determine the sign of the difference between the two values. Pizarri and Snook (2003, 2004) make a related point: it is necessary to utilize male and female fitness, rather than arbitrary phenotypic traits, if experimental approaches to testing SAC are to be useful.

Measuring fitness accurately is not child's play, to say the least. It is trite but true that the costs and benefits to the female must be measured under "ecologically realistic" rather than artificial conditions, if one wishes to make arguments concerning why some traits and not others occur in present-day organisms (Cordero & Eberhard 2003). Unfortunately, precise measurements of both direct and indirect payoffs in the field are extremely difficult to obtain; they are impossible in model species such as *Drosophila melanogaster* and *Tribolium castaneum* in which the natural habitat(s) are not even known. There is no guarantee that the balance of gains and losses under captive conditions is a reliable indicator of the balance under natural conditions. For instance, even such an "obvious" cost to females as reduced life span (Miller & Pitnick 2003) is not necessarily selectively important, if females in nature do not survive long enough to reap all of the benefits of an increased potential life span. For *Drosophila* flies, for example, one may

need to quantify the effects that a bewildering array of factors in nature, such as variations in limitations in oviposition substrate, nutrients in different types of food for larvae and adults, survival rates of adults, rates of parasitism of larvae and pupae at different population densities, microorganisms and secondary compounds present in different types of food that could influence larval and pupal survival, and densities of males and females that affect male–male competition and also female mating frequency. It is likely that there are interactions between some factors of this sort (Eady et al. 2007), making accurate analysis even more difficult.

Attempts to solve this "ecological realism" problem by using strains that have spent many generations in captivity (Orteiza et al. 2005) are problematic, because adaptations to captive environments are likely to be only partial. This is illustrated by a strain of *D. melanogaster* that has been used for sexual selection studies for several hundred generations in captivity (Orteiza et al. 2005). The rearing protocol for this strain has been to use eggs laid after the adult female was over two weeks old to raise the offspring for the subsequent generation of flies (Orteiza et al. 2005). This constitutes intense selection against oviposition early in the female's life. Nevertheless the females of this strain continue to lay many eggs during the first two weeks of their adult lives. If the females were truly adapted to this new selective environment, they would not lay eggs until reaching two weeks of age.

The point is that these (and thus other) female reproductive processes cannot be assumed to be finely adjusted to conditions in captivity, even in captive strains. Therefore measurements of direct and indirect female gains and losses from responding to male manipulations cannot be assumed to indicate the balance between gains and losses that occurred when these responses evolved. In summary, tests involving experimental evolution can be (and often are) very sophisticated technically, but nevertheless only relatively crude in their theoretical implications for the SAC–CFC controversy.

Acknowledgments I thank Janet Leonard and Alex Cordoba for inviting me to write this chapter, Carlos Cordero, Nick Tatarnic, Hojun Song, Marie Djernaes, Alfredo Peretti, and Rudolf Meier for access to unpublished work, Y. Kamimura and P. Schmid-Hempel for permission to quote personal communications, Santosh Jagadeeshan for comments on one section, and David Hosken and Rafael Lucas Rodriguez for comments on the entire chapter. My research was supported financially by the Smithsonian Tropical Research Institute and the Universidad de Costa Rica.

REFERENCES

Abalos, J. W. & Baez, E. C. 1966. Las arañas del género *Latrodectus* en Santiaqgo del Estero. Revista de la Facultad de Ciencias Exactas, Físicas y Naturales de la Universidad Nacional de Córdoba (Argentina), 27(3–4), 1–30.

Agnarsson, I., Coddington, J. A. & Knoflach, B. 2007. Morphology and evolution of cobweb spider male genitalia (Araneae, Theridiidae). Journal of Arachnology 35, 334–395.

Aisenberg, A. & Eberhard, W. G. 2009. Female cooperation in plug formation in a spider: effects of male copulatory courtship. Behavioral Ecology doi:10.1093/beheco/arp 117.

Alexander, R. D. 1964. The evolution of mating behaviour in arthropods. Symposium of the Royal Entomological Society of London, 2, 78–94.

Alexander, R. D., Marshall, D. C. & Cooley, J. R. 1997. Evolutionary perspectives on insect mating. Pp. 4–31 In Choe, J. C. & Crespi, B. J. (eds.) Mating Systems in Insects and Arachnids. Cambridge: Cambridge University Press.

Amsel, H. G., Gregor, F. & Reisser, H. (eds) 1965–2000. Microlepidoptera Palaearctica. Vienna: Goerg Fromme.

Andersson, M. 1994. Sexual Selection. Princeton, N.J.: Princeton University Press.

Anthes, N. & Michiels, N. K. 2007. Precopulatory stabbing, hypodermic injections and unilateral copulations in a hermaphroditic sea slug. Biology Letters, 3, 121–124.

Arnqvist, G. 1998. Comparative evidence for the evolution of genitalia by sexual selection. Nature, 393, 784–786.

Arnqvist, G. 2004. Sexual conflict and sexual selection: lost in the chase. Evolution, 58, 1383–1388.

Arnqvist, G. 2006. Sensory exploitation and sexual conflict. Philosophical Transactions of the Royal Society B, 361, 375–386.

Arnqvist, G. & Danielsson, I. 1999. Copulatory behavior, genital morphology, and male fertilization success in water striders. Evolution, 53, 147–156.

Arnqvist, G. & Rowe, L. 2002a. Antagonistic coevolution between the sexes in a group of insects. Nature, 415, 787–789.

Arnqvist, G. & Rowe, L. 2002b. Correlated evolution of male and female morphologies in water striders. Evolution, 56, 936–947.

Arnqvist, G. & Rowe, L. 2005. Sexual Conflict. Princeton, NJ: Princeton Univ. Press.

Arnqvist, G. & Thornhill, R. 1998. Evolution of animal genitalia: patterns of phenotypic and genotypic variation and condition dependence of genital and non-genital morphology in a water strider (Heteroptera: Gerridae: Insecta). Genetic Research, 71, 193–212.

Arnqvist, G., Thornhill, R. & Rowe, L. 1997. Evolution of animal genitalia: morphological correlates of fitness components in a water strider. Journal of Evolutionary Biology, 10, 613–640.

Arrow, G. 1951. Horned Beetles. The Hague: Dr. Junk.

Aubertin, D. 1933. Revision of the genus *Lucilia* R.-D. (Diptera, Calliphoridae). Linnean Journal of Zoology, 38, 389–436.

Baena, M. L. & Eberhard, W. G. 2007. Appearances deceive: female "resistance" behaviour in a sepsid fly is not a test of the male's ability to hold on. Ethology Ecology and Evolution, 19, 27–50.

Battin, T. 1993. The odonate mating system, communication, and sexual selection. Boletin Zoologia, 60, 353–360.

Beani, L., Giusti, F., Mercati, D., Lupetti, P, Paccagnini, E., Turrilazzi, S. & Dallai, R. 2005. Mating of *Xenos vesparum* (Rossi) (Strepsiptera, Insecta) revisited. Journal of Morphology, 265, 291–303.

Bertin, A. & Fairbairn, D. 2005. One tool, many uses: precopulatory sexual selection on genital morphology in *Aquarius remigis*. Journal of Evolutionary Biology, 18, 949–961.

Birkhead, T. 1996. In it for the eggs. Nature, 383, 772.

Bohme, W. 1983. The Tucano Indians of Colombia and the iguanid lizard *Plica plica*: ethnological, herpetological and ethological implications. Biotropica, 15, 148–150.

Bonduriansky, R. 2007. Sexual selection and allometry: a critical reappraisal of the evidence and ideas. Evolution 61, 838–849.

Bonduriansky, R. & Day, T. 2003. The evolution of static allometry in sexually selected traits. Evolution, 57, 2450–2458.

Brennan, P. L. R., Prum, R. O., McCracken K. G., Sorenson, M. D., Wilson, R. E., & Birkhead, T. R. 2007. Coevolution of male and female genital morphology in waterfowl. PLOS 2(5), e418.

Briceño, R. D. & Eberhard, W. G. in press. Experimental modifications of male genitalia confirm cryptic female choice theory of genital evolution. Journal of Evolutionary Biology

Briceño, R. D., Eberhard, W. G., & Robinson, A. S. 2007. Copulation behavior of *Glossina pallidipes* (Diptera: Muscidae) outside and inside the female, and genitalic evolution. Bulletin of Entomological Research 97, 471–488.

Brindle, A. 1976. The Dermaptera of Dominicana. Smithsonian Contributions to Zoology, 63, 1–25.

Callahan, P. s. & Chapin, J. B. 1960. Morphology of the reproductive system and mating in two representative members of the family Noctuidae, Pseudaletia unipuncta and Peridroma margaritosa, with comparison to Heliothis zea. Annals of the Entomological Society of America 53, 763–782.

Carayon, J. 1966. Traumatic insemination and the paragenital system. Pp 81–167 In: Usinger, R. (ed.) Monograph of Cimicidae. Philadelphia: Thomas Say Foundation 7, Entomological Society of America.

Carayon, J. 1972. Caracteres systematiques et classification des Anthocoridae (Hem.). Annal Societe Entomologique Francia (N.S.), 8, 309–349.

Cardé, R. T., Shapiro, A. M. & Clench, H. K. 1970. Sibling species in the eurydice group of Lethe (Lepidoptera: Satryidae). Psyche, 77, 70–103.

Chapman, R. F. 1969. Insects Structure and Function. London: English University Press. 1–819.

Chapman, T., Arnqvist, G., Banham, J. & Rowe, L. 2003. Sexual conflict. Trends in Ecology and Evolution, 18, 41–47.

Chitwood, B. G. & Chitwood, M. B. 1974. Introduction to Nematology. Baltimore: University Park Press.

Coddington, J. A. 1987. The shape of things that come. Cladistics, 3, 196–198.

Coddington, J. A. 1990. Ontogeny and homology in the male palpus of orb-weaving spiders and their relatives, with comments on phylogeny (Araneoclada: Araneoidea, Deinopoidea. Smithsonian Contributions to Zoology, 496, 1–52.

Coe, M. J. 1969. The anatomy of the reproductive tract and breeding in the spring haas, *Pedestes surdaster* larvalis Hollister. Journal of Reproduction and Fertility, Supplement 6, 159–174.

Cordero, C. & Eberhard, W. G. 2003. Female choice of sexually antagonistic male adaptations: a critical review of some current research. Journal of Evolutionary Biology, 16, 1–6.

Cordero, C. & Eberhard W. G. 2005. Interaction between sexually antagonistic selection and mate choice in the evolution of female responses to male traits. Evolutionary Ecology, 19, 111–122.

Cordero-Rivera, A., Andrés, J. A., Córdoba-Aguilar, A. & Utzeri, C. 2007. Postmating sexual selection: allopatric evolution of sperm competition mechanisms and genital morphology in calopterygid damselflies (Insecta: Odonata). Evolution, 58, 349–359.

Córdoba-Aguilar, A. 1999. Male copulatory sensory stimulation induces female ejection of rival sperm in a damselfly. Proceedings of the Royal Society of London, B, 266, 779–784.

Córdoba-Aguilar, A. 2002. Sensory trap as the mechanism of sexual selection in a damselfly genitalic trait (Insecta: Calopterygidae). American Naturalist 160, 594–601.

Córdoba-Aguilar, A. 2005. Possible coevolution of male and female genital form and function in a calopterygid damselfly. Journal of Evolutionary Biology, 18, 132–137.

Córdoba-Aguilar, A. 2009. Seasonal variation in genital and body size, sperm displacement ability, female mating rate and male harassment in two calopterygid damselflies (Odonata: Calopterygidae). Biological Journal of the Linnean Society, 96, 815–829.

Crean, C. S. & Gilburn, A. 1998. Sexual selection as a side-effect of sexual conflict in the sea-weed fly, *Coelopa ursina* (Diptera: Coelopidae). Animal Behaviour, 56, 1405–1410.

Danielsson, I. & Askenmo, C. 1999. Male genital traits and mating interval affect male fertilization success in the water strider *Gerris lacustris*. Behavioral Ecology and Sociobiology, 46, 149–156.

Darwin, C. 1871. The Descent of Man and Selection in Relation to Sex. Reprinted. New York: Modern Library.

Dixson, A. F. 1987. Obserrvations on the evolution of the genitalia and copulatory behavior in male primates. Journal of Zoology, London, 213, 423–443.

Dixson, A. F. 1998. Primate sexuality. Oxford, U.K.: Oxford University Press.

Dodson, G. 2000. In Fruit flies (Tephritidae): Phylogeny and Evolution of Behavior (Ed. by M. Aluja & A. Norrbom), pp. Del Ray FL: CRC Press.

Dominick, R. B., Ferguson, D. C., Franclemont, J. G., Hodges, R. W. & Munroe, E. G. (eds) 1971–1998. The Moths of America North of Mexico. London: E. W. Classey Limited and RBD Publications.

Dybas, L. K. & Dybas, H. S. 1981. Coadaptations and taxonomic differentiation of sperm and spermathecae in featherwing beetles. Evolution 35, 168–174.

Eady, P., Hamilton, L. & Lyons, R. E. 2007. Copulation, genital damage and early death in *Callosobruchus maculatus*. Proceedings of the Royal Society of London B, 274, 247–252.

Eberhard, W. G. 1985. Sexual Selection and Animal Genitalia. Cambridge, MA: Harvard Univ. Press.

Eberhard, W. G. 1990. Genitalic courtship in *Acmaeodera impluviata* (Coleoptera: Buprestidae). Journal of the Kansas Entomological Society, 63, 345–346.

Eberhard, W. G. 1993a. Copulatory courtship and genital mechanics of three species of *Macrodactylus* (Coleoptera, Scarabeidae, Melolonthinae). Ecology Ethology and Evolution, 5, 19–63.

Eberhard, W. G. 1993b. Evaluating models of sexual selection by female choice: genitalia as a test case. American Naturalist, 142, 564–571.

Eberhard, W. G. 1994. Evidence for widespread courtship during copulation in 131 species of insects and spiders, and implications for cryptic female choice. Evolution 48, 711–733.

Eberhard, W. G. 1996. Female Control: Sexual Selection by Cryptic Female Choice. Princeton, NJ, Princeton University Press.

Eberhard, W. G. 2000. Sexual behavior in the medfly, *Ceratitis capitata*. In: Fruit Flies (Tephritidae): Phylogeny and Evolution of

Behavior (Ed. by M. Aluja & A. Norrbom), pp. 457–487. Del Ray, FL: CRC Press.

Eberhard, W. G. 2001a. The functional morphology of species-specific clasping structures on the front legs of male *Archisepsis* and *Palaeosepsis* flies (Diptera, Sepsidae). Zoological Journal of the Linnean Society, 133, 335–368.

Eberhard, W. G. 2001b. Species-specific genitalic copulatory courtship in sepsid flies (Diptera, Sepsidae, *Microsepsis*). Evolution, 55, 93–102.

Eberhard, W. G. 2001c. Genitalic behavior during copulation in *Hybosciara gigantea* (Diptera: Sciaridae) and the evolution of species-specific genitalia. Journal of the Kansas Entomological Society, 74, 1–9.

Eberhard, W. G. 2001d. Multiple origins of a major novelty: moveable abdominal lobes in male sepsid flies (Diptera: Sepsidae) and the question of developmental constraints. Evolution and Development 3, 206–222.

Eberhard, W. G. 2002. Physical restraint or stimulation? The function(s) of the modified front legs of male *Archisepsis diversiformis* (Diptera, Sepsidae). Journal of Insect Behavior, 15, 831–850

Eberhard, W. G. 2003. Sexual behavior of male *Themira minor* (Diptera, Sepsidae), and movements of the male's sternal lobes and genitalic surstyli. Canadian Entomologist, 135, 569–581.

Eberhard, W. G. 2004a. Male-female conflicts and genitalia: failure to confirm predictions in insects and spiders. Biological Reviews, 79. 121–186.

Eberhard, W. G. 2004b. Rapid divergent evolution of sexual morphology: comparative tests of sexually antagonistic coevolution and traditional female choice. Evolution, 58, 1947–1970.

Eberhard, W. G. 2004c. Apparent stimulatory function of species-specific male genitalic setae in *Aelurus septentrionalis* (Hymenoptera: Tiphiidae). Journal of the Kansas Entomological Society, 77, 837–839.

Eberhard, W. G. 2005. Sexual morphology of male *Sepsis cynipsea* (Diptera: Sepsidae): lack of support for sexually antagonistic coevolution and lock and key hypotheses. Canadian Entomologist, 137, 551–565.

Eberhard, W. G. 2006. Sexually antagonistic coevolution in insects is associated with only limited morphological diversity. Journal of Evolutionary Biology, 19, 657–681.

Eberhard, W. G. 2009. Static allometry and animal genitalia. Evolution, 63, 48–66.

Eberhard, W. G. & Briceño, R. D. 1983. Chivalry in pholcid spiders. Behavioral Ecology and Sociobiology, 13, 189–195.

Eberhard. W. G. & Gelhaus, J. 2009. Genitalic stridulation in a male tupulid fly. Revista de Biología Tropical, 57(Suppl.1), 251–256.

Eberhard, W. G. & Pereira, F. 1995. The process of intromission in the medfly, Ceratitis capitata (Diptera, Tephritidae). Psyche, 102, 101–122.

Eberhard, W. G. 1996. Female Control: Sexual Selection by Cryptic Female Choice. Princeton, NJ, Princeton University Press.

Eberhard, W. G. & Pereira, F. 1996. Functional morphology of male genetic surstyli in the dungflies Archisepsis diversiformis and A. ecalcarata (Diptera: Sepsidae). Journal of the Kansas Entomological Society, 69, 43–60.

Eberhard, W. G. & Ramirez, N. 2004. Functional morphology of the male genitalia of four species of Drosophila: failure to confirm both lock and key and male–female conflict predictions. Annals of the Entomological Society of America, 97, 1007–1017.

Eberhard, W. G. & Gelhaus, J. 2009. Genitalic stridulation during copulation in a species of crane fly, Tipula (Bellardina) sp. (Diptera: Tipulidae). Revista de Biologia Tropical 57 (Suppl. 1), 251–256.

Eberhard, W. G., Huber, B. A., Rodriguez, R. L., Briceo, R. D., Salas, I., & Rodriguez, V. 1998. One size fits all? Relationships between the size and degree of variation in genitalia and other body parts in twenty species of insects and spiders. Evolution, 52, 415–431.

Edvardsson, M. & Canal, D. 2006. The effects of copulation duration in the bruchid beetle Callosobruchus maculatus. Behavioral Ecology, 17, 430–434.

Emlen, D. 1997. Alternative reproductive tactics and male dimorphism in the horned beetle Onthophagus acuminatus (Coleoptera: Scarabeidae). Behavioral Ecology and Sociobiology, 41, 335–341.

Enrodi, S. 1985. The Dynastinae of the World. Boston: Dr. Junk.

Exline, H. & Levi, H. W. 1962. American spiders of the genus Argyrodes (Araneae: Theridiidae). Bulletin of the Museum of Comparative Zoology, 127, 75–203.

Fennah, R. G. 1945. The Cixiini of the Lesser Antilles. Proceedings of the Biological Society of Washington, 58, 133–146.

Ferris, G. F. & Usinger, R. L. 1939. The family Polyctenidae (Hemiptera: Heteroptera). Microentomology, 4, 1–50.

Fisher, R. A. 1930. The Genetical Theory of Natural Selection. 2nd Edn. New York, NY: Dover.

Forbes, W. T. M. 1941. Does he stridulate? (Lepidoptera: Eupterotidae). Entomological News, 52, 79–82.

Förster, M., Gack, C. & Peschke, K. 1998. Morphology and function of the spermatophore in the rove beetle, Aleochara curtula (Coleoptera: Staphylinidae). Zoology, 101, 34–44.

Forster, W. & Wohlfahrt, T. A. 1952–1981. Die Schmetterlinge Mitteleuropas. Stuttgart: Franckh'sche Velagshandlung.

Geist, V. 1966. The evolution of horn-like organs. Behaviour, 27, 175–214.

Ghiselin, M. in press. Darwin's view. In: The Evolution of Primary Sexual Characters in Animals. (Ed. by Leonard, J. & Cordoba-Aguilar, A.) pp. ...

Gilligan, T. M. & Wenzel, J. W. 2008. Extreme intraspecific variation in Hystrichophora (Lepidoptera: Tortricidae) genitalia—questioning the lock-and-key hypothesis. Annales Zoologici Fennici, 45.

Gwynne, D. T. & Edwards, E. D. 1986. Ultrasound production by genital stridulation in Syntonarcha iriastis (Lepidoptera: Pyralidae): long distance signaling by male moths? Zoological Journal of the Linnean Society, 88, 363–376.

Hedin, M. 1997. Speciational history in a diverse clade of habitat-specialized spiders (Araneae: Nesticidae: Nesticus): inferences from geographic-based sampling. Evolution, 51, 1929–1945.

Höbel, G. & Gerhardt, H. C. 2003. Reproductive character displacement in the acoustic communication system of green tree frogs (Hyla cinera). Evolution, 57, 894–904.

Holman, L. & Snook, R. R. 2006. Spermicide, cryptic female choice, and the evolution of sperm form and function. Journal of Evolutionary Biology 19, 1660–1670.

Holland, B. & Rice, W. R. 1998. Chase-away sexual selection: antagonistic seduction versus resistence. Evolution, 52, 1–7.

Hosken, D. J. & Stockley, P. 2004. Sexual selection and genital evolution. Trends in Ecology and Evolution, 19, 87–93.

Hosken, D. J., Meyer, E. P., & Ward, P. I. 1999. Internal female reproductive anatomy and genitalic interactions during copula in the yellow dung fly, *Scathophaga stercoraria* (Diptera: Scathophaga). Canadian Journal of Zoology, 77, 1975–1983.

Hosken, D. J., Minder, A. M., & Ward, P. I. 2005. Male genital allometry in Scathophagidae (Diptera). Evolutionary Ecology, 19, 501–515.

House, C. M. & Simmons, L. W. 2003. Genital morphology and fertilization success in the dung beetle *Onthophagus taurus*: an example of sexually selected male genitalia. Proceedings of the Royal Society of London B, 278, 447–455.

House, C. M. & Simmons, L. W. 2007. No evidence for condition dependent expression of male genitalia in the dung beetle *Onthophagus taurus*. Journal of Evolutionary Biology, 20, 1322–1332.

Howell, G. I. & Herberstein, M. E. in preparation. Mirror-image genital dimorphism in *Ciulfina* praying mantids (Mantodea: Litergusidae).

Huber, B. A. & Eberhard, W. G. 1997. Courtship, copulation and genital mechanics in Physocyclus globosus (Araneae, Pholcidae). Canadian Journal of Zoology 74, 905–918.

Howell, G. I., Winnick C., Tregenza, T. & Herberstein, M. E. 2009. Genital shape correlates with sperm transfer success in the praying mantis Ciulfina klassi (Insecta: Mantodea). Behavioral Ecology and Sociobiology

Huber, B. A. 2003. Rapid evolution and species-specificity of arthropod genitalia: fact or artifact? Organisms Diversity and Evolution, 3, 63–71.

Huber, B. A. 2004. The significance of copulatory structures in spider systematics. In: Biosemiotik-praktische Anwendung und Konsequenzen fur die Einzelwissenschaften. (Ed. by J. Schult), pp. 89–100. Berlin: VWB Verlag.

Huber, B. A. & Pérez Gonzalez, A. 2001. Female genital dimorphism in a spider (Araneae: Pholcidae). Journal of Zoology, London, 255, 301–304.

Huemer, P., Karsholt, O. & Lyneborg, L. 1996. Microlepidoptera of Europe. Stenstrup, Denmark: Apollo Books.

Huxley, J. 1932. Problems of Relative Growth (republished 1972). New York: Dover Publications.

Ingram, K. K., Laamanen, T., Puniamoorthy, N. & Meier, R. 2008. Lack of morphological coevolution between male forelegs and female wings in *Themira* (Sepsidae: Diptera: Insecta). Biological Journal of the Linnean Society, 93, 227–238.

Jacob, A., Ganterbein, G., Braunwalder, M. E., Nentwig, W. & Kropf, C. 2004a. Complex male genitalia (hemispermatophores) are not diagnostic for cryptic species in the genus *Euscorpius* (Scorpiones: Euscorpiidae). Organism Diversity and Evolution, 4, 59–72.

Jacob, A., Ganterbein, G., Braunwalder, M. E., Nentwig, W. & Kropf, C. 2004b. Morphology and function in male genitalia (spermatophore) in *Euscorpius italicus* (Euscorpiidase, Scorpiones): complex spermatophore structures enable safe sperm transfer. Journal of Morphology, 260, 72–84.

Jagadeeshan, S. & Singh, R. 2006. A time-sequence functional analysis of mating behaviour and genital coupling in *Drosophila*: role of cryptic female choice and male sex-drive in the evolution of male genitalia. Journal of Evolutionary Biology, 19, 1058–1070.

Jamnbeck, H. 1965. The *Culicoides* of New York State (Diptera: Ceratopogonidae). New York State Museum of Science Service Bulletin, 399, 1–154.

Jocqué, R. 1998. Female choice, secondary effect of "mate check"? A hypothesis. Belgian Journal of Zoology, 128, 99–117.

Johnson, N. F. 1995. Variation in male genitalia of *Merope tuber* Newman (Mecoptera: Meropeidae). Journal of the Kansas Entomological Society, 68, 224–233.

Jordan, K. 1896. On mechanical isolation and other problems. Novitates Zoologicas, 3, 426–525.

Jordan, K. 1905. Der Gegensatz zwischen geographische und nicht-geographische Variation. Zeitschrift Wissenschaftlichen Zoologie, 83, 151–210.

Kamimura, Y. 2007. Twin intromittent organs of *Drosophila* for traumatic insemination. Biology Letters 3, 401–404.

Kamimura, Y. & Matsuo, Y. 2001. A "spare" compensates for the risk of destruction of the elongated penis of earwigs (Insecta: Dermaptera). Naturwissenschaften 88, 468–471.

Kawano, K. 2003. Character displacement in stag beetles (Coleoptera: Lucanidae). Annals of the Entomological Society of America, 96, 503–511.

Kodric-Brown, A., Sibly, R. M. & Brown, J. H. 2006. The allometry of ornaments and weapons. Proceedings of the National Academy of Sciences, U.S.A., 103, 8733–8738.

Koeniger, G. 1983. Die Entfernung des Begattungzeichens bei der Mehrfachpaarung der Bienenkonigen. Allegemeine Deutsche Imker (Aug.), 244–245.

Kokko, H., Brooks, R. Jennions, M., & Morley, J. 2003. The evolution of mate choice and mating biases. Proceedings of the Royal Society, London B, 270, 653–664.

Kotrba, M. 1995. The internal female genital organs of *Chaetodiposis* and *Diasemopsis* (Diptera: Diopsidae) and their stystematic relevance. Annals of the Natal Museum 36, 147–159.

Kotrba, M. 2006. The internal female reproductive tract of *Campichoeta* Macquart, 1835 and *Diastata* Meigen, 1830 (Diptera, Schizophora). Studia Dipterologica 13, 309–315.

Kraus, O. 1966. Isolationsmechanismen und Genitalstrukturen bei wirbellosen Tieren. Zoologische Anzeiger, 181, 22–38.

Kumashiro, M., Tsuji, Y. & Sakai, M. 2006. Genitalic autogrooming: a self-filling trash collection system in crickets. Naturwissenschaften, 93, 92–96.

Kunze, L. 1959. Die Functionsanatomischen Grundlagen der Kopulation Zwerzikaden, untersucht an *Euscelis plebejus* (Fall.) und einigen Typhlocybinen. Detsche Entomologische Zeitschrift, 6, 322–387.

Leonard, J. & Cordoba-Aguilar, A. ... The evolution of primary sexual characters: theory. In: The Evolution of Primary Sexual Characters in Animals. (Ed. by Leonard, J. & Cordoba-Aguilar, A.), pp. ...

Lewis, C. T. & Pollock, J. N. 1975. Engagement of the phallosome in blowflies. Journal of Entomology (A), 49, 137–147.

Liebherr, J. K. 1992. Phylogeny and revision of the *Platynus degallieri* species group (Coleoptera: Carabidae: Platini). Bulletin of the American Museum of Natural History, 214, 1–115.

Lloyd, J. E. 1979. Mating behavior and natural selection. Florida Entomologist, 62, 17–23.

Lloyd, J. E. 1997. Firefly mating ecology, selection and evolution. In: The Evolution of Mating Systems in Insects and Arachnids (Ed. by J. Choe & B. Crespie) pp. 184–92. Cambridge: Cambridge University Press.

Lüpold, S., McElligott, A. G. & Hosken, D. J. 2004. Bat genitalia: allometry, variation, and good genes. Biological Journal of the Linnean Society, 83, 497–507.

Markow, T. A. & O'Grady, P. M. 2005. Evolutionary genetics of reproductive behavior in *Drosophila*: connecting the dots. Annual Review of Genetics 39, 263–291.

Mayr, E. 1963. Animal Species and Evolution. Cambridge, MA: Harvard University Press.

McAlpine, D. K. 1988. Studies in upside-down flies (Diptera: Neurochaetidae). Part II. Biology, adaptations, and specific mating mechanisms. Proceedings of the Linnean Society of New South Wales, 110, 59–82.

McPeek, M. A., Shen, L., Torrey, J. Z., & Farid, H. 2008. The tempo and mode of three-dimensional morphological evolution in male reproductive structures. American Naturalist 171, E138–E178.

Mendez, V. 2002. Comportamiento sexual y dinámica de población de *Leucauge marina* (Araneae: Tetragnathidae). MSc Thesis, Universidad de Costa Rica.

Merrett, D. J. 1989. The morphology of the phallosome and accessary gland material transfer during copulation in the blowfly, *Lucilia cuprina* (Insecta, Diptera). Zoomorphology, 109, 359–366.

Michiels, N. 1998. Mating conflicts and sperm competition in simultaneous hermaphrodites. In: Sperm Competition and Sexual Selection. (Ed. by T. Birkhead & A. P. Møller), pp. 219–254. New York: Academic Press.

Mikkola, K. 1992. Evidence for lock-and-key mechanisms in the internal genitalia of the *Apamea* moths (Lepidoptera, Noctuidae). Systematic Entomology, 17, 145–153.

Mikkola, K. 2008. The lock-and-key mechanisms of the internal genitalia of the Noctuidae (Lepidoptera): How are they selected for? European Journal of Entomology 105, 13–25.

Miller, G. T. & Pitnick, S. 2002. Sperm–female coevolution in Drosophila. Science 298, 1230–1233.

Miller, G. T. & Pitnick, S. 2003. Functional significance of seminal receptacle length in *Drosophila melanogaster*. Journal of Evolutionary Biology, 16, 114–126.

Minder, A. M., Hosken, D. J. & Ward, P. I. 2005. Coevolution of male and female reproductive characters across the Scathophagidae (Diptera). Journal of Evolutionary Biology, 18, 60–69.

Møller, A. P. 1998. Sperm competition and sexual selection. In: Sperm Competition and Sexual Selection. (Ed. by T. Birkhead & A. P. Moller) Pp. 55–90. New York: Academic.

Moore, A. G., Gowaty, P. A. & Moore, P. J. 2003. Females avoid manipulative males and live longer. Journal of Evolutionary Biology, 16, 523–530.

Moreno-Garcia, M. & Cordero, C. 2008. On the function of male genital claspers in Stenomacra marginella (Heteroptera: Largidae). Journal of Ethology, 26, 255–260.

Morrow, E. H. & Arnqvist, G. 2003. Costly traumatic insemination and a female counter-adaptation in bed bugs. Proceedings of the Royal Society of London B, 270, 2377–2381.

Mühlhäuser, C. & Blanckenhorn, W. 2002. The costs of avoiding matings in the dung fly Sepsis cynipsea. Behavioral Ecology, 13, 359–365.

Mutanen, M. 2005. Delimitation of difficulties in species splits: a morphometric case study on the Euxoa tritici complex (Lepidoptera, Noctuidae). Systematic Entomology, 30, 632–643.

Mutanen, M. & Kaitala, A. 2006. Genital variation in a dimorphic moth Selenia tetralunaria (Lepidoptera, Geometridae). Biological Journal of the Linnean Society, 87, 297–307.

Mutanen, M. & Pretorius, E. 2007. Subjective visual evaluation versus traditional and geometric morphometrics in species delimitation: a comparison of moth genitalia. Systematic Entomology, 32, 371–386.

Mutanen, M., Kaitala, A. & Monkkonen, M. 2006. Genital variation within and between three closely related Euxoa moth species: testing the lock-and-key hipótesis. Journal of Zoology, 268, 109–119.

Nessler, S., Uhl, G. & Schneider, J. 2007. A non-sperm transferring genital trait under sexual selection: an experimental approach. Proceedings of the Royal Society of London B, 274, 2337–2341.

Ohno, S., Hoshizaki, S., Ishikawa, Y., Tatsuki, S. 2003. Allometry of male genitalia in a lepidopteran species Ostrinia laatipennis (Lepidoptera: Crambridae). Applied Entomology and Zoology, 38, 313–319.

Orteiza, N., Linder, J. E. & Rice, W. R. 2005. Sexy sons from remating do not recoup the direct costs of harmful male interactions in the Drosophila melanogaster laboratory system. Journal of Evolutionary Biology 18, 1315–1323.

Ortiz, P. G. 2002. Historia natural, sitios de apareamiento, comportamiento sexual y posible funcion de la alimentación nupcial en Ptilosphen variolataus (Diptera: Micropezidae). Masters thesis, Universidad de Costa Rica, San José.

Ossiannilsson, F., Russell, L. M. & Weber, H. 1970. Homoptera. In: Taxonomist's Glossary of Genitalia in Insects. (Ed. by S. L. Tuxen), pp. 179–189. Darien, CON: S-H Service Agency.

Otronen, M. 1990. Mating behavior and sperm competition in the fly, Dryomyza anilis. Behavioral Ecology and Sociobiology, 26, 349–356.

Otte, D. & Stayman, K. 1979. Beetle horns: some patterns in functional morphology. In: Sexual Selection and Reproductive Competition in Insects. (Ed. by M. Blum & N. Blum), pp. 259–292. New York: Academic Press.

Palestrini, C., Rolando, A. & Laiolo, P. 2000. Allometric relationships and character evolution in Onthophagus taurus (Coleoptera: Scarabaeidae). Canadian Journal of Zoology, 78, 1199–1206.

Paraq, A., Bennett, N. C., Faulkes, C. G. & Bateman, P. W. 2006. Penile morphology of African mole rats (Bathyergidae): structural modifications in relation to mode of ovulation and degree of sociality. Journal of Zoology, London, 270, 323–329.

Parker, G. A. 1970. Sperm competition and ints evolutionary consequences. Biological Reviews, 45, 525–567.

Parker, G. A. 1972. Reproductive behavior of Sepsis cynipsea (L.) (Diptera: Sepsidae). I. A preliminary analysis of the reproductive strategy and its associated behaviour patterns. Behaviour, 41, 172–206.

Parker, G. A. 1984. Sperm competition and the evolution of animal mating strategies. In: Sperm Competition and the Evolution of Animal Mating Systems. (Ed. by R. L. Smith) pp. 2–60. New York: Academic Press.

Parker, G. A. 2005. Sexual conflict over mating and fertilization: an overview. Philosophical

Transactions of the Royal Society B, 361, 235–259

Patterson, H. E. H. 1982. Perspective on speciation by reinforcement. South African Journal of Science, 78, 53–57.

Patterson, B. D. & Thaeler, C. S. 1982. The mammalian baculum: hypotheses on the nature of bacular variability. Journal of Mammology, 63, 1–15.

Peretti, A. 2003. Functional morphology of spematophores and female genitalia in bothriurid scorpions: genital courtship, coercion and other possible mechanisms. Journal of Zoology, London, 261, 135–153.

Peretti, A., Eberhard, W. G. & Briceño, R. D. 2006. Copulatory dialogue: female spiders sing during copulation to influence male genitalic movements. Animal Behaviour, 72, 413–421.

Phelan, P. L. 1997. Evolution of mate-signaling in moths: phylogenetic considerations and predictions from the asymmetric tracking, hypothesis. In: Mating Systems in Insects and Arachnicds. (Ed. by J. Choe & B. Crespie), pp. 240–256. Cambridge University Press, Cambridge, U.K.

Pizarri, T. & Snook, R. 2003. Sexual conflict and sexual selection: chasing away paradigm shifts. Evolution, 57, 1223–1236.

Pizarri, T. & Snook, R. R. 2004. Sexual conflict and sexual selection: measuring antagonistic coevolution. Evolution, 58, 1389–1393.

Pizzo, A., D. Mercurio, C. Palestrini, A. Roggero & A. Rolando. 2006a. Male differentiation patterns in two polyphenic sister species of the genus Onthophagus Latreille, 1802 (Coleoptera: Scarabeidae): a geometric morphometric approach. Journal of Zoology 44, 54–62.

Polihronakis, M. 2006. Morphometric analysis of intra-specific shape variation in male and female genitalia of Phyllophaga hirticula (Coleoptera: Scarabeidae: Melolonthinae). Annals of the Entomological Society of America, 99, 144–150.

Reinhardt, K., Harney, E., Naylor, R., Gorb, S. & Siva-Jothy, M. T. 2007. Female-limited polymorphism in the copulatory organ of a traumatically inseminating insect. American Naturalist 170, 931–935.

Rensch, B. 1960. Evolution Above the Species Level. New York, NY: Columbia University Press.

Rice, W. R. & Chippendale, A. K. 2001. Intersexual ontogenetic conflict. Journal of Evolutionary Biology, 41, 685–693.

Richards, O. W. 1927. The specific characters of the British humblebees (Hymenoptera). Transactions of the Royal Entomological Society of London, 75, 233–265.

Richards, O. W. 1978. The Social Wasps of the Americas. London: British Museum (Natural History).

Robertson, H. M. & Paterson, H. E. H. 1982. Mate recognition and mechanical isolation in Enallagma damselfiies (Odonata: Coenagrionidae). Evolution, 36, 243–250.

Robinson, J. V. & Novak, K. L. 1997. The relationship between mating system and penis morphology in ischnuran damselflies (Odonata: Coenagrionidae). Biological Journal of the Linnean Society, 60, 187–200.

Robinson, M. H. & Robinson, B. 1980. Comparative studies of the courtship and mating behavior of tropical araneid spiders. Pacific Insects Monographs, 36, 1–218.

Robson, G. C. & Richards, O. W. 1936. The Variation of Animals in Nature. London: Longmans, Green & Co.

Rodriguez, V. 1993. Fuentes de variación en la precedencia de espermatozoides de Chelymorpha alternans Boheman 1854 (Coleoptera: Chrysomelidae: Cassidinae). Master's thesis, Universidad de Costa Rica.

Rodriguez, V., Windsor, D. M. & Eberhard, W. G. 2004. Tortoise beetle genitalia and demonstration of a sexually selected advantage for flagellum length in Chelymorpha alternans (Chrysomelidae, Cassidini, Stolaini). In: New Developments in the Biology of Chrysomelidae, (Ed. by P. Jolivet, J. A. Santiago-Blay & M. Schmitt), pp. 739–748. The Hague: SPB Academic Publishing.

Roeder, K. D. 1967. Nerve Cells and Insect Behavior. Harvard University Press, Cambridge, MA.

Roig-Alsina, A. 1993. The evolution of the apoid endophallus, its phylogenetic implications, and functional significance of the genital capsule (Hymenoptera, Apoidea). Bolletin Zool., 60, 169–183.

Ronkainen, K., Kaitala, A. & Huttenen, R. 2005. The effect of abdominal spines on female mating frequency and fecundity in a water strider. Journal of Insect Behavior, 18, 619–631.

Rovner, J. S. 1968. Territoriality in the sheet-web spider *Linyphia triangularis* (Clerck) (Araneae, Linyphiidae). Zeitschrift für Tierphychologie, 25, 232–242.

Rowe, L. 1994. The costs of mating and mate choice in water striders. Animal Behaviour, 48, 1049–1055.

Rowe, L. & Arnqvist, G. 2002. Sexually antagonistic coevolution in a mating system: comparative approaches to address evolutionary processes. Evolution, 56, 754–767.

Sauter, A., Brown, M. J. F., Baer, B., & Schmid-Hempel, P. 2001. Males of social insects can prevent queens from multiple mating. Proceedings of the Royal Society B, 268, 1449–1454.

Schmid-Hempel, P. & Schmid-Hempel, P. 2000. Female mating frequencies in *Bombus* species from Central Europe. Insectes Sociaux, 47, 36–41.

Schmitz, G., Reinhold, K & Wagner, P. 2000. Allometric relationship between genitalic size and body size in two species of mordellid beetles (Coleoptera: Mordellidae). Annals of the Entomological Society of America, 93, 637–639.

Schulte-Hostedde, A. & Alarie, Y. 2006. Morphological patterns of sexual selection in the diving beetle Graphoderus liberus. Evolutionary and Ecological Research 8, 891–901.

Schultz, W. W., van Andel, P., Sabelis, I., & Mooyaart, E. 1999. Magnetic resonance imaging of male and female genitals during coitus and female sexual arousal. BMJ 13, 1596–1600.

Schulmeister, S. 2001. Functional morphology of the male genitalia and copulation in lower Hymenoptera, with special emphasis on the Tenthredinoidea s. str. (Insecta, Hymenoptera, 'Symphyta'). Acta Zoologica (Stockholm), 82, 331–349.

Scudder, G. 1971. Comparative morphology of insect genitalia. Annual Review of Entomology, 16, 379–406.

Shapiro, A. M. & Porter, A. H. 1989. The lock-and-key hypothesis: evolutionary and biosystematic interpretation of insect genitalia. Annual Review of Entomology, 34, 231–245.

Simmons, L. W. 2001. Sperm Competition and its Evolutionary Consequences in the Insects. Princeton University Press, Princeton, NJ.

Simmons, L. W. & Siva-Jothy, M. 1998. Sperm competition in insects: mechanisms and the potential for selection. In: Sperm Competition and Sexual Selection (Ed. by T. Birkhead & A. P. Møller), pp. 341–434. New York: Academic Press.

Simmons, L. W., Stockley, P., Jackson, R. L. & Parker, G. A. 1996. Sperm competition or sperm selection: no evidence for female influence over paternity in yellow dung flies *Scathophaga stercoraria*. Behavioral Ecology and Sociobiology, 38, 199–206.

Sirot, L. K. 2003. The evolution of insect mating systems through sexual selection. Florida Entomologist, 86, 124–133.

Siva-Jothy, M. J. 2006. Trauma, disease and collateral damage; conflict in cimicids. Philosophical Transactions of the Royal Society B 361, 269–275.

Smith, R. L. 1984. Sperm Competition and the Evolution of Animal Mating Systems. New York: Academic Press.

Song, H. 2006. Systematics of the Cyrtacanthacridinae (Orthoptera: Acrididae) with a focus on the genus *Schistocerca* Stål 1873: Evolution of locust phase polymorphism and study of insect genitalia. PhD. Thesis, Ohio State University, Columbus, OH.

Song, H. & Bucheli, S. R. 2009. Comparison of phylogenetic signal between male genitalia and non-genital characters in insect systematic. Cladistics 26 (2010), 23–35.

Song, H. & Bucheli, S. R. Submitted. Rate of genital evolution is not necessarily rapid: composite nature and phylogenetic utility of insect genitalia. Evolution.

Sota, T. & Kobuta, K. 1998. Genital lock-and-key as a selective agent against hybridization. Evolution, 52, 1507–1513.

Stockley, P. 2002. Sperm competition risk and male genital anatomy: comparative evidence for reduced duration of female sexual receptivity in primates with penile spines. Evolutionary Ecology, 16, 123–137.

Stockley, P. & Preston, B. T. 2004. Sperm competition and diversity in rodent copulatory behavior. Journal of Evolutionary Biology, 17, 1048–1057.

Stoks, R., Nystrom, J. L., May, M. L., & McPeek, M. A. 2005. Parallel evolution in ecological and reproductive traits to produce cryptic dameslfly species across the Holarctic. Evolution, 59, 1976–1988.

Stutt, A. D. & Siva-Jothy, M. T. 2001. Traumatic insemination and sexual conflict in the bed bug *Cimex lecularius*. Proceedings of the National Academy of Sciences, U.S.A., 98, 5683–5687.

Takami, Y. 2003. Experimental analysis of the effect of genital morphology on insemination success in the ground beetle *Carabus insulicola* (Coleoptera Carabidae). Ethology Ecology and Evolution 15, 51–61.

Tanabe, T. & Sota, T. 2008. Complex copulatory behavior and the proximate effect of genital and body size differences on mechanical reproductive isolation in the millipede genus *Parafontaria*. American Naturalist 171, 692–699.

Tanabe, T., Katakura, H. & Mawatari, S. F. 2001. Morphological difference and reproductive isolation: morphometrics in the millipede *Parafontaria tonominea* and its allied forms. Biological Journal of the Linnean Society, 72, 249–264.

Tatarnic, N. J., Cassis, G. & Hochuli, D. F. 2005. Traumatic insemination in the plant bug genus *Coridromius* Signoret (Heteroptera: Miridae). Biology Letters, 2, 58–61.

Tatsuta, H. & Akimoto, S.-I. 1998. Sexual differences in the pattern of spatial variation in brachypterous grasshopper *Podisma sapporensis* (Orthoptera: Podisminae). Canadian Journal of Zoology, 76, 1450–1455.

Tatsuta, H., Mizota, K. & Akimoto, S.-I. 2001. Allometric patterns of heads and genitalia in the stag beetle *Lucanus maculifemoratus* (Coleoptera: Lucanidae). Annals of the Entomological Society of America, 94, 462–466.

Taylor, C. A. & Knouft, J. H. 2006. Historical influences on genital morphology among sympatric species: gonopod evolution and reproductive isolation in the craydfish genus *Orconectes* (Cambaridae). Biological Journal of the Linnean Society. 89, 1–12.

Thornhill, R. 1982. Cryptic female choice and its implications in the scorpionfly *Harpobittacus nigriceps*. American Naturalist 122, 765–788.

Tinkham, E. R. & Renz, D. C. 1969. Notes on the bionomics and distribution of the genus *Stenopalmatus* in central California with description of a new species. Pan-Pacific Entomologist, 45, 4–14.

Turgeon, J., Stoks, R., Thum, R. A., Brown, J. M. & McPeek, M. A. 2005. Simultaneous Quaternary radiations of three damselfly clades across the Holarctic. American Naturalist 165, E78-E107.

Tuxen, L. 1970. A Taxonomist's Glossary of Genitalia of Insects. Darien, CON: S-H Service Agency.

Usinger, R. 1966. Monograph of Cimicidae. Philadelphia: Thomas Say Foundation 7, Entomological Society of America.

Usami, T., Yokoyama, J., Kubota, K., & Kawata, M. 2006. Genitali lock-and-key system promotes premating isolation by mate preference. Biological Journal of the Linnean Society 87, 145–154.

Vanacker, D., Vanden Borre, J., Jonckheere, A., Maes, L., Pardo, S., Hendrickx, F. & Maelfait, J.-P. 2003. Dwarf spiders (Erigoninae, Linyphiidae, Araneae): good candidates for evolutionary research. Belgian Journal of Zoology, 133, 143–149.

Verrell, P. A. 1992. Primate penile morphologies and social systems: further evidence for an association. Folia Primatologica, 59, 114–120.

Vitalievna, N. M. 1995. Seasonal variation in the male genitalia of *Plagodis dolabraria* (Linneaus, 1759) (Lepidoptera, Geometridae). Atalanta, 26, 311–313.

von Helverson, D. & von Helverson, O. 1991. Pre-mating sperm removal in the bushcricket *Metaplastes ornatus* Ramme 1931 (Orthoptera, Tettigonoidea, Phaneropteridae). Behavioral Ecology and Sociobiology, 28, 391–396.

Waage, J. K. 1979. Dual function of the damselfly penis: sperm removal and transfer. Science, 203, 916–918.

Ward, P. I. 1983. The effects of size on the mating behaviour of the dung fly *Sepsis cynipsea*. Behavioral Ecology and Sociobiology, 13, 75–80.

Ware, A. & Opell, B. D. 1989. A test of the mechanical isolation hypothesis in two similar spider species. Journal of Arachnology, 17, 149–162.

Wenninger, E. J. & Averill, A. L. 2006. Influence of body and genital morphology on relative male fertilization success in oriental beetle. Behavioral Ecology, 17, 656–663.

Werner, M. & Simmons, L. W. 2008. The evolution of male genitalia: functional integration of genital sclerites in the dung beetle *Onthophagus taurus*. Biological Journal of the Linnean Society 93, 257–266.

Whitman, D. W. & Loher, W. 1984. Morphology of male sex organs and insemination in the grasshopper *Taeniopoda eques* (Burmeister). Journal of Morphology, 179, 1–12.

Whitney, N. M., Pratt, H. L. & Carrier, J. C. 2004. Group courtship, mating behaviour and siphon sac function in the whitetip reef shark, *Triaenodon obesus*. Animal Behaviour, 68, 1435–1442.

Wickler, W. 1966. Ursprung und biologische Deutung des Genitalprasentierens mannlicher Primaten. Zeitschrift fürTierpsychologie 23, 422–437.

Wood, D. M. 1991. Homology and phylogenetic implications of male genitalia in Diptera. The ground plan. In: Proceedings of the Second International Congress of Dipterology (Ed. by Weismann, Orszagh, & Pont, A.), pp. 255–284. The Hague.

5

Killing Time

A Mechanism of Sexual Conflict and Sexual Selection

PATRICIA ADAIR GOWATY AND STEPHEN P. HUBBELL

INTRODUCTION

In Darwin's (1871) monumental book on sexual selection, he catalogued sex-limited secondary sexual characters that he argued evolved by sexual selection rather than natural selection. Yet, the great naturalist went to great lengths to exclude a curious set of traits from consideration. In chapter after chapter he put aside prehensile organs of males as ones for which "… it is scarcely possible to distinguish the effects of natural and sexual selection" (p. 257). Again and again, he mentioned traits that are restricted to males but unlikely to affect competition with same-sex rivals. Darwin included among these traits organs to seize females once found (p. 331) or to "prevent her escape" including the traumatic attachment to the female of male tentacles by sucker discs of some cephalopods (p. 325), the modified antenna of some of the lower crustaceans into an "elegant, and sometimes wonderfully complex, prehensile organ" (p. 330), the pincers of some male crustaceans (p. 331) that males use to "seize with impunity" females before they have molted their hard shells and to carry females about until mating can occur (p. 331). He listed as not worthy of consideration as intrasexually selected the anterior segment of Diplopoda males that are "modified into prehensive hooks which serve to

secure the female" (p. 340); "the sickle-shaped jaws" of sand-wasps that the "exceedingly ardent" males use to seize their partners (p. 342). He left out of consideration "the appendages at the tip of the tail, which are modified in an almost infinite variety of curious patterns to enable (male dragon-flies) to embrace the neck of the female" (p. 344). He remarked about insects in general that sex differences abound and "oftener in the males possessing diversified contrivances for retaining the females when found. But", he went on, "we are not here much concerned with sexual differences of these kinds" (p. 418). Further on, he dismisses from consideration the claspers of the chimaeroid fishes that "serve to retain the female" (volume 2, page 1); the claws on the front-feet of mud-turtles that are twice the size of the females and are used when the sexes unite, and the "prehensile claws" on the fore-legs of some salamanders and newts, which "no doubt aid the male in his eager search and pursuit of the female". In short, Darwin put organs for catching, seizing, restraining, holding, and preventing the escape of females aside as unlikely due to advantage acquired over rivals, even though these organs were seemingly important to gamete transfer, and even though similar organs were most often absent in females (but see, p. 332). He concluded, "… we may safely infer from the many singular contrivances possessed by the males, such as great jaws, adhesive cushions, spines, elongated legs, and others., for seizing the female ... that there is some difficulty in the act" (Darwin 1871, p. 421).

One cannot but wonder why such organs of prehension so rarely evolved in females if they had the symmetric advantage in gamete transfer implied by natural selection. If these organs are not associated with intrasexual competition, why has natural selection not worked as often on females as males to produce convenient anchoring organs? And, why would such organs so often be prehensile, bending, and responsive to opposing forces?

Indeed, could it be that the "difficulty in the act" is caused by females trying to get away? Could it be that females resist some males?

A PROBLEM FOR DARWIN THEN AND FOR EVOLUTIONARY THEORY NOW

Reading Darwin (1871) leaves one with the impression that he either did not have an explanation for these traits or that he strategically set them aside. He said he was setting them aside because he could not clearly attribute them to either sexual or natural selection. We find it hard to believe that Darwin did not imagine sexual conflict as a potential outcome of male ardor and female "coyness", i.e., as a mechanism of sexual selection as later scholars so readily did (e.g., Trivers 1972; Smuts & Smuts 1993), and because he describes many examples of sexual conflict especially in the chapters on the descent of man. Thus, we think it possible that Darwin put aside consideration of the traits that interest us here, because these male "contrivances" for holding, seizing, and grasping females sometimes hurt and killed these same females. Remember, Darwin's main goal with the 1871 volume was to assuage his critics who said that bizarre and elaborate male traits could not evolve by natural selection because these traits decreased the survival probability of their bearers. Surely Darwin would have muddied the waters of his defense had he attempted to explain why males would hurt (Johnstone & Keller 2000) and sometimes kill, through violent seizure or restraint, the very females they were trying to mate.

Indeed, it remains a general problem in evolutionary biology to understand why males are aggressive to females in so many organisms, why in some species males force copulate females or violently seize females, and sometimes hurt them before or after mating or even use "traumatic seizure and restraint". Why does sexual conflict occur when sexual reproduction is a fundamentally collaborative act (Eberhard 1996, 2005).

Most attempts to address these questions usually start under the assumption that, in general, selection has acted so that males are indiscriminate and females choosy (Darwin 1871; Trivers 1972; Parker et al. 1972; Parker & Simmons 1996). However, there is abundant evidence now that in many species, even those with asymmetries in gamete sizes and parental investment (Berglund 1994, 1995; Berglund & Rosenqvist 1993), that both sexes are sometimes choosy and both sexes sometimes indiscriminate (Bonduriansky 2001; Drickamer et al. 2000, 2003; Gowaty 1998; Gowaty et al. 2003a, 2002). That males, not just females, may be choosy in all species is interestingly problematic here because organs of traumatic insemination, ardent seizure and holding do seem to be limited in most taxa to males (Eberhard 1996) and it is easy to imagine their evolution if all males are indiscriminate. Is it possible to reconcile universal pre-mating assessment of fitness outcomes by males with the evolution of sex-associated organs for holding and seizing females?

Much of the contemporary discussion of sexual conflict is about the costs of sexually antagonistic alleles (Arnqvist & Rowe 2005), those alleles advantageous in one sex that, when expressed in the other sex, are costly to their bearers. In contrast, we cast our consideration of possible reasons for organs of seizure and restraint not in terms of antagonistic alleles, but in terms of sexually antagonistic selection pressures, which could, but probably do not, result in sexually antagonistic alleles. Sexually antagonistic selective pressures include those social and ecological problems created by individuals of one sex that individuals of the other sex have to solve to survive and reproduce (Gowaty 1996, 1997, 2002; Gowaty & Buschhaus 1998). We think about the problems of male seizure and restraint of females in terms of the back and forth, moment by moment changes in social forces of flexibly expressed and potentially adaptive coercion and resistance to coercion.

SEXUAL CONFLICT RESULTS FROM SEXUALLY ANTAGONISTIC SELECTION PRESSURES

Given sexually antagonistic selection pressures, several hypotheses exist to explain the contrivances that males use to seize and restrain females. The CODE hypothesis (Gowaty & Buschhaus 1998) is one. In the context of forced copulation, it says that aggression creates a dangerous environment for females, so that females trade social monogamy with a male for protection from his aggressive relatives or friends. A problem with this hypothesis is that the traits that Darwin set aside occur in species besides those that are socially monogamous. Another explanation is that ardent and violent seizing of females could increase the aggressor's probability of immediate insemination of the seized female, but this explanation does not explain why females would use the sperm (Eberhard 1996; Gowaty & Buschhaus 1998). Yet another possible explanation is that holding on or hurting one's mate decreases the likelihood that she will subsequently mate with the aggressor's rivals (Smuts & Smuts 1993; Smuts 1992; Johnstone & Keller 2000). A problem with this idea is that hurting one's mate also may reduce the probability that she will successfully reproduce with the aggressor, or once mated, actually use the aggressor's sperm to fertilize her eggs, or once her eggs are fertilized, be able or willing to invest in them further. A fourth possibility is that contrivances for seizing and holding females function in cryptic female choice (Eberhard 1996), that is, physiological mechanisms during copulation that give females additional information unavailable to them in pre-touching assessments of potential mates. And, a fifth possibility is that these contrivances function as communicative devices that allow males to transmit and females to receive communications, and vice versa (Baena & Eberhard 2007). These last two ideas remain interesting. Nevertheless, a problem in coercive mating, as it is in cryptic female choice or in information transfer, is whether females will use the inseminated sperm of a given male.

There are good reasons to expect that females may eject or kill, rather than use, the sperm of males who coerce copulations. Granted, before Eberhard's (1996) *Female Control*, some investigators imagined females as passive receptacles for male sperm rather than as the architects of sperm competition

(Gowaty 1994, 1997). Later, Gowaty and Bushhaus (1998) predicted that female waterfowl, in which forced copulation is notorious, facultatively use the secretions from cells in the second compartment of their cloacae to denature the sperm of males that force-copulate them. This hypothesis was suggested by the resemblance of the cells lining the cloacae to gut cells that secrete hydrochloric acid. Similarly, like others, we expect that the secretions in the lining of the reproductive tracts of most females may be able to denature sperm, as they may in *Drosophila arizonae* (Kelleher et al. 2007). And, most important to this discussion, we expect that females are facultatively able to regulate their secretions to denature or nurture sperm. Investigations (Knowles & Markow 2001; Pitnick et al. 1997; Markow & Ankney 1988; Gromko et al. 1984; Gromko & Markow 1993) of the insemination reaction of some *Drosophila* also suggest the possibility that the female components of the insemination reaction may hurt or kill sperm under some conditions, but under others may nurture sperm, guarding sperm until females use it, store it, or dump it. Thus, here we assume that females are not passive sperm receptacles and that females may do more than benignly sort among and store the sperm of inseminating males.

Given the likelihood of female resistance to male coercive attempts, and given the possibility that females kill the sperm of some males but not others, begs, in turn, the question of why males attempt to coerce females in the first place? Do all males attempt to coerce females? Is there between-male variation in the use or timing of potentially coercive contrivances? Is male coercion something that some males do all the time, or that all males might attempt some of the time? Have these potentially coercive "contrivances" evolved in tandem with physiological mechanisms of female resistance? Is there between-female variation in the use or timing of potentially resistant physiological responses to male coercive tactics? Are coercion and resistance to coercion adaptively flexible options for individuals?

As we now realize sexual conflict can result in differences in fitness among male rivals (Smuts & Smuts 1993) if some males are better able to manipulate or control females' reproductive decisions than other males (Sih et al. 2002). Furthermore, the idea that females would resist such manipulation and control is now familiar (Gowaty 1997; Gowaty & Buschhaus 1998; Forstmeier 2004; Jormalainen & Merilaita 1995; Moore et al. 2003), and just as variation in coercion may affect variances in male fitness, so can variation in resistance affect variances in female fitness. Thus, like coercion, resistance evolves under within-sex fitness variation.

In this chapter:

1. We describe a new hypothesis, the killing time hypothesis[1] (KTH) to explain these contrivances for seizure and holding of females.
2. We briefly describe a key element in the KTH, the switch point theorem (Gowaty & Hubbell 2009), which solves analytically for a focal individual the fraction of potential mates in the population who are acceptable as mates. We used the switch point theorem in the context of DYNAMATE (Gowaty & Hubbell, 2005), an individual based simulation model, to determine the dynamic reproductive decisions of individuals in a population under demographic stochasticity (see also Appendix 2).
3. We discuss results from a typical suite of runs of DYNAMATE, and we then focus on encounters in which sexual conflict over mating occurred.
4. We discuss the quantitative predictions of the KTH.
5. We end by discussing tests able to reject the KTH, if it does not apply.

THE KILLING TIME HYPOTHESIS FOR MALE ORGANS OF SEIZURE AND PREVENTION OF FEMALE ESCAPE

We hypothesize that the benefit that favored the evolution of male organs for seizing and restraining females is that they manipulate females' subsequent reproductive decisions. When coercive males *specifically reduce female time available for mating and reproduction*, females are more likely to use the sperm of males who these females previously assessed as unacceptable. We did not solve the switch point theorem (Gowaty & Hubbell 2009) or

design DYNAMATE (Gowaty & Hubbell 2005) specifically to study sexual conflict; rather sexual conflict and the KTH emerged from the rules of these models.

Assumptions of the KTH

Assumptions of the KTH include: (1) individuals of both sexes assess the likely fitness consequences of mating with a particular partner before subsequent reproductive decisions. We emphasize that we assume that selection has acted so that individuals — both males and females—make assessments of potential fitness outcomes before they have touched, something that the switch point theorem has proven should be adaptive (Gowaty & Hubbell, 2009) and that empirical studies have shown is the case in a variety of animals (Anderson et al. 2007; Drickamer et al. 2003; Gowaty and Hubbell 2005). (2) Chance effects on individual encounter probabilities, e, individual survival probabilities s, and their latencies, l (tables 5.1 and 5.2) affect time available for mating (Appendix A: Gowaty & Hubbell 2009); and chance effects on e, s, and l are sufficient to determine seasonal and lifetime variance in mating success (LVMS), not different from variances that usually are attributed to sexual selection (Gowaty & Hubbell 2005; Hubbell & Johnson 1987; Sutherland 1985; Snyder & Gowaty 2007). (3) Individuals initially assess fitness outcomes from a potential mating in terms of offspring viability and the number of offspring that are likely to reach reproductive age (Anderson et al. 2007; Bluhm & Gowaty, 2004b; Drickamer et al. 2000; Gowaty et al. 2003a; Moore et al. 2003). (4) Although it is clear that ecological, life-history, and demographic circumstances affect the relative power of one sex to control and/or manipulate the reproductive decisions of the other (Gowaty 1997), here we make the simplifying assumption that there are no intrinsic or extrinsic sex differences in power. (5) Females are able to kill or sequester the sperm of individual males. (6) Individuals who reproduce with partners they initially assess as conferring lower fitness than

---

1. To "kill time" is an idiomatic expression that goes back to the eighteenth century (Oxford English Dictionary). It is usually defined as "passing time aimlessly". Thus, we use it here in a different sense, as a mechanism that coercive individuals may use to manipulate or control the reproductive decisions of mates or rivals. In our usage "killing time" means to use up another's time in a way that levies opportunity costs on potential mates or rivals. Thus, we do mean that we hypothesize that coercive individuals decrease or eliminate time their potential mates or rivals have for mating and reproduction as a manipulative way to affect their subsequent reproductive decisions.

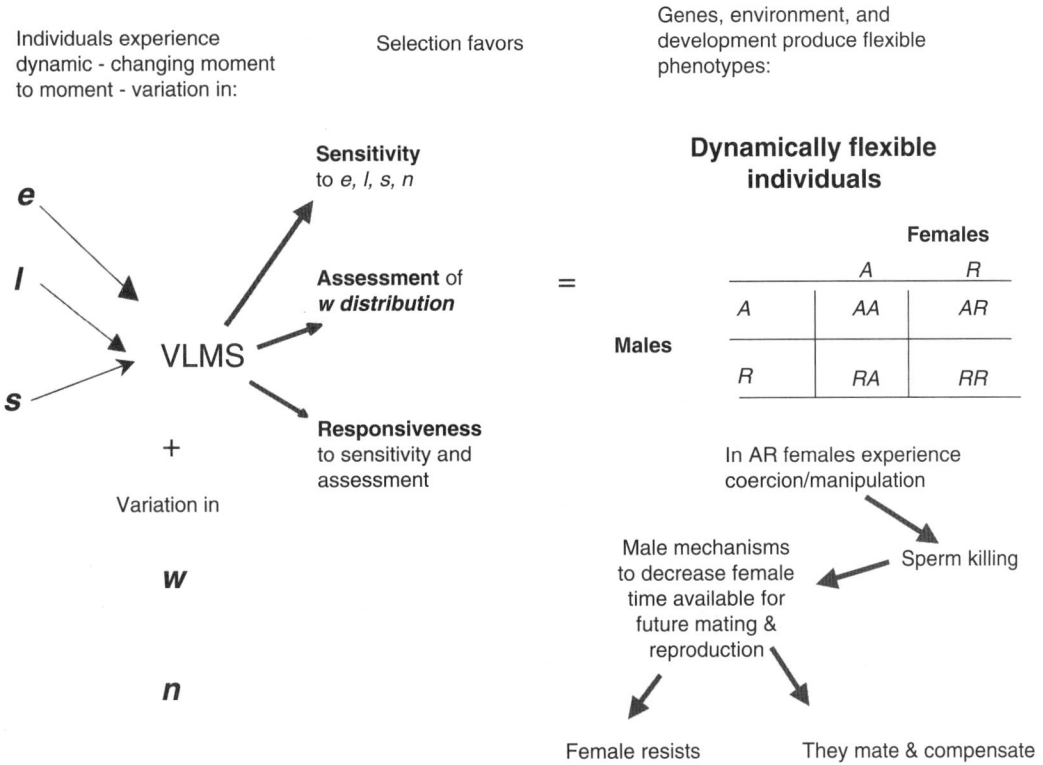

FIGURE 5.1 A schematic drawing of a scenario for the evolution of adaptively flexible individuals and their reproductive. See table 5.1 for definitions of terms and text for a discussion of the killing time hypothesis.

other potential mates compensate for expected deficits in offspring viability (Gowaty 2008; Gowaty et al. 2007; Byers and Waits 2006). (6) Compensation may extract survival costs from individual breeders (Anderson et al. 2007; Bluhm and Gowaty 2004b; Drickamer et al. 2000; Gowaty et al. 2003a; Moore et al. 2003).

A Reasonable Scenario for the Evolution of Adaptively Flexible Individuals

We hypothesize that because time available for mating is dynamically changing under demographic stochasticity (individuals dying, entering nonreceptivity), selection will favor individuals who are sensitive to these changes, and who make reproductive decisions that are weighted both by stochastic effects on lifetime mating success and the fitness that would be conferred, w, by mating with alternative

potential mates (Gowaty & Hubbell 2005, 2009). See also tables 5.1, 5.2 and Appendix A for definitions of fitness that would be conferred and the fitness distribution. Thus, we logically inferred that individuals are favored who can respond flexibly moment to moment to changing social and demographic environments to make adaptive reproductive decisions, such as whether to accept or reject an encountered potential mate. Figure 5.1 is a schematic, conceptual statement of the KTH.

The switch point theorem (Gowaty and Hubbell 2009) is the quantitative description of the time available for mating model describing how the reproductive decisions of flexible focal individuals should vary under time-varying values of e, s, l, n, and w (tables 5.1, 5.2 and Appendix A). All else equal there is more time available for mating for individuals with higher e and s, and shorter l (table 5.2). The average effects of varying e, s, and l on the fraction of acceptable mates are listed in table 5.2.

TABLE 5.1   Definitions of model parameters and frequently used terms

	Definition
Parameters of the model	
e	Individual encounter probability
s	Individual survival probability
l	Time interval from the onset of one copulation to receptivity to the next
n	The number of potentially mating opposite sex individuals
w	The fitness that would be conferred by mating with a given encountered potential mate
W	Fitness at f^*
f^*	Equilibrium number of acceptable mates which maximizes fitness
Other terms	
Fitness distribution or w distribution	The distribution of fitnesses of potential mates in the population. Conceptually this distribution is from a matrix of fitness conferred for all males and all females in the population if each mated with every opposite sex individual without carry-over effects.
Beta distribution	A function producing distributions of different shapes
VMS	Variance in mating success
LVMS	Lifetime variance in mating success
VRS	Variance in reproductive success
LVRS	Lifetime variance in reproductive success
KTH	Killing Time Hypothesis
Opportunity cost	Time lost during which a focal could have been searching for, encountering, and mating with potential mates that would have conferred higher fitness
Components of fitness	Number of mates, fecundity, productivity, offspring viability
Fecundity	Number of eggs laid or offspring born
Productivity	Number of offspring who survive to reproductive age
Egg-to-adult viability	Proportion of eggs that survive to reproductive age; a proxy for offspring health
AA	In an encounter of potential mates the male accesses the female as acceptable and the female accesses the male as acceptable
AR	In an encounter of potential mates the male accesses the female as acceptable but the female accesses the male as unacceptable (she rejects)
RA	In an encounter of potential mates the male accesses the female as unacceptable (he rejects) but the female accesses the male as acceptable
RR	In an encounter of potential mates the male accesses the female as unacceptable and the female accesses the male as unacceptable (both reject)

TABLE 5.2   How variation in e, s, and l affect time available for mating (Hubbell & Johnson, 1987; Gowaty & Hubbell, 2005, 2009) and individual switch points

Parameter	Effects on time available for mating	Effects on switch points*
s	Variation in s affects all states that individuals may pass through in the model; longer s = more time and lower s = less time available for mating and reproduction	All else equal: Higher s = lower switch point (fewer acceptable potential mates) Lower s = higher switch point (more acceptable potential mates)
e	Variation in e affects time in search; lower e = longer search time and thus, less time available and higher e = shorter search time and thus, more time available for mating and reproduction	All else equal: Higher e = lower switch point (fewer acceptable potential mates) Lower e = higher switch point (more acceptable potential mates)
l	Variation in l represents individuals' opportunity costs relative to same-sex competitors; longer l = less time available for mating and reproduction, while shorter l = more time available for mating and reproduction	All else equal: Longer l = lower switch point (fewer acceptable potential mates) shorter l = higher switch point (more acceptable potential mates)

* Switch points are for focal individuals and expressed in terms of the ranks of potential mates, they are the rank at which focals switch from accepting to rejecting potential mates; lower ranks represent better fitness conferred than higher ranks (best fitness conferred is from the potential mate ranked 1; worst fitness conferred is from potential mate ranked n, where n is the number of potential mates).

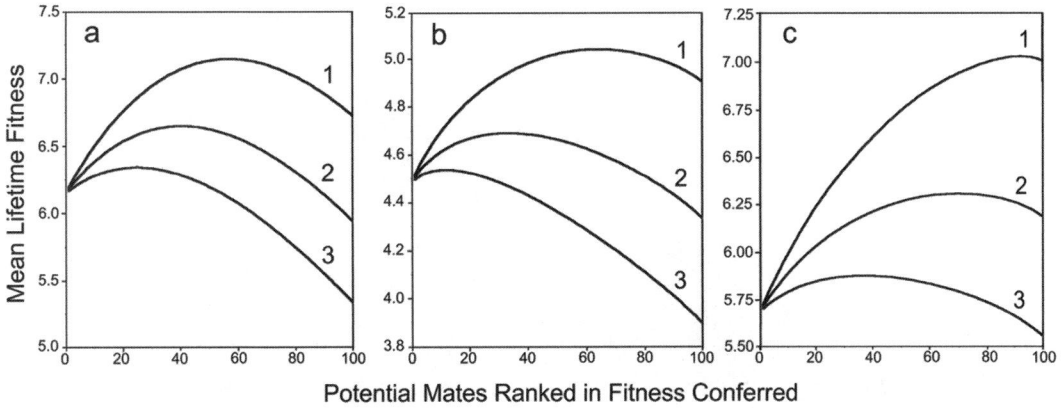

FIGURE 5.2 The switch point threshold between acceptable and unacceptable potential mates is the peak in each curve, which depends on five parameters: s, l, e, n, and w *distribution* (tables 5.1 and 5.2 and see appendix A). Here we hold s, e, and n constant ($e = 0.2$, $s = 0.97$, and $n = 100$) while varying l and the w distribution. Curves labeled 1 indicate individuals with an l of one; curves labeled 2 indicate individuals with l of two; and curves labeled 3 indicate individuals with l of three. In all of these cases, increasing l reduces the fraction of acceptable mates (i.e., reduces the reward for the decision rules that say accept more). Potential mates are ranked on the x axis from highest conferred fitness (rank 1) to lowest conferred fitness (rank 100). The points on the curve are the stochastic average lifetime relative fitness for individuals following a given accept/reject rule (accept x, reject $n–x$) for a given set of parameter values, e, s, l, n, and w distribution. Thus for each graph, each point on each curve represents the average lifetime fitness for individuals following the decision rule of accept x, reject $n–x$, where x = rank of potential mates. Thus, comparisons of the fitness rewards yields the decision rule for a given set of parameter values that maximizes average individual lifetime relative fitness, the switch point, f^*. Under demographic stochasticity, f^* is a dynamic fitness optimum. For example, in graph a, to maximize average lifetime relative fitness, individuals with $l = 1$ should accept any potential mate with rank 1 through 59 as encountered (i.e., acceptance does not depend on encounter sequence), after which potential mates with ranks 60 through 100 should be rejected (also independent of order of encounter). Graph (a) shows results for a w distribution of beta (1, 1), which is the highest variance fitness distribution, (b) a w *distribution* of beta (4, 7), which is a left-skewed distribution with more low fitnesses conferred than high fitnesses conferred, and c. a w *distribution* of beta (7, 4), which is a right-skewed distribution with more high than low fitnesses conferred. Note that in graphs (a) and (c) the highest mean lifetime relative fitness is in excess of 7, while for graph (b). the highest mean lifetime relative fitness is around 5, an effect here entirely of the shape of the w *distribution* (tables 5.1 and 5.2).

On the x axes of figure 5.2 are the potential mates ranked in fitness conferred, from the best fitness rank at 1 to worst fitness rank at n, which in this case is 100. This axis represents a series of rules for the fraction of mates acceptable, so that for rank 10 the rule is accept all potential mates through rank 10 and reject all mates with rank $n–10$. The y axis represents the instantaneous average lifetime relative fitness of the focal individual if it accepts all potential mates with ranks to the left of the corresponding rank on the x axis, and rejects all potential mates to the right of the rank on the x axis. This function reaches a peak, W, the switch point fitness,

which maximizes the average lifetime relative fitness of focal individuals experiencing specified variation in e, s, l, n, and w distribution. Any other switch point results in lower average lifetime relative fitness.

The Opportunity for Sexual Conflict

We ran DYNAMATE to simulate, under variation in e, s, l, w distribution, and n, how often males and females would, on encounter, both accept each other (from now on accept–accept or AA); the male

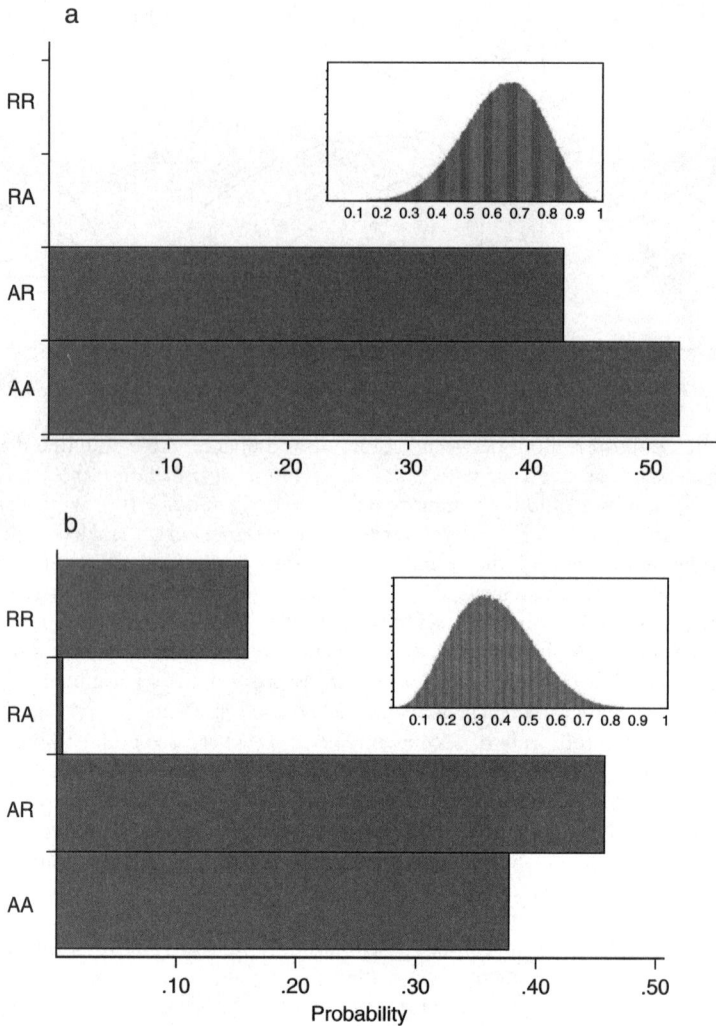

FIGURE 5.3 The distribution of pre-touching assessments and mating decisions during encounters of focal males and females in search for mates. DYNAMATE (table 5.1; appendix B) determines the reproductive decisions of each individual as they experience dynamically changing values of e, s, l, and n under demographic stochasticity using the switch point theorem (table 5.1; appendix A). DYNAMATE then tallied the cases in which encountered potential mates both accepted (AA); the male accepted but the female rejected (AR); the male rejected and the female accepted (RA), or both rejected (RR) matings, under initial $e = 0.99$ and $s = 0.99$ for individuals of both sexes. In each ensemble of runs the only initial sex differences were that male l = one; and female l = 20. In both ensemble sets n = 200, but in one set w *distributions* of the populations were beta 7.4 (a) and in the other beta 4.7 (b). The insets show the w *distributions*. In these runs males seldom find females unacceptable because male l is shorter than female l, but both males and females find more potential mates unacceptable when the w-distribution is left skewed (b) than right skewed (a). In left-skewed (b) distributions compared to right-skewed (a) distributions pre-touching assessments include more individuals who found the encountered potential mate unacceptable.

would accept, but the female reject (from now on accept–reject or AR); the male would reject and the female accept (from now on reject–accept or RA); and both the male and female would reject (from now on reject–reject, RR). The opportunity for sexual conflict over mating occurs only in cases AR or RA. To illustrate the approach, we show the results of two ensemble runs of 1000 populations each (figure 5.3a, b) with identical e, s, l, and n; the two runs differed only in their w distributions. In figure 5.3a most encounters were AA. In figure 5.3b about 38% of encounters were AA. In these populations, conflict over mating is not universal. Moreover, even when AA outcomes were less common as in the runs in figure 5.3b, AA matings yielded significantly higher fitness than matings in which one or both of the encountered potential mates rejected (results to be reported elsewhere). In the current runs, many more encounters were AR than RA, because males had fewer opportunity costs from any mating than females (males had lower l).

Similar asymmetries to those in figure 5.3 in AR and RA frequencies arise whenever males have more time for mating; that is, when they have higher encounter e or longer s than females (to be reported elsewhere). Likewise, when females have shorter l, and greater e and/or s than males, while all else is equal, RA encounters are more common than AR. The AR encounters, in which males accept but females reject, are of most interest in our discussion of the contrivances of males Darwin left out of his discussion of sexual selection.

Using the Switch Point Theorem to Predict Which Sex "Wins" the Pre-Mating Contest

Because we assumed no sex differences in power, in other words, we assumed there was no within or between sex variation in vulnerability to control of reproductive decisions by opposite sex individuals, we used fitness differentials from the switch point graphs (figure 5.4) to predict which sex would win the between-sex pre-mating behavioral contest, that is, whether mating would or would not occur. For example, consider the outcome of AR matings under a w distribution of beta (7,4), when the initial $e = 0.2$, $s = .97$, with male l = one and female l = three. Figure 5.4a, b shows that in these cases, if the pair mated, the mean lifetime loss of fitness for the female would be much larger than the relatively

small gain in fitness for the male. Then we used the relative fitness gains and losses to measure the relative strength of the male's attempt to manipulate or force the mating and the female's attempt to resist the mating. Whenever a female would have lost more fitness than the male would have gained from a mating, DYNAMATE determined that females won so the mating did not take place. Thus, in the AR encounter, the female resisted successfully, so that the pair did not mate. The outcomes of AR encounters in these pairs that differed only in their w distributions, 30% in figure 5.4a and 37% in figure 5.4b, of AR females successfully resisted mating. However, in the remaining AR encounters, males were successful in coercing females into mating.

The question we then asked was what was the force of selection on coerced females' post-mating physiological responses?

The Switch Point Theorem Predicts the Direction of Post-Mating Selection

Relative to non-coerced females, the decrement in fitness for coerced females will be a selection pressure favoring physiological mechanisms that decrease the opportunity cost from the coerced mating. In response, one would expect that counter-selection would also act on males to increase female "commitment" to the mating such as mechanisms to rescue sperm from deadly effects of female physiology, or those that would increase female time commitments to the current mating so dramatically that a good option for continued reproductive success may be collaborating with—using the sperm of—the coercing male (figure 5.1). The KTH predicts mechanisms of males, which females previously assessed as unacceptable, can use to coerce female commitment to reproduction with them. The mechanisms are ones that use up females' time.

Increasing Female Opportunity Costs Manipulate Females into Using the Sperm of Previously Unacceptable Males

Decreasing a female's instantaneous s, reduces her time available for mating (table 5.2), and is very likely the most effective way to impose opportunity costs on females (Gowaty & Hubbell 2009).

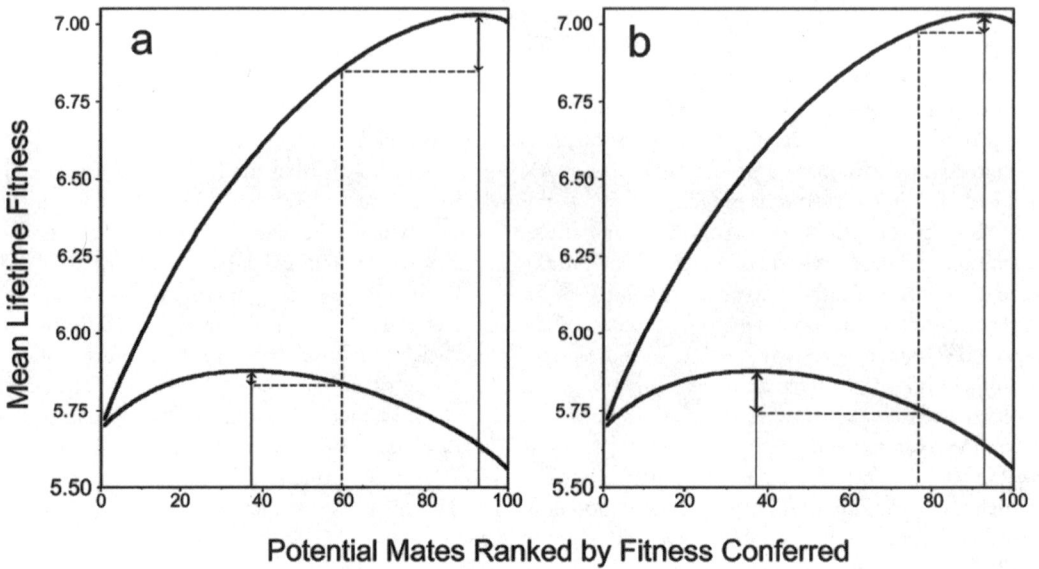

Potential Mates Ranked by Fitness Conferred

FIGURE 5.4 Two switch point graphs for AR encounters, when a male assesses a female as acceptable, but the female assesses a male as unacceptable. The graphs allow inferences of the outcomes (did they mate or not) and subsequent reproductive decisions. See figure 5.2 for a description of switch point graphs. The top curve in each graph represents the switch point curve for males; the bottom curve shows the switch point curve for females. The switch point theorem (Figure 5.2; appendix A) calculates for each focal individual the fraction of acceptable mates in the population, and proves that average lifetime fitness is maximized by mating with any combination of mates up to the rank of potential mates at which the peak is achieved and rejecting those from the peak rank to n (in this case 100). We infer therefore that on average individuals experiencing a given set of parameter values will enhance their fitness by accepting potential mates from rank one to the peak rank. On the graphs, the best fitness rank is 1; the worst fitness rank is 100. In both graphs males have l of one (the high curves) and the females have l of three (the low curve). In both (a) and (b) $e = 0.20$; $s = 0.97$, $n = 100$; and W = beta (7, 4) for both males and females. In (a) and in (b) females find any male ranked 1 through 38 acceptable (the first vertical line on each graph); and males find any female ranked 1 through 96 acceptable (the third vertical line on each graph); that is, rank 38 for females and rank 96 for males corresponds with the peak in average lifetime relative fitness for individuals experiencing parameter values like theirs, and thus the ranks corresponding to the peaks represent the switch points for these individuals. The encountered potential mates assess each other as rank 60 in (a) and 78 in (b) (note that we had each rank the other at the same rank for ease of drawing). Now, we consider whether it is the males or the females that gain or lose more average lifetime relative fitness by mating with the encountered potential mate. The double-sided arrows show the relative gain and loss for each male and female. In (a) the fitness gain from the mating would be greater for the male than the fitness loss for the female, so we called the male the "winner" in this encounter, so that they mated. In (b) the fitness gain for the male was less than the fitness loss for the female, so we assigned the female the "winner" of the encounter, so that they did not mate.

Decreasing instantaneous s may affect females' preexisting physiological pathways (West-Eberhard 1984) and thereby induce females to commit their reproductive resources to the sperm of males they previously assessed as unacceptable. All else equal, decreasing female s, would move her switch point to higher ranks (with the best rank being 1) so that a larger fraction of potential mates becomes acceptable (figure 5.2). In light of the time available for mating hypothesis (Gowaty & Hubbell 2009) the anti-intuitive idea that males would "harm their mates" makes sense.

Similarly, when males reduce female e, they not only directly affect the opportunities other males have for mating with her, but they also steal from females time available for mating (table 5.2). Decreasing female e produces an opportunity cost to females, and like opportunity costs associated with s, will move her switch point to higher ranks (so that a greater fraction of potential mates is acceptable) and may be sufficient to induce females to commit reproductive resources to the sperm of males that the manipulated female previously assessed as unacceptable.

If there is a reasonable likelihood that males previously assessed as unacceptable would encounter these females again, the KTH predicts that males will manipulatively increase the fraction of potential mates acceptable to her using mechanisms that decrease female l, which will move her switch point to higher ranks, that is, those with lower fitness. However, if they are unlikely to meet again, a manipulative male who increased female l would move her switch point to lower fitness ranks, thereby decreasing the fraction of potential mates acceptable to her (Gowaty & Hubbell 2009).

Males who manipulate female time available for mating are using up their own time available for mating, so one expects that such males will be sensitive to and able to respond to information from females about how long they have to hang on in order to move her range of acceptable males to their rank. If males hang on too long, they will impose further opportunity costs on themselves. We are currently studying male give-up time in coercive matings.

A sensitivity analysis (Gowaty & Hubbell 2009) of the switch point theorem demonstrated that changes in W with respect to changes in l, e, and s are such that $dW/ds > dW/de > dW/dl$ (Gowaty & Hubbell 2009), thus, when all else is equal, selection will favor coercive males who manipulate female instantaneous s before manipulating female e, and they should manipulate e before l. But, each of these options has costs for males; we are currently studying variation in the costs to males of different ways of manipulating females.

But whether males hang on or not, there is a rub, for if we are correct, females who males manipulate—by decreasing s, e, and/or l—into mating will often lose out in the fitness contests with other females who successfully avoid mating and reproduction with males they initially assessed as unacceptable.

The Evolutionary Benefit of Collaboration Between Resisted Males and Coerced Females

If it is the case that mating with rejected partners reduces offspring viability as it does in *Drosophila pseudoobscura*, mice, ducks, and fish (see review in Gowaty et al. 2007; Gowaty 2008), selection should act on parents to enhance the likelihood that their offspring from resisted matings survive to reproductive age by attempting to compensate for deficits in offspring viability (Bluhm and Gowaty 2004a,b; Byers & Waits 2006; Drickamer et al. 2000, 2003; Gowaty 2008; Gowaty et al. 2007). The compensation hypothesis predicts that both parents will be under selection to flexibly compensate, even when only one of them resists. Compensatory mechanisms previously described include the facultative production of more eggs in an attempt to expose any latent or under-represented variation in the genotypes of parents, or the production of eggs with higher variance in expressed genes than would be predicted by parental genotypes, as well as investment of resources that are likely to enhance offspring health and well-being (Anderson et al. 2007; Bluhm and Gowaty 2004a; Byers and Waits 2006; Gowaty et al. 2007; Navara et al. 2004, 2006). Post intromittant mechanisms of enhanced male investment (Knowles et al. 2004, 2005) may be through nutritive components in the sperm (Snook and Markow 1996; Pitnick et al. 1991). We have wondered if very large sperm tails (Pitnick and Markow, 1994) might not be a kind of yolk that could nourish embryos or deliver male-derived defensive antibiotics or antivirals to zygotes. If males are collaborating with females to produce healthy offspring, males also may contribute nutrients or other protective products to their zygotes and to their mates (Knowles et al. 2004).

Some Predictions of the KTH

The KTH like its parent model the switch point theorem is prediction rich; we emphasize only a few here:

1. When power asymmetries between the sexes are absent, the outcomes of sexual conflict over mating depend on the fraction of optimal fitness gained or lost in an RA or AR matings.

2. All else equal, if there are sex differences in l, the value of coercion for the sex with shorter l is greater than for the sex with longer l.

3. Similarly, all else equal, sex differences in e or s increase the value of coercion for individuals with lower e or lower s.

4. Populations in which l is the same for males and females, but males have a lower e, or lower s, will have more AR than RA matings than when systematic sex differences are absent.

5. Nonetheless, sexual conflict over mating is not universal.

6. Matings between individuals from AA encounters have offspring with higher egg-to-adult survival than matings in which one or both individuals reject.

7. Behavioral and physiological mechanisms that decrease overall time available for future matings for females are more exaggerated in AR assessment types than AA; for example, lengthened "courtship", or other mechanisms decreasing female survival or encounter probabilities may exploit pre-existing physiological pathways that induce females to commit more eggs to the sperm of a coercive male.

8. In AR and RA matings, one partner has assessed the likely fitness outcome as less advantageous than with other potential mates, thus, one or both of the partners will attempt to compensate (figure 5.1) for offspring viability deficits.

9. Thus, mechanisms of behavioral and physiological cooperation often will result when individuals are manipulated into matings they assessed in pre-touching encounters as unacceptable.

How to Reject the KTH as an Explanation for Organs for Seizing and Holding Females

Because the KTH is based on a theorem (Gowaty & Hubbell 2009), its predictions are likely to hold if its assumptions are met. Thus, we recommend that those interested in empirical tests of the KTH, evaluate whether the assumptions hold in their test species. For example, one might ask, what if there are no signs of pre-touching assessment in arenas designed to eliminate within-sex interactions and between-sex manipulation or force? In such species, the KTH would not hold. We suspect that when experimental evidence suggests that there are

no pre-touching assessments that cryptic female choice may more readily explain organs of holding and seizure, and we suspect that it is in such species that sperm killing and sequestration would be most likely to be easily observed. However, we continue to wonder who those species are given that the pre-touching behavior of individuals of so many species reliably indicates fitness conferred by alternative potential mates (Gowaty 2008).

Suggested Tests

Because an important assumption of the KTH is that individuals assess likely fitness outcomes before expressing the predicted flexibility in individual reproductive decisions, we recommend that investigators experimentally evaluate whether pre-touching behavioral assessments predict fitness outcomes. One can do this in two steps by evaluating individuals' pre-touching behavioral preferences in arenas in which touching and mating cannot take place followed by experimental breeding trials in which focal individuals are confined randomly with the potential mate it preferred or the one it did not. We suspect that in insects and other small organisms, human observers may be blind, deaf, or otherwise insensitive to signs of rejection, and that whenever investigators place multiple females and males in close proximity in a closed vial, for example, that males may take such signs as a green light and females may be trapped. This may be especially so if subjects are insects in which females show acceptance by being still and resist by just flying away (Gromko & Markow 1993; Gromko et al. 1984).

Once pre-touching assessments have been evaluated under conditions in which mating cannot take place, one can further evaluate the fitness outcomes of pre-touching assessments. Investigators might then further characterize behavior during and after copulation, tests to measure females' physiological responses to sperm and expression profiling among individuals as a function of their assessment and subsequent pairing types (e.g., AA, AR, RA, or RR). If we are correct about induced behavior, the expression profiles and/or the translational profiles, of these four assessment types will be different. We stress that sorting out differences in assessment types under sexual conflict over mating will require evaluation of differences in behavior and physiology when focal females and males are mated to those they behaviorally assess as acceptable or not in controlled pre-touching trials, as in studies in which within-sex competitive displays or between-sex force are controlled or eliminated. The hypothesis also predicts that translational proteomic profiling associated with both focal male and focal female subjects will reveal within-species variation in induced physiology whenever individuals are coerced or manipulated into mating. The species Darwin (1871) left out of the discussion and which meet the assumptions of the KTH would seem to be likely candidates for testing the KTH.

APPENDIX A. THE SWITCH POINT THEOREM (GOWATY & HUBBELL 2009)

We used a similar analytical approach to Hubbell and Johnson (1987) to solve the general case of continuous variation in fitness conferred by alternative potential mates in order to predict the fraction of potential mates that a focal individual would find acceptable to mate over its entire life. The analytical solution (Gowaty & Hubbell, 2009) computes the lifetime fitness of focal individuals that accept or reject a certain number of individuals of the opposite sex as potential mates as they move through an absorbing Markov chain. The transition probabilities between states in the Markov chain are functions of the focal individual's encounter probability, e, with any receptive potential mate, p_i, the probability that the receptive mate is individual i, the probability of survival for one time unit, s, and the duration of the latency period, l.

Equation (1) is the general solution for the lifetime fitness of an individual accepting (mating if encountered) f mating types and rejecting $n - f$ mating types out of a total of n potential mating types for different probabilities p_i of encountering the ith mating type. Mating types are ranked in

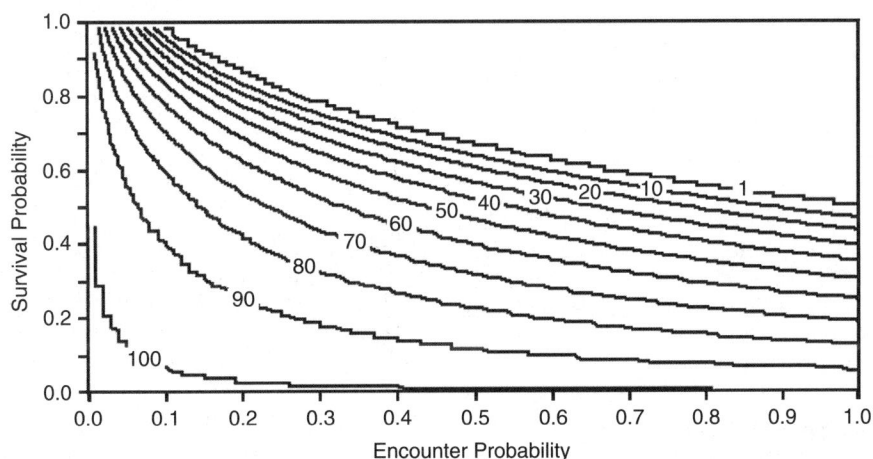

FIGURE 5.5 A contour plot showing the fraction of potential mates acceptable for a focal individual in a population with a w distribution of beta (1, 1) and $l = 10$, experiencing changing values of e and s. The top contour line represents the contour of 1% of potential mates acceptable; the next 10% of potential mates acceptable, and so on to the bottom contour of 100% acceptable mates. For an AR assessment in which a female had a survival probability of 0.6 and a manipulative male decreased her e with potentially mating individuals from 0.7 to 0.6, the fraction of potential mates acceptable to her would increase from 1% to between 10 and 20%. For an AR assessment in which a female had $e = 0.3$ and a manipulative male decreased her s from 0.99 to 0.7, the fraction of potential mates acceptable to her would increase from 1% to about 30%. For an RA assessment, where males assess a female as unacceptable, a female may likewise manipulate his switch point increasing the percentage of acceptable potential mates using mechanisms that reduce his e, s, or l.

fitness such that $w_1 \geq w_2 \geq \ldots \geq w_n$. Parameters are: e, probability of encountering a potential mate; s, probability of surviving one time unit; l, the number of time units in reproductive latency (time from onset of one copulation to receptivity to remating), and w_i is the fitness conferred by mating with the ith mating type or individual. If the different mating types are simply individuals that are potential mates, then $p_i = 1/n$ and equation (1) simplifies to equation (2). To find the maximum lifetime fitness for an individual, calculate equation (1) or (2) for values of f ranging from 1 to n.

$$W\left\{\sum\nolimits_f M|SA\right\}$$
$$= \frac{es^2\left(\sum_{i=1}^{f} p_i w_i\right)}{(1-s)+es\left[1-s^{l+1}\left(\sum_{i=1}^{f} p_i\right)-\sum_{i=f+1}^{n} p_i\right]} \quad (1)$$

$$W\left\{\sum\nolimits_f M|SA\right\} = \frac{es^2 \sum_{i=1}^{f} w_i}{(1-s)n + esf\left[1-s^{l+1}\right]} \quad (2)$$

The switch point is thus a threshold in the fitness of focal individuals that specifies the fraction of potential mates that the focal individual should accept or reject over its lifetime in order to optimize its lifetime fitness. The choosing individual should accept potential mates with fitness above the threshold, and should reject potential mates with fitness below this threshold.

It is important to emphasize that the switch point theorem specifies the instantaneous optimum switch point for the accept–reject threshold, which specifies the optimum rule that individuals should follow moment to moment as e, s, and l and the number of potential mates n change with the current social and physical environment. For example, switch points will change as some individuals die or enter latencies, which results in changes in the distribution of fitness of available potential mates. So, as these parameters change, the position of an individual's curve in fitness space will be time-varying, and thus their fitness threshold for switching from accepting to rejecting potential mates will also change. Because of the time-varying optimum, it is possible that a given individual will sometimes accept and sometimes reject even the same potential mate, depending on the current circumstances. Such flexible mating behavior of individuals will maximize lifetime fitness and therefore theoretically will always evolve.

For the purposes of this chapter it is sufficient to know that on average holding all else equal, shorter l, lower e, and lower s, for many w *distributions* increases the fraction of potential mates that are acceptable to a focal individual.

APPENDIX B. PROGRAM DYNAMATE

DYNAMATE is an individually based simulation model (each individual in a population is followed until death) that keeps track of individual behavior, mating frequencies, fitness conferred under constantly changing social and demographic variation, including when individuals are removed from the population because of latency or death (see figure 5.6). DYNAMATE can compare individuals with identical starting variables in populations; in DYNAMATE the probabilities of e, s, and l vary and change as individuals die or are removed from the mating pool by reproductive latency, and it tallies the reproductive decisions of focal individuals under the rules of the switch point theorem already described. And, if individuals mate under conflict (i.e, one accepts and the other rejects); and, assuming that neither sex has greater vulnerability to control than the other sex, DYNAMATE calculates the opportunity cost in fitness for those who reject but are manipulated into mating anyway, and fitness gains for those who successfully coerce. Thus, DYNAMATE assigns a relative fitness gain or loss to an individual of mating or not when one individual rejects and the other accepts during an encounter.

Acknowledgments We thank the editors for inviting our contribution to this volume and for their patience as we produced it. We thank those who read and commented on the manuscript including J.P. Drury, Brant C. Faircloth, Graham Pyke, William Eberhard, Theresa Markow, and the many who have listened to us talk about the switch point theorem and DYNAMATE since 2001 when we first began our collaboration. Neither of us could have done this work without the other, and we thank each other for our mutual—but not constant—patience and usual goodwill despite various roadblocks to continued work. PAG asked the motivating questions that stimulated our work. SPH solved the switch point theorem, and

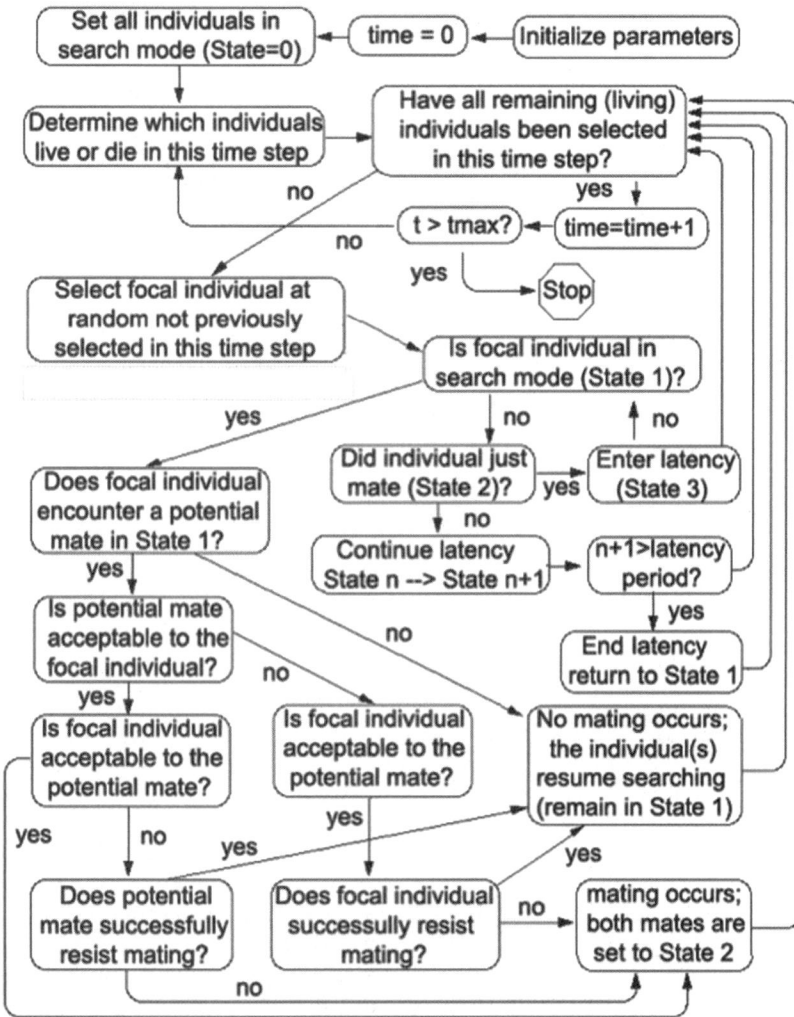

FIGURE 5.6 Flowchart of program DYNAMATE.

SPH and PAG worked together on the creation of DYNAMATE; PAG analyzed runs of the switch point theorem and DYNAMATE to obtain results displayed here. SPH drew several of the figures, and PAG wrote the manuscript. We acknowledge support from NSF grant "Beyond Bateman" that has allowed us to carry out experimental tests for chance effects on mating success variances.

REFERENCES

Anderson, W. W., Kim, Y.-K. & Gowaty., P. A. (2007) Experimental constraints on female and male mate preferences in *Drosophila pseudoobscura* decrease offspring viability and reproductive success of breeding pairs. Proceedings of the National Academy of Sciences of the United States of America, 104, 4484–4488.

Arnqvist, G. & Rowe, L. (2005) Sexual Conflict, Princeton, Princeton University Press.

Baena, M. L. & Eberhard, W. G. (2007) Appearances deceive: female "resistance" behaviour in a sepsid fly is not a test of male ability to hold on. Ethology, Ecology & Evolution, 19, 27–50.

Berglund, A. (1994) The operational sex-ratio influences choosiness in a pipefish. Behavioral Ecology, 5, 254–258.

Berglund, A. (1995) Many mates make male pipefish choosy. Behaviour, 132, 213–218.

Berglund, A. & Rosenqvist, G. (1993) Selective males and ardent females in pipefishes. Behavioral Ecology and Sociobiology, 32, 331–336.

Bluhm, C. K. & Gowaty, P. A. (2004a) Reproductive compensation for offspring viability deficits by female mallards, *Anas platyrhynchos*. Animal Behaviour, 68, 985–992.

Bluhm, C. K. & Gowaty, P. A. (2004b) Social constraints on female mate preferences in mallards *Anas platyrhynchos* decrease offspring viability and mother's productivity. Animal Behaviour, 68, 977–983.

Bonduriansky, R. (2001) The evolution of male mate choice in insects: a synthesis of ideas and evidence. Biological Reviews, 76, 305–339.

Byers, J. A. & Waits, L. (2006) Good genes sexual selection in nature. Proceedings of The National Academy of Sciences of the United States of America, 103, 16343–16345.

Darwin, C. (1871) The Descent of Man, and Selection in Relation to Sex, London, J. Murray.

Drickamer, L. C., Gowaty, P. A. & Holmes, C. M. (2000) Free female mate choice in house mice affects reproductive success and offspring viability and performance. Animal Behaviour, 59, 371–378.

Drickamer, L. C., Gowaty, P. A. & Wagner, D. M. (2003) Free mutual mate preferences in house mice affect reproductive success and offspring performance. Animal Behaviour, 65, 105–114.

Eberhard, W. G. (1996) Female Control: Sexual Selection by Cryptic Female Choice, Princeton, Princeton University Press.

Eberhard, W. G. (2005) Evolutionary conflicts of interest: are female sexual decisions different? American Naturalist, 165, S19–S25.

Forstmeier, W. (2004) Female resistance to male seduction in zebra finches. Animal Behaviour, 68, 1005–1015.

Gowaty, P. A. (1994) Architects of sperm competition. Trends in Ecology and Evolution 9, 160–161

Gowaty, P. A. (1996) Battles of the sexes and origins of monogamy. In Black, J. L. (Ed.) Black Partnerships in Birds. Oxford, Oxford University Press.

Gowaty, P. A. (1997) Sexual dialectics, sexual selection, and variation in mating behavior. In Gowaty, P. A. (Ed.) Feminism and Evolutionary Biology: Boundaries, Intersections, and Frontiers. New York, Chapman Hall.

Gowaty, P. A. (1998) Mate choice: overview. In Knobil, E. & Neill, J. D. (Eds) Encyuclopedia of Reproduction. San Diego, Academic Press.

Gowaty, P. A. (2002) Power asymmetries between the sexes, mate preferences, and components of fitness. In Travis, C. (Ed.) Sex, Aggression, and Power. Cambridge, Massachusetts, MIT Press.

Gowaty, P. A. (2008) Reproductive compensation. Journal of Evolutionary Biology, 21, 1189–1200.

Gowaty, P. A. & Buschhaus, N. (1998) Ultimate causation of aggressive and forced copulation in birds: female resistance, the code hypothesis, and social monogamy. American Zoologist, 38, 207–225.

Gowaty, P. A. & Hubbell, S. P. (2005) Chance, time allocation, and the evolution of adaptively flexible sex role behavior. Journal of Integrative and Comparative Biology, 45, 931–944.

Gowaty, P. A. & Hubbell, S. P. (2009) It's about time: chance, flexible individuals, and sexual selection. Proceedings of the National Academy of Sciences of the United States of America, 106, Suppl. 1, 10017–10024.

Gowaty, P. A., Steinichen, R. & Anderson, W. W. (2002) Mutual interest between the sexes and reproductive success in *Drosphila pseudoobscura*. Evolution, 56, 2537–2540.

Gowaty, P. A., Drickamer, L. C. & Holmes, S.-S. (2003a) Male house mice produce fewer offspring with lower viability and poorer performance when mated with females they do not prefer. Animal Behaviour, 65, 95–103.

Gowaty, P. A., Steinechen, R. & Anderson, W. W. (2003b) Indiscriminate females and choosy males: within- and between-species variation in *Drosophila*. Evolution, 57, 2037–2045.

Gowaty, P. A., Anderson, W. W., Bluhm, C. K., Drickamer, L. C., Kim, Y. K. & Moore, A. J. (2007) The hypothesis of reproductive compensation and its assumptions about mate preferences and offspring viability. Proceedings of the National Academy of

Sciences of the United States of America, 104, 15023–15027.

Gromko, M. H. & Markow, T. A. (1993) Courtship and remating in field population of *Drosophila*. Animal Behviour, 45, 253–262.

Gromko, M. H., Newport, M. E. A. & Kortier, M. G. (1984) Sperm dependence of female receptivity to remating in *Drosophila melanogaster*. Evoluion, 38, 1273–1282.

Hubbell, S. P. & Johnson, L. K. (1987) Environmental variance in lifetime mating success, mate choice, and sexual selection. American Naturalist, 130, 91–112.

Johnstone, R. A. & Keller, L. (2000) How males can gain by harming their mates: sexual conflict, seminl toxins, and the cost of mating. American Naturalist, 156, 366–377.

Jormalainen, V. & Merilaita, S. (1995) Female resistance and duration of mate-guarding in three aquatic peracarids (Crustacea). Behavioral Ecology and Sociobiology, 36, 43–48.

Kelleher, E. S., Swanson, W. J. & Markow, T. A. (2007) Gene duplication and adaptive evolution of digestive proteases in *Drosophila arizonae* female reproductive tracts. Plos Genetics, 3, 1541–1549.

Knowles, L. L. & Markow, T. A. (2001) Sexually antagonistic coevolution of a postmating-prezygotic reproductive character in desert *Drosophila*. Proceedings of the National Academy of Sciences of the United States of America, 98, 8692–8696.

Knowles, L. L., Hernandez, B. B. & Markow, T. A. (2004) Exploring the consequences of postmating-prezygotic interactions between the sexes. Proceedings of the Royal Society of London Series B—Biological Sciences, 271, S357–S359.

Knowles, L. L., Hernandez, B. B. & Markow, T. A. (2005) Nonantagonistic interactions between the sexes revealed by the ecological consequences of reproductive traits. Journal of Evolutionary Biology, 18, 156–161.

Markow, T. A. & Ankney, P. F. (1988) Insemination reaction in *Drosophila* found in species whose males contribute material to oocytes before fertilization. Evolution, 42, 1097–1101.

Moore, A. J., Gowaty, P. A. & Moore, P. J. (2003) Females avoid manipulative males and live longer. Journal of Evolutionary Biology, 16, 523–530.

Navara, K. J., Hill, G. E. & Mendonca, M. T. (2004) Females deposit more yolk androgens into eggs sired by less attractive males: a contradiction to the differential allocation hypothesis? Integrative and Comparative Biology, 44, 611–611.

Navara, K. J., Hill, G. E. & Mendonca, M. T. (2006) Yolk androgen deposition as a compensatory strategy. Behavioral Ecology And Sociobiology, 60, 392–398.

Parker, G. A. & Simmons, L. W. (1996) Parental investment and the control of sexual selection: predicting the direction of sexual competition. Proceeding of the Royal Society, London, Series B, 315–321.

Parker, G. A., Baker, R. R. & Smith, V. G. F. (1972) The origin and evolution of gamete dimorphism and the male–female phenomenon. Journal of Theoretical Biology, 36, 529–553.

Pitnick, S. & Markow, T. A. (1994) Male gametic strategies: sperm size, testes size, and the allocation of ejaculate among successive mates by the sperm-limited fly *Drosophila pachea* and its relatives. American Naturalist, 143, 785–819.

Pitnick, S., Markow, T. A. & Riedy, M. F. (1991) Transfer of ejaculate and incorporation of male-derived substances by females in the *Nonoptera* species group (Diptera, Drosophilidae). Evolution, 45, 774–780.

Pitnick, S., Spicer, G. S. & Markow, T. (1997) Phylogenetic examination of female incorporation of ejaculate in *Drosophila*. Evolution, 51, 833–845.

Sih, A., Lauer, M. & Krupa, J. J. (2002) Path analysis and the relative importance of male–female conflict, female choice and male–male competition in water striders. Animal Behaviour, 63, 1079–1089.

Smuts, B. (1992) Male aggression against women: an evolutionary perspective. Human Nature, 3, 1–44.

Smuts, B. & Smuts, R. W. (1993) Male aggression and sexual coercion of females in nonhuman primates and other mammals: evidnce and theoretical implications. Advances in the Study of Behavior, 22, 1–63.

Snook, R. R. & Markow, T. A. (1996) Possible role of nonfertilizing sperm as a nutrient source for female *Drosophila pseudoobscura* Frolova (Diptera: Drosophilidae). Pan-Pacific Entomologist, 72, 121–129.

Snyder, B. F. & Gowaty, P. A. (2007) A reappraisal of Bateman's classic study of intra-sexual selection. Evolution, 61, 2457–2468.

Sutherland, W. J. (1985) Chance can produce a sex difference in variance in mating success and account for Bateman's data. Animal Behaviour, 33, 1349–1352.

Trivers, R. L. (1972) Parental investment and sexual selection. In Campbell, B. (Ed.) Sexual Selection and the Descent of Man. Chicago, Aldine.

West-Eberhard, M. J. (1984) Sexual selection, competitive communication and species-specific signals in insects. In Lewis, T. (Ed.) Insect Communication. New York, Academic Press.

Two

PRIMARY SEXUAL CHARACTERS IN SELECTED TAXA

6

Gamete Release and Spawning Behavior in Broadcast Spawning Marine Invertebrates

KATIE E. LOTTERHOS AND DON R. LEVITAN

INTRODUCTION

The release of eggs and sperm into the environment for external fertilization is common (Giese & Kanatani 1987) and thought to be the ancestral mating strategy (Wray 1995). Most phyla have at least some species that release sperm into the environment (free spawning or spermcasting) and a majority have representatives that also release eggs (broadcast spawning) for external fertilization (Levitan 1998a).

A common feature of broadcast spawning taxa is a lack of sexually dimorphic adult characters (Strathmann 1990; Levitan 1998a). The rare instances of adult morphological differences among sexes occur in pair-spawning species and include body size differences or specialized appendages for clasping females (Levitan 1998a). The general absence of sexual dimorphism has been used as evidence that broadcast spawning invertebrates are not under sexual selection (Darwin 1871).

In contrast to this view on the nature of sexual selection based on sexual dimorphism, recent studies documenting the rapid evolution of gamete recognition proteins in external fertilizing species have invoked sexual selection and sexual conflict as the driving force of positive selection (Palumbi 1999; Swanson & Vacquier 2002; Haygood 2004). Only recently have data been collected on the patterns of reproductive variance and the intensity of sexual selection in male and female broadcast spawners

(Levitan 2004, 2005a, b; Levitan & Ferrell 2006; Levitan 2008). The results suggest that sexual selection can be intense in these species and that the nature of sexual selection is dependent on the distribution and abundance of individuals.

Given the nature of external fertilization, where the competition among individuals for mates is played out among gametes in the water column, it is not surprising that the primary sexual characteristics targeted by selection are associated with the spawning behaviors that mediate gamete competition and the traits of the gametes themselves. The influence of sperm availability and sexual selection on gamete traits has already received some attention (Levitan 1998b, 2006). Below we briefly summarize the relationship between gamete traits and fertilization and direct readers to the relevant literature. The remainder of the chapter is focused on the less studied question of how sexual selection might influence the evolution of the duration, release rate, and timing of gamete release. Because these features determine the size, shape, and duration of an individual's gamete distribution, it defines the extent to which one individual can interact with mates and mate competitors.

Gamete Traits

There is emerging evidence that selection caused by too few or too many sperm might have influenced the evolution of both sperm and eggs (Levitan 2002a;

Marshall & Keough 2003; Levitan & Ferrell 2006). Gamete traits that do best under sperm-limited conditions are those that either enhance the likelihood of sperm–egg collisions or increase the probability of fertilization given a collision. Higher rates of sperm–egg collisions can be a result of increasing egg cell size (Levitan 1993, 1996; Marshall & Keough 2003), associated structures that might capture sperm (Farley & Levitan 2001; Podolsky 2001, 2002), and chemical sperm attractants (Miller 1985; Riffell et al. 2004). Increasing the likelihood of fertilization can also be influenced by the properties of the egg surface that determine the fraction of sperm collisions that result in fertilization (Vogel et al. 1982; Levitan 1993). A potential cost to producing eggs too easy to fertilize can be an increase in the risk of heterospecific fertilizations (Levitan 2002b) or developmental failure caused by multiple sperm fusing with the egg surface (Styan 1998; Franke et al. 2002; Levitan 2004; Levitan et al. 2007). Eggs presenting a smaller sperm target, or a more restricted set of acceptable sperm, appear able to successfully block excess sperm (Levitan et al. 2007).

Sperm may also be under selection based on different levels of sperm availability. Sperm limited conditions may select for longer-lived sperm, as sperm must be viable for longer periods before encountering eggs, while sperm competitive conditions may select for more rapid sperm as unfertilized eggs are an ephemeral resource (Levitan 1993, 2000; Kupriyanova & Havenhand 2002). Empirical and theoretical examinations of how egg size might influence collision rates have caused for a reexamination of the evolution of anisogamy (eggs and sperm) from an isogamous condition (Levitan 1996; Dusenbery 2000; Randerson & Hurst 2001; Bulmer & Parker 2002; Podolsky 2004; Bode & Marshall 2007).

Spawning Behavior

Because of the difficulty associated with predicting the timing and location of spawning for most broadcast spawning taxa, there is a paucity of information on the timing (relative to mates), spawning duration (length of time gametes are released), and frequency (number of times an individual spawns in a season, or reproductive bout) of gamete release. Much of what is known is based on scattered reports of spawning behavior in nature or in the laboratory and what can be gleaned by experimentally inducing individuals to spawn via chemical or temperature stimulants. In this chapter, we are interested in how individuals parcel out their gametes during release and the effect this may have on male–female differences in time of spawning. What are the advantages and disadvantages of blasting out gametes for a short duration or dribbling out gametes for longer intervals? Does the optimal strategy depend on mate or competitor densities or environmental factors such as water flow? Do males and females pursue different spawning strategies?

Most theoretical treatments of the mechanics of spawning assume that individuals spawn as a plume. Plumes are a steady release of gametes that, in principle, establish a constant gradient of gamete concentration from the point of release. However, when gamete release is of short duration, a constant gradient may not be established and a time-dependent concentration gradient is established as gametes disperse from a source. Acknowledging that gamete release is an ephemeral process and that gamete concentration gradients have an ebb and flow is a necessary prerequisite for examining the consequences of different spawning strategies.

In this chapter we first survey different patterns of gamete release among taxa. We then discuss experimental manipulations of gamete release to examine how different release strategies influence male and female reproductive success under a variety of demographic conditions. Then, we introduce a model of turbulent diffusion from a point source that can be used to estimate gamete concentrations for different spawning durations. We expand this model to a two-dimensional population to explore how advection, male density, and male spawning duration affect the time when females choose to initiate spawning. Then, using comparative data from the literature, we examine the evidence for predictions from the model. Finally, we discuss how the effectiveness of these strategies may vary with the level of sperm competition, population density, and ambient water flow.

PATTERNS OF GAMETE RELEASE

Because our ability to predict the timing and location of spawning in marine populations is limited, and spawning is often nocturnal, many observations of natural spawning events are on lone individuals or a small portion of the population. In particular, although it is common to report the time of spawning in relation to some environmental variable (i.e., sunset or day in a lunar cycle), spawning duration is not commonly measured. For this section, we reviewed the literature in order to gain insight into the different strategies for gamete release in broadcast spawners.

We concentrate on broadcast spawning species, in which sperm and eggs are released into the water column. Spermcasting species are organisms in which males release sperm into the surrounding water, but females that retain eggs (e.g., some sponges—Reiswig 1970; byrozoans and ascidians—Yund & McCartney 1994; gorgonians—Lasker 2006) may have a very different dynamic (Bishop 1998). Females that retain eggs on or inside their surface are able to integrate sperm concentration in the seawater over the life of the egg. High fertilization rates have been observed as a result of this strategy (Phillippi et al. 2004; Lasker 2006). The dynamics of broadcast spawning species are different because the concentrations of sperm and eggs become diluted in the water column, and fertilization rates will decrease with decreasing concentrations. Therefore, the rate and timing of gamete release will have a large effect on fertilization rates of broadcast spawners.

The physiological act of spawning places gametes into the environment. The dispersal and dilution of those gametes can be influenced by the viscosity of the spawned material, adult morphology and posture, and the interaction of these factors with water flow. In polychaetes, sperm appear to advect away from adults in a plume, while eggs can form strings or clumps that retain on the female for seconds to minutes before being released (Thomas 1994a). Thomas (1994b) has suggested that adults may increase the viscosity of spawned material when flow and turbulence are high to reduce the rapid dilution of gametes. Enhanced residence times of eggs on the female's surface would allow eggs to sample the water column for sperm passing by. Under some conditions this might lead to higher levels of fertilization compared to eggs released into the water column that drift with a particular parcel of water.

Adults can also influence how flow interacts with gamete dispersal through movement patterns and posture. Many species climb objects prior to spawning or assume a posture that places their gonopores into the water column (Pearse 1979; McEuen 1988; Hendler 1991; Babcock et al. 1992; Minchin 1992; Stekoll & Shirley 1993).

Releasing gametes under the higher flow conditions above the surface boundary layer would increase the rate at which they would be advected away from adults (Yund & Meidel 2003). This suggests that not all species behave in a way that would enhance gamete retention; in some situations gamete dispersal may be advantageous.

Differences in viscosity, behavior, morphology and habitat may explain why in some taxa, gametes seem to disperse immediately into the water column, while in other taxa they may reside for short periods before being advected away. Observations of gametes being released directly into the water column include cnidarians and echinoderms (Minchin 1992; van Veghel 1994; Himmelman et al. 2008). In some cases, notably temperate sea urchins, eggs and sperm pool on the aboral surface for several seconds to minutes before being advected off the adult surface (Minchin 1992; Levitan 2002a; Himmelman et al. 2008). Sea urchins may retain gametes longer because their spines might create a boundary layer around the adults; however, the tropical sea urchin *Diadema antillarum*, which has extremely long spines, spawns gametes directly into the water column (DRL personal observation), suggesting that viscosity or expulsion rate may also influence dispersal rate. The temperate sea urchins that often retain gametes also live in wave swept environments, and gametes appear to lift off at greater rates during the change in water direction associated with wave action (DRL personal observation), which is when flow is accelerating and presumably highly turbulent. Laboratory experiments of gamete fertilization of sea urchins in mild to moderate unidirectional flow, where changes in velocity are eliminated, indicate that eggs are slowly released off of females for several hours and that a proportion of those eggs are fertilized during this period (Yund & Meidel 2003). But because individuals can spawn for several hours (Levitan 2002a), these experiments cannot distinguish between long residence times (the length of time a gamete remains on the surface) and long spawning times.

The spectrum of gamete release observed in a variety of taxa ranges from organisms that release their gametes in a near instantaneous release ("puff"), to organisms that spurt gamete batches or bundles over a period of time ("pulse"), to organisms that continuously release gametes from several minutes to over nearly an hour or longer ("plume"). Because measurements of spawning duration are not the main focus of most studies, the observations we cite are based on natural spawning in the field or laboratory and usually only for a few individuals (table 6.1).

A majority of observations are from organisms that release gametes continuously. Continuous spawners can release gametes for just a few seconds to several minutes to an hour or more (table 6.1, figure 6.1a, b). The advantage of releasing gametes for an extended period of time is that if synchrony in the population is low, the probability of finding a mate is increased. The cost of this behavior is a reduction in the concentration of gametes

TABLE **6.1** Spawning observations of marine invertebrate species that release gametes continuously (as a puff or plume) or in pulses. If the sexes exhibited different strategies, or only one sex was observed spawning, it is noted below

Species	Taxa (Phylum: Class: Order)	Spawning Behavior	Reference
Dryodora glandiformis, Bolinopsis vitrea, Pleurobrachia bachei	Ctenophora: Tentaculata	**Pulse:** Sperm are released in bursts over a 5 minute period, followed by bursts of egg release over 5–10 minutes (hermaphrodite).	Pianka 1974
Macrorynchia philippina	Cnidaria: Plumulariida: Aglaopheniidae	**Both:** Spawning of swimming medusiods lasted 1–2 minutes. Males released sperm continuously and females released batches of 1–4 eggs with each contraction.	Bourmaud & Gravier-Bonnet 2004
Heteractis magnifica	Cnidaria: Anthozoa: Actiniaria	**Pulse:** One female was observed releasing 1 to 10 eggs at a time over a period of 1 hour.	Babcock et al. 1992
Montastraea annularis complex	Cnidaria: Anthozoa: Scleractinia	**Continuous:** Synchronous release of buoyant egg-sperm bundles across the entire coral head under a minute.	Szmant 1986; Levitan et al. 2004
Montastraea cavernosa	Cnidaria: Anthozoa: Scleractinia	**Pulse:** Sperm expelled from male colonies as repeated plumes; female colonies rapidly released eggs.	Gittings et al. 1992
Acropora aspera group	Cnidaria: Anthozoa: Scleractinia	**Continuous:** Colonies spawned over a 15–20 minute period.	Van Oppen et al. 2002
Stephanocoenia intersepta	Cnidaria: Anthozoa: Scleractinia	**Continuous:** Male colonies release sperm over a period of 4–5 minutes, and females release eggs over 2–3 minutes.	Vize et al. 2005
Diploria strigosa	Cnidaria: Anthozoa: Scleractinia	**Continuous:** Eggs released over an 8 hour period, sperm release not noted.	Wyers et al. 1991
Pseudoplexaura porosa	Cnidaria: Anthozoa: Gorgonacea	**Continuous:** Males spawn for 60–90 minutes and females spawn for 25–40 minutes.	Coma & Lasker 1997
Phragmatopoma californica	Annelida: Polychaeta	**Pulse:** Males release a series (1–10) of sperm clouds or clumps into the water column over a 1–15 second period. **Continuous:** A female released eggs for 193 seconds.	Thomas 1994b
Spirobranchus giganteus	Annelida: Polychaeta	**Continuous:** Males and females spawned for approximately 15 minutes.	Babcock et al. 1992
Aspidosiphon fischeri (formerly: Paraspidosiphon fischeri)	Sipuncula: Phascolosomatidea: Aspidosiphoniformes	**Pulse:** A male spawned seven times in the laboratory over a period of 40 minutes (females not noted).	Rice 1975

TABLE **6.1** *(Cont.)*

Species	Taxa (Phylum: Class: Order)	Spawning Behavior	Reference
Phascolion cryptum (formerly: *Phascolion cryptus*)	Sipuncula: Sipunculidea: Golfingiiformes	**Pulse:** Released short intermittent spurts of sperm over 15 minutes (females not noted).	Rice 1975
Glottidia spp. and *Terebratalia* spp.	Brachiopoda	**Continuous:** Eggs were shed for an hour to several hours (males not noted).	Long & Stricker 1991
Mopalia lignosa	Mollusca: Polyplacophora: Neoloricata	**Pulse:** Males observed to release sperm in spurts lasting 3–5 minutes at 5–15 minute intervals. **Continuous:** Females released a steady stream of eggs.	Pearse 1979
Lepidochitona cinereus	Mollusca: Polyplacophora: Neoloricata	**Continuous:** Females released eggs over about 2.5 hours (males not noted).	Pearse 1979
Crassostrea spp.	Mollusca: Bivalvia: Ostreoida	**Pulse:** In females eggs are expelled in batches by contractions of adductor muscle (males not noted).	Andrews 1979
Hyotissa hyotis, Chama spp., *Arca* spp.	Mollusca: Bivalvia	**Pulse:** Gametes shed intermittently over a period of 15 minutes.	Babcock et al. 1992
Tridacna gigas	Mollusca: Bivalvia: Veneroida	**Pulse:** Males released sperm in contractions spaced 2–5 min apart, lasting for 1–2 hours. Female spawning not noted.	(Babcock et al. 1992)
Acanthaster planci	Echinodermata: Asteroida: Valvatida	**Continuous:** Spawning in both sexes lasted for about 30 minutes to an hour.	Babcock et al. 1992; Babcock et al. 1994
Linkia laevigata	Echinodermata: Asteroida: Valvatida	**Continuous:** Gamete release lasted 15–30 minutes.	Babcock et al. 1992
Amphioplus abditus	Echinodermata: Ophiuroidea: Ophiurida	**Pulse:** Release of gametes was intermittent over 20 minutes.	Hendler 1991
Gorgonocephalus eucnemis	Echinodermata: Ophiuroidea: Ophiurida	**Continuous:** Spawned for 1.5 hours.	Hendler 1991
Ophiothrix angulata	Echinodermata: Ophiuroidea: Ophiurida	**Continuous:** In females all eggs were shed "explosively" in one session.	Hendler 1991
Ophiothrix orstedii	Echinodermata: Ophiuroidea: Ophiurida	**Pulse:** Gametes ejected in 5 second bursts alternating with 30 second intervals of repose.	Hendler 1991
Ophiura robusta	Echinodermata: Ophiuroidea: Ophiurida	**Continuous:** A mass spawn lasted 30 minutes.	Himmelman et al. 2008
Ophiopholis aculeata	Echinodermata: Ophiuroidea: Ophiurida	**Continuous:** A mass spawn lasted 20 minutes.	Himmelman et al. 2008
Strongylocentrotus droebachiensis	Echinodermata: Echinoidea: Echinoida	**Continuous:** Individuals were observed to spawn for at least one hour.	Himmelman et al. 2008
Strongylocentrotus fransicanus	Echinodermata: Echinoidea: Echinoida	**Continuous:** Males were observed to spawn for 140 minutes and females for 60 minutes.	Levitan 2002a
Diadema antillarum	Echinodermata: Echinoidea: Echinoida	**Continuous:** Urchins were observed to spawn for at least 2 hours.	Randall et al. 1964

Continued

TABLE 6.1 *(Cont.)*

Species	Taxa (Phylum: Class: Order)	Spawning Behavior	Reference
Cucumaria lubrica	Echinodermata: Holothuroidea: Dendrochirotida	**Continuous:** In males, sperm exit the gonopore in bundles over a period of 2 or more hours. Females began to spawn 2–7 hours after males, and released eggs for 1–4 hours.	McEuen 1988
Cucumaria miniata	Echinodermata: Holothuroidea: Dendrochirotida	**Continuous:** Males released sperm for 3–6 hours. **Pulse:** Females released egg packets every 4–10 minutes, over a period of 0.75–4 hours.	McEuen 1988
Psolus chitonoides	Echinodermata: Holothuroidea: Dendrochirotida	**Continuous:** Males spawned for 1–3 hours, females spawned for 1–1.5 hours.	McEuen 1988
Pentamera populifera	Echinodermata: Holothuroidea: Dendrochirotida	**Pulse:** Males emitted sperm in pulses lasting 2.75–3 minutes, as many as 5 pulses were visible. **Continuous:** Females released eggs for 1–2 hours.	McEuen 1988
Molpadia intermedia	Echinodermata: Holothuroidea: Molpadiida	**Pulse:** In males, 1–2 second long puffs of sperm spurted out at intervals of 0.7–6.5 min. **Continuous:** A female was observed to release all eggs in an explosive burst.	McEuen 1988
Isostichopus fuscus	Echinodermata: Holothuroidea: Aspidochirotida	**Continuous:** Females released gametes on average for 48 min (± 18 min St. Err.) and males for 62 min (± 18 min St. Err.).	Mercier et al. 2007
Stephanometra spp.	Echinodermata: Crinoidea: Comatulida	**Pulse:** Several short bursts lasting 2–3 seconds.	Babcock et al. 1992
Lamprometra klunzingeri	Echinodermata: Crinoidea: Comatulida	**Continuous:** Spawning for both sexes lasts only about 25 seconds.	Holland 1991
Oxycomanthus japonicus	Echinodermata: Crinoidea: Comatulida	**Continuous:** Females spawned within 5 minutes.	Holland 1991

(compared to if all were released at once). This might influence reproductive success in females if optimal sperm concentrations are not reached and in males if high concentrations are needed to assure paternity during male competition. The advantage of releasing gametes in an explosive burst, is that gamete concentrations are not reduced (compared to if they were released over a period of time). However, this requires high synchrony with mates because gamete concentrations are more ephemeral.

Broadcasting spawning corals are an interesting case because they are hermaphroditic, and package gametes into egg–sperm bundles. In the *Montastraea annularis* complex, these bundles are released almost simultaneously across the entire coral head, where they float to the surface and break apart for fertilization (Szmant 1986; Levitan et al. 2004). Since self-fertilization is not evident in these species, this strategy ensures initially high concentrations of both types of gametes, but apparently is only successful if population synchrony is high (Levitan et al. 2004). Not surprisingly, spawning in these coral populations is highly predictable and can be calculated from lunar calendars and sunset times (Szmant 1986; Levitan et al. 2004).

Some animals may spawn for a length of time, but not do so continuously as in pluming organisms. For example, some invertebrates release gametes for several minutes, stop for a period of time,

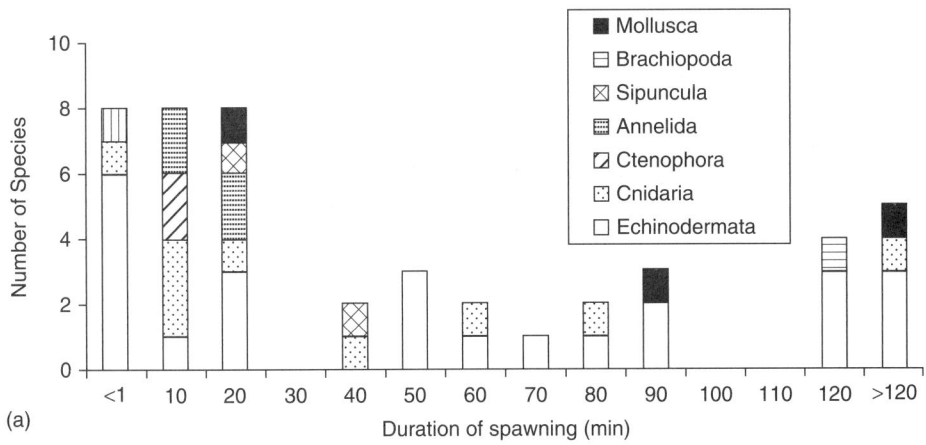

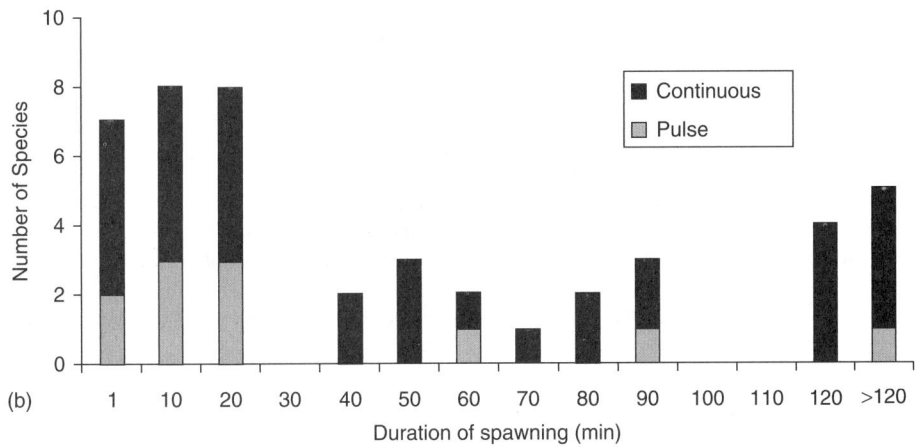

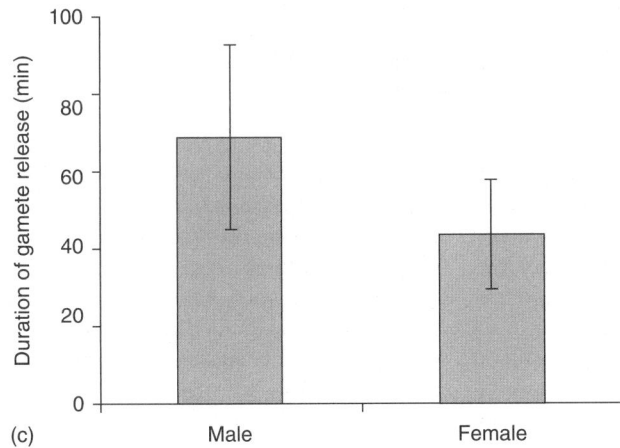

FIGURE 6.1 Results from a survey on gamete release observations in nature. Males and females were counted as separate within a species. (a) Histogram of spawning duration organized taxonomically. (b) Same histogram organized by type of gamete release (continuous or in pulses). (c) Average spawning duration for both males and females within a species, from studies in which both sexes were reported. Means are significantly different based on a paired *t*-test.

and then start releasing again (table 6.1). This may last for an hour or more. Similarly, other invertebrates may release gametes in quick, successive spurts that last anywhere from less than a second to over a half an hour. Whether the difference between individuals releasing the same number of gametes continuously or in pulses is related to fertilization success or is a result of physiological constraints has yet to be investigated.

We used the studies cited in table 6.1 to examine whether duration of gamete release exhibited taxonomic patterns, release patterns, or sex differences. For this analysis, males and females were counted as separately within a species if reported, and the midpoint was used if a range of spawning duration was reported. Only species for which total duration was reported were used. A wide range of behaviors are noted, even with a taxonomic group (e.g., Echinodermata, figure 6.1a). In order to examine whether there was a difference between continuous and pulsing spawners, we plotted the duration of gamete release against the number of species (figure 6.1b). On average, pulsers ($n = 11$, mean = 32.5 ± 13.9 min) tended to have a shorter spawning duration than organisms that spawn continuously ($n = 34$, mean = 67.1 ± 16.2 min, means marginally nonsignificant by t-test assuming unequal variances, $p = 0.057$). Within a species, we tested if there were significant differences between males and females in their spawning duration ($n = 13$, data was only used if male and female duration was observed in the same study). While the duration of spawning between males and females within a species was highly correlated (Pearson's correlation coefficient = 0.94), females spawned for a significantly less duration than males (figure 6.1c, one-tailed paired-t-test, t-stat = 2.17, $p = 0.025$). This supports the observation that females tend to start spawning after males and both sexes finish at the same time (Hamel & Mercier 1996; Levitan 2002a). Overall, observations of spawning durations in nature are spread throughout a wide range of times, but there may be considerable bias toward shorter times because they are easier to quantify. Notably, there appears to be no taxonomic pattern, which suggests that phylogenetic constraints do not limit these behaviors and that contemporary selective pressures (perhaps associated with demography or water flow characteristics) might drive spawning durations.

To date there is not enough data to make general conclusions about how spawning duration may vary within and between individuals. For instance, individuals may increase the rate of spawning in response to cues from conspecifics, which has been occasionally observed (Levitan 1998a; Himmelman et al. 2008). The duration of gamete release, along with the advective environment and the number of gametes released per unit time will determine the size and extent of the gamete cloud (Denny & Shibata 1989), and thus interactions with conspecifics of the opposite sex. In the next section, we outline experimental data that examines the consequences of variation in spawning times. Then, we use a model to explore how different spawning durations affect the distribution of gametes, and how this affects female choice in when to spawn.

EXPERIMENTAL MANIPULATIONS

Sea urchins provide a good model for mechanistic studies of gamete interactions because spawning is easy to manipulate, by KCl injection, and gamete performance can then be examined through controlled artificial releases in either the laboratory or field. In addition, adults can be manipulated to spawn at different times in the field, allowing examination of the costs and benefits of synchronous and asynchronous spawning behavior.

Although artificial induction of spawning removes the natural decision over when and where to spawn, there is some evidence that induced spawning events can closely mimic natural spawning events in terms of gamete release, aggregative behavior and fertilization rates in at least some species. Experimentally induced spawning events in which all individuals within a 5×5 meter area were injected with KCl, and immediately placed back in their original position had the same fertilization rates and nearest neighbor distances compared to naturally spawning sea urchins at those same densities (Levitan 2002a). This indicates, for at least some sea urchin species, that gamete release rates, gamete quality, and adult aggregative behavior are not markedly different during natural and induced spawning events.

Syringe Release Experiments

Gametes from the sea urchin *Strongylocentrotus franciscanus* were collected from adults and released via syringes to examine how subtle differences in the timing and position of gamete release influenced male and female fertilization success in and out of sperm competition (figure 6.2). Male success in the presence of competition was determined using microsatellite markers to assign paternity. Eggs and sperm were released in a 5-second burst at the same release rate as natural gamete release (Levitan 2005b). Sperm from either one or two males were released before and/or after egg release in 20-second intervals to examine the consequences of spawning early or late in the presence or absence of a competing male.

In the first set of treatments, all gametes were released into the center of the same parcel of seawater (which drifted downstream with the natural current), such that all subsequent releases were in the center of the gamete cloud of the initial gamete release. This mimicked individuals in a downstream linear arrangement, where gametes from one individual would pass over another individual, who would release gametes into that gamete cloud and then the pooled gamete cloud would potentially pass over a third individual for a similar release. The time differences between gamete releases (20 seconds) could also be viewed as the distance downstream between individuals; the distance gametes might drift and disperse in 20 seconds.

Two sets of treatments released gametes in the same relative position (figure 6.2a,b). Over both treatments, female success did not depend on whether male or female gametes were released first, but a higher proportion of eggs were fertilized when two males competed, because the total sperm released doubled (figures 6.2a,b, 6.3a). In the first

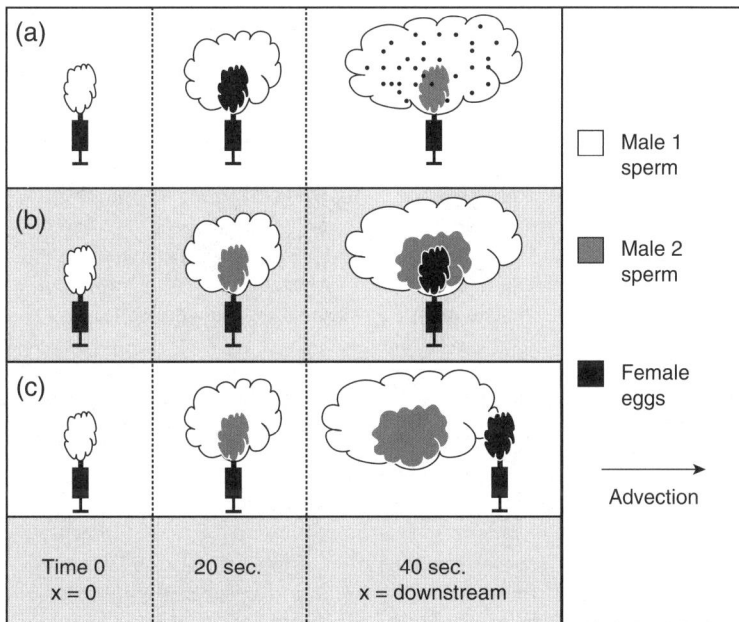

FIGURE **6.2** Schematic of the three experiments that examined the reproductive success of different sex-specific spawning behaviors under male competition (Levitan 2005b). The order of gamete release is presented from left to right, color-coded by male 1 (white), male 2 (grey), and the female (black). Gamete clouds were followed downstream as they dispersed to mimic a spawning event. (a) Spawning in same location with one male first: sperm from male 1 is released 20 seconds prior to egg release and male 2 was released 20 seconds post-egg release. Results from this experiment are presented in figure 6.3b. (b) Spawning in same location with both males first: sperm from male 1 is released 40 seconds prior to egg release, sperm from male 2 is released 20 seconds prior to egg release, and eggs are released in the center of the sperm cloud. Results from this experiment are presented in figure 6.4a. (c) Spawning at different locations with both males first: sperm from male 1 is released 40 seconds before egg release, sperm from male 2 is released 20 seconds before egg release, and eggs are released 1 m from the sperm cloud. Results from this experiment are presented in figure 6.4b.

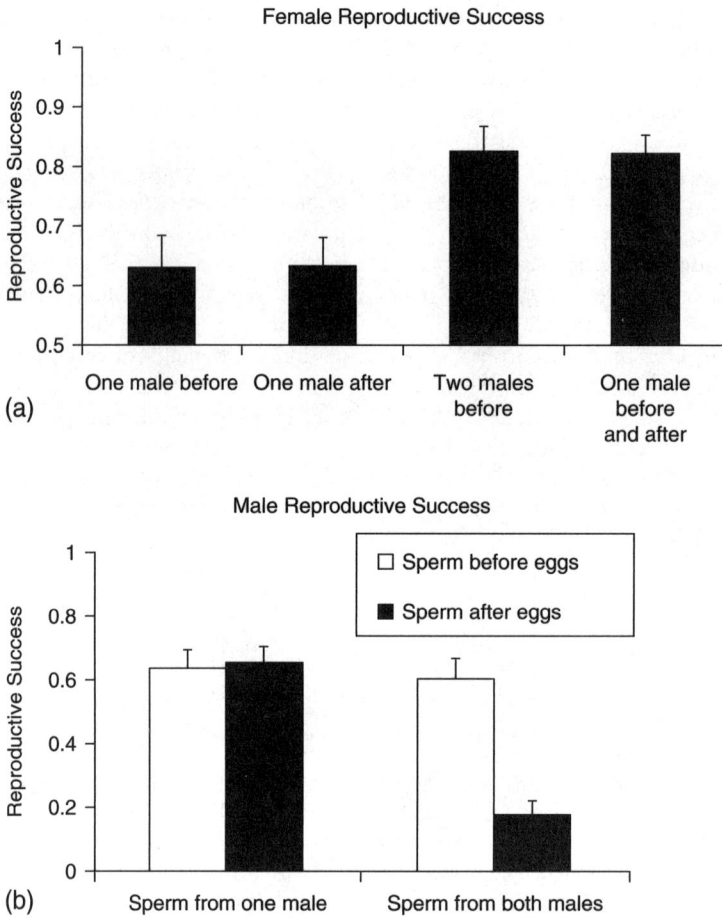

FIGURE **6.3** Male and female reproductive success as a function of synchrony and sperm competition. (a) Female reproductive success was higher when sperm from two males were released, but it was not dependent on spawning order (figure 6.2 a, b). (b) In the absence of competition, male reproductive success was the same if sperm was released before or after females. Sperm from the "one male" treatment tested each male in independent trials. In the presence of competition, the sperm released before the eggs were released garnered an advantage. For details see Levitan (2005b).

set of treatments, one male spawned before and one male after the female (figure 6.2a). When only one male released sperm, there was no difference in male success when sperm was released before or after egg release (figure 6.3b), but in competition, males that spawned after females lost to males that spawned before females (figures 6.2a, 6.3b). In the second set of treatments, both males spawned before the female (figure 6.2b). Males that spawned in closer synchrony with females (20 seconds prior to females as opposed to 40 seconds prior to females) garnered more fertilizations, especially so when in direct competition (figures 6.2b, 6.4a).

This set of trials indicated that when males compete, they should be selected to spawn just prior to females, assuming that they are physically close to females.

A third set of treatments considered spatial variance, such that subsequent gamete releases were one meter from the center of the gamete cloud (figure 6.2c). These different scenarios attempted to capture the potential trade-off of a male spawning earlier before female release and producing a more diffuse gamete cloud that covered a larger spatial area compared to a later spawning male that produced a more concentrated gamete cloud with

Sperm released in center

(a)

Sperm released off center

(b)

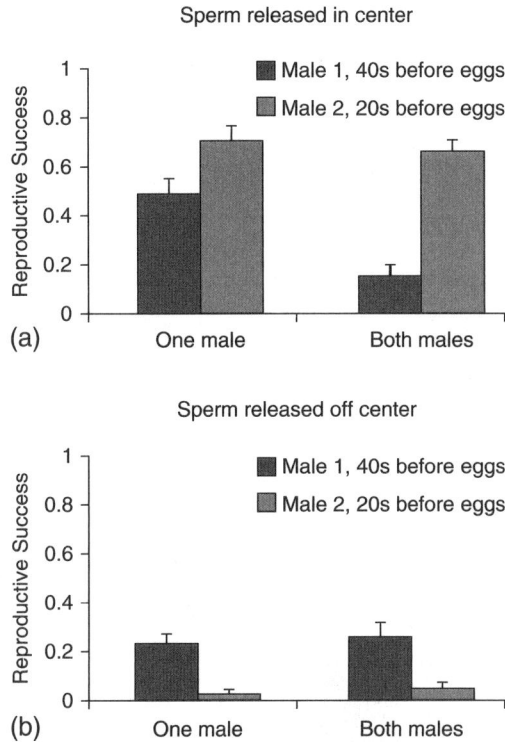

FIGURE **6.4** Male reproductive success when sperm were released 40 or 20 seconds before eggs were released. Eggs were released either (a) in the center of the sperm cloud or (b) 1 m from the center of the sperm cloud. Sperm from the "one male" treatments tested each male in independent trials. The male that released first had higher success when eggs were released away from the sperm cloud, but lower success if eggs were released in the center of the sperm cloud. For details see Levitan (2005b).

limited spatial coverage. In these trials that examined spatial variation in spawning, early spawning males had an advantage over later spawning males that were more synchronous with females. The sperm from early spawning males had the time to spread over a larger spatial area and could fertilize eggs and outcompete more synchronous males whose sperm was more concentrated over a smaller spatial area (figures 6.2c, 6.4b). These results suggest that males are selected to spawn before females, and the time difference between male and female spawning may depend on the spatial distribution of individuals and the degree of sperm competition (Levitan 2005b).

Gamete Release from Induced Spawning Experiments

The releases of gametes in the above experiments were in short puffs and the timing differences were staged at 20 second intervals. This provides a measure of the relative consequences of subtle differences in spawning times, but spawning durations in marine invertebrates can often extend for an hour or longer (table 6.1). To test the patterns noted above over longer spawning durations, individual sea urchins were induced to spawn at different intervals to determine how early or late initiation of male spawning influenced male reproductive success over a range of male to female distances and water flow (figure 6.5). Males placed in the center of the experiment were initiated to spawn 20 minutes, 10 minutes and 5 minutes before eggs were collected from spawning females. Females were placed between 0.5 and 3.5 meters from the group downstream of spawning males. Parentage of collected embryos from each female was determined from microsatellite loci and male reproductive success was examined as a function of how early he initiated spawning and the distance between the male and female (figure 6.5). The results indicated that early spawning males sired an equal number of

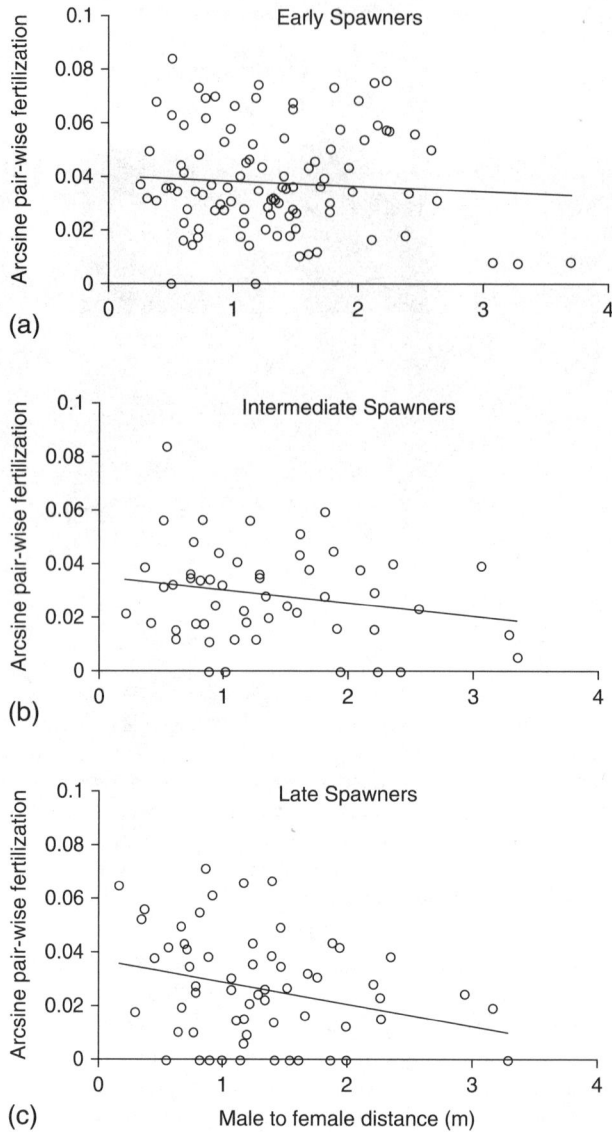

FIGURE **6.5** Fertilization success of a particular male–female pair (determined by microsatellite paternity assignment) as a function of the distance between them, for early, intermediate, and late spawning males. Significance of advection and male–female distance on fertilization success was tested with a multiple regression. (a) Early-spawning did not have a significant effect for male–female distance ($P < .05$). (b) For intermediate spawners this effect was marginally significant and (c) late-spawning males had a significant effect of male–female distance. For late spawners, greater distance between males and females resulted in lower levels of fertilization. For details see Levitan (2005b).

offspring independent of female distance, while late spawning males had decreasing paternity with female distance. This shows that early spawning males were able to more evenly cover the full spawning array than late spawning males.

These results from both short and long duration of gamete release, suggest that for sessile or sedentary organisms, there may be a benefit for males to spawn over a protracted period prior to females releasing eggs, so that sperm have already permeated the environment before females release their eggs. Another benefit to early spawning males is that their sperm will potentially be exposed to a wider diversity of egg genotypes. Variation in sea urchin reproductive success is in part a function of intraspecific gametic compatibility (Palumbi 1999, Evans & Marshall 2005, Levitan & Ferrell 2006). Field experiments with this same species of sea urchin indicate that even in the face of wide variation in spatial positioning during spawning events, male and female reproductive success could be predicted by their gamete recognition protein genotype (Levitan & Ferrell 2006).

The cost of spawning over a protracted period is that sperm are less concentrated at the point of release, assuming an inverse relation between spawning duration and the rate of gamete release. A potential cost to spawning too early is either missing the spawning event entirely or potential loss of gamete function caused by aging. Egg longevity is on the scale of several hours (Pennington 1985), while sperm longevity is dependent on sperm concentration (Levitan et al. 1991). Sperm, while concentrated remain inactive and can survive for hours to days, particularly at cold temperatures (Chia & Bickell 1983). Once advected into the water there appears to be a linear decline in sperm half-life with dilution, such that sperm at the boundary of being dense enough to fertilize any eggs have half-lives of around 5 to 10 minutes (Levitan 1993). While spawning events can certainly last longer than 5 minutes, the residence time of a particular cohort of sperm over the spawning event, once they are advected off the males and become diluted in the water column, is likely to be fairly short (on the scales of seconds to minutes at typical flow velocities; Levitan 2002a). When advection is low, bottom topography is complex (thus retaining gametes), or populations are sparse, sperm longevity may become increasingly important. Interestingly, of the three species that inhabit this environment in British Columbia, there is a correlation between sperm longevity and average population density; the species living at lowest densities has the greatest longevity.

To date, empirical data have been focused on why males may want to spawn sooner than or in synchrony with females, based on competition with other males. This data suggests that for females, the number of spawning males can influence reproductive success and thus spawning behavior. In order to explore the timing for females to initiate spawning given particular conditions, we introduce a model of gamete dispersal when releases are of finite duration rather than a constant plume.

MODELING FEMALE CHOICE IN A MALE POPULATION

Although there is a need for continued experimental studies to further examine the costs and benefits of different spawning strategies as a function of flow, spatial scale, and adult mobility, simulation models provide another tool for making predictions and generating hypotheses. Here we introduce an advection–diffusion model to explore how female behavior is predicted to be affected by male spawning duration, male density, and advection. This model estimates how a group of particles released from a point source diffuses through time (Box 6.1). We integrate this model over the spawning duration of an organism in order to understand how plumes are established and dissipate in space and time.

The type of model that has traditionally been used to explore fertilization success in marine broadcast spawners is the time-average plume model (Denny & Shibata 1989). After gametes have been released for a sufficient period of time, the plume model describes the spatial concentration gradient of gametes as a function of distance from the spawning organism. This model is independent of time, and has been used to model organisms that release gametes over long periods of time, such as sea urchins and seastars (Babcock et al. 1994; Levitan & Young 1995; Claereboudt 1999; Meidel & Scheibling 2001; Metaxas et al. 2002; Lundquist & Botsford 2004; Lauzon-Guay & Scheibling 2007).

The plume model, however, cannot be used to describe how concentrations change when an organism initiates spawning, or when an organism ceases to spawn. It can only be used to describe concentrations once a gradient has been established. As advection decreases, we can expect that it will take a longer period of time for a constant concentration to be established at downstream locations. Therefore, the plume model is not useful for exploring the reason for sexual differences in spawning, or how varying spawning duration in males can affect downstream sperm concentrations.

In order to understand how plumes are established at a downstream mate when an organism starts spawning, we introduce a model that describes

BOX **6.1**   A Model of Turbulent Diffusion from a Point Source

The equation describing the concentration of particles released from a point source can be obtained from classic texts on turbulent diffusion (Csanady 1973; Okubo 1980). The equation for two-dimensional diffusion from a point source is:

$$C(x,y,t) = \frac{Q}{(2\pi)^{1/2}\sigma_x\sigma_y} \exp\left[-\frac{(x-\alpha t)^2}{2\sigma_x^2} - \frac{y^2}{2\sigma_y^2}\right]$$

where C is the concentration at an x, y location at time t; Q is the total number of gametes released at the source; and σ is advection, which is assumed to be in the x-direction. The parameters σ_x and σ_y describe the variance in the spread of the particle cloud, which increases non-linearly with time:

$$\sigma_i = (2tD_i)^{1/2}$$

where t is time and D_i is the diffusion constant in the given direction.

This model estimates how a cloud of particles that are released from a point will diffuse through space and time. Therefore, in order to extend this model to explore different spawning durations, we integrate the point source model over every second for the duration of spawning. It is in this case essentially a plume model, but unlike the plume model it is able to describe the establishment and decay of a plume at a downstream mate as a function of time. It is therefore useful for examining the effect of different spawning durations. Since the point source model is integrated over the number of seconds an organism spawns for, the parameter Q is calculated as the rate of gamete release (the total number of gametes divided by the number of seconds an organism spawns).

The model is limited in that it assumes gametes are released independently. If gametes are held together by mucoid substances in clumps, fertilization models may be significantly underestimating fertilization success for free-spawners (Thomas 1994b). Additionally, the model assumes that the advective environment is homogeneous, and does not estimate the effect of instantaneous turbulent structures such as vortexes. Theoretical models predict that fertilization success may become enhanced if egg and sperm filaments become trapped in these instantaneous structures (Crimaldi & Browning 2004; Crimaldi et al. 2006). However, the model presented here is practical because it can be used to simulate spawning in a population of organisms, and it predicts the average concentration of gametes as a function of distance and time.

the time-dependent diffusion of a point source (Box 6.1). For the purposes of this exercise, we assume that our population is spread out on a two-dimensional plane, on which diffusion occurs. The question we use this model to address is: given a particular pattern of spawning in an upstream male population (i.e., advection, density, and spawning duration), when will females initiate spawning? Laboratory flume experiments with unidirectional flow indicate that most fertilization happens near or at the female (Yund & Meidel 2003). Therefore, the model assumes that female behavior is based on sperm concentration at the female, at the time when she can get the highest proportion of her eggs fertilized.

In the model, when sperm is saturating (above the sperm threshold for 100% fertilization), females will initiate spawning at the time this threshold is reached. Although this threshold varies within and among species and is influenced by water flow (Levitan et al. 1991; Levitan 1998b), we use a threshold of 10^5 sperm per mL, since empirical research has shown that this is the sperm concentration by which a female can have 100% of her eggs fertilized (Levitan 2002b). If sperm concentrations in a modeled reproductive bout do not reach the 10^5 threshold, (i.e., if sperm is limiting), we assume that females initiate spawning when sperm concentration has reached 90% of the maximum.

We modeled a male population of varying natural densities (1, 2, 3, 4, 6, 8, 12, 16, 24, 32, or 48 males in a 4-m² area) in a square area and placed a female at the downstream central edge of this male population. Stochasticity in the model results from the random placement and start time of sperm release for all males in each replicate. Organisms are assumed to be sessile and do not move while they are releasing gametes. The model was parameterized with data from sea urchins, and each male in the population had a spawning duration of 100 minutes (typical of sea urchins). The spread of spawning initiation in the male population was normally distributed, with 95% of males beginning to spawn within approximately 45 minutes, which is typical for sea urchins (figure 6.6; Levitan 2002a). Since the magnitude of advection can affect how quickly plumes get established, and therefore female spawning behavior, we modeled advections of 0.01, 0.02, and 0.05 m/s. For all trials, we kept diffusion coefficients constant at $D_x = 0.006$ and $D_y = 0.0006$. We choose these values because they are within the range of those found in marine environments (Koehl et al. 1993; K. Lotterhos unpublished data), and it is typical for diffusion in the direction of advection (x-axis) to be an order of magnitude larger (Csanady 1973). In order to explore how spawning duration in the male population affects female choice, we also modeled the

same set of parameters with each male having the same gamete release rate, but a shorter spawning duration of 10 minutes.

For each set of parameters, the response variable (timing of female initiation) was averaged for at least 20 replicate runs. We report the average time of spawning initiation in the female, in relation to the average time of spawning in the male population. We would like to emphasize that a negative value of female spawning initiation does not mean that females spawn before all males, but only before the male average.

MODEL RESULTS

In all scenarios modeled, it always benefited females to spawn after at least some males had initiated spawning, since it took time for advection to deliver sperm from the males to the females. However, in the highest density scenarios, the females are predicted to begin spawning before a large proportion of males have initiated spawning, since in these situations enough males have already released sperm to fertilize 100% of a female's eggs. We define the time for a female to initiate spawning as (a) when sperm concentrations reach 10^5 sperm per mL, and a female can have 100% of her eggs fertilized, or (b) when sperm concentrations are limiting

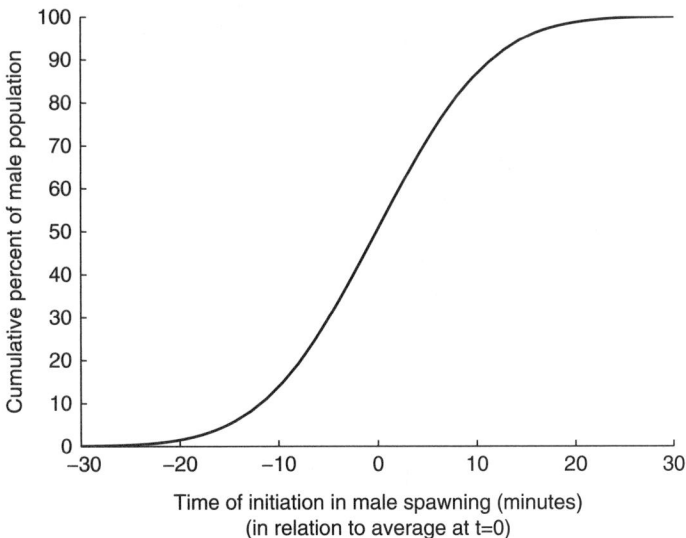

FIGURE **6.6** Cumulative distribution of the times males initiate gamete release in the model.

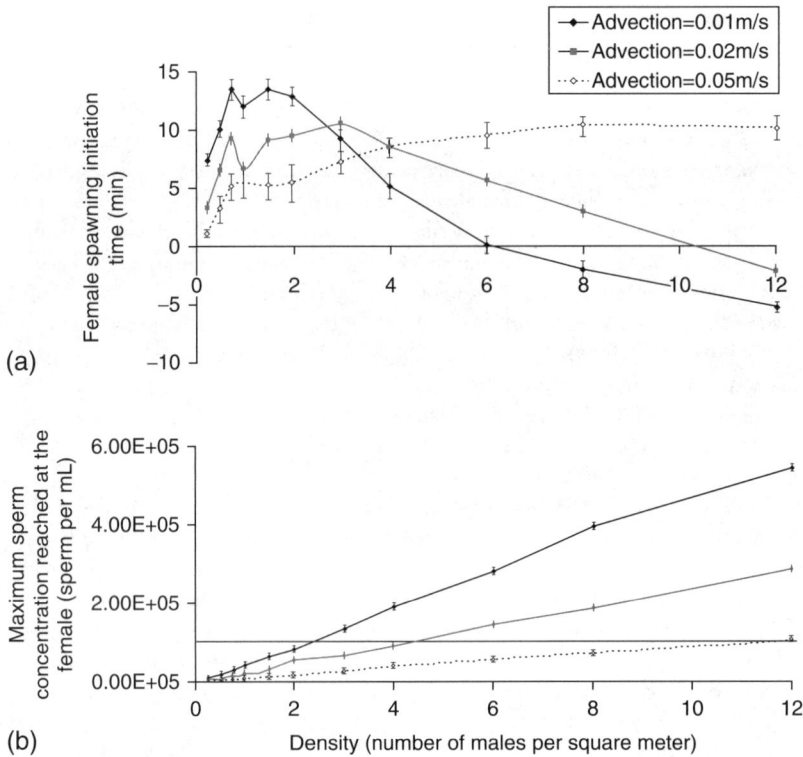

FIGURE **6.7** Model results for male spawning duration of 100 minutes. (a) Female spawning initiation is reported in relation to the average spawning in the male population, given conditions of advection and male density. (b) The maximum sperm concentration reached at the female as a function of advection and male density. The sperm threshold at which a female can get 100% of her eggs fertilized (10^5 sperm/mL) is shown by the black horizontal line.

(less that 10^5 sperm per mL), sperm concentrations are at 90% of the maximum sperm concentration reached at that trial.

When male spawning duration is long (100 minutes), female spawning initiation is the result of an interaction between male density and advection (figure 6.7a). Depending on advection, sperm can be limiting at a range of densities (below the 10^5 threshold; figure 6.7b). Inside this window of sperm-limitation, as density increases, females are predicted to wait longer for more males to spawn. This result is counter-intuitive to the notion that females should begin spawning earlier than males as density increases. This is because the amount of time males release gametes is longer than the initiation of male spawning in the population, so that the earliest males are still releasing sperm as the later males begin gamete release. However, depending on advection, eventually a density is reached when sperm becomes saturating (at the

10^5 sperm per mL threshold). Under sperm saturation, as male density increases females are predicted to initiate spawning earlier. Therefore, with increasing male density, females are predicted to delay spawning as long as sperm are limiting, and then spawn progressively sooner when sperm is saturating. The density at which this occurs is determined by advection, because it affects the rate at which sperm accumulates over the female. Higher advection shifts the transition to saturating sperm conditions (10^5 sperm/mL) to higher male densities (figure 6.7b).

We also explored how decreasing male spawning duration affects female timing in spawning (figure 6.8a, for male spawning duration of 10 minutes). Under a shorter male spawning duration, females are predicted to be more synchronous with the male population for a range of densities and advections. The decrease in female delay is caused by the relationship between male start times and duration.

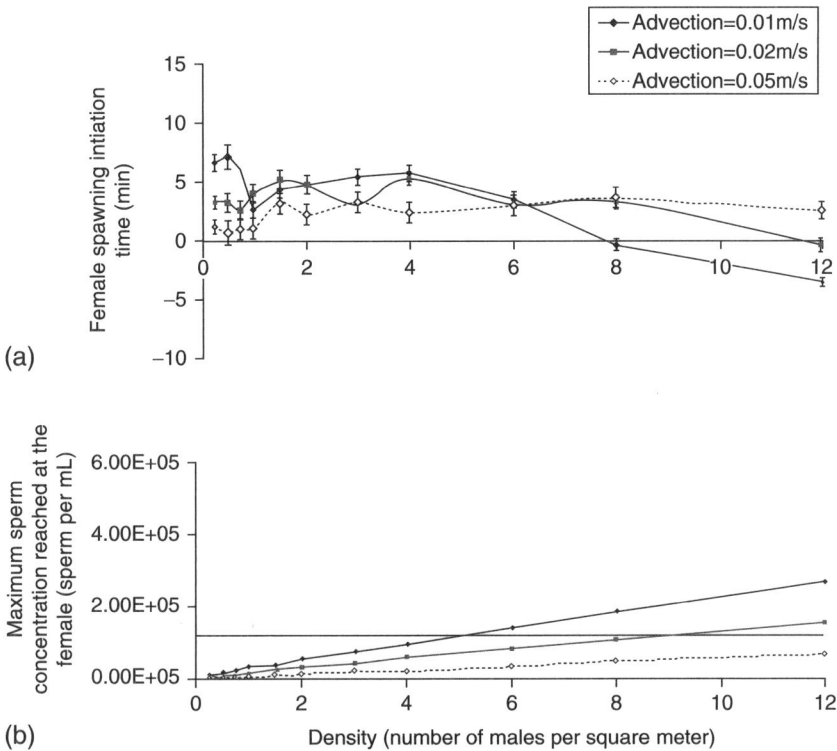

FIGURE **6.8** Model results for male spawning duration of 10 minutes. (a) Female spawning initiation is reported in relation to the average spawning in the male population, as a function of male density and advection. (b) Maximum sperm concentration reached at the female as a function of advection and male density. The sperm threshold at which a female can get 100% of her eggs fertilized (10^5 sperm/mL) is shown by the light grey horizontal line.

While start times vary among males, the female cannot wait until all males have initiated (as in the case of long male spawning duration), because by that time the early spawning males have finished spawning. This same relationship between start time and duration also results in increased variability within and among replicates. The interaction between advection and male density is more ambiguous in this case because sperm concentrations are near the threshold for a higher range of male densities (figure 6.8b).

Overall, this model suggests that there is a non-linear effect of spawning density on male–female synchrony and this effect is mediated by water flow. In addition, the model predicts a positive relationship between male spawning duration and the delay between male and female spawning times. However in all modeled scenarios females were predicted to initiate spawning after at least some males initiated spawning.

COMPARING THE MODEL TO EMPIRICAL DATA

The model above suggests that decreasing male spawning duration results in females becoming more synchronous with the male population. We wanted to test the generality of this result for broadcast spawning animals, but data on spawning duration and sexual differences in spawning initiation are scant and difficult to collect. The only data available for testing this question is from an extensive study on reproduction of siphonous green algae in the order Bryopsidales (Clifton 1997; Clifton & Clifton 1999).

Although this volume is focused on animal taxa, several species of algae are broadcast spawners that share characteristics with animal broadcast spawners, and can provide insight into processes of fertilization, sperm limitation, and sexual selection in the sea. Clifton's data is ideal for this question, because individuals release gametes for between 5 and 35 minutes, depending on the species. Thus green algae constitute a continuum of pluming durations. Because of the extensive survey performed by Clifton and colleagues for 20 months of nearly continuous daily monitoring, their dataset on green algae can be used as a reliable source to test the relationship between spawning duration and sexual differences in timing between males and females.

In all the dioecious green algae reported by Clifton (1997), males initiated spawning before the females. We plotted the average time difference between males and females as a function of the midpoint in spawning duration for all 15 dioecious species. There was a significant positive relationship between spawning duration and sexual differences in timing between males and females (figure 6.9; Adjusted R-squared = 0.49; F-statistic = 14.46 on 1 and 13 DF; p-value = 0.002).

We examined whether our model could be used to predict the differences in timing between males and females observed in Clifton's data, given a particular spawning duration in the male population. We modeled Clifton's data using an advection of 0.01 m/s, a population of 16 males in a 4-m^2 square plot, the midpoint of spawning duration observed for each algal species, and an arbitrary fixed total number of gametes for each species (10,000 gametes), such that increases in spawn duration was compensated for by a decrease in spawning rate (sperm/second). At these sperm release rates, densities, and water flows, sperm were always limiting in these simulations. The male population was modeled with 95% of males initiating spawning within a 25 minute period, which was typical of the species in Clifton's study (Clifton & Clifton 1999). Adjusting the parameter values could give better or worse fits to the empirical data, here we simply show a qualitatively similar response (figure 6.9). Most interesting would be examination of deviations from the linear predictions for each species and what factors, such as density, release rate, flow or gamete traits might explain these deviations.

CONCLUSIONS AND FUTURE DIRECTIONS

Patterns of gamete release vary among species and likely reflect differences in the selective pressures associated with successful fertilization. Experimental evidence suggests that sex differences in the timing of spawning can be influenced by the distribution and abundance of competitors and mates. At low densities selection on males will favor earlier male spawning times, because males that spawn earlier are able to spread out sperm over a greater spatial area. At low densities selection on females will

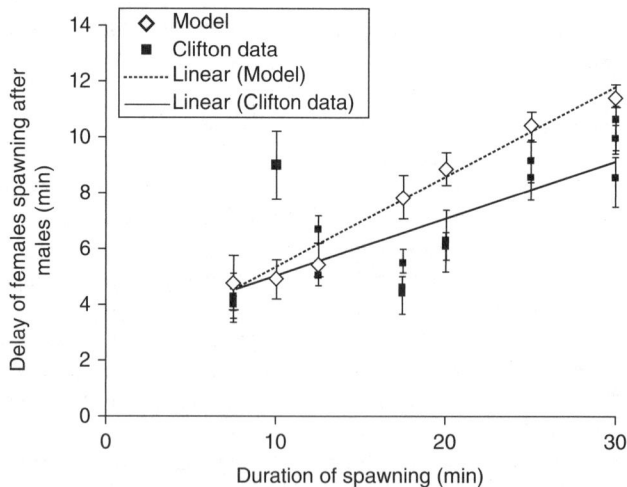

FIGURE 6.9 The relationship between spawning duration and the delay between male and female spawning. Empirical data in the green algae is presented in dark squares (Clifton and Clifton 1999); results from the model are presented in white diamonds. A significant positive relationship is observed in both the empirical and theoretical data (see text).

favor later female spawning times, because females should wait for sperm to accumulate near them as additional males join the spawning event. However, increasing mate densities will select for tighter synchrony between male and female spawning, because asynchronous males are outcompeted by more synchronous males, and sperm will accumulate above females in a shorter period of time.

Our modeling represents an early exploration into the costs and benefits of packaging gametes into long or short duration releases and how this packaging can interact with sex differences in the timing of spawning. The modeling predicts that when males have a long spawning duration, the time a female chooses to spawn is a result of a strong interaction between density and advection. Females will increasingly delay spawning with increasing density until sperm becomes saturating, at which point female delay begins to decrease. The male density at which sperm becomes saturating is larger at higher advections, because increasing advection results in lower sperm concentration at the female. As male spawning duration decreases, this interaction between density and advection becomes weaker, and female delay is shorter for a range of densities. Generally, these results predict a reduced time difference between the sexes in organisms with short spawning durations and high density populations. Comparisons of the model with data from broadcast spawning algae suggest that this approach might be fruitful, particularly when expectations fail to match predictions. More complex models and models that consider male and female strategies simultaneously (e.g., game theory) would be worth developing.

In this chapter, we documented various spawning behaviors and focused on spawning duration and its relation to sex difference in spawn time. However, this exploration is meant to stimulate additional experimental and theoretical studies. These studies need to be placed into the context of natural spawning observations. Spawning observations are still rarely reported in the literature and often these reports are not detailed enough to test the various hypotheses that might determine the costs and benefits of variation in spawning behavior or insights into how these behaviors evolve. Our hope is that this chapter motivates more quantitative measures of spawning behaviors and experiments on how various strategies influence the fertilization success of males and females.

REFERENCES

Andrews, J. D. 1979. Pelecypoda: Ostreidae. Pages 293–342 in A. C. Giese and J. S. Pearse, editors. Reproduction of Marine Invertebrates Vol. 5. Academic Press, New York.

Babcock, R., C. Mundy, J. Keesing, and J. Oliver. 1992. Predictable and unpredictable spawning events—*in-situ* behavioral data from free-spawning coral-reef invertebrates. Invertebrate Reproduction and Development 22: 213–228.

Babcock, R. C., C. N. Mundy, and D. Whitehead. 1994. Sperm diffusion-models and *in-situ* confirmation of long-distance fertilization in the free-spawning asteroid *Acanthaster planci*. Biological Bulletin 186: 17–28.

Bishop, J. D. D. 1998. Fertilization in the sea: are the hazards of broadcast spawning avoided when free-spawned sperm fertilize retained eggs? Proceedings of the Royal Society of London Series B—Biological Sciences 265: 725–731.

Bode, M. and D. J. Marshall. 2007. The quick and the dead? Sperm competition and sexual conflict in sea. Evolution 61: 2693–2700.

Bourmaud, C. and N. Gravier-Bonnet. 2004. Medusoid release and spawning of *Macrorynchia philippina* (Kirchenpauer, 1872—Cnidaria, Hydrozoa, Aglaopheniidae). Hydrobiologia 530–31: 365–372.

Bulmer, M. G. and G. A. Parker. 2002. The evolution of anisogamy: a game-theoretic approach. Proceedings of the Royal Society of London Series B—Biological Sciences 269: 2381–2388.

Chia, F. S. and L. R. Bickell. 1983. Echinodermata. Pages 545–620 in K. G. Adiyodi and R. G. Adiyodi, editors. Reproductive Biology of Invertebrates. Vol II. Spermatogenesis and Sperm Function. Wiley, New York.

Claereboudt, C. 1999. Fertilization success in spatially distributed populations of benthic free-spawners: A simulation model. Ecological Modelling 121: 221–233.

Clifton, K. E. 1997. Mass spawning by green algae on coral reefs. Science 275: 1116–1118.

Clifton, K. E. and L. M. Clifton. 1999. The phenology of sexual reproduction by green algae (Bryopsidales) on Caribbean coral reefs. Journal of Phycology 35: 24–34.

Coma, R. and H. R. Lasker. 1997. Effects of spatial distribution and reproductive biology on in situ fertilization rates of a broadcast-spawning invertebrate. Biological Bulletin 193: 20–29.

Crimaldi, J. P. and H. S. Browning. 2004. A proposed mechanism for turbulent enhancement of broadcast spawning efficiency. Journal of Marine Systems 49: 3–18.

Crimaldi, J. P., J. R. Hartford, and J. B. Weiss. 2006. Reaction enhancement of point sources due to vortex stirring. Physical Review E 74: 0163071–0163074.

Csanady, G. T. 1973. Turbulent Diffusion in the Environment. R. Reidel Publishing Company, Boston.

Darwin, C. 1871. The Decent of Man and Selection in Relation to Sex. J. Murray, London.

Denny, M. W. and M. F. Shibata. 1989. Consequences of surf-zone turbulence for settlement and external fertilization. American Naturalist 134: 859–889.

Dusenbery, D. B. 2000. Selection for high gamete encounter rates explains the success of male and female mating types. Journal of Theoretical Biology 202: 1–10.

Evans, J. P. and D. J. Marshall. 2005. Male-by-female interactions influence fertilization success and mediate the benefits of polyandry in the sea urchin *Heliocidaris erythrogramma*. Evolution 59: 106–112.

Farley, G. S. and D. R. Levitan. 2001. The role of jelly coats in sperm–egg encounters, fertilization success, and selection on egg size in broadcast spawners. American Naturalist 157: 626–636.

Franke, E. S., R. C. Babcock, and C. A. Styan. 2002. Sexual conflict and polyspermy under sperm-limited conditions: In situ evidence from field simulations with the free-spawning marine echinoid *Evechinus chloroticus*. American Naturalist 160: 485–496.

Giese, A. C. and H. Kanatani. 1987. Maturation and spawning. Pages 251–329 in A. C. Giese, J. S. Pearse, and V. B. Pearse, editors. Reproduction of Marine Invertebrates Vol. IX: Seeking Unity in Diversity. Blackwell Scientific/Boxwood Press, Palo Alto/ Pacific Grove, CA.

Gittings, S. R., G. S. Boland, K. J. P. Deslarzes, C. L. Combs, B. S. Holland, and T. J. Bright. 1992. Mass spawning and reproductive viability of reef corals at the East Flower Garden Bank, northwest Gulf of Mexico. Bulletin of Marine Science 51: 420–428.

Hamel, J. F. and A. Mercier. 1996. Gamete dispersion and fertilisation success of the sea cucumber *Cucumaria frondosa*. Beche-de-mer Information Bulletin 8: 34–40.

Haygood, R. 2004. Sexual conflict and protein polymorphism. Evolution 58: 1414–1423.

Hendler, G. 1991. Ophiuroida. Pages 356–513 in J. S. Pearse and V. B. Pearse, editors. Reproduction in Marine Invertebrates, Vol. 6. Boxwood Press.

Himmelman, J. H., C. P. Dumont, C. F. Gaymer, C. Vallieres, and D. Drolet. 2008. Spawning synchrony and aggregative behaviour of cold-water echinoderms during multi-species mass spawnings. Marine Ecology-Progress Series 361: 161–168.

Holland, N. D. 1991. Echinodermata: Crinoidea. Pages 247–300 in A. C. Giese, J. S. Pearse, and V. B. Pearse, editors. Reproduction of Marine Invertebrates Volume VI. The Boxwood Press, Pacific Grove, CA.

Koehl, M. A. R., T. M. Powell, and G. Dairiki. 1993. Measuring the fate of patches in the water: larval dispersal. Pages 50–60 in S. A. Levin, T. M. Powell, and J. H. Steele, editors. Lecture Notes in Biomathematics: Patch Dynamics XIII.

Kupriyanova, E. and J. N. Havenhand. 2002. Variation in sperm swimming behaviour and its effect on fertilization success in the serpulid polychaete *Galeolaria caespitosa*. Invertebrate Reproduction & Development 41: 21–26.

Lasker, H. R. 2006. High fertilization success in a surface-brooding Caribbean gorgonian. Biological Bulletin 210: 10–17.

Lauzon-Guay, J. and R. E. Scheibling. 2007. Importance of spatial population characteristics on the fertilization rates of sea urchins. Biological Bulletin 212: 195–205.

Levitan, D. R. 1993. The importance of sperm limitation to the evolution of egg size in marine invertebrates. American Naturalist 141: 517–536.

Levitan, D. R. 1996. Effects of gamete traits on fertilization in the sea and the evolution of sexual dimorphism. Nature 382: 153–155.

Levitan, D. R. 1998a. Chapter 6: Sperm limitation, gamete competition, and sexual selection in external fertilizers. Pages 173–215 Sperm Competition and Sexual Selection. Academic Press Ltd.

Levitan, D. R. 1998b. Does Bateman's principle apply to broadcast-spawning organisms? Egg traits influence in situ fertilization rates among congeneric sea urchins. Evolution 52: 1043–1056.

Levitan, D. R. 2000. Sperm velocity and longevity trade off each other and influence fertilization in the sea urchin *Lytechinus variegatus*. Proceedings of the Royal Society of London Series B—Biological Sciences 267: 531–534.

Levitan, D. R. 2002a. Density-dependent selection on gamete traits in three congeneric sea urchins. Ecology 83: 464–479.

Levitan, D. R. 2002b. The relationship between conspecific fertilization success and reproductive isolation among three congeneric sea urchins. Evolution 56: 1599–1609.

Levitan, D. R. 2004. Density-dependent sexual selection in external fertilizers: Variances in male and female fertilization success along the continuum from sperm limitation to sexual conflict in the sea urchin *Strongylocentrotus franciscanus*. American Naturalist 164: 298–309.

Levitan, D. R. 2005a. The distribution of male and female reproductive success in a broadcast spawning marine invertebrate. Integrative and Comparative Biology 45: 848–855.

Levitan, D. R. 2005b. Sex-specific spawning behavior and its consequences in an external fertilizer. American Naturalist 165: 682–694.

Levitan, D. R. 2006. The relationship between egg size and fertilization success in broadcast-spawning marine invertebrates. Integrative and Comparative Biology 46: 298–311.

Levitan, D. R. 2008. Gamete traits influence the variance in reproductive success, the intensity of sexual selection, and the outcome of sexual conflict among congeneric sea urchins. Evolution 62: 1305–1316.

Levitan, D. R. and D. L. Ferrell. 2006. Selection on gamete recognition proteins depends on sex, density, and genotype frequency. Science 312: 267–269.

Levitan, D. R. and C. M. Young. 1995. Reproductive success in large populations—empirical measures and theoretical predictions of fertilization in the sea biscuit *Clypeaster rosaceus*. Journal of Experimental Marine Biology and Ecology 190: 221–241.

Levitan, D. R., M. A. Sewell, and F. S. Chia. 1991. Kinetics of fertilization in the sea urchin *Strongylocentrotus franciscanus*: interaction of gamete dilution, age, and contact time. Biological Bulletin 181: 371–378.

Levitan, D. R., H. Fukami, J. Jara, D. Kline, T. M. McGovern, K. E. McGhee, C. A. Swanson, and N. Knowlton. 2004. Mechanisms of reproductive isolation among sympatric broadcast-spawning corals of the *Montastraea annularis* species complex. Evolution 58: 308–323.

Levitan, D. R., C. P. terHorst, and N. D. Fogarty. 2007. The risk of polyspermy in three congeneric sea urchins and its implications for gametic incompatibility and reproductive isolation. Evolution 61: 2007–2014.

Long, J. A. and S. A. Stricker. 1991. Brachiopoda. Pages 47–85 in A. C. Giese, J. S. Pearse, and V. B. Pearse, editors. Reproduction of Marine Invertebrates Volume VI. The Boxwood Press, Pacific Grove, CA.

Lundquist, C. J. and L. W. Botsford. 2004. Model projections of the fishery implications of the Allee effect in broadcast spawners. Ecological Applications 14: 929–941.

Marshall, D. J. and M. J. Keough. 2003. Sources of variation in larval quality for free-spawning marine invertebrates: Egg size and the local sperm environment. Invertebrate Reproduction and Development 44: 63–70.

McEuen, F. S. 1988. Spawning behaviors of northeast Pacific sea cucumbers (Holothuroidea, Echinodermata). Marine Biology 98: 565–585.

Meidel, S. K. and R. E. Scheibling. 2001. Variation in egg spawning among subpopulations of sea urchins *Strongylocentrotus droebachiensis*: a theoretical approach. Marine Ecology Progress Series 213: 97–110.

Mercier, A., R. H. Ycaza, and J.-F. Hamel. 2007. Long-term study of gamete release in a broadcast-spawning holothurian: predictable lunar and diel periodicities. Marine Ecology Progress Series 329: 179–189.

Metaxas, A., R. E. Scheibling, and C. M. Young. 2002. Estimating fertilization success in marine benthic invertebrates: a case study with the tropical sea star *Oreaster reticulatus*. Marine Ecology Progress Series 226: 87–101.

Miller, R. L. 1985. Demonstration of sperm chemotaxis in Echinodermata—Asteroidea, Holothuroidea, Ophiuroidea. Journal of Experimental Zoology 234: 383–414.

Minchin, D. 1992. Multiple species, mass spawning events in an Irish sea lough: the effect of temperature on spawning and recruitment of invertebrates. Invertebrate Reproduction and Development 22: 229–238.

Okubo, A. 1980. Diffusion and Ecological Problems: Mathematical Models. Springer-Verlag, Berlin.

Palumbi, S. R. 1999. All males are not created equal: Fertility differences depend on gamete recognition polymorphisms in sea urchins. Proceedings of the National Academy of Sciences of the United States of America 96: 12632–12637.

Pearse, J. S. 1979. Polyplacophora. Pages 27–85 in A. C. Giese, J. S. Pearse, and V. B. Pearse, editors. Reproduction in Marine Invertebrates Vol. 5. Academic Press.

Pennington, J. T. 1985. The ecology of fertilization of echinoid eggs—the consequences of sperm dilution, adult aggregation, and synchronous spawning. Biological Bulletin 169: 417–430.

Phillippi, A., E. Hamann, and P. O. Yund. 2004. Fertilization in an egg-brooding colonial ascidian does not vary with population density. Biological Bulletin 206: 152–160.

Pianka, H. D. 1974. Ctenophora. Pages 201–266 in A. C. Giese and J. S. Pearse, editors. Reproduction of Marine Invertebrates. Academic Press, New York and London.

Podolsky, R. D. 2001. Evolution of egg target size: An analysis of selection on correlated characters. Evolution 55: 2470–2478.

Podolsky, R. D. 2002. Fertilization ecology of egg coats: physical versus chemical contributions to fertilization success of free-spawned eggs. Journal of Experimental Biology 205: 1657–1668.

Podolsky, R. D. 2004. Life-history consequences of investment in free-spawned eggs and their accessory coats. American Naturalist 163: 735–753.

Randall, J. E., R. E. Schroeder, and W. A. Starck. 1964. Notes on the biology of the echinoid Diadema antillarum. Caribbean Journal of Science 4: 421–433.

Randerson, J. P. and L. D. Hurst. 2001. A comparative test of a theory for the evolution of anisogamy. Proceedings of the Royal Society of London Series B—Biological Sciences 268: 879–884.

Reiswig, H. M. 1970. Porifera—sudden sperm release by tropical Demospongiae. Science 170: 538–539.

Rice, M. E. 1975. Sipuncula. Pages 67–128 in A. C. Giese and J. S. Pearse, editors. Reproduction of Marine Invertebrates. Academic Press, New York.

Riffell, J. A., P. J. Krug, and R. K. Zimmer. 2004. The ecological and evolutionary consequences of sperm chemoattraction. Proceedings of the National Academy of Sciences of the United States of America 101: 4501–4506.

Stekoll, M. S. and T. C. Shirley. 1993. In-situ spawning behavior of an Alaskan population of pinto abalone, Haliotis kamtschatkana (Jonas, 1845). Veliger 36: 95–97.

Strathmann, R. R. 1990. Why life histories evolve differently in the sea. American Zoologist 30: 197–207.

Styan, C. A. 1998. Polyspermy, egg size, and the fertilization kinetics of free-spawning marine invertebrates. American Naturalist 152: 290–297.

Swanson, W. J. and V. D. Vacquier. 2002. Reproductive protein evolution. Annual Review of Ecology and Systematics 33: 161–179.

Szmant, A. M. 1986. Reproductive ecology of Caribbean reef corals. Coral Reefs 5: 43–53.

Thomas, F. I. M. 1994a. Physical properties of gametes in 3 sea urchin species. Journal of Experimental Biology 194: 263–284.

Thomas, F. I. M. 1994b. Transport and mixing of gametes in three free-spawning Polychaete Annelids: Phragmatopoma californica (Fewkes), Sabellaria cementarium (Moore), and Schizobranchia insignis (Bush). Journal of Experimental Marine Biology and Ecology 179: 11–27.

Van Oppen, M. J. H., B. L. Willis, T. Van Rheede, and D. J. Miller. 2002. Spawning times, reproductive compatibilities and genetic structuring in the Acropora aspera group: evidence for natural hybridization and semi-permeable species boundaries in corals. Molecular Ecology 11: 1363–1376.

van Veghel, M. L. J. 1994. Reproductive characteristics of the polymorphic Caribbean reef building coral Montastrea annularis.1. Gametogenesis and spawning behavior. Marine Ecology Progress Series 109: 209–219.

Vize, P. D., J. A. Embesi, M. Nickell, P. D. Brown, and D. K. Hagman. 2005. Tight temporal consistency of coral mass spawning at the Flower Garden Banks, Gulf of Mexico, from 1997–2003. Gulf of Mexico Science 1: 107–114.

Vogel, H., G. Czihak, P. Chang, and W. Wolf. 1982. Fertilization kinetics of sea-urchin eggs. Mathematical Biosciences 58: 189–216.

Wray, G. A. 1995. Evolution of larvae and developmental modes. Pages 412–448 in L. McEdward, editor. Ecology of Marine Invertebrate larvae. CRC, Boca Raton.

Wyers, S. C., H. S. Barnes, and S. R. Smith. 1991. Spawning of hermatypic corals in Bermuda —a pilot-study. Hydrobiologia 216: 109–116.

Yund, P. O. and M. A. McCartney. 1994. Male reproductive success in sessile invertebrates - competition for fertilizations. Ecology 75: 2151–2167.

Yund, P. O. and S. K. Meidel. 2003. Sea urchin spawning in benthic boundary layers: Are eggs fertilized before advecting away from females? Limnology and Oceanography 48: 795–801.

7

Prosobranchs with Internal Fertilization

ALAN N. HODGSON

INTRODUCTION: WHAT IS A PROSOBRANCH?

The Gastropoda is one of the most diverse groups of metazoans with estimates of the number of species ranging between ~37,000 to ~150,000 (Bieler 1992; Ponder & Lindberg 1997; Kay et al. 1998; Aktipis et al. 2008). For much of the twentieth century three subclasses of gastropods (Prosobranchia, Opisthobranchia, Pulmonata) were recognized, with the Prosobranchia being further divided into the Archaeogastropoda, Mesogastropoda, and Neogastropoda. The re-examination of gastropod systematics by Golikov and Starobogatov (1975), along with the discovery of new higher taxa from deep sea habitats, especially hydrothermal vents and methane seeps, stimulated considerable research on gastropod anatomy (macro and micro) and systematics. It became apparent that the gastropod divisions *sensu* Thiele (1929–1935) were not phylogenetically sound, and investigators have since used detailed morphology (including ultrastructure), along with fossil information and molecular data sets to present new views of gastropod phylogeny (e.g., Haszprunar 1988a; Lindberg 1988; Bieler 1992; Tillier et al. 1994; Ponder & Lindberg, 1997; Winnepenninckx et al. 1998; Colgan et al. 2000, 2003; Harasewych & McArthur 2000; Wagner 2001; McArthur & Harasewych 2003; Grande et al. 2004; Geiger & Thacker 2005; Nakano & Ozawa 2007; Kano 2008; also see Chapters 9 to 15 in Ponder & Lindberg 2008).

Unfortunately the phylogenies that have been produced using different data sets do not always agree, and higher gastropod systematics are still equivocal. Ponder and Lindberg (1996, 1997) proposed that the Gastropoda could be split into two subclasses, the Eogastropoda and Orthogastropoda, although this division is yet to be fully supported by molecular data (Colgan et al. 2003). The Eogastropoda has one Order, the Patellogastropoda. Within the Orthogastropoda there are five main Superorders; Vetigastropoda (which possibly includes vent taxa including the Neomphalina and Lepetelloidea; also see Kano 2008, for discussion on vetigastropod systematics), Cocculiniformia, Neritimorpha (= Neritopsina), Caenogastropoda and Heterobranchia (includes the former subclasses Opisthobranchia and Pulmonata as well as some groups, e.g., Pyramidelloidea and Omalogyroidea, once considered prosobranchs). Whilst gastropod phylogeny remains unresolved, these superorders have largely been supported by the recent combined morphological and molecular phylogenetic analysis of Aktipis et al. (2008) (figure 7.1). Thus with the exception of the Heterobranchia, the "prosobranchs" are scattered among these higher taxa. In short, the Prosobranchia (and its traditional orders Archaeogastropoda, Mesogastropoda, and Neogastropoda) is now recognized as being paraphyletic, not a valid taxonomic unit, and at best a grade of organization. For the purposes of this review, however, the term prosobranch has been retained and the only group not considered here is the Heterobranchia.

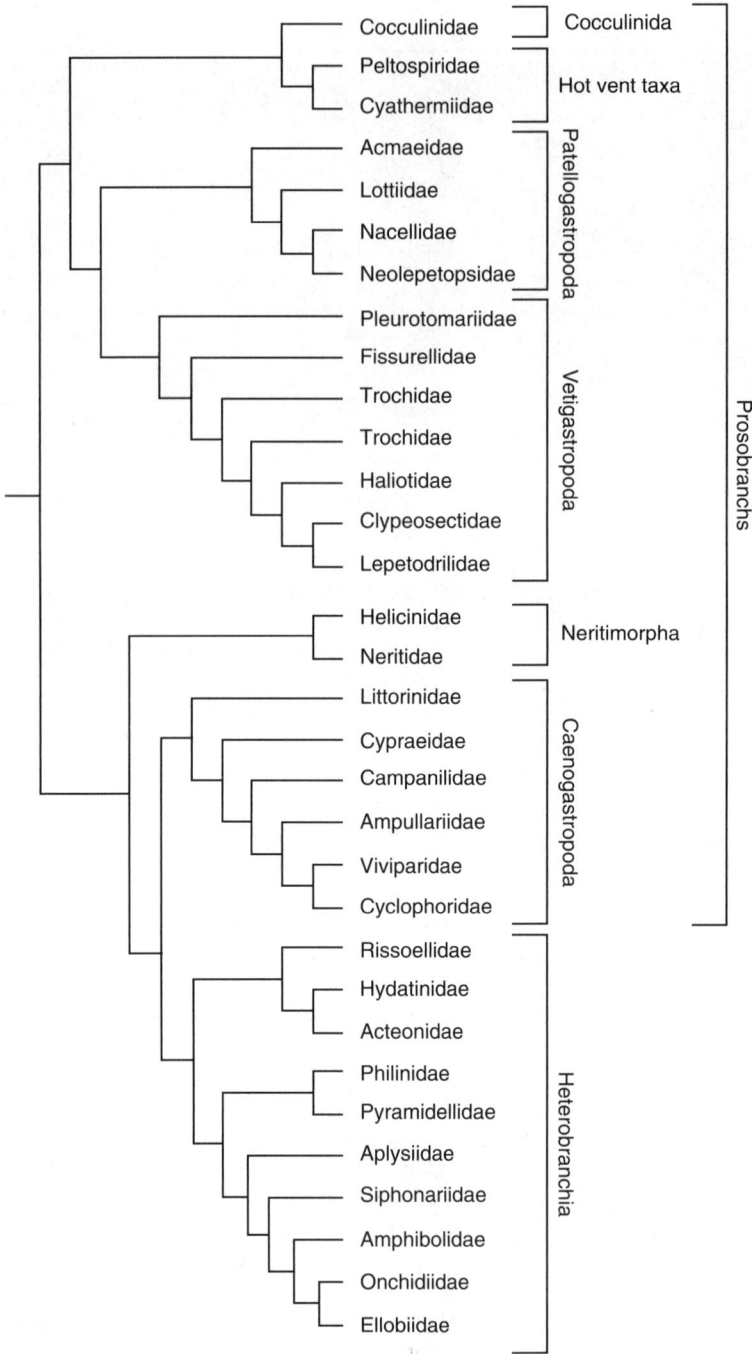

FIGURE 7.1 Phylogeny of the Gastropoda based on morphological and molecular data (modified after Aktipis et al. 2008).

The prosobranchs constitute just over half of the Gastropoda (53–54%) (Bieler 1992; Fretter et al. 1998), which means that there must be between 20,000 and 75,000 species. Fretter and Graham (1994: p. 16) define a prosobranch as "Those gastropods with anterior mantle cavity, ctenidia anterior to the heart, and a twisted visceral loop". The group have radiated into most aquatic habitats and there are even terrestrial representatives (a few species of Neritoidea [Neritimorpha], Littorinoidea, and Cyclophoroidea [Caenogastropoda]). It is therefore not surprising that prosobranchs are morphologically and physiologically diverse, and this is reflected in many aspects of their reproductive biology including their sex characteristics.

MODES OF FERTILIZATION, THE ANCESTRAL CONDITION AND REPRODUCTIVE ANATOMY

Fertilization of prosobranch eggs occurs in one of three "environments", in the surrounding water (external fertilization), in the female reproductive tract (true internal fertilization), and less commonly in the mantle cavity (which can be considered a type of internal fertilization as it requires a mechanism of sperm transfer). The majority of the Patellogastropoda and Vetigastropoda (e.g., Fissurellidae, Haliotidae, Trochoidea, Pleurotomariidae), have separate sexes (gonochoristic), external fertilization, and pelagic larvae. Males and females tend to be of a similar size probably because sperm competition is absent (Levitan 1998), and reproductive systems are very simple, consisting of a single mesodermal gonad and gonoduct through which gametes enter the environment via the right kidney and renal opening (Webber 1977; Voltzow 1994; Fretter & Graham 1994). In patellogastropods the gonad lies between the foot and the visceral mass, whereas in vetigastropods with a coiled shell, the gonad is adjacent to the digestive gland on the columellar side of the visceral mass. A copulatory organ and accessory glands are usually absent. In some deep sea representatives, however, the neck of males may be elaborated and serve to direct sperm (Hickman 1992) or a penis may be present (Kano 2008). In one or two trochacean genera (e.g., *Gibbula*, *Calliostoma*) the distal region of the female duct is glandular, with eggs being deposited in its mucous secretions

(Webber 1977; Hickman 1992). These are, however, relatively minor anatomical modifications.

Are external fertilization and a simple reproductive anatomy the ancestral (primitive) condition in prosobranchs? The traditional view is that they are, because patellogastropods are considered the basal gastropod clade (Ponder & Lindberg 1997; Colgan et al. 2000; McArthur & Harasewych 2003) (figure 7.1). Furthermore, copulatory organs and seminal receptacles are considered derived and to have evolved relatively recently in vetigastropods (Quinn 1983; Kano 2008). There are, however, challenges to the majority view. Buckland-Nicks and Scheltema (1995) have, on the basis of a study of sperm morphology of the Neomeniomorpha, a primitive molluscan group, argued that internal fertilization was the ancestral condition of Bilateria and molluscs. They postulated that "perhaps the internally fertilizing Neritimorpha with their primitive ctenidia and simple spiral shell are, in fact, closer to the stem Gastropoda than the docoglossan limpets?" Furthermore, some recent phylogenies (e.g., see Remigio & Hebert 2003; Yoon & Kim 2007) show that the Neritimorpha are basal gastropods. If these authors are correct, then internal fertilization would be the ancestral gastropod condition. Lindberg (2008), however, observes that the phylogenetic placement and relationships of the Patellogastropoda and Neritimorpha to other gastropods, are not resolved.

Internal fertilization, and fertilization within the mantle cavity, has resulted in anatomically variable and complex reproductive systems. This is due to the different problems that internal fertilization presents to the two sexes. Males have had to evolve mechanisms of sperm transfer, and overcome challenges of the pre-fertilization environment in the female reproductive tract (Buckland-Nicks 1998). Females must receive and store sperm, and many internal fertilizers have encapsulated larval development, requiring the development of accessory structures that provide embryo nutrition and protection. Internal fertilization is found in one or two species of acmaeid limpet (Lottiidae: Patellogastropoda) (Golikov & Kussakin 1972; Lindberg 1983 cited in Ponder & Lindberg 1997), many deep sea and vent Vetigastropoda (e.g., Clypeosectidae, Skeneidae, Lepetodrilidae, Peltospiridae, Neomphalidae, Seguenzioidea; Quinn 1983, 1991; Haszprunar 1987, 1988b, 1989b) (figure 7.2), all Caenogastropoda, Neritimorpha, and Cocculinoidea (Fretter et al. 1998). Whilst most taxa with internal

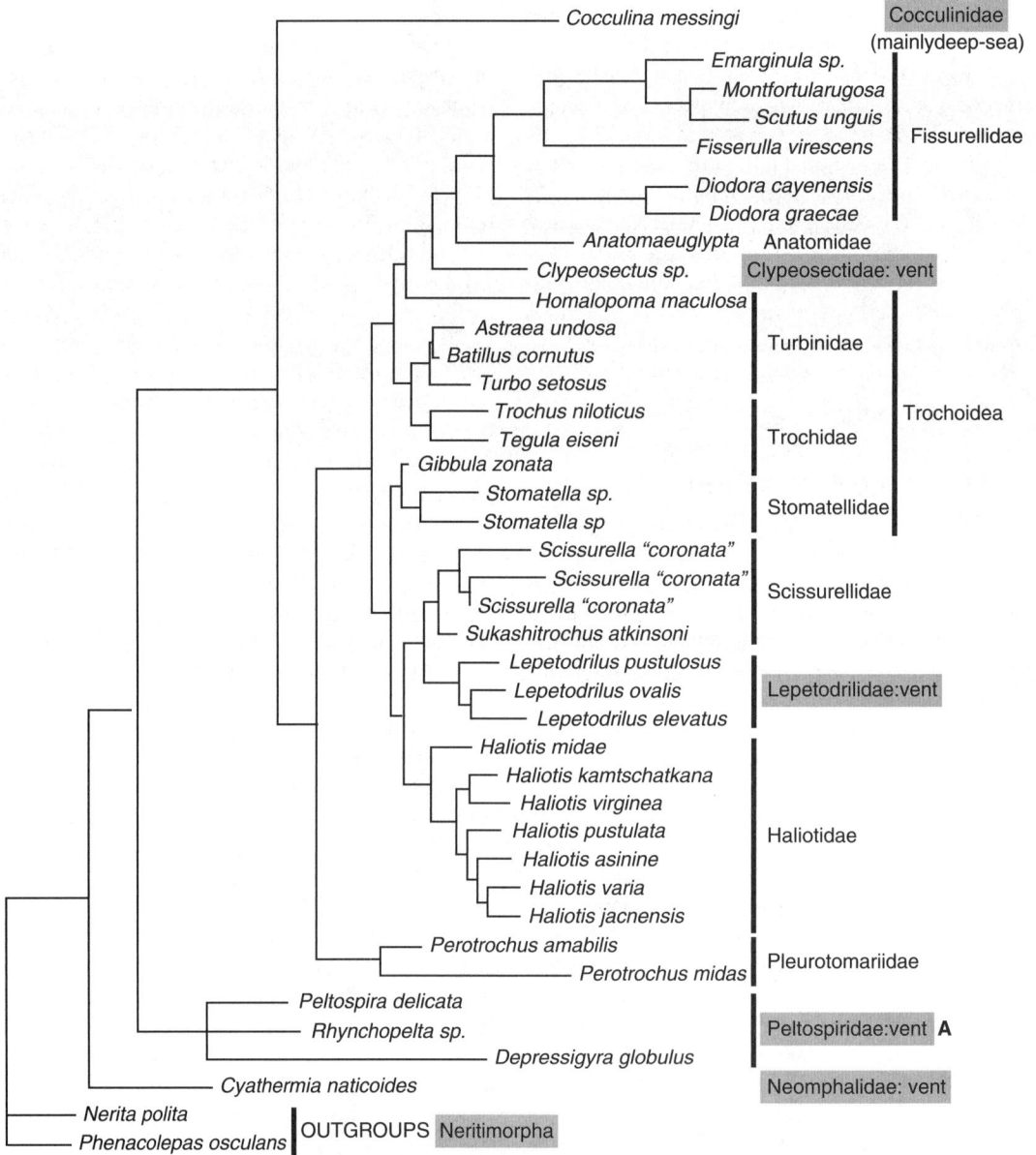

FIGURE 7.2 Phylogeny of the Vetigastropoda based on molecular data (modified from Geiger and Thacker 2005). Taxa highlighted in grey have internal fertilization or fertilization in the mantle cavity (Lepetodrilidae). A = aphallic. Note that both aphallic and phallic taxa can be found in the Neritimorpha.

fertilization are gonochoristic (97% of 2,100 genera; Heller 1993), there are some notable hermaphroditic exceptions, for example, Calyptraeidae (Caenogastropoda) and cocculiniform limpet groups (see Webber 1977 for review of early literature; Haszprunar 1988c; Hoagland & Ponder 1998; Chen & Soong 2000; Collin 2003).

The gross anatomy (and tissue structure) of the reproductive systems of prosobranchs with internal fertilization has been described and well reviewed by a number of authors (e.g., Houston 1976, 1985, 1990; Webber 1977; Fretter 1984, 1989; Houbrick 1987; Haszprunar 1988a,b,c; Fretter & Graham 1994; Voltzow 1994; Strong 2003; see also all sections in chapter 15, Beesley et al. 1998). Therefore a brief overview only is presented here.

The Male Reproductive System

Male prosobranchs with internal fertilization have a single testis (consisting of numerous acini) that lies alongside, and can ramify through, the digestive gland in the coiled visceral hump (Voltzow 1994). Sperm are passed into a ciliated coelomic gonoduct (or testis duct) of mesodermal origin that

functions in part as a seminal vesicle in most species. This region may be greatly swollen and convoluted (e.g., many neritimorphs and higher caenogastropods, Houston 1976; Demaintenon 2001; Simone 2003; Strong 2003) (figure 7.3A), or elaborated into a series of pockets (as in the architaenoglossan *Pomacea canaliculata*; Andrews 1965). In addition to storing autosperm, the seminal vesicle also plays a role in gamete phagocytosis (Ponder 1972, 1973; Houston 1976; Buckland-Nicks & Chia 1976; Purchon 1977; Fretter 1984; Fretter & Graham 1994). The seminal vesicle region leads into the ectodermal pallial gonoduct via the renal vas deferens, the two often being separated by a muscular sphincter which controls the release of sperm (Houston 1976; Purchon 1977; Voltzow 1994). In the majority of prosobranchs the pallial gonoduct is closed (figure 7.3B) (Fretter & Graham 1994), but in a number of diverse caenogastropod families (e.g., Calyptraeidae, Littorinidae, Cypraeidae, Capulidae, Cerithiidae, Naticidae [neogastropod]; figure 7.4) and the Neritidae (Neritimorpha), is an open ciliated groove in the mantle (figure 7.3A). Typically there is a prostate gland associated with the pallial duct,

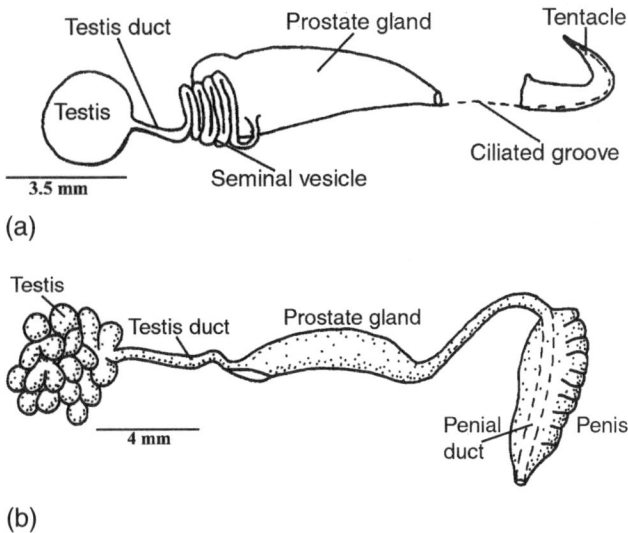

FIGURE 7.3 Examples of two male reproductive systems: (a) with an open pallial gonoduct in the form of a ciliated groove that continues along the length of a tentacle that acts as a copulatory organ (*Neritina latissima*; Neritimorpha). Note that in some caenogastropods the pallial duct may be open for all or part of its length; (b) closed pallial gonoduct and penial tube (*Acanthina angelica*; Caenogastropoda). (a) modified after Houston 1990 and (b) after Houston 1976.

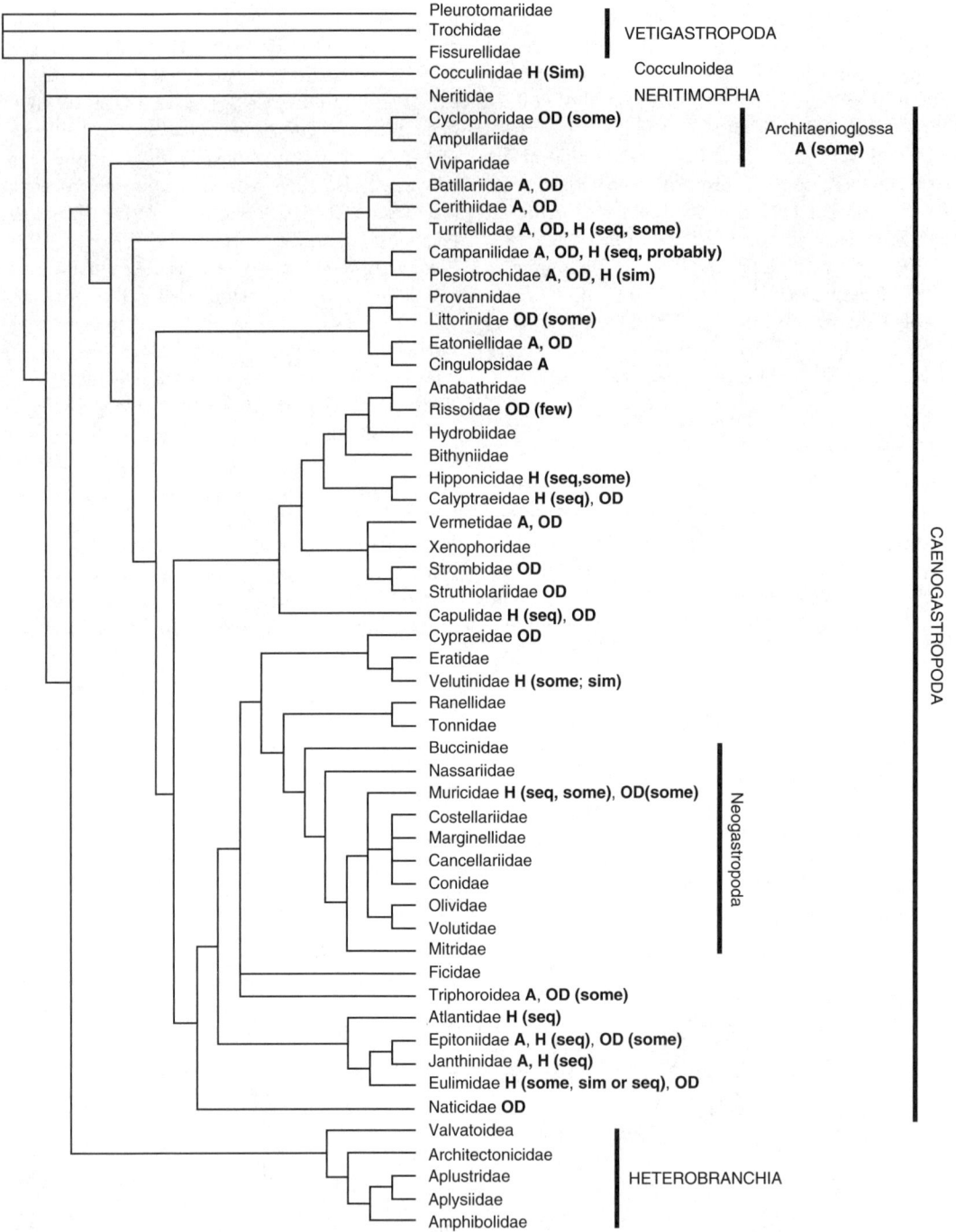

FIGURE 7.4 Phylogeny of the Caenogastropoda based on morphological data (modified from Ponder et al. 2008) with some reproductive features added. Note that not all caenogastropod families are included. A = aphallic; H = hermaphroditic; OD = open genital duct; seq = sequential; Sim = simultaneous.

although in some calyptraeids (Fretter & Graham 1994) and neogastropods (e.g., *Columbella fuscata*; Houston 1976) this gland is absent. The prostate may be divided into regions (Runham 1988; Kano & Kase 2002) that presumably produce different secretions. In species that form spermatophores, the prostate and anterior region of the pallial gonoduct can contribute to their production (Hadfield & Hopper 1980; Runham 1988; Robertson 1989), although details on the process of spermatophore formation are generally lacking (Glaubrecht & Strong 2004).

In most prosobranchs the contents of the prostate and the sperm are discharged through a ciliated tube, to a penis that is anterior in position (figure 7.3B). In some taxa, however, the sperm from the pallial gonoduct are transported to the penis in a ciliated groove. Some notable (and often abundant) unrelated taxa (figures 7.1 and 7.4) are aphallic (e.g., Vetigastropoda: Peltospiridae; Caenogastropoda: Cerithiidae, Turritellidae, Capulidae, Epitoniidae, Etoniellidae, Cingulopsidae, Campanilidae, Plesiotrochidae, Hipponicidae, Janthinidae, Melaniidae, Vermetidae) (Fretter 1984, 1989; Fretter et al. 1998; Kano & Kase 2003). In addition,

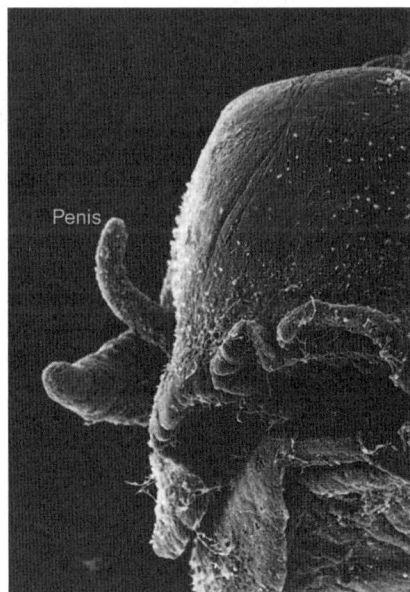

FIGURE 7.5 Scanning electron microscope image of the anterior of *Erginus* sp. (Patellogastropoda) showing the presence of a penis (unpublished photograph courtesy of J. Buckland-Nicks).

both phallic and aphallic species can be present within a taxon, for example, Neritimorpha (Sasaki 1998; Sasaki et al. 2006).

In patellogastropods and vetigastropods that have internal fertilization, the copulatory organ is formed from different tissues in the different taxa. In internally fertilizing Lottiidae it is formed from cephalic tissue (Golikov & Kussakin 1972; Scott & Kenny 1998). *Erginus* (Lottiidae) for example has a true penis (figure 7.5) with a ciliated penial groove, rather like that of some caenogastropods (Buckland-Nicks personal communication). In the vetigastropod taxa, Clypeosectidae and Seguenziidae, the copulatory organ is formed from epipodial tissue (Haszprunar 1989a; Quinn 1991), whereas in the Lepetodrilidae and Gorgolepetidae, from the left oral lappet (Fretter 1988). However, in *Gorgoleptis emarginatus* and *G. patulus* (Lepetodrilacea) the putative penis arises from the left oral region (McLean 1988; Fretter 1988). In the Skeneidae (Trochoidea) the penis is formed from the propodium (Warén & Bouchet 1993; Kano 2008). In many hydrothermal vent vetigastropods one or both cephalic tentacles may be modified as a copulatory organ. In the Neomphalidae the enlarged left cephalic tentacle acts as a penis (Fretter et al. 1981; McLean 1981, 1990; Fretter 1984, 1989; Israelsson 1998) although in *Melanodrymia aurantiaca*, both tentacles are used (Haszprunar 1989b). In most cocculiniform limpets (Haszprunar 1988c; Leal & Harasewych 1999; Strong et al. 2003) and the vetigastropod *Choristella* (Haszprunar 1998) the right cephalic tentacle is a copulatory organ (the bathysciadiids have a penis close to the right cephalic tentacle). In *Cocculina* the penis is a structure that arises from the right side of the foot or neck region (Haszprunar 1987; Sasaki 1998). This suggests that internal fertilization has been independently derived in vetigastropod taxa, and that different groups have found different solutions to sperm transfer.

In neritimorphs the penis is formed from cephalic tissue, although in *Bathynerita naticoidea* and *Shinkailepas* spp. (Neritimorpha, Phenacolepadidae) the right cephalic lappet forms a simple tapering penis (Warén & Bouchet 1993; Sasaki et al. 2003).

The penis of most caenogastropods, which is muscular and non-invaginable, is pedal in origin, innervated by the pedal ganglia and located behind the right cephalic tentacle at the base of the head (figure 7.6). In a few taxa, for example Anabathridae (figure 7.4) and their close relatives the Embandidae, the coiled penis is situated in the middle of the head, and is innervated by the cerebral ganglia. This suggests that a penis has arisen at least twice in caenogastropods.

The penis of prosobranchs is often a relatively simple muscular organ, round to oval in cross-section, with either a closed penial vas deferens (most taxa) or an open sperm groove, for example the caenogastropod *Littorina littorea* (figure 7.6). In some taxa the penis is elaborate in structure and shape (figure 7.7). For example in caenogastropods the penis may have a terminal or sub-terminal papilla (Houston 1976; Harasewych 1984; Simone 2003) which is considered a derived condition. In strombids the tip is spade-shaped, and some species have an accessory pad and auxiliary projections

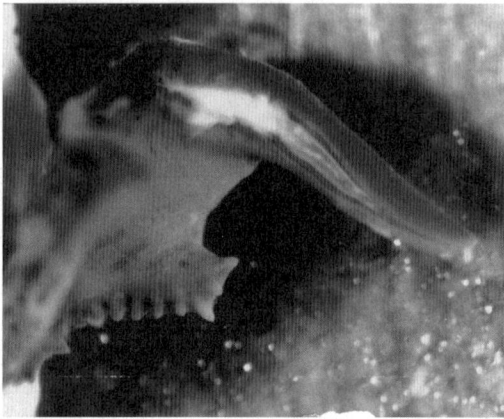

(a)

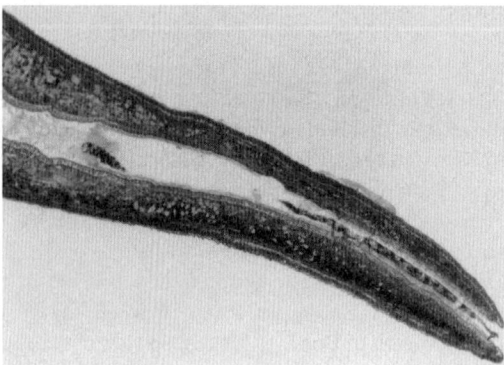

(b)

FIGURE 7.6 (a) Extended cephalic penis of *Littorina littorea* (Caenogastropoda) showing sperm in the dorsal groove. (b) Section of the penial filament with sperm inside groove (from Buckland-Nicks et al. 1998).

(Reed 1995a). Bithyniid (Bithyniidae) penes have an accessory lobe that is glandular (Fretter et al. 1998). That of littorinids has glands (Reid 1986) that produces a visco-elastic secretion that has been suggested to bind the penis in place during copulation (Buckland-Nicks & Worthen 1993; Buckland-Nicks et al. 1999). The mucous secreting glands of the distal portion of the penis of strombids may have a similar function (Reed 1995a). In littorinids the penis delivers sperm into the bursa copulatrix, and whilst Reid (1989) found no correlation between the length of the relaxed penis and position of the bursa, Buckland-Nicks et al. (1999) noted that in four species of littorinid, there was a correlation with the extended penis. Penis length, however, can vary within a species, being influenced by body size and the reproductive cycle (Stroben et al. 1996; Barroso & Moreira 1998; Ramon & Amor 2002). Penis morphology can also vary between sister species, for example in littorinids (Reid 1989), strombids (Reed 1995a), and calyptraeids (Simone 2002; Collin 2003) (figure 7.7).

The Female Reproductive System

The genital system of female prosobranchs is more elaborate than that of the males. This is because it not only produces eggs, it is also the site of fertilization, and therefore has structures for receiving and storing sperm (the bursa copulatrix and seminal receptacle or receptaculum seminalis). Furthermore, in many taxa the females protect the fertilized eggs with coverings and capsules that are secreted by glands (albumen and capsule) associated with the reproductive system.

Prosobranchs have a single ovary (similar in position to the testis) with an oviduct that is modified along its length. In some taxa the pallial oviduct is open (figure 7.8A), and in most aphallic species there is a long opening to this duct. Phallic species tend to have a closed pallial oviduct (Fretter et al. 1998), but there are exceptions (e.g., Cingulopsidae). During mating, sperm or spermatophores are deposited in the bursa copulatrix (e.g., Houston 1990; Reed 1995b), which therefore acts as a temporary site of allosperm storage (Paterson et al. 2001). From here sperm move to the seminal receptacle, which may be a widened part of the oviduct, or a blind ending pouch that is connected to the oviduct (figure 7.8). Within the seminal receptacle the sperm are orientated with their heads embedded in the epithelial cells of the wall (Houbrick

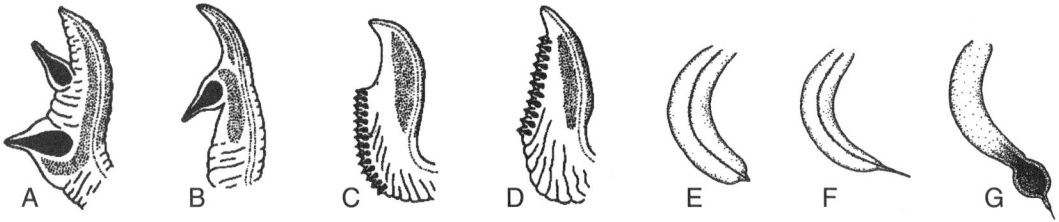

FIGURE 7.7 Examples of penis morphology from different species of caenogastropods, (A to D) littorinids, and (E to G) calyptraeids (A to D from Reid 1989; E to G from Collin 2003).

1973, 1993; Giusti & Selmi 1985; Sasaki 1998; Buckland-Nicks et al. 1999). In pleurocerids (Caenogastropoda) sperm can be stored over winter (Dillon 2000), and littorininds can store sperm for between three months (Paterson et al. 2001) and one year (Erlandsson & Johannesson 1994). In *Viviparus ater* (Viviparidae) fertile sperm can be stored for up to two years (Trüb 1990 cited in Oppliger et al. 2003). Part of the seminal receptacle, however, can ingest spermatozoa (Houston 1985) and in some the receptacle is connected to a gametolytic gland, which may be in the form of a sac-like structure or blind-ending tubules that ingest sperm (Runham 1988; Fretter & Graham 1994). Not all caenogastropods, however, have two sperm storage organs. The seminal receptacle has been lost in the Pomatopsidae and some Lacuninae (Littorinidae), sperm being stored in either the renal oviduct (Reid 1989) or ovary (Buckland-Nicks & Darling 1993).

Glandular regions associated with the reproductive system include an albumen gland and a capsule gland (figure 7.8). The former can take the form of an enlargement of the oviduct or a pouch that opens into the oviduct or as part of the capsule gland. These glands are composed of a number of cell types that produce mucoid and protein secretions (Runham 1988; Fretter & Graham 1994). In brooding species the pallial oviduct may be modified as a brood chamber (Houbrick 1988).

Whilst the majority of prosobranchs have a single opening to the female reproductive system, in the Neritimorpha the number of openings varies both within and between taxa. Two openings (diauly) is most common (figure 7.8B), but examples of monauly and triauly can be found (Strong 2003; Sasaki et al. 2006). In diaulic taxa, one opening leads to the bursa and seminal receptacle, and the second leads to the oviduct and its glands (Houston

1990; figure 7.8B). The two components are joined posteriorly by a duct. In triaulic neritids, the third aperture is an enigma, as its function is not known.

An unusual feature of the female reproductive system of some ampullariids, littorinids and rissoids (three very distantly related caenogastropod taxa, figure 7.4) is the presence of a vestigial or rudimentary penis, which lies near the anus (Thiriot-Quiévreux 1982; Reid 1986; Keawjam 1987; Takada 2000). A pseudopenis has also been

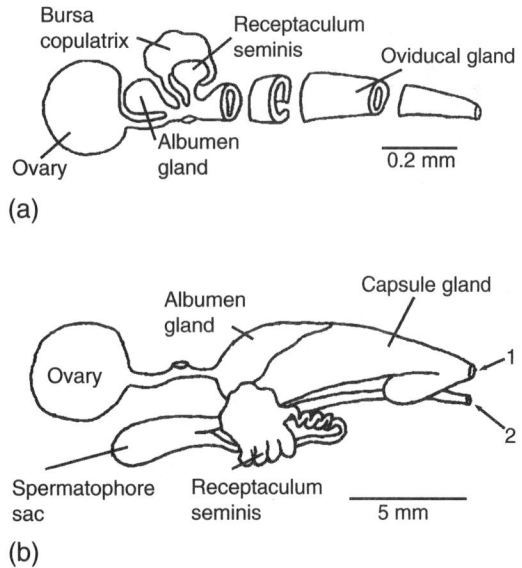

(a)

(b)

FIGURE 7.8 Examples of female reproductive systems: (a) with an open pallial oviduct (*Seila assimilata*; Caenogastropoda); (b) diaulic system of *Nerita funiculata* (Neritimorpha); 1 = nidamental opening, 2 = genital opening (diagrams modified after Houston 1985 (a) and 1990 (b)).

described in some hydrobiid snails (Arconada & Ramos 2002). In the ampullariids both sexes begin to develop a copulatory structure, but when the ovary first appears its growth is arrested (Andrews 1964). Laboratory experiments on littorinids have shown that in males, hormones produced from the pedal ganglia and right ocular tentacle stimulate growth of the penis in males (Reid 1986). Takeda (2000) has proposed a "steroid hormone theory" to explain the development of accessory sex organs in prosobranchs.

Simultaneous and Sequential Hermaphrodites

There are only a few prosobranchs that are simultaneous hermaphrodites (see Fretter 1984 for review; figure 7.4) and it is likely that this trait arose independently in those higher taxa in which it occurs (Fretter et al. 1998). The majority of the deep sea Cocculiniformia (Cocculinoidea and Lepetelloidea) are simultaneous hermaphrodites (Haszprunar 1987, 1988b,c, 1998; Strong & Harasewych 1999), as well as some lamellariids (Velutinidae), eulimids (Eulimidae) and plesiotrochids (all caenogastropod taxa; figure 7.4). Haszprunar (1988c) has proposed that hermaphroditism is the most plesiomorphic condition in the Cocculiniformia. Whether this is true for other taxa is unclear.

The reproductive system of the Cocculiniformia is relatively simple, with sperm and eggs being produced simultaneously in different regions of the ovotestis, or in a separate ovary and testis

(figure 7.9). The gonoduct(s) are ciliated and may or may not be glandular. Transferred sperm are stored in one or more seminal receptacles (Haszprunar 1988c, 1998; Strong & Harasewych 1999). By contrast in species of *Velutina* (Velutinidae, Caenogastropoda) there is a single gonad but there are separate male and female ducts (Wilson 1998). The male duct has a prostate and the female a separate bursa copulatrix, albumen and capsule gland (see figure 6 in Fretter 1984).

Consecutive protandric hermaphroditism (to date protogyny has not been observed) in caenogastropods has been reported from disparate taxa (Hoagland 1978; Collin 2000; Morton & Jones 2001; Richter & Luque 2004) (see figure 7.4). In addition a few species of taxa that are primarily gonochoristic, for example the vermetid *Serpulorbis squamigerus* (Hadfield 1966 cited in Bieler & Hadfield 1990), the littorinid *Mainwaringia rhizophila* (Reid 1986), the vitrinellid *Cyclostremiscus beauii* (Bieler & Mikkelsen 1988), the turritellids *Vermicularia spirata* and *Gazameda gunni* (Bieler & Hadfield 1990), and the assimineid *Rugapedia androgyna* (Fukuda & Ponder 2004) are also hermaphrodites. Although the majority of prosobranchs are gonochoristic (Heller 1993), Bieler & Hadfield (1990) suggest that the number of taxa that are sequential hermaphrodites may be underestimated.

Because protandry is found in unrelated prosobranch groups, it suggests that sequential hermaphroditism has evolved independently in these taxa. For some protandry is probably a reproductive

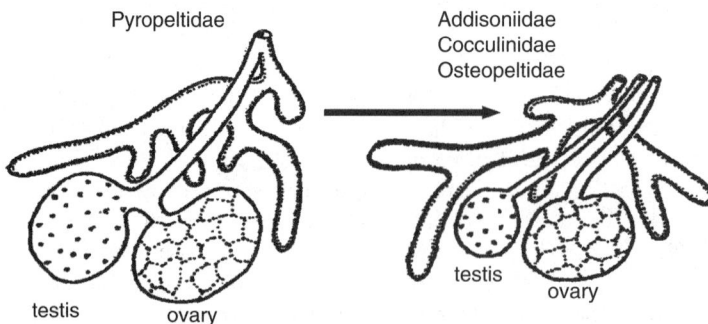

FIGURE 7.9 Two arrangements of the reproductive systems of the simultaneous, hermaphroditic Cocculiniformia. The arrow indicates the proposed evolutionary pathway and in the most apomorphic condition (Choristellidae) the sexes are separate (diagram not presented). (Diagrams are modified from Haszprunar 1988c).

solution either to low densities (Bieler & Mikkelsen 1988), or to a sessile existence (Bieler & Hadfield 1990) as is the case in the Calyptraeidae (e.g., Hoagland 1978; Collin 2000). Coralliophilids also tend to be sessile or have low mobility as adults (Richter & Luque 2004). Whether protandric hermaphroditism is a derived or primitive feature is equivocal and may depend on the group. Slavoshevskaya (1984), for example, has proposed that hermaphroditism was the ancestral condition in the Rissoacea, and that gonochorism is therefore derived. By contrast Wright (1988) and Simone (2002) have proposed that hermaphroditism in calyptraeids is derived, their ancestors being gonochoristic.

In protandric species, sex change can be influenced by the gender of neighboring individuals (see Coe 1938a,b; Hoagland 1978; Warner et al. 1996; Chen et al. 1998; Morton & Jones 2001; Soong & Chen 2003; Richter & Luque 2004), the presence of a large female delaying sex change in others (Soong & Chen 1991; Collin 1995; Warner et al. 1996). By contrast in *Calyptrea chinensis*, *Crepidula adunca*, *C. convexa* (Calyptraeidae) and the vitrinellid *Cyclistremiscus beauii*, sex change happens at a specific age or size (Wyatt 1961; Bieler & Mikkelsen 1988; Chen et al. 1998). Both these mechanisms must trigger changes in those steroid hormones that are thought to control the development of accessory sex organs in prosobranchs (Takeda 2000). In *Crepidula fornicata* these are produced by the cerebral ganglia of the central nervous system (Fretter & Graham 1994). When the masculinizing hormones are no longer produced, ovarian development occurs (Joosse & Geraerts 1983; Wright 1988). Further evidence of the important role of hormones in the expression of sexual characters has been provided by the now well documented effects of endocrine disruptors (e.g., TBT and TPT) on snails. Low levels of these organotin pollutants can cause male sex characters (e.g., penis and vas deferens) to develop in females, as well as the initiation of spermatogenesis (e.g., see Gibbs et al. 1987; Oehlmann et al. 1991; Horiguchi et al. 2006). This phenomenon is particularly prevalent in some neogastropod families (e.g., Muricidae, Buccinidae, Nassariidae) (Fioroni et al. 1991; Horiguchi et al. 1997, 2006), and in some populations impairment of reproductive success results in population decline or even local extinction (Bryan et al. 1986; Gibbs & Bryan 1986, 1996; Horiguchi et al. 2006).

Sex change in some prosobranchs occurs during a reproductive resting stage (Richter & Luque 2004). The extent of morphological change during sex change varies. In *Epitonium* (Epitoniidae), changes are slight as the glandular pallial duct is open in both sexes, and males are aphallic. During the transition to the female condition a receptaculum develops (Fretter 1984). The morphological change in calyptraeids is more substantial. The pallial duct, which is an open groove in males, closes in females and develops glandular walls that form the albumen and capsule glands as well as a seminal receptacle (Fretter 1984). In *Calyptraea morbida* the penis length decreases in size with an increase in female size so that large females have a vestigial penis only (Chen & Soong 2000). Similarly rudiments of a penis remain after male to female sex change in other protandric caenogastropods (Hoagland 1978; Warén 1983; Reid 1986; Bieler & Mikkelsen 1988; Bieler & Hadfield 1990; Soong & Chen 1991; Collin 2000; Richter & Luque 2004).

THE MALE GAMETE AND GAMETE PACKAGES— STRUCTURE AND DIVERSITY

Spermatozoa and Spermatozeugmata

The fertilizing spermatozoa (euspermatozoa) of prosobranchs with true internal fertilization are highly modified haploid cells known as introsperm (Rouse & Jamieson 1987). In addition to the euspermatozoa, however, many prosobranchs also produce non-fertilizing sperm (paraspermatozoa). The two sperm types are formed alongside one another in the testis but develop from distinct lineages of germ cells (Hodgson 1997; Buckland-Nicks et al. 1999). Detailed morphological studies on both sperm types have not only provided insights into fertilization biology, but have also been used to resolve taxonomic and phylogenetic problems, a topic that is discussed later. The literature on the morphology of prosobranch sperm is considerable (e.g., see Nishiwaki 1964; Giusti & Selmi 1982; Maxwell 1983; Kohnert & Storch 1984a,b; Koike 1985; Healy 1983, 1986, 1988, 1996a,b; Hodgson 1995; Buckland-Nicks 1998 for reviews), and a brief overview only of the structure of each sperm type is presented here.

Introsperm—Despite the immense variability in the size and fine structure of prosobranch eusperm, they all have a number of features in common. All are long and filiform, with a cylindrical to rod-shaped nucleus that is capped by a conical acrosome. Posterior to the nucleus is a centriolar complex from which the single axoneme emerges. The axoneme is surrounded by a modified mitochondrial sleeve (elongated mid-piece) posterior to which is a glycogen piece and end-piece (figure 7.10). An annulus is present at the junction of the mid-piece and glycogen piece in all groups except the Neritimorpha (Healy 1988, 2001) and Lepetodrilidae (figure 7.11) (Hodgson et al. 1997).

In a few prosobranchs sperm are probably deposited in the mantle cavity of the female where fertilization may occur. The euspermatozoa produced are known as ent-aquasperm. These spermatozoa have a much simpler morphology when compared to introsperm, yet are not as simple in structure as the sperm from species with true external fertilization (ect-aquasperm). For example whilst the sperm of the hydrothermal vent species *Lepetodrilus fucensis* (Lepetodrilidae) has a free axoneme and lacks a glycogen piece (features typical of aquasperm, Healy 1988), the sperm have an elongated nucleus, modified basal plate, derived centriolar complex and flagellum with a distal accessory sheath (Hodgson et al. 1997) (figure 7.11). These are features that are not associated with true ect-aquasperm. To date ent-aquasperm have been described from a few vetigastropods only (Lepetodrilidae, Hodgson et al. 1997; possibly *Sinezona* sp. (Scissurellidae), Healy 1990a; Skeneiformia, Healy & Ponder cited in Hodgson et al. 1997).

Paraspermatozoa—Paraspermatozoa have been described from a variety of prosobranch taxa, including the Neritimorpha, Vetigastropoda, and Caenogastropoda (Nishiwaki 1964; Melone et al. 1980; Giusti & Selmi 1982; Healy 1986, 1988, 1990b; Buckland-Nicks 1998). Because parasperm occur in these distantly related clades, sperm heteromorphism has probably evolved more than once, and has a long evolutionary history within prosobranchs. In the caenogastropods parasperm are very common and reach their greatest morphological diversity (figure 7.12), with some species producing more than one type of parasperm (Buckland-Nicks et al. 1982). Buckland-Nicks and Hodgson (2005) suggest that within the Caenogastropoda at least, parasperm appeared in a common

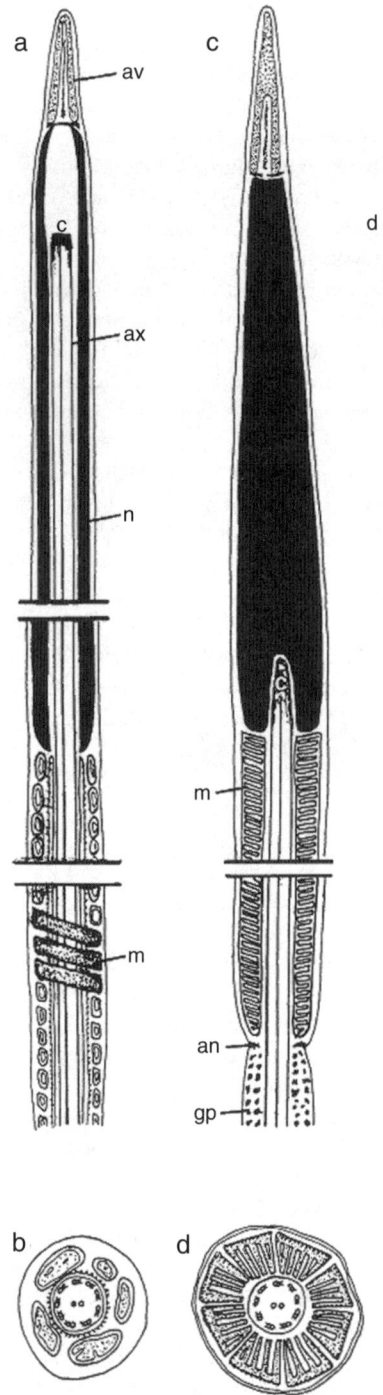

FIGURE 7.10 Examples of prosobranch introsperm in LS (a, c) and TS through the mid-piece (b, d). av, acrosomal vesicle; an, annulus; ax, axoneme; gp, glycogen piece; m, modified mitochondria; n, nucleus (from Healy 2001).

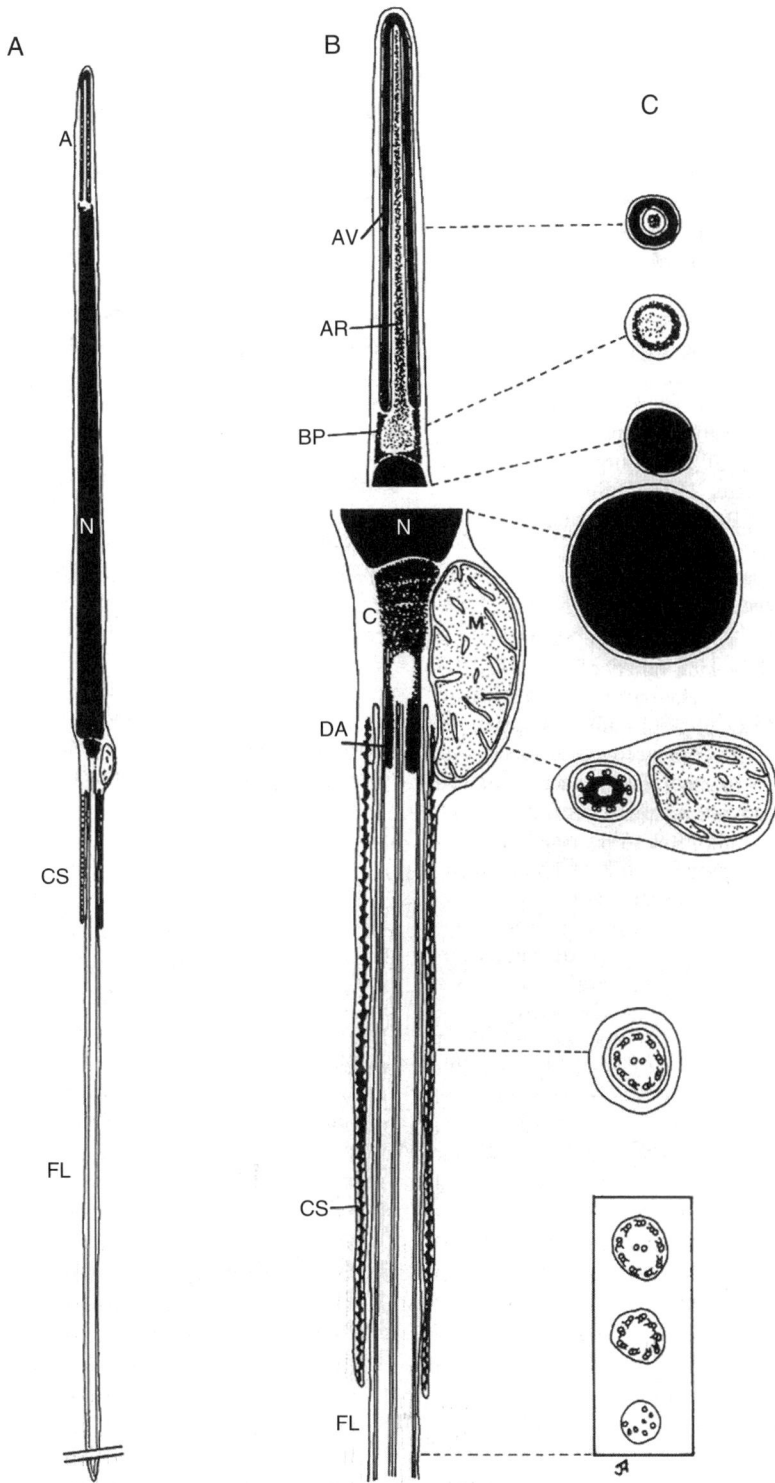

FIGURE 7.11 Entaquasperm of *Lepetodrilus fucensis* in longitudinal (A, B) and transverse (C) sections. A, acrosome; AV, acrosomal vesicle; AX, axial rod; BP, basal plate; C, centriolar complex; CS, cytoplasmic sheath; DA, distal accessory sheath; FL, flagellum; M, mitochondrion of midpiece; N, nucleus (from Hodgson et al. 1997).

ancestor, prior to the divergence of this group. Parasperm are also considered a plesiomorphic feature of littorinids (Warén & Hain 1996). Although prosobranch parasperm are morphologically diverse, the majority lack any nuclear material (apyrene) or have a nuclear remnant (oligopyrene), lack an acrosome (exceptions are Cerithioidea and Campaniloidea; Healy & Jamieson 1981; Buckland-Nicks & Hodgson 2005) and are multi-flagellate. Many also possess large amounts of storage product (Tochimoto 1967; Buckland-Nicks 1998; Buckland-Nicks & Hodgson 2005).

The function and evolution of all prosobranch parasperm continues to intrigue researchers. This is in part due to their variability in size and shape, as well as motility, but largely a result of not being able to track their fate once transferred to the female. Some parasperm undoubtedly play a role in eusperm transportation, for example in the aphallic Cerithiopsidae and Janthinidae. In these, and some other taxa, large numbers of euspermatozoa become attached to the tail region of the large multi-axonemed parasperm to form a spermatozeugma (figure 7.12D,E,F,I). It is possible that as a unit they swim faster than an individual sperm would (Buckland-Nicks 1998). Spermatozeugmata, however, are also produced by some phallic taxa, and they probably do not have a transportation role. In *Littorina littorea* and *Fusitriton oregonensis* the sperm of the spermatozeugmata separate into eusperm and parasperm by the time the ejaculate reaches the end of the penis (Buckland-Nicks et al. 1999; Buckland-Nicks & Tompkins 2005). Using homogenates from the prostate gland, Buckland-Nicks et al. (1999) were able to show that the alkaline prostate fluid can cause the eusperm and parasperm to separate. Buckland-Nicks (1998) has suggested that spermatozeugmata might break through sperm plugs to rival males an ability to deposit fertilizing sperm, and it is interesting to note that in some cerithiodeans the parasperm possess an anterior acrosome-like structure (Buckland-Nicks & Hodgson 2005).

Many taxa, however, produce parasperm and eusperm that do not form physically connected swimming units, and in a number of species the parasperm are immotile (Buckland-Nicks 1998). In those species in which the composition of the seminal receptacle has been studied, only eusperm have been found (e.g., *Cerithium muscarum*; Houbrick 1973). It is unlikely that these parasperm would play a role in transportation of the eusperm to the female or within the female tract. Furthermore, in a number of taxa the eusperm and parasperm are encased in a spermatophore (see below). One suggestion for the function of such parasperm (as well as some carrier parasperm) is that they act as nutritional nuptial gifts (Hanson et al. 1952; Reed 1995a; Buckland-Nicks 1998). This is because the head region of parasperm contains numerous glycoprotein vesicles (Tochimoto 1967; Buckland-Nicks 1998; Buckland-Nicks & Hodgson 2005), which are clearly transferred to the female during mating. Parasperm have been observed to be digested in the bursa copulatrix (Dembski 1968) and in *Strombus* spp. to disintegrate within two hours of their introduction into the female (Reed 1995a). Such gifts may facilitate egg production. The game theory models of Kura and Yoda (2001), however, predict that nutritional benefits will only be found in species where multiple matings are rare. Dembski (1968) also suggested that the breakdown products of the parasperm, along with prostate secretions may activate eusperm motility. Alternatively parasperm may play a role in sperm competition (discussed in a following section) or as countermeasures against spermicides as recently shown for one sperm type in *Drosophila* (Holman & Snook 2008).

Spermatophores

The males of a number of species from about 21 prosobranch families examined to date, package their sperm for delivery into spermatophores (see Roberston 1989 for earlier literature; Jamieson & Newman 1989; Nakano & Nishiwaki 1989; Bieler & Hadfield 1990; Houston 1990; Houbrick 1991a,b, 1992, 1993; Zehra & Perveen 1991; Kennedy 1995; Sasaki 1998; Dillon 2000; Kano & Kase 2002; Glaubrecht & Strong 2004; Calvo & Templado 2005). Thus spermatophores are found in disparate taxa from diverse habitats suggesting that they have evolved independently a number of times. Spermatophores, which can be complex in structure (e.g., in vermetids), and large (up to 3 cm long in the neritimorph *Neritina reclivata*; Andrews 1936) can contain both eusperm and parasperm. Spermatophores are another evolutionary solution to sperm transfer and internal fertilization in aphallic prosobranchs such as the Cerithiidae, Vermetidae, and possibly the Peltospiridae (Hadfield & Hopper 1980; Warén personal communication cited in Israelsson 1998). The actual transfer of spermatophores, however, has been observed in a

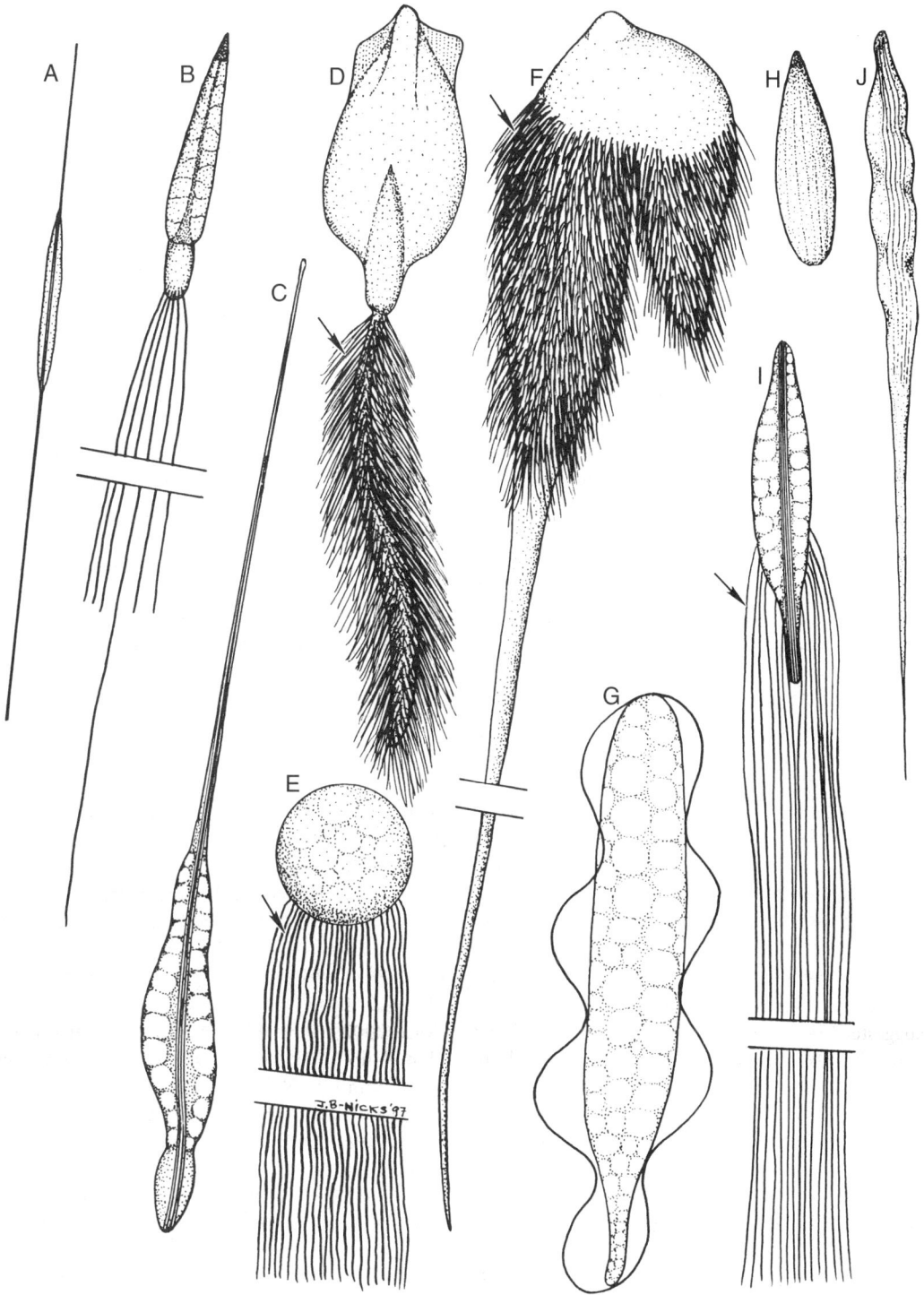

FIGURE 7.12 Diagrammatic examples of the major forms of parasperm of prosobranchs. (A) *Nerita*; (B) *Cerithium*; (C) *Serpulorbis*; (D) *Janthina*; (E) *Littorina*; (F) *Epitonium*; (G) *Strombus*; (H) *Conus*; (I) *Fusitrion* carrier; (J) *Fusitriton* lancet sperm. Arrows show eusperm attached to parasperm to form spermatozeugmata (D, E, F, I) (from Buckland-Nicks 1998).

few species of cerithioideans and vermetids only. In the caenogastropods *Cerithium muscarum* and *Modulus modulus* the male releases the spermatophore close to the entrance to the mantle cavity of female. After the spermatophore breaks down (which can take up to 20 minutes) the sperm are drawn into the mantle cavity by cilia (Houbrick 1973). In *M. modulus* spermatophores are drawn into the spermatophore receptacle by ciliary currents (Houbrick 1980) where they eventually break down. In other cerithioideans, spermatophore transfer might occur by water currents (Fretter 1989; Calvo & Templado 2005) although *Turritella communis* engages in pseudo-copulatory behavior (Kennedy 1995). Hadfield and Hopper (1980) suggest that spermatophore production was an important pre-adaptation for vermetids to become sessile, and in those species that are not gregarious, the morphologically complex inflated pelagic spermatophores are released into the water where they are rapidly captured by the large mucous nets of females (Hadfield and Hopper 1980; Calvo & Templado 2005).

Spermatophores, however, are not unique to aphallic species, being produced by phallic members of the Neritimorpha (aquatic taxa only; Robertson 1989) and Heteropoda.

COPULATION, SPERM COMPETITION AND PATERNITY

In animals that have internal fertilization, mutiple matings by females are now known to be common. Females may then 'select' the sperm from a number of rival males (see Birkhead & Møller 1998 for review and Eberhard 2000 for a discussion on 'sperm choice' vs. 'cryptic female choice') and offspring can be sired by a number of males.

In comparison to other pair mating gastropods, detailed studies on prosobranch mating are few. Nevertheless these few studies have shown that the females of some species will, over a short period of time, mate with several males (Dembski 1968; Erlandsson & Johannesson 1994; Staub & Ribi 1995; Reed 1995b; Baur 1998; Oppliger et al. 2003; Yusa 2004), and that this can result in multiple paternity (Gaffney & McGee 1992; Warner et al. 1996; Paterson et al. 2001; Oppliger et al. 2003; Walker et al. 2007). There are several possible advantages that females gain from being polyandrous. First it ensures that sufficient sperm are received.

Because some males can copulate with more than one female over a short period of time, it is possible that later-mated females receive a depleted ejaculate. Second, using sperm from more than one male may increase the genetic diversity of her offspring. Third, it increases the chances of receiving viable sperm, as some mates could have been sterile. Finally, having multiple mates could allow the female to choose the "fittest" sperm. Whether this form of post-copulatory selection and resultant sperm competition occurs in prosobranchs is not known, but female *Littorina scabra* have been seen dumping sperm that are possibly unwanted (Buckland-Nicks 1998). For some species at least, multiple mating is beneficial to females in that they produce more offspring than those of the same size that mate with one male only (Oppliger et al. 2003). How widespread this is within prosobranchs is not known.

One of the ways by which males may promote their paternity is by the production of paraspermatozoa (see also previous discussion on the transportation role of parasperm). Recently, it has been found that in the caenogastropod *Viviparus ater*, as the risk of sperm competition increases, males produce relatively more parasperm that are also longer (Oppliger et al. 1998, 2003). Furthermore, those males with longer parasperm sired a greater proportion of offspring (Oppliger et al. 2003). How parasperm achieve this is still unknown. They may simply block the sperm of rival males. For example, the lancet parasperm of *Fusitriton* help form a sperm plug in the bursa copulatrix (Buckland-Nicks 1998). Parasperm could also suppress the desire of the female to mate with another male, as has been found in insects (Silbergleid et al. 1984) or as previously mentioned counteract spermicides in the female tract.

Competition to sire offspring, therefore, can be intense and has lead males to develop other behaviours to promote their paternity. Mate guarding and prolonging copulation are ways by which males could achieve this. Observations on prosobranch pre- and post-copulatory mate guarding, however, are rare; Gibson (1964) has observed pre-copulatory guarding in *Littorina planaxis* and Bradshaw-Hawkins and Sander (1981) reported post-copulatory guarding in *Strombus pugilus*. In some prosobranchs copulation can last from one to several hours (Gibson 1964; Struhsaker 1966; Houbrick 1973; Houston 1976; Martel et al. 1986; Reed 1995a; Albrecht et al. 1996; Dillon 2000; Paterson et al. 2001; Nieman & Lively 2005).

Lengthy copulation not only means that the female genital tract is blocked, but it may also ensure that there is time for sperm to be transferred from the bursa copulatrix to the seminal receptacle, which is deep in the oviduct. In some littorinids this region is out of reach of the penial filament, and therefore sperm are in no danger of physical removal by another male's penis (Buckland-Nicks & Chia 1990; Paterson et al. 2001). To date, however, there is no evidence that the penis of any male prosobranch can remove another's sperm. The recent study of paternity in *Pomacea canaliculata* (Yusa 2004), however, has shown that in this species at least, when a female is mated with two males, offspring were sired by the second. This suggests that the sperm of the second male displaced that of the first and may explain why males of this species remain in copula for up to 18 hours (Albrecht et al. 1996).

REPRODUCTIVE CHARACTERS, CLUES TO PROSOBRANCH PHYLOGENY AND SYSTEMATICS

Because the reproductive systems of gastropods with internal fertilization can be anatomically complex, they have the potential to provide valuable characters for phylogenetic analyses. Very few characters derived from gross reproductive anatomy, however, have been used to help unravel higher phylogenetic relationships within the Gastropoda. Of the 25 reproductive characters used by Ponder and Lindberg (1997) in their phylogenetic analyses, only five were from reproductive anatomy. By contrast 17 were characters derived from sperm morphology (discussed below). One of the reasons for this is that many anatomical features (e.g., bursa copulatrix, seminal vesicle, oviducal glands, open/closed gonoducts) are not homologous (Fretter et al. 1998; Sasaki 1998), and have undoubtedly evolved independently in different groups. For example, Purchon (1977) suggested that in prosobranchs closed genital ducts were derived from open ones. In caenogastropods whilst basal architaenoglossans can have closed ducts, the open condition is found in many basal taxa (e.g., Batillariidae, Cerithiidae, and Turritellidae) (figure 7.4). In such taxa males are also aphallic, an indication that this is the primitive condition in caenogastropods. Open genital ducts, however, are also present in taxa with more derived features (e.g., Calyptraeidae

and Cypraeidae; figure 7.4), suggesting that it is secondarily derived in some lineages. Furthermore, both open and closed genital ducts occur within some caenogastropod families (e.g., Littorinidae, Muricidae, Epitoniidae; figure 7.4). Reid (1989) considers open ducts an apomorphic feature of littorinids. A further example is the penis, which in the majority of gastropods is situated adjacent to the right cephalic tentacle. Anatomical and developmental studies, however, have shown that there is variation in the origin and innervation of prosobranch penes. In most it is pedally innervated, but examples of innervation from the right pleural ganglion and cerebral ganglion can be found. Ponder et al. (2008) suggest that in the hypsogastropods (= all caenogastropods except the archaetaenoglossa and cerithioideans) the cephalic penis has been independently derived in several taxa. Furthermore, the penes of architaenioglossans also have independent origins, from the mantle in the Ampullarioidea, the right tentacle in the Viviparoidea, and a cephalic origin in the Cyclophoroidea (Ponder et al. 2008). Kano (2008) also proposes that in those minute and deep-sea vetigastropod taxa with internal fertilization, the penis (and seminal receptacle) has evolved independently several times. This is thought to be a result of size and/or the need to ensure fertilization in habitats where population numbers are very low (Kano 2008). Robertson (1989) also concluded that because spermatophores occur in very disparate prosobranch groups, they cannot be used in higher systematic studies.

Although seemingly of little value at a higher taxonomic level, a number of studies have used characters from reproductive anatomy in their phylogenetic investigations of specific prosobranch taxa. Houbrick (1980) suggested that unique features of the pallial gonoducts of the Modulidae separated this family from other cerithiaceans. Sasaki (1998) produced a phylogeny of the archaeogastropods in which 10 of the 93 morphological characters were from reproductive features, and Strong (2003) used nine reproductive characters (out of a total of 64) in her phylogeny of the Caenogastropoda. In their molecular phylogenetic study of the neritimorphs, Kano et al. (2002) suggested that the only morphological characters that were diagnostic of clades, were reproductive ones.

At a lower taxonomic level anatomical reproductive characters have been used to varying degrees in the generation of phylogenetic trees of some muricoideans (Harasewych 1984—3 of 15 characters), cerithiodeans (Houbrick 1987—2 of 21 in one analysis and 4 of 25 in a second; 1988—13 of 58; 1993—7 of 21); littorinids (Reid 1989—26 of 53), Rapaninae (Kool 1993—6 out of 18), calyptraeids (Simone 2002—16 of 112; Collin 2003—17 of 79) and cocculinoideans (Strong et al. 2003—5 of 31).

Features of the reproductive system have long been used to separate gastropod taxa at the species and generic level. For example Reid (1989) notes that for littorinids at both a generic and specific level, "the shape of the penis is the single most important taxonomic character" and the female and male genitalia are useful for generic and species differentiation in the hydrobiids (Runham 1988; Arconada & Ramos 2002; Haase 2003). Whilst reproductive characters have been of great value at a generic level in the cocculinoideans, Strong et al. (2003) suggest that once more taxa are studied, the structure of their copulatory organs may be very informative phylogenetically.

The diversity of animal genitalia has been regarded as being brought about by sexual selection, and genitalia are often species specific (Eberhard 1985, 1996). Male genitalia in particular appear to be subject to rapid evolution through sexual selection (Eberhard 1993). As in other animals, variation in gastropod penis morphology may have arisen to prevent inter-specific mating through morphological or sensory incompatibility. Unfortunately there are very few studies that have examined prosobranch genitalia with this in mind. Species of *Strombus* have been shown to possess unique penes, and interspecific copulation can occur, but such matings do not result in offspring (Reed 1995a,b). Indiscriminate interspecific copulations have also been recorded in littorinids (Ito & Wada 2006) and morphological incompatibility of the penis with the female duct of another species may prevent fertilizations.

Haszprunar (1988c) found that although the genital system of the Cocculiformia is simple, there are morphological characters that can be used for classification at the generic level, and that an evolutionary trend in their genital system is apparent. Haszprunar (1988c) proposed that hermaphroditism is the most plesiomorphic state in these gastropods. In the Lepetellidae, Pyropeltidae, and Pseudococculinidae the ovotestis has quite distinct male and female regions with a single gonoduct and common genital opening. A slightly more advanced condition is seen in the Osteopeltidae, Cocculinellidae,

and Addisoniidae, which all have a separate male and female gonad and two gonoducts and reproductive openings. The most advanced condition is in the Choristellidae which are gonochoristic. Thus the evolution of gonochorism in the Cocculinoidea would appear to be secondary.

Studies of gastropod eusperm and parasperm morphology and genesis have provided valuable insights into taxonomic and systematic relationships (e.g., Healy 1983, 1986, 1987, 1988, 1995, 1996b, 2001; Hodgson 1995). For example sperm morphology supported the suggestion that the Omalogyroidea and Pyramidelloidea are not prosobanch taxa and should be placed within the Heterobranchia (Healy 1988). Healy (1995) also concluded, that based on sperm morphology the Architectonicoidea are associated with the Heterobranchia and not prosobranchs (figure 7.4). As mentioned previously, the majority of reproductive characters used in Ponder and Lindberg's phylogenetic analysis of gastropods were derived from sperm morphology.

Although Robertson (1989) concluded that spermatophore morphology could not help to resolve systematic relationships between higher prosobranch taxa, Hadfield and Hopper (1980) suggested that spermatophores could provide species specific structures, a conclusion supported by Glaubrecht and Strong (2004). Whether such species specificity, along with male genital morphology is a result of sexual selection remains to be determined (Glaubrecht & Strong 2004).

Although ovarian morphology and oogenesis can also help resolve systematic and phylogenetic questions (Eckelbarger 1994), very little detailed work has been undertaken on gastropods (Hodgson & Eckelbarger 2000), an avenue of research that could prove valuable in the future. By contrast egg size and egg capsule morphology has been used in systematics at a generic and specific level (e.g., see Hoagland 1986; Reid 1989; Mak 1995).

CONCLUDING REMARKS

The majority of pair mating prosobanch gastropod taxa are gonochoristic, with hermaphroditism occurring in a number of unrelated groups. Whilst prosobranchs have reproductive systems with a reasonably common anatomical plan, wide variation in reproductive and genital morphology exists within and between taxa, and a number of similar reproductive features (e.g., penis) have often arisen independently. At a higher taxomonic level it is therefore difficult to see phylogenetic trends in anatomy. An exception to this is sperm structure, the study of which has contributed greatly to understanding prosobranch systematics. By contrast at the generic and species level, reproductive and genital morphology, as well as gamete and egg mass structure, can provide valuable taxonomic and phylogenetic characters.

Whether variation in reproductive and genital anatomy is a result of natural, or sexual selection (as shown for other invertebrates; Eberhard 1985; Møller 1998), or a combination of the two, remains equivocal, although it seems unlikely that differences in sperm morphology at a species level would be a result of natural selection. Studies on sexual selection in prosobranchs have tended to concentrate on mating behavior to see whether size-assortative mating occurs (e.g., Erlandsson & Johannesson 1994; Staub & Ribi 1995; Johnson 1999) and other detailed investigations on sexual selection are now required. If we are to have a greater appreciation of the selective forces operating on prosobranch genitalia detailed studies of the female reproductive tract with respect to male genital morphology, and copulation and the fate of eusperm and parasperm, along with mating experiments using molecular markers to study paternity are required, and could prove to be very enlightening.

REFERENCES

Aktipis, S.W., Giribet, G., Lindberg, D.L. & Ponder, W.F. 2008. Gastropoda. In: Phylogeny and Evolution of the Mollusca (Ed. W.F. Ponder & D.R. Lindberg). pp. 201–237. Berkeley: University of California Press.

Albrecht, E.A., Carreño, N.B. & Castro-Vazquez, A. 1996. A quantitative study of copulation and spawning in the South American apple-snail, *Pomacea canaliculata* (Prosobranchia, Ampullariidae). The Veliger, 39, 142–147.

Andrews, E.A. 1936. Spermatophores of the snail *Neritina reclivata*. Journal of Morphology, 60, 191–209.

Andrews, E.B. 1964. The functional anatomy and histology of the reproductive system of some pilid gastropod molluscs. Proceedings of the Malacological Society of London, 36, 121–140.

Andrews, E.B. 1965. The functional anatomy of the mantle cavity, kidney and blood system of some pilid gastropods (Prosobranchia). Journal of Zoology, 146, 70–94.

Arconada, B. & Ramos, M.A. 2002. *Spathogyna*, a new genis for *Valvata* (?*Tropodina*) *fezi* Altimira, 1960 from eastern Spain: another case of pseudohermaphroditism in the Hydrobiidae (Gastropoda). Journal of Molluscan Studies, 68, 319–327.

Barroso, C.M. & Moreira, M.H. 1998. Reproductive cycle of *Nassarius reticulatus* in the Ria de Aveiro, Portugal: implications for imposex studies. Journal of the Marine Biological Association of the United Kingdom, 78, 1233–1246.

Baur, B. 1998. Sperm competition in molluscs. In: Sperm Competition and Sexual Selection (Ed. T.R. Birkhead & A. P. Møller). pp. 225–305. London: Academic Press.

Beesley, P.L., Ross, G.J.B. & Wells, A. 1998. Mollusca: The Southern Synthesis. Fauna of Australia. Vol. 5, Chapter 15. Prosobranchia. CSIRO Publishing, Melbourne, Part B viii pp. 565–1234.

Bieler, R. 1992. Gastropod phylogeny and systematics. Annual Review of Ecology and Systematics, 23, 311–338.

Bieler, R. & Hadfield, M.G. 1990. Reproductive biology of the sessile gastropod *Vermicularia spirata* (Cerithioidea: Turritellidae). Journal of Molluscan Studies, 56, 205–219.

Bieler, R. & Mikkelsen, P.M. 1988. Anatomy and reproductive biology of two western Atlantic species of Vitrinellidae, with a case of protandrous hermaphroditism in the Rissoacea. Nautilus, 102, 1–29.

Birkhead, T. R. & Møller, A.P. 1998. Sperm Competition and Sexual Selection. London: Academic Press.

Bradshaw-Hawkins, V.I. & Sander, F. 1981. Notes on the reproductive biology and behaviour of the West-Indian fighting conch, *Strombus pugilis* Linnaeus in Barbados, with evidence of mate guarding. The Veliger, 24, 159–164.

Bryan, G.W., Gibbs, P.E., Hummerstone, L.G. & Burt, G.R. 1986. The decline of the gastropod *Nucella lapillus* around the south-west of England: evidence for the effect of tributyltin from anti-fouling paints. Journal of the Marine Biological Association of the United Kingdom, 66, 611–640.

Buckland-Nicks, J. 1998. Prosobranch parasperm: sterile germ cells that promote paternity? Micron, 29, 267–280.

Buckland-Nicks, J. & Chia, F-S. 1976. Fine structural observations of sperm resorption in the seminal vesicle of a marine snail *Littorina scutulata* (Gould, 1849). Cell and Tissue Research, 172, 503–515.

Buckland-Nicks, J.A. & Chia, F-S. 1990. Egg capsule formation and hatching success in the marine snail *Littorina sitkana*. Philosophical Transactions of the Royal Society, London B, 326, 159–176.

Buckland-Nicks, J. & Darling, P. 1993. Sperm are stored in the ovary of *Lacuna* (*Epheria*) *variegata* (Carpenter, 1864) (Gastropoda: Littorinidae). Journal of Experimental Zoology, 267, 624–627.

Buckland-Nicks, J. & Hodgson, A.N. 2005. Paraspermatogenesis of cerithioidean snails: retention of an acrosome and nuclear remnant. Journal of Morphology, 264, 314–326.

Buckland-Nicks, J. & Scheltema, A. 1995. Was internal fertilization an early innovation of early Bilateria? Evidence from sperm structure of a mollusc. Proceedings of the Royal Society of London, Biological Sciences, 261, 11–18.

Buckland-Nicks, J. & Tompkins, G. 2005. Paraspermatogenesis in *Ceratostoma foliatum* (Neogastropoda): confirmation of programmed nuclear death. Journal of Experimental Zoology, 303, 723–741.

Buckland-Nicks, J. & Worthen, G.T. 1993. Functional morphology of the mammiliform penial glands of *Littorina saxatilis* (Gastropoda). Zoomorphology, 112, 217–225.

Buckland-Nicks, J., Williams, D., Chia, F-S. & Fontaine, A. 1982. Studies on the polymorphic spermatozoa of a marine snail. I—genesis of the apyrene sperm. Biology of the Cell, 44, 305–314.

Buckland-Nicks, J., Bryson, I., Hart, L. & Partridge, V. 1999. Sex and a snail's sperm: on the transport, storage and fate of dimorphic sperm in Littorinidae. Invertebrate Reproduction and Development, 36, 145–152.

Calvo, M. & Templado, J. 2005. Spermatophores of three Mediterranean species of vermetid gastropods (Caenogastropoda). Journal of Molluscan Studies, 71, 301–311.

Chen, M-H & Soong, K. 2000. Sex change in the hat snail, *Calyptraea morbida* (Reeve) (Gastropoda: Calyptraeidae): An analysis of substratum, size and reproductive characteristics. The Veliger, 43, 210–217.

Chen, M.-H., Yang, Y.-W. & Soong, K. 1998. Preliminary observations on change of sex by the coral inhabiting snail *Coralliophila violacea* (Lamarck) (Gastropoda: Coralliophilidae). Journal of Experimental Marine Biology and Ecology, 230, 207–212.

Coe, W.R. 1938a. Influence of association on the sexual phases of gastropods having protandric consecutive sexuality. Biological Bulletin, 75, 274–285.

Coe, W.R. 1938b. Conditions influencing change of sex in molluscs of the genus *Crepidula*. Journal of Experimental Zoology, 77, 401–424.

Colgan, D.J., Ponder, W.F. & Eggler, P.E. 2000. Gastropod evolutionary rates and phylogenetic relationships assessed using partial 28S rDNA and histone H3 sequences. Zoological Scripta, 29, 29–63.

Colgan, D.J., Ponder, W.F., Beacham, E. & Macaranas, J.M. 2003. Gastropod phylogeny based on six segments from four genes representing coding or non-coding and mitochondrial or nuclear DNA. Molluscan Research, 23, 123–148.

Collin, R. 1995. Sex, size and position: a test of models predicting size at sex change in the protandrous gastropod *Crepidula fornicata*. American Naturalist, 146, 815–831.

Collin, R. 2000. Sex change, reproduction, and development of *Crepidula adunca* and *Crepidula lingulata* (Gastropoda: Calyptraeidae). The Veliger, 43, 24–33.

Collin, R. 2003. The utility of morphological characters in gastropod phylogenetics: an example from the Calyptraeidae. Biological Journal of the Linnean Society, 78, 541–593.

Demaintenon, M.J. 2001. Analysis of reproductive system ontogeny and homology in *Nassarius vibex* (Gastropoda: Buccinidae: Nassariinae). Journal of Molluscan Studies, 67, 37–49.

Dembski, W.J. 1968. Histochemische Untersuchungen über Funkton un Verbleib eu- und oligopyrener Spermien von *Viviparus contectus* (Millet 1813) (Gastropoda: Prosobranchia). Zeitschrift für Zellforschung und Mikroskopische Anatomie, 89, 151–179.

Dillon, R.T. Jr. 2000. The Ecology of Freshwater Molluscs. Cambridge: Cambridge University Press.

Eberhard, W.G. 1985. Sexual Selection and Animal Genitalia. Cambridge MA: Harvard University Press.

Eberhard, W.G. 1993. Evaluating models of sexual selection: genitalia as a test case. The American Naturalist, 142, 564–571.

Eberhard, W.G. 1996. Female Control: Sexual Selection by Cryptic Female Choice. Princeton: Princeton University Press.

Eberhard, W.G. 2000. Criteria for demonstrating postcopulatory female choice. Evolution, 54, 1047–1050.

Eckelbarger, K.J. 1994. Diversity of metazoan ovaries and vitellogenic mechanisms: implications for life history theory. Proceedings of the Biological Society of Washington, 107, 193–218.

Erlandsson, J. & Johannesson, K. 1994. Sexual selection on female size in a marine snail, Littorina littorea (L.). Journal of Experimental Marine Biology and Ecology, 181, 145–157.

Fioroni, P., Oehlmann, J. & Stroben, J. 1991. The pseudohermaphroditism of prosobranchs: morphological aspects. Zoologischer Anzeiger, 226, 1–26.

Fretter, V. 1984. Prosobranchs. In: The Mollusca. V. 7. (Ed. A.S. Tompa, N.H. Verdonk & J.A.M. van den Biggelaar), pp. 1–45. Orlando: Academic Press.

Fretter, V. 1988. New archaeogastropod limpets from hydrothermal vents, superfamily Lepetodrilacea. Part 2: Anatomy. Philosophical Transactions of the Royal Society of London, B, 319, 33–82.

Fretter, V. 1989. The anatomy of some new archaeogastropod limpets (Superfamily Peltospiracea) from hydrothermal vents. Journal of Zoology, 218, 123–169.

Fretter, V. & Graham, A. 1994. British Prosobranch Molluscs. Their Functional Anatomy and Ecology. London: Ray Society Publications.

Fretter, V., Graham, A. & McClean, J.H. 1981. The anatomy of the Galapagos Rift limpet, Neomphalus fretterae. Malacologia, 21, 337–361.

Fretter, V., Graham, A., Ponder, W.F. & Lindberg, D.R. 1998. Prosobranchia Introduction. In: Mollusca: The Southern Synthesis. Fauna of Australia. Vol. 5. (Ed. P.L. Beesley, G.J.B. Ross & A. Wells), pp. 605–638. Melbourne: CSIRO Publishing, Part B viii.

Fukuda, H. & Ponder, W.F. 2004. A protandric assimineid gastropod: Rugapedia androgyna n. gen. and n. sp. (Mollusca: Caenogastropoda: Rissooidea) from Queensland, Australia. Molluscan Research, 24, 75–88.

Gaffney, P.M. & McGee, B. 1992. Multiple paternity in Crepidula fornicata. The Veliger, 35, 12–15.

Geiger, D.L. & Thacker, C.E. 2005. Molecular phylogeny of Vetigastropoda reveals non-monophyletic Scissurellidae, Trochoidea and Fissurelloidea. Molluscan Research, 25, 47–55.

Gibbs, P.E. & Bryan, G.W. 1986. Reproductive failure in populations of the dog-whelk, Nucella lapillus, caused by imposex induced by tributyltin from antifouling paints. Journal of the Marine Biological Association of the United Kingdom, 66, 767–777.

Gibbs, P.E. & Bryan, G.W. 1996. TBT-induced imposex in neogastropod snails: masculization to mass extinction. In: Tributyltin: Case Study of an Environmental Contaminant (Ed. S.J. De Mora), pp. 212–236. Cambridge: Cambridge University Press.

Gibbs, P.E., Bryan, G.W., Pascoe, P.L. & Burt, G.R. 1987. The use of the dog whelk, Nucella lapillus, as an indicator or tributyltin (TBT) contamination. Journal of the Marine Biological Association of the United Kingdom, 67, 507–523.

Gibson, D.G. III. 1964. Mating behaviour in Littorina planaxis Philippi (Gastropoda: Prosobranchia). Veliger, 7, 134–139.

Giusti, F. & Selmi, M.G. 1982. The atypical sperm in the prosobranch molluscs. Malacologia, 22, 171–181.

Giusti, F. & Selmi, M.G. 1985. The seminal receptacle and sperm storage in Cochlostoma montanum (Issl) (Gastropoda: Prosobranchia). Journal of Morphology, 184, 121–133.

Glaubrecht, M. & Strong, E.E. 2004. Spermatophores of thalassoid gastropods (Paludomidae) in Lake Tanganyika, East Africa, with a survey of their occurrence in Cerithioidea: functional and phylogenetic implications. Invertebrate Biology, 123, 218–236.

Golikov, A.N. & Kussakin, O.G. 1972. Sur la biologie de la reproduction des Patelles de la famille Tecturidae (Gastropoda: Docoglossa) et sur la position systématique de ses subdivisions. Malacologia, 11, 287–294.

Golikov, A.N. & Starobogatov, Y.I. 1975. Systematics of prosobranch gastropods. Malacologia, 15, 185–232.

Grande, C., Templado, J., Cervera, J.L. & Zardoya, R. 2004. Molecular phylogeny of Euthyneura (Mollusca: Gastropoda). Molecular Biology and Evolution, 21, 303–313.

Haase, M. 2003. A new spring snail of the genus *Graziana* (Caenogastropoda: Hydrobiidae) from Switzerland. Journal of Molluscan Studies, 69, 107–112.

Hadfield, M.G. & Hopper, C.N. 1980. Ecological and evolutionary significance of pelagic spermatophores of vermetid gastropods. Marine Biology, 57, 315–326.

Hanson, J., Randall, J.T. & Bayely, S.T. 1952. The microstructure of the spermatozoa of the snail *Viviparus*. Experimental Cell Research, 3, 65–78.

Harasewych, M.G. 1984. Comparative anatomy of four primitive muricacean gastropods: implications for trophonine phylogeny. American Malacological Bulletin, 3, 11–26.

Harasewych, M.G. & McArthur, A.G. 2000. A molecular phylogeny of the Patellogastropoda (Mollusca: Gastropoda). Marine Biology, 137, 183–194.

Haszprunar, G. 1987. The anatomy of *Addisonia* (Mollusca, Gastropoda). Zoomorphology, 106, 269–278.

Haszprunar, G. 1988a. On the origin and evolution of major gastropod groups. With special reference to the Streptoneura. Journal of Molluscan Studies, 54, 367–441.

Haszprunar, G. 1988b. Anatomy and affinities of pseudococculinid limpets (Mollusca, Archaeogastropoda). Zoologica Scripta, 17, 161–179.

Haszprunar, G. 1988c. Comparative anatomy of cocculiniform gastropods and its bearing on archaeogastropod systematics. Malacological Review Supplement, 4, 64–84.

Haszprunar, G. 1989a. New slit limpets (Scissurellacea and Fissurellacea) from hydrothermal vents. Part 2. Anatomy and relationships. Contributions in Science, Number 408, 1–17.

Haszprunar, G. 1989b. The anatomy of *Melanodrymia aurantiaca* Hickman, a coiled archaeogastropod from the East Pacific hydrothermal vents (Mollusca, Gastropoda). Acta Zoologica, 70, 175–186.

Haszprunar, G. 1998. Superorder Cocculiniformia. In: Mollusca: The Southern Synthesis. Fauna of Australia. Vol. 5, Part B (Ed. P.L. Beesley, G.J.B. Ross & A. Wells), pp. 653–656. Melbourne: CSIRO Publishing: viii.

Healy, J.M. 1983. Ultrastructure of euspermatozoa of cerithiacean gastropods (Prosobranchia: Mesogastropoda). Journal of Morphology, 178, 57–75.

Healy, J.M. 1986. Ultrastructure of paraspermatozoa of cerithiacean gastropods (Prosobranchia: Mesogastropoda). Helgoländer Meeresuntersuchungen, 40, 177–199.

Healy, J.M. 1987. Spermatozoon ultrastructure and its bearing on gastropod classification and evolution. Australian Zoologist, 24, 108–113.

Healy. J.M. 1988. Sperm morphology and its systematic importance in the Gastropoda. Malacological Review Supplement, 4, 251–266.

Healy, J.M. 1990a. Sperm structure in the scissurellid gastropod *Sinezona* sp. (Prosobanchia: Pleurotomarioidea). Zoologica Scripta, 19, 189–193.

Healy, J.M. 1990b. Euspermatozoa and paraspermatozoa in the trochid gastropod *Zalipais laseroni* (Trochoidea: Skeneidae). Marine Biology, 105, 497–507.

Healy, J.M. 1995. Sperm and spermiogenic ultrastructure in the Mathildidae with a review of sperm morphology and its systematic importance in the Architectonicoidea (Gastropoda). Journal of Molluscan Studies, 61, 361–373.

Healy, J.M. 1996a. Molluscan sperm ultrastructure: correlation with taxonomic units within the Gastropoda, Cephalopoda and Bivalvia. In: Origin and Evolutionary Radiation of the Mollusca (Ed. J. Taylor), pp. 99–113. Oxford: Oxford University Press.

Healy, J.M. 1996b. Euspermatozoan ultrastructure in *Bembicium auratum* (Gastropoda): comparison with other caenogastropods especially Littorinidae. Journal of Molluscan Studies, 62, 57–63.

Healy, J.M. 2001. Spermatogenesis and oogenesis. In: The Biology of Terrestrial Molluscs (Ed. G.M. Baker), pp. 357–382. Wallingford, U.K: CABI Publishing.

Healy, J.M. & Jamieson, B.G.M. 1981. An ultrastructural examination of developing and mature paraspermatozoa in *Pyrazus ebeninus*

(Mollusca, Gastropoda, Potamididae). Zoomorphology, 98, 101–119.

Heller, J. 1993. Hermaphroditism in molluscs. Biological Journal of the Linnaean Society, 48, 19–42.

Hickman, C.S. 1992. Reproduction and development of trochacean gastropods. The Veliger, 35, 245–272.

Hoagland, E. 1978. Protandry and the evolution of the environmentally-mediated sex change: a study of the Mollusca. Malacologia, 17, 365–391.

Hoagland, E. 1986. Patterns of encapsulation and brooding in the Calyptraeidae (Prosobranchia: Mesogastropoda). American Malacological Bulletin, 4, 173–183.

Hoagland, K.E. & Ponder, W.F. 1998. Superfamily Calyptraeoidea. In: Mollusca: The Southern Synthesis. Fauna of Australia. Vol. 5, Part B (Ed. P.L. Beesley, G.J.B. Ross & A. Wells), pp. 772–774. Melbourne: CSIRO Publishing viii.

Hodgson, A.N. 1995. Spermatozoal morphology of Patellogastropoda and Vetigastropoda (Mollusca: Prosobranchia). Memoirs Muséum National d'Histoire Naturelle, 166, 167–177.

Hodgson, A.N. 1997. Paraspermatogenesis in gastropod molluscs. Invertebrate Reproduction and Development, 31, 31–38.

Hodgson, A.N. & Eckelbarger, K.J. 2000. Ultrastructure of the ovary and oogenesis in six species of patellid limpets (Gastropoda: Patellogastropoda) from South Africa. Invertebrate Biology, 119, 265–277.

Hodgson, A.N., Healy, J.M. & Tunnicliffe, V. 1997. Spermatogenesis and sperm structure of the hydrothermal vent prosobranch gastropod Lepetodrilus fucensis (Lepetodrilidae, Mollusca). Invertebrate Reproduction and Development, 31, 87–97.

Holman, L. & Snook, R. 2008. A sterile sperm caste protects brother fertile sperm from female-mediated death in Drosophila pseudoobscura. Current Biology, 18, 292–296.

Horiguchi, T., Shiraishi, H., Shimizu, M. & Morita, M. 1997. Imposex in sea snails, caused by organotin (tributyltin and triphenyltin) pollution in Japan: a survey. Applied Organometallic Chemistry, 11, 451–455.

Horiguchi, T., Kojima, M., Hamada, F., Kajikawa, A., Shiraishi, H., Morita, M. & Shimizu, M. 2006. Impact of tributyltin and triphenyltin on ivory shell (Babylonia japonica) populations. Environmental Health Perspectives, 114, 13–19.

Houbrick, R. S. 1973. Studies on the reproductive biology of the genus Cerithium (Gastropoda: Prosobranchia) in the Western Atlantic. Bulletin of Marine Science, 23, 875–904.

Houbrick, R.S. 1980. Observations on the anatomy and life history of Modulus modulus (Prosobranchia: Modulidae). Malacologia, 20, 117–142.

Houbrick, R.S. 1987. Anatomy, reproductive biology, and phylogeny of the Planaxidae (Cerithiacea: Prosobranchia). Smithsonian Contributions to Zoology, 445, 1–57.

Houbrick, R.S. 1988. Cerithioidean phylogeny. Malacological Review Supplement, 4, 88–128.

Houbrick, R.S. 1991a. Systematic review and functional morphology of the mangrove snails Terebralia and Telescopium (Potamididae; Prosobranchia). Malacologia, 33, 289–338.

Houbrick, R.S. 1991b. Anatomy and systematic placement of Faunus Montfort 1810 (Prosobranchia: Melanopsidae). Malacological Review, 24, 35–54.

Houbrick, R.S. 1992. Monograph of the genus Cerithium Bruguière in the Indo-Pacific (Cerithiidae: Prosobranchia). Smithsonian Contributions to Zoology, 501, 1–211.

Houbrick, R.S. 1993. Phylogenetic relationships and generic review of Bittiinae (Prosobranchia: Cerithioidea). Malacologia, 35, 261–313.

Houston, R.S. 1976. The structure and function of neogastropod reproductive systems with special reference to Columbella fuscata Sowerby 1832. The Veliger, 19, 27–46.

Houston, R.S. 1985. Genital ducts of Cerithiacea (Gastropoda: Mesogastropoda) from the Gulf of California. Journal of Molluscan Studies, 51, 183–189.

Houston, R.S. 1990. Reproductive systems of neritimorph archaeogastropods from the Eastern Pacific, with special reference to Nerita funiculata Menke, 1851. The Veliger, 33, 103–110.

Israelsson, I. 1998. The anatomy of Pachyderma laevis (Archaeogastropoda: 'Peltospiridae'). Journal of Molluscan Studies, 64, 93–109.

Ito, A. & Wada, S. 2006. Intrasexual copulation and mate discrimination in a population of Nodilittorina radiata (Gastropoda: Littorinidae). Journal of Ethology, 24, 45–49.

Jamieson, B.G.M. & Newman, L.J. 1989. The phylogenetic position of the heteropod Atlanta gaudichaudi Souleyet (Mollusca,

Gastropoda), a spermatological investigation. Zoologica Scripta, 18, 269–278.

Johnson, L.J. 1999. Size assortative mating in the marine snail *Littorina neglecta*. Journal of the Marine Biological Association of the United Kingdom, 79, 1131–1132.

Joosse, J. & Geraerts, P.M. 1983. Endocrinology. In: The Mollusca Volume 4 (Ed. A.S.M. Saleuddin & K.M. Wilbur), pp. 318–406. New York: Academic Press.

Kano, Y. 2008. Vetigastropod phylogeny and a new concept of Seguenzioidea: independent evolution of copulatory organs in the deep-sea habitats. Zoologica Scripta, 37, 1–21.

Kano, Y. & Kase, T. 2002. Anatomy and systematics of the submarine-cave gastropod *Pisulina* (Neritopsina: Neritiliidae). Journal of Molluscan Studies, 68, 365–384.

Kano, Y. & Kase, T. 2003. Systematics of the *Neritilia rubida* complex (Gastropoda: Neritiliidae): three amphidromous species with overlapping distributions in the Indo-Pacific. Journal of Molluscan Studies, 69, 273–284.

Kano, Y., Chiba, S. & Kase, T. 2002. Major adaptive radiation in neritopsine gastropods estimated from 28S rRNA sequences and fossil records. Proceedings of the Royal Society of London. Series B, 269, 2457–2465.

Kay, E.A., Wells, F.E. & Ponder, W.F. 1998. Class Gastropoda. In: Mollusca: The Southern Synthesis. Fauna of Australia. Vol. 5 (Ed. P.L. Beesley, G.J.B. Ross & A. Wells), pp. 566–568. Melbourne: CSIRO Publishing, viii.

Keawjam, R. 1987. The apple snails of Thailand: aspects of comparative anatomy. Malacological Review, 20, 69–90.

Kennedy, J.J. 1995. The courtship, pseudo-copulation behaviour and spermatophore of *Turritella communis* Risso 1826 (Prosobranchia: Turritellidae). Journal of Molluscan Studies, 61, 421–434.

Kohnert, R. & Storch, V. 1984a. Vergleichend-ultrastrukturelle Untersuchungen zur Morphologie eupyrener Spermien der Monotocardia (Prosobranchia). Zoologische Jahrbucher, 111, 51–93.

Kohnert, R. & Storch, V. 1984b. Elektronmickroskopische Untersuchungen zur Spermiogenese der eupyrenen Spermien der Monotocardia (Prosobranchia). Zoologische Jahrbucher, 112, 1–32.

Koike, K. 1985. Comparative ultrastructural studies on the spermatozoa of the Prosobranchia (Mollusca: Gastropoda). Science Report of the Faculty of Education, Gunma University, 34, 33–153.

Kool, S.P. 1993. Phylogenetic analysis of the Rapaninae (Neogastropoda: Muricidae). Malacologia, 35, 155–259.

Kura, T. & Yoda, K. 2001. Can voluntary nutritional gifts in seminal flow evolve? Journal of Ethology, 19, 9–15.

Leal, J.H. & Harasewych, M.G. 1999. Deepest Atlantic molluscs: hadal limpets (Mollusca, Gastropoda, Cocculiniformia) from the northern boundary of the Caribbean Plate. Invertebrate Biology, 118, 116–136.

Levitan, D.R. 1998. Sperm limitation, gamete competition, and sexual selection in external fertilizers. In: Sperm Competition and Sexual Selection (Ed. T.R. Birkhead & A.P. Møller), pp. 175–217. San Diego, Academic Press.

Lindberg, D.R. 1988. The Patellogastropoda. Malacological Review Supplement, 4, 35–63.

Lindberg, D.R. 2008. Patellogastropoda, Neritimorpha, and Cocculinoidea. In: Phylogeny and Evolution of the Mollusca (Ed. W.F. Ponder & D.R. Lindberg), pp. 271–296. Berkeley, University of California Press.

Mak, Y.M. 1995. Egg capsule morphology or five Hong Kong rocky shore littorinids. Hydrobiologia, 309, 53–59.

Martel, A., Larrivée, D.H. & Himmelman, J.H. 1986. Behaviour and timing of copulation and egg-laying in the neogastropod *Buccinum undatum* L. Journal of Experimental Marine Biology and Ecology, 96, 27–42.

Maxwell, W.L. 1983. Mollusca. In: Reproductive Biology of Invertebrates. Volume 2 (Ed. K.G. Adiyodi & R.G. Adiyodi), pp. 275–342. New York: Wiley.

McArthur, A.G. & Harasewych, M.G. 2003. Molecular systematics of the major lineages of the Gastropoda. In: Molecular Systematics and Phylogeography of Mollusks (Ed. C. Lydeard & D.R. Linberg). pp. 140–160. Washington and London: Smithsonian Books.

McLean, J.H. 1981. The Galapagos Rift limpet *Neomphalus*: Relevance to understanding evolution of a major Paleozoic–Mesozoic radiation. Malacologia, 21, 291–336.

McLean, J.H. 1988. New archaeogastropod limpets from hydrothermal vents, superfamily Lepetodrilacea. 1. Systematic descriptions.

Philosophical Transactions of the Royal Society London Series B, 318, 1–32.

McLean, J.H. 1990. A new genus and species of neomphalid limpet from the Mariana vents with a review of current understanding of relationships among the Neomphalacea and Peltospiracea. Nautilus, 104, 77–86.

Melone, G., Donin, D.L.L. & Cotelli, F. 1980. The paraspermatic cell (atypical spermatozoon) of the Prosobranchia: a comparative ultrastructural study. Acta Zoologica, 61, 191–201.

Møller, A.P. 1998. Sperm competition and natural selection. In: Sperm Competition and Sexual Selection (Ed. T.R. Birkhead & A.P. Møller), pp. 55–90. San Diego, Academic Press.

Morton, B. & Jones, D.S. 2001. The biology of *Hipponix australis* (Gastropoda: Hipponicidae) on *Nassarius pauperatus* (Nassariidae) in Princess Royal Harbour, Western Australia. Journal of Molluscan Studies, 67, 247–255.

Nakano, D. & Nishiwaki, S. 1989. Anatomical and histological studies on the reproductive system of *Semisulcospira libertina* (Prosobranchia: Pleuroceridae). Venus, 48, 263–273.

Nakano, T. & Ozawa, T. 2007. Worldwide phylogeography of limpets of the order Patellogastropoda: molecular, morphological and palaeontological evidence. Journal of Molluscan Studies, 73, 79–99.

Nieman, M. & Lively, C.M. 2005. New Zealand snails (*Potamopyrgus antipodarum*) persist in copulating with asexual and parasitically castrated females. The American Midland Naturalist, 154, 88–86.

Nishiwaki, S. 1964. Phylogenetic study on the type of the dimorphic spermatozoa in Prosobranchia. Science Reports of Tokyo Kyoiku Daigaku. B, 11, 237–275.

Oehlmann, J., Stroben, E. & Fioroni, P. 1991. The morphological expression of imposex in *Nucella lapillus* (Linnaeus) (Gastropoda: Muricidae). Journal of Molluscan Studies, 57, 375–390.

Oppliger, A., Hosken, D.J. & Ribi, G. 1998. Snail sperm production characteristics vary with sperm competition. Proceedings of the Royal Society of London B, 265, 1527–1534.

Oppliger, A., Naciri-Graven, Y., Ribi, G. & Hosken, D.J. 2003. Sperm length influences fertilization success during sperm competition in the snail *Viviparus ater*. Molecular Ecology, 12, 485–492.

Paterson, I.G., Partridge, V. & Buckland-Nicks, J. 2001. Multiple paternity in *Littorina obtusata* (Gastropoda, Littorinidae) revealed by microsatellite analyses. Biological Bulletin, 200, 261–267.

Ponder, W.F. 1972. The morphology of some mitriform gastropods with special reference to their alimentary and reproductive systems (Neogastropoda). Malacologia, 11, 295–342.

Ponder, W.F. 1973. The origin and evolution of the Neogastropoda. Malacologia, 12, 295–338.

Ponder, W.F. & Linberg, D.R. 1996. Gastropod phylogeny—challenges for the 90's. In: Origin and evolutionary radiation of the Mollusca (Ed. J.D. Taylor), pp. 135–154. Oxford: Oxford University Press.

Ponder, W.F. & Lindberg, D.R. 1997. Towards a phylogeny of gastropod molluscs: an analysis using morphological characters. Zoological Journal of the Linnean Society, 19, 83–265.

Ponder, W.F. & Lindberg, D.R. (Eds) 2008. Phylogeny and Evolution of the Mollusca. Berkeley: University of California Press.

Ponder, W.F., Colgan, D.J., Healy, J.M., Nützel, A., Simone, L.R.L. & Strong, E.E. 2008. Caenogastropoda. In: Phylogeny and Evolution of the Mollusca. (Ed. W.F. Ponder & D.R. Lindberg), pp. 331–383. Berkeley, University of California Press.

Purchon, R.D. 1977. The Biology of the Mollusca. Oxford: Pergamon Press.

Quinn, J.F. Jr. 1983. A revision of the Seguenziacea Verrill, 1884 (Gastropoda: Prosobranchia). 1. Summary and evaluation of the superfamily. Proceedings of the Biological Society of Washington, 96, 725–757.

Quinn, J.F. Jr. 1991. Systematic position of *Basilissopsis* and *Guttula*, and a discussion of the phylogeny of the Seguenzioidae (Gastropoda: Prosobranchia). Bulletin of Marine Science, 49, 575–598.

Ramón, M. & Amor, M.J. 2002. Reproductive cycle of *Bolinus brandaris* and penis and genital duct size variations in a population affected by imposex. Journal of the Marine Biological Association of the United Kingdom, 82, 435–442.

Reed, S.E. 1995a. Reproductive anatomy and biology of the genus *Strombus* in the Caribbean: I. Males. Journal of Shellfish Research, 14, 325–330.

Reed, S.E. 1995b. Reproductive anatomy and biology of the genus *Strombus* in the

Caribbean: II. Females. Journal of Shellfish Research, 14, 331–336.

Reid, D.G. 1986. *Mainwaringia* Nevill, 1885, a littorinid genus from Asiatic mangrove forests, and a case of protandrous hermaphroditism. Journal of Molluscan Studies, 52, 225–242.

Reid, D.G. 1989. The comparative morphology, phylogeny and evolution of the gastropod family Littorinidae. Philosophical Transactions of the Royal Society of London. Series B, 324, 1–110.

Remigio, E.A. & Hebert, P.D.N. 2003. Testing the utility of partial COI sequences for phylogenetic estimates of gastropod relationships. Molecular Phylogenetics and Evolution, 29, 641–647.

Richter, A. & Luque, A.A. 2004. Sex change in two Mediterranean species of Coralliophilidae (Mollusca: Gastropoda: Neogastropoda). Journal of the Marine Biological Association of the United Kingdom, 84, 383–392.

Robertson, R. 1989. Spermatophores of aquatic non-stylommatophoran gastropods: A review with new data on *Heliacus* (Architectonicidae). Malacologia, 30, 341–364.

Rouse, G.W. & Jamieson, B.G.M. 1987. An ultrastructural study of the spermatozoa of the polychaetes *Eurothoe complanata* (Amphinomidae), *Clymenella* sp. and *Micromaldane* sp. (Maldanidae), with a definition of sperm types in relation to reproductive biology. Journal of Submicroscopic Cytology, 19, 573–584.

Runham, N.W. 1988. Mollusca. In: Reproductive Biology of Invertebrates. Vol. III. Accessory Sex Glands (Eds. K.G. & R.G. Adiyodi), Chapter 8, pp. 113–188. Chichester: John Wiley.

Sasaki, T. 1998. Comparative anatomy and phylogeny of the recent Archaeogastropoda (Mollusca: Gastropoda). The University Museum, The University of Tokyo, Bulletin No. 38, 1–224.

Sasaki, T., Okutani, T. & Fujikura, K. 2003. New taxa and new records of patelliform gastropods associated with chemoauthosynthesis-based communities in Japanese waters. The Veliger, 46, 189–210.

Sasaki, T., Okutani, T. & Fujikura, K. 2006. Anatomy of *Shinkailepas myojinensis* Sasaki, Okutani & Fujikura, 2003 (Gastropoda: Neritopsina). Malacologia, 48, 1–26.

Scott, B.J. & Kenny, R. 1998. Superfamily Neritoidea. In: Mollusca: The Southern Synthesis. Fauna of Australia. Vol. 5, Part B (Ed. P.L. Beesley, G.J.B. Ross & A. Wells, A), pp. 694–702. Melbourne: CSIRO Publishing viii.

Silberglied, R.E., Shepherd, J.G. & Dickinson, J.L. 1984. Eunuchs: The role of apyrene sperm in Lepidoptera? American Naturalist, 123, 255–265.

Simone, L.R.L. 2002. Comparative morphological study and phylogeny of representatives of the superfamily Calyptraeoidea (including Hipponicoidea) (Mollusca, Caenogastropoda). Biotropica Neotropica, 2, 1–137.

Simone, L.R.L. 2003. Revision of the genus *Benthobia* (Caenogastropoda, Pseudolividae). Journal of Molluscan Studies, 69, 245–262.

Slavoshevskaya, L.V. 1984. An analysis of the organisation of the Rissoacea (Mollusca, Gastropoda). Zoologichesky Zhurnal, 63, 361–372. (In Russian with English abstract)

Soong, K. & Chen, J.L. 1991. Population structure and sex-change in the coral-inhabiting snail *Coralliophila violacea* at Hsiao-Liechiu, Taiwan. Marine Biology, 111, 81–86.

Soong, K. & Chen, M-H. 2003. Sex expression of an immobile coral-inhabiting snail, *Quoyula monodonta*. Marine Biology, 143. 351–358.

Staub, R. & Ribi, G. 1995. Size-assortative mating in a natural population of *Viviparus ater* (Gastropoda: Prosobranchia) in Lake Zürich, Switzerland. Journal of Molluscan Studies, 61, 237–247.

Stroben, E., Oehlmann, U., Schulte-Oehlmann, U. & Fioroni, P. 1996. Seasonal variations in the genital ducts of normal and imposex-affected prosobranchs and its influence on biomonitoring indexes. Malacological Review, Supplement 6, 173–184.

Strong, E.E. 2003. Refining molluscan characters: morphology, character coding and a phylogeny of the Caenogastropoda. Zoological Journal of the Linnean Society, 137, 447–554.

Strong, E.E. & Harasewych, M.G. 1999. Anatomy of the hadal limpet *Macleaniella moskalevi* (Gastropoda, Cocculinoidea). Invertebrate Biology, 118, 137–148.

Strong, E.E., Harasewych, M.G. & Haszprunar, G. 2003. Phylogeny of the Cocculinoidea (Mollusca, Gastropoda). Invertebrate Biology, 122, 114–125.

Struhsaker, J.W. 1966. Breeding, spawning, spawning periodicity and early development in the Hawaiian *Littorina*; *L. pintado* (Wood), *L. picta* (Philippi) and *L. scabra* (Linné). Proceedings of the Malacological Society of London, 37, 137–166.

Takada, N. 2000. Development of a penis from the vestigial penis in the female apple snail *Pomacea canaliculata*. Biological Bulletin, 199, 316–320.

Thiele, J. 1929–1935. Handbuch der Systematischen Weichtierkunde. Gustav Fischer: Jena Vol. 1 vi pp. 1–778; Vol.2 v pp. 779–1134.

Thiriot-Quievreux, C. 1982. Donnée sur la biologie sexuelle des Rissoidae (Mollusca: Prosobranchia). International Journal of Invertebrate Reproduction, 5, 167–180.

Tillier, S., Maselot, M., Guerdox, J. & Tillier, A. 1994. Monophyly of major gastropod taxa tested from partial 28S rRNA sequences, with emphasis on Euthyneura and hot-vent limpets Peltospiroidae. The Nautilus, Supplement 2, 122–140.

Tochimoto, T. 1967. Comparative histochemical study on the dimorphic spermatozoa of the Prosobranchia with special reference to polysaccharides. Science Reports of Tokyo Kyoiku Daigaku. B, 13, 75–109.

Voltzow, J. 1994. Gastropoda: Prosobranchia. In: Microscopic Anatomy of Invertebrates (Ed. F.W. Harrison and A.J. Kohn), Volume 5: Mollusca I, pp. 111–252. New York: Wiley-Liss.

Wagner, P.J. 2001. Gastropod phylogenetics: progress, problems, and implications. Journal of Palaeontology, 75, 1128–1140.

Walker, D., Power, A.J., Sweeney-Reeves, M. & Avise, J.C. 2007. Multiple paternity and female sperm usage along egg-case strings of the knobbed whelk, *Busycon carica* (Mollusca; Melongenidae). Marine Biology, 151, 53–61.

Warén, A. 1983. A generic revision of the family Eulimidae. Journal of Molluscan Studies. Supplement 13, 1–95.

Warén, A. & Bouchet, P. 1993. New records, species, genera, and a new family of gastropods from hydrothermal vents and hydrocarbon seeps. Zoologica Scripta, 22, 1–90.

Warén, A. & Hain, S. 1996. Description of Zerotulidae fam. Nov. (Littorinoidea), with comments on an Antarctic littorinid gastropod. The Veliger, 39, 277–334.

Warner, R.R., Fitch, D.L. & Standish, J.D. 1996. Social control of sex change in the shelf limpet, *Crepidula norrisiarum*: size-specific responses to local group composition. Journal of Experimental Marine Biology and Ecology, 204, 155–167.

Webber, H.H. 1977. Gastropoda: Prosobranchia. In: Reproduction of Marine Invertebrates IV (Ed. A.C. Giese & J.S. Pearse), pp. 1–97. New York: Academic Press.

Wilson, B. 1998. Superfamily Velutinoidea. In: Mollusca: The Southern Synthesis. Fauna of Australia. Vol. 5, Part B (Ed. P.L. Beesley, G.J.B. Ross & A. Wells), pp. 786–790. Melbourne: CSIRO Publishing viii.

Winnepenninckx, B., Steiner, G., Backeljau, T. & De Wachter, R. 1998. Details of gastropod phylogeny inferred from 18S rRNA sequences. Molecular Phylogenetics and Evolution, 9, 55–63.

Wright, W.G. 1988. Sex change in the Mollusca. Trends in Ecology and Evolution, 3, 137–140.

Wyatt, H.V. 1961. The reproduction, growth and distribution of *Calyptraea chinensis* (L.). The Journal of Animal Ecology, 30, 283–302.

Yoon, S.H. & Kim, W. 2007. 18S ribosomal DNA sequences provide insight into the phylogeny of patellogastropod limpets (Mollusca: Gastropoda). Molecules and Cells, 23, 64–71.

Yusa, Y. 2004. Inheritance of colour polymorphism and the pattern of sperm competition in the apple snail *Pomacea canaliculata* (Gastropoda: Ampullariidae). Journal of Molluscan Studies, 70, 43–48.

Zehra, I. & Perveen, R. 1991. Studies on the breeding season, egg capsule and early larval development of *Theliostyla albicilla* (Linne, 1758) from Karachi coast. Pakistan Journal of Zoology, 23, 35–38.

8

Opisthobranchs

ÁNGEL VALDÉS, TERRENCE M. GOSLINER, AND MICHAEL T. GHISELIN

INTRODUCTION

Opisthobranchs are a diverse group of gastropod mollusks that display an evolutionary trend towards the reduction and loss of the shell in the adult state. The great majority of species are marine, but some freshwater examples have been described. Opisthobranchs have colonized all marine ecosystems, including some extreme environments such as the deep sea, hydrothermal vents, and cold seeps. Pelagic and interstitial species have also been documented. Many species display bright external color patterns associated with the possession of chemical defenses (Cimino & Ghiselin 2009). The great majority of opisthobranchs are simultaneous hermaphrodites (Ghiselin 1966; Gosliner 1985, 1994; Schmekel 1985). In most species the male system matures first, but the animals are not functionally protandric, for they both receive and donate sperm (Gosliner 1994). Only highly modified and derived acochlidiaceans, adapted to interstitial and freshwater habitats, display secondary gonochorism. This condition appears to have developed independently in different lineages of acochlidiaceans (Gosliner 1994; Neusser et al. 2006).

For a relatively small number of species (6,000 according to the estimates of Gosliner & Draheim 1996), opisthobranch mollusks display a vast range of hermaphroditic reproductive configurations and very complex structures. These configurations have been well studied and have served as the basis for discussion on the phylogenetic relationships of opisthobranchs (Ghiselin 1966; Gosliner 1985, 1991, 1994; Schmekel 1985; Mikkelsen 1996).

Different lines of evidence suggest that a great deal of this variation is due to the parallel acquisition and loss of reproductive structures in different lineages (Gosliner & Ghiselin 1984; Mikkelsen 1996). Reconstructing the history of reproductive morphology change in opisthobranchs has broad implications for understanding the evolutionary diversification of this group of mollusks, and offers opportunities to study the action of sexual selection in the evolution of primary sexual reproductive characteristics.

Ghiselin (1966) suggested that the hermaphroditic system of opisthobranchs is the result of the gradual superimposition of "prosobranch" male and female systems. He identified a series of inefficient features in this hypothetical merged reproductive system (various functions may interfere with each other), and argued that the subsequent transformation of the system lead to solutions for those inefficiencies. For example, according to Ghiselin (1966), in the hypothetical merged reproductive system the storage of endogenous sperm in the ampulla interferes with the passage of eggs, and the movement of eggs could eject the sperm. More important, because in the merged system endogenous and exogenous sperm and eggs must move through the undivided gonoduct guided by ciliary currents and grooves, there is a risk of misdirection causing self-fertilization or loss of gametes. Different lineages evolved a variety of solutions (in most cases involving splits of genital ducts) so the effects of the inefficiencies are decreased or the inefficiencies no longer exist. Modifications from one kind of adaptation to solve a particular inefficiency to another are not likely because they would not increase fitness once the inefficiency is overcome. Ghiselin (1966) also proposed that the simplest of all solutions is the formation of three closed, separated tubes one for each of the three major functions of the original undivided duct—movement of endogenous sperm, movement of exogenous sperm, and movement of eggs. Although this configuration has never been achieved, a variety of independent splits of the pallial gonoduct have increased the efficiency of the system in several groups of opisthobranchs. Since the publication of Ghiselin's (1966) work, a series of phylogenetic hypotheses for the Opisthobranchia have been proposed, allowing for testing of Ghiselin's model and tracking of the evolutionary steps that increased the efficiency of the system in different lineages of opisthobranchs (Gosliner 1985; Schmekel 1985; Mikkelsen 1996).

The phylogenetic relationships within opisthobranch mollusks remain uncertain and it is unclear whether they constitute a monophyletic group. They are currently considered to be derived heterobranch gastropods related to pulmonates (Dayrat & Tillier 2002; Wägele & Klussmann-Kolb 2005; Vonnemann et al. 2005; Grande et al. 2004; Dinapoli & Klussmann-Kolb 2010). Alternative phylogenetic analyses that are discussed at the end of this chapter provide contradictory interpretations of the relationships within the Opisthobranchia (figure 8.1). For the purposes of this chapter we will consider Opisthobranchia in its traditional sense, as including a series of lineages of mainly marine hermaphroditic gastropods displaying an evolutionary trend towards the reduction or complete loss of the shell. Members of this group include the "Cephalaspidea" (headshield slugs, bubble shells, and relatives), from which the Acteonidae should probably be removed, the Anaspidea (sea hares), the Sacoglossa (sap-sucking slugs), the pelagic Pteropoda, including Gymnosomata (sea angels) and Thecosomata (sea butterflies), the interstitial Acochlidiacea, the highly diverse Nudibranchia, and the traditional paraphyletic "Notaspidea" (side-gill slugs), which is now divided into the Pleurobranchoidea (which is sister to nudibranchs forming the Nudipleura), and the Tylodinoidea, a basal opisthobranch clade of uncertain relationships.

In this chapter we review some of these transformations in light of newly available information derived from the introduction of new techniques, phylogenetic hypotheses, and a wealth of information published during the last decade.

THE REPRODUCTIVE SYSTEM OF OPISTHOBRANCHS

Generalities

The hermaphrodite reproductive systems of opisthobranchs show considerable variation and specialization. The general physiology of the reproductive system features production of gametes, transfer and maturation of gametes in various organs, copulation, fertilization, and production of the egg mass. The most basal opisthobranch reproductive system hypothesized by Ghiselin (1966) would include all the necessary organs to conduct all the functions of both sexes (figure 8.2).

Authors recognize three main organizational configurations in the reproductive system of opisthobranchs: monaulic, diaulic and triaulic. Some of these modes are present in different lineages. In many cases rampant homoplasy has produced independent evolution of nearly identical arrangements, especially between Sacoglossa and Nudipleura.

In both the male and female parts of the reproductive system there are two distinct regions, the gonad region and the gonoduct region (figure 8.2). The gonad has a coelomic origin and it is involved in the production of male and female gametes that subsequently enter the gonoduct (where copulation, fertilization, and egg mass production take place). In most hermaphroditic species the male and female acini of the gonad are interdigitated and not clearly separated. Often the gonad is superimposed on or around the digestive gland and these two organs are indistinguishable without the aid of histological techniques. In species in which the male and female acini are separated from the female acini, the latter are situated peripherally around a central male acinus (Gosliner, 1994). Only some acochlidiaceans appear to have a distinct ovary and testis.

The gonoduct consists of two distinct regions, the proximal region of coelomic, mesodermal origin

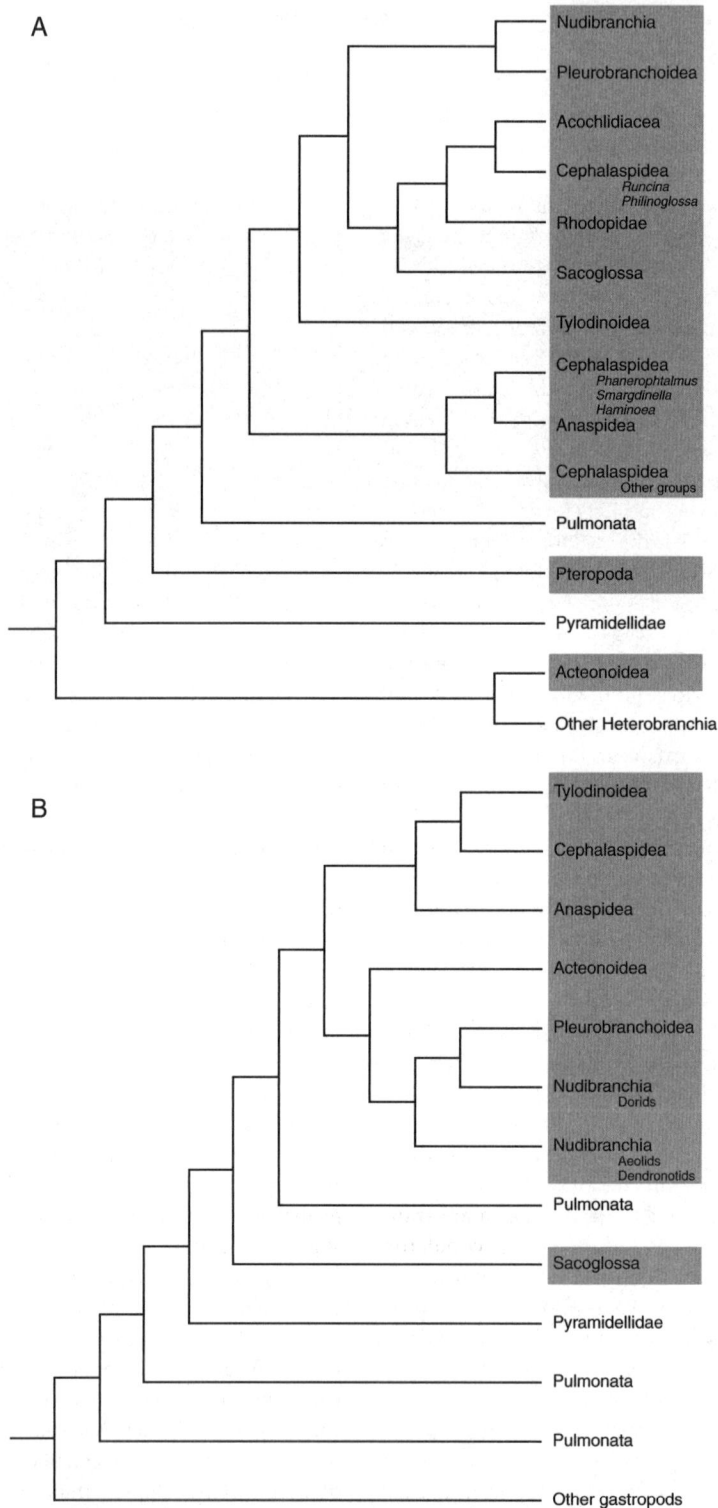

FIGURE 8.1 Phylogenetic hypotheses of relationships within the traditional Opisthobranchia highlighted in grey, showing this group to be paraphyletic. (A) Morphological phylogeny by Wägele & Klussmann-Kolb (2005), Acteonidae needs to be excluded to make Opisthobranchia monophyletic; (B) Molecular phylogeny by Grande et al. (2004), Sacoglossa needs to be excluded to make Opisthobranchia monophyletic.

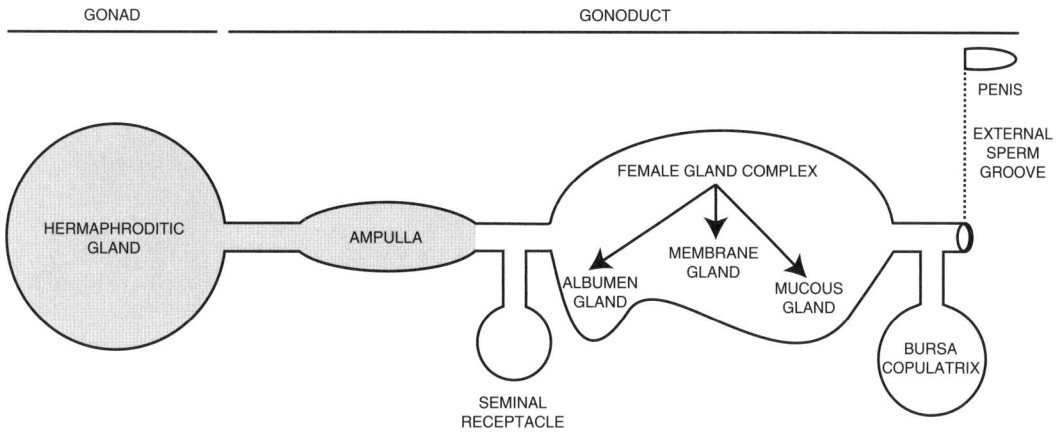

FIGURE **8.2** Ancestral hermaphrodite reproductive system hypothesized by Ghiselin (1966) with the arrangement of the different organs. Organs shaded are of coelomic, mesodermal origin; organs with no shade are of ectodermal origin.

and the pallial distal region of ectodermal origin (figure 8.2). In most species the coelomic gonoduct (also called the hermaphroditic duct), is shared between the male and female systems. The hermaphroditic duct contains a muscular swelling called the ampulla, which stores endogenous sperm and transports gametes between the gonad and the pallial gonoduct. Ghiselin (1966) hypothesized that the ampulla originated in ancestor male systems and later became part of the common gonoduct in hermaphrodites. In the ampulla, self-fertilization is prevented by ciliary movement of eggs around sperm or because the sperm are not fully activated at that point (Gosliner 1994).

The male gonoduct is often referred to as the deferent duct and includes a prostate and usually a single copulatory organ. In basal clades the deferent duct is not closed off and the sperm are transported between the gonoduct to the copulatory organ by an external ciliated sperm groove (Beeman 1970). In more derived groups, such as many sacoglossans and all nudipleurans, sperm transport is fully internalized. The prostate is rich in protein secretory cells. In most groups the prostate surrounds a portion of the deferent duct, but in derived sacoglossans and nudipleurans it consist of a series of glandular organs connected to the deferent duct. In members of the Umbraculidae and seahares a distinct prostatic gland has not been observed. In the latter, glands within the pallial gonoduct function like a prostate (Gosliner 1994). In monaulic opisthobranchs a prostate gland is generally associated

with the cephalic penis. This condition may or may not be homologous to the prostatic gland of other more derived taxa, which is associated with the deferent duct.

The copulatory apparatus of the male system contains an eversible or protrusible penis that may have different types of spines and copulatory glands (figures 8.3 and 8.4). In the plesiomorphic condition the penis is situated on the right side of the head and connected to the external sperm groove. This arrangement is virtually identical to that found in caenogastropods, some pulmonates, and other heterobranchs. In acteonids, the penis is situated at the opening of the mantle cavity, next to the opening of the female organs (figures 8.5B and 8.6A). This arrangement, in which there is no external sperm groove, is a unique morphology of acteonids, probably independently derived. The penis possesses a ciliated groove that transports sperm although in many cases it has been transformed into a closed tube encapsulated into the deferent duct. Such a closed, muscular duct allows more rapid movement of the semen. It is not clear whether these different penial structures are homologous. The region of the deferent duct that contains the penis is often called the ejaculatory portion since it often includes a highly muscularized region that likely facilitates seminal discharge. In some cephalaspideans, such as *Runcina* and *Phanerophthalmus* there is a seminal bulb associated with the penis. These animals produce spermatophores. This configuration permits rapid impregnation.

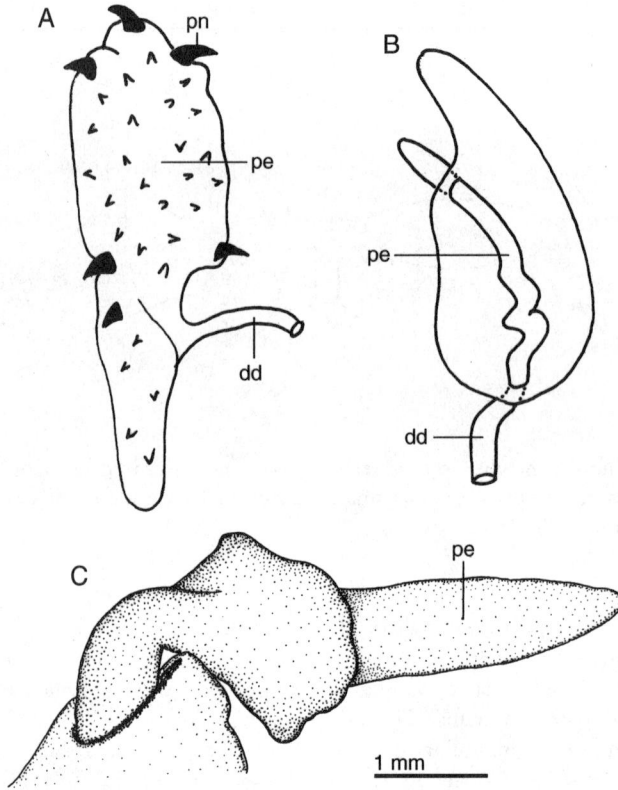

FIGURE 8.3 Penial morphology. (A) *Rictaxis* sp., showing penial spines; (B) *Hydatina zonata*; (C) *Bathydoris aioca*. Abbreviations: dd, deferent duct; pe, penis; pn, penial spine.

Derived opisthobranchs may have accessory copulatory organs such as penial spines that assist copulation. The presence of copulatory spines is particularly important in species with hypodermic insemination such as several species of sacoglossans (figure 8.4F). Also commonly found are penial hooks (figures 8.3A and 8.4A), that probably provide anchorage during penetration (Valdés 2004) and penial glands, accessory glands and accessory spines (Gosliner 1994) whose functions are poorly understood.

In species of dorid nudibranchs there is a wide range of variation of the penial morphology and armature. The plesimorphic condition is a protrusible, unarmed penis found in species of *Bathydoris* (figure 8.3C; Valdés 2002). In the rest of the dorids the penis is eversible (figures 8.4A–C). The penis contains penial hooks in virtually all species of "phanerobranch" dorids (a paraphyletic group of dorids that cannot retract the gill inside of the mantle) and porostome dorids (the radula-less dorids).

In cryptobranch dorids (dorids with a retractable gill), penial hooks are present in at least some members of *Aldisa*, *Discodoris*, *Platydoris*, *Gargamella*, *Baptodoris*, *Cadlina*, *Chromodoris*, and other genera. The homology of these spines is uncertain. It either represents a plesiomorphic condition that has been lost multiple times or represents multiple independent acquisitions. Species of *Taringa* have a trumpet-shaped penial cuticle (figure 8.4B). Members of *Jorunna*, *Goslineria*, *Paradoris*, and *Pharodoris* have copulatory spines (figures 8.4D, E) and copulatory glands whose function is not fully understood (see Valdés & Gosliner 2001; Valdés 2001).

The female part of the gonoduct consists of sperm-containing structures and secretory organs. There are usually two sperm-containing structures: a distal bursa copulatrix or gametolytic gland and a proximal seminal receptacle. The bursa copulatrix is normally a thin walled structure that receives

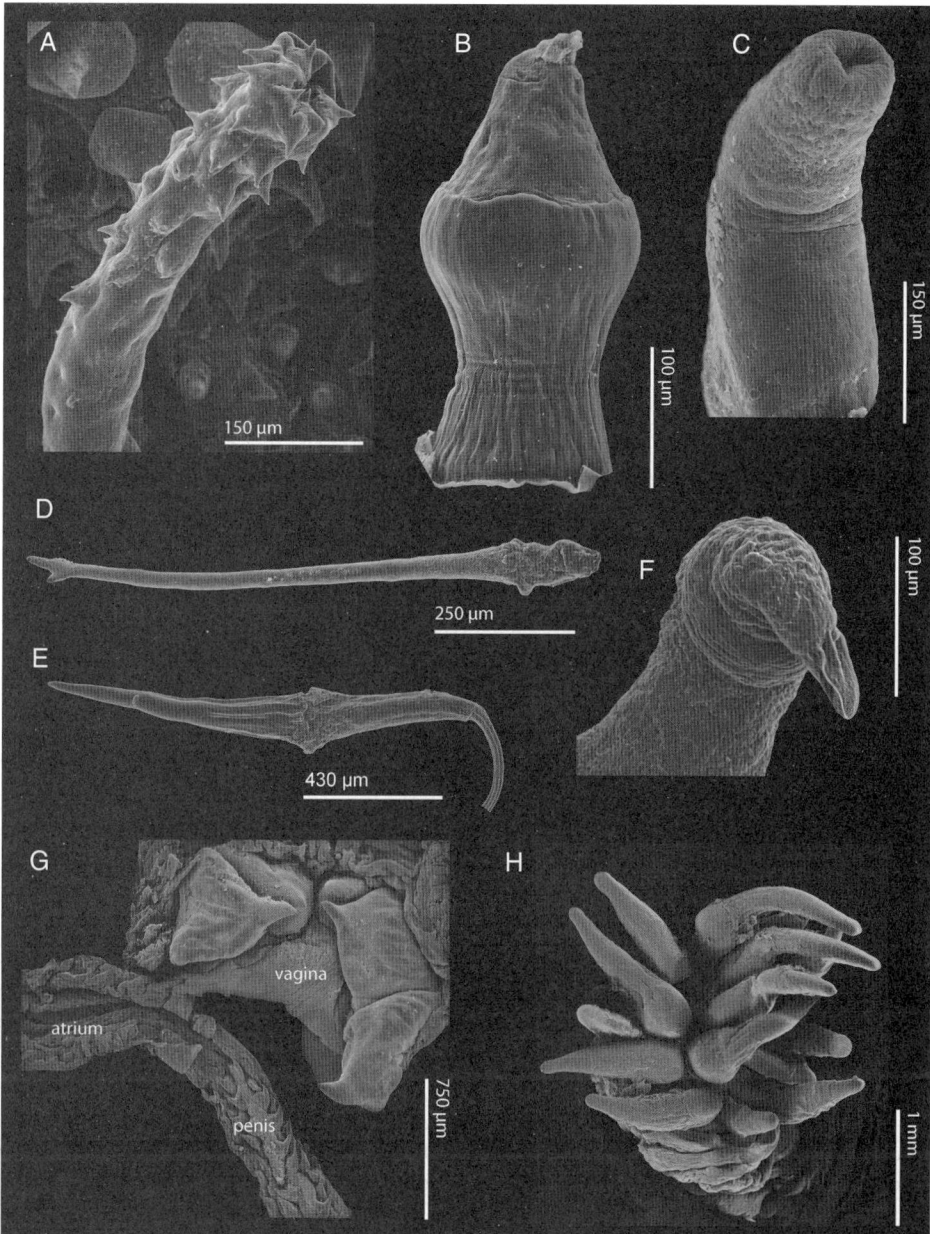

FIGURE **8.4** Scanning electron micrographs of penial and vaginal morphology. (A) Penis of *Baptodoris cinnabarina*, showing penial spines; (B) Penis of *Taringa telopia*, showing the trumpet shaped penial cuticle; (C) Penis of *Doris pseudoargus*, lacking spines; (D) Copulatory spine of *Pharodoris philippinensis*, showing a bifid tip; (E) Copulatory spine of *Asteronotus cespitosus*, showing a simple tip; (F) Penis of *Elysia tuca*, showing an apical penial spine; (G) Atrium, vagina and penis of *Gargamella immaculata*, showing penial and vaginal spines. (H) Female copulatory organ of *Cylichnium* sp.

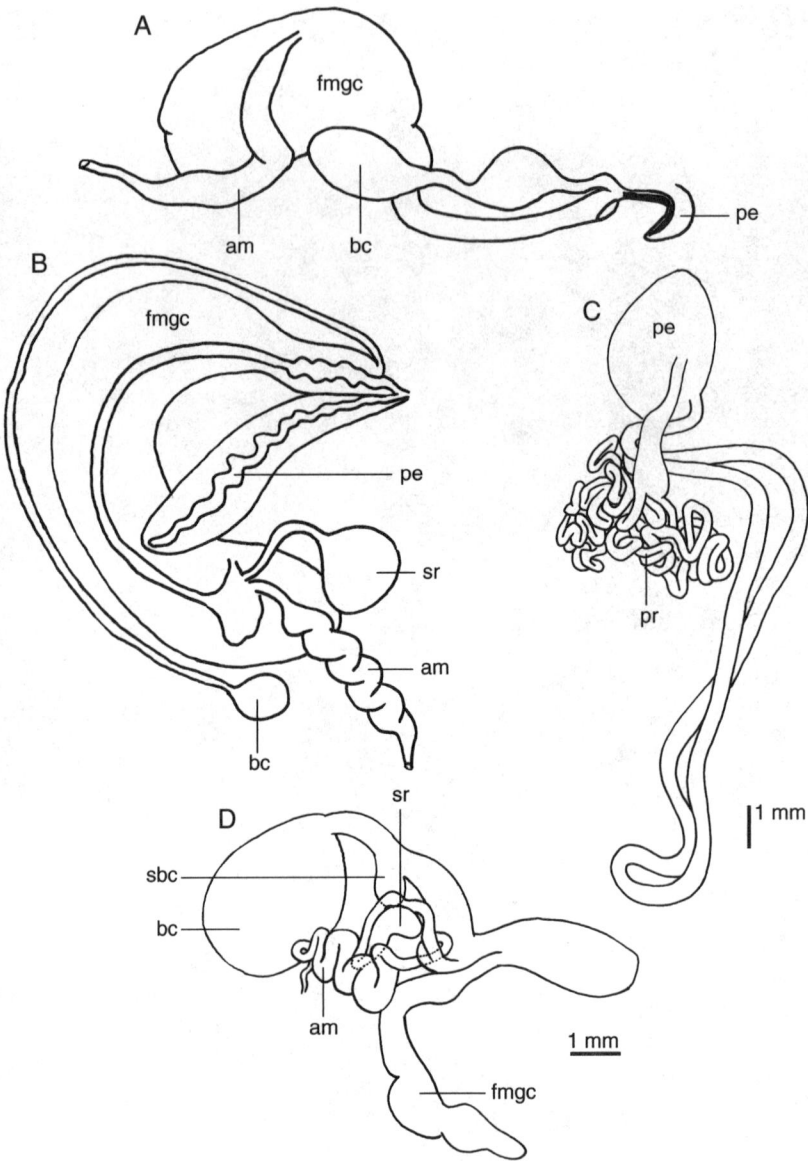

FIGURE **8.5** Reproductive systems of several basal opisthobranchs. (A) Monoaulic system of *Tylodina fungina*; (B) Monoaulic system of *Pupa* sp., with the penis situated at the opening of the mantle cavity; (C–D) Oodiaulic system of *Philine* spp., (C) is the male portion and (D) is the female and hermaphroditic portions. Abbreviations: am, ampulla; bc, bursa copulatrix; fmgc, female gland complex; pe, penis; pr, prostate; sbc, secondary bursa copulatrix; sr, seminal receptacle.

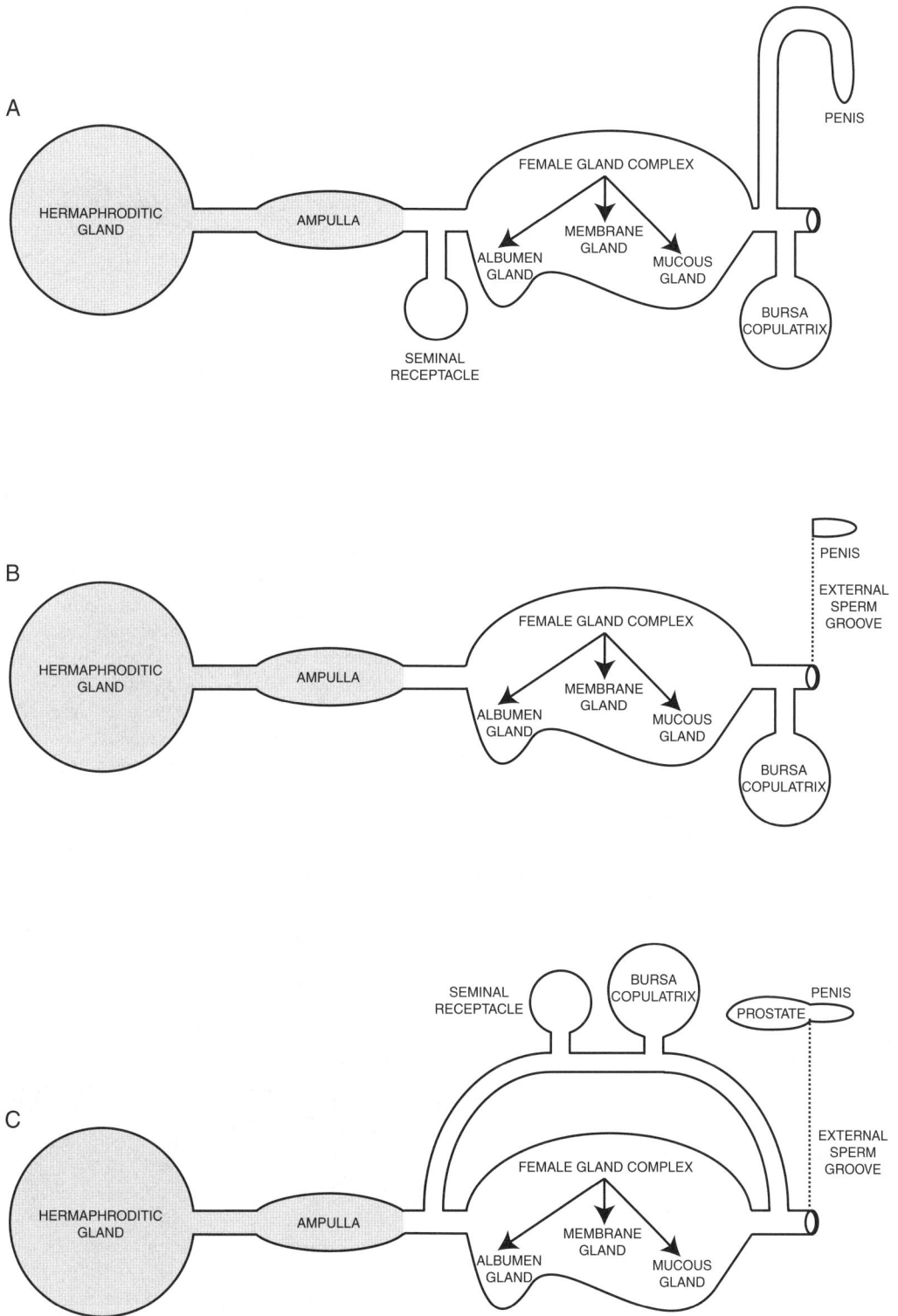

FIGURE 8.6 Schematic representations of different reproductive systems. (A) Monoaulic system as found in *Pupa* sp.; (B) Monoaulic system as found *Tylodina fungina*; (C) Oodiaulic system as found in some species of *Philine*.

sperm after copulation. The wall of the bursa copulatrix is lined with columnar endothelial cells, including both secretory and resorptive cells (Schmekel 1971). It may contain a layered mixture of granules, spermatozoa in various stages of breakdown, amorphous materials, and yellow-red to dark-red oil droplets (Beeman 1970). The secretory cells contain large vacuoles and scattered mitochondria whereas the resorptive cells have numerous microvilli on the edge of the lumen and contain abundant mitochondria on the inside of the microvillar surface. Closer to the base is a region occupied largely by smooth endoplasmic reticulum.

The alternative names bursa copulatrix and gametolytic gland for this organ are suggestive of different functions. Whereas bursa copulatrix indicates a receiving and storage function, gametolytic gland suggests a sperm digestion function. It appears that the bursa copulatrix may carry on both functions. Brandriff and Beeman (1973) found a high level of elaboration of cell types, extensive areas of cytoplasmic RNA, and vigorous excretory activity in the bursa copulatrix of *Aplysia* and *Phyllaplysia*, which suggest a role for this organ in providing food by digesting sexual materials. Alternatively, the interaction of the bursa copulatrix with the bag cells of the nervous system is probably involved in coordinating ovulation of the individual depending on the availability of exogenous sperm (Brandriff and Beeman 1973). Although Mikkelsen (1996) hypothesized that a bursa and a gametolytic gland are not homologous structures, all previous and subsequent workers have considered them homologous but perhaps having developed structural and physiological modifications and distinct functions.

In most monaulic systems the bursa copulatrix is situated near the gonopore; this is the plesiomorphic state (figures 8.5B and 8.6A, B). In more derived groups with diaulic and triaulic systems the bursa copulatrix is located more proximally, near the seminal receptacle. In many taxa the bursa copulatrix has been lost and only a seminal receptacle is present (Gosliner 1994). For instance, in several groups of aeolid nudibranchs, phylogenetically basal species have both a bursa copulatrix and a seminal receptacle, and the more derived species of each clade have only a seminal receptacle. In other groups, all members of each clade lack a bursa copulatrix. Conversely, some taxa such as many Dendronotina may lack a receptaculum but possess a bursa.

The seminal receptacle is normally muscular and its function is related to storage and possible release of exogenous sperm at the time of fertilization. Spermatozoa can be stored in the seminal receptacle for extended periods of time. The ultrastructure of this organ has been studied by Schmekel (1971) who showed that the outer wall is highly muscular whereas the endothelium consists of columnar cells. Each cell contains large vacuoles, a few small mitochondria and Golgi complexes. Hadfield and Switzer-Dunlap (1984) stated that a seminal receptacle is probably present in all opisthobranch groups except for acochlidiaceans. As mentioned above, in many groups only one of the two female sperm-containing structures is present, and its identification is based on its proximal or distal position. However, some cases are difficult to determine. Derived acteonids of the genera *Acteon*, *Rictaxis*, and *Microglyphis* possess only a distal seminal receptacle. This organ contains oriented sperm in the duct, similar to that found in the pouch of the seminal receptacle of other gastropods. The main difference between acteonids and most other opisthobranchs is that in acteonids the oriented spermatozoa are not found in the pouch itself. Also, the seminal receptacle of acteonids contains a brownish liquid similar to that found in the bursa copulatrix of other opisthobranchs. More research is required to determine the homology of this organ to either the seminal receptacle or bursa copulatrix of other opisthobranchs. In at least some species of the genera *Haminoea*, *Atys*, *Runcina*, *Diaphana*, *Bulla*, *Hancockia*, *Doto*, *Tritonia*, *Lomanotus*, *Melibe*, *Janolus*, and *Armina* only a bursa copulatrix is present and the seminal receptacle appears to have been lost.

The female gonoduct opens through a copulatory tube called the vagina. Numerous species have different types of vaginal glands and armature that seem to be involved in copulation (figure 8.4G). Species of *Cylichnium* possess a complex female copulatory organ composed of densely packed papillae at the female opening (figure 8.4H). Vaginal spines and cuticular linings are present in some nudibranchs such as *Baptodoris*, *Platydoris*, and *Gargamella* (Garovoy et al. 1999; Dorgan et al. 2002; Valdés & Gosliner 2001).

Monaulic Systems

The ancestral "one tube" type is characterized by the presence of only one (hermaphroditic) gonoduct.

This configuration is present in several lineages with a well-developed shell as well as in groups with shell internalization (Ghiselin 1966; Gosliner 1994). Gonochoric acochlidiaceans are also monaulic.

This configuration (figures 8.6A, B) is characterized by having a single preampullar hermaphroditic gonoduct that expands into an ampulla, which functions as a short-term storage vesicle for endogenous sperm. The ampulla then connects into the postampullar duct, which leads to the pallial portion of the gonoduct. Near this point is the seminal receptacle, which stores exogenous sperm for long periods of time. The hermaphroditic duct then enters the female gland complex, where the fertilized eggs are covered with secretions. The female gland complex opens into the gonopore where it joins the duct of the bursa copulatrix, which stores exogenous sperm right after copulation and also likely has a gametolytic function. From the gonopore, a ciliated sperm grove carries endogenous sperm to the non-protrusible penis. This arrangement is the most plesiomorphic and only persists in *Tylodina* (figures 8.5A and 8.6B), but there are several major modifications of this ancestral monaulic reproductive systems found in other basal opisthobranchs (Gosliner 1981a). For instance, some species of *Ringicula*, and all members of Bullacea, Philinacea, Diaphanidae, Thecosomata, Gymnosomata, Phillinoglossa, and Acochlidiacea have a basic monaulic system, but the penis is protrusible, located in the cephalic region, and generally has a prostatic gland at its proximal end. Two other species of *Ringicula* have either an incomplete separation of the female and male ducts or a complete separation, becoming an androdiaulic system (see below).

Diaulic Systems

This configuration is characterized by a split hermaphroditic gonoduct and has evolved several times independently. Two main organizations of diaulic systems are observed, oodiaulic (figure 8.6C) and androdiaulic (figure 8.7A), which have distinct phylogenetic origins.

In androdiaulic reproductive systems there is a separate male duct terminating at the penis (figures 8.7A and 8.8A); the other duct serves as both an oviduct and vagina. Acteonids, many sacoglossans, some pleurobranchids, and some nudibranchs possess androdiaulic reproductive systems. However in acteonids the penis is non-protrusible.

In sacoglossans, pleurobranchids, and nudibranchs there is an additional division of the genital ducts to include a separate insemination or vaginal duct. Gosliner (1991) described details of the anatomy of androdiaulic reproductive systems that suggest that this configuration has evolved at least four times independently within acteonids, *Ringicula*, sacoglossans and the Nudipleura clade. Schmekel (1970) recognized two different types of androdiaulic systems that she identified as diauly I and II. In diauly I, the bifurcation of the hermaphroditic duct into the deferent duct and the oviduct lies near the genital aperture, as does the branching of the glandular duct from the oviduct. Diauly II differs by the more proximal location of the bifurcation of the hermaphroditic duct. The seminal receptacle, when it is present, is located either near or far from the genital opening in diauly I and II, respectively. According to García-Gómez et al. (1990) some aeolid nudibranchs show intermediate characteristics between the two types and could be classified as diauly II based on the bifurcation of the hermaphroditic duct and diauly I based on the position of the seminal receptacle. Based on the descriptions of the reproductive systems in androdiaulic nudibranchs, García-Gómez et al. (1990) concluded that in aeolids with diauly II, the seminal receptacle is more proximal than the bifurcation of the glandular branch of the oviduct (figure 8.9D–E), while in the species with diauly I, the seminal receptacle is more distal than this branch of the oviduct and, therefore, is located more proximal to the genital aperture.

In oodiaulic reproductive systems there is a separate duct composed of the female glands through which the eggs traverse (figure 8.5D and 8.6C). Rudman (1978) applied the term oodiaulic to describe the branching of the oviduct in some species of philinacean opisthobranchs. This configuration is functionally and phylogenetically different from true oodiaulic systems because the eggs do not appear to traverse through the duct, but rather receive the secretory products. Additionally, the branching of the duct in the Philinacea includes the membrane and albumen glands but not the mucous gland (Gosliner 1980). Oodiaulic systems are found in all members of the Anaspidea, including the most basal clade members of the genus *Akera*.

Triaulic Systems

The most derived configuration consists in independent, subsequent splits of the hermaphroditic

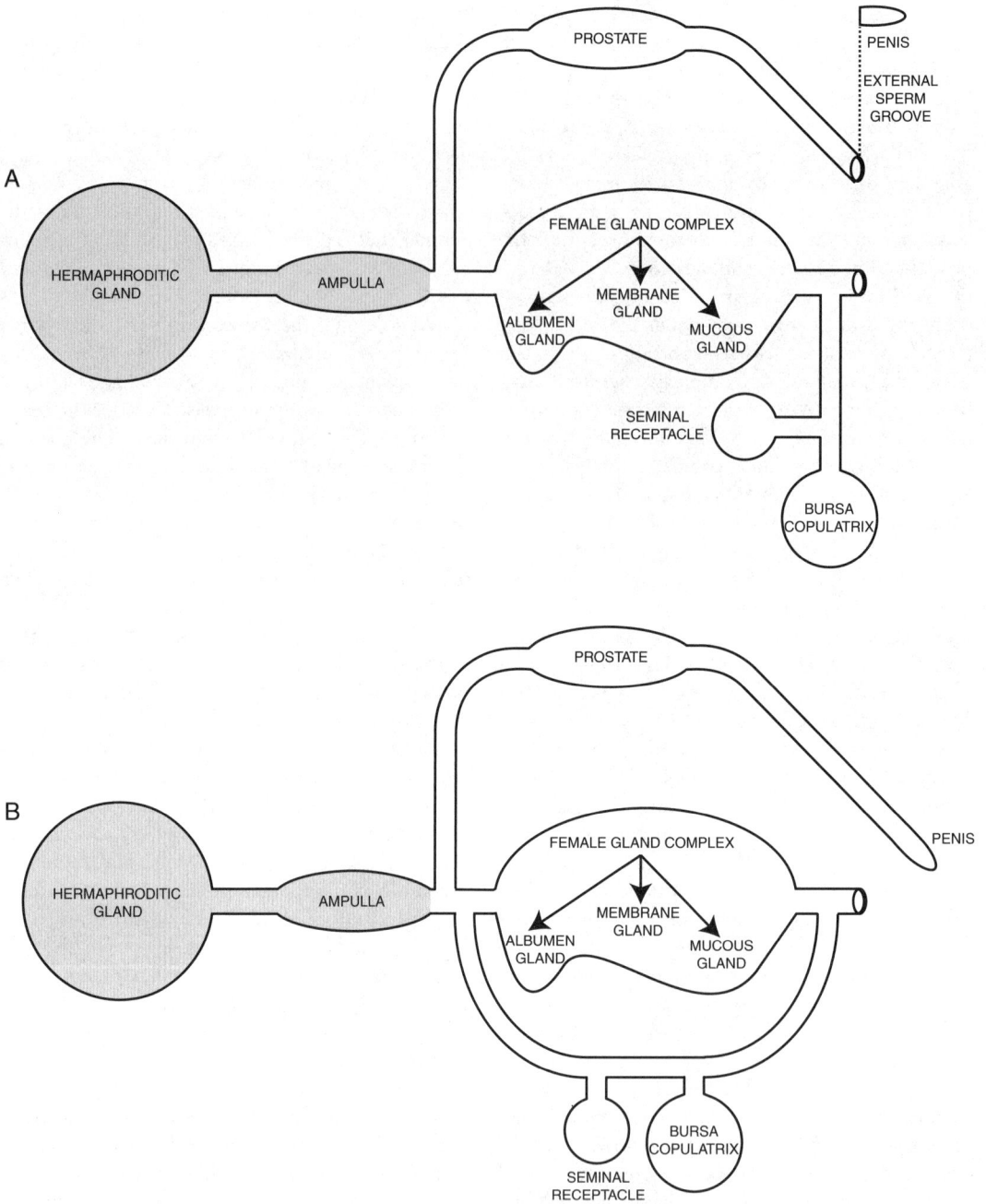

FIGURE 8.7 Schematic representations of different reproductive systems. (A) Androdiaulic system as found in *Ascobulla californica*; (B) Generic triaulic system with two reproductive openings.

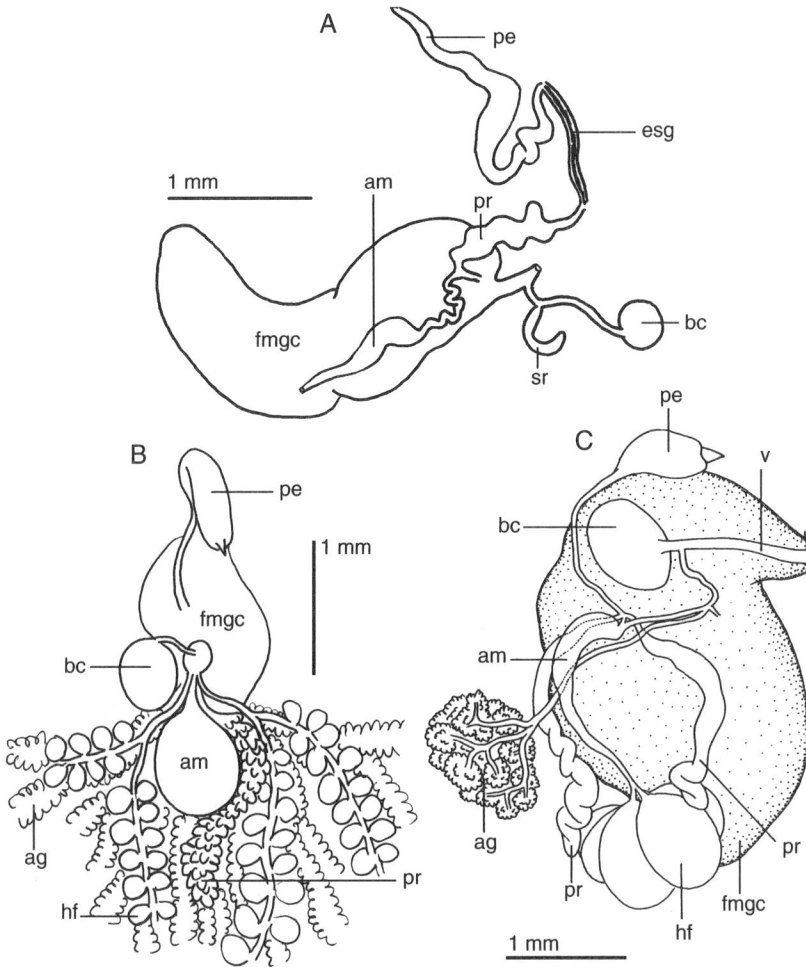

FIGURE **8.8** Reproductive systems of sacoglossans. (A) Androdiaulic system of *Ascobulla californica*; (B) Triaulic system of *Elysia cauze*; (C) Triaulic system of *Cyerce antillensis*. Abbreviations: ag, albumen gland; am, ampulla; bc, bursa copulatrix; esg, external sperm groove; fmgc, female gland complex; hf, hermaphroditic follicle; pe, penis; pr, prostate; sr, seminal receptacle; v, vagina.

gonoduct producing a diverse range of morphological modes. Several distinct types of reproductive system have been called triaulic (figure 8.7B). These likely reflect distinct evolutionary histories (Cervera et al. 2000).

According to Jensen (1996) numerous non-shelled sacoglossans have a separate vaginal opening that in some cases continues with a separate duct conducting the exogenous sperm to the seminal receptacle, forming a triaulic system (figures 8.8B, C). However, the arrangement of the vaginal opening and vaginal duct differs widely among the sacoglossans, and Jensen (1996) argued that these

structures are probably not homologous in the various lineages. Species of *Thuridilla* have a dorsal vagina, usually with a separate opening. In *Elysia viridis* the latero-dorsal vaginal opening connects to the mucous gland (Sanders-Esser 1984), whereas in *Elysia leucolegnote* it connects through the pericardium to the duct of the seminal receptacle (Jensen 1990), and in *Elysia australis* the vaginal duct connects directly to the seminal receptacle (Jensen & Wells 1990). In *Cyerce* the vaginal duct connects to the duct of the seminal receptacle (Sanders-Esser 1984) and in limapontiids the separate vaginal opening connects to the seminal

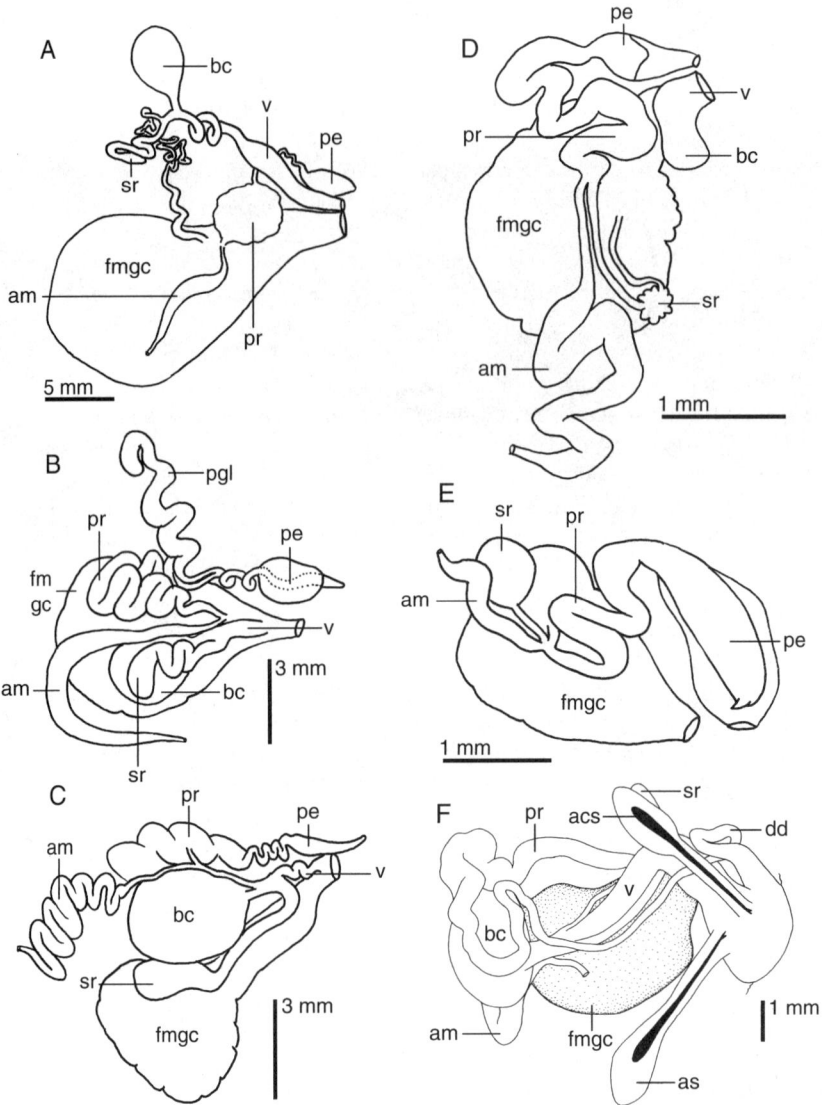

FIGURE 8.9 Reproductive systems of Nudipleura. (A) Triaulic system of *Bathyberthella antarctica*, showing the oviduct entering into the female gland complex and a completely differentiated vagina; (B) Triaulic system of *Berthella sideralis*, showing the oviduct entering into the female gland complex and lacking a differentiated vaginal opening; (C) Triaulic system of *Pleurobranchus areolatus*, showing the oviduct connecting with the bursa copulatrix and seminal receptacle and lacking a differentiated vagina; (D) Diaulic system of *Flabellina iodinea*; (E) Diaulic system of *Phidiana lynceus*, lacking a differentiated bursa copulatrix; (F) Traulic system of *Pharodoris diaphora*, showing large copulatory spines. Abbreviations: acs, accessory copulatory spine; am, ampulla; as, accessory sac; bc, bursa copulatrix; dd, deferent duct; fmgc, female gland complex; pe, penis; pgl, penial gland; pr, prostate; sr, seminal receptacle; v, vagina.

receptacle, where present, or to the fertilization region (Gascoigne 1976).

Cervera et al. (2000) described three distinct patterns of division of reproductive systems of pleurobranchids producing three different triaulic arrangements. In *Pleurobranchea*, *Pleurobranchella*, *Euselenops*, *Anidolyta*, and *Pleurobranchus* the hermaphroditic duct divides into the deferent duct and an oviduct that is connected to the seminal receptacle and the bursa copulatrix (figure 8.9C). The exogenous sperm storage organs are connected to the female glands mass only at the common genital atrium. While this is technically a triaulic system, it is very different from the arrangement found in *Berthella*, *Berthellina*, *Pleurehdera*, and *Tomthompsonia*, in which the oviduct enters the female gland complex and then exits at the nidamental gonopore (figure 8.9B). There it meets the opening of a separate vaginal duct that joins with a distinct bursa copulatrix and seminal receptacle. In *Polictenidia*, *Bathyberthella*, and *Parabathyberthella* the oviduct also enters directly into the female glands mass, but an additional uterine duct joins the seminal receptacle and bursa copulatrix with the female gland complex in addition to having a vaginal opening adjacent to the nidamental gonopore (figure 8.9A) and this arrangement is identical to that found in dorid nudibranchs (figure 8.9F). The condition found in *Pleurobranchea*, *Pleurobranchella*, *Euselenops*, *Anidolyta*, and *Pleurobranchus* is probably plesiomorphic since it is more similar to the arrangement found in the more basal, monoaulic *Tylodina*.

In dorid nudibranchs, basal members currently classified in *Bathydoris* are androdiaulic whereas the rest of the dorids, including a derived species of *Bathydoris*, are triaulic. Most species of the aeolids have diaulic reproductive systems, but in *Bajaeolis bertschi* there is a separate vaginal duct and it is triaulic. In *Janolus* several basal members are androdiaulic and more derived ones are triaulic (Gosliner 1981b). It appears that triaulic systems have evolved at least three different times independently in nudibranchs.

The Female Glands

This is a complex of glands consisting of different arrangements of glandular tissue involved in egg-mass formation. They are also referred to as nidamental glands (Klussmann-Kolb 2001). In many cases the different glandular components of the female glands are indistinguishable with gross morphological examination and their study requires histological sectioning. Traditionally, in opisthobranchs, as well as in pulmonates (De Jong Brink 1969; Plesh et al. 1971) and pyramidellids (Fretter & Graham 1949), the female gland complex is considered to be composed of a proximally situated albumen gland, a membrane gland, and a distal mucous gland. Ghiselin (1966) proposed that the ancestor of all opisthobranchs possessed three-part female glands, which would constitute the plesiomorphic state.

The albumen gland is the area of the female glands mass in closest proximity to the point where the hermaphroditic duct enters the complex. It produces a layer of albumen that first surrounds the fertilized ova. The albumen nourishes the developing embryo. In derived sacoglossans the albumen gland consists of numerous diffuse branches (Marcus & Marcus 1970) rather than a compact, discrete mass as in basal sacoglossans and other opisthobranchs (Klussmann-Kolb 2001).

Klussmann-Kolb (2001) considered that an albumen gland is present in all groups of opisthobranchs except for pleurobranchoids, nudibranchs, and the cephalaspideans *Haminoea*, *Chelidonura*, and *Philinopsis* in which the most proximal of the glands would be a capsule gland, and those taxa would lack a true albumen gland. At the same time Klussmann-Kolb (2001) recognized that albumen and capsule glands are homologous because of their shared proximal position, similar structure and histology, similar mode of secretion, and similar histochemical properties.

Because albumen and capsule glands seem to be homologous and share the same function, we consider more appropriated to refer to them with the same name. Thus, they are here referred to as albumen gland.

Albumen glands are likely present in all opisthobranchs with possibly some exceptions. For instance, Ghiselin (1964) observed that at least some eolids appear not to have secretions consistent with the presence of an albumen gland. In *Hermissenda crassicornis* the membrane (see below) is laid down over the fertilized eggs themselves, which appear not to have an albumen layer around them (Ghiselin 1964). After the membrane layer has been deposited the egg then shrinks, producing a space between it and the membrane. This seems to indicate that in some groups the albumen gland could have been lost, but further research is necessary to clarify this point.

The ultrastructure of the albumen gland has been described by Klussman-Kolb (2001) as having the shape of a sac or a tube. The secretory products are mostly packed in the form of large round or elliptically shaped vesicles and have a homogeneous or amorphous structure. The glandular cells contain active nuclei and large areas of endoplasmic reticulum and Golgi complexes in the basal part of the cell. The supporting cells have short cilia. Albumen layers, when they are present, can be detected by means of specific stains, vital dies and other methods (Ghiselin 1964).

In virtually all species of opisthobranchs one or more eggs are enclosed in a thin capsule named membrane. Ghiselin (1966) suggested the term membrane gland to characterize the portion of the female gland complex that produces the membrane. This was suggested to eliminate the ambiguity of the term "capsule" that has been used to refer to the albumen gland and the membrane gland in opisthobranchs, and to the portion of the nidamentary gland mass that produces the coriaceous capsules in terrestrial pulmonates, which are probably homologous with the mucous gland of opisthobranchs. Rudman (1972, 1978) suggested that a membrane gland is absent in species of *Philine* and *Odontoaglaja*, but Gosliner (1980) confirmed the presence of a membrane gland in several species of Aglajidae and Klussmann-Kolb (2001) showed the presence of a membrane gland in all main opisthobranch lineages.

Ghiselin (1964) examined and dissected several species of opisthobranchs during the process of egg masses formation. He observed that it is physically necessary for the eggs with any surrounding albumen to pass over the tissue that secretes the membrane. Therefore the series of eggs must traverse the cavity of the membrane gland.

Opisthobranch eggs are commonly sheathed with a single egg per membrane, but many species in different lineages have multiple eggs per membrane (Hurst 1967; Gosliner 1981b). Larger animals tend to have more eggs surrounded by a single membrane. For instance, some species of *Aplysia* may have more than 100 eggs per membrane (Usiki 1970). The egg membranes of opisthobranchs are held together by thin strings called chalazae—this is a derived characteristic shared by several heterobranchs including most euthyneurans.

The ultrastructure of the membrane gland has been described by Klussman-Kolb (2001) as tubular and often coiled. It comprises a small glandular area. The secretory products are normally heterogeneous packed into small mucous vesicles. The supporting cells usually have long cilia.

The mucous gland constitutes the largest portion of the female gland complex. It produces the gelatinous matrix that protects the eggs and holds them together. Evidence indicates that this gelatinous sheath provides protection against low salinity (Woods & DeSilets 1997) and provides chemical cues for larval metamorphosis (Gibson & Chia 1994). Generally, the mucous gland is lobate, with some basal species of Aglajidae and Philinidae having more than one lobe, which appears to be the plesiomorphic condition for these groups (Rudman 1974; Gosliner 1980).

The ultrastructure of the mucous gland has been described by Klussman-Kolb (2001) as having the shape of a large tube with mostly wide coils. The secretions can take several different textures and consistencies. The main difference with other female glands is the absence of endoplasmic reticulum in the glandular cells. The supporting cells normally have small cilia.

Klussman-Kolb (2001) described additional glands of the female gland complex including the oviducal gland, the spermoviduct gland and the atrial gland that receive different names in different groups of opisthobranchs but can all be referred to as the adhesive region. One or several of these glands are found in members of Cephalaspidea, Anaspidea, Sacoglossa, and Nudibranchia.

Several species of nudibranchs posses a vestibular gland in addition to the female gland complex. The vestibular gland is attached to the distal oviduct of and its function is currently unknown. This gland is probably related to reproduction, and possibly to egg mass formation. Detailed investigation of the vestibular gland of *Dendrodoris nigra* by Klussmann-Kolb & Brodie (1999) revealed the presence of symbiotic bacteria aligned between the microvilli of the glandular cells. Identical looking bacteria were also found in the mucous layers of its egg masses. The number of bacteria in the egg mass increased during development. These bacteria are not external contaminants but are actively reproducing and are stored within the nudibranch for an apparently functional but presently unknown reason. Several hypotheses have been proposed to explain their function, including a possible role in the breakdown of the egg mass mucous coating and egg capsule, a protective function preventing the egg mass from colonization by other organisms, or a nutritional source for the larvae (Klussmann-Kolb & Brodie 1999).

The female gland complex may also play a role in the transfer of defensive secondary metabolites into the egg mass. The presence of defensive chemicals in the gonad and egg mass of several opisthobranch species has been reported (e.g., Cimino et al. 2005). Metabolites from the gonad deter predation on the eggs and larvae (Cimino & Ghiselin 1999). Metabolites from the albumen gland inhibit growth of bacteria (Kamiya et al. 1984).

STUDY OF THE REPRODUCTIVE SYSTEM OF OPISTHOBRANCHS— MODERN TECHNIQUES AND LIMITATIONS

Historical Background

In the last few years there have been a number of attempts to reconstruct the phylogenetic history of opisthobranchs. Most authors recognized the importance of reproductive structures for phylogenetic reconstruction, and therefore numerous studies have dealt with the identification of reproductive homologies using different techniques. However, the identification of organs has been proven difficult and subjective and authors disagree on the criteria used to determine the identity of particular structures. The wide variation of the reproductive morphologies between and within lineages and the numerous cases of the independent gain and loss of organs have exacerbated this problem. In the sections below we will discuss the different techniques and criteria used for the identification of reproductive organs and the implications for a broad understanding of the evolution of the reproductive system in opisthobranchs.

Gross Morphology

The most commonly used technique. It is based on the principle that reproductive structures can be identified and homologies between structures of different taxa can be determined by the relative position of the organs as well as their overall morphology.

Many reproductive structures have an external appearance that allows a "quick and dirty" identification of the organ. For example, the ampulla is normally an elongate, muscular organ that appears shiny externally. In species with a seminal receptacle and a bursa copulatrix, these organs are often distinguished by the external gross morphology, the bursa copulatrix being generally larger and having thin walls, whereas the seminal receptacle is normally smaller and muscular.

The position of the organs is also a commonly used indication of the identity of a reproductive organ. The ampulla, female glands, penis, and vagina are often easy to identify based on their positions in the reproductive system, which indicate their function. However, the identification of other organs such as exogenous sperm vesicles is often more complicated. The relative position of the bursa copulatrix and the seminal receptacle has been used for the identification of these two organs, because the seminal receptacle is often more proximal than the bursa copulatrix. Also the receptaculum often contains embedded sperm, is thick-walled and may have a shiny appearance while the bursa is usually thin-walled with a dull appearance. In monaulic and diaulic systems the bursa copulatrix is commonly distal and opens directly in the gonopore through the vagina. The seminal receptacle is often connected near to the postampullar duct or the female gland in a much more proximal position. However, in some cases the seminal receptacle connects directly into the mid-region of the vagina or it is absent. In a typical triaulic arrangement the seminal receptacle connects to the bursa copulatrix by a duct that inserts near the vaginal insertion in the latter. However, in variations of this system present in chromodorid nudibranchs, the seminal receptacle emerges from the vagina at a more distal position than the bursa copulatrix.

It is possible that the organs referred to as "seminal receptacle" in different lineages of opisthobranchs are not homologous. This idea is based on of the wide range of variation in the position and morphology of these organs between different taxa, the differences in their position relative to that of the bursa copulatrix, and the numerous cases in which a seminal receptacle has apparently been lost. On the contrary, the bursa copulatrix is consistent in position and gross morphology and likely homologous across lineages. However, this view has been challenged by histological studies that aim to identify organs based on ultrastructure, under the assumption that these techniques would provide a more objective criterion for determining homologies and ultimately to identify reproductive organs (Schmekel 1971; Wägele & Willan 2000).

Histology

Gross morphological studies are often supplemented with histological observations, which allow for the identification of secretory products using stains, including histochemical ones. The study the ultrastructural of reproductive organs has been suggested to offer advantages over traditional morphological examinations (Wägele & Willan 2000; Klussman-Kolb 2001). Schmekel (1971) proposed for the first time an identification system of exogenous sperm vesicles based on histology, the seminal receptacle being composed of a folded wall of indistinct cells surrounded by a thick muscular layer and containing sperm oriented perpendicular to the walls, and the bursa copulatrix composed of apocrine-secreting cells and containing degraded exogenous sperm. According to Wägele and Willan (2000), both organs as defined by Schmekel (1971) are only found in dorid nudibranchs. The presence of a bursa copulatrix in aeolid nudibranchs reported by Gosliner (1994) has been challenged by Wägele and Willan (2000) as impossible to determine because of lack of histological information.

The use of histological data for the identification of organs is hampered by the same problems and limitations as traditional observation of gross morphology. Variation in cellular components of homologous structures is poorly known. For example, assumption that the presence of certain cell types in an organ is a definitive evidence of homology has obvious problems of circularity. Homologous organs may or may not have the same function, morphology, and ultrastructure; they are homologous because they are derived from the same organ in their common ancestor, not because they are similar (Ghiselin 2005). Undoubtedly, the ultrastructure of an organ provides information about its function and possible origin, but is by no means a definitive answer to the problem of organ identification; it may reflect a set of newly acquired functions of the organ rather than its evolutionary history. Only detailed studies of development, tissue differentiation, and genetics could potentially provide definitive answers to this problem.

Physiology

Understanding how the reproductive organs function is a powerful tool to determine homologies even in the absence of other data. However, due to the high degree of homoplasy and subsequent acquisition and loss of several organs, this exercise of functional comparison must be done with caution. In this context, the relative position of an organ may be a good proxy for determining its function and therefore allows for estimates of homology, as it is a reflection of the developmental pathways and genetic diversity required to produce that organ. Although the seminal receptacles of different species of opisthobranchs may not be homologous, the bursa copulatrix (defined as the thin-walled organ connected proximally to the vagina) of most species shares a similar position, morphology and function, and it is likely a homologous organ across most taxa. It is also likely that other reproductive organs such as the prostate, female glands, penis, and vagina are homologous across many taxa.

THE EVOLUTION OF THE REPRODUCTIVE SYSTEM OF OPISTHOBRANCHS

Current Phylogenetic Hypotheses

Different lines of evidence conflict on whether opisthobranch gastropods constitute a polyphyletic complex of lineages of simultaneous hermaphrodites or a monophyletic group. Dayrat and Tillier (2002) proposed a morphological phylogenetic hypothesis for the Opisthobranchia in which the relationships between members of this group are not resolved. These authors found that the clade Euthyneura includes Pulmonata and at least 10 opisthobranch clades of unresolved relationships (Thecosomata, Gymnosomata, Acochlidioidea, Pyramidelloidea, Runcinoidea, Cephalaspidea, Sacoglossa, Umbraculoidea, Pleurobranchoidea, and Nudibranchia). Wägele and Klussmann-Kolb (2005), also based on morphological evidence, proposed a phylogenetic hypothesis in which opisthobranchs are a monophyletic group when the Actenoidea and the pelagic Pteropoda are removed.

Recent molecular phylogenies also failed to produce consistent results with one another and with morphological data. Vonnemann et al. (2005)

based on 18S and 28S gene sequences produced a phylogenetic hypothesis for the Opisthobranchia in which the monophyly of this group is relatively well supported. Acteonidae would be basal to the Nudipleura but relatively derived within the Opisthobranchia, which contradicts most morphological phylogenies produced to date, but would be supported by the presence of androdiaulic reproductive systems in several basal members of Nudipleura and Acteonidae. Also, in the hypothesis of Vonnemann et al. (2005) Tylodinoidea, along with Acochlidiacea, Anaspidea, and Cephalaspidea are basal, unresolved lineages, suggesting a possible relationship between clades with an external sperm groove.

Grande et al. (2004) produced another phylogenetic hypothesis for the Opisthobranchia based on mitochondrial genes that also indicates that Acteonoidea are basal to Nudipleura. However, in this hypothesis the Sacoglossans would not share an immediate common ancestor with other opisthobranchs.

Malaquias et al. (2008), further investigated the phylogeny of cephalaspidean opisthobranchs employing three genes and greater taxon sampling. They redefined a monophyletic Cephalaspidea to exclude Acteonacea and Runcinacea. Sacoglossa, Anaspidea, Gymnosomata, and Thecosmata are all more closely related to Cephalaspidea than to Acteonacea. Gymnosomes and thecosomes form a monophyletic Pteropoda. At this point in time, the phylogenetic relationships of opisthobranchs are not fully resolved, but greater taxon and gene sampling has yielded more consistent results that are promising.

Homoplastic Events

The basic organization of the reproductive system of opisthobranch gastropods appears to have been the subject of independent transformations often leading to parallel evolution of analogous arrangements. The lack of consensus between different phylogenetic hypotheses and particularly between morphological and molecular data, prevents specialists from being able to analyze the modifications of reproductive characters, structures, and functions throughout the evolution of opisthobranchs. In some cases it would seem like reversals to more structurally plesiomorphic conditions have occurred. According to molecular and morphological phylogenetic analysis the monaulic Tylodina is not a basal opisthobranch, but rather a derived

group either sister to the Cephalaspidea (figure 8.1B) or sister to a larger clade containing nudibranchs, sacoglossans, and other groups (figure 8.1A). In all cases more basal groups have either diaulic or even triaulic reproductive systems, which seems to indicate that the monaulic condition in Tylodina is secondarily evolved. In groups for which we have a good understanding of the basic phylogenetic relationships such as the cephalaspideans, anaspideans, and Sacoglossa, independent splits of the pallial gonoduct that increase the efficiency of the system have been documented. For instance, as mentioned above, Jensen (1996) shown that several sacoglossans have independently evolved a separate vaginal opening that in some cases continues with a separate duct conducting the exogenous sperm to the seminal receptacle, forming a triaulic system. All members of the Anaspidea have oodiaulic systems that are superficially similar to those found in some cephalaspideans such as Philine, but most likely evolved independently and display functional differences.

Changes in Reproductive Morphology—Structural Implications

Shell reduction in opisthobranchs is the consequence of a shift from mechanical to chemical defense (Faulkner & Ghiselin 1983). Reduction and loss of the shell have led to restructuring of the body with a tendency to displacement of the respiratory apparatus and some organs associated with it toward the rear end of the body (detorsion). Lengthening of the sperm groove in some cephalaspideans is a maladaptive side effect of the displacement of the gonopore to the rear (Gosliner 1981a). There has also been an untwisting of the nervous system and cephalization in the form of ganglia being concentrated at the front end of the body (Gosliner 1981a). Chemical defense is also largely responsible for the adaptive radiation of opisthobranchs, many of which derive both metabolites and nutriment from the food (others synthesize defensive metabolites *de novo*). The anatomy of the digestive tract is closely adapted to the properties of the food. Because there is no close connection between the mode of feeding and the exigencies of reproduction, evolutionary trends in the two systems are not strongly correlated except as a result of common ancestry. Resource availability does influence such factors as effective population density; however, food should have an indirect effect on reproductive strategies. The reproductive system is rather autonomous in its evolutionary trends and tendencies. There is no necessary connection between anagenesis in one part of the body and another. Acteonaceans have the most plesiomorphic arrangements of mantle organs and the central nervous system; yet they have evolved a fairly derived androdiaulic reproductive system (Gosliner 1981a). There may, however, be contingent correlations between trends in different parts of the body. The transformation from monaulic forms with separate male and female organs into diaulic or triaulic, compacted forms has been accompanied by substantial modifications in body structure and rearrangement of the other organs. The animals tend to be more detorted and more highly cephalized.

Changes in Reproductive Morphology—Functional Implications

The diversity of reproductive system organizations has important functional implications in copulation and reproductive behavior. There are a number of novel reproductive behaviors throughout the phylogeny of this group, including hypodermic insemination or formation of reproductive chains, which are possible due to particular morphological characteristics in several lineages. Some of the features that have evolved are evidently the result of sexual selection, and therefore are secondary, not primary, sex characters.

At the same time opisthobranchs display a wide variation in the morphology of the reproductive system, they also show remarkable variation in reproductive behavior. Simultaneous hermaphroditism is the most common reproductive mode, and this seems to be an advantage in species living at low effective population densities (Ghiselin 1969). In the great majority of species individuals copulate in pairs by placing their bodies with their anterior right sides against each other. The penis of each individual enters the vagina of the partner simultaneously and sperm are exchanged at the same time. Copulation is in some cases accompanied by premating behavior (e.g., Rutowski 1983; Angeloni 2003). In some species of facelinid aeolids the cerata adjacent to the gonopore are modified to grasp the partner during copulation.

Several species of sacoglossans display hypodermic injection of sperm anywhere on the partner's body and this is possible because of the presence of a penial spine. This is usually a reciprocal event within a mating pair. Schmitt et al. (2007) described the mating behavior of *Elysia timida* that includes a combination of a long series of hypodermic transfers followed by a short phase with standard insemination into the vagina. In both phases the two mating individuals show a high degree of transfer symmetry and synchrony. The evolution of hypodermic insemination can be traced by observation of mating pairs or the documentation of the presence of hollow penial spines in different species of sacoglossans. According to the phylogenetic hypothesis of the Sacoglossa proposed by Jensen (1996) and her comprehensive documentation of anatomical features within this group, it is likely that hypodermic insemination has evolved twice and/or has been lost several times. Figure 8.10 shows the presence of penial spines in different lineages of sacoglossans. Both shelled and non-shelled clades contain members with hypodermic insemination, but this feature is much more widespread in non-shelled clades. Within shelled groups only *Berthelinia*, *Volvatella*, and *Asocobulla* display this condition. It is possible that most shelled groups have lost the penial spines, as has happened in some non-shelled groups such as *Hermaea* or *Stiliger*, or that this condition evolved at least twice in the Sacoglossa, as it seems to be supported by the absence of penial spines in *Sohgenia*, one of the most basal non-shelled sacoglossans and by the morphological differences between the spines found in different groups. For example, as summarized by Jensen (1996), limapontiids have a short, curved penial spines, whereas *Caliphyilla* has a long, whip-like extension, called a "flagellum" situated at the tip of the penis and armed with a minute spine, and *Cyerce* and *Mourgona* have long, almost straight penial spines. Regardless of how many times hypodermic insemination has evolved in sacoglossans, it has evolved at least twice in opisthobranchs, as some members of the nudibranch genus *Polycera* also display this behavior (Rivest 1984). Hypodermic insemination has broad implications for sexual selection in hermaphrodites, for instance it may increase sexual conflict by allowing the sperm donor to avoid the sperm digesting organs of the recipient and directly fertilize the eggs (Michiels & Newman 1998). However, Angeloni (2003) studied the mating behavior of *Alderia* and found that

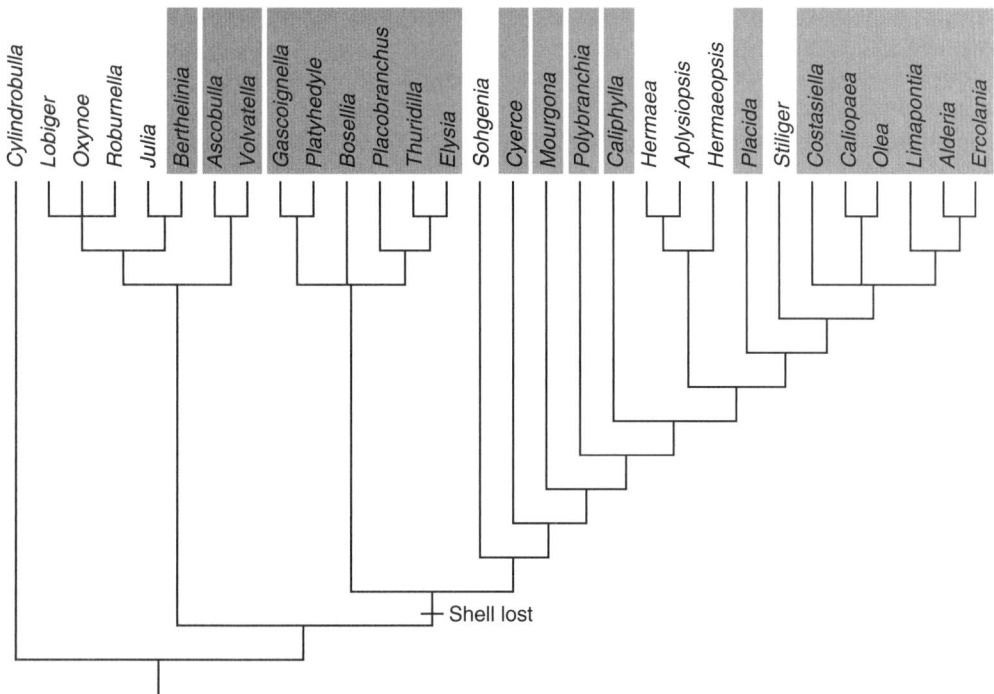

FIGURE 8.10 Phylogenetic hypothesis of the Sacoglossa extracted and simplified from Jensen (1996). Taxa highlighted in grey contain at least one species that displays hypodermic insemination.

injections of sperm all over the recipient's body resulted in fertilization, thus there does not seem to be any advantage for the sperm donor in this regard. Another possible evolutionary advantage of hypodermic insemination is that it allows for quick mating and avoidance of precopulatory behaviors as well as the need for consent by the recipient (Trowbridge 1995). Angeloni's (2003) research showed that *Alderia* displays precopulatory behavior and that brief insemination events did not result in fertilization. Only reciprocal, longer mating events resulted in egg fertilization. The reasons for the evolution of hypodermic insemination remain elusive.

The deposition of spermatophores is a variation of hypodermic insemination that has been described for several species, including cephalaspideans (Ghiselin 1963), thecosomes (Lalli & Wells 1978), nudibranchs (Tardy 1966; Haase & Karlsson 2000), and acochlidiaceans (Haase & Wawra 1996). The phylogenetic distance between these groups clearly indicates that this feature has evolved at least four times independently, possibly more, as both dorid and aeolid nudibranchs display

this condition. There is also substantial variation in the reproductive modes of species using spermatophores. For example, in cephalaspideans (Ghiselin 1963) and thecosomes (Lalli & Wells 1978) the spermatophores are deposited into the recipient's reproductive system, close to the genital opening. In gonochoristic acochlidioideans, the spermatophores attach to the body wall of the female (Haase & Wawra 1996). In the nudibranch *Aeolidiella glauca*, the penises of both partners are protruded simultaneously and stroke each other's backs depositing a spermatophore onto the notum (Haase & Karlsson 2000). Sperm enters the recipient through histolysis and reaches the seminal receptacle by an unknown mechanism. More recently, Haase & Karlsson (2003) found that individuals of *A. glauca* avoid mating with potential partners already carrying a spermatophore received during a previous mating. Thus, evolution of spermatophore deposition in opisthobranchs could be (at least in some cases) related to avoidance of sperm competition.

Several species of cephalaspideans and sea hares maintain seasonally high-density populations and their reproductive behavior is often modified. Leonard & Lukowiak (1984a,b, 1985), Michiels et al. (2003), and Anthes and Michiels (2005) described an alternation of sex roles during copulation in *Navanax inermis* and *Chelidonura sandrana*. Alternation of sex roles has implications for sexual conflict derived from sperm trading and sperm competition (Anthes & Michiels 2005). In both cases reciprocity of copulation seems to occur regularly indicating positive selection for the male function.

In sea hares, because of the long distance between the penis opening and the gonopore, species can greatly modify their mating positions. Many species of *Aplysia* form chains that may contain as many as 20 mating individuals. The posterior individual in the chain acts only as a male, the anterior individual acts only as a female, and all animals in between act as female to one partner and male to a second partner. The formation of these chains is stimulated by the production of several powerful water-borne sex pheromones released during egg-laying called attractins (Cummins et al. 2004; Painter et al. 1991, 2004). It also appears that these pheromones are functional across taxa and attractins extracted from *Aplysia californica* stimulate sexual activity in *Aplysia brasiliana*.

The evolution of penial spines and other anchorage systems in dorid nudibranchs has been, in some cases, related to particular functional needs. For example, Valdés (2004) argued that the presence of a vaginal cuticular lining in *Platydoris* and *Otinodoris* may offer a more stable substrate for the anchoring penial hooks to attach, because they appear to penetrate the hard surface of the cuticle. *Otinodoris* and *Platydoris* include species that reach large sizes, considerably larger than most other groups of dorid nudibranchs. The presence of the vaginal cuticular lining may provide an advantage for larger individuals, probably related to the forces involved in the potential separation of larger and heavier individuals during copulation.

Changes in Reproductive Morphology—Systematic and Phylogenetic Implications

In general terms, there is little correlation between changes in the organization of the reproductive system and changes in the body organization, digestive system, and the process of shell internalization. Thus, the organization of the reproductive system

is a fundamental tool for understanding opisthobranch evolution and reconstructing phylogenetic relationships within this group.

In the Aglajidae and Gastropteridae, Anthes et al. (2008) found a correlated evolution between morphological, behavioral and ecological traits. For instance, higher proportions of simultaneous, reciprocal (rather than unilateral) matings, where associated with larger penises and smaller seminal receptacles, but also with lower degrees of gregariousness. This seems to indicate that species that evolved smaller groups have also evolved larger ejaculates and reciprocal copulations to maximize the reproductive efficiency of sexual encounters (Anthes et al. 2008). This scenario is consistent with theoretical work on the evolution of hermaphroditism, in which gonochorism would be a disadvantage in species with low-density populations (Ghiselin 1966).

CONCLUSIONS AND SUGGESTIONS FOR FUTURE RESEARCH

While considerable progress has been made in the study of reproduction of opisthobranchs over the last decades, there are still significant limitations to our studies. Perhaps the greatest limitation still remains the lack of a robust phylogeny of the Opisthobranchia, other Euthyneura and other heterobranch gastropods. Considerable progress has been made in recent molecular phylogenetics, and it is encouraging that better taxon sampling seems to resolve rather than compound these issues. Exciting behavioral studies of sperm competition and reproductive modes (Anthes & Michiels 2005) have shed new light on the evolution of reproductive behavior and its relationship to copulatory structures. Evolutionary developmental studies are also likely to provide significant advances in our understanding of the relationship between genetics and developmental pathways. Despite all of the likely advances in novel fields of reproductive, genetic, behavioral, and developmental studies, only robust phylogenetic studies will provide the evolutionary road map for understanding the direction of evolutionary change. These phylogenies will also permit us to understand the number of independent instances of acquisition of evolutionary novelties, which is a critical issue in understanding reproductive evolution that is so well reflected in the tremendous variation of reproductive features and modes manifested in opisthobranch gastropods.

Acknowledgments We are grateful to Janet Leonard for her invitation to write this chapter and her helpful suggestions and constructive criticisms throughout the editorial process. Two anonymous reviewers also made constructive criticisms on the manuscript. This project has been supported by the US National Science Foundation through the PEET grant "Phylogenetic systematics of the Nudibranchia" (DEB-0329054) to T.M. Gosliner and A. Valdés. The SEM work was conducted at the Natural History Museum of Los Angeles County facility supported by the NSF MRI grant DBI-0216506.

REFERENCES

Angeloni, L. 2003. Sexual selection in a simultaneous hermaphrodite with hypodermic insemination: body size, allocation to sexual roles and paternity. Animal Behaviour 66: 417–426.

Anthes, N. & Michiels, N.K. 2005. Do "sperm trading" simultaneous hermaphrodites always trade sperm? Behavioral Ecology 16: 88–195.

Anthes, N., Schulenburg, H. & Michiels, N.K. 2008. Evolutionary links between reproductive morphology, ecology and mating behavior in opisthobranch gastropods. Evolution 62: 900–916.

Beeman, R.D. 1970. The anatomy and fuctional morphology of the reproductive system in the opisthobranch mollusk, *Phyllaplysia taylori* Dall, 1900. The Veliger 13: 1–31.

Brandriff, B. & Beeman, R.D. 1973. Observations on the gametolytic gland in the anaspidean opisthobranchs *Phyllaplysia taylori* and *Aplysia californica*. Journal of Morphology 141: 395–410.

Cervera, J.L., Gosliner, T.M., García-Gómez, J.C. & Ortea, J.A. 2000. A new species of Berthella Blainville, 1824 (Opisthobranchia: Notaspidea) from the Canary Islands (Eastern Atlantic Ocean), with a re-examination of the phylogenetic relationships of the Notaspidea. Journal of Molluscan Studies 66: 301–311.

Cimino, G. & Ghiselin, M.T. 1999. Chemical defence and evolutionary trends in biosynthetic capacity among dorid nudibranchs (Mollusca: Gastropoda: Opisthobranchia). Chemoecology 9: 187–207.

Cimino, G., & Ghiselin, M.T. 2009. Chemical defense and the evolution of opisthobranch gastropods. Proceedings of the California Academy of Sciences 60: 175-422.

Cimino, G., Fontana, A., Cutignano, A. & Gavagnin, M. 2005. Biosynthesis in opisthobranch mollusks: General outline in the light of recent use of stable isotopes. Phytochemistry Reviews 3: 285–307.

Cummins, S.F., Nichols, A.E., Rajarathnam, K. & Nagle, G. T. 2004. A conserved heptapeptide sequence in the waterborne attractin pheromone stimulates mate attraction in *Aplysia*. Peptides 25: 185–189.

Dayrat, B. & Tillier, S. 2002. Evolutionary relationships of euthyneuran gastropods (Mollusca): A cladistic re-evaluation of morphological characters. Zoological Journal of the Linnean Society 135: 403–470.

De Jong Brink, M. 1969. Histochemical and electron microscope observations on the reproductive tract of *Biomphalaria glabrata* (*Australorbis glabratus*), intermediate host of *Schistosoma mansoni*. Zeitschrift für Zellforschung und mikroskopische Anatomie 102: 507–542.

Dinapoli, A. & Klussmann-Kolb, A. 2010. The long way to diversity – Phylogeny and Evolution of the Heterobranchia (Mollusca: Gastropoda). Molecular Phylogenetics and Evolution 55.

Dorgan, K.M., Valdés, A. & Gosliner, T.M. 2002. Phylogenetic systematics of the genus *Platydoris* (Mollusca, Nudibranchia, Doridioidea) with descriptions of six new species. Zoologica Scripta 31: 271–319.

Faulkner, D.J. & Ghiselin, M.T. 1983. Chemical defense and the evolutionary ecology of dorid nudibranchs and some other opisthobranch gastropods. Marine Ecology Progress Series 13: 295–301.

Fretter, V. & Graham, A. 1949. The structure and mode of life of the Pyramidellidae, parasitic opisthobranchs. Journal of the Marine Biological Association of the UK 28: 493–532.

García-Gómez, J.C., Cervera, J.L. & García, F.J. 1990. Description of *Eubranchus linensis* new species (Nudibranchia), with remarks on diauly in nudibranchs. Journal of Molluscan Studies 56: 585–593.

Garovoy, J.M., Valdés, A. & Gosliner, T.M. 1999. Two new species of *Gargamella* from South

Africa (Mollusca, Nudibranchia). Proceedings of the California Academy of Sciences 51: 245–257.

Gascoigne, T. 1976. The reproductive systems and classification of the Stiligeridae (Opisthobranchia: Sacoglossa). Journal of the Malacological Society of Australia 3: 157–172.

Ghiselin, M.T. 1963. On the functional and comparative anatomy of *Runcina setoensis* Baba, an opisthobranch gastropod. Publications of the Seto Marine Biological Laboratory 11: 389–398.

Ghiselin, M.T. 1964. Reproductive Function and the Evolution of Opisthobranch Gastropods. Ph.D. Dissertation, Stanford University.

Ghiselin, M.T. 1966. Reproductive function and the phylogeny of opisthobranch gastropods. Malacologia 3: 327–378.

Ghiselin, M.T. 1969. The evolution of hermaphroditism among animals. Quarterly Review of Biology 44: 189–208.

Ghiselin, M.T. 2005. Homology as a relation between parts of individuals. Theory in Biosciences 124: 91–103.

Gibson, G. D. & Chia, F.-S. 1994. A metamorphic inducer in the opisthobranch *Haminaea callidegenita*: Partial purification and biological activity. Biological Bulletin 187: 133–142.

Gosliner, T.M. 1980. Systematics and phylogeny of the Aglajidae (Opisthobranchia: Mollusca). Zoological Journal of the Linnean Society 68(2): 325–360.

Gosliner, T.M. 1981a. Origins and relationships of primitive member of the Opisthobranchia (Mollusca: Gastropoda). Biological Journal of the Linnean Society 16: 197–225.

Gosliner, T.M. 1981b. The South African Janolidae (Mollusca: Nudibranchia) with the descriptions of a new genus and two new species. Annals of the South African Museum 86: 1–42.

Gosliner, T.M. 1985. The aeolid nudibranch family Aeolidiidae (Gastropoda: Opisthobranchia) from the tropical southern Africa. Annals of the South African Museum 95: 233–267.

Gosliner, T.M. 1991. Morphological parallelism in opisthobranch gastropods. Malacologia 32: 313–327.

Gosliner, T.M. 1994. Gastropoda: Opisthobranchia. In: Harrison, F.W. & Kohn, A.J. (eds.), Microscopic Anatomy of Invertebrates, Vol. 5: Mollusca I, 253–355. Wiley-Liss, New York.

Gosliner, T. M. & Draheim, R. 1996. Indo-Pacific opisthobranch gastropod biogeography: how do we know what we don't know. American Malacological Bulletin 12: 37–43.

Gosliner, T.M. & Ghiselin, M.T. 1984. Parallel evolution in opisthobranch gastropods and its implications for phylogenetic methodology. Systematic Zoology 33: 255–274.

Grande, C. Templado, J., Cervera, J.L. & Zardoya, R. 2004. Molecular phylogeny of Euthyneura (Mollusca: Gastropoda). Molecular Biology and Evolution 21: 303–313.

Haase, M. & Karlsson, A. 2000. Mating and the inferred fuction of the genital system of the nudibranch, *Aeolidiella glauca* (Gastropoda: Opisthobranchia: Aelidioidea). Invertebrate Biology 119: 287–298.

Haase, M. & Karlsson, A. 2003. Mate choice in a hermaphrodite: You won't score with a spermatophore. Animal Behaviour 67: 287–291.

Haase, M. & Wawra, E. 1996. The genital system of *Acochlidium fijiense* (Opisthobranchia: Acochlidioidea) and its inferred function. Malacologia 38: 143–151.

Hadfield, M. & Switzer-Dunlap, M. 1984. Opisthobranchs. In: Tompa, A. & Verdonk, N. (eds.), The Mollusca, Vol. 7, 209–350. Academic Press, New York.

Hurst, A. 1967. The egg masses and veligers of thirty northeast Pacific opisthobranchs. The Veliger 9: 225–288.

Jensen, K.R. 1990 Three new species of Ascoglossa (Mollusca, Opisthobranchia) from Hong Kong, and a description of the internal anatomy of *Costasiella pallida* Jensen, 1985. In: Morton, B. (ed.), Proceedings of the Second International Marine Biological Workshop: The Marine Flora and Fauna of Hong Kong and Southern China. Pp. 419–432. Hong Kong University Press, Hong Kong.

Jensen, K.R. 1996. Phylogenetic systematics and classification of the Sacoglossa (Mollusca, Gastropoda, Opisthobranchia). Philosophical Transactions of the Royal Society of London, (B) Biological Sciences 351: 91–122.

Jensen, K.R. & Wells, F.E. 1990. Sacoglossa (=Ascoglossa) (Mollusca, Opisthobranchia) from southern Western Australia. In: Wells, F.E., Walker, D.I., Kirkman, H. & Lethbridge, R. (eds.), Proceedings of the Third International Workshop: The Marine Biology of the Albany Area. Pp. 297–331. Western Australian Museum, Perth, Australia.

Kamiya, H., Muramoto, K. & Ogata, K. 1984. Antibacterial activity in the egg masses of a sea hare. Experientia 40: 947–949.

Klussmann-Kolb, A. 2001. Comparative investigation of the genital systems in the Opisthobranchia (Mollusca: Gastropoda) with special emphasis on the nidamental glandular system. Zoomorphology 120: 215–235.

Klussmann-Kolb, A. & Brodie, G. 1999. Internal storage and production of symbiotic bacteria in the reproductive system of a tropical marine gastropod. Marine Biology 133: 443–447.

Lalli, C.M. & Wells, F.E. 1978. Reproduction in the genus *Limacina* (Opisthobranchia: Thecosomata). Journal of Zoology 186: 95–108.

Leonard, J.L. & Lukowiak, K. 1984a. Male–female conflict in a simultaneous hermaphrodite resolved by sperm trading. American Naturalist 124: 282–286.

Leonard, J.L. & Lukowiak, K. 1984a. An ethogram of the sea slug *Navanax inermis* (Opisthobranchia, Gastropoda). Zeitschrift für Tierpsychologie 65: 327–345.

Leonard, J.L. & Lukowiak, K. 1985. Courtship, copulation and sperm-traiding in the sea slug, *Navanax inermis* (Opisthobranchia: Cephalaspidea). Canadian Journal of Zoology 63: 2719–2729.

Malaquias, M., Mackenzie-Dodds, J., Gosliner, T.M., Bouchet, P. & Reid, D.G. 2008. A molecular phylogeny of the Cephalaspidea *sensu lato* (Gastropoda Euthyneura): Architectibranchia redefined and Runcinacea reinstated. Zoologica Scripta 38: 23–41.

Marcus, Er. & Marcus, Ev. 1970. Opisthobranchs from Curaçao and faunistically related regions. Studies on the fauna of Curaçao and other Caribbean Islands 33: 1–129.

Michiels, N.K. & Newman, L.J. 1998. Sex and violence in hermaphrodites. Nature 391: 647.

Michiels, N.K., Raven-Yoo-Heufes, A. & Kleine Brockmann, K. 2003. Sperm trading and sex roles in the hermaphroditic opisthobranch sea slug *Navanax inermis*: Eager females or opportunistic males? Biological Journal of the Linnean Society 78: 105–116.

Mikkelsen, P.M. 1996. The evolutionary relationships of Cephalaspidea s.l. (Gastropoda: Opisthobranchia): a phylogenetic analysis. Malacologia 37: 375–442.

Neusser, T.P., Hess, M., Haszprunar, G. & Schrödl, M. 2006. Computer-based three-dimensional reconstruction of the anatomy of *Microhedyle remanei* (Marcus, 1953), an interstitial acochlidian gastropod from Bermuda. Journal of Morphology 267: 231–247.

Painter, S.D., Chong, M.G., Wong, M.A., Gray, A., Cormier, J.G. & Nagle, G.T. 1991. Relative contributions of the egg layer and egg cordon to pheromonal attraction and the induction of mating and egg-laying behaviour in *Aplysia*. Biological Bulletin 181: 81–94.

Painter, S.D., Cummins, S.F., Nichols, A.E., Akalal, D.G., Schein, C.H., Braun, W., Smith, J.S., Susswein, A.J., Levy, M., de Boer, P.A., ter Maat, A., Miller, M.W., Scanlan, C., Milberg, R.M., Aweedler, J.V. & Nagle, G.T. 2004. Structural and functional analysis of *Aplysia* attractins, a family of water-borne protein pheromones with interspecific attractiveness. Proceedings of the National Academy of Sciences 101: 6929–6933.

Plesh, B., De Jong Brink, M. & Boer, H.H. 1971. Histological and histochemical observations of the reproductive tract of the hermaphrodite pond snail *Lymnaea stagnalis* (L.). Netherlands Journal of Zoology 21: 180–201.

Rivest, B.R. 1984. Copulation by hypodermic injection in the nudibranchs *Palio zosterae* and *P. dubia* (Gastropoda, Opisthobranchia). Biological Bulletin 167: 543–554.

Rudman, W.B. 1972. The genus *Philine* (Opisthobranchia, Gastropoda). Proceedings of the Malacological Society of London 40: 171–187.

Rudman, W.B. 1974. A comparison of *Chelidonura*, *Navanax* and *Aglaja*, with other genera of the Aglajidae (Opisthobranchia: Gastropoda). Zoological Journal of the Linnean Society 54: 185–212.

Rudman, W.B. 1978. A new species and genus of the Aglajidae and the evolution of the philinacean opisthobranch molluscs. Zoological Journal of the Linnean Society 62: 89–107.

Rutowski, R.L. 1983. Mating and egg mass production in the aeolid nudibranch Hermissenda crassicornis (Gastropoda: Opisthobranchia). Biological Bulletin 165: 276–285.

Sanders-Esser, B. 1984. Vergleichende Untersuchungen zur Anatomie und Histologie der vorderen Genitalorgane der Ascoglossa (Gastropoda: Euthyneura). Zoologische Jahrbücher, Abteilung für Anatomie und Ontogenie der Tiere 111: 135–243.

Schmekel, L. 1970. Anatomie der Genitalorgane von Nudibranchiern (Gastropoda Euthyneura). Pubblicazioni della Stazione Zoologica di Napoli 38: 120–217.

Schmekel, L. 1971. Histologie und Feinstruktur der Genitalorgane von Nudibranchiern (Gastropodoa, Euthyneura). Morphologie und Ökologie der Tiere 69: 115–183.

Schmekel, L. 1985. Aspects of the evolution within the opisthobranchs. In: Truman, E.R. & Clarke, M.R. (eds.), The Mollusca, Vol. 10: 221–267. Academic Press, London.

Schmitt, V., Anthes, N. & Michiels, N.K. 2007. Mating behaviour in the sea slug *Elysia timida* (Opisthobranchia, Sacoglossa): hypodermic injection, sperm transfer and balanced reciprocity. Frontiers in Zoology 4: 17.

Tardy, J. 1966. Spermatophores chez quelques espèces d'Aeolidielles des côtes européennes (Gastéropodes, Nudibranches). Comptes Rendus des Séances de la Société de Biologie et de ses Filiales 160: 369–371.

Trowbridge, C.D. 1995. Hypodermic insemination, oviposition, and embryonic development of a pool-dwelling ascoglossan (=sacoglossan) opisthobranch: *Ercolania felina* (Hutton, 1882) on New Zealand shores. The Veliger 38: 203–211.

Usuki, I. 1970. Studies on the life history of the Aplysiae and their allies in the Sado district of the Japan Sea. Science Reports of Niigata University, Series D (Biology) 7: 91–105.

Valdés, A. 2001. Deep-sea dorid nudibranchs (Mollusca, Opisthobranchia) from the tropical West Pacific, with descriptions of two new genera and nineteen new species. Malacologia 43: 237–311.

Valdés, A. 2002. Phylogenetic systematics of "*Bathydoris*" s.l. Bergh, 1884 (Mollusca, Nudibranchia), with the description of a new species from New Caledonian deep waters. Canadian Journal of Zoology 80: 1084–1099.

Valdés, A. 2004. Morphology of the penial hooks and vaginal cuticular lining of some dorid nudibranchs (Mollusca, Opisthobranchia). American Malacological Bulletin 18: 49–53.

Valdés, A. & Gosliner, T.M. 2001. Systematics and phylogeny of the caryophyllidia-bearing dorids (Mollusca, Nudibranchia), with the description of a new genus and four new species from Indo-Pacific deep waters. Zoological Journal of the Linnean Society 133: 103–198.

Vonnemann, V., Schrödl, M., Klussmann-Kolb, A. & Wägele, H. 2005. Reconstruction of the phylogeny of the Opisthobranchia (Mollusca: Gastropoda) by means of 18S and 28S rRNA gene sequences. Journal of Molluscan Studies 71: 113–125.

Wägele, H. & Klussmann-Kolb, A. 2005. Opisthobranchia (Mollusca, Gastropoda) – more than just slimy slugs. Shell reduction and its implications on defence and foraging. Frontiers in Zoology 2: 3.

Wägele, H. & Willan, R. C. 2000. Phylogeny of the Nudibranchia. Zoological Journal of the Linnean Society 130: 83–181.

Woods, H.A. & Desilets, R.L. Jr. 1997. Egg-Mass gel of *Melanochlamys diomedea* (Bergh) protects embryos from low salinity. Biological Bulletin 193: 341–349.

9

Basommatophoran Gastropods

PHILIPPE JARNE, PATRICE DAVID, JEAN-PIERRE POINTIER, AND JORIS M. KOENE

INTRODUCTION

Basommatophoran snails essentially comprise all the pulmonate gastropods living in freshwater. The Basommatophoran order is monophyletic and encompasses five families: Acroloxidae, Chilinidae, Lymnaeidae, Physidae, and Planorbidae (including the Ancylidae)—the position of the Glacidorbidea and Amphibolidae remains discussed (see Bouchet & Rocroi, 2005; Klussmann-Kolb et al. 2008). Their phylogenetic relationships based on molecular evidence are given in figure 9.1. The Siphonaridae, classically considered as primitive Basommatophorans, are probably basal to the whole Pulmonata clade (Klussmann-Kolb et al. 2008) and will not be considered here. From the Index to Organism Names (ION in ISI database; www.isiknowledge.com), after correction for misspellings and redundancies, we get estimates of 125 genera and 900 species. Strong et al. (2008) provided a lower figure, with about 500 species. However both numbers are probably over-estimates of the actual number of genera and species, because fossil species are included in ION lists and many species names are synonyms. The latter is mainly caused by species descriptions based on shell morphology alone (while this variation is largely due to environmental and developmental variation). For example, *Physa* (*Physella*) *acuta* has been described as *heterostropha*, *integra*, *borbonica*, among others. Up to a hundred names have been given to species in the *Physa* genus, whereas the actual number probably lies between 15 and 20

(Wethington & Lydeard 2007). The real number of Basommatophoran species might therefore be closer to *ca.* 300. A plethora of genus names has also been used, with substantial variation over time (spanning over a century). As a consequence it is difficult to compare information among authors and papers. Within the Basommatophora, the families Lymnaeidae, Physidae, and Planorbidae represent 87% of the genera and 91% of the species (ION database). These include several genera of medical importance, serving as intermediate hosts of parasites that affect humans (*Lymnaea*, *Bulinus* and *Biomphalaria*). Unsurprisingly, most research including on reproduction has focused on these three families.

The Basommatophoran order is essentially made up of freshwater species, with a limited tolerance to salinity, and a distribution that ranges from the Equator to the Polar circles. Several species have extremely large distributions, sometimes as a result of recent extensions. A striking example of invasive species is *P. acuta*, now being cosmopolitan. Other species are restricted to very few sites (e.g., *Bulinus camerunensis*; see Brown 1994). Overall Basommatophora occupy a wide variety of habitats. For example, some species are able to live in very unstable ponds in which water is available for only a few weeks per year. Habitat size also ranges from very small ditches and ponds to large rivers and lakes, including human-made habitats such as irrigated areas. A consequence of occupying unstable habitats is wide variation in population size and recurrent bottlenecks, which is presumably

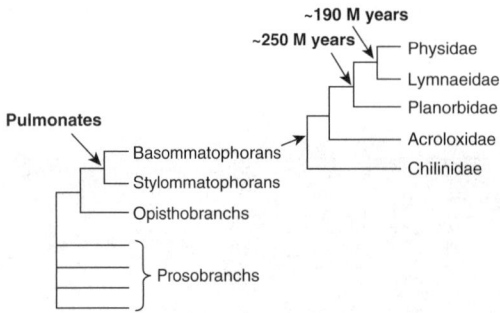

FIGURE **9.1** Phylogenetic relationships among Basommatophoran families (on the right) based on the molecular analysis (one mitochondrial and two nuclear genes) of Walther et al. (2006). The Planorbidae include the Planorbidae and the Ancylidae. The position of Amphibolidae and Glacidorbidea is not known. The sister group of the Basommatophora is the Stylommatophora, or land snails, together forming the pulmonates gastropods. The other large groups of gastropods are the Opisthobranchs and the Prosobranchs. Branch lengths are not at scale. Dating based on molecular data and the fossil record (C. Albrecht, personal communication).

a primary factor in the evolution of mating systems (Jarne & Städler 1995).

Much about the ecology, population genetics and dynamics, snail communities and distribution areas has been reviewed in Brown (1994) and Dillon (2000). Most species are diploid, and the few exceptions are all found within the *Bulinus* genus (Brown 1994). Generally, these snails have short life cycles and are relatively easy to maintain and breed, which allows for extensive control over external factors (e.g., food, light, and temperature) as well as experimental manipulations.

Basommatophora are simultaneous hermaphrodites and thus male and female at the same time. Male and female gametes are produced in a single hermaphroditic organ. Fertilization is internal, and cross-fertilization requires copulation. Self-fertilization is generally possible (reviews in Geraerts & Joosse 1984; Jarne et al. 1993; Jordaens et al. 2007). The reproductive morphology is rather complex and will be described in the next section. We will try to identify variability at various taxonomic levels as well as possible constraints or trade-offs in order to predict where selection is likely to play a role. A good deal of information is available on the nervous and endocrine control of

reproduction, and was essentially derived from the model species *Lymnaea stagnalis*. This will be described in the third section in connection with the structures previously described. The hermaphroditic nature of Basommatophora opens fascinating questions with regard to the evolution of reproduction described in the penultimate section, especially the evolution of hermaphroditism, self-fertilization, and sex polymorphisms. Perhaps even more interesting are the perspectives on sexual selection and sex allocation: the inter-gender conflict in gonochoric species translates into a double conflict, within and among individuals. These questions will be dealt with in the final main section.

STRUCTURAL AND FUNCTIONAL REPRODUCTIVE MORPHOLOGY

General Morphology of the Reproductive System

The general morphology and function of reproductive organs of several Basommatophoran species are well known, and a schematic diagram of the reproductive system is presented in figure 9.2 (partly based on the reviews of Geraerts & Joosse 1984; Jarne et al. 1993; Jordaens et al. 2007). The system can be divided into a hermaphroditic, male and female part. This general division is overall similar to that observed in both the sister-group Stylommatophorans (see Baur, this book; figure 10.1) and the Opisthobranchs, the other large group of hermaphroditic Gastropods (see Valdés et al., this book). The main difference lies in Stylommatophorans having a single gonopore and the Opisthobranchs displaying more variability with male and female parts varying from monaulic to triaulic. In Basommatophorans, the general structure of reproductive organs is known down to the genus and species level, since there traits are used for delimiting species. However, the description of the organs at the cellular and physiological levels is largely based on *L. stagnalis*.

Gametes and Hermaphroditic Organs

Gametes of both genders are produced in a single organ called the ovotestis. The ovotestis is divided into acini, each with an efferent duct, that join in a

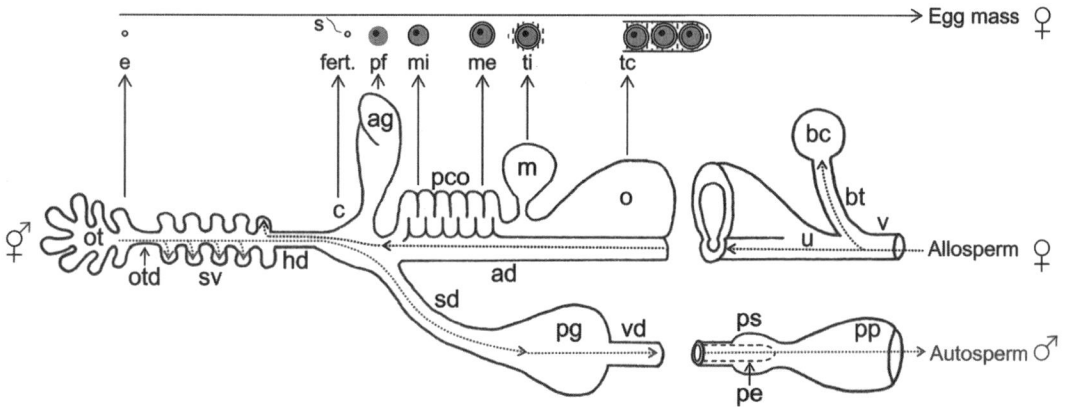

FIGURE 9.2 General morphology of reproductive tracts and gamete movement in Basommatophora. The reproductive system is divided in a hermaphroditic part (ot, otd, sv, hd, and c), a female part (ag, pco, ad, m, o, u, bc, bt, and v) and a male part (sd, pg, vd, pe, ps, and pp). The grey dotted lines and arrows indicate the route that the own autosperm take after being produced in the ovotestis and subsequently when they are transferred to a partner. The black dotted lines and arrows indicate the fate of sperm received from a partner (allosperm) within the female tract. The upper part of the figure indicates the fate of eggs from their production to egg capsules (abbreviations of different components are indicated), and arrows indicate in which gland each step takes place. More details in text. Abbreviations: ad, allosperm duct; ag, albumen gland; bc, bursa copulatrix (previously called spermatheca); bt, bursa tract (or pedunculus); c, carrefour (or fertilization pouch); e, egg cell (oocyte); fert., fertilization; hd, hermaphroditic duct; m, muciparous gland (or oviduct pouch); me, membrana externa; mi, membrana interna; o, oothecal (nidamental) gland; ot, ovotestis; ot, ovotestis duct; pco, pars contorta (or oviduct); pe, penis; pf, perivitellin fluid; pg, prostate gland; pp, preputium; ps, penis sheath; s, sperm cell (spermatozoid); sd, sperm duct; sv, seminal vesicles; tc, tunica capsulis; ti, tunica interna; u, uterus; v, vagina; vd, vas deferens.

central lumen (Joosse & Reitz 1969). Each acinus produces both oocytes (eggs) and spermatozoa (sperm). Starting from the germinal epithelial ring of each acinus, the Sertoli cells, which are the sperm nurse cells, form an epithelium that divides each acinus into a spermatogenic and oogenic compartment (De Jong-Brink 1969; Joosse & Reitz 1969). In other words, spatial specialization in a given gender occurs at the acinus level, but not at the whole organ level.

In the ovotestis, oocytes are continually produced during the reproductive season and several stages can be distinguished in oogenesis (e.g., De Jong-Brink & Geraerts 1982). The oogonia are first apposed and then enveloped by follicle cells (oocyte nurse cells). Next the developing oocyte forms clefts with the follicle cell, after which the oocyte matures into a ripe oocyte. Ripe oocytes are stored in the lumen of the ovotestis until ovulation.

Spermatogenesis starts with spermatogonia that develop into spermatocytes. Subsequently, five different stages of spermatid differentiation can be distinguished (De Jong-Brink et al. 1977). This differentiation occurs on Sertoli cells, and the late spermatids secrete the superfluous cytoplasm into these cells. The cytoplasmic bridges between the spermatids break and the now mature spermatozoa are spermiated (Rigby 1982). Basommatophora have only one type of sperm, fertilizing spermatozoa (also euspermatozoa or eupyrene; Healy 1983) with considerable length variation among species. For example, measurements range from 365 µm in *P. acuta* (Brackenbury & Appleton 1991) to 690 µm in *L. stagnalis* (Rigby 1982). Their structure has been described by, for example Healy (1983) and does not seem to include any remarkable features.

Ripe spermatozoa exit the ovotestis via the exiting duct of the lumen and are stored in the seminal vesicles (figure 9.2). From there they are either used for transfer to a partner, or resorbed once their quality starts degrading (e.g., Joosse et al. 1968). After copulation it will take a few days for the seminal vesicles to be refilled, as shown in

Bulinus globosus (Rudolph 1983) and *L. stagnalis* (De Boer et al. 1997a).

The hermaphroditic duct, leading away from the seminal vesicles, arrives at a junction where the female and male tracts separate. This junction is referred to as the carrefour or fertilization pouch (figure 9.2). Sperm storage has long been assumed to occur in this fertilization pouch (Geraerts & Joosse 1984), which is why it is sometimes labelled as receptaculum seminis. However, this suggestion seems an extrapolation from the situation in Stylommatophora rather than based on direct evidence. The latter have a well-defined allosperm storage organ, called spermatheca, attached to their fertilization pouch (Baur, this book). Tomé and Ribeiro (1998) did not find any evidence for sperm storage in their detailed histological study of the carrefour region of the freshwater snail *Biomphalaria tenagophila*. Also, in *L. stagnalis* the carrefour is neither used nor suited for allosperm storage, but is the ultimate site where fertilization can take place before egg packaging starts (Plesch et al. 1971; Koene et al. 2008).

The above indicates that the site of allosperm storage remains elusive, while Basommatophorans are able to store sperm from partners for several days to months (Jarne et al. 1991; Koene et al. 2008). Moreover, the amount of allosperm stored will be much smaller than that of autosperm. Hence, besides the question about the site of allosperm storage, how allosperm outcompete autosperm in outcrossing species is still unclear. Such priority of allosperm in the fertilization process may be due to a difference in activation and/or capacitation of auto- and allosperm, but the mechanism remains to be discovered.

Female Organs

Depending on the species, the oviduct can be further divided into several female structures (from albumen gland to vagina; figure 9.2). Most are involved in egg packaging along the following sequence, essentially known from studies in *L. stagnalis* (figure 9.2). After fertilization, the albumen gland first adds galactogen-rich perivitellin fluid to each egg (Plesch et al. 1971; Wijsman 1989). Each egg is then enveloped in two membranes, the membrana interna and externa, secreted by the posterior and anterior part of the pars contorta. The muciparous gland secretes a mucus that fuses the eggs together, the tunica interna, after which the tunica capsulis

surrounds the whole egg mass in the oothecal gland (Plesch et al. 1971). The egg mass is then deposited via the female gonopore and stuck to the substrate. The palium gelatinosum aids in sticking the mass to the substrate, and may be secreted by the glandular cells found in the gonopore (Plesch et al. 1971). The tract from the oothecal gland to the junction with the bursa tract is called the uterus, beyond this junction towards the gonopore the tract is called the vagina. The female tract ends in the female gonopore, which lies behind the separate male gonopore.

Sperm from a partner (allosperm; figure 9.2) are initially received in the vaginal duct, uterus, and part of the oothecal gland (Loose & Koene 2008). From there, a small portion of the sperm is transported along the allosperm duct, which is actually not a proper duct but rather an evaginated ridge running along the oviduct (therefore often called the spermoviduct). The bulk of allosperm, however, is not stored but ends up in the bursa copulatrix, which digests sperm. This gametolytic organ is sometimes confusingly-enough called the spermatheca or seminal receptacle; given its function such terms should be avoided.

Male Organs

From the carrefour region, the male system starts as a thin sperm duct that gradually widens as it reaches the prostate gland (figure 9.2). During copulation, autosperm is transported thought this duct and seminal fluid is added when the sperm pass through the prostate gland. The semen is then further transported via the vas deferens towards the penis. The penis lies within a penis sheath and is supported by a much larger preputium which ends in the male gonopore. The preputium is also attached with a number of preputium retractor muscles. The preputium and penis together are often referred to as the penial complex or phallus, which usually has a more or less externally-visible preputial gland (e.g., *L. stagnalis*; Plesch et al. 1971) and sometimes a penial stylet (e.g., *Gyraulus percarinatus*; Paraense 2003). The penial complex is presumably the most intricate part of the reproductive organs. The function of some parts remains elusive. For example, we have no proper explanation for the presence of penial stylets, although a comparison with the love darts found in Stylommatophora (Koene & Schulenburg 2005) or with traumatic insemination organs (e.g., Siva-Jothy 2006) might be illuminating.

Ontogeny

We know very little about the ontogeny of the reproductive tracts, and the most recent descriptions date back to Fraser (1946) in *L. stagnalis*. De Larambergue (1939) description for *Bulinus truncatus* focuses on the development of the male part, but overall provides a very similar description. Fraser (1946) distinguished three primary anlagen. The first of these primordia of the reproductive system appears after six days of development in the egg. This has a mesodermal origin and forms the hermaphroditic part of the reproductive tract, the ovotestis. At around the same time, originating from the lining of the mantle cavity, an ectodermal anlage starts to develop into the entire female tract, most of the male tract, and the hermaphroditic duct. Already within one day the latter joins the developing ovotestis. At this time the bursa copulatrix and its duct originate by splitting off from the developing female tract. The female gonopore also originates from the female tract, which simply grows through the body wall.

The future male gonopore is the origin of the third anlage, which appears at day 13 of development. This starts as an ectodermal evagination of the surface epithelium just posterior to the right tentacle (*i.e.*, the location of the male gonopore), and develops into the penis, preputium, and anterior part of the vas deferens. Once the first mature spermatozoa start appearing the hermaphroditic duct develops the pockets that become the seminal vesicles, which store the ripe sperm.

Understanding the ontogenesis of the genital tracts should help to evaluate possible evolutionary scenarios. For example, organs of different origin might evolve independently more easily than organs of the same origin. This could provide an explanation for phally polymorphism, that is, the presence/absence of the penial complex independently of the rest of the reproductive tract. In both *B. truncatus* and *L. stagnalis*, the penial complex indeed has a very distinct ectodermal origin, while the other parts of the reproductive system are mostly of mesodermal origin and all develop simultaneously (De Larambergue 1939; Fraser 1946).

Protandry, the expression of male prior to female reproduction, can be considered as a reflection of ontogeny, and has been reported in some Basommatophoran species. For example, Wethington and Dillon (1997) mentioned that individuals begin reproducing as male prior to reproducing as female in *P. acuta*. This is certainly an aspect that calls for more attention, since the evolution of protandry and of simultaneous hermaphroditism can be understood under the same conceptual framework (Charnov 1982).

Variation

Variation in primary sexual characters (PSCs) among Basommatophoran species has been used for taxonomic and/or medical purposes, but we know of no comprehensive picture at the scale of the whole group. A fairly large body of literature therefore describes PSCs, often including realistic drawings made using a *camera lucida*. Surprisingly enough, a good part of this literature is of little use, because of poor-quality drawings or uncertain scale, and mostly concerns New World species. In a first attempt to propose a more formal framework for describing variation in PSCs, we characterized the various organs and glands using a series of characters (presence/absence, size, shape; see table 9.1). This was possible in 62 species (2 Physidae, 13 Lymnaeidae, 47 Planorbidae including 25 *Biomphalaria* species; the full dataset is available from the authors upon request). A proper analysis of such data requires using a comparative approach in order to place the evolution of PSCs into a phylogenetic framework. Such a perspective has already been taken in terrestrial snails (Koene & Schulenburg 2005), but we are not aware of any such work in Basommatophorans. Thirty-two of these species have also been included in molecular phylogenies (2 Physoidea, 8 Lymnaeoidea, 22 Planorboidea including 14 *Biomphalaria* species; see DeJong et al. 2001; Morgan et al. 2002; Remigio 2002; Bargues et al. 2003; Walther et al. 2006; Jørgensen et al. 2007; Wethington & Lydeard 2007). However, conducting a comparative analysis is not possible because the available phylogenies are (i) either at family, or at lower levels, with no single phylogeny including all the main genera, even for the three main families; (ii) based on different marker genes, making it impossible to build a super-tree; and (iii) often incomplete within families or within genera.

We therefore resorted to a simpler, more descriptive analysis based on the meristic characters described in table 9.1. There is no striking Bauplan variation within the Basommatophorans with

TABLE 9.1 Discrete and continuous characters that can be used to characterize the various parts of reproductive tracts in Basommatophora. Variation was retrieved from the description of 62 species from the main three families (more details in text and in figure 9.4)

Section	Organ/gland	Characteristics
Hermaphroditic	Ovotestis	General shape: spherical or linear
		Type: diverticulated or acinous/lobular
		If diverticulated: number of diverticula
		If diverticula: number of bifurcations per diverticulum
	Ovotestis duct	Length between ovotestis and seminal vesicles
	Seminal vesicles	Type: tubular, with short lobes, with elongated lobes
		Size
	Carrefour	Area
Female	Albumen gland	Area
	Pars contorta	Type: straight, somewhat contorted, very contorted
		Length (between carrefour and muciparous gland)
	Muciparous gland	Present or absent
	Nidamental gland and uterus	Length (between muciparous gland and vagina)
	Bursa copulatrix	Length and width
	Canal of bursa copulatrix	Length
Male	Sperm duct (SD)	Length (between carrefour and prostate gland)
	Prostate gland	Type: thickening of SD, folded and linear, diverticulated and spherical with diverticula joining on a single point, diverticulated and linear
		If presence of diverticula: size of diverticula
		Area
	Vas deferens	Length (from prostate gland to beginning of penial complex)
	Penial complex	Preputium sheath: length and width
		Prepution: length and width
		Appendices (presence/absence): preputial gland on preputium (see *Physa*), preputial organ with pipe (see *Helisoma*), flagellum (see *Drepanatrema*), stylet (see *Gyraulus*), penial gland

regard to the general morphology of the reproductive systems. Shell chirality therefore does not seem to be influential here. Variation essentially concerns organs, glands and ducts shape, size or structure (e.g., number of lobes), or the occurrence of some minor structures (oviduct and vaginal gland). Qualitative, discrete variation in the organization and/or shape of the ovotestis, oviduct, prostate and penis complex mainly appears at the genus level within the Planorbidae, and among families (figures 9.3 and 9.4). There is also more resemblance between Lymnaeidae and Physidae than with Planorbidae, which is consistent with phylogenetic relationships (figure 9.1). The ovotestis of Planorbidae is made of well-individualized lobules (or diverticula) while that of Lymnaeidae and Physidae is not (figures 9.3 and 9.4). The oviduct has a relatively straight, elongated shape in most Planorbidae, but a contorted one in Physidae and Lymnaeidae. Within the Planorbidae, two groups of genera can be distinguished based on prostate structure (spherical *vs.* linear; figure 9.4). Some drawings also distinguish an oviduct and/or a vaginal gland, while others do not. It is not clear whether this reflects truly different types, because these so-called glands vary in size from barely visible to very salient, with no clear demarcation from the oviduct and vagina. It remains unclear whether this reflects genuine (e.g. temporal) variation in organ shape. Perhaps the most striking variation is in the structure of the penial complex, which is basically similar in the main three families, but exhibits a variety of distinctive annexes in particular genera or species, especially in the Planorbidae (figure 9.4). These include preputial glands (*P. acuta*), preputial organs acting as a "hold-fast" mechanical aid in copulation (*Helisoma*; Abdel-Malek 1952), flagellae, that is paired tubular structures attached to the penis sheath (*Drepanotrema, Plesiophysa*) and penial glands (*Plesiophysa*). Perhaps one of the most striking traits is the presence of chitinous stylets in the penis of *Gyraulus* species.

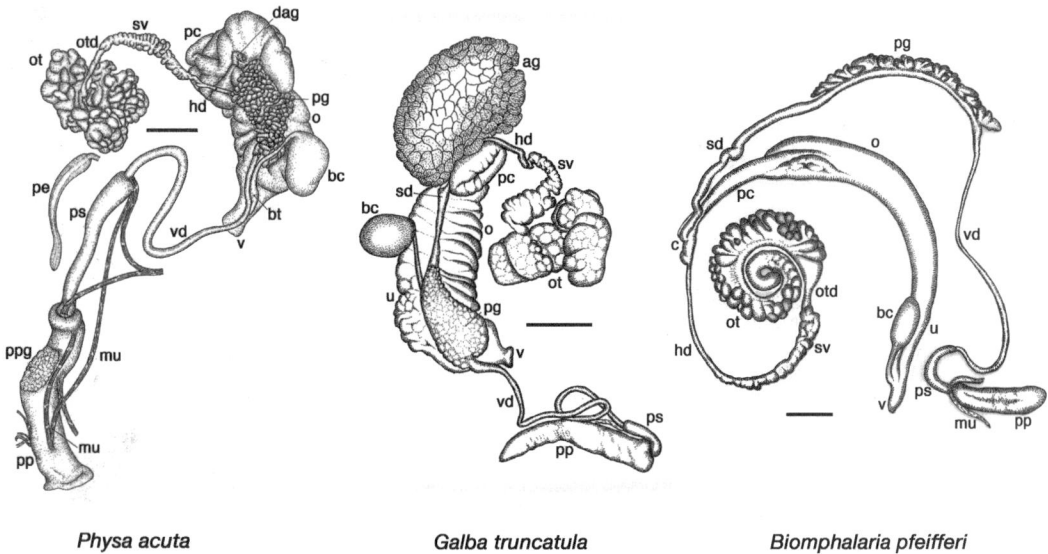

Physa acuta   *Galba truncatula*   *Biomphalaria pfeifferi*

FIGURE **9.3** Reproductive tracts in one species from each of the main Basommatophoran families (*Physa acuta* for the Physidae, *Galba* (*Lymnaea*) *truncatula* for the Lymnaeidae, and *Biomphalaria pfeifferi* for the Planorbidae). mu, muscles; ppg, preputial gland; dag, duct of albumen gland; other legends as in figure 9.2. Scale bars = 1 mm. The albumen gland has been removed in *P. acuta* and *B. pfeifferi* to obtain a better view of other organs. In *P. acuta* the penis is depicted on the left side of the penis sheath. Drawing of *P. acuta* is from Paraense & Pointier (2003). Other drawings are from authors.

We also observed lots of quantitative variation in the relative proportion of various organs (results not shown; table 9.1). The bulk of variation in duct length (hermaphroditic duct, oviduct, and vas deferens) seems constrained by the general shape of individuals: species with a large number of whorls and flat shells (Planorbidae such as *Biomphalaria*, *Drepanotrema*, *Gyraulus*, and *Helisoma*) have much longer ducts than species with rounded shells (Physidae, Lymnaeidae, and some Planorbidae such as *Bulinus*). However, other less constrained aspects, such as the length ratio of penis sheath to preputium and width/length ratios of both penis sheath and preputium, are highly variable among species within genera and often form the basis of accurate delimitation of species (see, e.g., Jackiewicz 1993). Of course a more formal analysis of this quantitative variation would be required.

A detailed example of anatomical variation at the genus level is available for the male part in European Lymnaeoidae (Jackiewicz 1993). A gradient of complexity in the prostate gland and penis complex was observed among species, and Jackiewicz (1993) used both to determine limits among subgenera and to build a tentative phylogeny, based on anatomical affinities. She assumed that evolution proceeds from simple (plesiomorphic) to complex (apomorphic) characters. As an example, the ancestral state of the prostate was assumed to be simple with no internal folds, as in *Omphiscola glabra*, and evolved towards an increasing number of folds (figure 9.5). A similar scenario was assumed for the number and folding patterns of the preputium, and for the area surrounding its opening near the penis sheath (presence and complexity of papilla, the so-called sarcobelum). The evolution of these traits can now be mapped on molecular phylogenies (Remigio 2002; Bargues et al. 2003, 2004; figure 9.5) which substantially differ from the morphological phylogeny. The *Radix* clade is monophyletic in both the molecular and morphological phylogenies. However, *O. glabra* is no longer basal in the molecular phylogenies, and the *Stagnicola* subgenus is no longer valid given that some species group with American species (*Catascopia*) while others fall within *Lymnaea sensu stricto*. As a consequence, the ancestral prostate likely had internal folds, and the ancestral preputium had two folds in this group. Simplification presumably occurred independently in the *G. truncatula* and *O. glabra*

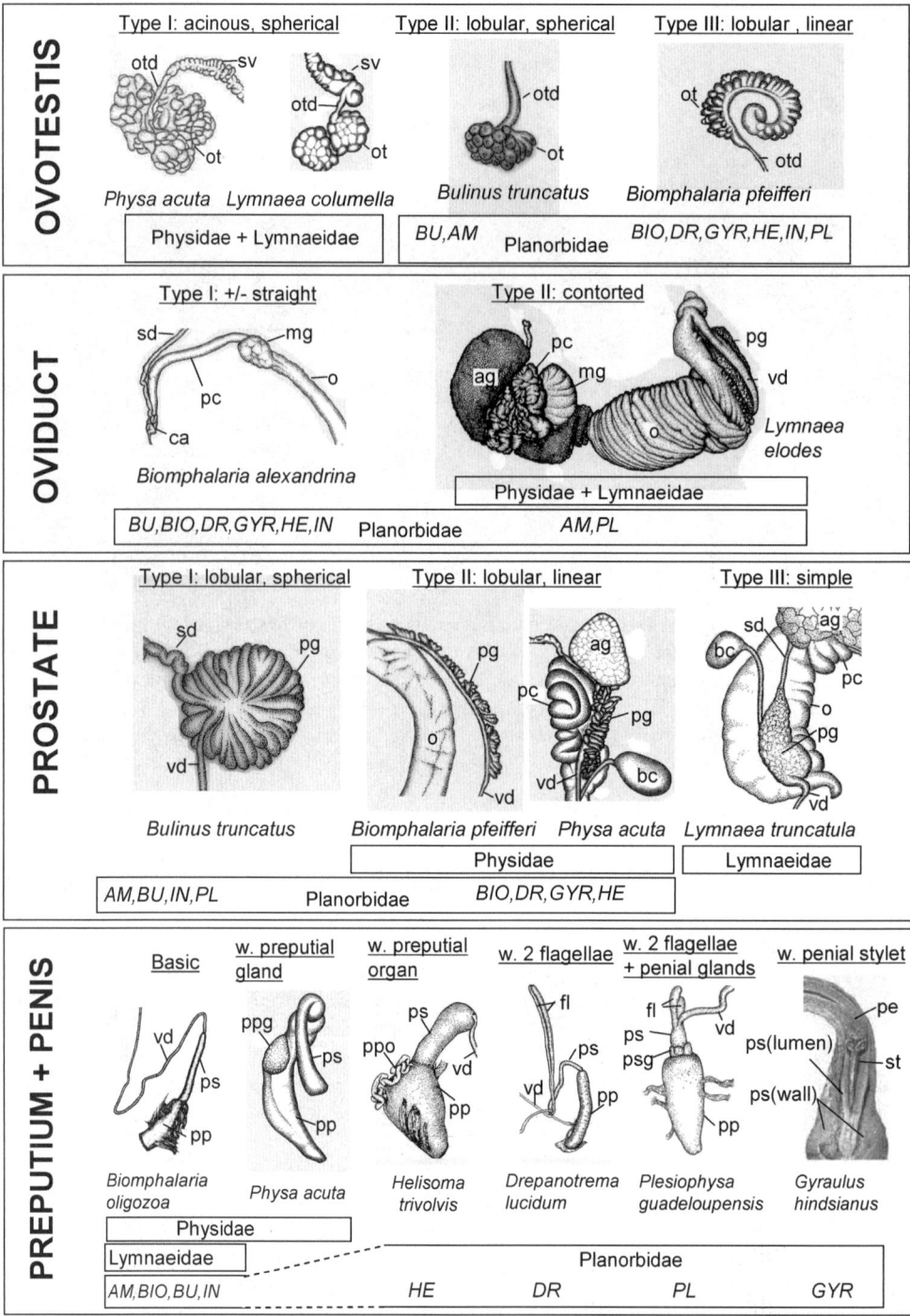

FIGURE 9.4 Family and genus-level variation in various parts of the reproductive tracts in representative genera of the three large Basommatophoran families. The families are indicated at the bottom of each panel, and the dotted lines in the 'Preputium + penis' panel means that no Planorbidae has a preputial gland. The genera studied include *Physa* (Physidae), *Lymnaea s.l.* (Lymnaeidae), and the following Planorbidae genera: *Amerianna* (AM), *Biomphalaria* (BIO), *Bulinus* (BU), *Drepanotrema* (DR), *Gyraulus* (GYR), *Helisoma* (HEL), *Indoplanorbis* (IN), *Pleisiophysa* (PL) for which we found appropriate representations of genital anatomy in the literature. Character states are illustrated by representative examples, not drawn to scale. fl, flagellae; ppg, preputial gland; ppo, preputial organ; psg, penis sheath glands; st, penis stylet; other legends as in figure 9.2. w. = with.

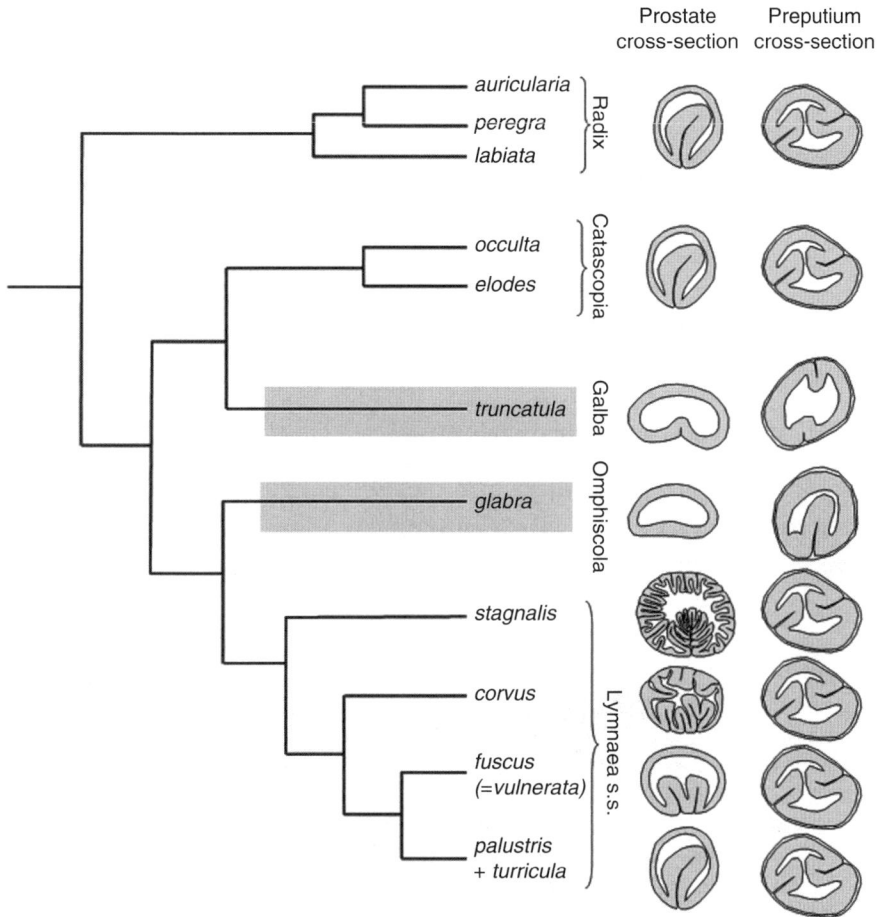

FIGURE 9.5 Evolution of prostate and preputium complexity in Lymnaeidae. A tentative and partial phylogeny of the genus *Lymnaea* (focusing mostly on European species) was reconstucted after Remigio (2002) and Bargues et al. (2003, 2004). Subgenera are as in Bargues et al. (2003). Prostate and preputium cross-sections are redrawn after Jackiewicz (1993). Shading indicates taxa known to be highly selfing.

lineages while others became more complex (such as *Lymnaea s.s.*). Interestingly both *G. truncatula* and *O. glabra* are small species, living in temporary, unstable habitats such as flooded grass fields, and are the only European species known to have very high self-fertilization rates. This suggests a functional link between self-fertilization and the reduction of the internal surfaces of the organs involved in male copulatory behavior. This initial analysis can be much improved, for example by using a more precise phylogeny, and by extending it to other groups.

Variation in PSCs also occurs within species, and its most striking—discrete—expression is aphally, a situation in which the penial complex or phallus fails to develop in some individuals. This will be considered later in more detail. Developmental differences due to different mating regimes, resulting in different prostate sizes, have been reported (Koene & Ter Maat 2004; Koene 2006). Nothing more seems known about discrete PSC variation. Intra-individual variation has been reported in relation to seasonal environmental changes, sexual activity, and/or parasitic infection. Examples are differences in prostate gland size due to time since mating (e.g., De Boer et al. 1997a), and differences in albumen gland size due to time since egg laying (e.g., De Jong-Brink & Geraerts

1982; Koene et al. 2006). Quite clearly, much has to be done here to understand the selective pressures acting on PSCs.

NERVOUS AND ENDOCRINE CONTROL OF REPRODUCTION

Egg laying

Female Neuro-Endocrine Organs and Substances

In *L. stagnalis*, the main neuro-endocrine center that controls female reproduction is the bilateral Caudo-Dorsal Cell cluster (CDCs) in the central nervous system (figure 9.6). CDCs release peptides into the blood that induce egg laying (Ter Maat et al. 1988). Eleven of the thirteen identified peptides,

including the egg laying hormone CDCH, are encoded on the same gene (figure 9.6; Vreugdenhil et al. 1988; Jiménez et al. 2004). The Dorsal Bodies (DBs; figure 9.6) are a second bilateral endocrine organ important for female development and oogenesis. Their action is mediated via an unidentified hormone that increases egg production, and stimulates growth and development of female accessory sex organs (Geraerts & Joosse 1975).

Cells morphologically similar to CDCs and DBs have been found in several Planorbidae and Lymnaeoidae (Roubos & Van der Ven 1987). Extracts containing CDCH also induce egg laying in different Lymnaeoidae, but cross-species tests show genus-specificity (Dogterom & Van Loenhout 1983). The foregoing indicates that a comparison of the CDCH gene within the Basommatophora should be at least as interesting as the comparison made with the distantly-related Opisthobranch

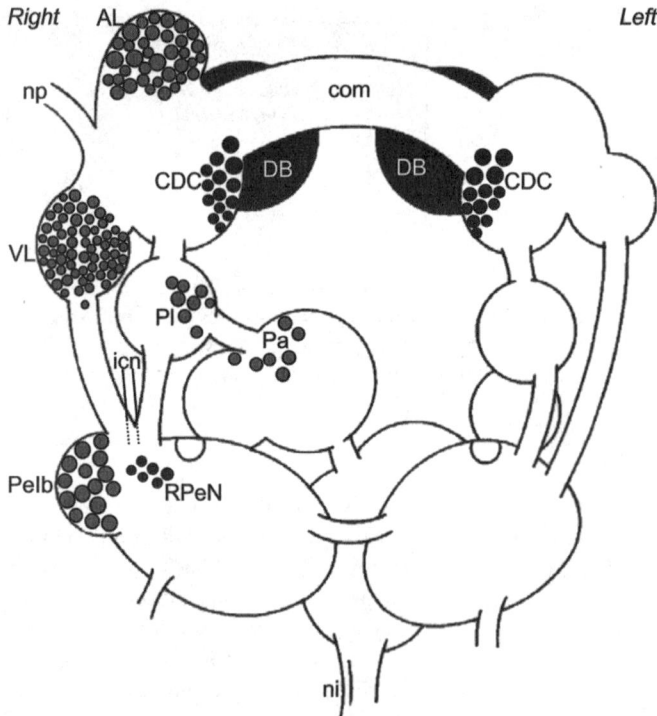

FIGURE 9.6 Schematic drawing of the ganglia of the central nervous system seen in *Lymnaea stagnalis* from the ventral side (front side is on top of drawing). The grey and black areas show, respectively the neuronal clusters that are involved in male and female reproduction. Abbreviations: AL, anterior lobe; CDC, caudo-dorsal cells; com, cerebral commissure; DB, dorsal bodies; icn, inferior cervical nerve; ni, nervus intestinalis; np, nervus penis; Pa, parietal cluster; PeIb, pedal Ib cluster; Pl, pleural cluster; RPeN, right pedal neurons; VL, ventral lobe.

Aplysia californica which revealed remarkable evolutionary conservation (Vreugdenhil et al. 1988).

Female Behavior

Most Basommatophora lay their eggs in egg masses that they fix to the substrate. A number of the behavioral components described below are found in several Basommatophoran species, but are mainly based on *L. stagnalis*. The key peptide for egg laying is CDCH, because its injection into the blood suffices to trigger egg laying (Ter Maat et al. 1987). In spontaneous egg laying all peptides encoded on the CDCH gene are released during the *resting phase* (figure 9.6; Ter Maat et al. 1986). The ovulation hormone CDCH inhibits the right pedal motor neurons (figure 9.6), which is probably why locomotion stops during this phase. CDCH also initiates ovulation, and this release of ripe eggs is followed by fertilization and packaging by the female accessory glands.

The passing of eggs through the tract is sensed by ciliated cells that send nervous signals to the central nervous system via the intestinal nerve (figure 9.6), triggering the *turning phase* (figure 9.7A). During this phase shell turning and locomotion are initiated by the increased activity of the motor neurons that project into the inferior cervical nerves (figure 9.6) that control the columellar muscle, involved in shell turning (Hermann et al. 1997). Buccal rasping is performed to clean the substrate for proper attachment of the egg mass and continues as long as eggs pass through the female tract (Ferguson et al. 1993). *Oviposition* starts when the egg mass emerges from the female gonopore and is fixed to the substrate (figure 9.7A). When the egg mass has been deposited, the animal crawls along the mass (*inspection phase*, figure 9.7A) before leaving it behind.

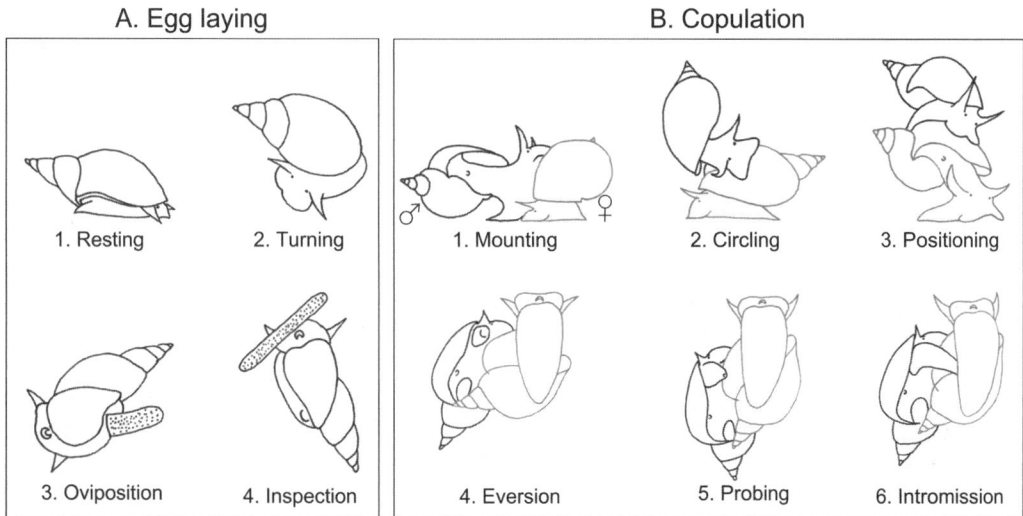

FIGURE **9.7** Egg laying and copulation in *Lymnaea stagnalis*. The drawings show the different stages of both reproductive behaviors. The egg laying behavior is redrawn from Ter Maat et al. (1987). (A) Egg laying starts with a resting phase of on average 40 min, during which locomotion stops, the shell is held still and slightly pulled forward over the tentacles and no rasping occurs. During the turning phase, lasting around 60 min, locomoting starts again, the shell is turned back and forth by 90°, and the surface is rasped. Oviposition usually lasts about 10 min during which the animal continues rasping, but stops shell turning and hardly moves. During inspection the animal crawls along the mass for up to 30 min without rasping or shell turning. (B) Copulation starts with mounting on the shell. For some 5 min, the sperm donor then performs counterclockwise circling, first towards the shell's tip and then towards its margin. It then takes on average 17 min from positioning to intromission, during which eversion of and probing with the preputium occur. Once intromission is reached an average insemination takes 35 min. The sperm donor is displayed in black, the sperm recipient in grey.

Courtship and Copulation

Male Neuro-Endocrine Organs and Substances

In *L. stagnalis*, the main centres controlling male reproductive behavior are situated on the right side of the central nervous system (figure 9.6), just as the male organs and gonopore. The main clusters of neurons are the right anterior lobe, right ventral lobe, right pedal Ib cluster, and some more dispersed cells in the right parietal and pleural ganglia (De Boer et al. 1996). Neurons in all these clusters project into the penial nerve, the only nerve innervating the male copulatory organs. Work in *Helisoma trivolvis* revealed very similar neuronal clusters (Young et al. 1999). The bilateral anterior lobes (figure 9.6) of the cerebral ganglia usually show striking right–left asymmetry. In dextral species, the right lobe is bigger (Koene et al. 2000) and in sinistral species it is the left one (e.g., *B. glabrata*; Lever et al. 1965), in concordance with the location of the male organs. The peptide APGWamide is expressed in virtually all right anterior lobe neurons and in some on the left. APGW is also found in *B. truncatus* (De Lange & Van Minnen 1998) and actually plays a key role in male copulatory behavior of gastropod molluscs in general (De Lange & Van Minnen 1998; Koene et al. 2000). APGW is usually co-expressed in anterior lobe cells with one of the other involved neuropeptides (De Lange et al. 1998). The ventral lobe (figure 9.6) is very prominent in the right cerebral ganglion of a dextral species like *L. stagnalis*. FMRFamide, as well as several other peptides encoded on the same exon, are expressed in the ventral lobe neurons (De Lange et al. 1998).

Male Behavior

Mating behavior is rather consistent across investigated species, as for example in *Bulinus octoploidus* (Rudolph & White 1979), *P. acuta* (DeWitt 1991), and *L. stagnalis* (Van Duivenboden & Ter Maat 1985). Courtship behaviour starts with mounting, circling, and positioning on the shell (figure 9.7B). Once the animal has found the proper position, the partially everted preputium becomes visible as a white bulge behind the right tentacle (left in sinistral species; figure 9.7B). *Eversion of the preputium* requires relaxation of the male gonopore and preputium retractor muscles. Many endocrine substances affect contractions of these muscles.

In *L. stagnalis* injection of APGW into the blood causes penial eversion (De Boer et al. 1997b). In *B. glabrata*, the presence of FMRF or serotonin-uptake inhibitors (like prozac) in the water cause eversion (e.g., Fong et al. 2005). The different muscle layers of the preputium, probably controlled by the ventral lobe and pleural and parietal motor neurons and their peptides, allow for the fine-tuned movements essential for finding the female gonopore (De Lange et al. 1998).

During *probing*, the fully-everted preputium probes under the lip of the partner's shell in search of the female gonopore (figure 9.7B). This most likely requires a sensory mechanism ensuring the correct position of the preputium prior to eversion and intromission of the penis. The most obvious candidates for this are the sensory neurons at the distal tip of the preputium, which could also control minor positional adjustments necessary for penis intromission (De Lange et al. 1998).

Courtship ends, and copulation starts, with *intromission* of the penis into the female gonopore (figure 9.7B). Once intromission is reached, sperm and seminal fluid are transferred into the vaginal duct of the recipient. Semen includes prostaglandins, possibly the egg laying hormone CDCH (Van Minnen et al. 1992), and several other peptides (Koene, Nagle & Ter Maat, unpublished data). These substances may be used to activate and nourish sperm but could potentially also manipulate partners, as detailed later. The vas deferens then shows rhythmic contractions that start in the pacemaker area near the prostate gland (De Lange et al. 1998). Several peptides seem involved in the regulation of these contractions (El Filali et al. 2006). Interestingly, the amino-acid sequence of one of the involved peptides, Conopressin, resembles vertebrate Vasopressin/Oxytocin, which are also involved in ejaculation (Van Kesteren et al. 1995).

Once insemination has finished the penis and preputium are retracted. Penis retraction is probably regulated by relaxation of its supporting preputium muscles and a decrease in hydrostatic pressure (De Jong-Brink 1969). For the much larger preputium, retraction seems to require the preputium retractor muscles, because when this set of muscles is cut retraction becomes impossible (Ter Maat & Koene, unpublished results). Their contraction is induced and modulated by many substances, most importantly APGW relaxes them and antagonizes the contractile effect of serotonin (e.g., Fong et al. 2005). This contributes to eversion and retraction of the male copulatory organ.

As with the identified egg laying peptides, several genes producing the peptides involved in the male function have been identified in *L. stagnalis*. However, so far these have not been compared between different species. This would be extremely interesting, especially given that the use of some of these substances seems highly conserved in distantly-related gastropods (Koene et al. 2000).

Motivation to Mate

Female Sexual Drive

Most freshwater pulmonates perform one sexual role within a copulation. Because sperm recipients are usually rather inactive when being mounted, the general assumption has long been that they are continually receptive as females (Van Duivenboden & Ter Maat 1985). However, animals may not always be prepared to receive sperm, and this may especially be the case when self-fertilization is favored, a rather common situation among Basommatophora. Studies that specifically looked at recipient behavior are limited to two species, *Physa gyrina* and *P. acuta* (*heterostropha*), and found that individuals have ways of discouraging a mating partner once it has mounted the shell, for example by swinging the shell back-and-forth or biting the phallus (DeWitt 1996; Wethington & Dillon 1996). Similar behaviors are found in *L. stagnalis* (Hoffer & Koene, unpublished data) and such potential insemination-avoidance behaviors could be used as indicators of the willingness to be inseminated. This has already been done successfully in studies on *P. acuta,* which revealed that such behaviors become more pronounced in matings between kin, indicating that mate recognition may be used to avoid inbreeding (Facon et al. 2006; McCarthy & Sih 2008). In addition, one might expect such behaviors to be more pronounced in species that preferentially self-fertilize. Female drive also seems to be influenced by the infection status and genetic resistance to trematode parasites in *B. glabrata* (Webster & Gower 2006). Its physiological basis, especially in connection with the regulation of egg laying, is still unknown.

Male Sexual Drive

Freshwater snails are not always motivated to donate sperm, and sexual isolation is often used to increase mating activity for experiments (reviewed in Koene & Ter Maat 2005). If both individuals are motivated to mate as males, the individual that was sexually isolated longest will act as male first; afterwards, role alternation can take place so that both individuals get to mate in both roles sequentially (Van Duivenboden & Ter Maat 1985; Wethington & Dillon 1996; Koene & Ter Maat 2005). The increase in size of the prostate gland in *L. stagnalis*, which produces the seminal fluid, is the permissive trigger for performing the male role (De Boer et al. 1997a). As a consequence, this species only mates as a male when enough seminal fluid is present (Koene & Ter Maat 2005).

The increase in size of the prostate gland is detected in the central nervous system via a small branch of the penial nerve (figure 9.6; De Boer et al. 1997a), and may be mediated by neurons in the connective tissue surrounding the prostate gland (De Lange et al. 1998; De Lange & Van Minnen 1998). Interestingly, cutting this nerve results in complete elimination of the male function (De Boer et al. 1997a). It is likely that a similar regulation of male motivation occurs in other Basommatophora, but this awaits investigation.

An additional factor that seems important for male motivation is the partner's identity. After copulating once in the male role, thus having partially depleted the prostate gland, individuals are able to inseminate again, but only do so when the partner they encounter is novel (Koene & Ter Maat 2007). The latter is in accordance with predictions from sperm competition theory, as is the finding that more sperm are donated to virgin individuals (Loose & Koene 2008). Moreover, as with female drive, male motivation might be affected by partner kinship and infection status (Facon et al. 2006; Webster & Gower 2006; McCarthy & Sih 2008).

THE EVOLUTION OF REPRODUCTIVE SYSTEMS

Mating Systems

Long-Term Stability of Hermaphroditism

Separate sexes (dioecy or gonochorism) is presumably the ancestral state in Gastropods, and hermaphroditism has evolved recurrently (Heller 1993). All Pulmonate species—Basommatophorans and Stylommatophorans—are hermaphroditic (e.g., Jordaens et al. 2007), meaning that hermaphroditism has been stable in this group for the last 300 to 400 million years. This impressive temporal stability, which can be found in other animal groups like the trematodes (Jarne & Auld 2006), certainly calls for an explanation. It is unlikely that this is associated with the colonization of freshwater habitats; numerous counter-examples could indeed be found in marine Gastropods, as well as in other animal groups. It is also difficult to envisage explanations relating to demography or population structure, since Basommatophorans show wide variation in this respect. Our most general theory of the transition between hermaphroditism and separate sexes would probably be that hermaphroditism is maintained in Basommatophorans as a result of optimal sex allocation between male and female functions (Charnov 1982; Greeff & Michiels 1999). However, this is of little help when it comes to understanding what constrains sex allocation. A functional explanation might be that this transition involves several steps that might not be easily completed. One reason is that the female and male tracts are not fully separated (figure 9.2), as they are in other hermaphroditic animals and in Angiosperms. In the latter situation, evolving separate sexes can be envisioned through a two-step process with gynodioecy as an intermediate step (Charlesworth & Charlesworth 1978). In Basommatophorans, at least four steps are required since the evolution of females from an hermaphrodite requires both the loss of the male organ and of the male part of the gonad (and a symmetrical evolution for males). Our current knowledge of both the ontogeny and molecular basis of reproductive tracts is far too limited to be more specific, and this is certainly a field in which much needs to be done. Genes involved in the transition between separate sexes and hermaphroditism have been characterized in *Caenorhabditis* (Braendle & Félix 2006), and this might be a fruitful avenue for snail research.

Is Selfing Associated to Specific Traits?

Self-fertilization and its evolution is arguably the most widely studied topic with regard to mating systems in Basommatophorans. As a consequence our understanding of the evolution of selfing rates is most advanced in this animal group (Jarne & Auld 2006). Basommatophorans can indeed both self- and cross-fertilize (Geraerts & Joosse 1984; Jarne et al. 1993). However the distribution of selfing rates among species is U-shaped; in other words, selfing rates at the species level are either very high, or very low in the *ca.* 20 species in which thorough studies have been conducted based on genetic markers (Jarne & Auld 2006; authors' unpublished data). *L. stagnalis* might be the only exception with populations exhibiting a wide range of selfing rates (Puurtinen et al. 2007). The evolution of selfing rate has not been set in a phylogenetic framework in Basommatophora, but we do know that the mating system can switch from one state to another within families and genera (Escobar 2008; see figure 9.5), and both highly inbreeding and outbreeding species are found in the genera *Bulinus*, *Biomphalaria*, *Lymnaea* (*sensu lato*), and *Physa*. Two closely related pairs of species of particular interest are *B. glabrata* (outcrosser)/*B. pfeifferi* (selfer) and *P. acuta* (outcrosser)/*Aplexa (Physa) marmorata* (selfer), because they are the best candidates for studying evolutionary transitions in selfing rates and their consequences. In both cases, selfing is the derived conditions (Escobar 2008).

Several hypotheses have been proposed to explain the evolution of selfing rate (reviewed in Jarne & Charlesworth 1993; Goodwillie et al. 2005). Most genetic models posit that the intrinsic genetic advantage of selfers is counter-balanced by inbreeding depression. These models predict an association between high selfing rates and low inbreeding depression, and vice versa. This idea, however, does not account for why residual selfing (respectively outcrossing) is maintained in outcrossing (respectively selfing) species. The reproductive assurance hypothesis might be of help here: when the probability of finding a partner is low, it might be worth waiting for some time before switching to selfing. This is the essence of the waiting time model (Tsitrone et al. 2003a, b). A review of empirical work in Basommatophora (13 species) indicates that low selfing rates are associated with high inbreeding depression and long waiting time, as expected from theory (Escobar 2008). Evidently, other selective forces might be involved as well (e.g., local adaptation), but this still awaits a thorough theoretical as well as empirical treatment.

Botanists have long recognized that the evolution of selfing is associated with the evolution of specific reproductive traits (e.g., morphological). In freshwater snails, this has been referred to as a 'selfing syndrome' (Doums et al. 1996).

We already mentioned the association between selfing rate, inbreeding depression, and waiting time. Outcrossing is also associated with much more frequent copulations than selfing (Tian-Bi et al. 2008). Unfortunately, it is not yet possible to conduct a thorough analysis of the relationship between selfing rate and PSCs, because the number of species in which both aspects have been quantified is rather limited. The comparison between *Lymnaea* species suggests a functional link between increased selfing and reduction of the internal surfaces of the penial complex (figure 9.5). The reduction in allocation to the male function as a consequence of increased selfing rates (Charnov 1982) has not been tested, neither the idea that substances involved in sexual selection and sperm competition are not expressed in selfing species.

Life Without a Phallus

Aphally has evolved recurrently in Pulmonates, but seems limited in Basommatophora to a few species of the *truncatus/tropicus* group of the genus *Bulinus* (Doums et al. 1998). Phally polymorphism is the only discrete variation in reproductive tracts that has been studied in detail, and much of what we know has been derived from *B. truncatus*. The frequency of aphallic snails is high in natural populations of this species, though no purely aphallic population has been detected. Aphally has a complex basis involving both genetic and environmental factors, and can be thought of as a binary trait with complex genetic determination (Ostrowski et al. 2000). A general (threshold) model has been proposed to fit this situation, but much has to be done for a more complete understanding of the genetical and physiological bases of aphally. This is of importance, because the production of 'aphallic mutants' might provide hints about how reproductive tracts evolve in Basommatophorans. As mentioned earlier, the phallus differs in ontogenetic origin from the other parts of the reproductive tract. This coincides with research showing that phally status is determined early in life, before egg hatching (Ostrowski et al. 2002), and suggests that aphallic individuals should readily occur in natural populations.

Another evident, open question is how phally polymorphism is maintained in natural populations. One possible explanation is derived from the theory of sex allocation: resources that are not invested into male organ development, maintenance

and use might be diverted towards fecundity or other functions related to fitness (Doums et al. 1998). Strikingly, even extremely controlled experiments (e.g., in terms of genotypes) failed to detect a difference in life-history traits between aphallic and regular, euphallic snails (Ostrowski et al. 2003). A possible explanation is that these experiments were conducted on isolated individuals, and thus did not fully account for costs associated with the male function. This also calls for deeper knowledge on how energy is allocated to the two sexual functions.

Sex Allocation and Sexual Selection

Sexual selection favours traits that enhance reproductive success and can result in sexual conflicts between partners: what is a good for one partner might be harmful to the other. Such conflicts can lead to extravagant and costly behaviors and traits (Arnqvist & Rowe 2005). This issue has attracted the attention of evolutionary biologists for decades, and has been approached experimentally essentially in gonochoric species with work on, for example, sperm competition and sexual conflicts (review in Andersson 1994; Arnqvist & Rowe 2005). In contrast, still little is known about sexual selection in hermaphroditic species. One reason is that it has long been thought that sexual selection could not act in hermaphrodites (see Charnov 1982; Leonard 2006). Recent studies have proven that various expressions of sexual selection do occur in hermaphroditic animals, including sperm competition, fast evolution of PSCs or elaborate courtship (Koene & Schulenburg 2005; Anthes et al. 2006; Koene et al. 2007). It is true though that sexual selection should be considered from a different perspective in hermaphrodites, even when self-fertilization is not possible. The reason is that each individual is both male and female, and selection favors equal sex allocation in both genders. Sexual selection among individuals is therefore strongly associated to sex allocation within individuals.

From theoretical work on sexual selection and sex allocation in hermaphrodites (Charnov 1982; Leonard 2006) arises one key-question: is there a preferred gender, and what kinds of conflicts are generated by such preference? Theoretical approaches have been driven by the idea that there is actually a preferred gender, but some recent studies suggest that preference may vary over time as a function of a variety of parameters, including previous sexual experience, ecological conditions and infection status (Facon et al. 2006; Webster & Gower 2006; Koene et al. 2007). For example, large individuals may tend to prefer the female role (Ohbayashi-Hodoki et al. 2004), while individuals that have been isolated for some times prefer the male role (De Boer et al. 1997a; Facon et al. 2007). Moreover, the mechanisms underlying these processes have remained largely unexplored. We highlight here the most important aspects.

Sex Allocation

Sex allocation is the way in which individuals allocate their resources to their male and female functions. This is an issue that has been studied in plants, but not much has been done in animals, and freshwater snails are no exception. The theory predicts equal sex allocation in the male and female functions (Charnov 1982), but approaching sex allocation experimentally is not easy, especially in animals. Fixed costs of PSCs (e.g., building of organs) might be evaluated through weight or nitrogen contents (Koene et al. 2006), but the continuous production of gametes and fluids in the breeding season, sometimes over several years, and repeated courtship makes such costs hard to estimate. A related issue is how resources can be redirected from one function to another. Some hints can be derived from experimental work in *L. stagnalis*: cutting the nerve registering the filling-state of the prostate gland results in complete elimination of male mating behavior. In pairs this treatment roughly doubles egg production. This may indicate that the resources that are no longer allocated to the male function are reallocated to the female function (De Visser et al. 1994; Koene 2006). Alternatively, given that animals were paired within treatments, the non-copulants may have laid more because suppression of egg laying was absent (Van Duivenboden et al. 1985).

Courtship

Attracting partners and courtship behavior are essential aspects of sexual selection and sex allocation. Not much is known on how partners detect each other, especially when population density is low, but chemical signals must be involved. Courtship has already been described in some Basommatophoran species. Mating via shell mounting by one individual seems to be the rule in these snails, although occasional simultaneously reciprocal matings have been reported (reviewed in Jordaens et al. 2007). It is unclear why other mating positions have not evolved, especially given that a face-to-face position is observed in a large number of terrestrial Pulmonates. Moreover, for the latter group it has been proposed that shell shape (high- or low-spired shells) and mating positions (shell mounting and face-to-face) are correlated (Davison et al. 2005). If the aforementioned is true, and given the wide variation in shell shapes among Basommatophorans, we might also expect variation in mating position. This clearly requires a comparative analysis at the scale of Pulmonates, or even Gastropods, and more detailed studies of mating behavior, especially in poorly studied Basommatophoran genera. More generally, it should be very fruitful to compare the details of mating behavior within and among species and relate this to the selfing rate, the number of partners, and the structure of the reproductive tracts. Our analysis already reveals wide variation in penial complex morphology.

Copulating, Manipulating Partners and Sperm Competition

The prime function of copulation certainly is to transfer sperm, but this also provides an excellent opportunity to manipulate sexual partners through seminal products. The impact of receipt of one or more ejaculates has been reported in a number of studies. For example, compared to snails that receive one ejaculate, virgins start laying eggs significantly later in outcrossing species. The difference in the onset of egg laying between these two treatments has been defined above as the waiting time. This accelerated onset of egg laying could be caused by the presence of allosperm, but it has also been suggested that a bioactive substance (*i.e.*, allohormone: Koene & Ter Maat 2001) may be involved (Koene & Ter Maat 2004; Koene et al. 2006). That the frequency of semen receipt is also important beyond the onset of egg laying is clear, but whether this results in decreased or increased egg laying is still undecided (respectively Van Duivenboden et al. 1985; Koene et al. 2006). The presence of semen (sperm and/or seminal fluid) can result in resource sub-optimal allocation in the recipient, hinting at a potential sexual conflict between the recipient and the donor over investment in eggs (reviewed in Koene 2006).

Basommatophoran snails are able to store sperm from partners, hence there is scope for sperm competition. We already mentioned that there is much to do to determine the location of allosperm storage, the physiological basis of interaction between different allosperm, and to explain why allosperm outcompete autosperm in outcrossing species but not in selfing species. It is possible that those seminal substances mentioned above play a role. The few studies documenting the outcome of sperm competition in simple situations all found roughly equal paternity for the first and second donor (e.g., Koene et al. 2008).

Egg Laying as an Extension of PSCs

Investment in the female function seems to amount mainly in the choice of a suitable egg laying site and the provisioning of eggs. There is only one example of parental care in Basommatophorans: two *Protancylus* species retain their egg strings underneath their mantle until hatching (Albrecht & Glaubrecht 2006). In all other species the structure and size of capsules seems therefore of primary importance. The capsules are composed of mucopolysacharides, polysaccharide–protein complexes and mucoproteins, and these different types of secretions can also be recognized in the different parts of the female tract (e.g., Duncan, 1975). As explained earlier, the egg mass consists of a number of different layers produced by the female glands, and the number of layers surrounding each egg differs across species (Bondesen 1950). Only in the Physidae and Lymnaeidae are the eggs enveloped by a second external membrane (produced by the anterior muciparous gland; figure 9.2), which is reflected in the complexity of the female tract (e.g., Duncan 1975). It is unknown whether mothers transfer substances to their eggs and egg capsules beyond nutriments, for example as protection against microbes.

Other important differences between species are the number and position of eggs in egg capsules (e.g., Bondesen 1950). The Physidae and Lymnaeidae lay masses containing several layers of eggs, while the Planorbidae masses contain a single layer of eggs and are thus much flatter in shape and more adherent to egg laying substrates (e.g., Bondesen 1950). There is also large variation in the number of eggs per capsule among species, genera and families. Capsule size also seems generally smaller in selfing than in outcrossing

individuals from outcrossing species (e.g., Jarne et al. 1991).

Within species, egg laying seems to be influenced by a number of factors. For example, adult *L. stagnalis* can lay egg capsules at a frequency of at least one per week, and such capsules contain between 50 and 150 eggs depending on the individual's body size (Koene et al. 2007). Another determinant of number of eggs per capsule is the time since last oviposition, because ripe oocytes accumulate in the ovotestis (Ter Maat et al. 1983) and perivitellin fluid accumulates in the albumen gland (Koene et al. 2006). *L. stagnalis* seems to have a preference for laying eggs during the day, but other species have been found to predominantly lay eggs during the night (*Planorbarius corneus* and *H. trivolvis*; Cole 1925). The laying frequency also varies widely among species (Bondesen 1950; Dogterom & Van Loenhout 1983). Other factors that modify egg laying are short light cycles, starvation, old age, dirty water (Koene et al. 2007) and parasitic infection (Bayne & Loker 1987). For example, egg laying in *L. stagnalis* can be triggered by a transfer from dirty to clean water. Although this is known as the clean water stimulus, the effect is actually caused by a combination of clean water, clean surface, and higher oxygen (Ter Maat et al. 1983). In sum, frequency of laying and number of eggs per capsule are affected by numerous internal and external factors, which needs to be kept in mind when studying egg laying.

CONCLUSIONS AND PERSPECTIVES

This overview indicates that there is still much to do for describing PSCs and their variation, as well as for providing functional and evolutionary explanations to this variation. We would like to propose some paths along which our knowledge might progress.

- Perhaps the most immediate goal would be to get a broader phylogenetic picture about the structure of reproductive tracts. Given the rather low number of Basommatophoran species, getting an appropriate anatomical description of these tracts for all species (or at least one for each genus) does not seem out of reach. This certainly calls for taxonomical clarification in some groups.

- If we are able to simultaneously build a phylogeny of Basommatophora (again a reasonable objective), this would allow for conducting a comparative analysis of PSCs, as done for example in Stylommatophora (Koene & Schulenburg 2005). This is of prime importance for determining the direction of PSC evolution, eventually in relation to characteristics such as selfing rate or relative allocation to various life-history traits. Quite clearly, a similar comparative perspective could be taken with substances involved in reproduction (e.g., neuropeptides) and would allow for cross-linking this with detailed morphological and behavioral data. In some instances, the genes involved in reproduction have begun to be pinpointed and comparison could start here. However, a comparative analysis of such genes will be greatly facilitated when the first fully sequenced Basommatophoran genome will be complete. *B. glabrata*, a biomedical model species because it is the intermediate host for the parasite that causes bilharziasis, is currently being sequenced. Given that *P. acuta* and *L. stagnalis* are both important model species, for respectively evolution and neuro-endocrinology, their genome sequences should be next in line.
- Vexing questions about reproductive tracts should be vigorously tackled, including the sites of allosperm storage, of self-fertilization and cross-fertilization, the role of several male accessory organs like stylets and preputial glands, the influence of seminal fluids on sexual partners, and the processes by which allosperm outcompete autosperm in outcrossing species (and not in selfing species).
- Beyond advocating a broad phylogenetic approach, we propose to focus some efforts on a few species, in such a way that the three main families are represented. *P. acuta* is an obvious candidate, because of its broad geographic distribution and relatively short lifecycle under laboratory conditions. It might be studied in concert with the closely related *A. marmorata* which, contrary to *P. acuta*, is a selfer. The *B. glabrata/B. pfeifferi* pair might be the counterpart in Planorbidae. *L. stagnalis* should remain the star of neurological and physiological studies while such knowledge should also be extrapolated to other species and combined with evolutionary and ecological approaches.

These are the basic assets upon which we will be able to address the evolutionary questions mentioned above on firmer ground. The stability of hermaphroditism, the evolution of selfing, of sexual polymorphisms and of sex allocation, as well as the intensity of sexual selection, remain big questions in evolutionary biology. Since Basommatophora are easily manipulated in the laboratory, a good part of empirical answers to these questions could come from this group of animals. Another extremely important question that was not addressed here is speciation. There is ample literature linking PSCs and reproductive isolation (see Coyne & Orr 2004). However essentially nothing has been done in Basommatophora. In connection to this, self-fertilization has been proposed as a mechanism favoring speciation (Antonovics 1968), and this might be evaluated in Basommatophora.

Acknowledgements The authors thank J. Leonard for inviting them to write this chapter and (especially) for her patience, J. Auld, J. Leonard, A. Wethington and an anonymous referee for critical reading, C. Albrecht for discussions and the French "Centre National de la Recherche Scientifique" (PICS program) for supporting exchanges that led to this chapter.

REFERENCES

Abdel-Malek, E. T. 1952. The preputial organ of snails in the genus *Helisoma* (Gastropoda: Pulmonata). American Midland Naturalist, 48, 94–102.

Albrecht, C. & Glaubrecht, M. 2006. Brood care among Basommatophorans: a unique reproductive strategy in the freshwater limpet snail *Protancylus* (Heterobranchia : Protancylidae), endemic to ancient lakes on Sulawesi, Indonesia. Acta Zoologica, 87, 49–58.

Andersson, M. 1994. Sexual Selection. Princeton: Princeton University Press.

Anthes, N., Putz, A. & Michiels, N. K. 2006. Sex role preferences, gender conflict and sperm trading in simultaneous hermaphrodites: a new framework Animal Behaviour, 72, 1–12.

Antonovics, J. 1968. Evolution in closely adjacent plant populations. V. Evolution of self-fertility. Heredity, 23, 219–238.

Arnqvist, G. & Rowe, L. 2005. Sexual Conflict. Princeton: Princeton University Press.

Bargues, M. D., Horák, P., Patzner, R. A., Pointier, J.-P., Jackiewicz, M., Meier-Brook, C. & Mas-Coma, S. 2003. Insights into the relationships of Palearctic and Nearctic lymnaeids (Mollusca: Gastropoda) by rDNA ITS-2 sequencing and phylogeny of Stagnicoline intermediate host species of *Fasciola hepatica*. Parasite, 10, 243–255.

Bargues, M. D., Artigas, P., Jackiewicz, M., Pointier, J.-P. & Mas-Coma, S. 2004. Ribosomal DNA ITS-1 sequence analysis of European Stagnicoline Lymnaeidae (Gastropoda). Heldia 6, 57–68.

Baur, B. 2010. Stylommatophoran gastropods. In: The Evolution of a "Primary" Sexual Characters In Animals (Ed. by Cordoba-Aguilar, A. & Leonard, J.L.), pp. 197–217. Oxford University Press.

Bayne, C. J. & Loker, E. S. 1987. Survival within the snail host. In: The Biology of Schistosomes—From Genes to Latrines (Ed. by Rollinson, D. & Simpson, A. J. G.), pp. 321–346: Academic Press.

Bondesen, P. 1950. Egg capsules of river limpet snails: Material for experimental biology. Science 2, 603–605.

Bouchet, P. & Rocroi, J.-P. 2005. Classification and nomenclator of gastropod families. Malacologia, 47, 1–397.

Brackenbury, T. D. & Appleton, C. C. 1991. Morphology of the mature spermatozoon of *Physa acuta* (Drapernaud, 1801) (Gastropoda, Physidae). Journal of Molluscan Studies, 57, 211–218.

Braendle, C. & Félix, M.-A. 2006. Sex determination: ways to evolve a hermaphrodite. Current Biology, 16, R468–R471.

Brown, D. S. 1994. Freshwater Snails of Africa and their Medical Importance. London: Taylor & Francis Ltd.

Charlesworth, B. & Charlesworth, D. 1978. Model for evolution of dioecy and gynodioecy. American Naturalist, 112, 975–997.

Charnov, E. L. 1982. The Theory of Sex Allocation. Princeton: Princeton University Press.

Cole, W. H. 1925. Egg laying in two species of *Planorbis*. American Naturalist, 59, 284–286.

Coyne, J. A. & Orr, H. A. 2004. Speciation. Sunderland: Sinauer Associates.

Davison, A., Wade, C. M., Mordan, P. B. & Chiba, S. 2005. Sex and darts in slugs and snails (Mollusca: Gastropoda: Stylommatophora). Journal of Zoology, 267, 329–338.

De Boer, P. A. C. M., Jansen, R. F. & Ter Maat, A. 1996. Copulation in the hermaphroditic snail *Lymnaea stagnalis*: A review. Invertebrate Reproduction and Development, 30, 167–176.

De Boer, P. A. C. M., Jansen, R. F., Koene, J. M. & Ter Maat, A. 1997a. Nervous control of male sexual drive in the hermaphroditic snail *Lymnaea stagnalis*. Journal of Experimental Biology, 200, 941–951.

De Boer, P. A. C. M., Ter Maat, A., Pieneman, A. W., Croll, R. P., Kurokawa, M. & Jansen, R. F. 1997b. Functional role of peptidergic anterior lobe neurons in male sexual behavior of the snail *Lymnaea stagnalis*. Journal of Neurophysiology, 78, 2823–2833.

DeJong, R. J., Morgan, J. A. T., Paraense, W. L., Pointier, J.-P., Amarista, M., Ayeh-Kumi, P. F. K., Babiker, A., Barbosa, C. S., Brémond, P., Canese, A. P., de Souza, C. P., Dominguez, C., File, S., Gutierrez, A., Incani, R. N., Kawano, T., Kazibwe, F., Kpikpi, J., Lwambo, N. J. S., Mimpfoundi, R., Njiokou, F., Poda, J. N., Sene, M., Velasquez, L. E., Yong, M., Adema, C. M., Hofkin, B. V., Mkoji, G. M. & Loker, E. S. 2001. Evolutionary relationships and biogeography of *Biomphalaria* (Gastropoda: Planorbidae) with implications regarding its role as host of the human bloodfluke, *Schistosoma mansoni*. Molecular Biology and Evolution, 18, 2225–2239.

De Jong-Brink, M. 1969. Histochemical and electron microscope observations on the reproductive tract of *Biomphalaria glabrata* (*Australorbis glabratus*), intermediate host of *Schistosoma mansoni*. Zeitschrift für Zellforschung und Mikroskopische Anatomie, 102, 507–542.

De Jong-Brink, M. & Geraerts, W. P. M. 1982. Oogenesis in gastropods. Malacologia, 22, 145–149.

De Jong-Brink, M., Boer, H. H., Hommes, T. G. & Kodde, A. 1977. Spermatogenesis and the role of sertoli cells in the freshwater snail *Biomphalaria glabrata*. Cell and Tissue Research, 181, 37–58.

De Lange, R. P. J. & Van Minnen, J. 1998. Localization of the neuropeptide APGWamide in gastropod molluscs by in situ hybridization and immunocytochemistry. General and

Comparative Endocrinology, 109, 166–174.

De Lange, R. P. J., Joosse, J. & Van Minnen, J. 1998. Multi-messenger innervation of the male sexual system of *Lymnaea stagnalis*. Journal of Comparative Neurophysiology, 390, 564–577.

De Larambergue, M. D. 1939. Etude de l'autofécondation chez les gastéropodes pulmonés: recherches sur l'aphallie et la fécondation chez *Bulinus (Isidora) contortus*. Bulletin Biologique de la France et de la Belgique 73, 19–231.

De Visser, J. A. G. M., Ter Maat, A. & Zonneveld, C. 1994. Energy budgets and reproductive allocation in the simultaneous hermaphrodite pond snail, *Lymnaea stagnalis* (L.): a trade-off between male and female function. American Naturalist, 144, 861–867.

DeWitt, T. J. 1991. Mating behavior of the freshwater pulmonate snail, *Physa gyrina*. American Malacological Bulletin, 9, 81–84.

DeWitt, T. J. 1996. Gender contests in a simultaneous hermaphrodite snail: a size-advantage model for behaviour. Animal Behaviour, 51, 345–351.

Dillon, R. J. J. 2000. The Ecology of Freshwater Molluscs. Cambridge: Cambridge University Press.

Dogterom, G. E. & Van Loenhout, H. 1983. Specificity of ovulation hormones of some basommatophoran species studied by means of iso- and heterospecific injections. General and Comparative Endocrinology, 52, 121–125.

Doums, C., Viard, F., Pernot, A.-F., Delay, B. & Jarne, P. 1996. Inbreeding depression, neutral polymorphism, and copulatory behavior in freshwater snails: A self-fertilization syndrome. Evolution, 50, 1908–1918.

Doums, C., Viard, F. & Jarne, P. 1998. The evolution of phally polymorphism. Biological Journal of the Linnean Society, 64, 273–296.

Duncan, C. J. 1975. Reproduction. In: Pulmonates (Ed. by Fretter, V. & Peake, J.), pp. 309–365. London: Academic Press.

El Filali, Z., Van Minnen, J., Liu, W. K., Smit, A. B. & Li, K. W. 2006. Peptidomics analysis of neuropeptides involved in copulatory behavior of the mollusk *Lymnaea stagnalis*. Journal of Proteome Research, 5, 1611–1617.

Escobar, J. S. 2008. Consanguinité et évolution phénotypique—Systèmes de reproduction,

valeur sélective et histoire de vie chez des mollusques Pulmonés. PhD thesis. Université Montpellier II.

Facon, B., Ravigné, V. & Goudet, J. 2006. Experimental evidence of inbreeding avoidance in the hermaphroditic snail *Physa acuta*. Evolutionary Ecology, 20, 395–406.

Facon, B., Ravigné, V., Sauteur, L. & Goudet, J. 2007. Effect of mating history on gender preference in the hermaphroditic snail *Physa acuta*. Animal Behaviour, 74, 1455–1461.

Ferguson, G. P., Pieneman, A. W., Jansen, R. T. & Ter Maat, A. 1993. Neuronal feedback in egg-laying behaviour of the pond snail *Lymnaea stagnalis*. Journal of Experimental Biology, 178, 251–259.

Fong, P. P., Olex, A. L., Farrell, J. E., Majchrzak, R. M. & Muschamp, J. W. 2005. Induction of preputium eversion by peptides, serotonin receptor antagonists, and selective serotonin reuptake inhibitors in *Biomphalaria glabrata*. Invertebrate Biology, 124, 296–302.

Fraser, L. A. 1946. The embryology of the reproductive tract of *Lymnaea stagnalis appressa* Say. Transactions of the American Microscopical Society 65, 279–298.

Geraerts, W. P. M. & Joosse, J. 1975. Control of vitellogenesis and of growth of female accessory sex organs by the dorsal body hormone (DBH) in the hermaphroditic freshwater snail *Lymnaea stagnalis*. General and Comparative Endocrinology, 27, 450–464.

Geraerts, W. P. M. & Joosse, J. 1984. Freshwater snails (Basommatophora). In: The Mollusca. Reproduction (Ed. by Tompa, A. S., Verdonk, N. H. & van den Biggelaar, J. A. M.), pp. 142–208. London: Academic Press.

Goodwillie, C., Kalisz, S. & Eckert, C. G. 2005. The evolutionary enigma of mixed mating in plants: occurrence, theoretical explanations, and empirical evidence. Annual Review of Ecology, Evolution and Systematics, 36, 47–79.

Greeff, J. M. & Michiels, N. K. 1999. Sperm digestion and reciprocal sperm transfer can drive hermaphrodite sex allocation to equality. American Naturalist, 153, 421–430.

Healy, J. M. 1983. An ultrastructural study of Basommatophoran spermatozoa (Mollusca, Gastropoda). Zoologica Scripta, 12, 57–66.

Heller, J. 1993. Hermaphroditism in Mollusks. Biological Journal of the Linnean Society, 48, 19–42.

Hermann, P. M., De Lange, R. P. J., Pieneman, A. W., Ter Maat, A. & Jansen, R. F. 1997. Role of neuropeptides encoded on CDCH-1 gene in the organization of egg-laying behavior in the pond snail, *Lymnaea stagnalis*. Journal of Neurophysiology, 78, 2859–2869.

Jackiewicz, M. 1993. Phylogeny and relationships within the European species of the family Lymnaeidae (Gastropoda: Pulmonata: Basommatophora). Folia Malacologica, 5, 61–95.

Jarne, P. & Auld, J. R. 2006. Animals mix it up too: the distribution of self-fertilization among hermaphroditic animals. Evolution, 60, 1816–1824.

Jarne, P. & Charlesworth, D. 1993. The evolution of the selfing rate in functionally hermaphrodite plants and animals. Annual Review of Ecology and Systematics, 24, 441–466.

Jarne, P. & Städler, T. 1995. Population genetic structure and mating system evolution in freshwater pulmonates. Experientia, 51, 482–497.

Jarne, P., Finot, L., Delay, B. & Thaler, L. 1991. Self-fertilization versus cross-fertilization in the hermaphroditic freshwater snail *Bulinus globosus*. Evolution, 45, 1136–1146.

Jarne, P., Vianey-Liaud, M. & Delay, B. 1993. Selfing and outcrossing in hermaphrodite freshwater gastropods (Basommatophora): where, when and why. Biological Journal of the Linnean Society, 49, 99–125.

Jiménez, C. R., Ter Maat, A., Pieneman, A., Burlingame, A. L., Smit, A. B. & Li, K. W. 2004. Spatio-temporal dynamics of the egg-laying-inducing peptides during an egg-laying cycle: a semi-quantitative matrix-assisted laser desorption/ionization mass spectrometry approach. Journal of Neurochemistry, 89, 865–875.

Joosse, J. & Reitz, D. 1969. Functional anatomical aspects of the ovotestis of *Lymnaea stagnalis*. Malacologia 9, 101–109.

Joosse, J., Boer, M. H. & Cornelisse, C. J. 1968. Gametogenesis and oviposition in *Lymnaea stagnalis* as influenced by γ-irradiation and hunger. Symposia of the Zoological Society of London, 22, 213–235.

Jordaens, K., Dillen, L. & Backeljau, T. 2007. Effects of mating, breeding system and parasites on reproduction in hermaphrodites: pulmonate gastropods (Mollusca) Animal Biology, 57, 137–195.

Jørgensen, A., Jørgensen, L. V. G., Kristensen, T. K., Madsen, H. & Stothard, J. R. 2007.

Molecular phylogenetic investigations of *Bulinus* (Gastropoda: Planorbidae) in Lake Malawi with comments on the topological incongruence between DNA loci. Zoologica Scripta, 36, 577–585.

Klussmann-Kolb, A., Dinapoli, A., Kuhn, K., Streit, B. & Albrecht, C. 2008. From sea to land and beyond—New insights into the evolution of euthyneuran Gastropoda (Mollusca). BMC Evolutionary Biology, 8, 57.

Koene, J. M. 2006. Tales of two snails: sexual selection and sexual conflict in *Lymnaea stagnalis* and *Helix aspersa*. Integrative and Comparative Biology, 46, 419–429.

Koene, J. M. & Schulenburg, H. 2005. Shooting darts: co-evolution and counter-adaptation in hermaphroditic snails. BMC Evolutionary Biology, 5, 25.

Koene, J. M. & Ter Maat, A. 2001. "Allohormones": a class of bioactive substances favoured by sexual selection. Journal of Comparative Physiology A, 187, 323–326.

Koene, J. M. & Ter Maat, A. 2004. Energy budgets in the simultaneously hermaphroditic pond snail, *Lymnaea stagnalis*: a trade-off between growth and reproduction during development. Belgian Journal of Zoology, 134, 41–45.

Koene, J. M. & Ter Maat, A. 2005. Sex role alternation in the simultaneously hermaphroditic pond snail *Lymnaea stagnalis* is determined by the availability of seminal fluid. Animal Behaviour, 69, 845–850.

Koene, J. M. & Ter Maat, A. 2007. Coolidge effect in pond snails: male motivation in a simultaneous hermaphrodite. BMC Evolutionary Biology, 7, 212.

Koene, J. M., Jansen, R. F., Ter Maat, A. & Chase, R. 2000. A conserved location for the central nervous system control of mating behaviour in gastropod molluscs: Evidence from a terrestrial snail. Journal of Experimental Biology, 203, 1071–1080.

Koene, J. M., Montagne-Wajer, K. & Ter Maat, A. 2006. Effects of frequent mating on sex allocation in the simultaneously hermaphroditic great pond snail (*Lymnaea stagnalis*). Behavioral Ecology and Sociobiology, 60, 332–338.

Koene, J. M., Montagne-Wajer, K. & Ter Maat, A. 2007. Aspects of body size and mate choice in the simultaneously hermaphroditic pond snail *Lymnaea stagnalis*. Animal Biology, 57, 247–259.

Koene et al. 2009. The fate of received sperm in the reproductive tract of a hermaphroditic snail and its implications for fertilisation Evolutionary Ecology, 23, 533–543.

Leonard, J. L. 2006. Sexual selection: lessons from hermaphrodite mating systems. Integrative and Comparative Biology, 46, 349–367.

Lever, J., De Vries, C. M. & Jager, J. C. 1965. On the anatomy of the central nervous system and location of neurosecretory cells in *Australorbis glabratus*. Malacologia 2, 219–230.

Loose, M. J. & Koene, J. M. 2008. The effect of body weight, insemination duration and rearing condition on sperm transfer in a simultaneous hermaphrodite. Invertebrate Biology, In press.

Loose & Koene. 2008. Sperm transfer is affected by mating history in the simultaneously hermaphroditic snail *Lymnaea stagnalis* Invertebrate Biology, 127, 162–167.

McCarthy, T. M. & Sih, A. 2008. Relatedness of mates influences mating behaviour and reproductive success of the hermaphroditic snail *Physa gyrina*. Evolutionary Ecology Research, 10, 77–94.

Morgan, J. A. T., DeJong, R. J., Jung, Y. H., Khallaayoune, K., Kock, S., Mkoji, G. M. & Loker, E. S. 2002. A phylogeny of planorbid snails, with implications for the evolution of *Schistosoma* parasites. Molecular Phylogenetics and Evolution, 25, 477–488.

Ohbayashi-Hodoki, K., Ishihama, F. & Shimada, M. 2004. Body size-dependent gender role in a simultaneous hermaphrodite freshwater snail, *Physa acuta*. Behavioral Ecology, 15, 976–981.

Ostrowski, M. F., Jarne, P. & David, P. 2000. Quantitative genetics of sexual plasticity: the environmental threshold model and genotype-by-environment interaction for phallus development in the snail *Bulinus truncatus*. Evolution, 54, 1614–1625.

Ostrowski, M.-F., Jarne, P., Berticat, O. & David, P. 2002. Ontogenetic reaction norm for binary traits: the timing of phallus development in the snail *Bulinus truncatus*. Heredity, 88, 342–348.

Ostrowski, M. F., Jarne, P. & David, P. 2003. A phallus for free? Quantitative genetics of sexual trade-offs in the snail *Bulinus*

truncatus. Journal of Evolutionary Biology, 16, 7–16.

Paraense, W. L. 2003. A bird's eye survey of Central American planorbid molluscs. Memorias do Institute Oswaldo Cruz, 98, 51–67.

Paraense, W. L. & Pointier, J. P. 2003. *Physa acuta* Draparnaud, 1805 (Gastropoda: Physidae): a studies of topotypic specimens. Memorias do Institute Oswaldo Cruz, 98, 513–517.

Plesch, B., De Jong-Brink, M. & Boer, H. H. 1971. Histological and histochemical observations on the reproductive tract of the hermaphrodite pond snail *Lymnaea stagnalis* (L.). Netherlands Journal of Zoology, 21.

Puurtinen, M., Knott, K. E., Suonpää, S., Nissinen, K. & Kaitala, V. 2007. Predominance of outcrossing despite low apparent fitness costs of self-fertilization. Journal of Evolutionary Biology, 20, 901–912.

Remigio, E. A. 2002. Molecular phylogenetic relationships in the aquatic snail genus *Lymnaea*, the intermediate host of the causative agent of fascioliasis: insights from broader taxon sampling. Parasitology Research, 88, 687–696.

Rigby, J. E. 1982. The fine structure of differentiating spermatozoa and Sertoli cells in the gonad of the pond snail, *Lymnaea stagnalis*. Journal of Molluscan Studies, 48, 111–123.

Roubos, E. W. & Van der Ven, A. M. H. 1987. Morphology of neurosecretory cells in basommatophoran snails homologous with egg-laying and growth-hormone producing cells of *Lymnaea stagnalis*. General and Comparative Endocrinology, 67, 7–23.

Rudolph, P. H. 1983. Copulatory activity and sperm production in *Bulinus* (*Physopsis*) *globosus* (Gastropoda: Planobidae). Journal of Molluscan Studies, 49, 125–132.

Rudolph, P. H. & White, J. K. 1979. Egg laying behaviour of Bulinus octoploidus Burch (Basommatophora: Planorbidae). Journal of Molluscan Studies 45, 355–363.

Siva-Jothy, M. T. 2006. Trauma, disease and collateral damage: conflict in cimicids. Philosophical Transactions of the Royal Society B, 361, 269–275.

Strong, E. E., Gargominy, O., Ponder, W. F. & Bouchet, P. 2008. Global diversity of gastropods (Gastropoda; Mollusca) in freshwater. Hydrobiologia, 595, 149–166.

Ter Maat, A., Lodder, J. C. & Wilbrink, M. 1983. Induction of egg-laying in the pond snail *Lymnaea stagnalis* by environmental stimulation of the release of ovulation hormone from the caudo-dorsal cells. International Journal of Invertebrate Reproduction 6, 239–247.

Ter Maat, A., Dijcks, F. A. & Bos, N. P. A. 1986. In vivo recordings of neuroendocrine cells (Caudo-Dorsal Cells) in the pond snail. Journal of Comparative Physiology, 158, 853–859.

Ter Maat, A., Van Duivenboden, Y. A. & Jansen, R. F. 1987. Copulation and egg-laying behavior in the pond snail. In: Neurobiology: Molluscan Models (Ed. by Boer, H. H., Geraerts, W. P. M. & Joosse, J.), pp. 255–261. Oxford, New York: North-Holland Publishing Company Amsterdam.

Ter Maat, A., Geraerts, W. P. M., Jansen, R. F. & Bos, N. P. A. 1988. Chemically mediated positive feedback generates long-lasting discharge in a molluscan neuro-endocrine system. Brain Research, 438, 77–82.

Tian-Bi, T. N., N'Goran, K. E., N'Guetta, S.-P., Matthys, B., Sangare, A. & Jarne, P. 2008. Prior selfing and the selfing syndrome in animals: an experimental approach in the freshwater snail *Biomphalaria pfeifferi*. Genetics Research, 90, 61–72.

Tomé, L. A. & Ribeiro, A. F. 1998. The functional organization of the carrefour in the reproductive tract of *Biomphalaria tenagophila* (Mollusca, Planorbidae). Invertebrate Reproduction and Development, 34, 25–33.

Tsitrone, A., Duperron, A. & David, P. 2003a. Delayed selfing as an optimal mating strategy in preferentially outcrossing species: theoretical analysis of the optimal age at first reproduction in relation to mate availability. American Naturalist, 162, 318–331.

Tsitrone, A., Jarne, P. & David, P. 2003b. Delayed selfing and resource reallocations in relation to mate availability in the freshwater snail *Physa acuta*. American Naturalist, 162, 474–488.

Valdés, A., Gosliner, T. M. & Ghiselin, M. T. 2010. Opisthobranchs. In: The Evolution of a «Primary» Sexual Characters In Animals (Ed. by Córdoba-Aguilar, A. & Leonard, J. L.): Oxford University Press, pp. 148–172.

Van Duivenboden, Y. A. & Ter Maat, A. 1985. Masculinity and receptivity in the hermaphrodite pond snail, *Lymnaea stagnalis*. Animal Behaviour, 33, 885–891.

Van Kesteren, R. E., Smit, A. B., De Lange, R. P., Kits, K. S., Van Golen, F. A., Van Der Schors, R. C., De With, N. D., Burke, J. F. & Geraerts, W. P. M. 1995. Structural and functional evolution of the vasopressin/oxytocin superfamily: vasopressin-related conopressin is the only member present in *Lymnaea*, and is involved in the control of sexual behavior. Journal of Neurosciences, 15, 5989–5998.

Van Minnen, J., Schallig, H. D. F. H. & Ramkema, M. D. 1992. Identification of putative egg-laying hormone containing neuronal systems in gastropod molluscs. General and Comparative Endocrinology, 86, 96–102.

Vreugdenhil, E., Jackson, J. F., Bouwmeester, T., Smit, A. B., Van Minnen, J., Van Heerikhuizen, H., Klootwijk, J. & Joosse, J. 1988. Isolation, characterization, and evolutionary aspects of a cDNA clone encoding multiple neuropeptides involved in the stereotyped egg-laying behavior of the freshwater snail *Lymnaea stagnalis*. Journal of Neurosciences, 8, 4184–4191.

Walther, A. C., Lee, T., Burch, J. B. & Ó Foighil, D. 2006. *E Pluribus Unum*: A phylogenetic and phylogeographic reassessment of *Laevapex* (Pulmonata: Ancylidae), a North American genus of freshwater limpets. Molecular Phylogenetics and Evolution, 40, 501–516.

Webster, J. P. & Gower, C. M. 2006. Mate choice, frequency dependence, and the maintenance of resistance to parasitism in a simultaneous hermaphrodite. Integrative and Comparative Biology, 46, 407–418.

Wethington, A. R. & Dillon, R. T. 1996. Gender choice and gender conflict in a non-reciprocally mating simultaneous hermaphrodite, the freshwater snail *Physa*. Animal Behaviour, 51, 1107–1118.

Wethington, A. R. & Dillon, R. T. J. 1997. Selfing, outcrossing and mixed mating in the freshwater snail *Physa heterostropha*: lifetime fitness and inbreeding depression. Invertebrate Biology, 116, 192–199.

Wethington, A. R. & Lydeard, C. 2007. A molecular phylogeny of Physidae (Gastropoda : Basommatophora) based on mitochondrial

DNA sequences. Journal of Molluscan
Studies, 73, 241–257.

Wijsman, T. C. M. 1989. Glycogen and galactogen
in the albumin gland of the fresh-water snail
Lymnaea stagnalis—effects of egg laying,
photo period and starvation. Comparative
Biochemistry and Physiology A, 92, 53–59.

Young KG, Chang JP, Goldberg, JI. 1999 :
Gonadotropin-releasing hormone neuronal
system of the freshwater snails *Helisoma
trivolvis* and *Lymnaea stagnalis*:
Possible involvement in reproduction.
Journal of Comparative Neurology, 404,
427–437.

10

Stylommatophoran Gastropods

BRUNO BAUR

INTRODUCTION

The snails and slugs grouped in the subclass Pulmonata constitute one of the three subclasses of Gastropoda, the other two being the Prosobranchia and Opisthobranchia. The subclass Pulmonata has two major subdivisions, the orders Basommatophora and Stylommatophora. Stylommatophoran gastropods are a large and highly diverse group, probably exceeding 30,000 species. They occur in a wide variety of terrestrial habitats such as river embankments, grasslands, soil, leaf litter in forest, exposed cliff walls, stone deserts, decaying wood and trees (Solem 1984). Stylommatophoran gastropods are hermaphrodites (see below), have a lung inside the mantle cavity, two pairs of tentacles, eyes at the tips of the long pair of tentacles, and a usually coiled shell. Stylommatophorans show a great diversity in life-history traits, including "primary" sexual characters. There is abundant information on the morphology of reproductive organs in the literature, primarily as a result of systematic studies (for reviews see Duncan 1975; Tompa 1984; Nordsieck 1985; Runham 1988; Luchtel et al. 1997; Barker 2001; Gomez 2001). However, detailed functional information is not so plentiful and restricted to a few model species and thus not representative of the phylogenetic diversity in this animal group.

The stylommatophorans are a group of gastropods that moved from aquatic to terrestrial environments (Little 1990). This required significant adaptations in all possible life processes including reproduction, which is characterized by internal fertilization, direct development by means of cleidoic eggs, and often elaborate courtship behavior (Tompa 1984).

My focus is on primary sexual characters in stylommatophoran gastropods. I review characters which determine reproductive success, and reflect on, for which of these characters and selective events stylommatophoran gastropods ought to serve as particularly suitable research models for simultaneous hermaphrodites. Most interestingly, the review shows that in stylommatophorans any separation of characters into "primary" sexual characters (i.e., those that have evolved through natural selection on reproductive efficiency) and "secondary" sexual characters (i.e., those that have evolved through sexual selection) is difficult. Almost all reproductive characters seem to be shaped by both natural and sexual selection in this group of hermaphrodites. Exceptions might be dart shooting and accessory sexual organs (see below: Auxiliary copulatory organs), which are sexually selected serving primary in competition with other mates. By compiling the existing information, this review should stimulate future studies on sexual traits in stylommatophoran gastropods.

REPRODUCTIVE BIOLOGY

Stylommatophorans are simultaneous hermaphrodites, albeit with the gonad predominantly in the male phase initially, and predominantly in the female phase towards the end of the reproductive cycle in some species (Gomez 2001). Cross-fertilization is prevalent, but self-fertilization is widespread, particularly in species with tiny shells (e.g., *Vallonia pulchella* (Whitney 1938); *Punctum pygmaeum* (Baur 1987)) and in slugs (South 1992). The frequency of selfing varies greatly among species and even among populations (Heller 2001). In some species, it is rare, in others it occurs occasionally, and still in others self-fertilization occurs regularly. Self-fertilization has evolved in several phylogenetically independent lines (Heller 1993).

Courtship and mating duration ranges from a few to more than 36 hours in terrestrial gastropods and thus often exceed the period favourable for locomotor activity (conditions of high air humidity; Lind 1973, 1976; Jeppesen 1976; Chung 1987). During courtship and copulation terrestrial gastropods are exposed to severe water loss and more susceptible to predation than single adults (Pollard 1975). In most species actual intromission and sperm transfer is rather short compared with the extended courtship (Tompa 1984). In many species, however, courtship and mating behavior is very complex, demanding a high amount of coordination between partners. Hence, interspecific differences in mating behavior might cause effective reproductive barriers. For example, the pronounced behavioral differences between *Deroceras rodnae* and *D. praecox* and the unsuccessful interspecific mating attempts indicate the occurrence of a prezygotic barrier (Reise 1995). Individuals of many species transfer spermatozoa packed in a spermatophore, whose shape and size are species-specific and thus of taxonomic significance. Other species deliver spermatozoa as an unpacked sperm mass.

Hermaphroditic land snails would greatly enhance their reproductive success by choosing large mates because female fecundity (number of clutches, clutch size, and egg size) is positively correlated with shell size (Wolda 1963; Baur 1988a; Baur & Raboud 1988). However, mating has been reported to be random with respect to shell size in *Cepaea nemoralis* (Wolda 1963), *Arianta arbustorum* (Baur 1992a), and *Succinea putris* (Jordaens et al. 2005) and with respect to shell colour and banding pattern in *C. nemoralis* (Schilder 1950; Schnetter 1950; Lamotte 1951; Wolda 1963). In *Achatina fulica*, which is protandrous, adults capable of producing both sperm and eggs, were more favored as mating partners than young adults which

produce only sperm (Tomiyama 1996). Size-assortative mating has been observed among old adults of *A. fulica* (Tomiyama 1996). In contrast to the species listed above, *A. fulica* shows indeterminate growth.

Mate-choice tests with *A. arbustorum* from geographically isolated populations in Sweden and Switzerland revealed that snails preferred to mate with individuals from their population of origin, and pairs involving snails from two distant Swiss populations showed a reduced fertility, indicating effects of outbreeding depression (Baur & Baur 1992). In general, mating between closely related individuals can incur substantial fitness costs (i.e., inbreeding depression). However, individuals of *A. arbustorum* mated randomly with respect to the degree of relatedness, indicating a lack of inbreeding avoidance by selective mating (B. Baur & A. Baur 1997). Snails that mated with full-sibs did not differ in number of eggs, hatching success of eggs, or number of offspring produced from those mated with unrelated conspecifics. In another population of *A. arbustorum*, Chen (1993) found that eggs of inbred snails showed a lower hatching success (30.4%) than those of outbred snails (48.5%). Furthermore, inbred offspring reared in the garden had a higher mortality rate than outbred offspring reared in the same environment, but no difference was found when offspring from both groups were kept in the laboratory. This result supports the hypothesis that cross-fertilization in simultaneous hermaphrodites is maintained by inbreeding depression. It also shows that the extent of negative inbreeding effects varies between populations and environments in which the snails are kept.

The two basic modes of reproductive strategies in Stylommatophora comprise semelparity, in which animals reproduce during one season only, after which they die; and iteroparity, in which animals reproduce during several seasons (Heller 2001). Semelparity is uncoupled from the annual cycle and thus the life span of many semelparous species extends to more than 1 year. Heller (2001) analyzed life-history data comprising species in 35 genera, of which 15 genera are semelparous and 20 iteroparous. This sample amounts to approximately 2% of the estimated 1700 genera of terrestrial gastropods.

The life history of terrestrial gastropods is highly dependent on climate; the animals aestivate when it is too hot and hibernate when it is too cold. For example, *Cristataria genezarethana*, a rock-dwelling species in Mediterranean habitats, spends 95–98% of its life time within crevices, emerging to the surface of the rock during brief periods of high air humidity to feed and mate (Heller & Dolev 1994). In this species, maturity is reached within about 11 years and individuals live for at least 16 years. In general, the life span of Stylommatophora ranges from several months to 20 years (in *Helix pomatia* some individuals may reach an age of 40 and more years; Heller 1990).

GENITAL MORPHOLOGY AND FUNCTION

Stylommatophoran gastropods exhibit a great diversity in their reproductive system, reflecting their phylogeny (Barker 2001). The different functions of the reproductive system include:

1. production of ova and sperm;
2. storage and transport of mature gametes in a suitable medium;
3. structural and physiological roles in courtship and copulation;
4. transfer of endogenous sperm (autosperm) to the mating partner's reproductive duct;
5. reception of exogenous sperm (allosperm);
6. supplying a site and proper medium for fertilization of ova;
7. covering the zygote with nutritive and protective layers;
8. oviposition; and
9. resorption of remnant and excess reproductive products (Gomez 2001).

The terminology of the morphology of the gastropod reproductive tract is often confusing, partly due to the use of the same term for different structures. For simplicity, I use descriptive terms throughout this chapter (figure 10.1).

Gonad and Gonoduct

Stylommatophorans have only a single gonad, the ovotestis, which produces both oocytes and spermatozoa (figure 10.1). The gonad, located among the lobes of the digestive gland toward the posterior part of the body, consists of numerous acini containing both male and female germ cells (South 1992). In most species, the male germ cells appear to differentiate and mature earlier in the life

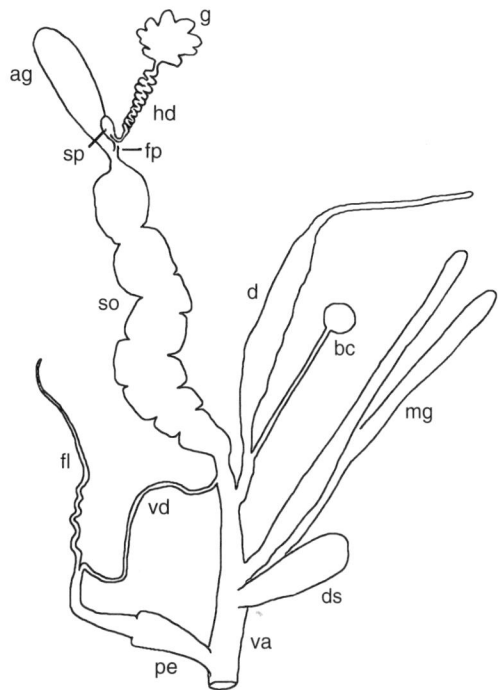

FIGURE 10.1 Schematic drawing of the reproductive morphology of a stylommatophoran gastropod with one dart and a diverticulum. ag, albumen gland; bc, bursa copulatrix; d, diverticulum; ds, dart sac; e, epiphallus; fl, flagellum; fp, fertilization pouch; g, gonad; hd, hermaphroditic duct; mg, mucous glands; pe, penis; so, spermoviduct; sp, spermatheca; va, vagina; vd, vas deferens. The carrefour consists of the spermatheca and the fertilization pouch.

cycle than the female germ cells (Duncan 1975; Luchtel et al. 1997). The ovotestis opens to a gonoduct (= hermaphrodite duct). When released, both male and female gametes pass along the hermaphroditic duct. Thereafter, they follow separate paths (figure 10.1). The hermaphrodite duct varies in complexity between higher taxa (Duncan 1975).

In many species, autosperm are stored in the seminal vesicle of the hermaphrodite duct throughout the year (Lind 1973). Phagocytosis of autosperm by the hermaphrodite duct epithelium has been reported in *Helix pomatia* and *Oxychilus cellarius* (Rigby 1963). Sperm can be expelled from the hermaphrodite duct at times other than copulation to be eventually digested (as are foreign sperm) in the bursa copulatrix (see below).

Carrefour

In stylommatophoran gastropods, the carrefour includes structures for allosperm storage (the spermatheca or female sperm-storage organ), and for oocyte fertilization and the coating of zygotes with the albumen layer (Gomez 2001). The fertilization of oocytes occurs in a specialized region of the carrefour, which has the form of a pouch in most species. Secretory cells occur in the walls of the fertilization chamber; their secretions are thought to provide a medium for gamete fusion (Gomez et al. 1991).

Sperm received (allosperm) travel through the spermoviduct to the spermatheca (figure 10.1). They reach the carrefour within 4 h of copulation in the slug *Deroceras reticulatum* (Runham & Hogg 1992). In *Helix pomatia* and *Cornu aspersum* (formerly *Helix aspersa*), only 0.02–0.1% of the allosperm that are transferred reach the storage organ; the majority of them within 12 h of copulation (Lind 1973; Rogers & Chase 2001). In the spermatheca allosperm are stored for long periods. Viable allosperm have been found up to 4 years after the last copulation in the sperm-storage organ of stylommatophoran gastropods (see below: Sperm competition).

There is an enormous variability in the structure and morphology of the carrefour in stylommatophorans. For example, the carrefour is not divided into separate spermatheca and fertilization chamber in *Trigonephrus gypsinus* (Brinders & Sirgel 1992). *Oxychilus draparnaudi* and *Bradybaena fruticum* have a single spermathecal tubule beside the fertilization chamber (Flasar 1967; Bojat et al. 2001a). In *Succinea putris* two spermathecal tubules occur (Rigby 1965), and 34 tubules have been recorded in the spermatheca of *Drymaeus papyraceus* (van Mol 1971). There is also a considerable within-species variation in the number of spermathecal tubules (e.g., 3–5 in *H. pomatia*, Lind 1973; 2–9 in *Arianta arbustorum*, Haase & Baur 1995; Baminger & Haase 1999; 4–19 in *Cornu aspersum*, Koemtzopoulos & Staikou 2007). The blind-ending tubules unite to a common duct, which opens into the fertilization chamber. In *A. arbustorum*, the musculature surrounding the spermathecal tubules is arranged in a complex three-dimensional network (Bojat et al. 2001b, c). If there were a selective activation of the muscles of each tubule (which has not yet been examined), this would allow the animal to expel sperm stored in single tubules and thus promotes a selective fertilization of eggs. The ciliation of the common duct is probably responsible for the distribution of incoming sperm among the tubules. The spermatheca is expandable and can accommodate more sperm than would be expected from the initial volume (Beese & Baur 2006).

As a consequence of the large intraspecific variation in the number of spermathecal tubules, different individuals might have different possibilities to store allosperm from more than one mating partner. Mixing of sperm from different mates would be more likely in a less structured spermatheca, whereas a large number of tubules would allow better separation of spermatozoa from different mates. Comparing female and male reproductive traits in six populations of *A. arbustorum*, Beese et al. (2006a) found an association between spermatheca volume and the number of sperm transferred. This suggests that post-copulatory mechanisms drive a correlated evolution between sperm characteristics and female reproductive traits in hermaphroditic gastropods.

A variety of adaptive explanations have been proposed to explain the diversity of female sperm-storage organs (Beese et al. 2009). One hypothesis claims that the differentiation of sperm-storage organs is dictated by demands of sperm storage capacity arising from differences in animal longevity and/or egg productivity, or by selection for functional design to match sperm morphology in order to efficiently store and utilize sperm (Pitnick et al. 1999). Females which live long or produce multiple clutches in consecutive years may require more specialized organs to provide nourishment or protection (e.g., through anchoring the sperm inside the storage organ) to maintain the viability of sperm (Smith & Yanagimachi 1990). Consequently, the evolution of sperm-storage organs should be coupled with life history. Moreover, female reproductive morphology is presumably associated with habitat specificity, because of adaptations of the life-history traits to local conditions. The evolution of female morphology may also simply track sperm length that evolves due to selection independent of female sperm stores (Pitnick et al. 1999), which might result in evolutionary correlations between the length of sperm-storage organs and sperm length documented in several gonochoristic animals (Presgraves et al. 1999).

Another widely supported hypothesis is that postcopulatory sexual selection has played an important role in the evolution of this trait, due to the potential influence of female sperm stores on the extent of non-random paternity (Eberhard 1996). A prerequisite for sexual selection via sperm competition is that the sperm of two or more males coexist within the reproductive tract of the female at the time of fertilization (Parker 1970). In the past few years, increasing attention has been paid to the possibility that females of many species are active not only in precopulatory choice but also in controlling the processes of sperm storage and use (Eberhard 1996; Birkhead & Møller 1998). The presence of storage organs may allow females to maintain viable sperm from multiple mates and thus selectively bias the fertilization success of sperm in relation to male behavior (Siva-Jothy & Hooper 1995) or male genotype (Ward 1998). Males always try to monopolize females (Chapman et al. 2003). Females, however, may benefit from increased within- and between-male variance in sperm traits in their reproductive tract (Jennions & Petrie 2000). The resulting male–female conflict over sperm use could have favored the evolution of adaptations in the female that control the events after copula and, vice-versa, counter-adaptations by the male that manipulate sperm storage processes (Rice & Holland 1997). The adaptations often involve harmful behavior and might lead to perpetual antagonistic co-evolution between certain traits or the evolution of new traits (Lessells 2006), resulting in increased inter-sexual specializations. The presence of female sperm-storage organs should therefore be linked with the presence of complex or peculiar reproductive traits. Moreover, diverse mating systems that impose different levels of selection pressure on postcopulatory processes are expected to covary with the presence of sperm-storage organs and their complexity.

Beese et al. (2009) examined morphologically the presence and complexity of the spermathecae in the carrefour in 47 species of stylommatophoran gastropods and used partial 28 rDNA sequences to reconstruct a molecular phylogeny for these species. The phylogenetic reconstruction supported several gains and losses of the spermathecae in stylommatophorans indicating rapid evolutionary changes, which could have gone jointly with explosive radiations of families during Mesozoic and the Late Cretaceous/Early Tertiary. Moreover, a complex spermatheca was associated with the occurrence of love darts or any kind of auxiliary copulatory organ (see below), the presence of a long flagellum at the penis and cross-fertilization as the predominant mating system. However, the results of Beese et al. (2009) also suggest associations of carrefour complexity with body size, reproductive strategy (semelparity vs. iteroparity), reproductive mode (oviparity vs. ovoviviparity), and with habitat type.

Albumen Gland

The albumen gland is a compound tubular gland that produces albumen or perivitelline fluid for the egg. In stylommatophorans, the gland increases in size with sexual maturation of the animal (Gomez 2001). The number of eggs that can be produced at any one time appears to depend on the size of the albumen gland (Tompa 1984). Secretory cells in the albumen gland contain large amounts of galactogen (Duncan 1975).

Spermoviduct, Free Oviduct and Vagina

In Stylommatophora, the male and female gametes follow separate pathways from the carrefour along a common duct (= spermoviduct), and separate male and female ducts then diverge at the distal end of the spermoviduct. In many species, the lumen of the spermoviduct is incompletely divided into two grooves (Luchtel et al. 1997). These grooves are lined by ciliated and secretory cells and are unequal in size, with a larger female groove and a narrow male groove (South 1992). The female groove produces supporting layers of the eggs, while the male groove produces seminal fluid (Gomez 2001). As the egg descends along the oviductal channel, the perivitelline membrane, the jelly or organic matrix of the inner egg cover, and the outer egg cover are sequentially deposited (Bayne 1968). In many species, the calcium content of the outer egg cover increases gradually as the egg passes along the oviductal channel (Tompa 1984). In ovoviviparous species (see below: Ova), the distal portion of the oviductal gland, together with the adjoining free oviduct, functions as a uterus for brooding of young after their hatching from eggs retained in the female duct (Tompa 1979). Both the free oviduct and vagina have a thick muscular wall, which may be attributed to their role in copulation and oviposition (Gomez 2001).

During copulation, sperm masses or spermatophores containing spermatozoa are transferred in many species into the vagina of the partner. Following sperm transfer the partners may quickly separate (Reise 2007). In a variety of species, however, there is a period of immobility (e.g., lasting 0.5–9 hours in *Helix pomatia*; Lind 1976), during which the spermatophore is transported in the reproductive tract of the recipient towards the bursa copulatrix, where it is eventually digested (see below). During this period sperm leave the spermatophore. Depending on the location of the spermatophore in the female reproductive tract, sperm may reach the spermatheca (female sperm-storage site) or they may be transported into the bursa copulatrix where they are eventually digested.

Vas Deferens, Epiphallus and Spermatophore

The vas deferens is a ciliated, narrow and partly folded duct, which functions to transport autosperm (figure 10.1). Peristalsis in the wall, together with ciliary action, contribute to the movement of seminal fluids along the duct, including their expulsion during mating (Runham 1988). When present, the epiphallus is usually a highly muscular organ, with the lumen larger and more folded than the vas deferens lumen. In a variety of stylommatophorans spermatophores are formed by the epiphallus (head filament and sperm container) and flagellum (= epiphallic caecum; tail) during copulation. The spermatophore is largely composed of secretory material containing glycosaminoglycans and mucoproteins (Mann 1984). In *Arianta arbustorum*, spermatophore formation is initiated more or less synchronously in mating partners a few minutes after penis intromission (Baminger & Haase 2001). The part of the spermatophore that contains the sperm increases in size until shortly before the spermatophore is transferred (approximately 90 min after penis intromission). Growth and final size of the spermatophore, however, are not adjusted between the mating partners.

Spermatophores have a species-specific shape and surface structure, which is of taxonomical significance (Baur 1998). Spermatophores may be smooth, elaborately spined, calcified or uncalcified (Tompa 1984). In *Helix pomatia*, the spermatophore is 6–8 cm long and consists of a distinctive tip, a body (sperm container), and a long tail (Meisenheimer 1907). In species with spermatophores, snails form a single spermatophore at each mating and exchange it reciprocally.

The adaptive significance of the spermatophore in stylommatophorans with well-developed copulation organs and internal fertilization is unclear. In *H. pomatia*, sperm leave the spermatophore body through the spermatophore tail in the stalk of the bursa copulatrix and migrate into the spermatheca (Lind 1973). The spermatophore and any remaining sperm are digested later in the bursa copulatrix. Lind (1973) suggested that the function of the spermatophore is to ensure that a number of sperm can migrate into the oviduct and reach the spermatheca without coming into contact with the digesting bursa copulatrix (see below). Thus, the significance of the peculiar way of transferring sperm may be to allow only the most active sperm to pass to the spermatheca and thus can be considered as a means to mitigate sperm selection of the recipient (which still might occur in the spermatheca).

Penis

Penial morphology of Stylommatophora is highly variable and species-specific (Barker 2001; Reise 2007). It has been suggested that the penis is the prime species recognition character in mating and, particularly in copulation success (Gomez 2001).

The penis of stylommatophorans is a muscular organ that is everted at copulation and is typically inserted into the genital atrium and vagina of the mate. The basic structure of the penis is a tube consisting of a non-ciliated, non-glandular epithelium, surrounded by a thick muscular wall with inner circular and outer longitudinal layers (Gomez 2001). The contraction of the penis wall effects the hydrostatic pressure necessary to evert the penis, while a retractor muscle affects retraction. The epithelium lining the lumen of the penis is often folded transversely, and several types of raised ridges or folds, known as pilasters and spines, may function as stimulator or hold-fast surfaces during copulation (Baur 1998). Special glandular regions can be present in the penis wall of some species. For example, the appending penial gland of slugs of the genus *Deroceras* consists of one or more finger-like appendages located at the end of the penis. In *Deroceras gorgonium*, the penial gland is particularly large, consisting of a huge bundle of branched processes (Reise et al. 2007). In most *Deroceras* species, the appending glands are everted during copulation and spread on the partner's upper body wall transferring a secretion (Reise et al. 2007). The gland of *D. gorgonium*, however, was also spread underneath the partner's body. It is suggested that this secretion has a similar manipulative function as those involved in dart shooting (see below: Auxiliary copulatory organs).

In those stylommatophoran species that have a penis, the animals insert it simultaneously or sequentially into the partners's genital pore during mating. Some stylommatophorans, however, have external sperm exchange, by which sperm is deposited on the mate's everted penis without intromission (Emberton 1994). Several slug species exhibit spectacular aerial matings (Gerhardt 1933; Chace 1952; Falkner 1992). Copulating pairs are hanging on thick mucus ropes suspended from trees or vertical walls with everted penes. The penes entwine and exchange sperm at their tips, completely outside of the body. In these species the penis is often remarkably long in relation to body length. For example, the uncoiled penis reaches a length of 60 cm in the 12–15 cm long slug *Limax corsicus* and 85 cm in the 13–15 cm long *Limax redii* (Falkner 1990). In other species with external sperm exchange, courtship and mating are usually performed on horizontal surfaces such as old leaves. The mating partners are side by side and exchange sperm masses from penis to penis, for example in *Deroceras rodnae* (Reise 1995).

Other stylommatophoran species, especially slugs, normally lack penes altogether; mating in these species may be accomplished by pressing the genital pores together directly, or by using non-homologous penis-like structures (derived from other parts of the terminal genitalia) for sperm transfer (Tompa 1984).

Phally polymorphism refers to a male genital polymorphism, in which two or three sexual morphs co-occur. The penis can be reduced or absent. These animals are referred to as hemiphallic and aphallic, respectively. The female reproductive organs, however, are always fully developed. Euphallic individuals, in contrast, have fully developed male and female reproductive organs. Hemiphallic and aphallic individuals cannot transfer sperm to mating partners; they can theoretically only reproduce uniparentally (i.e., by parthenogenesis or self-fertilization) or by outcrossing as females, whereas euphallic individuals can reproduce by outcrossing as male or female, as well as by selfing. Aphallic and hemiphallic individuals have been reported in numerous species of basommatophoran and stylommatophoran gastropods (Watson 1923; Pokryszko 1987; Doums et al. 1998;

Jordaens et al. 1998; Leonard et al. 2007). In most studies the proportion of aphallic individuals varies widely among populations. In the rock-dwelling land snail *Chondrina clienta*, the frequency of aphally varied from 52–99% in 23 natural populations in Sweden (Baur et al. 1993) and from 1–89% in 21 populations of *C. avenacea* in Switzerland (Baur & Chen 1993). Jordaens et al. (1998) recorded frequencies of hemiphallic individuals of 81–100% in 17 European populations of *Zonitoides nitidus*, with differences between geographic regions (Belgium, Germany, and Sweden).

Phally polymorphism evolved at least 13 times independently in basommatophoran and stylommatophoran gastropods, each time with euphally as the ancestral condition (Schrag & Read 1996). The determination of aphally, however, is still unclear. In the basommatophoran snail *Bulinus*, breeding experiments proved that both genetic and environmental factors may play a role in phally expression (see chapter of Jarne et al.). In *C. clienta*, phally expression can be influenced by environmental conditions (Baur et al. 1993). In the slug *Deroceras laeve*, the development of male reproductive organs is inhibited by low temperatures and/or exposure to light (Nicklas & Hoffmann 1981).

An increase in the frequency of uniparental reproduction, most probably self-fertilization, is assumed in populations with large proportions of aphallic or hemiphallic individuals. In fact, the lack of heterozygotes in otherwise polymorphic *C. clienta* and *Z. nitidus* populations indicates a uniparental breeding system (Baur & Klemm 1989; Jordaens et al. 1998). With increasing number of aphallics or hemiphallics in a population the extent of sperm competition may also decrease.

In the slug *Deroceras laeve*, aphally appears to be a character entirely different from that in all other species (Pokryszko 1987; Jordaens et al. 2006). This species has external sperm transfer between intertwined penes and the penis is needed for sperm receipt. Hence, aphallic and hemiphallic individuals are unable to receive sperm. These individuals are therefore restricted to reproduce uniparentally (Reise & Hutchinson 2002). This means that the selfing rate of aphallic and hemiphallic individuals is one and that there is no frequency-dependent selection on aphally and therefore polymorphism cannot be maintained. Genetic drift and possibly directional selection will then ultimately lead to the fixation of aphallics or euphallics, depending on which of the two phally morphs has the higher relative fitness. Hence, there are only two stable states in *D. laeve*, purely aphallic and purely euphallic populations (Jordaens et al. 2006). Interestingly, however, in contrast to sex allocation theory, aphallic individuals of *D. laeve* did not reallocate resources from the lost male function towards the female function, nor towards any life-history trait (Jordaens et al. 2006).

Sperm exchange is usually reciprocal in stylommatophorans species with reciprocal intromission or external sperm transfer (Baur 1998; Baur et al. 1998). However, unidirectional sperm transfer may occur. Reise (1995) observed that during three out of 15 apparently normal copulations of *Deroceras rodnae* only one of the partners transferred a sperm mass.

Diverticulum and Bursa Copulatrix

The bursa tract diverticulum, when present, is a blind-ended tube, which is especially long in several species (e.g., in *Eobania vermiculata*; Tompa 1984). As a lateral continuation of the lower part of the bursa duct, the diverticulum is widespread in the Stylommatophora and apparently plesiomorphic (Barker 2001; Koene & Schulenberg 2005). The diverticulum is specifically positioned relative to the bursa duct opening (figure 10.1). During mating, it functions as the site of spermatophore uptake (Barker 2001). Within the lumen of the diverticulum of *Arianta arbustorum*, the spermatophore wall is dissolved or at least partly broken down. The digested material is taken up by epithelial cells and accumulated in cells of the connective tissue (Beese et al. 2006b). In *Helix pomatia*, the length of the diverticulum is highly variable; in some individuals it may even be reduced or entirely lacking (Hochpoechler & Kothbauer 1979; van Osselaer & Tursch 2000). The origin of the diverticulum is not clear. It has been suggested that the separation of the allospermiduct led to the evolution of the diverticulum (Hochpoechler & Kothbauer 1979; Visser 1981). However, there is no convincing evidence for this hypothesis.

The so-called bursa copulatrix (= gametolytic gland; Tompa 1984) is a sacculate reservoir, commonly connected to the female reproductive system via a thin duct. Sperm received (allosperm) travel up the spermoviduct to reach the spermatheca, where they are stored until fertilization (Lind 1973). The vast majority of allosperm (99.98% in *Cornu aspersum*; Rogers & Chase 2001), however, is

transferred into the bursa copulatrix. The function of the bursa copulatrix is the extracellular digestion and subsequent resorption of excess gametes (primary allosperm) and other reproductive products, such as secretions from the albumen gland, oviductal glands, seminal channel, and remnants of the spermatophore (Németh & Kovacs 1972; Els 1978; Gomez et al. 1991; Beese et al. 2006b). The stalk of the bursa copulatrix can exhibit strong peristaltic waves (Lind 1973).

In stylommatophorans with a sperm digesting organ connected to the female part of the reproductive tract, there is—at least theoretically—an opportunity for sperm selection by the female function of the hermaphrodite (cryptic female choice; cf. Eberhard 1996). In *A. arbustorum*, the length of the diverticulum shows a positive allometry and a high phenotypic variation compared to snail size, which suggests that the diverticulum is under directional sexual selection (Beese et al. 2006b). It has been suggested that the diverticulum in *A. arbustorum* has evolved in response to selection pressures imposed by divergent evolutionary interests between male and female function (Beese et al. 2006b). Indeed, a comparative study across stylommatophoran species indicates counter-adaptations between presence, relative length and placement of the diverticulum and the flagellum length (Koene & Schulenburg 2005).

Auxiliary Copulatory Organs

The reproductive system of Stylommatophora is plesiomorphically equipped with an auxiliary copulatory organ that plays an active role during mating (Gomez 2001). This auxiliary organ has been thought to facilitate reciprocal copulation (Nordsieck 1985) and mutual exchange of male gametes (Tompa 1984). More recently, its potential role in sperm competition has begun to be explored. Reflecting the great diversity in morphology, numerous terminologies have been applied to the auxiliary organ and its components (for reviews see Tompa 1984; Barker 2001).

For many stylommatophorans, the auxiliary copulatory organ comprises a tubular gland opening through a prominent papilla into the penis (Gomez 2001). During copulation, the papilla is protruded from the genital pore and pressed against the partner's body or genitalia, and may even be introduced into the mate's genitalia. This activity can be accompanied by expulsion of secretory material from the gland of the auxiliary copulatory organ.

In many stylommatophorans, the papilla of this auxiliary organ is equipped with a sharp, calcified or chitinous dart within a so-called dart sac (Davison et al. 2005). One or several glands, in a compact glandulous mass or elongate tubules, open to the sac. The dart is used to pierce the body of the mating partner during courtship. Even though darts may wound a partner, the elaborate structure of the dart apparatus suggests that it serves some adaptive function.

Dart shooting is best studied in *Cornu aspersum* (for a review see Chase 2007). Dart shooting occurs when a snail quickly everts the basal tubercle of the dart sac out of its everted genitals. The dart is never propelled through the air, because it is firmly attached by its base to the tubercle until it is lodged in the partner's tissue. Occasionally, the dart does not hit the partner. A new dart is produced within 5–6 days after dart shooting. Chase and Vaga (2006) found that in mating *C. aspersum* each dart was shot independently, and each animal appeared to be interested only in getting off the best possible shot, probably one that penetrates deeply near the genital pore. The outcomes of the dart shots affect neither the probability that the courtship will culminate in copulation nor the size of the ensuing sperm donation (Adamo & Chase 1988; Chase & Vaga 2006). The dart transfers a substance that induces conformational changes in the female reproductive tract of the recipient, closing off the entrance to the gametolytic bursa copulatrix and thus most likely reduce sperm digestion (Koene & Chase 1998; Chase & Blanchard 2006). Thus, successful dart shooting more than doubles the number of donated sperm that are stored by the recipient (Rogers & Chase 2001), and it significantly increases the relative paternity when a dart shooter competes with an unsuccessful shooter (Landolfa et al. 2001; Rogers & Chase 2002). In the Helicidae (e.g., *C. aspersum* and *H. pomatia*), the dart is bladed and shot once into the partner, where it stays behind the skin (Chase 2007). In some species (*Euhadra subnimbosa*, *Helminthoglypta* spp.), however, mating snails stab their partner repeatedly with the same dart (Koene & Chiba 2006).

These findings indicate a function for the dart shooting behavior in the survival and storage of allosperm. Thus, an indirect cost might be the partial loss of control over fertilization by the female function. It follows that any defense against this cost should be expressed in the sperm-receiving organs, but not in courtship behavior or sperm transfer. Indeed, Koene & Schulenburg (2005) found evidence for the coevolution of the dart apparatus and the bursa tract diverticulum. Their phylogenetic analyses revealed that the length of the diverticulum increases as the size and complexity of the dart apparatus increases. One interpretation of these findings is that the female function is responding defensively to male-function manipulation by means of the dart (Chase & Vaga 2006). Dart shooting is perhaps the best studied sexually selected behavior in stylommatophorans.

Ova

The majority of stylommatophorans are oviparous with eggs laid singly or in batches; embryogenesis occurs after oviposition. All oviparous stylommatophorans thus far examined deposit individual eggs (i.e., each ovum is surrounded by its own egg shell or distinct jelly layer) and not egg masses or capsules, such as occur in the freshwater pulmonates and in most marine prosobranchs (Tompa 1976). Eggs of stylommatophorans are cleidoic; they contain all the nutrients and trace elements needed for a successful embryonic life and have direct development (Tompa 1980).

Parental investment may often be critical to the survival and growth of young, but the larger the investment per offspring, the lower the number of offspring that can be produced. Several models have been developed to predict the optimal size of offspring under different environmental conditions. Although the models make different predictions, these are based on the assumption that egg size is a reliable measure of the amount and quality of resources invested in each offspring (i.e., larger eggs are supposed to contain more organic material).

The among-species variation in egg size is pronounced ranging from 0.5 mm (in *Punctum pygmaeum* with an adult shell width of 1.5 mm; Baur 1989) to 50 mm (in *Strophocheilus (Borus) popelairianus* with a shell length of 15–23 cm; Bequaert 1948). Within species, egg size of *Arianta arbustorum* varies both among populations and seasons, as do protein concentration of eggs and clutch size (A. Baur & B. Baur 1997, 1998). Within clutch, however, eggs vary little in size (A. Baur 1994). Protein concentrations in freshly-laid eggs range from 14.2% of their dry weight in *Helix pomatia* (Alyakrinskaya 1981) to 25.5% in

A. arbustorum (A. Baur 1994) and 38.8% in *Sphincterochila boissieri* (Yom-Tov 1971).

Stylommatophoran eggs contain calcium carbonate which is used for the calcification of the embryonic shell and for the deposition of calcium reserves in the first calcium cells which differentiate during the embryonic life (Fournié & Chétail 1984). Developing embryos resorb calcium from their egg shells (Tompa 1976), and in some species the hatching young eat their own egg shell and those of unhatched siblings (B. Baur 1992b, 1994a).

In several stylommatophoran species hatchling size is positively correlated with egg size (e.g., in *Cepaea nemoralis* and *A. arbustorum*; Wolda 1963; Baur 2007). Thus, a larger parental investment in single eggs results in larger hatchlings. Large hatchlings, in turn, may enjoy an enhanced survivorship compared to small hatchlings. For example, large hatchlings of *Strophocheilus oblongus* more frequently survived immediate posthatching starvation than small ones kept under identical conditions (Tompa 1984). However, hatchlings of *A. arbustorum* emerging from larger eggs have also a longer developmental period than those from smaller eggs (Baur 2007). Because the resources available to gastropods for gamete production are finite, the animals may produce either few large eggs, or many small ones. In many stylommatophoran species, egg size is negatively correlated with the number of eggs produced (Heller 2001).

There is a large within-species variation in number and size of egg clutches, mainly influenced by the size of the parent snail and by environmental factors such as intra- and interspecific competition, and seasonality in climate (Wolda 1967; Wolda & Kreulen 1973; Baur 1990; Baur & Baur 1990). Snails of minute size produce only a few eggs at any time, and deposit them singly (e.g., six eggs in *Punctum pygmaeum* during an average life span of 170 days; Baur 1989). Larger gastropods produce many more eggs in their life: *Deroceras reticulatum* up to 500 eggs (Carrick 1938), *A. arbustorum* 800 eggs (Baur & Raboud 1988) and *Vaginulus borellians* over 1300 eggs (Runham & Hunter 1970).

As a modification of simple oviparity, the eggs may be retained for periods of different length within the female reproductive tract, resulting in a shorter time from laying to hatching (hereafter called egg retention). In ovoviviparous species, eggs are retained within the female reproductive tract for the entire embryonic period. Hatching may occur just after oviposition, or young may hatch

from the egg inside the female reproductive tract followed by birth (B. Baur 1994b). In viviparous species, there is a transfer of nutritional material from the parent to the developing embryo, which is retained in the female reproductive tract until extrusion as free living young. Variations in pattern of egg retention and development are virtually continuous, but a division into oviparous species, species with egg retention, ovoviviparous and viviparous species is generally accepted (Solem 1972).

Egg-retaining snails can keep their eggs for any (longer) period of time in the female reproductive tract after they are formed. If the conditions for oviposition become favorable (e.g., the soil is moist or soft enough to allow a hole to be excavated for a nest), then the snails release their eggs immediately, as any oviparous snail would do. For example, *Limicolaria martensiana* retains its eggs when it aestivates during the dry season in Central Africa (Owiny 1974). At the beginning of the rainy season, eggs and young are immediately deposited, ensuring them the best prospects of survival.

In ovoviviparous species the eggs are arranged in succession in the reproductive tract. The egg shell becomes resorbed by the parent or is consumed by the embryo, which uses the calcium carbonate to build up its own shell. Thus, ovoviviparity is an extreme form of egg retention (the young hatch in the reproductive tract). In some species, reproduction can either be oviparous, egg-retaining or ovoviviparous (Owen 1965; Peake 1978). For example, individuals of *Lacinaria biplicata* are usually ovoviviparous, but under favorable environmental conditions they lay eggs with well-developed embryos (Falkner 1990). Partulidae are ovoviviparous snails that occur on the Society Islands. In these snails the egg shell is resorbed by the parent before birth (Murray & Clarke 1966).

There are few reports on viviparity in stylommatophorans. In *Tekoulina pricei*, which occurs in the Cook Islands, 5–7 embryos of increasing size were found in the uterine oviduct (Solem 1972). However, details on the mechanism of nutrition are unknown.

SPERM AND EJACULATE CHARACTERISTICS

Stylommatophoran sperm are characterized by a small head with a very small but often complex nucleus (Thompson 1973). The nucleus consists of

densely packed chromatin and often has a pronounced helical surface sculpturing. The large
interspecific variation in sperm morphology is frequently used as a taxonomic character (Thompson
1973; Healy 1988, 1996; Luchtel et al. 1997).

Parivar (1981) described sperm dimorphism in
the slug *Arion ater* involving two different types of
sperm, eupyrene and apyrene. The most obvious
difference between the two sperm types is the lack
of an acrosome in the apyrene form. Apyrene sperm
are not capable of fertilizing eggs. Luchtel et al.
(1997) suggested that apyrene sperm may have
some physiological or endocrinological role in the
gonad. There is, however, no further evidence for
sperm dimorphism in any other stylommatophoran
species. Yet, abnormal sperm (lack of acrosomes,
modified centrioles, multiple flagella) occur in large
numbers in several terrestrial slug species (e.g.,
Milax gagates and *Deroceras agrestis*; South 1992).
The function of abnormal sperm still remains to be
clarified in Stylommatophora.

Much interest has been focused on theory concerning the significance of the size, number and
quality of sperm (e.g., Parker 1982; Parker & Begon
1993; Pizzari & Birkhead 2002). The size of sperm
may influence their power and swimming speed as
well as longevity because of changes in the energetic demands of longer or shorter flagella. In taxa
with sperm storage organs, sperm length may determine the ability to reach the storage organs
first and to move to the ovum from the storage
organs once ovulation takes place. However,
assuming a fixed resource budget, smaller sperm
may allow males to produce more gametes, which
may be adaptive if sperm compete numerically
(Parker 1982). Confounding variables, such as the
morphology and biochemistry of the female reproductive tract, might also affect sperm form and
function.

Information on the size of spermatozoa is summarized in Thompson (1973). Spermatozoa of
stylommatophorans are among the longest of
the molluscs (e.g., 850 μm in *Helix pomatia* and
1140–1400 μm (of which the head accounts for
only 10 μm) in *Hedleyella falconeri*). Minoretti and
Baur (2006) examined variation in sperm length in
individuals of *Arianta arbustorum* from four natural populations in Switzerland. Sperm are monomor-
phic in this species. Like other stylommatophorans,
A. arbustorum produces extremely long sperm.
Independent of adult shell size, sperm length

differed among populations (mean values of four
populations: 878, 898, 913, and 939 μm) and—to
a minor extent—even among individuals within
populations. Individual snails showed consistent
sperm length in successive matings. Mean sperm
length of an individual, however, was not correlated with the number of sperm delivered in a spermatophore. No further information on intraspecific
and interindividual variation in sperm length is
available for any other stylommatophoran species.

Sperm number, in some cases, is an important
determinant for achieving successful fertilization in
sperm competition (Birkhead & Møller 1998).
Theoretical models and empirical evidence from
various studies suggest that, fundamentally, numerical superiority is an adaptive strategy for sperm
competition (Birkhead & Møller 1998). In *Succinea
putris*, individuals with different mating history
transferred between 188,000 and 6,392,000 sperm
to their partners (Jordaens et al. 2005), whereas
in individuals of *A. arbustorum* that copulated
for the first time the number of sperm delivered
ranged between 803,000 and 3,969,000 (Baur et al.
1998). However, only a small portion of the
sperm transferred may reach the female sperm-
storage organ of the mating partner (see above:
Carrefour). After a successful copulation individuals needed 8–21 days to replenish their sperm
reserves (Locher & Baur 1999; Hänggi et al. 2002).
Snails from different populations of *A. arbustorum*
differed in the number of sperm delivered (Baminger
et al. 2000).

A synthesis of the available literature on gonochoristic animals indicates that sperm-quality traits
(proportion of live, morphologically normal spermatozoa, and motility of spermatozoa) affect fertilization success and that they are important in both
sperm competition and cryptic female choice (Snook
2005). Mean sperm velocity in *A. arbustorum* was
neither influenced by the shell size of the snails, nor
did it differ between two populations (Minoretti &
Baur 2006). However, mean sperm velocity differed among individual snails (range 52–112 μm/s).
Furthermore, the percentage motility and longevity
of sperm differed between snails from the two
populations, but were not affected by shell size. No
correlations were found between length, velocity,
percentage motility, and longevity of sperm. Thus,
individual snails differed in sperm quality. This
interindividual variation may partly explain differences in fertilization success (see below).

SPERM COMPETITION AND CRYPTIC FEMALE CHOICE

Male Aspects

Sperm competition is the competition between the spermatozoa from two or more males to fertilize the eggs of a single female during one reproductive cycle (Parker 1970). Evidence for promiscuity and multiple paternity in broods is available for several stylommatophoran species. Individuals of *Helix pomatia*, *Cornu aspersum*, *Cepaea nemoralis*, and *Arianta arbustorum* have been observed to mate repeatedly with different partners in the course of a reproductive season resulting in multiple-sired broods (Wolda 1963; Murray 1964; Baur 1988b; Lind 1988; Fearnley 1996). Viable allosperm indicated by fertilized eggs have been found 108 days after the last copulation in the tropical snail *Limicolaria flammea* (Egonmwan 1990), 520 days in *L. martensiana* (Owiny 1974), 341 days in *Achatina fulica* and 476 days in *Macrochlamys indica* (Raut & Ghose 1979) and 4 years in *Cepaea nemoralis* (Duncan 1975). However, sperm viability is not a simple function of time. In *A. fulica* and *M. indica*, the viability of sperm stored is influenced by the length of the aestivation period (Raut & Ghose 1982).

Direct evidence for sperm competition in stylommatophorans is so far scarce, exceptions are *C. aspersum* (see above: Auxiliary copulatory organs), *Succinea putris* and *A. arbustorum* (B. Baur 1994c; Rogers & Chase 2002; Jordaens et al. 2005; Chase & Vaga 2006). Paternity analysis in broods of wild-caught *A. arbustorum* indicated that at least 63% of the snails used sperm from two or more mates for the fertilization of their eggs (B. Baur 1994c). Sperm precedence is the differential sperm usage from consecutive matings (mating order effect). It is typically measured as the proportion of eggs fertilized by the second of two mates (the P_2 value). Sperm precedence (P_2) in doubled-mated *A. arbustorum* was influenced by the time between the two matings when the mating delay exceeded 70 days (one reproductive season). In the first brood of snails that mated twice within 70 days, P_2 averaged 0.34, indicating precedence of sperm from the first mate (B. Baur 1994c). In contrast, P_2 averaged 0.76 in broods of snails that remated in the following season, indicating a decreased viability of sperm from the first mate. Analysis of long-term sperm utilization in 23 snails that laid 3–9 batches over

2 years revealed striking differences among individuals. Five snails (21.7%) exhibited precedence of sperm from the first mate throughout, eight snails (34.8%) showed precedence of sperm from the second mate throughout, whereas 10 snails (43.5%) exhibited sperm mixing in successive batches. This indicates that different mechanisms might be involved in creating the observed inter-individual variation in sperm precedence.

Female Aspects

Until recently, most research concentrated on male aspects of sperm competition in gonochoristic animals. In the past few years, there has been increasing interest in the possibility that females influence the outcome of sperm competition by cryptic female choice and selective sperm use (Eberhard 1996). Females might be able to discriminate between and differentially utilize the sperm of different males, a process referred to as "sperm choice" (Birkhead 1998). There are broad and narrow definitions of "sperm choice"; some authors make it synonymous with "cryptic female choice" (see Eberhard 2000; Kempenaers et al. 2000; Pitnick & Brown 2000). Cryptic female choice has been defined as nonrandom paternity biases resulting from female morphology, physiology or behavior that occur after coupling (Pitnick & Brown 2000). This definition ascribes to sperm choice any biases in paternity owing to the way females handle sperm, regardless of the specific mechanism or evolutionary causes, and regardless of proximate control. The only relevant consideration for this definition is whether a female-mediated process generates sexual selection on males. A general problem with cryptic female choice is that it is difficult to rule out the direct influence from males (e.g., Edvardsson & Arnqvist 2000).

In the context of cryptic female choice, the sites of sperm storage (spermatheca), fertilization and sperm digestion (bursa copulatrix) are of major interest. The sites of sperm storage were examined in *Arianta arbustorum* that remated successfully. In some snails a part of the spermathecal tubules was filled with spermatozoa, while in other animals no sperm were found in the spermatheca (Haase & Baur 1995). In these snails sperm were found exclusively in the sperm-digesting bursa copulatrix. This suggests that the female reproductive system of *A. arbustorum* may be able to control fertilization by a selective digestion of sperm from certain mating partners (cf. Eberhard 1991, 1996).

The morphology of the sperm storage organ (spermatheca) may also influence the outcome of sperm competition in stylommatophorans, as shown in insects (Simmons & Siva-Jothy 1998; see above: Carrefour). Bojat and Haase (2002) assessed the amount of allosperm stored in the spermatheca of *A. arbustorum* in relation to the structure of the spermatheca (number of spermathecal tubules) in 18 individuals that had copulated once. Snails differed in patterns of sperm storage: two individuals used 100% of their spermathecal tubules, two used 80%, three 75%, two 66.7%, one 50%, two 40%, three 33.3%, two 25%, and one used 20%. The main tubule always contained sperm (51–100% of the total amount of sperm stored, i.e., more than all lateral tubules combined). The amount of sperm stored was not correlated with the volume of the received spermatophore. However, the amount of sperm stored was positively correlated with the number of spermathecal tubules. This suggests that the female function of the receiver controls the number of sperm stored.

Baminger and Haase (1999) examined whether the variation in number of spermathecal tubules and the amount of allosperm stored are influenced by the risk of sperm competition, as indicated by the local density of adult *A. arbustorum* in six natural populations in the Eastern Alps, Austria. The number of spermathecal tubules ranged from two to nine. However, snails from the six populations did not differ in either the mean number of spermathecal tubules or the cumulative length of the tubules. Individuals from different populations did not differ in the amount of sperm stored, neither was the amount of sperm stored correlated with population density. Similarly, no correlation was found between the number of spermathecal tubules or the cumulative length of all tubules and the local density of five *C. aspersum* populations in Greece (Koemtzopoulos & Staikou 2007). This suggests that the risk of sperm competition does not affect the number of spermathecal tubules. However, it is still not known whether individuals in high-density populations store allosperm from a larger number of different mating partners than those in low-density populations.

The variation in genitalia size could be related to the intensity of sexual selection. Baminger and Haase (2000) tested this hypothesis by examining the variability of the distal genitalia involved in spermatophore production, reception, and manipulation in adult individuals of *A. arbustorum* from six natural populations. The intensity of sexual selection was estimated by measuring local population density. The size of the genitalia was unexpectedly inversely related to population density, probably because of an increased inhibitory effect of snail mucus (i.e., an effect of intraspecific competition). Patterns of variation of female and male characters did not differ. However, the influence of sexual selection on genitalia size and variance could not be unambiguously determined.

Sexual Conflict

Mating conflicts arise because males are generally interested in mating frequently and promiscuously, whereas females prefer to be selective. Conflicts occur after insemination because the male is interested in having all his sperm survive so that a maximum number can fertilize eggs. The female, however, benefits from mating with other males that will cause the displacement of sperm, or she may digest some of his sperm to gain energy. Sexual conflict can also occur in hermaphrodites because the male and female functions retain their separate interests even though they are united in the same individual (Michiels 1998). Sexual conflict might be manifested in several ways, but mating conflict is especially prominent in hermaphrodites because two individuals sometimes attempt to perform the same sexual role (Leonard 1991; Michiels 1998).

Hermaphroditic individuals in a population would benefit from mating primarily in the more fitness-enhancing sexual role, leading to a conflict of interest between two prospective mating partners (Charnov 1979). Gamete trading might have evolved to resolve the sexual conflict in simultaneous hermaphrodites (Leonard 1991). The gamete-trading model is based on the premise that the preferred role for a simultaneous hermaphrodite will be the one that controls fertilization. In particular, this model predicts that when the female function controls fertilization, the mating system will be based on sperm trading.

Baur et al. (1998) examined whether individuals of *Arianta arbustorum* adjust sperm release according to the potential risk of sperm competition incurred with a virgin or nonvirgin mating partner and whether sperm trading occurs in mating pairs. In controlled mating trials, focal snails were allowed to copulate with virgin or nonvirgin partners to simulate a different risk of sperm competition in a given mating. The number of sperm transferred was not related to the mating history of the partner. This indicates that individuals of *A. arbustorum* are not able to adjust sperm expenditure to the mating history of the partner. Furthermore, individuals of *A. arbustorum* did not respond to experimentally increased cues from conspecifics, which were assumed to mimic a high risk of sperm competition by delivering more sperm (Locher & Baur 2000a).

Sexual conflict may play a key role in the evolution of dart shooting in Stylommatophora (reviewed in Chase 2007). The dart increases allosperm storage and paternity, probably via the transfer of an allohormone that inhibits sperm digestion (see Auxiliary copulatory organs). An interspecific comparison of dart-possessing gastropods revealed coevolution between darts and spermatophore-receiving organs that is consistent with counter-adaptation against an allohormone manipulation (Koene & Schulenburg 2005). Furthermore, sexual conflict may occur in the allocation of reproductive resources to the male and female function in stylommatophorans (Locher & Baur 2000b, 2002).

CONCLUSIONS AND SUGGESTIONS FOR FUTURE RESEARCH

Primary sexual characters are those fundamental for successful reproduction. This review provides insight into the enormous variation in reproductive characters and patterns in stylommatophoran gastropods and indicates potential consequences for fertilization and sperm competition. Selection acting on primary characters can be complex, especially in simultaneous hermaphrodites, because selection on the female role may also affect the male role within the same individual (and vice-versa). In this group of hermaphrodites almost all reproductive characters appear to be shaped by both natural and sexual selection. Furthermore, there is increasing evidence that intersexual counter-adaptations may drive correlated reproductive character evolution in stylommatophoran gastropods (Davison et al. 2005; Koene & Schulenburg 2005; Beese et al. 2006b). However, there is also evidence that life-history traits and habitat specificity have potentially influenced the evolution of reproductive morphology (Beese et al. 2009).

Life-history traits and habitat specificity should be integrated more frequently into studies of reproductive trait divergence. Several stylommatophoran species may be well-suited for studies on the evolution of reproductive traits, sexual conflicts and sperm competition. For example, the adaptive significance of variation in sperm characters such as length and swimming velocity is still not known. Careful studies of the morphology of the female reproductive tract with respect to allosperm storage and mating experiments using molecular markers for paternity analyses should prove to be particularly rewarding. There are many topics that remain largely unexplored and there is much to be learned in this most interesting animal group.

Acknowledgements   I thank Anette Baur, Kathleen Beese, Janet Leonard, and two anonymous reviewers for constructive comments on the manuscript. Financial support was received from the Swiss National Science Foundation.

REFERENCES

Adamo, S. A. & Chase, R. 1988. Courtship and copulation in the terrestrial snail, *Helix aspersa*. Canadian Journal of Zoology, 66, 1446–1453.

Alyakrinskaya, I. O. 1981. Egg nutrient content in gastropods. *Doklaidi Akademia Nauk SSSR*, 260, 245–248.

Baminger, H. & Haase, M. 1999. Variation in spermathecal morphology and amount of sperm stored in populations of the simultaneously hermaphroditic land snail *Arianta arbustorum*. Journal of Zoology, 249, 165–171.

Baminger, H. & Haase, M. 2000. Variation of distal genitalia in the simultaneously hermaphroditic land snail *Arianta arbustorum* (Pulmonata, Stylommatophora) caused by sexual selection? Biological Journal of the Linnean Society, 71, 599–613.

Baminger, H. & Haase, M. 2001. Spermatophore formation in the simultaneously hermaphroditic land snail *Arianta arbustorum* (Pulmonata: Stylommatophora: Helicidae). Netherlands Journal of Zoology, 51, 347–360.

Baminger, H., Locher, R. & Baur, B. 2000. Incidence of dart shooting, sperm delivery, and sperm storage in natural populations of the simultaneously hermaphroditic land snail *Arianta arbustorum*. Canadian Journal of Zoology, 78, 1767–1774.

Barker, G. M. 2001. Gastropods on land: phylogeny, diversity and adaptive morphology.

In: The Biology of Terrestrial Molluscs (Ed. by G. M. Barker), pp. 1–146. Wallingford: CABI Publishing.

Baur, A. 1994. Within- and between-clutch variation in egg size and nutrient content in the land snail *Arianta arbustorum*. Functional Ecology, 8, 581–586.

Baur, A. & Baur, B. 1997. Seasonal variation in size and nutrient content of eggs of the land snail *Arianta arbustorum*. Invertebrate Reproduction and Development, 32, 55–62.

Baur, A. & Baur, B. 1998. Altitudinal variation in size and composition of eggs in the land snail *Arianta arbustorum*. Canadian Journal of Zoology, 76, 2067–2074.

Baur, B. 1987. The minute land snail *Punctum pygmaeum* (Draparnaud) can reproduce in the absence of a mate. Journal of Molluscan Studies, 53, 112–113.

Baur, B. 1988a. Population regulation in the land snail *Arianta arbustorum*: Density effects on adult size, clutch size and incidence of egg cannibalism. Oecologia, 77, 390–394.

Baur, B. 1988b. Repeated mating and female fecundity in the simultaneously hermaphroditic land snail *Arianta arbustorum*. Invertebrate Reproduction and Development, 14, 197–204.

Baur, B. 1989. Growth and reproduction of the minute land snail *Punctum pygmaeum* (Draparnaud). Journal of Molluscan Studies, 55, 383–387.

Baur, B. 1990. Seasonal changes in clutch size, egg size and mode of oviposition in *Arianta arbustorum* (L.) (Gastropoda) from alpine populations. Zoologischer Anzeiger, 225, 253–264.

Baur, B. 1992a. Random mating by size in the simultaneously hermaphroditic land snail *Arianta arbustorum*: experiments and an explanation. Animal Behaviour, 43, 511–518.

Baur, B. 1992b. Cannibalism in gastropods. In: *Cannibalism*: Ecology and Evolution among Diverse Taxa (Ed. by M. A. Elgar & B. J. Crespi), pp. 102–127. Oxford: University Press.

Baur, B. 1994a. Interpopulation variation in propensity for egg cannibalism in the land snail *Arianta arbustorum*. Animal Behaviour, 48, 851–860.

Baur, B. 1994b. Parental care in terrestrial gastropods. Experientia, 50, 5–14.

Baur, B. 1994c. Multiple paternity and individual variation in sperm precedence in the simultaneously hermaphroditic land snail *Arianta*

arbustorum. Behavioral Ecology and Sociobiology, 35, 413–421.

Baur, B. 1998. Sperm competition in molluscs. In: Sperm Competition and Sexual Selection (Ed. by T. R. Birkhead & A. P. Møller), pp. 255–305. London: Academic Press.

Baur, B. 2007. Reproductive biology and mating conflict in the simultaneously hermaphroditic land snail *Arianta arbustorum*. American Malacological Bulletin, 23, 157–172.

Baur, B. & Baur, A. 1990. Experimental evidence for intra- and interspecific competition in two species of rock-dwelling land snails. Journal of Animal Ecology, 59, 301–315.

Baur, B. & Baur, A. 1992. Reduced reproductive compatibility in the land snail *Arianta arbustorum* from distant populations. Heredity, 69, 65–72.

Baur, B. & Baur, A. 1997. Random mating with respect to relatedness in the simultaneously hermaphroditic land snail *Arianta arbustorum*. Invertebrate Biology, 116, 294–298.

Baur, B. & Chen, X. 1993. Genital dimorphism in the land snail *Chondrina avenacea*: frequency of aphally in natural populations and morph-specific allocation to reproductive organs. The Veliger, 36, 252–258.

Baur, B. & Klemm, M. 1989. Absence of isozyme variation in geographically isolated populations of the land snail *Chondrina clienta*. Heredity, 63, 239–244.

Baur, B. & Raboud, C. 1988. Life history of the land snail *Arianta arbustorum* along an altitudinal gradient. Journal of Animal Ecology, 57, 71–87.

Baur, B., Chen, X. & Baur, A. 1993. Genital dimorphism in natural populations of the land snail *Chondrina clienta* and the influence of the environment on its expression. Journal of Zoology, 231, 275–284.

Baur, B., Locher, R. & Baur, A. 1998. Sperm allocation in the simultaneously hermaphroditic land snail *Arianta arbustorum*. Animal Behaviour, 56, 839–845.

Bayne, C. 1968. Histochemical studies on the egg capsules of eight gastropod molluscs. Proceedings of the Malacological Society of London, 38, 199–212.

Beese, K. & Baur, B. 2006. Expandable spermatheca influences sperm storage in the simultaneously hermaphroditic snail *Arianta arbustorum*. Invertebrate Reproduction and Development, 49, 93–101.

Beese, K., Beier, K. & Baur, B. 2006a. Coevolution of male and female reproductive traits in a simultaneously hermaphroditic land snail. Journal of Evolutionary Biology, 19, 410–418.

Beese, K., Beier, K. & Baur, B. 2006b. Bursa tract diverticulum in the hermaphroditic land snail *Arianta arbustorum* (Stylommatophora: Helicidae): Morphology, function, and evolutionary implications. Journal of Morphology, 267, 940–953.

Beese, K., Armbruster, G. F. J., Beier, K. & Baur, B. 2009. Evolution of female sperm storage organs in the carrefour of stylommatophoran gastropods. Journal of Zoological Systematics and Evolutionary Research, 47, 49–60.

Bequaert, J. 1948. Monography of the Strophocheilidae, a neotropical family of terrestrial mollusks. Bulletin of the Museum of Comparative Zoology Harvard, 100, 1–210.

Birkhead, T. R. 1998. Cryptic female choice: Criteria for establishing female sperm choice. Evolution, 52, 1212–1218.

Birkhead, T. R. &. Møller A. P. (Eds). 1998. Sperm Competition and Sexual Selection. London: Academic Press.

Bojat, N. C. & Haase, M. 2002. Sperm storage in the simultaneously hermaphroditic land snail, *Arianta arbustorum*. Journal of Zoology, 258, 497–503.

Bojat, N. C., Sauder, U. & Haase, M. 2001a. Functional anatomy of the sperm storage organs in Pulmonata: The simple spermatheca of *Bradybaena fruticum* (Gastropoda, Stylommatophora). Zoomorphology, 121, 243–255.

Bojat, N. C., Sauder, U. & Haase, M. 2001b. The spermatheca in the land snail, *Arianta arbustorum* (Pulmonata: Stylommatophora): Muscle system and potential role in sexual selection. Invertebrate Biology, 120, 217–226.

Bojat, N. C., Sauder, U. & Haase, M. 2001c. The spermathecal epithelium, sperm and their interactions in the hermaphroditic land snail *Arianta arbustorum* (Pulmonata, Stylommatophora). Zoomorphology, 120, 149–157.

Brinders, E. M. & Sirgel, W. F. 1992. The morphology and histology of the genital system of *Trigonephrus gypsinus* and *Trigonephrus latezonatus* (Gastropoda: Pulmonata). Annals of the University Stellenbosch, 3, 1–27.

Carrick, R. 1938. The life-history and development of *Agriolimax agrestis*, the grey field slug. Transactions of the Royal Society of Edinburgh, 59, 563–597.

Chace, L. M. 1952. The aerial mating of the great slug. Discovery, 13, 356–359.

Chapman, T., Arnqvist, G., Bangham, J. & Rowe, L. 2003. Sexual conflict. Trends in Ecology and Evolution, 18, 41–47.

Charnov, E. L. 1979. Simultaneous hermaphroditism and sexual selection. Proceedings of the National Academy of Sciences U.S.A., 76, 2480–2484.

Chase, R. 2007. The function of dart shooting in helicid snails. American Malacological Bulletin, 23, 183–189.

Chase, R. & Blanchard, K. 2006. The snail's love-dart delivers mucus to increase paternity. Proceedings of the Royal Society of London, Series B, 273, 1471–1475.

Chase, R. & Vaga, K. 2006. Independence, not conflict, characterizes dart-shooting and sperm exchange in a hermaphroditic snail. Behavioral Ecology and Sociobiology, 59, 732–739.

Chen, X. 1993. Comparison of inbreeding and outbreeding in hermaphroditic *Arianta arbustorum* (L.) (land snail). Heredity, 71, 456–461.

Chung, D. J. D. 1987. Courtship and dart shooting behavior of the land snail *Helix aspersa*. The Veliger, 30, 24–39.

Davison, A., Wade, C. M., Mordan, P. B. & Chiba, S. 2005. Sex and darts in slugs and snails (Mollusca: Gastropoda: Stylommatophora). Journal of Zoology, 267, 329–338.

Doums, D., Viard, F. & Jarne, P. 1998. The evolution of phally polymorphism. Biological Journal of the Linnean Society, 64, 273–296.

Duncan, C. J. 1975. Reproduction. In: Pulmonates, Vol.1 (Ed. by V. Fretter & J. Peake), pp. 309–365. London: Academic Press.

Eberhard, W. G. 1991. Copulatory courtship and cryptic female choice in insects. Biological Review, 66, 1–31.

Eberhard, W. G. 1996. Female Control: Sexual Selection by Cryptic Female Choice. Princeton: University Press.

Eberhard, W. G. 2000. Criteria for demonstrating postcopulatory female choice. Evolution, 54, 1047–1050.

Edvardsson, M. & Arnqvist, G. 2000. Copulatory courtship and cryptic female choice in red flour beetles *Tribolium castaneum*. Proceedings of the Royal Society of London, Series B, 267, 1–5.

Egonmwan, R. I. 1990. Viability of allosperm in the garden snail *Limicolaria flammea*, Muller

(Gastropoda: Pulmonata). Bioscience Research Communications, 2, 87–92.

Els, W. J. 1978. Histochemical studies on the maturation of the genital system of the slug *Deroceras laeve* (Pulmonata, Limacidae), with special reference to the identification of mucosubstances secreted by the genital tract. Annals of the University Stellenbosch, 1, 1–116.

Emberton, K. C. 1994. Polygyrid land snail phylogeny: external sperm exchange, early North American biogeography, iterative shell evolution. Biological Journal of the Linnean Society, 52, 241–271.

Falkner, G. 1990. Binnenmollusken. In: *Weichtiere.* Europäische Meeres- und Binnenmollusken (Ed. by R. Fechter & G. Falkner), pp. 112–278. Munich: Mosaik Verlag.

Falkner, G. 1992. Grandioser Seilakt zu nächtlicher Stunde: Paarung des Tigerschnegels. In: Die grosse Bertelsmann Lexikothek, Naturenzyklopädie Europas, Vol. 6 (Ed. by J. H. Reichholf & G. Steinbach), pp. 282–283. Munich: Mosaik Verlag.

Fearnley, R. H. 1996. Heterogenic copulatory behaviour produces non-random mating in laboratory trials in the land snail *Helix aspersa* Müller. Journal of Molluscan Studies, 62, 159–164.

Flasar, I. 1967. Der innere Bau der Befruchtungstasche bei *Oxychilus draparnaudi* (Beck) und die Geschichte ihrer Entdeckung und Erforschung bei anderen Pulmonaten. Acta Societatis Zoologicae Bohemoslovacae, 31, 150–158.

Fournié, J. & Chétail, M. 1984. Calcium dynamics in land gastropods. American Zoologist, 24, 857–870.

Gerhardt, U. 1933. Zur Kopulation der Limaciden. I. Mitteilung. Zeitschrift für Morphologie und Ökologie der Tiere, 27, 401–450.

Gomez, B. J. 2001. Structure and functioning of the reproductive system. In: The Biology of Terrestrial Molluscs (Ed. by G. M. Barker), pp. 307–330. Wallingford: CABI Publishing.

Gomez B. J., Angulo, E. & Zubiaga, A. 1991. Ultrastructural analysis of the morphology and function of the spermatheca of the pulmonate slug *Arion subfuscus*. Tissue and Cell, 23, 357–365.

Haase, M. & Baur, B. 1995. Variation in spermathecal morphology and storage of spermatozoa in the simultaneously hermaphroditic land snail *Arianta arbustorum* (Gastropoda: Pulmonata: Stylommatophora). Invertebrate Reproduction and Development, 28, 33–41.

Hänggi, C., Locher, R. & Baur, B. 2002. Intermating interval and number of sperm delivered in the simultaneously hermaphroditic land snail *Arianta arbustorum* (Pulmonata: Helicidae). The Veliger, 45, 224–230.

Healy, J. M. 1988. Sperm morphology and its systematic importance in the Gastropoda. Malacological Review, Supplement, 4, 251–266.

Healy, J. M. 1996. Molluscan sperm ultrastructure: correlation with taxonomic units within the Gastropoda, Cephalopoda and Bivalvia. In: Origin and Evolutionary Radiation of the Mollusca (Ed. by J. D. Taylor), pp. 99–113. Oxford: University Press.

Heller, J. 1990. Longevity in molluscs. Malacologia, 31, 259–295.

Heller, J. 1993. Hermaphroditism in molluscs. Biological Journal of the Linnean Society 48, 19–42.

Heller, J. 2001. Life history strategies. In: The Biology of Terrestrial Molluscs (Ed. by G. M. Barker), pp. 413–445. Wallingford: CABI Publishing.

Heller, J. & Dolev, A. 1994. Biology and population dynamics of a crevice-dwelling landsnail, *Cristataria genezarethana* (Clausiliidae). Journal of Molluscan Studies, 60, 33–46.

Hochpoechler, F. & Kothbauer, H. 1979. Triaulie bei Heliciden (Gastropoda). Zur phylogenetischen Bedeutung des Bursa copulatrix Divertikels. Zeitschrift für Zoologie, Systematik und Evolutionsforschung, 17, 281–285.

Jennions, M. D. & Petrie, M. 2000. Why do females mate multiply? A review of the genetic benefits. Biological Reviews, 75, 21–64.

Jeppesen, L. L. 1976. The control of mating behaviour in *Helix pomatia* L. (Gastropoda: Pulmonata). Animal Behaviour, 24, 275–290.

Jordaens, K., Backeljau, T., Ondina, P., Reise, H. & Verhagen, R. 1998. Allozyme homozygosity and phally polymorphism in the land snail *Zonitoides nitidus* (Gastropoda, Pulmonata). Journal of Zoology, 246, 95–104.

Jordaens, K., Pinceel, J. & Backeljau, T. 2005. Mate choice in the hermaphroditic land snail *Succinea putris* (Stylommatophora: Succineidae). Animal Behaviour, 70, 329–337.

Jordaens, K., Van Dongen, S., Temmerman, K. & Backeljau, T. 2006. Resource allocation in a simultaneously hermaphroditic slug with phally polymorphism. Evolutionary Ecology, 20, 535–548.

Kempenaers, B., Foerster, K., Questiau, S., Robertson, B. C. & Vermeirssen, E. L. M. 2000. Distinguishing between female sperm choice versus male sperm competition: A comment on Birkhead. Evolution, 54, 1050–1052.

Koemtzopoulos, E. & Staikou, A. 2007. Variation in spermathecal morphology is independent of sperm competition intensity in populations of the simultaneously hermaphroditic land snail *Cornu aspersum*. Zoology, 110, 139–146.

Koene, J. M. & Chase, R. 1998. Changes in the reproductive system of the land snail *Helix aspersa* caused by mucus from the love dart. Journal of Experimental Biology, 201, 2313–2319.

Koene, J. M. & Chiba, S. 2006. The way of the Samurai snail. American Naturalist, 168, 553–555.

Koene, J. M. & Schulenberg, M. 2005. Shooting darts: Co-evolution and counter-adaptation in hermaphroditic snails. BMC Evolutionary Biology, 5, 25.

Lamotte, M. 1951. Recherches sur la structure génétique des populations naturelles de *Cepaea nemoralis*. Bulletin biologique de la France et de la Belgique, Supplement 35, 1–239.

Landolfa, M. A., Green, D. M. & Chase, R. 2001. Dart shooting influences paternal reproductive success in the snail *Helix aspersa* (Pulmonata, Stylommatophora). Behavioral Ecology, 12, 773–777.

Leonard, J. L. 1991. Sexual conflict and the mating systems of simultaneously hermaphroditic gastropods. American Malacological Bulletin, 9, 45–58.

Leonard, J. L. Westfall, J. A. & Pearse, J. S. 2007. Phally polymorphism and reproductive biology in *Ariolimax (Ariolimax) buttoni* (Pilsbry and Vanatta, 1896) (Stylommatophora: Arionidae). American Malacological Bulletin, 23, 121–135.

Lessells, C. M. 2006. The evolutionary outcome of sexual conflict. Philosophical Transactions of the Royal Society of London, Series B, 361, 301–317.

Lind, H. 1973. The functional significance of the spermatophore and the fate of spermatozoa in the genital tract of *Helix pomatia* (Gastropoda: Stylomatophora). Journal of Zoology, 169, 39–64.

Lind, H. 1976. Causal and functional organization of the mating behaviour sequence in *Helix*

pomatia L. (Pulmonata, Gastropoda). Behaviour, 59, 162-202.

Lind, H. 1988. The behaviour of *Helix pomatia* L. (Gastropoda, Pulmonata) in a natural habitat. Videnskablige meddelelser fra Dansk naturhistorisk forening i København, 147, 67–92.

Little, C. 1990. The Terrestrial Invasion: An Ecophysiological Approach to the Origins of Land Animals. Cambridge: University Press.

Locher, R. & Baur, B. 1999. Effects of intermating interval on spermatophore size and sperm number in the simultaneously hermaphroditic land snail *Arianta arbustorum*. Ethology, 105, 839–849.

Locher, R. & Baur, B. 2000a. Sperm delivery and egg production of the simultaneously hermaphroditic land snail *Arianta arbustorum* exposed to an increased sperm competition risk. Invertebrate Reproduction and Development, 38, 53–60.

Locher, R. & Baur, B. 2000b. Mating frequency and resource allocation to male and female functions in the simultaneous hermaphrodite land snail *Arianta arbustorum*. Journal of Evolutionary Biology, 13, 607–614.

Locher, R. & Baur, B. 2002. Nutritional stress changes sex-specific reproductive allocation in the simultaneously hermaphroditic land snail *Arianta arbustorum*. Functional Ecology, 16, 623–632.

Luchtel, D. L., Martin, A. W., Deyrup-Olsen, I. & Boer, H. H. 1997. Gastropoda: Pulmonata. In: Microscopic Anatomy of Invertebrates. Vol. 6B: Mollusca II (Ed. by F. W. Harrison & A. J. Kohn), pp. 459–718. New York: Wiley-Liss.

Mann, T. 1984. Spermatophores. Development, Structure, Biochemical Attributes and Role in the Transfer of Spermatozoa. Berlin: Springer Verlag.

Meisenheimer, J. 1907. Biologie, Morphologie und Physiologie des Begattungsvorganges und der Eiablage von *Helix pomatia*. Zoologisches Jahrbuch, Abteilung Systematik und Ökologie, 25, 461–502.

Michiels, N. K. 1998. Mating conflicts and sperm competition in simultaneous hermaphrodites. In: Sperm Competition and Sexual Selection (Ed. by T. R. Birkhead & A. P. Møller), pp. 219–254. London: Academic Press.

Minoretti, N. & Baur, B. 2006. Among- and within-population variation in sperm quality in the simultaneously hermaphroditic land snail

Arianta arbustorum. Behavioral Ecology and Sociobiology, 60, 270–280.

Murray, J. 1964. Multiple mating and effective population size in *Cepaea nemoralis*. Evolution, 18, 283–291.

Murray, J. & Clarke, B. C. 1966. The inheritance of polymorphic shell characters in *Partula* (Gastropoda). Genetics, 54, 1261–1277.

Németh, A. & Kovacs, J. 1972. Ultrastructure of the epithelial cells of seminal receptacle in the snail *Helix pomatia* with special reference to the lysosomal system. Acta Biologica Academiae Scietiarum Hungaricea, 23, 299–308.

Nicklas, N. L. & Hoffmann, R. J. 1981. Apomictic parthenogenesis in a hermaphroditic terrestrial slug, *Deroceras laeve* (Müller). Biological Bulletin, 160, 123–135.

Nordsieck, H. 1985. The system of the Stylommatophora (Gastropoda), with special regard to the systematic position of the Clausiliidae: Importance of the excretory and genital systems. Archiv für Molluskenkunde, 116, 1–24.

Owen, D. F. 1965. A population study of an equatorial land snail, *Limicolaria martensiana* (Achatinidae). Proceedings of the Zoological Society of London, 144, 361–382.

Owiny, A. M. 1974. Some aspects of the breeding biology of the equatorial land snail *Limicolaria martensiana* (Achatinidae: Pulmonata). Journal of Zoology, 172, 191–206.

Parivar, K. 1981. Spermatogenesis and sperm dimorphism in land slug *Arion ater* L. (Pulmonata, Mollusca). Zeitschrift für mikroskopische-anatomische Forschung Leipzig, 95, 81–92.

Parker, G. A. 1970. Sperm competition and its evolutionary consequences in the insects. Biological Review, 45, 535–567.

Parker, G. A. 1982. Why are there so many tiny sperm? Sperm competition and the maintenance of two sexes. Journal of Theoretical Biology, 96, 281–294.

Parker, G. A. & Begon, M. E. 1993. Sperm competition games: sperm size and number under gametic control. Proceedings of the Royal Society of London, Series B, 253, 255–262.

Peake, J. F. 1978. Distribution and ecology of the Stylommatophora. In: Pulmonates, Vol. 2A: Systematics, Evolution and Ecology (Ed. by V. Fretter & J. F. Peake), pp. 429–526. London: Academic Press.

Pitnick, S. & Brown, W. D. 2000. Criteria for demonstrating female sperm choice. Evolution, 54, 1052–1056.

Pitnick, S., Markow, T. A. & Spicer, G. S. 1999. Evolution of multiple kinds of female sperm-storage organs in *Drosophila*. Evolution, 53, 1804–1822.

Pizzari, T. & Birkhead, T. R. 2002. The sexually-selected sperm hypothesis: Sex-biased inheritance and sexual antagonism. Biological Reviews, 77, 183–209.

Pokryszko, B. M. 1987. On the aphally in the Vertiginidae (Gastropoda: Pulmonata: Orthurethra). Journal of Conchology, 32, 365–375.

Pollard, E. 1975. Aspects of the ecology of *Helix pomatia* L. Journal of Animal Ecology, 44, 305–329.

Presgraves, D. C., Baker, R. H. & Wilkinson, G. S. 1999. Coevolution of sperm and female reproductive tract morphology in stalk-eyed flies. Proceedings of the Royal Society of London, Series B, 266, 1041–1047.

Raut S. K. & Ghose, K. C. 1979. Viability of sperm in two land snails, *Achatina fulica* Bodwich and *Macrochlamys indica* Godwin-Austen. The Veliger, 21, 486–487.

Raut S. K. & Ghose, K. C. 1982. Viability of sperm in aestivating *Achatina fulica* Bodwich and *Macrochlamys indica* Godwin-Austen. Journal of Molluscan Studies, 48, 84–86.

Reise, H. 1995. Mating behaviour of *Deroceras rodnae* Grossu & Lupu, 1965 and *D. praecox* Wiktor, 1966 (Pulmonata: Agriolimacidae). Journal of Molluscan Studies, 61, 325–330.

Reise, H. 2007. A review of mating behavior in slugs of the genus *Deroceras* (Pulmonata: Agriolimacidae). American Malacological Bulletin, 23, 137–156.

Reise, H. & Hutchinson, J. M. C. 2002. Penis-bitting slugs: wild claims and confusions. Trends in Ecology and Evolution, 17, 163.

Reise, H., Visser, S. & Hutchinson, J. M. C. 2007. Mating behaviour in the terrestrial slug *Deroceras gorgonium*: is extreme morphology associated with extreme behaviour? Animal Biology, 57, 197–215.

Rice, W. R. & Holland, B. 1997. The enemies within: intergenomic conflict, interlocus contest evolution (ICE), and the intraspecific Red Queen. Behavioral Ecology and Sociobiology, 41, 1–10.

Rigby, J. E. 1963. Alimentary and reproductive systems of *Oxychilus cellarius*. Proceedings of the Zoological Society of London, 141, 311–359.

Rigby, J. E. 1965. *Succinea putris*, a terrestrial opisthobranch mollusc. Proceedings of the Zoological Society of London, 144, 445–487.

Rogers, D. W. & Chase, R. 2001. Dart receipt promotes sperm storage in the garden snail *Helix aspersa*. Behavioral Ecology and Sociobiology, 50, 122–127.

Rogers, D. W. & Chase, R. 2002. Determinants of paternity in the garden snail *Helix aspersa*. Behavioral Ecology and Sociobiology, 52, 289–295.

Runham, N. W. 1988. Mollusca. In: Reproductive Biology of Invertebrates, Vol. III, Accessory Sex Glands (Ed. by K. G. Adiyodi & G. Adiyodi), pp. 113–188. Chichester: John Wiley & Sons.

Runham, N. W. & Hogg, J. 1992. The pulmonate carrefour. Proceedings of the 9th International Malacological Congress Leiden, 303–308.

Runham, N. W. & Hunter, P. J. 1970. Terrestrial Slugs. London: Hutchinson.

Schilder, A. 1950. Die Ursachen der Variabilität bei *Cepaea*. Biologisches Zentralblatt, 69, 79–103.

Schnetter, M. 1950. Veränderungen der genetischen Konstitution in natürlichen Populationen der polymorphen Bänderschnecken. Verhandungen der Deutschen Zoologischen Gesellschaft, 13, 192–206.

Schrag, S. J. & Read, A. F. 1996. Loss of male outcrossing ability in simultaneous hermaphrodites: phylogenetic analyses of pulmonate snails. Journal of Zoology, 238, 287–299.

Simmons, L. W. & Siva-Jothy, M. 1998. Sperm competition in insects: Mechanisms and the potential for selection. In: Sperm Competition and Sexual Selection (Ed. by T. R. Birkhead & A. P. Møller), pp. 341–434. London: Academic Press.

Siva-Jothi, M. T. & Hooper, R. E. 1995. The disposition and genetic diversity of stored sperm in females of the damselfly *Calopteryx slendens xanthostoma* (Charpentier). Proceedings of the Royal Society of London, Series B, 259, 313–318.

Smith, T. T. & Yanagimachi, R. 1990. The viability of hamster spermatozoa stored in the isthmus of the oviduct: the importance of sperm-epithelium contact for sperm survival. Biology of Reproduction, 42, 450–457.

Snook, R. R. 2005. Sperm in competition: not playing by the numbers. Trends in Ecology and Evolution, 20, 46–53.

Solem, A. 1972. *Tekoulina*, a new viviparous tornatellinid land snail from Rarotonga, Cook Islands. Proceedings of the Malacological Society of London, 40, 93–114.

Solem, A. 1984. A world model of land snail diversity and abundance. In: World-wide Snails: Biogeographical Studies on Non-Marine Mollusca (Ed. by A. Solem & A. C. van Bruggen), pp. 6–22. Leiden: Brill.

South, A. 1992. Terrestrial Slugs: Biology, Ecology and Control. London: Chapman and Hall.

Thompson, T. E. 1973. Euthyneuran and other molluscan spermatozoa. Malacologia, 14, 167–206.

Tomiyama, K. 1996. Mate-choice criteria in a protandrous simultaneously hermaphroditic land snail *Achatina fulica* (Férussac) (Stylommatophora: Achatinidae). Journal of Molluscan Studies, 62, 101–111.

Tompa, A. S. 1976. A comparative study of the ultrastructure and mineralogy of calcified land snail eggs. Journal of Morphology, 150, 861–888.

Tompa, A. S. 1979. Oviparity, egg retention and ovoviviparity in pulmonates. Journal of Molluscan Studies, 45, 155–160.

Tompa, A. S. 1980. Studies on the reproductive biology of gastropods: part 3. Calcium provision and the evolution of terrestrial eggs among gastropods. Journal of Conchology, 30, 145–154.

Tompa, A. S. 1984. Land snails (Stylommatophora). In: The Mollusca, Vol. 7, Reproduction (Ed. by A. S. Tompa, N. H. Verdonk & J. A. M. van den Biggelaar), pp. 47–140. London: Academic Press.

van Mol, J.-J. 1971. Notes anatomiques sur les Bulimulidae (Mollusques, Gastéropodes, Pulmonés). Annales Societé Royale Zoologique de Belgique, 101, 183–226.

van Osselaer, C. & Tursch, B. 2000. Variability of the genital system of *Helix pomatia* L., 1758 and *H. lucorum* L., 1758 (Gastropoda: Stylommatophora). Journal of Molluscan Studies, 66, 499–515.

Visser, M. H. C. 1981. Monauly versus diauly as the original condition of the reproductive system of Pulmonata and its bearing on the interpretation of the terminal ducts. Zeitschrift für Zoologie, Systematik und Evolutionsforschung, 19, 59–68.

Ward, P. I. 1998. A possible explanation for cryptic female choice in the yellow dung fly, *Scathophaga stercoraria* (L.). Ethology, 104, 97–110.

Watson, H. 1923. Masculine deficiencies in the British Vertigininae. Proceedings of the Malacological Society of London, 15, 270–280.

Whitney, M. 1938. Some observations on the reproductive cycle of a common land snail, *Vallonia pulchella*: Influence of environmental factors. Proceedings of the Indiana Academy of Sciences, 47, 299–307.

Wolda, H. 1963. Natural populations of the polymorphic landsnail *Cepaea nemoralis* (L.). Archives Néerlandaises de Zoologie, 15, 381–471.

Wolda, H. 1967. The effect of temperature on reproduction in some morphs of the landsnail *Cepaea nemoralis* (L.). Evolution, 21, 117–129.

Wolda, H. & Kreulen, D. A. 1973. Ecology of some experimental populations of the landsnail *Cepaea nemoralis* (L.). II. Production and survival of eggs and juveniles. Netherlands Journal of Zoology, 23, 168–188.

Yom-Tov, Y. 1971. The biology of two desert snails *Trochoidea (Xerocrassa) seetzeni* and *Sphincterochila boissieri*. Israel Journal of Zoology, 20, 231–248.

11

An Ancient Indirect Sex Model

Single and Mixed Patterns in the Evolution of Scorpion Genitalia

ALFREDO V. PERETTI

INTRODUCTION

The evolutionary history of terrestrial scorpions dates back to their appearance in the middle Silurian (Rolfe 1985; Polis 1990; Dunlop et al. 2008). Traditionally, scorpions have been considered to be among the most basal arachnids (Weygoldt & Paulus 1979) but in many recent molecular phylogenies (e.g., Wheeler & Hayashi 1998) scorpions are in a distal position in the tree. Although their placement within the Arachnida remains controversial, scorpions are unquestionably monophyletic (Coddington et al. 2004). This arachnid order has 18 extant families (Prendini & Wheeler 2005) which are rather unique among terrestrial arthropods in many of their life history characteristics such as great ecological plasticity (i.e., they can adjust to distinctly different habitats including high and low humidity environments), pronounced intra- and interspecific aggressiveness, and the presence of a "ancient" form of sperm transfer (Polis 1990; Weygoldt 1990). After a ritualized and complex courtship, the male deposits a spermatophore on the soil, from which the female receives the sperm (Farley 2001). Once fertilization is accomplished, embryos undergo a highly specialized viviparous development that last from several months to well over a year, depending upon species (Polis & Sissom 1990). Like many other chelicerates with indirect sperm transfer, such as amblypygids, pseudoscorpi-

ons, uropygids, and some mites (Thomas & Zeh 1984; Weygoldt 1990; Proctor 1998), scorpions exhibit complex behavior patterns during courtship and associated with the sperm transfer, which serve to induce the female to pick up the sperm from the spermatophore (Polis & Sissom 1990; Benton 2001; Peretti 2003). The importance of female cooperation during mating, especially in the sperm transfer phase, is not yet well understood (Jacob et al. 2004a; Peretti & Carrera 2005).

Although the sperm transfer mechanism is conservative in its principal features (i.e., deposition of a sclerotized and tailed spermatophore and always with temporal pair formation), it is also diverse in other important respects in different species of the same family (e.g., great diversity in some parts of the spermatophore) (Hjelle 1990). From a historical perspective, it must be realized that sperm transfer in scorpions by a partially sclerotized spermatophore was originally reported, almost simultaneously, by authors in 1955–1956 (review in Francke 1979). Since then, spermatophores from 18 genera and six recent families have been described (Francke 1979; Peretti 2003). In this respect, our current knowledge of the detailed functional morphology of scorpion genitalia comes from only from recent studies (e.g., Francke 1979; Benton 1992a, 1993; Peretti 1992, 1996, 2003; Jacob et al. 2004a). Given that rapid and divergent evolution in male genitalia is one of the most widespread patterns of animal evolution (Eberhard 1985, 1993, 2001), scorpion spermatophores offer interesting new data to explain the evolution of male genitalia (Arnqvist 1998; Eberhard 2004a; Hosken & Stockley 2004), including those of other arachnids (see also the chapters in this book by Huber and Eberhard on spiders and Machado et al. on Opiliones).

The purpose of this chapter is to document interesting phenomena in scorpions, examining possible evolutionary transitions in some reproductive traits. The present review is divided in two sections, the first focuses on the functional morphology of spermatophores and associated female genitalia, especially in some families in which the available results offer the opportunity to test the power of hypotheses of natural and sexual selection to explain the observed patterns. In the second section, I will describe other male reproductive traits such as the fine structure of spermatozoa and the occurrence of different types of mating plugs. In both sections, evolutionary implications, especially associations with sperm competition, are discussed in detail.

FUNCTIONAL MORPHOLOGY OF SCORPION GENITALIA: MAIN PATTERNS AND POSSIBLE EVOLUTIONARY PRESSURES

Spermatophores: Origin and Morphology

In scorpions the sperm transfer sequence starts with the male bringing the female towards the deposited spermatophore while, as a rule, he continues grasping her pedipalps with his own pedipalps (Polis & Sissom 1990) (figure 11.1a). During the last part of the approach to the spermatophore, the female lowers her genital operculum, leaving her gonopore exposed. The spermatophores are formed in the paraxial organs of the male reproductive system, the right paraxial organ producing the right half of the spermatophore, hemispermatophore), and the left paraxial organ the left hemispermatophore (Francke 1979). These hemispermatophores join together just as the spermatophore emerges from the male gonopore (Hjelle 1990; Farley 2001). There are two main types of spermatophores: "flagelliform" and "lamelliform" (Francke 1979). The flagelliform spermatophore occurs in the

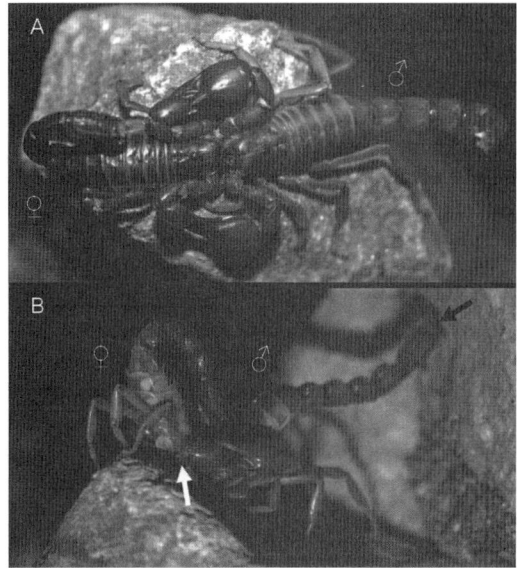

FIGURE 11.1 Example of a mating in lamelliform scorpions. Courtship and sperm transfer in *Bothriurus burmeisteri*: (A) During the courtship the male grasps the female's pedipalps with his own pedipalps. (B) Male pushing of the female during the initial part of the sperm transfer. As a result of this behaviour the lamella of the spermatophore moves down and compresses the trunk (white arrow), unfolding the capsule inside the female's genital atrium. To move his body forward appropriately, the male presses the end of his fifth metasoma segment against a stone (black arrow) (photos courtesy of Patricia Carrera).

families Buthidae and Microcharmidae (figure 11.2) while the lamelliform type appears in most other scorpion families (Stockwell 1989; Prendini & Wheeler 2005). Flagelliform spermatophores are relatively simple: they have an apical filament, the flagellum, connecting the spermatophore with the male genital region during mating. By pushing the female backwards, the spermatophore pivots on its base, and gets compressed as it passes through the arc. The gel in the trunk acts like a plunger in a syringe and the sperm is ejected, this process lasts two to four seconds (Shulov & Amitai 1958; Peretti 1991; Benton 1992a, 2001). In contrast, lamelliform spermatophores are more complex (figure 11.3a) and insemination can last several minutes (Polis & Sissom 1990). They bear a lever-shaped lamella, the bending of which causes

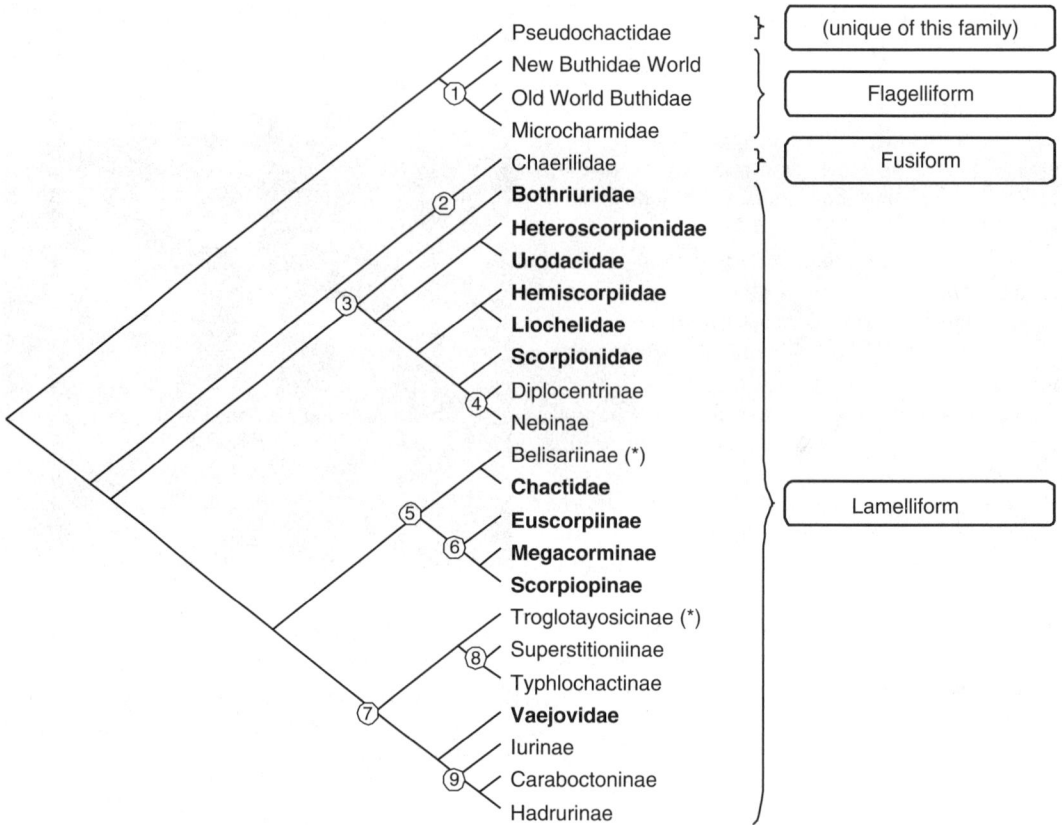

FIGURE 11.2 Types of spermatophores among scorpion families. The presence of a flagelliform spermatophore is restricted to the basal family Buthidae and to the closely-related family Microcharmidae. Complex spermatophore capsules (family or subfamily names in bold) have evolved independently among many unrelated families of lamelliform scorpions. Cladogram based on Coddington et al. (2004) (in this picture Ischnuridae is replaced by Liochelidae according with the change proposed by Fet & Bechly 2001). Abbreviations: (1) Buthoidea, (2) Chaeriloidea, (3) Scorpionoidea, (4) Diplocentridae, (5) Chactoidea, (6) Euscorpiidae, (7) Vaejovoidea, (8) Superstitioniidae, (9) Iurida; (*) Troglotayosicidae.

eversion of the capsule and ejects sperm into the female's atrium (Alexander 1959) (figure 11.1b). The capsule is a sclerotized region ornamented with lobes (e.g., basal and distal lobes), hooks, spines, tubercles, or other apophyses, and is traversed by the sperm duct (Francke 1979). At first glance, the mechanism of action of all scorpion lamelliform spermatophores seems to be similar (e.g., Angermann 1957 in Euscorpiidae; Francke 1979 in Diplocentridae and Vaejovidae; Maury 1968 in Bothriuridae) since the sperm transfer functions by a lever mechanism pressing the sperm into the female genital atrium (also called "chamber" or "vagina") (Jacob et al. 2004a). Although the flagelliform and lamelliform types are the most

common among scorpions, Stockwell (1989) proposed a third type, "fusiform", for the simple spermatophores of Chaerilidae and included it as a character state in his phylogenetic analysis. The fusiform spermatophore was also discussed by Prendini (2000). The term "fusiform" means that this type combines some features of both flagelliform and lamelliform spermatophores. In addition to these three types, the unique spermatophore of Pseudochactidae has been recently illustrated and described by Prendini et al. (2006). This spermatophore, while apparently allied to the flagelliform type, is unlike that of any other scorpion. Therefore, the situation is more complex than portrayed above. Unfortunately the mechanisms

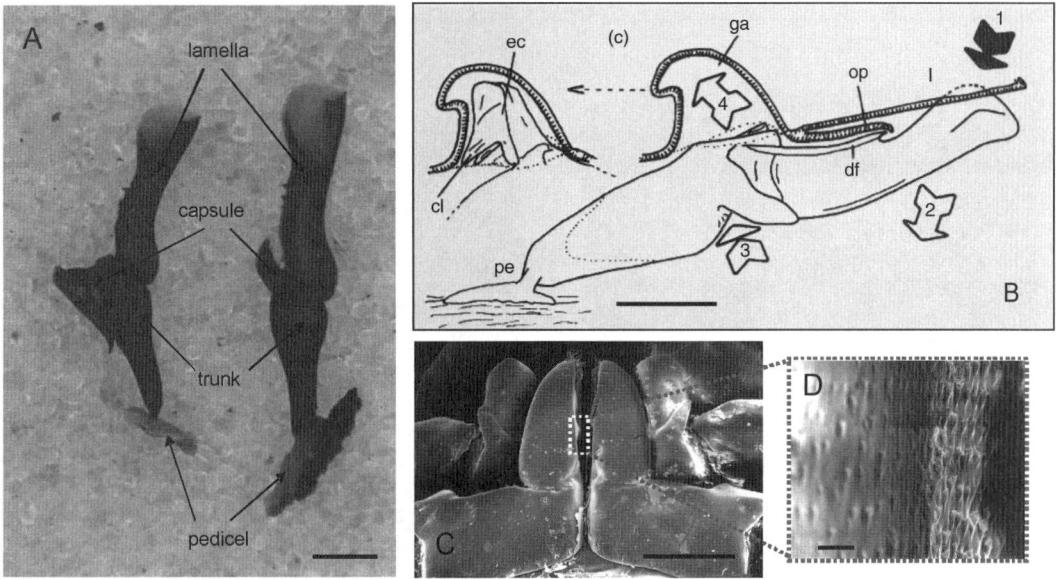

FIGURE 11.3 Examples of form and function in lamelliform spermatophores: (A) Principal parts of the spermatophore of *Bothriurus bonariensis* (Bothriuridae) in the pre-insemination stage (right) and post-insemination stage (left; note the everted capsule, with rest of sperm). (B) Copulatory mechanics (illustrated in the bothriurid *Timogenes elegans*). Flows show the sequence of pressures produced on the lamella by the female. The capsule is everted into the genital atrium. I, coxa of first pair of legs; cl: capsular lobes; dorsal fold; ec, everted capsule; ga, genital atrium; op, female genital operculum; pe, pedicel. (C) Anterior portion of the female's ventral face in *B. bonariensis* showing the space between the leg coxae (white square) of the first pair of legs in which the distal part of the lamella fit; (D) note on right a detail of the space showing many microspicules that facilitate the anchoring of lamella (Photos: A, C and D, Alfredo V. Peretti, unpublished; C, diagram from Peretti 2003). Scale bars: (A and B): 2 mm; (C): 1 mm; (D): 100 μm.

of action of the spermatophores of the Chaerilidae and Pseudochactidae have never been studied.

In this chapter I will refer especially to traits of the lamelliform type of spermatophore. Among the scorpions possesing this type, such as well studied examples belonging to the Euscorpiidae, there is a precise fit of the curved genital operculum into an "excavation" between the lamella and one basal lobe of the spermatophore (Jacob et al. 2004a). In two related bothriurids, *Timogenes* and the *Bothriurus bonariensis* group, there are some complementary structures that fit the spermatophore onto the female operculum. The most important is the dorsal fold (figure 11.3b) (Peretti 2003; Mattoni 2003). In all the lamelliform scorpions that have been studied, the intercoxal space of the female's legs functions as a "guideway" to receive and fit the lamella (figure 11.3c) from its basal to the distal part (Peretti 2003). In *Euscorpius italicus*, a longitudinal impression on the female's sternum serves a similar function (Jacob et al. 2004a). However, sexual dimorphism in the sternum morphology of these and other euscorpiids and in the intercoxal morphology of bothriurids is not apparent. It must be realized that all these anchoring mechanisms of the spermatophore against the female genitalia always require female cooperation.

Complexity of the Spermatophore Capsule

The most complex part of the spermatophore is the capsule (for an example see figure 11.4). Complex hemispermatophore capsules occcur in many families of lamelliform scorpions (figure 11.2) such as Bothriuridae (Maury 1980; Mattoni 2003),

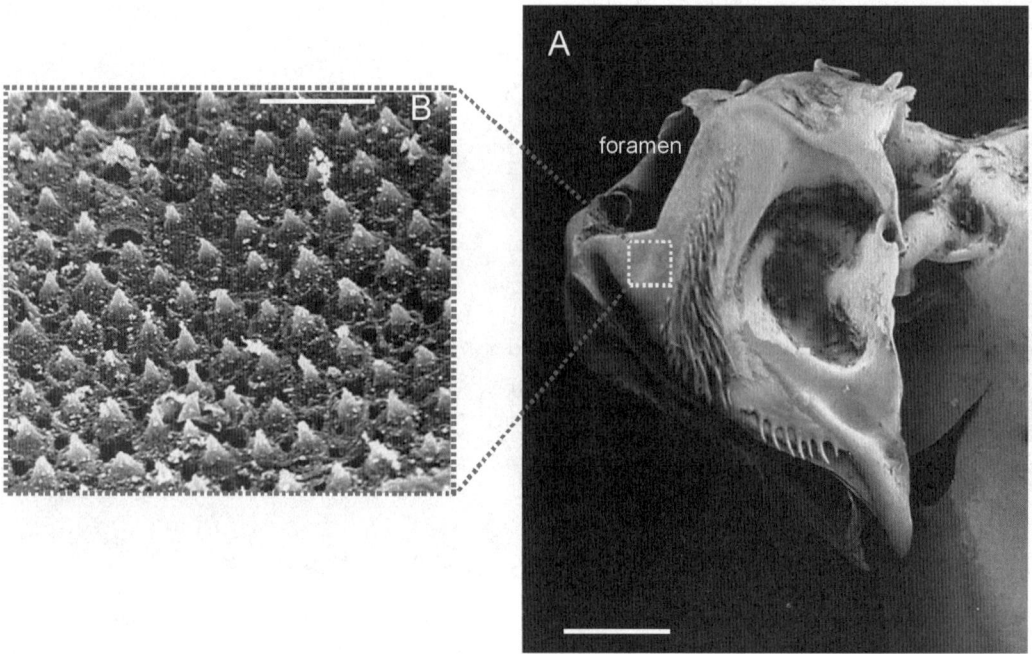

FIGURE 11.4 Example of a complex everted capsule: (A) Lateral view of the post-insemination capsule of the bothriurid *Brachistosternus ferrugineus*. Note the presence of rows of many large spines. (B) View of the microspines around the foramen (adapted from Peretti 2003). Scale bars: (A) 500 μm; (B) 10 μm.

Heteroscopionidae (Lourenco 1996), Urodacidae (Koch 1977), Hemiscorpiidae, Liochelidae (Monod & Lourenco 2005), Scorpionidae (Lamoral 1979), Chactidae, Euscorpiidae (Soleglad & Sissom 2001), and Vaejovidae (Stockwell 1989; Francke & González-Santillán 2006). Although hemispermatophores are commonly used in systematics, in general the morphology of used spermatophores is not. A very interesting point is that the everted capsules of used spermatophores offer the possibility of identifying structures which are generally hidden in the hemispermatophores, and they could show that genitalic diversity is larger than previously believed. For example, capsular eversion not only brings about a complete change in the orientation of the lobes, but also exposes other smaller structures, such as microspines.

Recently, the complexity and possible evolution of spermatophore capsule was investigated in some groups of the Bothriuridae and Euscorpiidae (Peretti 2003; Jacob et al. 2004a), in which the eversion of the capsule was experimentally induced to produce movements of the lamella similar to those that occur during normal copulations. In this interesting scenario of functional morphology, Jacob et al. (2004a) provided the most recent and detailed analysis of a spermatophore of the Euscorpiidae, using *Euscorpius italicus* as study animal. In this species the distal and basal lobes of the capsule hook into two cavities on the inner side of the female's genital operculum. In addition, in *Euscorpius* a so-called "crown-like structure" hooks into a membranous area in the genital atrium that moves backwards during sperm transfer, widening the gonopore. The sperm duct of the spermatophore is coated with numerous spicules on its outer side. These authors also presented evidences that the sclerotized dorsal and basal lobes of the spermatophore and the crown-like structures of the capsule function as a device for opening the female genital atrium providing that she does not reject the male at that moment. One thing is clear, this complex interaction of spermatophore and genital opening guarantees safe sperm transfer (the hypothesis supported by these authors, see below). All *Euscorpius* species investigated so far have very highly complex spermatophores with the sole exception of *E. flavicaudis*, where the capsular region is simple (Jacob et al. 2004a,b). The crown-like structures could be homologous to the horns (basal lobes) of the bothriurids, as both are situated just beneath the opening (foramen) of the sperm duct.

All species of the Bothriuridae have microspines in the apical region of the everted capsule, that contact the cuticular wall of the genital atrium which contains large pores with sensilla (Peretti 2003; Carrera 2008). The shapes of microspines vary among species of different genera. For instance, the tips of some *Bothriurus* species are blunter than those of *Timogenes* species while those in *Brachistosternus* species are thinner and more pointed at the tip. Each pressure on the lamella induces clear movements of the portion of the everted capsule containing the microspines rubbing the wall of the female atrium (Peretti 2003; P. Carrera, personal observation). Microspines are apparently common in other lamelliform spermatophores, such as those of the families Vaejovidae, Scorpionidae, and Chactidae (A.V. Peretti & C.I. Mattoni, unpublished data). None of these microstructures that rub the female's genital atrium detach from the capsule during or after the sperm transfer. Detachable structures of the spermatophore are associated with the presence of sclerotized mating plugs, a subject examined in the second section of this chapter.

Patterns of Allometry and Asymmetry of Spermatophores

Data from insects, birds, and mammals (Alatalo et al. 1988; Møller 1991; Andersson 1994; Møller & Pomiankowski 1993) show that sexually selected characters (in these cases, mostly visual signals) in general have rather significant phenotypic variation, which is evident as soon as one considers the high *cv* (coefficient of variation) values they have and the wide dispersion of data around the allometric line. To date, allometry and asymmetry of male and female body structures and of the spermatophores have only been analyzed in two species of the Bothriuridae, *Bothriurus bonariensis* and *Brachistosternus ferrugineus* (Peretti et al. 2001; A.V. Peretti & S. Outeda-Jorge, in preparation). More recently, representatives of the Vaejovidae and Diplocentridae have been analyzed (A.V. Peretti, unpublished data). In all these scorpions, the

highest values of phenotypic variation found for the body parts correspond to those that are involved in a very direct way in mating, especially the spine-shaped apophysis of the chela of the male pedipalp, the genital operculum, and the pectinal plate (Peretti et al. 2001). The spermatophore structures showing the highest *cv* values were the frontal crest and the lateral edge of the lamella and the height of the capsule. In this context, spermatophore structures that serve to fit the lamella to the female's intercoxal space (e.g., distal crests) could be subject to sexual selection through different forms of female choice (Andersson 1994; Eberhard 1996, 1997). In contrast, the lowest—and similar—*cv* values were found in "basic traits" of the spermatophore: the lengths of the lamina and the trunk, and in part the length of the capsular lobe. Moreover, these structures exhibit the lowest values of antisymmetry. This may be a consequence of the stabilizing pressure to which, primarily the first two structures, are subject because they form the *basic* regions in all the families that have lamelliform spermatophores. Because the size of the spermatophore stems from these parts, their phenotypic stabilities in the species will result in a mean size that would permit all males to be able "potentially" to inseminate any female independently, up to some extent, of the size of their genitalia and/or body. In fact, adult females show a higher (very significant) variation in size than males in the studied population, with a magnitude almost two and a half times larger than that of males. None of the genitalic characters of the studied scorpions showed positive or negative allometry (Peretti et al. 2001). Indeed, the basic "spermatophore-body" allometric line (male length–spermatophore length) was relatively low in these species. The low or moderate allometric values registered in spermatophores agree with those observed by Eberhard et al. (1998) in the genitalia of insects and spiders since they usually have also smaller values than other body traits.

Thornhill & Møller (1998) pointed out that developmental stability reflects the ability of individuals to cope with their environment during ontogeny given their genetic background. An inability to cope with environmental and genetic perturbations is reflected in elevated levels of fluctuating asymmetry and other measures of developmental instability. In scorpions all the structures of the spermatophore have antisymmetry which may be related to the way in which its parts are formed in the male reproductive system since there is simultaneous development of each hemispermatophore in two, not one, molds of tissue. This could produce larger differences between the structures of the two hemispermatophores than those that appear for the paired body characters, which might correspond to the presence of antisymmetry rather than fluctuating asymmetry (Peretti et al. 2001). Furthermore, it is possible that the two paraxial organs have a certain degreee of physiological independence (Peretti 2000). Alternatively, as an example of a possible evolutionary transition between types of asymmetry, the sister species of *B. bonariensis*, *B chacoensis*, has rather similar spermatophores with the exception of the existence of a long distal filament in the right capsular lobe. Some analyzed *B. bonariensis* males have, near the tip of the right capsular lobe, rudiments of different lengths of a filament that could be related with that existing in *B. chacoensis*, making it possible that they represent different evolutionary stages (A.V. Peretti & C.I. Mattoni, in preparation).

Clearly, this is a case of directional asymmetry, a phenomenon that has been observed in the left lobe and other capsular structures of *Brachistosternus* species (San Martin 1969; Maury 1975; Peretti 2003; Ojanguren-Affilastro 2005). Indeed, the greatest capsular complexity in scorpions is found in this genus. In general, the species have similar asymmetrical capsules: the left capsular lobe has the cylindrical apophysis and the external lamellar apophysis. In the right hemispermatophore the former is replaced by the internal lamellar apophysis. The internal lamellar apophysis has an area of many strong spines. The precise association between asymmetrical structures in the spermatophore and the female's genital atrium is not yet sufficiently understood. However, despite the fact that these structures remain pressing the wall of the basal region of the genital atrium after capsular eversion has finished, females withdraw from the spermatophore both before and after sperm transfer without any apparent difficulty. Furthermore, spines and apophyses are not structures that force the spermatophore into the female genital atrium (Peretti 2003, personal observation).

Female Genital Atrium: Form and Ultrastructure

In lamelliform scorpions, the female has a distinct ovoid and flexible genital atrium and two spermathecae (also termed "sperm receptacles") (figure 11.5) which increase in size with successive

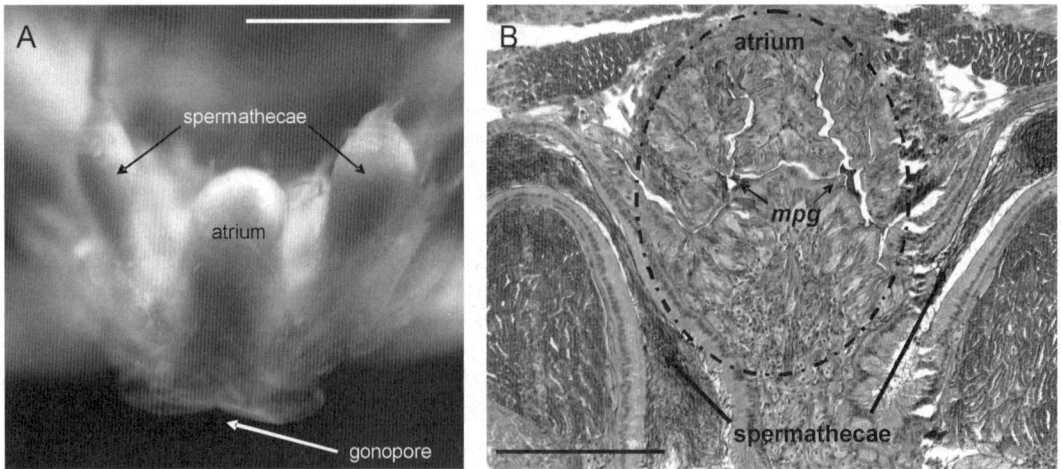

FIGURE 11.5 Example of the general form of a female's genital atrium and spermathecae in lamelliform scorpions. (A) Dorsal view in *Bothriurus asper*. (B) Sagital section of the anterior part of the female reproductive system in *B. flavidus*. Note the abundance of connective tissue, the cuticular wall covering the atrium, and the presence of a membranous mating plug (mpg) in the lumen. The spermathecae are filled with free sperm (photos: (A) from Mattoni & Peretti 2004; (B) Alfredo Peretti & Patricia Carrera unpublished). Scale bars: (A) 1 mm; (B) 500 μm.

inseminations (Peretti & Battán-Horenstein 2003; Volschenk et al., 2008). In contrast, in buthids the atrium is only represented by the union of the two enlarged and generally thin spermathecae. In some species with lamelliform spermatophores, the distal part of the atrium has a small chamber (Francke 1982). The genital atrium is always connected to the two spermathecae by a short duct. The sperm are mainly stored in the spermathecae, with some interspecific differences in the form in which spermatozoa are preserved (see the next section).

Unlike the thin wall of the atrium of buthid females, in the lamelliforms there is a well developed cuticular wall covering the epithelium of the atrium. The atrial epithellium of scorpions is characterized by prismatic glandular cells with microvilli and cells with intracellular canals (Lhajoui et al. 2004), some of which contain secretions. In fact, it was observed that among bothriurids and euscorpids all of the cuticular wall of the female genital atrium contains large numbers of small pores with thin ducts associated with epithelial secretory cells. The stratum that covers the atrium, just below the basement membrane of the epithelium, is a loose connective tissue. In many places, the musculature enters deeply into the connective tissue, reaching the base of the epithelium (Peretti 2003; Lhajoui et al. 2004; Carrera 2008). A very interesting point is that, in the Bothriuridae, small nerve fibers, ending in close contact with the atrial surface, where there are many larger pores containing a campaniform-like sensillum (Peretti 2003; Carrera 2008), project from the major ventral nerve cord, which passes very near the dorsal part of the atrium, However, the innervation of the female's genital atrium in other families has not yet been studied.

Evaluation of Natural and Sexual Selection Hypotheses

Natural Selection

The "safe sperm transfer" hypothesis. Clearly, the primary function of genitalia, including spermatophores, is to ensure sperm transfer. If we ignore this "obvious" point our final interpretation on the function of genitalic structures could be very biased. Stabilizing selection may determine that each species has low values of the coefficient of variation, the slope of the "body-spermatophore" allometric

line and antisymmetry in variables of the "basic" parts of spermatophore (i.e., length of the trunk and lamella) in order to ensure sperm transfer. Nevertheless, recently and more radically, natural selection has been suggested to be the principal force modeling the design of spermatophore morphology in scorpions (Jacob et al. 2004a). This possibility was summarized as the "sperm transfer improvement" or "safe sperm transfer" hypothesis (Peretti 2003; Jacob et al. 2004a). It must be realized that this idea was proposed by Grasshoff (1975) and Kraus (1984) who suggested, working on spider genitalia, that natural selection guides genital evolution (and its diversity) to improve sperm transfer success exclusively. In others words, complex genitalia evolved in order to maintain a complex coupling mechanism that safeguards sperm transfer.

In scorpions, the microspines and other structures of the capsule may also have a function in holding the spermatophore in the correct place to ensure that the sperm is expelled into the genital atrium. A variant of this idea would be a genitalic intersexual communication hypothesis (Peretti 2003) where, for example, stimuli caused by tubercles, etc. of the scorpion spermatophore would provide the female with information on whether sperm transfer is being performed correctly at that moment; thus she could better control her body movements. Jacob et al. (2004a), after evaluating the functional morphology of the spermatophore in *E. italicus* and other *Euscorpius* species, arrived at a dual conclusion since they observed that the numerous spicules of the sperm duct, among other capsule traits, could serve as a sealing mechanism. They concluded that "safeguarding sperm transfer" is one driving force for the evolution of male genital complexity in scorpions, but suggested this conclusion could be limited, because the hypothesis of Grasshoff and Kraus was restricted to explanation of the evolution of spider genitalia (see the chapter by Huber & Eberhard for an updated evaluation of this idea in spiders). Nevertheless, the safe sperm transfer hypothesis does not seem stronger than the other hypotheses since the controversial questions do not revolve around the function of the genitalia per se, but the function of the complex modifications of the male genitalia (which, in general, do not seem to be explicable in terms of sperm transfer per se). Perhaps the sperm transfer hypothesis is not an "alternative" to the other explanations but rather an "obvious complement" of them.

The lock-and-key hypothesis. This option predicts species-specific morphological complementarity between males and females (Shapiro & Porter
1989). Female genitalia should be capable of physically excluding the male genitalia of closely related
species but not those of their own species (Eberhard
1993, 1997). Although this hypothesis has been
rejected as an explanation of the diversification of
animal genitalia (Eberhard 1985, 1993; Arnqvist
1998), it is important to point out its principal
arguments for scorpion genitalia. Firstly, as was
observed in the genera *Bothriurus* (Peretti et al.
2001) and *Euscorpius* (Jacob et al. 2004b), character displacement does not exist in spermatophores
or female genitalia when closely related species of
scorpions are sympatric or parapatric. Indeed,
when the distributions of closely related species
overlap (where the risk of inter-specific pairings is
greater), there are incompatibilities in behavior
rather than in genital structure, and these occur
during the first stages of courtship (Le Pape &
Goyffon 1975; Polis & Sissom 1990; Peretti et al.
2001; Benton 2001).

Furthermore, the female genital atrium is in general soft and very uniform among species where
spermatophores are species-specific, and it could
not exclude cross-specific spermatophores (Peretti
2003; Jacob et al. 2004a). These data agree with
comparative studies carried out in other arthropods, which also failed to find support for a lock-
and-key genitalic mechanism (Eberhard 1985,
2001; Shapiro & Porter 1989; Arnqvist 1998). The
only case of genitalic incompatibility between
closely related species was observed in the bothriurid sister species *Bothriurus bonariensis* and
B. chacoensis. These have allopatric distributions
with possible zones of overlap in some regions of
central Argentina. Although sperm transfer was
complete in these matings, there is a certain degree
of lack of compatibility of both genitalia and behavior. This occurs principally because the capsular
lobes are larger in *B. bonariensis* which makes the
detachment of the used spermatophore from
the female's atrium during the post-insemination
stage difficult (Peretti & Acosta 1999; A.V. Peretti
& C.I. Mattoni, in preparation).

From Sexual Selection

The female choice hypothesis. Postcopulatory
female choice is the ability of a female to bias the
fertilization success of the males that copulate with

her (Eberhard 1996). In this context, copulatory
(genitalic or behavioral) courtship could be used by
males to alter the likelihood of females accepting,
storing or using a male's sperm (Eberhard 1996,
2002). The presence of genitalic copulatory courtship has been suggested for species of the
Bothriuridae and Euscorpiidae (Peretti 2003; Jacob
et al. 2004a); as mentioned above, groups with
complex capsular regions containing microspines,
spicule-coated sperm ducts and crown-like structures. In addition, new groundbreaking observations in species of more families, such as the
Vaejovidae and Chactidae (e.g., C.I. Mattoni &
E. Florez, unpublished data) found that microspines
are common structures and show a great diversity
in shapes and disposition on the capsule. Moreover,
the presence of nerve fibers close to the genital
atrium is clear in species of the Bothriuridae (Peretti
2003; A. V. Peretti & P. Carrera, unpublished data),
a phenomenon that has not yet been examined in
other scorpions. Although genitalic copulatory
courtship could be very widespread in some of the
Bothriuridae and Euscorpiidae, its frequency cannot
be exactly determined. However, one fact seems
clear: the large interspecific variability observed in
capsular structures (such as microspines aparently
used for internal courtship) but not in the female
genital atrium (i.e., its shape is uniform), agrees
with the predictions proposed by the sexual selection by female choice hypothesis for animal genitalia (Eberhard 1985).

One further hypothesis which could be taken
into account to explain some aspects of scorpion
genitalia is female choice by "mechanical fit"
(Eberhard 1985; Huber 1993, 1995, 1998; Huber
& Eberhard 1997). According to this hypothesis,
the female receives no sensory input and does
not exert "active" choice, but "passively" chooses
by the morphology of her genital organs and
associated body parts. Female genital morphology
would represent a female preference that discriminates between variants of the male genitalia by
the number of sperm transferred (Huber 1995).
This hypothesis might be acceptable for some lamellar structures that fit between the female's coxae
(e.g., the distal crest and the thin border of lamella)
and at the inner face of operculum (e.g., the dorsal
fold in *B. bonariensis* and *T. elegans*). These parts
are highly variable among species not only from
different genera but also within the same genus
(Maury 1980, 1982; Peretti 1993; Mattoni 2003).
Perhaps the greatest diversity appears in *Bothriurus*

(Maury 1980). The female contact zones are sclerotized, especially the space among the legs' coxae (Peretti 1993; Peretti, et al. 2001). There is no sexual dimorphism in the surface of the inner side of the coxae (C. Mattoni & A. Peretti, unpublished data). If female morphology is almost uniform, and they nevertheless fail to cooperate when contacted by non-conspecific lamella, then the choice is "active" rather than passive. If the female morphology differs (as is often the case in the female genitalia of spiders—Huber 1993, 1998; Eberhard 1996), then the choice could be "passive". For both versions of the female choice hypothesis, however, the mechanism should be cryptic female choice because so far occasional disrupting of mating by the female has been observed only in some bothriurid scorpions, such as *B. flavidus* (Peretti 1993, 1996). However, how cryptic female choice could act after sperm transfer is uncertain (Jacob et al. 2004a; Peretti 2003).

The sexual conflict hypothesis: genitalic coercion after capsule eversion. In scorpions, female cooperation is always necessary to allow intromission and males are not equipped to force the entrance of the spermatophore into the female genital atrium as would be the case if strong male–female conflict over the control of copulation was taking place (Alexander et al. 1997; Chapman et al. 2003; Arnqvist & Rowe 2005). In addition, scorpion females move their bodies in many different directions even while over the spermatophore. None of structures of the lamella or capsule prevents the movement of the female when she decides to redirect her body to interrupt insemination (Peretti 1996; Carrera 2008). In fact, scorpion females are able to display a wide range of types of "evasion" during courtship and sperm transfer (Polis & Sissom 1990; Benton 2001; Carrera et al., in press).

The only possible example of *genitalic coercion* was observed in *B. bonariensis* (Peretti 1992, 2003). In this species the majority of sperm are transferred immediately after the eversion of the capsule is completed, this transfer continues during the whole time that the female has the spermatophore capsule in her atrium. The long time that the everted capsule of the *B. bonariensis* spermatophore remains in the atrium could be necessary to push the sperm from the genital atrium to the sperm receptacles. Genitalic coercion may be produced by the strong anchoring mechanism of the two large capsular lobes of the spermatophore against the atrial wall

that prevents the female from interrupting the insemination (Peretti 2003). However, in a typical post-transfer sequence, the male of *B. bonariensis* helps to disengage the large capsular lobes by pushing her slowly backward. This occurs only once the insemination is complete, never before, and causes the spermatophore capsule to come free from the female atrium. Therefore, slight genital coercion in a late stage of the mating is not imcompatible with subsequent male help to the female.

On the Occurrence of Mixed Patterns

Lamelliform spermatophores do not fit a single selection pattern, since because of their morphological complexity, the different structures will be subject to different selective pressures. Sexual selection may have guided the evolution and divergence and/or convergence of some particular structures of the spermatophore (e.g., lamellar folds, capsular microspines), even though the allometric relationships of the basic parts remain under stabilizing selection.

If we accept the presence of parallel selective pressures (Kokko et al. 2003), avoiding endless discussions about which mechanism is "the best", then we will be able to form new questions to understand the evolution of animal genitalia. Thus, the help that the male of *B. bonariensis* provides to disengage the capsular lobes after sperm transfer is finished, may indicate that intersexual cooperation and strong conflict may co-occur in a single species but at different stages. In the general context of sexual conflict, the coexistence of evidence for different, and apparently opposing, hypotheses in a single species has also been observed in another arachnid with indirect sperm transfer, the solpugid *Oltacola chacoensis*, in which sexual coercion does not exclude luring behavior during a single copulation (Peretti & Willemart 2007). In this species mating involves vigorous male grasping performed with the claw-like chelicerae against the female gonopore and subsequent intense female shaking. However, this is not incompatible with the occurrence of copulatory courtship in the form of subtle but continuous stimulatory tapping on the female body.

On the other hand, some spermatophore structures may have more than one function. For example, well developed capsular lobes may function not only to ensure entry of the capsule into the atrium (avoiding sperm loss), but also to make interruption of the insemination process by the female difficult and to stimulate her. The same could be true for the microspines of the capsule which could serve to perform sexual stimulation, to communicate to the female the position of the evaginated capsule in the atrium, and to ensure sperm transfer. Co-occurrence of these apparently opposing strategies during sperm transfer has not been explored in depth and may suggest the possible occurrence of mixed behavioral evolutionary patterns in some groups. Also, one must realize that each structure or group of them might function in more than one context, so that the whole spermatophore is a unit as well as being a sum of "subunits" (Peretti 2003). Indeed, this trend is also found in other arachnids with indirect sperm transfer mediated by spermatophore such as amblypygids (Weygoldt 1999).

In addition, an involuntary bias could have often occurred in previous investigations on this subject, whereby the more conspicuous behavioral and morpho-genitalic data that are commonly associated with forced copulation or luring behavior are given in more detail, depending on the behavioral pattern and species, thus affecting our perception and interpretation of the facts (Peretti & Córdoba-Aguilar 2007).

OTHER REPRODUCTIVE TRAITS: STRUCTURE, EVOLUTIONARY PATTERNS AND ASSOCIATION WITH SPERM COMPETITION

Sperm and Associated Male and Female Traits

Testis and Accessory Glands

The scorpion testis basically consists of long tubules that are highly coiled in some lamelliforms, whereas the buthid testis consists of four non-coiled long tubules joined transversely. The anterior portion of the testis continues as the deferens ducts, which carries sperm to the seminal vesicle. Finally, the sperm passes from the seminal vesicle to the medial and apical part of the trunk of the hemispermatophore during the simultaneous process of formation and deposition of the spermatophore (Hjelle 1990; Farley 2001). Although some variation may be due to the amount of sperm contained, data from scorpions show that seminal vesicle size varies

with species. In the Bothriuridae there is no direct relationship between the length of hemispermatophore and length of seminal vesicle (Peretti & Battán-Horenstein 2003). The presence of an ampulla on the deferent duct may increase the available space for storage of spermatozoa. In some bothriurid species, the lack of an ampulla, as in *B. bonariensis* and *B. chacoenis*, could be offset by the larger seminal vesicle in these species.

Each paraxial organ is accompanied by the seminal vesicle and, depending on species, by one or more accessory gland. There is no consensus on the terminology applied to these glands (Hjelle 1990). In lamelliforms, there are generally medial and oval glands (Sissom 1990). Many studies on male reproductive system in buthids were published during the 1950s. For example, Alexander (1959) described the reproductive system morphology in some South African buthids, in particular, the oval and cylindrical accessory glands. This author observed that these two glands were located in the anterior and posterior portions of the seminal vesicle, respectively. Although their functions remain unknown, Alexander suggested that the oval gland may produce the adhesive substance that sticks the pedicel of the spermatophore to the soil (Alexander 1957). However, Bücherl (1956) suggested that this secretion might contain nutritive substances for spermatozoa. In some species the cylindrical gland may secrete a substance that forms a plug inside the genital atrium of inseminated females (see below) (Vachon 1953; Hjelle 1990).

Contrary to statements by Sissom (1990), there are male accessory glands (medial and oval) in the Bothriuridae. Vachon (1953) and Shulov and Amitai (1958) postulated that, in buthids, secretions of the male glands could be involved in formation of a gel-like plug that is usually found in the genital atrium of inseminated females (see below). However, more studies are necessary to confirm this possibility.

Scorpion Spermatozoa: Fine Morphology and Main Evolutionary Patterns

In all cases the spermatozoa of scorpions are filiform-flagellate (figure 11.6) divided into two basic parts, head and tail, the latter is subdivided in midpiece and endpiece (the free flagellum) (Alberti 1983, 2000). The filiform-flagellate organization of the scorpion spermatozoon is plesiomorphic, as

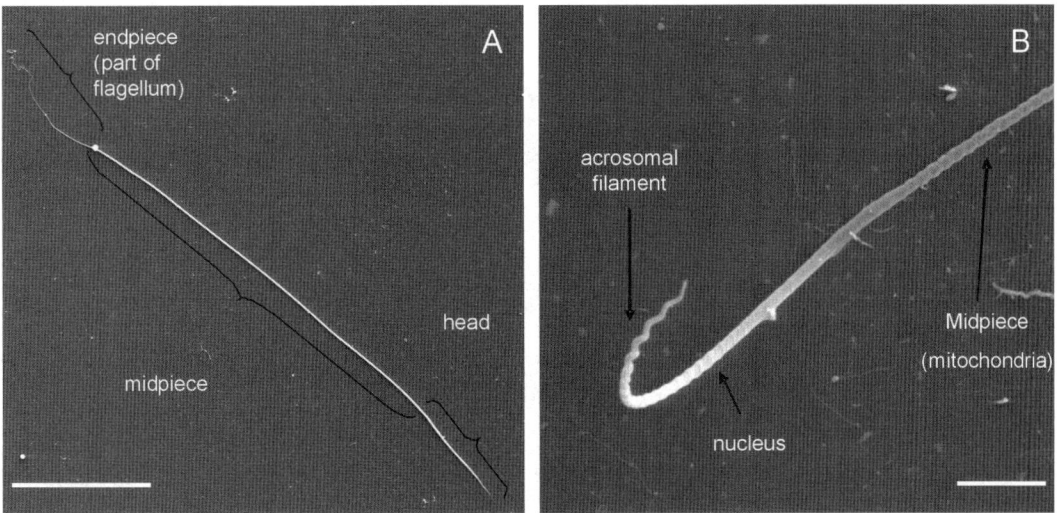

FIGURE 11.6 Main parts of a scorpion spermatozoon: (A) sperm of the buthid *Zabius fuscus*. (B) Anterior portion of the spermatozoon showing the position of some organelles. Note the helical arrangement of the acrosomal and mitochondrial regions, a typical pattern in Buthidae (photos: Anja Klann, Afredo Peretti & Peter Michalik unpublished). Scale bars: (A) 30 μm; (B) 3 μm.

compared to other Arachnida (Alberti 1983). For example, compared with a basal group of Chelicerata, the Xiphosura *Limulus*, the spermatozoa of scorpions present plesiomorphies such as an enlarged form, a free flagellum, and a coaxial arrangement of the centrioles (Cruz Landim & Ferreira 1973; Jespersen & Hartwick 1973; Alberti 1983). However, there are also certain synapomorphies such as a threadlike shape, a flagellar tunnel separating the flagellum from the mitochondrial sheath and modified elongated mitochondria (Alberti 1983, 2000). New fine structural studies of the spermatozoa of scorpions have been carried out on species of the Buthidae, Bothriuridae, Scorpionidae, Euscorpiidae, Troglotayosicidae, Vaejovidae, and Iuridae (Vignoli et al., 2008; Klann et al. 2007 in preparation).

There is great variation on sperm size in each family, ranging from 130 to 300 micrometers. The largest sperm are found in the iurid *Hadrurus arizonensis* (275 mu) (Jespersen & Hartwick 1973) whereas the smaller spermatozoon belong to the basal family Buthidae (e.g., in *Zabius fuscus* from 130 to 160 mu) (Peretti, personal observation). In the family Bothriuridae the mean size is 260 micrometers, varying from 155 in *Orobothriurus lourencoi* to 358 in *Centromachetes pococki*

(Vrech et al. 2007). Comparing all the species studied to date, the head, midpiece and endpiece represent, respectively, 16%, 56%, and 28% of the total sperm length. These proportions, in particular that of the sperm head, are stable within the order, without significant differences at the level of family or genus (A.V. Peretti, unpublished data). However, where there are important differences, they principally involve the relationship between the length of the midpiece and the length of the flagellum. In other words, the two characters are inversely correlated. Thus, in the iurid *H. arizonensis* the midpiece represents 64% of the sperm length which directly determines that the flagellum only occupies 27% (Jespersen & Hartwick 1973). On the other hand, there is an opposite situation in a bothriuridae, *Urophonius brachycentrus*, in which 40% of the sperm length is determined by the midpiece and 47% by the flagellum. Interestingly, in this species the number of spermatodesms transferred by the male is lower than that in species without sclerotized mating plugs (D. Vrech & A.V. Peretti, personal observation). As we can observe in table 11.1, increase in the total length of the scorpion spermatozoon is apparently determined by an increase in the length of its midpiece. However, further data sources are needed to confirm such possible relationship.

TABLE 11.1 Phenotypic correlation between some sperm traits from data on Buthidae (*Z. fuscus, T. bahiensis*), Bothriuridae (*T. elegans, U. brachycentrus*) and Iuridae (*H. arizonensis*) species. Negative: −; positive: +; no relationship: o

Correlated sperm traits	Correlation	*rs*	*P*
Sperm length × head length	o	0.50	0.39
Sperm length × midpiece length	+	0.90	0.037
Sperm length × flagellum length	o	0.50	0.39
Head × midpiece length	o	0.10	0.87
Head × flagellum length	o	0.000	0.99
Midpiece length × flagellum length	o	0.60	0.28

Furthermore, there is great diversity in the number and arrangement of components of the spermatozoon not only between families but also within the same family. Alberti (2000) remarked on the different variants in the morphology of the flagellar axoneme that have arisen from the plesiomorphic pattern of the central tubules 9×2+2 seen in Xiphosura and in some buthids. From this plesiomorphic state the number of central tubules is reduced to a single tubule (9×2+1) or may be absent (9×2+0). If we consider all the buthids studied so far, it is possible to see that the plesiomorphic pattern occurs in species of the "Old-World" group (Fet et al. 2003) such as *Androctonus amoreuxi* and *Buthus occitanus* (André 1963; Alberti 1983). In contrast, in buthids of the "New-World" group this pattern appears in *Zabius fuscus*, whereas reduction of the number of central tubules is evident in other species such as *Centruroides vittatus* (a single tubule) and *Tityus bahiensis* (none) (Hood et al. 1972; Cruz Landim & Ferreira 1973). However, the 9×2+1 pattern is very critical, since it is was not observed regularly and the sperms possessing this pattern of the central tubules could also be damaged or similar (Peter Michalik & Anja Klann, personal communication). Further studies needed to confirm the 9×2+1 arrangement as a regular axonemal pattern in scorpions. Lamelliform families show the derived axonemal pattern "9×2+0" (Hood et al. 1972; Alberti 1983, 2000;

Klann et al. 2007; Michalik & Mercati 2007; Vignoli et al., 2008; A. V. Peretti & D. Vrech, unpublished data). Buthids also show other characters that are not seen in the lamelliform families, such as the cork-screw shaped anterior part of the nucleus, the presence of only two mitochondria in the midpiece, and a lack of aggregation of mature spermatozoa into spermatodesms (Alberti 1983).

Another interesting characteristic from an evolutionary framework is the visible increase in the number of mitochondria that appears in the sperm midpiece of derived families. For example, the Buthidae have only two mitochondria while other scorpions investigated are characterized by spermatozoa with three (Bothriuridae—Peretti, personal observation) to six–eight mitochondria (Euscorpiidae; André 1959), four, five, and six being the predominant numbers in some families (Troglotayosicidae: ca. 4; Iuridae and Vaejovidae: ca. 5; Scorpionidae: 6 —reviewed in Alberti 1983; Michalik & Mercarti 2007; Vignoli et al., 2008). A very stable sperm trait, and closely associated with the presence of more than three mitochondria, is the presence of granular mitochondrial associated structures (MAS) (Jespersen & Hartwick 1973; Michalik & Mercati 2007; Vignoli et al., 2008). This is a spongework in the middle piece which shows structural affinities with smooth endoplasmic reticulum (Jespersen & Hartwick 1973). The function of the MAS is still obscure (Vignoli et al., 2008). An hypothesis is that the MAS could facilitate the transference of energy from the mitochondrion to the flagellum (Peter Michalik, personal communication).

On the other hand, as was mentioned before, aggregation of the spermatozoa into spermatodesms or sperm packages is a characteristic of all the lamelliform scorpions (Mattoni 2003; Peretti & Battán Horenstein 2003; Vrech et al. 2007; Vignoli et al., 2008). Nevertheless, the size and shape of the spermatodesms are highly variable among the families. For example, in the Bothriuridae the presence of long spermatodesms with elongated curved tips (the region containing the sperm heads) is common, whereas in the Diplocentridae the spermatodesms are coiled (Vrech et al. 2007; J. Contreras-Garduño, unpublished data). The diversity of the sperm aggregations in each family has not been explored in detail. In the Bothriuridae, preliminary studies have shown that the size and form of the spermatodesms differs between genera (Peretti & Battán-Horenstein 2003; Vrech et al. 2007).

The number of spermatozoa per spermatodesm is also variable among the few families in which this feature has been examined. In the Bothriuridae a spermatodesm contains approximately 220 spermatozoa while in the Iuridae the number is nearly 400 (Jespersen & Hartwick 1973; Vrech et al. 2007; D. Vrech, unpublished data).

Spermathecae and Sperm Maintenance

In bothriurid females the spermatozoa are completely unpackaged both within the female genital atrium and in the spermathecae 1 hour after insemination. The released spermatozoa remain in the atrium only 24 h in the Bothriuridae and 4 days in the Euscorpiidae (Peretti & Battán-Horenstein 2003; Lhajoui et al. 2004). After this period all sperms were located in the spermathecae. In some bothriurids, it is easy to induce sperm activation. For example, in species of *Timogenes* approximately 25% of the released spermatozoa extracted from the atrium and seminal receptacles start to move 30–50s after coming into contact with Ringer's saline. After the activation, the motile spermatozoa of bothriurids migrate to the spermathecae, possibly supported by peristaltic movements of the female atrium, which possesses a prominent muscular layer (Peretti 2003; Carrera 2008).

It is important to point out that, as in many other arthropods, insemination and fertilization in scorpions are not synchronous (Farley 2001). Therefore, selective mechanisms might trigger the sperm activation. Indeed, Alberti (1983) suggested that spermatozoa released from the male are possibly in a transport form, which is altered within the female into an active state through a capacitation process—a scenario also reported from other arachnid groups such as spiders (e.g., Alberti 2000; Michalik et al. 2005; Burger et al. 2006a, b). The initial activation in bothriurid females may occur as the spermatodesms enter the genital atrium from the everted spermatophore capsule and break to release the spermatozoa. However, a "stopping" mechanism seems to occur immediately after release since all spermatozoa extracted from the atrium showed no movement, except after immersion in saline solution. It seems probable that in the female sperm activation–inactivation may be mediated, at least in part, by substances secreted by epithelial secretory cells of the genital atrium

and spermatheca. It has been suggested in other arthropods that such female secretions may favor the survival of some spermatozoa over others (Eberhard 1996).

In bothriurids the spermatozoa are generally arranged close to each other and their anterior parts penetrating into distinct secretory areas of the epithelium of the spermathecae (figure 11.7) (Peretti & Battán-Horenstein 2003; Carrera 2008). The cytoplasm of these apical secretory cells contains many mitochondria and vacuoles. The residual sperms remain in the lumen of the spermathecae and are surrounded by a dense substance. The contact between spermatozoa and the secretory cells may indicate some role in nourishing and/or preserving of the sperm. However, this could be also just a kind of sperm phagocytosis by the epithelial cells. For example, in euscorpiid *E. carpathicus*, Lhajoui et al. (2004) observed hemocytes and spermathecal cells containing spermatozoa and suggested this could be an evidence for phagocytosis allowing sperm degradation. Secretions in the cytoplasm of these apical cells may be involved and should be studied further. For example, it would be very interesting to investigate if this type of phagocytosis also occurs in other scorpions and if there is a relationship between phagocytosis and sperm morphology (e.g., some sperm traits evolved for escape from phagocytic hemocytes). In other scorpions, only Kovoor et al. (1987) have reported that it may be possible for females of some species of *Centruroides* (Buthidae) to store sperm. In these buthids the spermatozoa were completely covered by glandular tissue.

Possible Evolutionary Pressures on the Sperm Traits

It is evident from the foregoing review that considerable variation among scorpion spermatozoa exists. Besides, our current knowledge on the diversity within each family is inadequate to determine whether this is lower than the, apparently very consistent, diversity existing between the families. Nevertheless, we can consider the main ideas suggested by previous authors, and, additionally, evaluate other options from the data reported here. First, considering the available data on sperm length, we can see there may be a strong phylogenetic constraint. For example, the shortest spermatozoa appear in the basal family Buthidae and the longest in the Iuridae, which is one of the more

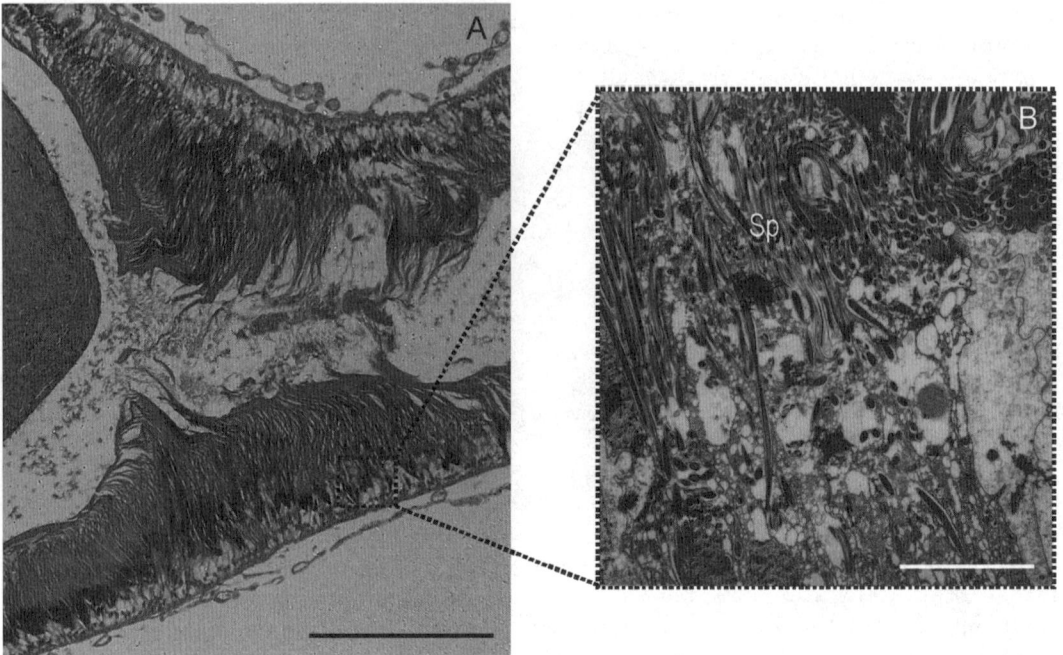

FIGURE 11.7 Sperm maintenance in the bothriurid scorpion *Brachistosternus ferrugineus*. General and detailed view showing the anterior parts of spermatozoa (Sp) inserted deeply into glandular portion of epithelium. The contact between fresh spermatozoa and the secretory cells may suggest nourishing and/or preserving of the sperm. Scale bars: (A) 100 μm; (B) 8 μm (photos: (A) Alfredo Peretti; unpublished; (B) adapted from Peretti & Battán-Horenstein 2003).

derived families. However, at an intrafamiliar level we also find appreciable diversity between some studied species. Bothriurids are very useful to illustrate this: within the genus *Bothriurus* there is a general tendency toward a shorter length in the most derived group of species: scorpions of the derived "Brazilian group" have shorter sperm that those of the more basal "*Andinobothriurus* group" (C.I. Mattoni & A.V. Peretti, unpublished data). It was recently determined that in species of the Bothriuridae spermatozoon length as well as spermatodesm length are not dependent variables of the male's size (Vrech et al. 2007). This interesting point has not yet been studied in other families. Regarding phenotypic correlations between sperm traits, it was observed that among the Bothriuridae the presence of long spermatozoa result in a lower number of spermatozoa per package (Peretti, personal observation; Vrech et al., in preparation). In another studied family, Iuridae, this pattern does not occur since they have long sperm and also large

spermatodesms (Jespersen & Hartwick 1973; D. Vrech, personal communication)

Among bothriurids, sperm length may be negatively associated with the presence of direct sperm competition (DSC). This suggestion is based on the fact that in species with a high frequency of female remating, and with more scramble competition between males (e.g., abundant vagrant males), the spermatozoa are significatively shorter than those of other species in which the female remates once per reproductive season (Peretti 1993; Peretti & Carrera 2005; P. Carrera, unpublished data). Interestingly, a similar pattern has been observed in other animals (Snook 2005), such as the cricket *Gryllus bimaculatus* (Gage & Morrow 2003). Such a pattern appears opposite to that shown by many other animals, in which direct DSC often promotes an increment in the sperm length (for a review in insects see Simmons, 2001). In this respect, the trend of some bothriurid scorpions toward shorter sperm in a context of DSC could be explained as a male strategy to optimize the transference of a higher number of spermatozoa per package (as said before, two inversely correlated sperm traits) and/ or to facilitate the preservation of the transferred sperm inside the spermathecae epithellium. Interestingly, in bothriurids the long spermatozoa (determined by a long flagellum) occur in species, such as *Urophonius brachycentrus*, in which males show an efficient mechanism to avoid DSC by blocking the female's gonopore with complex mating plugs once the insemination has ended (Carrera, 2008) (see below for plug diversity).

Some characteristics of scorpion spermatozoa may be affected by the structure and dimension of the female's genital tract, especially the spermathecae. As mentioned before, scorpion females posses two oval spermathecae between the genital atrium and ovariuterus. Like other arthropods with flagellate spermatozoa (reviewed in Simmons 2001; Snook 2005), a positive correlation between sperm length and size of the spermathecae is found in some scorpions. Examples of such patterns are the iurids *H. arizonesis* and *Caraboctonus keyserlingi*, where females maintain very long sperm in their two big spermathecae (Carrera 2008; Volschenk et al., 2008). Furthermore, both species also show "this gigantism" in other male reproductive traits, such as very large testes and seminal vesicles (Jespersen & Hartwick 1973; Vrech et al., 2007). In the Bothriuridae, however, increase in spermathecae size affects principally diameter instead of length (Peretti 2003; Carrera 2008; Volschenk et al., 2008), a fact associated with the presence of DSC (resulting in shorter sperm).

With regard to axonemal patterns, Hood et al. (1972) noted that in scorpions as well as other animals (Platyhelminthes, polychaetes, some insects and fishes) the absence of central tubules does not imply lack of sperm motility. However, Alberti (1983) has suggested that in scorpions the diversity of axonemal patterns reflects reduced importance of the motile apparatus in the spermatozoa, which are equipped with spermatophores as a mean for sperm transfer in their terrestrial habitats in contrast to ancestral aquatic lineages. Michalik & Mercati (2007) have noted that the 9×2+0 axonemal pattern present in the scorpionid *Opistophthalmus penrithorum* is responsible for a "helical" movement of the flagellum instead of the typical "lateral" movements shown by the plesiomorphic disposition 9+2 (Hood et al. 1972). Recently, helical movement was also observed in

another species with a 9+0 axoneme, the bothriurid *Timogenes elegans* (A.V. Peretti & D. Vrech, unpublished data). It is important to realize that in all these cases, independently of the axoneme type, flagellar movement is slow, never showing rapid undulations as in the spermatozoa of vertebrates (Gomendio et al. 1998; Pitcher et al. 2007). At our present state of knowledge the functional implications of these patterns of movements are not clear. However, helical movement seems to produce a lower velocity than lateral movement (the latter is more common among other animals with flagellar spermatozoa). In this context, when more data are available, it will be interesting to compare the motility of both scorpion axonemal patterns with the 9×2+2 pattern of insect flagellosperm (Werner & Simmons 2008). Among insects a double helical movement in which a small amplitude, high frequency wave is superimposed on a high amplitude, low frequency wave is common. Apparently, this pattern is an adaptation to movement within narrow genital ducts with the large wave increasing the contact with, and the small wave pushing the spermatozoon off, the duct wall (Werner & Simmons 2008). From this scenario in insects, it would be useful to explore whether a similar pattern of motility may occur in some scorpions as well as how it interacts with an intricate female reproductive tract, in which two ovoid and flexible spermathecae are followed by very long and narrow ducts of the ovary–uterus (Hjelle 1990; Volschenk et al., 2008). In summary, returning to Alberti's (1983) idea, the presence of sperm transfer by a sclerotized spermatophore could not explain why a rapid and/or differential velocity inside the long reproductive tract of scorpion females would not be necessary. Without doubt, this aspect is one of the most interesting for further studies in the group.

The high number of mitochondria in non-buthid scorpions may be explained in more than one way. One hypothesis would be that more mitochondria may produce faster spermatozoa, also counteracting the loss of axonemal tubules. An alternative hypothesis is that more mitochondria could produce better endurance of the spermatozoa during locomotion throughout the long ovariuterus of scorpion females (Hjelle 1990; Francke 1982; Warburg & Rosenberg 1990, 1992; Farley 2001). Further studies are needed to document such possible relationships. By comparing three scorpion families, the basal Buthidae, and the derived

Bothriuridae and Iuridae, we see that an increase in sperm length is due to increment in the size of the midpiece (A. V. Peretti & D. Vrech, unpublished data), the region in which the mitochondria are located. This fact may reinforce the idea supported for many animals with flagellate spermatozoa about the importance of this region for production of better velocity or endurance (Immler et al. 2007) instead of the head as in others (reviewed in Malo et al. 2006). As is clear in more derived families, more mitochondria are associated with the presence of MAS, a combination that might optimize the transfer of energy to the flagellum during sperm movement.

Finally, the presence of spermatodesms in lamelliform scorpions but not in the flagelliform Buthidae may be related to differences in the sperm transfer mechanism between these two types of spermatophore (Peretti & Battán-Horenstein 2003). For example, insemination lasts some minutes in lamelliforms but only takes seconds in flagelliforms so that a possible hypothesis is that the aggregation of spermatozoa protects the sperm cells against dehydration, but this protection is unnecessary in buthids. Therefore, aggregation of spermatozoa into spermatodesms in lamelliforms, and the exterior improvement of covering and compactation of the included sperm, could have arisen by pressures of natural selection directed to guarantee the preservation of the sperm in transfer during the longer insemination of these groups as compared with the "simple model" of buthids (Weygoldt 1990). However, pressures of sexual selection cannot be excluded, for example, in a scenario of sperm competition (e.g., by producing packages with more sperm in a context of DSC).

Evaluation of Traits Likely Associated with Avoiding DSC: "The Mating Plug Enigma"

The structure and function of genital or "mating" plugs have been explored in diverse animal groups, including many arthropods such as insects and spiders. In general, male genital plugs have been interpreted as a mechanism to prevent or reduce sperm competition (Parker 1984; Wigby & Chapman 2004). For example, when a male plug completely blocks the gonopore or genital atrium of an inseminated female, she will be deprived of the opportunity to select among the sperm of multiple males. However, there are few sources of evidence actually

showing that mating plugs prevent sperm competition (Elgar 1998; Birkhead & Møller 1998; Simmons 2001). In spiders the presence of one or more plugs blocking the female gonopore, internally and/or externally, is a widespread feature of inseminated females in many species (Austad 1984). For example, depending on the species, the mating plug can be formed by coagulation of the ejaculate on the external aperture of the spermathecae (Masumoto 1993; Méndez & Eberhard, unpublished data), or by a chitinous part of the male copulatory organ that breaks inside the spermathecal ducts (e.g., Kaston, 1970; Austad, 1984; Christenson, 1990). However, the effectiveness of these mating plugs in avoiding direct sperm competition is variable as inseminated females invariably remate, and, furthermore, new males have a chance to remove the external plugs (Masumoto 1991, 1993). In addition, internal chitinous plugs from the male copulatory organ do not impede new sperm from entering the spermathecae (Schneider et al., 2001). Scorpions are the other order of arachnids in which mating plugs have been often observed in inseminated females (Polis & Sissom, 1990), but doubts persist with regard to their origin and function (Mattoni & Peretti, 2004).

In this section I will review the subject in scorpions, considering the following aspects of each plug type: morphology, origin, position and duration within the female, effect on female remating, function, and evolutionary patterns.

Diversity of Mating Plugs: Complexity and Functional Morphology

Gel-like Mating Plugs In many species, immediately after the transfer of spermatozoa has finished (but with the spermatophore capsule still attached to the female), a gel-like mating plug is expelled from the spermatophore and enters directly into the female's genital chamber (Mattoni & Peretti 2004). Historically, the first term for the plug was 'spermatocleutrum' (Pavlovsky 1924), used for the whitish, shrunken mass found in the 'vaginal opening' of fertilized females. Spermatocleutrum was later used for many decades to refer to any type of plug found in the female's genital tract, independent of origin, structure, and function (Polis & Sissom 1990): for example, Varela (1961) in the bothriurid, *B. bonariensis*, Bücherl (1956), Shulov and Amitai (1958), and Probst (1972) in buthids, Smith (1966) in scorpionids, and Fox (1975) for vaejovids. However, more recently, spermatocleutrum had been mainly used only for the gel-like plugs in those families where sclerotized and more complex plugs have been found as well (e.g., Stockwell, 1989 in Vaejovidae and Urodacidae; Castelvetri & Peretti 1999 in Bothriuridae).

There are no comparative data on the fine differences between gel-like plugs of the mentioned families and their origin is uncertain (Shulov & Amitai 1958). Polis and Sissom (1990) suggested that it is secreted by the female sometime after mating, and is often lost prior to parturition. Indeed, as it has been observed in other arthropods, not only males but also females can play an active role in producing the plug in the female genital tract (Eberhard 1996; Mendez & Eberhard, unpublished data; A. Aisenberg, personal communication). This means that the presence of a plug in the female's gonopore does not mean that it was produced by males to avoid DSC. However, a female origin should not be generalized to all scorpions, because, in some bothriurids, gel-like plugs seem to be formed by a granular substance secreted by the medial accessory glands of the male's paraxial organs (Peretti 2003; Peretti & Battán-Horenstein 2003; Mattoni, unpublished data). These plugs are formed by the combination of the granules and hard elements (not spermatodesms or free spermatozoa) that can be observed inside a fresh spermatophore (Peretti 2003). In the Euscorpiidae, Lhajoui et al. (2004) observed the presence of abundant fibrillar material (absent inside the seminal vesicle of the male), apparently originating in the compaction of the microvilli detached from the epithelial cells of the spermathecae. These authors suggest this material may participate in the formation of the plug.

Typically, the gel-like plug appears in the center of the lumen of the female genital atrium and increases in size with further matings. Clear internal obstructions were observed only in female *B. bonariensis*, where a hard-amorphous plug has been observed to block the spermathecal ducts (Castelvetri & Peretti 1999; Peretti 2003). Whether males or females could remove this type of plugs has not been explored. There are no data on the changes that the gel-like plugs undergo inside the female during gestation. However, they look smooth when expelled at the beginning of parturition, before the emergence of the larvae (Varela 1961; Castelvetri & Peretti 1999).

Benton (1992b) observed that females of *Euscorpius flavicaudis* seem to accept just one mating per reproductive season. Some plugs (apparently of the gel-like type) have also been found in inseminated females of other *Euscorpius* (Jacob et al. 2004a; Jacob, unpublished data). It is difficult, however, to confirm whether the low sexual receptivity was because females had a plug and/or due to other factors. In summary, there is no consensus as to the function of gel-like plugs and even though their presence does not impede the intromission of a new spermatophore and ejaculate, in some cases it may make the localization of new sperm in the spermathecae more difficult (Castelvetri & Peretti 1999).

Sclerotized Mating Plugs Two terms have been used for sclerotized plugs: mating plug and genital plug (Mattoni & Peretti 2004); the former in the Vaejovidae (e.g., Stockwell 1989; Sissom 1993) and the latter in the Bothriuridae (Peretti 2003; Mattoni & Peretti 2004). One thing is clear, all the sclerotized plugs are produced by one or more definite cuticular parts of the male's spermatophore (Mattoni & Peretti 2004; Contreras-Garduño et al. 2006). There are two subtypes of sclerotized mating plugs: simple and complex plugs (figure 11.8).

Simple Plugs Filamental. So far this has been found as an autapomorphy of the bothriurid *B. bonariensis* (Mattoni, 2003). The filamental plug is formed by a thin filament extending from the capsular foramen of the spermatophore (figure 11.8.a) that is broken during sperm transfer and remains in the genital atrium, together with previous gel-like plugs (Peretti 2003). Each hemi-mating plugs is formed on the inner side of each basal lobe (hooks), fusing helically with each other during capsular eversion. Removal of filamental plugs by males and/or females after mating has not been observed. This type of plug is partially degraded during the gestation stage and expelled without difficult during parturition (Castelvetri & Peretti, 1999). Interestingly, post-mating sexual receptivity of the female is not affected. One example is that of *B. bonariensis* in which a majority of inseminated females (76% containing filamental and gel-like plugs) remate up to four times per reproductive season and only pregnant females are unreceptive (Castelvetri & Peretti 1999; Peretti 2003). This filamental plug does not affect the eversion process of a new spermatophore. However, because of its more internal position in the female

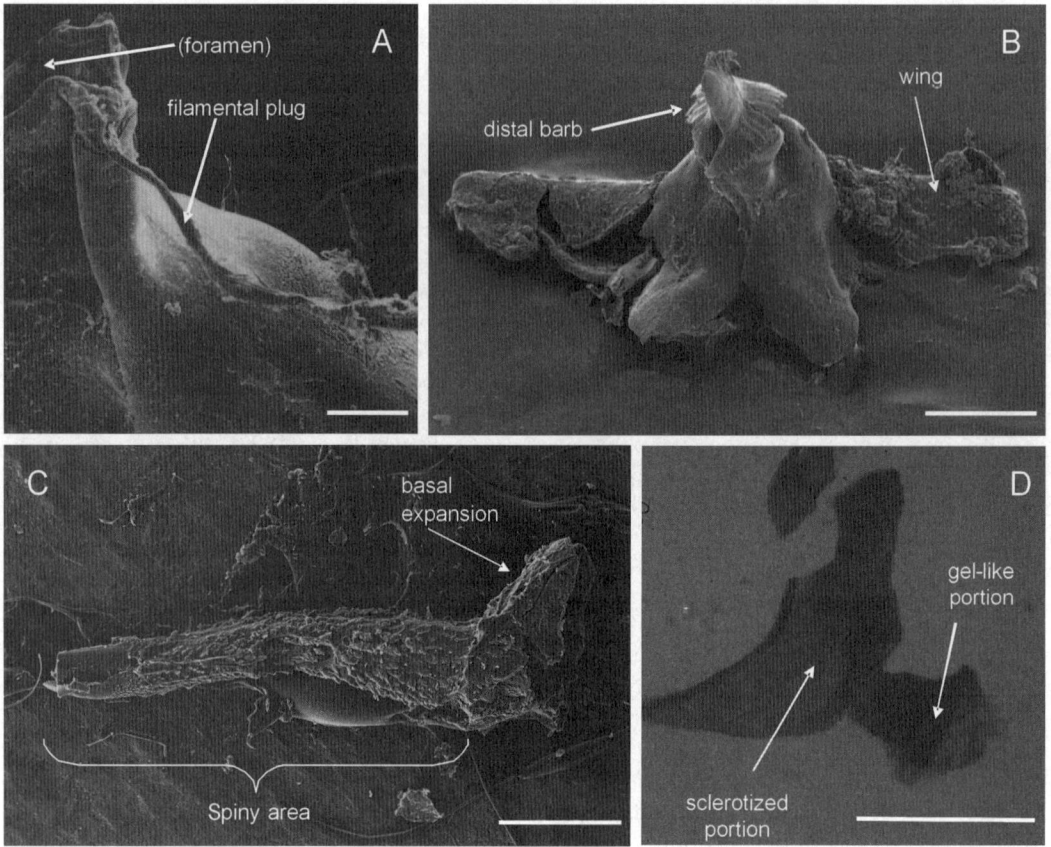

FIGURE 11.8 Four forms of sclerotized mating plugs in scorpions. (A) Simple plug: filamental plug of *B. bonariensis*. (B) Complex plug: distally barbed plug of *Vaejovis punctatus*. (C) Complex plug: cone-shaped plug of *B. asper*. (D) Mixed plug of *Phoniocercus sanmartini* (photos: (A) from Peretti 2003; (B) courtesy of Jorge Conteras-Garduño; (C) from Mattoni & Peretti 2004; (D) courtesy of Camilo Mattoni). Scale bar: (A, B, and C): 200 μm; D: 500 μm.

genital atrium, this type of plug can block the initial portion of the spermathecal ducts and ensure that the sperm of a second male is confined to the genital atrium (Castelvetri & Peretti 1999).

Membranous. This is a synapomorphy of the *flavidus* group of *Bothriurus* (Bothriuridae) (Mattoni 2003; Mattoni & Peretti 2004). The membranous genital plug occurs near the genital opening of inseminated females. The plug is formed by membranous components that are in close contact with the region of the spermatophore foramen. After the reproductive season, this plug suffers a progressive degradation in the female's genital atrium (Peretti, unpublished data) and does not affect her post-mating sexual receptivity (Peretti 1993; Peretti &

Battán-Horenstein 2003). The position of the plug only partially prevents the intromission of new spermatophores (Peretti 2003, unpublished data).

Complex Plugs Distally barbed. This is the most well-known example of a genital plug originating from a male scorpion and is found in inseminated females of several species of the family Vaejovidae (Stockwell 1989). In general, descriptions have relied only on one hemi-mating plug (one of the two chitinous structures—a part of the lobes—that will fuse during spermatophore deposition to form the mating plug; Stockwell 1989). These descriptions have usually been used in taxonomy (e.g., Stockwell 1989; Sissom & Stockwell 1992; Sissom 1993).

Each hemi-mating plug has a 'distal barb' (naked or with small-teeth) that enhances anchorage of the plug to the cuticular wall of female genital atrium once sperm transfer has concluded. The best known example are the mating plugs of *V. punctatus*. Contreras-Garduño et al. (2006) provide detailed data on its morphology and function. This plug completely blocks the female's genital atrium and gonopore. The basal portion of the plug has two wide expansions, the 'wings' (figure 11.8b), which remain covering the gonopore and two large spines that anchor in the most proximal part of the genital atrium. The apical portion of the plug shows a 'distal barb' consisting of a 'crown' of many small, sharp and posteriorly oriented (to the female's gonopore) spines. This apical portion anchors against the cuticular wall of the internal region of female's genital chamber. This type of plug cannot be displaced by pushing it to the interior of the female's genital chamber as the two large spines of its basal part are oriented towards the internal region of this space and fit strongly against the proximal portion of its cuticular wall. Moreover, the two 'wings' of the basal part of the plugs remain external to the gonopore, which would obstruct movement into the female's inside. Five phases of degradation of the mating plug were recognized. In the fifth phase (after more than seven months) each inseminated female has only little pieces of the mating plug, at a time in which these females had well developed embryos. The plug is completely degraded before the parturition season, so that the emergence of embryos is not impeded. None of the inseminated females—those bearing a complete or partially degraded mating plug according to the phase of degradation—accept a new mating. Although males started courtship in all cases, females always respond negatively. These data reported by Contreras-Garduño et al. (2006) are evidence that this type of mating plug is effective in preventing DSC because females did not remate until several months later. It seems that the complex anchoring mechanism prevents potential removal by the female.

Cone-shaped. In the Bothriuridae, a giant sclerotized plug appears in Bothriurus asper and its as yet undescribed sister species (Mattoni 2003). The giant spiny cone-shaped plug of B. asper (figure 11.8c) is the most complex plug found among the bothriurids. Given the large size and complex shape, the genital plug of the B. asper group is unique, not only among the Bothriuridae, but in the order Scorpiones (in part, this type of plug resembles the genital plug of the Vaejovidae). It is formed by fusion of the basal lobes of the hemispermatophore and completely fills the genital atrium of inseminated females. Fusion is complete in B. asper and the surface of the plug has many microspines that anchor it to the cuticular wall of the atrium. Despite this strong anchorage, the cuticular wall shows no evidence of lesions. The plug also blocks the two ducts that allow communication between the genital atrium and the spermathecae. As in the *V. punctatus* plug, in the B. asper group the plug also suffers progressive degradation inside the female: The color changes from brown to blackish and the plug becomes completely degraded in females with full developed embryos. Mattoni & Peretti (2004) analyzed only preserved material from museums (living specimens were not available). Inseminated females (i.e., those containing spermatozoa in the sperm receptacles) always had just one plug inside their genital tract. It was necessary to destroy the tissues of the female's genital chamber for removal of the inserted plug. In summary, all these data on size and structural complexity suggest that cone-shaped plugs are efficient male devices to prevent DSC.

Mixed plugs   In the family Urodacidae there is a big structure in the hemispermatophore lobes that has been referred as a hemi-mating plug (Stockwell 1989; Prendini 2000; Volschenk, personal observation) and that could have a similar function to the plugs described here for the complex sclerotized plugs. This mating plug would be a "mixed" subtype because of the combination of sclerotized and gelatinous components (Mattoni & Peretti 2004). An example is found in the bothriurid genus *Phoniocercus* (figure 11.8d) where the plug is formed by the fusion of some hemispermatophore detachable structures (associated with the basal lobe) with a hard sclerotized substance that resembles the gel-like plug (Mattoni, unpublished data). A similar plug subtype appears in inseminated females of *Urophonius* species, a genus related to *Phoniocercus* (Prendini 2000, 2003). In *U. brachycentrus* the inserted mixed plug completely blocks the gonopore and, similarly to the complex plugs above, prevents the intromission of spermatophores of further males. In this species female remating is only partially affected since inseminated females accept new males which, however, never deposit a spermatophore (Carrera 2008).

Main Evolutionary Patterns and Link Between Avoidance of DSC and Sexual Conflict

Given that the presence of a gel-like plug in the atrium of inseminated females of flagelliform and lamelliform scorpions is common (figure 11.9), Stockwell (1989) and Mattoni and Peretti (2004) postulated that this type represents a plesiomorphic character in the order. New comparative data

from the Bothriuridae confirm this supposition for the gel-like plug, which has been found in most species in which mating behavior and/or reproductive morphology have been examined (Carrera 2008; Mattoni, unpublished data). Therefore, this plesiomorphic (Mattoni & Peretti 2004) type of genital plug is usually shown by species that also contain other, more elaborate, genital plugs. Other types of genital plugs described here, represent apomorphies (Mattoni & Peretti 2004; Peretti 2007) of some *Bothriurus* species (table 11.2): filamental for *B. bonariensis*, membranous for the *B. flavidus* group, cone-shaped for the *B. asper* group, and the mixed plug subtype for the

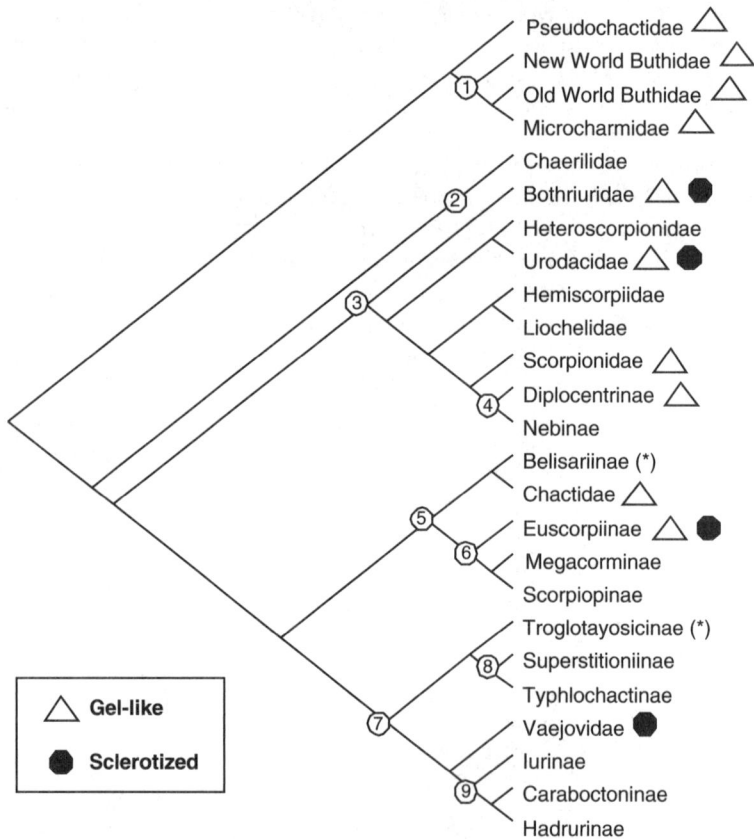

FIGURE 11.9 General occurrence of the two principal types of mating plugs within the Order Scorpiones. Sclerotized plugs from the spermatophore capsule have evolved independently among lamelliform families. Cladogram based on Coddington et al. (2004). Abbreviations: (1) Buthoidea, (2) Chaeriloidea, (3) Scorpionoidea, (4) Diplocentridae, (5) Chactoidea, (6) Euscorpiidae, (7) Vaejovoidea, (8) Superstitioniidae, (9) Iurida; (*) Troglotayosicidae.

clade *Phoniocercus-Urophoniu*. The complex morphology of the genital plug of the *B. asper* group and its origin from detachable basal lobes of the spermatophore are novel features in the Bothriuridae.

This structural pattern also occurs in the Vaejovidae and Urodacidae. This is very important because it is the best evidence that these functional and morphological similarities are evidently not completely homologous and must have evolved independently. Indeed, the conspicuous character "presence of a complex sclerotized plug" (Mattoni & Peretti 2004), with an apparently similar function of preventing DSC, shows little tendency to be lost, occurring in all members of the Urodacidae and most Vaejovidae (Stockwell 1989; Prendini 2000).

In relation to the mating plugs of the Vaejovidae, it is clear that we need more detailed morphological and functional data on mating plugs of other vaejovids to explore the presence or lack of alternative patterns. Stockwell (1989) realized that it is necessary to better define the phylogenetic relationships among the groups to permit outgroup comparisons. So the plesiomorphic state (Stockwell 1989) is determined to be the "smooth" condition of the plug by the functional criterion, i.e., a change from simple to complex. The "toothed" condition is considered apomorphic (Stockwell 1989). Examples of these two opposite patterns are the smooth distal barb of the mating plug in the genus *Serradigitus* and the genera *Syntropis* and *Vaejovis* in which the distal barb is provided with numerous teeth. In contrast, there are no mating plugs in the *V. mexicanus* group, possessing the plesiomorphic state, i.e., only a gel-like mating plug (Stockwell 1989; Contreras-Garduño, personal observation). The question remains as to which evolutionary forces are behind the whole variety of reported for hemi-mating plugs (e.g., presence or absence of some spines, shapes of the wings) (Stockwell 1989; Sissom 1992; E. González, unpublished data).

With regard to the mating plug as a male strategy to avoid DSC, this review indicates that postmating sexual receptivity is affected by the type of plug anchored inside the female (table 11.2). With sclerotized plugs females reject almost all new males while gelatinous plugs do not affect female behavior, at least not strongly. More comparative studies using species that show both extremes of mating plugs are necessary to test whether the morphological properties of plugs correlate with female remating probability. However, if we take into account the data on complex sclerotized plugs, we could say that the effectiveness of a plug transferred by the male in avoiding sperm competition may be a function of the anchoring devices that it bears and its effective blocking of the female's genital chamber and gonopore (Contreras et al. 2005). However, the case of *V. punctatus* also suggests that in this species the mechanical barrier imposed by the mating plug and male behavior may not be the only mechanism preventing female remating (Polis & Sissom 1990; Peretti & Carrera 2005). This is because, after mating plug removal, all inseminated females still rejected new males. Possibly, when the mating plug was removed, the female was damaged. This may have induced her to avoid any further copulation as a way to recover (Contreras-Garduño et al. 2006). Another important thing to realize, is that in none of the scorpion species have the males developed behavioral and/or morphological strategies to remove complex mating plugs. At a behavioral level, the high level of aggression shown by many reluctant females after a mating may impede new males from removing the plug by manipulation of the gonopore (e.g., with the pedipalps).

Contreras-Garduño et al. (2006) suggested that pregnancy may be another possible cause for the reduction in remating propensity in females blocked by complex plugs, since the presence of the plug may induce hormonal and behavioral changes which make females unreceptive. This may explain why females in *V. punctatus* in 3, 4, or 5 stage of degradation did not mate, but both after a few minutes, and one or two weeks after their first mating females did mate again, but only partially. There are alternative hypotheses, not related to sperm competition, to explain the presence of a plug in inseminated females (Contreras-Garduño et al. 2006) as been mentioned. One of them is that the presence of a mating plug produced by the male may serve to reduce loss and sperm desiccation in the gonopore. This idea could also be extended to any type of mating plug, not only for that of *V. punctatus* that these authors studied in detail. However, at least for this species as well as in other vaejovids and lamelliform scorpions with complex sclerotized plugs, this hypothesis does not explain the great complexity and diversity. Indeed, this idea could apply to any type of plug (from gel-like to hard and sclerotized) found inside the female, independently of whether it was produced by the male or the female. Perhaps, temporary gel-like plugs

TABLE 11.2 Current knowledge of the principal types of mating plugs in scorpions

Type of plug	Origin	In the female			Efficacy	Examples of species	References
		Position	Degradation	Effect			
Gel-like	Male acc. glands	Genital Atrium Spermathecae	Partial	No	Low?	*Tityus bahienis* (Buthidae)	Bücherl 1956; Stockwell 1989; Polis & Sissom 1990 Althaus et al. *in press*
	Mainly sperm	Female gonopore (full)	Partial	Yes	High	*Euscorpius italicus* (Euscorpiidae)	
Sclerotized Simple *Filamental*	Part of spermatophore	Genital Atrium (partial)	No?	No	Intermediate	*Bothriurus bonariensis* (Bothriuridae)	Peretti 2003
Membraneous	Part of spermatophore	Genital Atrium (partial)	Partial?	No	Intermediate?	*Bothriurus flavidus* (Bothriuridae)	Peretti 2003; Mattoni & Peretti 2004
Complex *Barbed*	Part of spermatophore	Genital Atrium (full)	Complete	Yes	High	*Vaejovis punctatus* (Vaejovidae)	Stockwell 1989; Contreras-Garduño et al. 2006
Cone-shaped	Part of spermatophore	Genital Atrium (full)	Complete	Yes?	High?	*Bothriurus asper* (Bothriuridae)	Mattoni & Peretti 2004
Mixed	Part of spermatophore	Genital Atrium (full)	Complete?	Yes?	High?	*Phoniocercus sanmatini* (Bothriuridae)	Mattoni & Peretti 2004

could have this adaptive value. Another alternative hypothesis is that the mating plug of *V. punctatus* serves to prevent the entry of pathogen agents into the gonopore and sperm. However, there is no supporting information about this and, again, the diversity reported from the hemi-mating plugs of vaejovids (e.g., in Stockwell 1989; Sissom 1993; E. González, personal communication) cannot be explained by this hypothesis.

A possible explanation for vaejovid sexual features, is that females may have resolved the conflict in their favor, are able to choose males on the basis of the males' sperm competition abilities (Eberhard 1996; Cordero & Eberhard 2003), and may be actually choosing males with better mating plugs (Contreras-Garduño et al. 2006). These authors proposed that even with this apparent form of male control of mating, female choice could still act if:

1. females 'screen' males before sperm transference;

2. the effectiveness of the mating plug is selected by females; and,
3. after copulation, there is selection for males that are better at searching for the few receptive females.

Nevertheless, there is a detail of the gel-like mating plugs in *B. bonariensis* that could be linked with sexual conflict: the mortality associated with the accumulation of plugs inside the atrium and the spermathecae when females remate several times (Castelvetri & Peretti 1999). At present it is not known if the female deaths were due to the plugs or to accumulation of toxic substances in the ejaculate (e.g., Chapman et al. 1995 in *Drosophila*) or to other causes. Indeed, some of the differences among complex mating plugs may be explained using a sexual conflict perspective in which males' and females' sexual interests are not similar (Contreras-Garduño et al. 2006). This perspective indicates that because of intense male–male competition, males will evolve different means to increase their fertilization success (Chapman 2006). This evolutionary pressure will make females mate suboptimally or with high costs for each mating (Arnqvist & Rowe 2005). For example, in scorpion species without mating plugs (e.g., the buthid *Zabius fuscus*, the bothriurid *Timogenes dorbignyi*) females are polyandrous, remating more than once per reproductive season (Polis & Sissom 1990; Peretti 1993). In contrast, the transfer of a complex sclerotized mating plug by the male clearly implies the imposition of "forced monoandry" for the inseminated female. Therefore, in scorpions, restriction of female mating activity by complex mating plugs may be considered an interesting new example of sexual conflict resulting from adaptations to sperm competition (Stockley 1997).

CONCLUSIONS AND PROSPECTS

Throughout this chapter, I have shown data suggesting it is unwise to assume that only a single pattern of selection exists in either the spermatophore or other reproductive characters of scorpions. The fact that spermatophore structures may have more than one function is such a real phenomenon that we should also think that many mixed patterns of selective pressures may be occurring. For example, the coexistence of evidence for different, and apparently opposing, hypotheses in a single species in which each structure, or group of them, might function in more than one context. This is not exclusive to scorpion genitalia. In spiders, Huber (1996) identified the presence of mixed and inconsistent patterns of selection in the genitalic traits of pholcids, and proposed viewing the genitalia as multifunctional organs.

On the other hand, Ghiselin's chapter (in this book) discusses this matter: in the section "furthering effective mating and fertilization", spermatophores are shown as examples of parts of the male's reproductive system that are primary sexual characters, produced by natural selection. However, as Huber (1996) pointed out, the morphologically complex genitalia of spiders and those of many arthropods, in which we can include scorpions, will often not be assignable to primary or secondary sexual traits. In this respect, he suggested also the substitution of a classification into "displaying" and "coupling" traits for the traditional classification of

sexual traits into primary and secondary sexual traits. Indeed, the presence of mixed patterns of evolution was also noted by Andersson (1994) for behavioral or morphological traits participating in sexual interactions that are probably favored by both sexual and natural selection. Based on Huber's statements, spermatophores and spermatozoa of scorpions may show both types of traits, each selected separately. Figure 11.10 summarizes some sexual characters of the scorpion spermatophore that were reviewed in the present chapter, and following Ghiselin's criteria, whether they should be considered as primary, secondary or tertiary (i.e., "mixed") characters.

Conclusions from current knowledge on genitalia of a few scorpions cannot be automatically extended to other characters and/or still unexamined groups of scorpions since generalizations may be misleading. It must be noted that our understanding of the diversity across different groups in functional morphology, still has several gaps. To homologize certain parts of the hemispermatophores across the whole order a "hemispermatophore homology project" has recently been started by scorpionologists from many regions of the world (Volschenk et al., in preparation).

There are many interesting areas waiting for research from a combined functional and evolutionary perspective. For example, it is necessary to determine whether the innervation of the deeper part of the epithelium of the female atrium could detect pressures exerted by the spermatophore capsule against the cuticular wall. Moreover, more studies are necessary on the detailed location of sensilla in the female's genital atrium in species of different families and on their correspondence with each part of the capsule, in order to confirm if spines and rough surfaces in the spermatophore always stimulate the female as genitalic copulatory courtship.

On the other hand, the absence, in the female atrium, of lesions, caused by the spiny surface of everted capsules or the detached complex plug, may be attributed to the presence, in bothriurids and vaejovids, of a folded atrial epithelium, covered by a thick cuticular wall. This interesting correlation may suggest sexually antagonistic coevolution. For example, this female's trait could be considered a counteradaptation selected by this type of coevolution to avoid injury during capsular evertion of the male's spermatophore. A similar situation was observed in seed beetles (Rönn et al. 2007), in

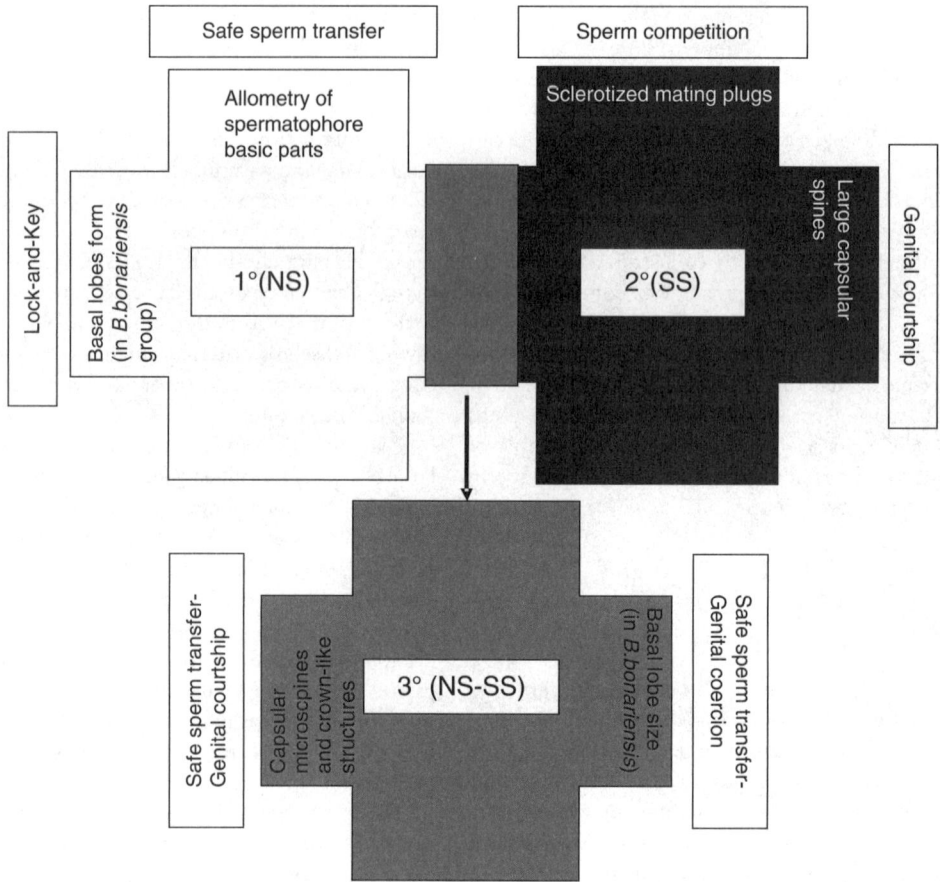

FIGURE 11.10 Diagrammatic comparison of some characters of scorpion spermatophores (using bothriurids as model) and their position as primary, secondary or tertiary sexual characters according to Ghiselin's criteria as to whether they owe their existence to natural selection, sexual selection or both, respectively.

which spinier male genitalia that could cause harm to females are associated with reinforcement of the copulatory tract (connective tissue, in this case).

Finally, further studies are needed to document the diversity and many functional aspects of scorpion spermatozoa, such as the relationship between motility and the size of some sperm traits. In fact, the presence of sperm transfer by a sclerotized spermatophore cannot explain why a rapid movement and/or endurance inside the long female reproductive tract would not be necessary. Without doubt, this is one of the most interesting questions for studies in the group. Sperm production (with the complementary evaluation of testis size) and its relationship with the risk of sperm competition in different lineages represents a promising area.

Studies on sperm competition may also benefit from examining the presence of a "differential" conservation of the sperm in the spermathecae in more species to evaluate the presence of cryptic female choice in the form of sperm selection.

Acknowledgments Patricia Carrera, Camilo Mattoni, David Vrech, Peter Michalik, Erich Volschnek, and Edmundo González let me read and liberally cite their work, in press or in preparation. Several other individuals also helped me with the literature, including Bill Eberhard, Gerd Alberti, Anja Klann, Alain Jacob, and Lorenzo Prendini. I especially thank Camilo Mattoni, David Vrech, Peter Michalik, Anja Klann, Alex Córdoba-Aguilar,

Janet Leonard, and three anonymous reviewers for providing helpful comments on the manuscript. This work was supported by CONICET, FONCYT and SECYT-UNC of Argentina.

REFERENCES

Alatalo, R. V., Höglund, J. & Lundberg, A. 1988. Patterns of variation in tail ornament size in birds. Biol. J. Linn. Soc., 34: 363–374.

Alberti, G. 1983. Fine structure of scorpion spermatozoa (*Buthus occitanus*; Buthidae, Scorpiones). J. Morphol., 177: 205–212.

Alberti, G. 2000. Chelicerata. In Adiyodi, K. G. & Adiyodi, R. G. (Eds) Reproductive Biology of the Invertebrates. Vol. IX, Part B (Progress in male gamete ultrastructure and phylogeny). Oxford & I B H Publis. Co., New Delhi. Pp. 311–388.

Alexander, A. J. 1957. The courtship and mating of the scorpion, *Opisthophthalmus latimanus*. Proc. Zool. Soc. Lond., 128: 529–544.

Alexander, A. J. 1959. Courtship and mating in the Buthid scorpions. Proc. Zool. Soc. Lond., 133: 145–169.

Alexander, R. D., Marshall, D. C. & Cooley, J. R. 1997. Evolutionary perspectives on insect mating. In: The Evolution of Mating Systems of Insects and Spiders. Choe, J. C. & Crespi, B. J. (Eds). Cambridge: Cambridge University Press. Pp. 4–31.

Andersson, M. 1994. Sexual Selection. Princeton University Press, Princeton, 599 pp.

André, J. 1959. Etude au microscope électronique de l'évolution du chondriome pendant le spermatogénése du scorpion *Euscorpius flavicaudis*. J. Ultra. Res., 2: 288–308.

André, J. 1963. Some aspects of specialization in sperm. In: General Physiology of Cell Specialization. Mazia, D. & Tyler, A. (Eds). New York: McGraw-Hill, Pp. 91–115.

Angermann, H. 1957. Über Verhalten, Spermatophorenbildung und Sinnesphysiologie von *Euscorpius italicus* Herbst und vertwandten Arten (Scorpiones, Chactidae). Z. Tierpsychol., 14: 276–302.

Arnqvist, G. 1998. Comparative evidence for the evolution of genitalia by sexual selection. Nature, 393: 784–786.

Arnqvist, G. & Rowe, L. 2002. Antagonistic coevolution between the sexes in a group of insects. Nature, 415: 787–789.

Arnqvist, G. and Rowe, L. (2005). Sexual Conflict. Princeton University Press, Princeton.

Austad, S. N. 1984. Evolution of sperm priority patterns in spiders. In: Sperm Competition and the Evolution of Animal Mating Systems (Smith, R. L., Ed.). Academic Press, Orlando. Pp. 223–249.

Benton, T. G. 1992a. Courtship and mating in *Leiurus quinquestriatus* (Scorpiones; Buthidae). In Arachnida. Cooper, J. E., Pearce-Kelly, P. & Williams, D. L. (Eds). London: Chiron Publications. Pp. 83–98.

Benton, T. G. 1992b. Determinants of male mating success in a scorpion. Anim. Behav. 43, 125–135.

Benton, T. G. 1993. The courtship behavior of the scorpion, *Euscorpius flavicaudis*. Bull. Br. Arachnol. Soc., 9: 137–141.

Benton, T. G. 2001. Reproductive biology. In Scorpion Biology and Research. Brownell, P. & Polis, G. A. (Eds). Oxford: Oxford University Press. Pp. 278–301.

Birkhead, T. R. & Møller, A. P. 1998. Sperm Competition and Sexual Selection. Academic Press, San Diego.

Bücherl, W. 1956. Escorpioes e Escorpionismo no Brasil Observacoes sobre o aparelho reprodutor masculino e o acasalamento de *Tityus bahiensis*. Mem. Inst. Butantan 27: 121–155.

Burger, M., Graber, W., Michalik, P. & Kropf, C. 2006a. *Silhouettella loricatula* (Arachnida, Araneae, Oonopidae): A Haplogyne spider with complex female genitalia. J. Morphol., 267: 663–677.

Burger, M., Michalik, P., Graber, W., Jacob, A., Nentwig, W. & Kropf, C. 2006b. Complex genital system of a haplogyne spider (Arachnida, Araneae, Tetrablemmidae) indicates internal fertilization and full female control over transferred sperm. J. Morphol., 267: 166–186

Castelvetri, S. & Peretti, A. V. 1999. Receptividad sexual y presencia de tapón genital en hembras de *Bothriurus bonariensis* (C. L. Koch) (Scorpiones, Bothriuridae). Revue Arachnol., 13: 15–23.

Carrera, P. 2008. Estudio comparado de los patrones de selección sexual en escorpiones y ácaros acuáticos: evaluación de hipótesis de conflicto sexual y cooperación intersexual. 345 pp., Doctoral Thesis, Fac. Cien. Exactas, Físicas y Nat., Univ. Nac. Córdoba, Argentina.

Carrera, P., Mattoni, C. & Peretti, A. V. 2009. Chelicerae as male grasping organs in scorpions: sexual dimorphism and associated behavior. Zoology. 112: 332–350.

Chapman, T. 2006. Evolutionary conflicts of interest between males and females. Curr. Biol., 16: 744–754.

Chapman, T., Liddle, L. F., Kalb, J. M., Wolfner, M F. & Partridge, L. 1995. Cost of mating in *Drosophila melanogaster* females is mediated by male accessory gland products. Nature, 373: 241–244

Chapman, T., Arnqvist, G., Bangham, J. & Rowe, L. 2003. Sexual conflict. Trends Ecol. Evol., 18: 41–47.

Christenson, T. E. 1990. Natural selection and reproduction: a study of the golden orb-weaving spider. In: Contemporary Issues in Comparative Psychology (Dewsbury, D. A., Ed.). Sinauer Associates Inc., Massachusetts. Pp. 149–174.

Coddington, J. A., Giribet, G., Harvey, M. S., Prendini, L., Walter, D. E. 2004. Arachnida. In: Cracraft, J. & Donoghue, M. (Eds). Assembling the Tree of Life. Oxford University Press, Oxford. Pp. 296–318.

Contreras-Garduño, J., Peretti, A. V. & Córdoba-Aguilar, A. 2006. Evidence that mating plug is related to null female mating activity in the scorpion *Vaejovis punctatus*. Ethology, 112: 152–163.

Cordero, C. & Eberhard, W. G. 2003. Female choice of antagonistic male adaptations: a critical review of some current research. J. Evol. Biol., 16: 1–6.

Cruz Landim, C. & Ferreira, A. 1973. The spermatozoa of *Tityus bahiensis* (Perty). Scorpion —Buthidae. Cytologia, 38: 187–194.

Dunlop, J. A., Tetlie, E. O. & Prendini, L. 2008. Reinterpretation of the silurian scorpion *Proscorpius osborni* (Whitfield): integrating data from palaeozoic and recent scorpions. Palaeontology, 51: 303–320.

Eberhard, W. G. 1985. Sexual Selection and Animal Genitalia. Cambridge: Harvard University Press.

Eberhard, W. G. 1993. Evaluating models of sexual selection: genitalia as a test case. Am. Nat. 142: 564–571.

Eberhard, W. G. 1996. Female Control: Sexual Selection by Cryptic Female Choice. Princeton: Princeton University Press.

Eberhard, W. G. 1997. Cryptic female choice in insects and spiders. In The Evolution of Mating Systems of Insects and Spiders. Choe, J. C. & Crespi, B. (Eds). Cambridge: Cambridge University Press. Pp. 32–57.

Eberhard, W. G. 2001. Species-specific genitalic copulatory courtship in sepsid flies (Diptera, Sepsidae, *Microsepsis*) and theories of genitalic evolution. Evolution, 55: 93–102.

Eberhard, W. G. 2004a. Male–female conflicts and genitalia: failure to confirm predictions in insects and spiders. Biol. Rev. Camb. Phil. Soc., 79: 121–186.

Eberhard, W. G. 2004b. Rapid divergent evolution of sexual morphology: comparative tests of sexually antagonistic coevolution and traditional female choice. Evolution, 58: 1947–1970.

Eberhard, W. G., Rodriguez, R. L., Huber, B. A., Briceño, R. D., Salas, I. & Rodriguez, V. 1998. One size fits all? Relationships between the size and degree of variation in genitalia and other body parts in 16 species of insects and spiders. Evolution, 52: 415–431.

Elgar, M. A. 1998. Sperm competition and sexual selection in spiders and other arachnids. In Sperm Competition and Sexual Selection. Birkhead, T. R. & Møller, A. P. (Eds) London: Academic Press. Pp. 307–340.

Farley, R. 2001. Structure, reproduction and development. In: Scorpion Biology and Research. Brownell, P. & Polis, G. A. (Eds). Oxford: Oxford University Press. Pp. 13–78.

Francke, O. F. 1979. Spermatophores of some North American scorpions (Arachnida, Scorpiones). J. Arachnol., 7: 19–32.

Francke, O. F. 1982. Parturition in scorpions (Arachnida, Scorpiones): A review of the ideas. Revue Arachnol., 4: 27–37.

Francke, O. F. & Gonález-Santillán, E. 2006. A new species belonging to the *Vaejovis punctipalpi* group (Scorpiones, Vaejovidae) from southern Mexico. J. Arachnol., 34: 586–591.

Fet, V. & Bechly, G. 2001. Opinion 2037 (Cases 3120 and 3120a). Liochelidae Fet & Bechly 2001 (1879) (Scorpiones): adopted as a valid substitute name of Ischnuridae Simon 1879 in order to remove homonymy with Ischnurinae Fraser 1957 (Insecta, Odonata). Bull. Zool. Nomen., 60: 159–161.

Fet, V., Gantenbeim B., Gromov, A. E., Lowe, G. & W. R. Lourenço. 2003. The first molecular phylogeny of Buthidae (Scorpiones). Euscorpius, (4): 1–10.

Fox, W.K. 1975. Bionomics of two sympatric scorpion populations (Scorpionida: Vaejovidae).

Ph. D. dissertation, Arizona State University, Tempe. 85 pp.

Gage, M. J. G. & Morrow, E. H. 2003. Experimental evidence for the evolution of numerous, tiny sperm via sperm competition. Curr. Biol., 13: 754–757.

Gomendio, S., Harcourt, A. H. & Roldán, E. R. S. 1998. Sperm competition in mammals. In: Sperm Competition and Sexual Selection. Birkhead, T. R. & Møller, A. P. (Eds) London: Academic Press. Pp. 667–756.

Grasshoff, M. 1975. Die Evolution komplizierter Kopulationsorgane—ein separater Adaptationsverlauf. Aufs. Reden senckenb. Naturf. Ges., 27: 61–68.

Hjelle, J. T. 1990. Anatomy and morphology. In: The Biology of Scorpions. Polis, G. A. (Ed.). Standford: Stanford University Press. Pp. 56–63.

Hood, R. D., Watson, O. F., Deason, T. R. & Benton, C. l. B. 1972. Ultrastructure of scorpion spermatozoa with atypical axonemes. Cytobios, 5: 162–177.

Hosken, D. J. & Stockley, P. 2004. Sexual selection and genital evolution. Trends Ecol. Evol., 19: 87–93.

Huber, B. A. 1993. Female choice and spider genitalia. Boll. Acc. Gioenia Sci. Nat., 26: 209–214.

Huber, B. A. 1995. Genital morphology and copulatory mechanics in *Anyphaena accentuata* (Anyphaenidae) and *Clubiona pallidula* (Clubionidae: Araneae). J. Zool. Lond., 235: 689–702.

Huber, B. A. 1996. Genitalia, fluctuating asymmetry, and patterns of sexual selection in Physocyclus globosus (Araneae: Pholcidae). Rev. Suisse Zool., 289–294.

Huber, B. A. 1998. Sexual selection in pholcid spiders (Araneae, Pholcidae): artful chelicerae and forceful genitalia. J. Arachnol., 27: 135–141.

Huber, B. A. & Eberhard, W. G. 1997. Courtship, genitalia, and genital mechanics in *Physocyclus globosus* (Araneae, Pholcidae). Canad. J. Zool., 74, 905–918.

Immler, S., Moore, H. D. M. Breed, W. G. & Birkhead, T. R. 2007. By hook or by crook? morphometry, competition and cooperation in rodent sperm. Plose One, 1, e170: 1–5.

Jacob, A., Gantenbein, I., Braunwalder, M. E., Nentwig, W. & Kropf, C. 2004a. Morphology and function of male genitalia (spermatophores) in *Euscorpius italicus* (Euscorpiidae, Scorpiones): complex spermatophores structures enable safe sperm transfer. J. Morphol., 260, 72–84.

Jacob, A., Gantenbein, I., Braunwalder, M. E., Nentwig, W. & Kropf, C. 2004b. Complex male genitalia (hemispermatophores) are not diagnostic for cryptic species in the genus *Euscorpius* (Scorpiones: Euscorpiidae). Org. Divers. & Evol., 4: 59–72.

Jespersen, A. & Hartwick, R. 1973. Fine structure of spermiogenesis in scorpions from the family Vejovidae. J. Ultraestruct. Res., 45: 366–383.

Kaston, B. J. 1970. Comparative biology of American black widow spiders. Trans. San Diego Soc. Nat. Hist. 16, 33–82.

Klann, A., Peretti, A. V., Vignoli, V., Talarico, G., Alberti, G., Carrera, P. & Michalik, P. 2007. Sperm diversity in scorpions (Scorpiones). Abstract Book 17th int. Cong. Arachnol., (Brazil): 57.

Koch, L. E. 1977. The taxonomy, geographic distribution and evolutionary radiation of australo-papuan scorpions. Rec. West. Aust. Mus., 5(2): 83–367.

Kokko, H., Brooks, R., Jennions, M. D. & Morley, J. 2003. The evolution of mate choice and mating biases. Proc. R. Soc. Lond. B, 270: 653–664.

Kovoor, J., Lourenco, W. & Muñoz-Cuevas, A. 1987. Conservation des espermatozoides dans les voies genitales des femelles et biologie de la reproduction des scorpions (Chélicérates). C. R. Acad. Sc. 304 Série III, 10: 259–264.

Kraus, O. 1984. Male spider genitalia: evolutionary changes in structure and function. Verh. Naturwiss. Ver. Hamburg (NF), 27: 373–382.

Lamoral, B. 1979. The scorpions of Namibia. Ann. Natal Mus., 23 (3): 497–784.

Le Pape, M. M. G. & Goyffon, M. 1975. Accouplement interspécifique suivi de parturition dans le genre *Androctonus* (Scorpionida, Buthidae). C. R. Acad. Sc. Paris, 280, Sér. D: 2005–2008.

Lhajoui, F., Lautié, N., Soranzo, L. & Stockmann, R. 2004. Ultra structure of female genital ducts of *Euscorpius carpathicus* (L) (Scorpiones, Euscorpiidae) after spermatophore uptake. Abstract book 16th Int. Cong. Arachnol. (Belgium): 245.

Lourenco, W. R. 1996. Faune de Madagascar (87): Scorpiones (Chelicerata, Scorpiones) 102 pp. Publ. Gov. Rep. Malgache.

Malo, A. F., Gomendio, M., Garde, J., Lang-Lenton, B., Soler, A. J. & Roldán, E. R. S. 2006. Sperm design and sperm function. Biol. Lett., 2: 246–249.

Masumoto, T. 1991. Males' visit to females' webs and female mating receptivity in the spider, *Agelena limbata* (Araneae: Agelenidae). J. Ethol., 9: 1–7.

Masumoto, T. 1993. The effect of the copulatory plug in the Funnel-web spider, Agelena limbata (Aranea: Agelenidae). J. Arachnol., 21: 55–59.

Mattoni, C. I. 2003. Patrones evolutivos en el género *Bothriurus* (Scorpiones, Bothriuridae): análisis filogenético. i–vii + 249 pp. Doctoral Thesis, Fac. Cien. Exactas, Físicas y Nat., Univ. Nac. Córdoba, Argentina.

Mattoni, C. I. & Peretti, A. V. 2004. The giant and complex genital plug of the *asper* group of *Bothriurus* (Scorpiones, Bothriuridae): morphology and comparison with other scorpion genital plugs. Zool. Anz., 243: 75–84.

Maury, E. A. 1968. Aportes al conocimiento de los escorpiones de la República Argentina I. Observaciones biológicas sobre *Urophonius* brachycentrus (Thorell, 1877) (Bothriuridae). Physis, 27: 407–418.

Maury, E. A. 1975. La estructura del espermató-foro en el género *Brachistosternus* (Scorpiones, Bothriuridae), Physis, C., 34: 179–182.

Maury, E. A. 1980. Usefulness of the hemiesper-matophore in the systematics of the scorpion family Bothriuridae. Proc. 8th Int. Arachnol. Congr. (Wien): 335–339. Wien, Verlag H. Egermann.

Maury, E. A. 1982. El género *Timogenes* Simon 1880 (Scorpiones, Bothriuridae). Rev. Soc. Entomol. Arg., 41: 23–48.

Michalik, P. & Mercati, D. 2007. Spermatozoa and spermiogenesis of *Opistophthalmus penrithorum* (Scorpioninae, Scorpionidae, Scorpiones). Abstract Book 17th Int. Cong. Arachnol., (Brazil): 137.

Michalik, P., Reiher, W., Suhm-Tintelnot, M., Coyle, F.A. & Alberti, G. 2005. The female genital system of the folding-trapdoor spider *Antrodiaetus unicolor* (Hentz, 1842) (Antrodiaetidae, Araneae)—an ultrastructural study of form and function with notes on reproductive biology of spiders. J. Morphol., 263: 284–309.

Møller, A. P. 1991. Fluctuating asymmetry in male sexual ornaments may reliably reveal male quality. Animal Behaviour, 40: 1185–1187.

Møller, A. P. & Pomiankowski, A. 1993. Fluctuating asymmetry and sexual selection. Genetica, 89: 267–279.

Monod, L. & Lourenco, W.R. 2005. Hemiscorpiidae (Scorpiones) from Iran, with description of two new species and notes on biogeography and phylogenetic relationships. Rev. Suisse Zool., 112: 869–941.

Ojanguren-Affilastro, A. A. 2005. Estudio monográfico de los escorpiones de la República Argentina. Rev. Ibérica de Aracnol., 11: 75–246.

Parker, G. A. 1984. Sperm competition and the evolution of animal mating strategies. In Sperm Competition and the Evolution of Animal Mating System. Smith, R. L. (Ed.). New York: Academic Press. Pp. 2–60.

Pavlovsky, E. N. 1924. Studies on the organization and development of scorpions. Quarterly J. Micros. Sci., 68: 615–640.

Peretti, A. V. 1991. Comportamiento de apareami-ento de *Zabius fuscus* (Thorell) (Scorpiones, Buthidae). Bol. Soc. Biol. Concepción (Chile), 62: 123–146.

Peretti, A. V. 1992. El espermatóforo de *Bothriurus bonariensis* (C. L. Koch) (Scorpiones, Bothriuridae). Bol. Soc. Biol. Concepción (Chile), 63: 125–138.

Peretti, A. V. 1993. Estudio de la biología reproductiva en escorpiones Argentinos (Arachnida, Scorpiones): un enfoque etológ-ico. Doctoral Thesis, Facultad de Ciencias Exactas, Físicas y Naturales, Universidad Nacional de Córdoba, Argentina.

Peretti, A. V. 1996. Comportamiento de transferencia espermática de *Bothriurus flavidus* Kraepelin (Scorpiones, Bothriuridae). Rev. Soc. Entomol. Arg., 55: 7–20.

Peretti, A. V. 2000. Existencia de cortejo en el campo de machos de *Bothriurus bonariensis* (Scorpiones, Bothriuridae) que carecen de un órgano paraxial. Rev. Soc. Entomol. Arg., 59: 96–98.

Peretti, A. V. 2003. Functional morphology of spermatophores and female genitalia in bothriurid scorpions: genital courtship, coercion and other possible mechanisms. J. Zool. Lond., 261: 135–153.

Peretti, A. V. & Acosta, L. E. 1999. Comparative analysis of mating in scorpions: the post-transfer stage in selected Argentinian bothriurids (Chelicerata, Scorpiones, Bothriuridae). Zool. Anz., 237: 259–265.

Peretti, A. V. & Battán-Horenstein, M. 2003. Comparative analysis of the male reproductive system in Bothriuridae scorpions: structures associated with the paraxial organ and sperm packages. Zool. Anz., 242: 21–31.

Peretti, A. V. & Carrera, P. 2005. Female control of mating sequences in the mountain scorpion *Zabius fuscus*: males do not use coercion as a response to unreceptive females. Anim. Behav. 69, 453–462.

Peretti, A. V. & Willemart, R. H. 2007. Sexual coercion does not exclude luring behavior in the climbing camel-spider *Oltacola chacoensis* (Solifugae, Ammotrechidae). J. Ethol., 25: 29–39.

Peretti, A. V. & Córdoba-Aguilar, A. 2007. On the value of fine-scaled behavioural observations for studies of sexual coercion. Ethol. Ecol. and Evol., 19: 77–86.

Peretti, A. V., Acosta, L. E. & Martínez, M. A. 2000. Comportamiento de apareamiento en tres especies de *Bothriurus* del grupo *prospicuus*: estudio comparado y su relación con *Bothriurus flavidus* (Scorpiones, Bothriuridae). Revue Arachnol., 13: 73–91.

Peretti, A. V., Depiante, L. & Battán-Horenstein, M. 2001. Allometry and asymmetry of body characters and spermatophores in *Bothriurus bonariensis* (C. L. Koch) (Scorpiones, Bothriuridae). In Scorpions 2001. In Memoriam Gary A. Polis. 345–355. Fet, V. & Selden, P. (eds). Brit. Arachnol. Soc., Burnham Beeches, Bucks. xi +404pp.

Pitcher, T. E., Rood, F.H. & Rowe, L. 2007. Sexual colouration and sperm traits in guppies. J. Fish Biol., 70: 165–177.

Polis, G. A. 1990. Introduction. In The Biology of Scorpions. Polis, G. A. (Ed.). Stanford: Stanford University Press. Pp. 1–8.

Polis, G. A. & Sissom, W. D. 1990. Life history. In The Biology of Scorpions. Polis, G. A. (Ed.). Stanford: Stanford University Press. Pp. 161–223.

Prendini, L. 2000. Phylogeny and classification of the superfamily Scorpionoidea Latreille 1802 (Chelicerata, Scorpiones): An exemplar approach. Cladistics, 16(1): 1–78.

Prendini, L. 2003. A new genus and species of bothriurid scorpion from the Brandberg Massif, Namibia, with a reanalysis of bothriurid phylogeny and a discussion of the phylogenetic position of *Lisposoma* Lawrence. Systematic Entomology, 28: 1–24.

Prendini, L. & Wheeler, W. C. 2005. Scorpion higher phylogeny and classification, taxonomic anarchy, and standards for peer review in online publishing. Cladistics, 21: 446–494.

Prendini, L., Volschenk, E.S. Maaliki, S. & Gromov, A.V. 2006. A 'living fossil' from Central Asia: The morphology of *Pseudochactas ovchinnikovi* Gromov, 1998 (Scorpiones: Pseudochactidae), with comments on its phylogenetic position. Zoologischer Anzeiger, 245: 211–248.

Probst, P. J. 1972. Zur fortoflanzungsbiologie und zur Entwicklung der Giftdrüsen beim scorpion *Isometrus maculatus* (DeGeer, 1778) (Scorpiones: Buthidae). Acta Trop., 29: 1–87.

Proctor, H. C. 1998. Indirect sperm transfer in arthropods: behavioral and evolutionary trends. Annu. Rev. Entomol., 43: 153–174.

Rolfe, W. D. 1985. Early invertebrate arthropods: A fragmentary record. Phil. Trans. R. Soc., B, 309: 207–218.

Rönn, J., Katvala, M. & Arnqvist, G. 2007. Coevolution between harmful male genitalia and female resistance in seed beetles. Proc. Natl. Acad. Sci. USA, 104: 10921–10925.

San Martin, P. R. 1969. Estudio sobre la compleja estructura del esqueleto esclerificado del órgano paraxil del género "*Brachistosternus*" (Bothriuridae, Scorpionida). Bol. Soc. Biol. Concepción (Chile), 41: 13–30.

Schneider, J. M., Thomas, M. L. & Elgar, M. A. 2001. Ectomised conductors in the golden orb-web spider, *Nephila plumipes* (Araneoidea): a male adaptation to sexual conflict?. Behav. Ecol. Sociobiol., 49: 410–415.

Shulov, A. & Amitai, P. 1958). On mating habits of three scorpions: *Leiurus quinquestriatus* (H. & E.), *Buthotus judaicus* E. Sim, and *Nebo hierochonticus* E. Sim. Arch. Inst. Pasteur d'Algérie, 38: 117–129.

Simmons, L. W. 2001. Sperm competition and its evolutionary consequences in the insects. i–xiii—434 pp., Princeton University Press, Princeton.

Shapiro, A. M. & Porter, A. H. 1989. The lock-and-key hypothesis: evolutionary and biosystematic interpretation of insect genitalia. Ann. Rev. Entomol., 34: 231–245.

Sissom, W. D. 1990. Systematics, biogeography, and paleontology. In: Polis, G. A. (Ed.) The Biology of Scorpions. Stanford University Press, Stanford. Pp. 64–160.

Sissom, W. D. 1993. A new species of *Vaejovis* (Scorpiones, Vaejovidae) from western Arizona, with supplemental notes on the male of *Vaejovis spicatus* Haradon. J. Arachnol., 21: 64–68.

Sissom, W. D. & Stockwell, S. A. 1992. The genus *Serradigitus* in Sonora Mexico (Scorpiones, Vaejovidae), with descriptions of four new species. Insecta Mundi, 5: 197–214.

Smith, G. T. 1966. Observations on the life history of the scorpion *Urodacus abruptus* Pocock (Scorpionida), and an analysis of its home sites. Aust. J. Zool., 14: 383–398.

Snook, R. R. 2005. Sperm in competition: not playing by the numbers. Trends Ecol. Evol., 20: 46–53.

Soleglad, M. E. & Sissom, W. D. 2001. Phylogeny of the family Euscorpiidae Laurie, 1896: a major revision. 25–111. In: Scorpions 2001. In Memorian Gary A. Polis. V. Fet & P. A. Selden (Eds).Brit. Arachnol. Soc., Burnham Beeches, Bucks. pp. xi +404.

Stockley, P. 1997. Sexual conflict resulting from adaptations to sperm competition. Trends Ecol. Evol., 12: 154–159.

Stockwell, S.A. 1989. Revision of the Phylogeny and Higher Classification of Scorpions (Chelicerata). Ph.D. Thesis, University of California, Berkeley, California. pp. 413.

Thomas, R. H. & Zeh, D. W. 1984. Sperm transfer and utilization strategies in arachnids: ecological and morphological contraints. In: R. L. Smith (Ed.) Sperm Competition and Evolution of Animal Mating Systems. Academic Press, Orlando. Pp. 180–221.

Thornhill, R. & Møller, A. P. 1998. The relative importance of size and fluctuating asymmetry in sexual selection. Behav. Ecol., 9: 546–551.

Vachon, M. 1953. Quelques aspects de la biologie des scorpions. Endeavour, 12: 80–89.

Varela, J. C. 1961. Gestación, nacimiento y eclosión de *Bothriurus bonariensis* var. *bonariensis* (Koch, 1842) (Bothriuridae,

Scorpiones). Rev. Fac. Hum. y Cien. de Montevideo, 19: 5–24.

Vignoli, V., Klann, A. & Michalik, P. 2008. Spermatozoa and sperm packages of the European troglophylous scorpion *Belisarius xambeui* Simon, 1879 (Troglotayosicidae, Scorpiones). Tissue & Cell, 40: 411–416.

Volschenk, E., Mattoni, C. I. & Prendini, L. 2008. Comparative anatomy of the mesosomal organs of scorpions (Chelicerata, Scorpiones), with implications for the phylogeny of the order. Zool. J. Linn. Soc., 154: 651–675.

Vrech, D., Peretti, A. V. & Mattoni, C. I. In press. Sperm packages morphology in scorpions and its relation to phylogeny. Zool. J. Linn. Soc.

Warburg, M. R. & Rosenberg, M. 1990. The morphology of the female reproductive system in the three scorpion species. Acta Zool. Fenninca, 190: 393–396.

Warburg, M. R. & Rosenberg, M. 1992. The reproductive system of female *Buthotus judaicus* (Scorpiones; Buthidae). Biol. Struc. Morphogen., 4: 33–37.

Werner, M. & Simmons, L. W. 2008. Insect sperm motility. Biol. Rev., 83: 191–208.

Weygoldt, P. 1990. Arthropoda—Chelicerata: sperm transfer. In: K. G. & R. G. Adiyodi (Eds) Reproductive Biology of Invertebrates. Vol IV, Part. B (fertilization, development and parental care). Oxford & I B H Publis Co., New Delhi. Pp. 77–119.

Weygoldt, P. 1999. Spermatophores and the evolution of female genitalia in whip spiders (Chelicerata, Amblypygi). J. Arachnol., 27: 103–116.

Weygoldt, P. & Paulus, H.F. 1979. Unteruschungen zur morphologie, taxonomie und phylogenie der Chelicerata. II. Cladogramme und die enfaltung der Chelicerata. Z. Zool. Syst. Evol., 17: 117–200.

Wheeler, W. C. & Hayashi, C. Y. 1998. The phylogeny of the extant Chelicerate orders. Cladistics, 14: 173–192.

Wigby, S. & Chapman, T. 2004. Sperm competition. Curr. Biol., 14: 100–103.

12

Spider Genitalia

Precise Maneuvers with a Numb Structure in a Complex Lock

WILLIAM G. EBERHARD AND BERNHARD A. HUBER

INTRODUCTION

The structures with which male spiders transfer sperm to females are unique among all animals in several respects. From the ancestral arachnid sperm transfer mechanism via an external spermatophore (Thomas & Zeh 1984), male spiders have evolved to transfer sperm to the female by using their pedipalps, a portion of the male's body that is not associated directly with his primary genitalia. Similar but independent evolution of "secondary" genitalia involving different male structures that subsequently come under sexual selection has occurred in several other arachnid orders (solifugids, ricinuleids, some pseudoscorpions, and mites—Kaestner 1968; Thomas & Zeh 1984; Alberti & Michalik 2004), as well as in other groups, such as odonate insects. The use of the pedipalps as secondary male intromittent genitalia is a unique synapomorphy of spiders, without convergence in any other arachnid, and without a single known reversal.

More important, however, is the fact that the portion of the male spider's pedipalp that has become specialized to receive and transfer sperm, the palpal "bulb", is apparently unique among all animal genitalia in that it lacks nerves, and thus also lacks sense organs and muscles (Eberhard & Huber 1998b). Much of this chapter will be dedicated to exploring the consequences of this lack of nerves for the evolution of spider genitalia. We will concentrate first on males because we believe that this lack of nerves in males has probably been largely responsible for many peculiarities of genitalic evolution in spiders, and then turn to females.

Despite the profound differences associated with a lack of nerves, spider genitalia clearly share the overall evolutionary trend seen in the male genitalia of many other groups, in being relatively distinct morphologically even among closely related species. This pattern of sustained, relatively rapid divergent evolution in genitalia was discovered long ago by spider taxonomists (summary, Comstock 1967), and their accumulated work constitutes a treasure chest of information on how spider genitalia have evolved. Although it is possible that the generality of this trend may be somewhat overestimated due to the possible bias of some taxonomists to recognize species mostly on the basis of genitalic differences (Huber 2003a; Song 2006) (see chapter 04 in this book), there is an independent indication that sustained rapid divergence has characterized genitalia in spiders. Male spider genitalia have such a diverse array of different sclerites that it has been very difficult to homologize them (e.g., Gering 1953: 33; Merrett 1963; Platnick 1975; Coddington 1990; Griswold et al. 1998, 2005; Agnarsson et al.

249

2007; Kuntner et al. 2008). Spider genitalia may even have a greater tendency toward qualitative rather than just quantitative changes than other traits (Huber 2003a).

Spider genitalia are an interesting "test" case for the various hypotheses that attempt to explain genitalic diversity, both in the sense of the same trend occurring in a different structure (the palpal bulb), and also because this structure has such strange characteristics (lacking nerves, muscles and sense organs).

MALE SPIDER GENITALIA

Spider Sperm and Sperm Transfer—the Male Perspective

Spiders and their closest relatives all transfer sperm in an inactive state, with the flagellum rolled up around the nucleus (Alberti 1990; Alberti & Michalik 2004). Within spiders, apparent vestiges of ancestral spematophores still occur. Sperm are packaged in small transfer units (coenospermia) in the Mesothelae and Mygalomorphae (Alberti 1990; Alberti & Michalik 2004; Michalik et al. 2004), taxa that are characterized by many plesiomorphic characters (figure 12.1). The more derived Araneomorphae mostly transfer sperm cells individually, and each sperm is surrounded by its own secretion sheath (cleistospermia).

In its simplest form, the genital bulb is a bulbous (pyriform) organ with no further subdivisions, as in many Mygalomorphae and Haplogynae (figures 12.1 and 12.2). Many other groups have evolved highly complex bulbs, however, consisting of a variety of sclerites that are connected by membranes (hematodochae) that can be inflated by hydraulic pressure and thus move the sclerites (most Entelegynae; figures 12.2–12.4). Inflation of hematodochal membranes that are twisted, folded irregularly, or composed of fibers of different elasticity can produce complex movements of sclerites (Osterloh 1922; Lamoral 1973; Grasshoff 1968; Loerbroks 1984; Huber 1993a, 2004b).The 'primitive' Mesothelae have moderately complex bulbs, suggesting that, after an early elaboration when palps evolved to transfer sperm, evolution has proceeded in both directions, towards simplification and towards higher complexity (Kraus 1978, 1984; Haupt 1979; Coddington 1990). The aberrant family Pholcidae, in which the male inserts a unique,

elaborate extension of the palpal segment just basal to the bulb deep into the female, and in which the female genitalia are also unusual in being largely membranous and lacking spermathecae, will be omitted from most of the discussions here.

Before sperm transfer, a male spider must charge his palps with sperm. The male constructs a small sperm web (which ranges from a single thread to an elaborate structure of silk lines), deposits a drop of sperm from his gonopore on the ventral surface of his abdomen, and takes the sperm up into his palpal bulb. The bulb contains a blind-ended, tube-like invagination (the sperm duct) that is formed by highly specialized cuticle; in most species the sperm duct is relatively rigid, is porous, and is surrounded by a glandular epithelium (Cooke 1970; Lopez & Juberthie-Jupeau 1985; Lopez 1987). During sperm uptake (induction), sperm is probably sucked into the sperm duct by removing the fluid that fills this duct through its rigid walls (presumably the epithelium imbibes the liquid); ejaculation is probably effected by the inverse mechanism of moving fluid into the lumen of the sperm duct through its walls (Cooke 1966; Juberthie-Jupeau & Lopez 1981; Lopez & Juberthie-Jupeau 1982, 1985; Lopez 1987). Other mechanisms must exist, however, because in some spiders the wall of the sperm duct lacks pores (Cooke 1970; Lopez 1987). Insertion and ejaculation can be surprisingly rapid in some species (< 5 s in *Argiope*—Schneider et al. 2005b; see Huber 1998), also leading one to wonder if secretion of these gland cells is the complete explanation. In mesothelid spiders the wall may be more flexible, and collapse under hemolymph

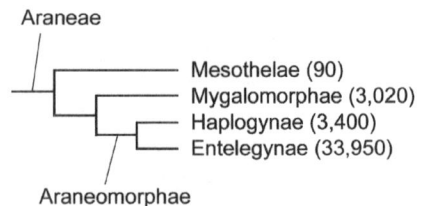

```
Araneae
   |
   |
   |_____ Mesothelae (90)
   |___ |_____ Mygalomorphae (3,020)
        |___ |___ Haplogynae (3,400)
             |___ Entelegynae (33,950)
        |
Araneomorphae
```

FIGURE 12.1 Simplified relationships among the major groups of spiders, with approximate numbers of known species (from Coddington and Levi 1991; Platnick 2007). By far the greatest number of species are in the more derived group Entelegynae, in which both male and female genitalia are also the most complex and diverse.

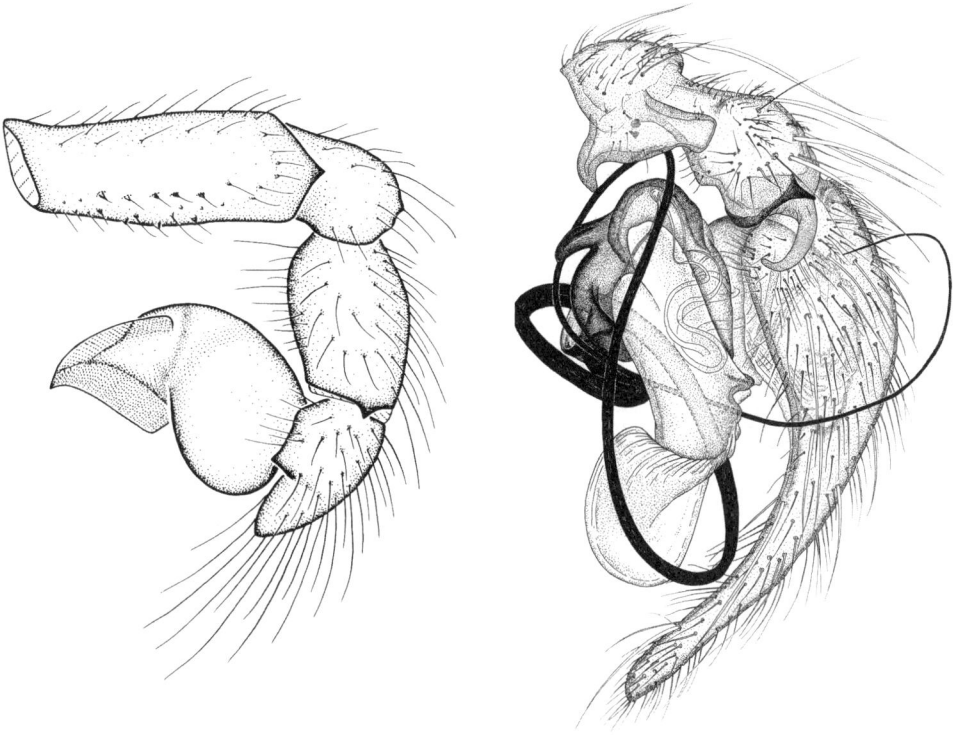

FIGURE 12.2 Male spider genitalia range from simple to extremely complex, but mapping of genital bulb complexity on cladograms suggests that medium complex bulbs are plesiomorphic, while very simple bulbs like that of *Segestrioides tofo* (left) and highly complex bulbs like that of *Histopona torpida* (right) are derived (from Platnick 1989; Huber 1994; with permission from AMNH and Blackwell).

pressure during ejaculation (Kraus 1984; Haupt 2003, 2004).

In a virgin male, the fluid that is pulled out of the duct would presumably have been produced when the bulb developed during the penultimate instar; in a nonvirgin, it could be the secretions that pushed sperm out during a previous ejaculation. The extraordinary complexity of the internal sperm ducts of the palps of some spiders (Coddington 1986; Sierwald 1990; Huber 1995b; Agnarsson 2004; Agnarsson et al. 2007; Kuntner 2007) suggests that this account is seriously incomplete; but to date no hypotheses to explain the function of this complex morphology are available. In theridiids, sperm duct trajectories vary greatly between genera, but are often constant within species and genera (Agnarsson et al. 2007). The fact that copulation does not always result in sperm transfer (Bukowski & Christenson 1997a; Schneider et al.

2005a, b) also suggests that additional, still un-appreciated processes may occur.

Movements of sperm once they have been deposited within the female have seldom been studied. Soon after a female's second copulation in the lycosid *Schizocosa malitiosa*, the ejaculates of the two males, which can be distinguished because the second male's sperm are still encapsulated while those of the first male are decapsulated, are already largely mixed in most parts of the spermatheca (Useta et al. 2007). The female of the tetragnathid *Leucauge mariana* has compound spermathecae, with one soft-walled chamber in which sperm are deposited and decapsulated, and two other rigid chambers to which decapsulated sperm then move (or are moved) (Eberhard & Huber 1998a). Similarly, the dysderid *Dysdera erythrina* has compound spermathecae with different glands hypothesized to function for short-term and

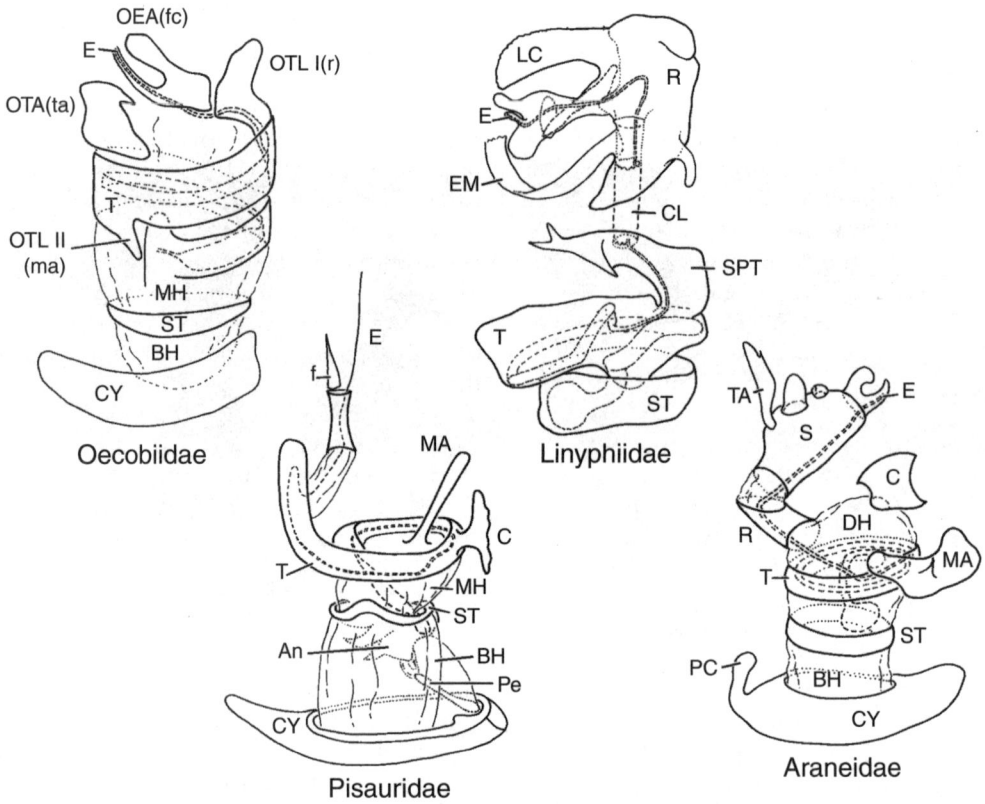

FIGURE 12.3 Schematic illustrations of palpal bulb designs in different groups of spiders, in which putatively homologous sclerites are labeled (from Coddington 1990) illustrating the diversity of sclerites and their arrangements. E indicates the embolus, the structure through whose tip sperm are transferred to the female, CY is the cymbium, the most distal segment of the palp that carries the genital bulb (with permission).

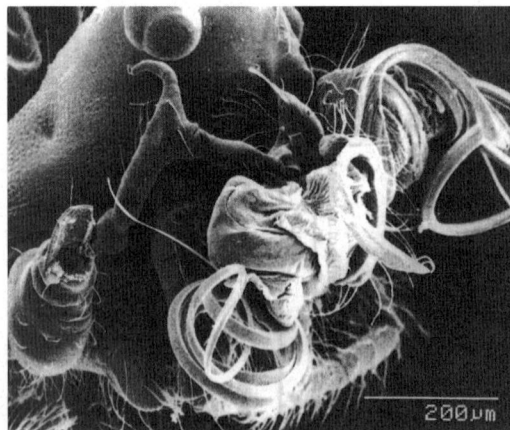

FIGURE 12.4 A male *Anapisona simoni* is partially hidden behind his elaborate, partially expanded genitalia, illustrating both the elaborate complexity often found in spider male genitalia and the appreciable material investment that they sometimes represent.

long-term storage (Uhl 2000). The second of two intromissions (one on each side) in the araneid *Micrathena gracilis* is twice as long as the first, and experimental manipulations showed that the prolongation of the second intromission did not influence the amount of sperm transferred, but did increase the amount of sperm stored from the first intromission (Bukowski & Christenson 1997a), suggesting that active female participation in sperm storage is induced by the male palp.

Evidence that Palpal Bulbs Lack Neurons

Histological studies using stains capable of differentiating nerve cells have failed to reveal any neurons in the bulbs of mature males (Osterloh 1922; Harm 1931; and Lamoral 1973 on six different families). Sections of the palp in a member of a seventh family (Linyphiidae) showed that a thin basal neck ("column") that connects more distal portions of the bulb with the rest of the bulb is made of solid cuticle, with only the sperm duct inside and no space for nerves (B. Huber, unpublished on *Neriene montana*). Ultrathin sections also failed to reveal nerves in the palpal bulb in yet another family (M. Suhm unpublished on *Amaurobius*, cited in Eberhard and Huber 1998b).

Additional, less direct histological data from many other species also suggest that palpal bulbs are not innervated. Glands in the bulb of *Amaurobius* lack both muscles and neurons to control the release of their products (Suhm et al. 1995). There are muscles that originate from the more proximal portions of the palp and insert at the base of the bulb in some spiders, but as Levi (1961) noted, no muscles have ever been seen within any palpal bulb. Sectioning studies showed that there were no muscles of any kind in the palpal bulbs in 76 genera of 56 different families in all major taxonomic groups (Huber 2004b).

In addition, external cuticular sense organs such as slit sensilla and setae (socketed epidermal bristles) appear to be completely lacking on palpal bulbs (figure 12.5; Eberhard & Huber 1998b; Berendonck & Greven 2005). The setae that are present on large areas of a spider's body, and that are innervated and function as tactile organs (Foelix 1985), are conspicuous by their absence in SEM micrographs of the bulbs of a large variety of groups (e.g., Kraus 1978; Opell 1979; Coddington 1986; Kraus & Kraus 1988; Griswold 1987, 1990, 1991,

1994, 1997; Hormiga 1994; Haupt 2003; Griswold et al. 2005; Bond & Platnick 2007; Miller 2007a).

Our earlier speculation (Eberhard & Huber 1998b) that the reason for the lack of nerves in the bulb is due to its developmental derivation from the palpal claw (e.g., Harm 1931) (the claw lacks neurons) is contradicted by the finding that both rudimentary claws and bulbs occur during bulb development in some spiders (Coddington 1990). The reason nerves are missing from palpal bulbs is not known. Perhaps both the bulb and the claw are derived from the same anlagen. Muscles attached to the base of the bulb are thought to be homologous with the levator and depressor muscles of the claw (Cooke 1970).

The portion of the palp just basal to the bulb, the cymbium, is not involved directly in sperm transfer, although it sometimes makes direct contact with the female during copulation. In contrast with the bulb, the cymbium is generally richly innervated and usually bears many setae (figure 12.2). Presumably there are sensors in the cymbium and or the membranes and muscle (if present) that unite the cymbium with the bulb, but they have apparently never been searched for. A male spider may thus have at least some information regarding the position of his bulb with respect to his cymbium during copulation. There is behavioral evidence of at least a crude sensitivity, as a male *Leucauge mariana* can apparently sense whether or not the

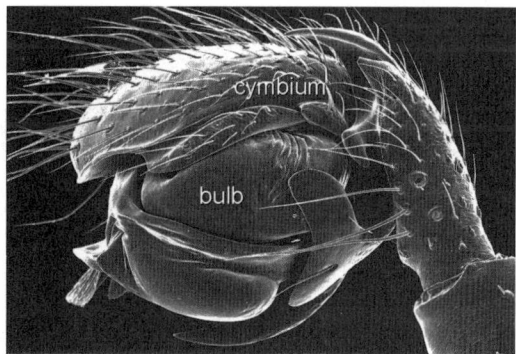

FIGURE 12.5 Distal palpal segments and genital bulb of a male linyphiid (*Triplogyna major*), showing the total absence of hairs on the bulb while hairs cover most of the cymbium and other palpal segments (from Miller 2007a; with permission from Blackwell).

structures at the tip of his bulb (the embolus and conductor) have entered the sperm droplet when he is taking up sperm into the bulb (Eberhard & Huber 1998a).

SEM photographs of palps (Silva 2003 on Ctenidae; Griswold et al. 2005 on several families; Miller 2007 on Linyphiidae) show that the cymbium, paracymbium, and tibial apophyses also sometimes have small regions lacking setae; apparently these areas contact the female epigynum during copulation. In contrast, there are abundant setae on the portion of the cymbium of *Leucauge mariana* that rests loosely on a featureless portion of the surface of the female abdomen away from the epigynum (the exact site varies) (Eberhard & Huber 1998a). Perhaps this loss of setae is an adaptation to fit more tightly with the female, or to avoid damage that would otherwise result from friction with the female during copulation. Additional groups need to be checked to see whether similar bald spots occur in other taxa, and whether areas lacking setae consistently contact the female. This pattern has a major implication. Males do not seem to be in urgent need of sensory information from the sites specialized to contact particular sites on the female.

The lack of innervation in the male intromittent genitalia of spiders is in clear contrast with other groups like mammals and insects. For instance the intromittent phallic organs and the associated genitalic structures arising nearby are provided with sense organs and muscles in many insects (Snodgrass 1935; Peschke 1978, 1979; Chapman 1998; Sakai et al. 1991; Schulmeister 2001).

Consequences for Males of Lack of Genitalic Innervation

Because of the lack of nerves in the palpal bulb, the challenges faced by a male spider attempting to copulate can be likened to those of a person attempting to adjust a complex, delicate mechanism in the dark, using an elongate, elaborately formed fingernail. A male spider is more or less "sensorily blind" when he attempts to perform the selectively all-important act of inseminating a female. Spider males are likely to have difficulty in achieving the proper alignment with both the external and internal portions of the female (which are often quite complex—see below). The only sensations it is reasonable to expect to be available to the male would be from more basal portions of his palp such

as the cymbium, the connections between the bulb and the cymbium, and from the hydraulic system (pressure changes, perhaps flow of fluid into the bulb?) that is involved in inflating the palpal hematodochae.

Apparent confirmation that male spiders have difficulty positioning their palps precisely with respect to the female comes from behavioral observations of possibly exploratory movements of the male's palp in the close vicinity of the female's copulatory openings, variously called "scraping" (Rovner 1971; Blest & Pomeroy 1978; Huber 1995b; Eberhard and Huber 1998a; Stratton et al. 1996), "stroking" (Bristowe 1926; Melchers 1963), "rubbing" (Montgomery 1903; Bristowe 1929), "scrabbling" (Robinson & Robinson 1980), "beating" (Robinson & Robinson 1973), "poking" (Whitehouse & Jackson 1984; Fromhage & Schneider 2006), "slapping" (Gering 1953) "fumbling" (Snow et al. 2006), "flubs" (Watson 1991), and "brushing" (Senglet 2004). Flubs are very widespread: they were reported in 40% of 151 species in 38 families in a survey study (Huber 1998). Some authors have concluded that these movements represent failed intromission attempts (Watson 1991); other non-exclusive hypotheses are that these movements represent exploration, or courtship stimulation of the female (Robinson 1982; Stratton et al. 1996; Eberhard 1996). The fact that male *Portia labiata* and *P. schultzi* scrape on one side, then scrape and insert on the other side (Jackson & Hallas 1986) implies that scraping in this species has a stimulatory function rather than being just a searching movement. Salticid and lycosid males trying to mate with females whose genitalia were experimentally sealed, scraped for prolonged periods (Rovner 1971), suggesting a searching function and at least crude sensory feedback. Fragmentary observations of male *Nephila edulis* withdrawing their palps from the already-inseminated side of female epigyna to shift to the other, non-inseminated side (Jones & Elgar 2008) also hints at sensitivity of some sort.

One solution to possible orientation problems would be to develop "preliminary locking" structures, either on the bulb or on the more basal, innervated palpal segments (cymbium, tibia, etc.), whose engagement with the female would require less precise alignment with her, but would provide a stable point of support to facilitate more precise alignment during subsequent stages of intromission that demand more precision. They might even enable the male to sense that such preliminary alignment had occurred, via sensations from the cymbium or its articulation with the palp. Preliminary locks, and sclerites specialized to produce locking of this sort between male and female are widespread in spiders (figure 12.6; van Helsdingen 1965 on the paracymbium of *Lepthyphantes*; Eberhard & Huber 1998a on the conductor of *Leucauge*; Melchers 1963 on the "retinaculum II" of *Cupiennius*; Loerbroks 1983, 1984 and Huber 1995a, b on the rta of various families; Stratton et al. 1996 on the median apophysis of *Schizocosa*). As the different positions of these structures and their widely separate taxa suggest, preliminary locking has probably evolved several times independently. Coupling is sometimes a multi-stage process. In *Agelenopsis*, the embolus engages the female, the hematodocha expands and couples the conductor to the female, and the embolus then enters the female (Gering 1953). For reasons that are not clear, some groups have lost palpal locking structures (e.g., some linyphiids lack a paracymbium, G. Hormiga personal communication; some lycosids lack a rta, Griswold 1993).

A second important consequence relates to the difficulty of fine motor control over a structure that lacks muscles. The male genitalia of spiders are moved only by more proximal muscles in the palp, and by internal pressure changes that result in inflation of the membranes between sclerites (hematodochae) within the bulb. Although there are very few studies concerning the degree of variability in the genitalic movements in spiders (or other animals for that matter; most studies of the functional anatomy of genitalia are unfortunately extremely typological), it seems likely that this type of movement mechanism results in less ability to make fine adjustments in movements compared with structures controlled by individual muscles, as in the genitalia of other groups. Spiders probably have some general control of movements during intromission, for instance of whether some hematodochae inflate while others do not, but there is probably little fine control; for instance, the sequence with which the one to three hematodochae of a bulb inflate seems to be fixed. In the tetragnathid *Leucauge mariana*, the movements of the palpal bulb prior to and following removal of a copulatory plug in the female showed no perceptible qualitative differences (Méndez & Eberhard, unpublished data).

The anatomical lack of nerves precludes direct sensory feedback from palpal bulbs, and some experimental manipulations of males (Rovner 1966,

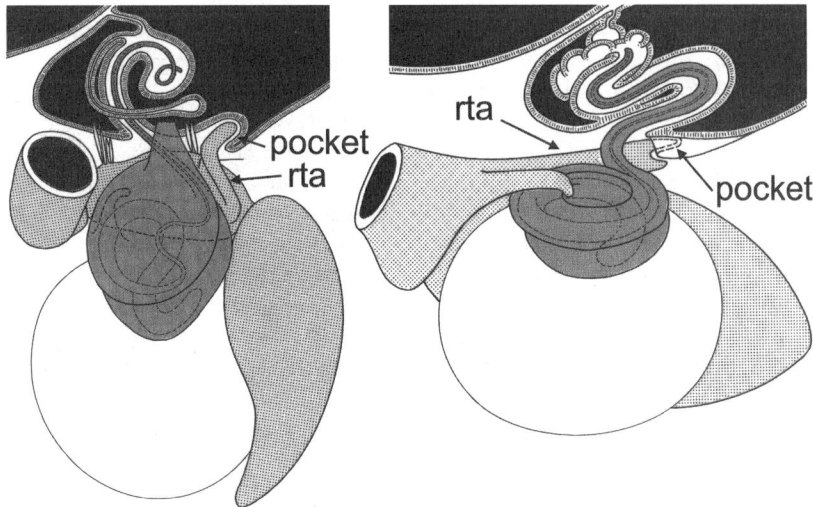

FIGURE 12.6 Females of the rta clade often provide the male with cooperative structures that facilitate coupling of his genitalia. In the cases shown here (left: *Anyphaena accentuata*; right: *Philodromus aureolus*), the females (black) provide pockets for the male retrolateral tibial apophyses (rta); palps in light gray, bulbs in dark gray (from Huber 1995a, b; with permission from Blackwell).

1967, 1971) and female *Rabidosa* (= *Lycosa*) *rabida* (Rovner 1971) reveal lack of propioceptive feedback from the palps. Nevertheless, some species show surprising discriminations. Male *Latrodectus hasselti* spiders tended to use the palp containing the greater amount of sperm first when they copulate (Snow & Andrade 2004). Proper positioning of the male of *L. rabida* on the female depended on feedback from the male palps, and when one side of the female's epigynum was artificially sealed, the male gradually decreased his attempts to insert his palp on that side (although males did perform some "pseudoinsertions" on the plugged side) (Rovner 1971). Removal of the bulb in this species reduced the usual male attempts to moisten the bulb in his mouth. Male *Argiope bruennichi* never attempted to copulate using the stump of an ablated palp (Nessler et al. 2007a). The mechanism(s) responsible for selective hematodochal inflations and the differences in their patterns (e.g., pulsating rather than sudden inflations in *Agelenopsis*) presumably involve differences in hydraulic pressure (Gering 1953). The extensive "cleaning" or "lubricating" of the palps by the male that is often associated with copulation (van Helsdingen 1965; Costa 1979; Lopez 1987) might result in softening of the membranes of the palp, causing them to become more flexible and thus to move palpal sclerites in particular ways (Gering 1953).

A further function in spiders, which appears to be much more common than in insects, is the use of some palpal sclerites to brace or push others, both during the process of preliminary locking and during subsequent orientation and deeper intromission. This bracing function appears to be widespread in spiders (Gering 1953; van Helsdingen 1971; Grasshoff 1973; Loerbroks 1983; Huber 1993a, 1995a; Costa & Pérez-Miles 1998; Eberhard & Huber 1998a; Knoflach & Pfaller 2004; Agnarsson et al. 2007). It is much less common in insects, and was not even included in the review of genital functions by Scudder (1971), or in a survey of functions documented in 43 species of Diptera (Eberhard 2004a). Still another related function not mentioned for insects but present in spiders is that of the "tethering membrane" of *Agelenopsis* which guides the movements of other sclerites (Gering 1953). Presumably these differences occur because the lack of muscles makes independent adjustments of the positions of different structures more difficult in spiders, and also because of their lack of sensory feedback during coupling.

There is a wealth of morphological variation with as yet unknown functions and even apparently paradoxical (e.g., the longitudinally split embolus in the theridiid *Anelosimus*, the extremely long coiled embolus that does not contain a sperm duct in the theridiid *Stemmops*—Agnarsson et al. 2007) that lies below the level of this necessarily superficial review.

FEMALE SPIDER GENITALIA

Sperm Storage and Fertilization

The female spider copulatory organ is closely associated with the gonopore on the ventral surface of the abdomen. Sperm are usually stored in separate internal receptacles ("spermathecae"). In the plesiomorphic "haplogyne" condition, sperm are introduced through the same opening that is used for oviposition (figure 12.7). The spermathecae of haplogyne spiders have only a single duct, through which sperm both enter and exit the receptacle (the "cul-de-sac morphology" of Austad 1984). In the derived, entelegyne condition, an "insemination" (or copulatory) duct which connects each spermatheca with the outside is used to introduce sperm into the spermatheca; and a separate "fertilization" duct, running from the receptacle to the uterus, is used to transfer sperm to the eggs (Wiehle 1967a; Cooke 1970). Austad (1984) called this two duct arrangement a "conduit" morphology, and proposed that haplogyne and entelegyne female morphology may influence sperm precedence patterns. In some entelegynes, however, both ducts enter the same end of the spermatheca, resulting in an effectively cul-de-sac design; in addition, there is no clear relationship between these designs and sperm use patterns (Uhl & Vollrath 1998; Uhl 2002). The conduit morphology could also affect sperm usage by promoting the evolution of copulatory plugs by males (see below). Cul-de-sac designs have evolved secondarily from conduits in two and perhaps four families (Dimitrov et al. 2007). Hypodermic insemination, which circumvents female ducts, has recently been discovered in one species (Rezác 2007).

The standard belief is that eggs are fertilized as they reach the portion of the oviduct near the mouth of the fertilization duct, but the discovery of fertilized eggs in the ovarian cavity of the theridiid *Achaearanea tepidariorum* (Suzuki 1995) indicates

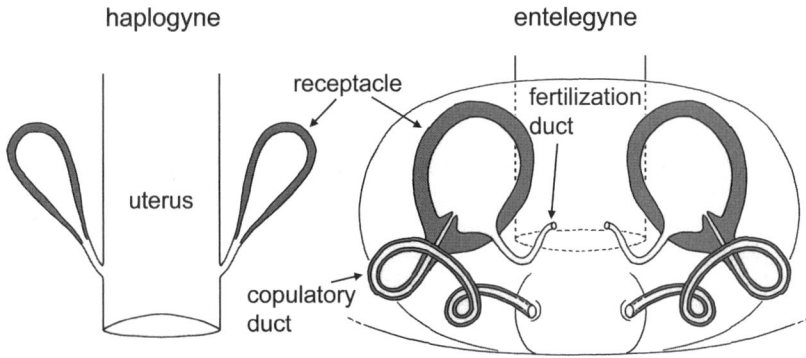

FIGURE 12.7 Female spider genitalia, schematic. In the haplogyne design, sperm enters and exits the receptacle (gray) through the same duct. In the entelegyne design, sperm enters through a copulatory (insemination) duct and exits through a fertilization duct (after Wiehle 1967b; with permission from Senckenberg Gesellschaft).

that sperm sometimes range more widely within the female (see also Burger et al. 2006a). The presence of a flap covering the opening of the fertilization duct into the oviduct in the nephilid *Nephila edulis* (Uhl & Vollrath 1998) also hints that sperm may move into the oviduct at times other than oviposition. The function of this flap is uncertain; it lacks muscles (G. Uhl, personal communication). It is not known whether similar flaps occur in other species.

While male spider genitalia are universally paired, the female genitalia vary. A few Mesothelae have a single spermatheca, and in some other "primitive" species the female has a pair of receptacles but the male can fill both with a single palp (Haupt 1979, 2003; Costa et al. 2000). Kraus (1978) suggested that the unpaired vulva of *Liphistius* is plesiomorphic, and that paired receptacles are derived; Forster (1980) and Forster et al. (1987) suggested that a bursal storage is plesiomorphic and that receptacles evolved several times independently. In at least one group (some tetragnathids) an unpaired sperm storage organ or area in the oviduct has been secondarily derived from paried spermathecae (Dimitrov et al. 2007). The finding of sperm in the ovarian cavity of other spiders suggests one possible, but untested explanation for the loss of spermathecae: males under sexual selection may have "short-circuited" female storage organs and introduced sperm directly into the oviduct. Perhaps this change was facultative at

first, as some unpaired sacs that might store sperm are present in other tetragnathids that still have spermathecae (Dimitrov et al. 2007).

In most araneomorphs (the majority of spiders— figure 12.1) the spermathecae are paired, and each must be inseminated separately. Almost universally each spermatheca is inseminated by the insertion of a different palp (von Helversen 1976). This makes it possible for females to influence insemination by interrupting copulation after a male has inseminated only one side ("hemicopulation") (e.g., Bukowski & Christenson 1997b). Detailed proof that such behavior can alter sperm precedence patterns was obtained in the theridiid *Latrodectus hasselti*. When two males were forced to inseminate a single spermatheca, there was strong first male precedence (mean 78.9% of the offspring). When, in contrast, males inseminated opposite spermathecae, the first male had no advantage (49.3% of the offspring). Because females control whether the first male inseminates one or both spermathecae, and because females often remate, a female can thus alter a first male's chances of obtaining paternity advantage (Snow & Andrade 2005).

It is theoretically possible that the female further influences paternity by favoring the use of sperm in one spermatheca over that in the other in fertilizing her eggs. Such a bias has never to our knowledge been demonstrated, and there is evidence that it does not occur in *Nephila* (e.g., Jones & Elgar 2008).

None of the male traits that Snow and Andrade (2005) measured in *L. hasselti* correlated with paternity success when each male inseminated a different spermatheca, but (as they note) their small sample size and the limited number of traits they measured mitigates against confident conclusions (Snow & Andrade 2005). A few species have many more spermathecae (up to about 100) (Forster & Platnick 1984); their significance (and even whether they all store sperm) is not known.

External Rigidity and Internal Complexity

The more external portions of entelegyne female genitalia are often strongly sclerotized and rigid. Associated with this trend to female rigidity is the fact that, in strong contrast to many other animal groups, the morphology of the female genitalia is very often species-specific in form. All of the rigid portions, including the epigynum, the ducts, and to a lesser extent the spermathecae themselves, show diverse forms. This tendency toward rigid species-specific female genitalia has been exploited by taxonomists for many years, and taxonomic descriptions of spider species usually include descriptions of both male and female genital morphology. There is thus a huge (and to date largely unexploited) accumulation of data on female genitalic morphology which can be used to check for evolutionary trends.

The more internal portions of the female genitalia are less well known; in at least some species they are very complex. Recent studies of haplogyne spiders revealed several moveable sclerites attached to muscles (Burger et al. 2003, 2006b; Burger 2007, 2008) (figure 12.8). Proposed functions include locking of one area of the female's reproductive tract closed, packaging a male's ejaculate in a secretion that prevents sperm mixing, and ejecting it from her body as a single mass (Burger et al. 2006b; Burger 2007, 2008). There are also muscles attached to female internal genitalia in other groups such as Antrodiaetidae (Michalik et al. 2005), Dysderidae (Cooke 1966; Burger & Kropf 2007), Pholcidae (Uhl 1994; Huber 2004a), Pisauridae (Carico & Holt 1964), and Theridiidae (Berendonck & Greven 2005) whose functions are poorly understood.

One generalization about female genitalic morphology is that the insemination and fertilization ducts of entelegyne spiders show quite different patterns of evolution (figure 12.9). The insemination

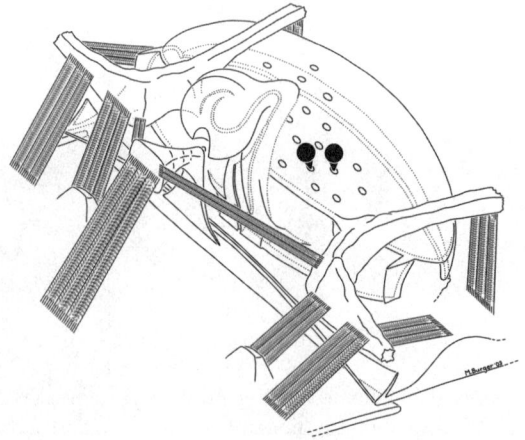

FIGURE **12.8** The complex array of muscles (shaded) and sclerites in the internal genitalia of the female haplogyne oonopid *Silhouettella loricatula* imply that the female plays an active role in sperm management in her body (from Burger et al. 2006b; with permission from Wiley).

duct, through which sperm enter the spermatheca, is usually much longer and more tortuously coiled than the fertilization duct, through which sperm leave the spermatheca to enter the oviduct and fertilize the eggs. In extreme cases insemination ducts are coiled in >15 loops. Fertilization ducts, in contrast, are generally simpler and shorter, usually running directly from the spermatheca to the oviduct (Eberhard 1996). The selection responsible for the elaboration of these two types of duct thus seems to be related not to the sperm themselves, but to the access that the sperm (or the male genitalia) have to the spermathecae. In some groups of Linyphiidae with long coiled insemination ducts or furrows, the male has a long thread-like embolus that is inserted into the coiled female tube and reaches the spermatheca (van Helsdingen 1969; Hormiga & Scharff 2005). Long emboli are also known to traverse long coiled insemination ducts in other families (Wiehle 1961; Abalos & Baez 1963; Uhl & Vollrath 1998; Jocqué 1991; Snow et al. 2006; Jäger 2005), in one extreme case, the theridiid *Kochiura aulica*, the embolus is three times the length of the male's entire body (Agnarsson et al. 2007). In some other species of Linyphiidae, in contrast, the insemination duct is very thin and the male genitalia do not enter the duct (Wiehle's "Anschluss-Embolus" group). In the

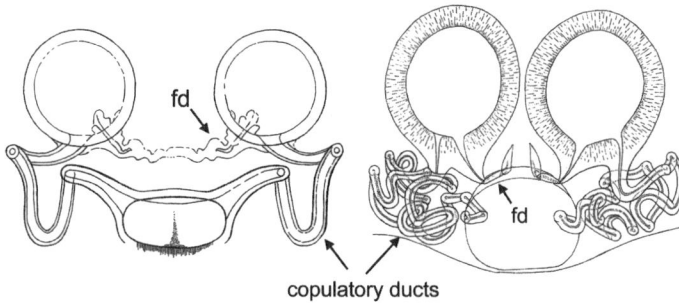

FIGURE 12.9 Female internal genitalia of two representatives of *Theridion* (Theridiidae) illustrate the longer, more tortuous ducts sperm need to traverse to enter storage sites. The species on the left has short copulatory and fertilization ducts (fd). The species on the right has highly elongated copulatory ducts, but the fertilizations ducts have remained short (from Wiehle 1967a; with permission from Senckenberg Gesellschaft).

linyphiid *Neriene* the embolus falls far short of reaching the spermatheca, and the long duct is actually an open groove. Prior to sperm transfer, this groove is filled with a substance (presumably produced by the female; B. Huber, unpublished data) through which the sperm are then pushed or sucked; a long duct that is inaccessible for the male could test his ability to push the sperm (or to induce the female to push/suck) rather than to insert his embolus deeply. Experimental manipulations of male palps in the theridiid *Latrodectus hasselti* showed that males obtained more paternity when they penetrated deep enough to ejaculate sperm directly into the spermatheca rather than in the insemination duct (Snow et al. 2006).

The insemination ducts and (especially) the spermathecal walls are often riddled with pores that connect the lumen with glandular ductules (e.g., Coyle et al. 1983; Uhl & Vollrath 1998). In some cases in which the insemination ducts are longer than the emboli, the part of the duct that is not traversed by the embolus is glandular and has been hypothesized to aid in sperm transport (Baum 1972). Glandular ducts also occur, however, in at least one species in which the embolus reaches the spermatheca (Uhl & Vollrath 1998 on *Nephila*). At present, we are nearly completely ignorant of the functions of the glands associated with spermathecae and their ducts. Products of these glands have been hypothesized to induce sperm to emerge from their membranous capsules ("decapsulate") (Eberhard & Huber 1998a), or nourish or otherwise maintain sperm. In addition, they could be responsible for sperm transport, causing the spermathecae to take up sperm by absorbing the liquid contents of the spermatheca, and/or to expel sperm by secreting into the lumen and thus displacing the sperm (Foelix 1996; but see Berendonck & Greven 2005). The usually rigid walls of entelegyne spermathecae and their ducts seem to rule out sperm transport by female muscular contractions.

Sensory Blindness of Contact Structures

A third, more surprising possible generalization about female spider genitalia is based on the large number of SEM micrographs in the taxonomic literature. Female genitalia (in particular the epigynum) generally lack setae, at least on the externally visible portions that are contacted by the male bulb during copulation, and thus lack possible tactile sense organs (Huber 1993a; Berendonck & Greven 2005) (figure 12.10). The abdominal cuticle of spiders is typically densely covered with setae, so the lack of setae on the epigynum, which may reduce damage due to abrasion with the male's genitalia, is a derived feature. It is less clear (because close-up SEM photos are needed, and taxonomic works generally do not provide such photos) whether epigyna also lack slit-sense organs that could sense stress in the cuticle. Epigyna are typically very dark and heavily sclerotized, however, and thus seem unlikely to be bent by the forces applied by male palps. Clearly there are exceptions (e.g., the atrium of *Linyphia triangularis* stretches during copulation—van Helsdingen 1969; the scape of many araneids is deflected during copulation—Grasshoff

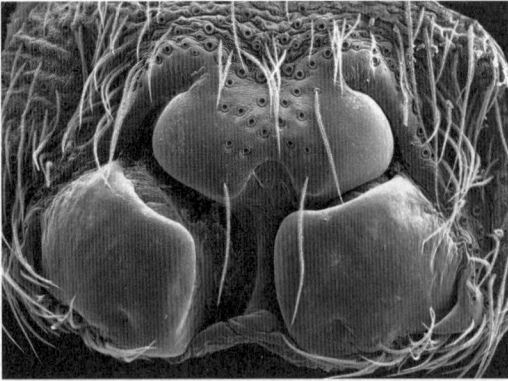

FIGURE 12.10 Mechanoreceptive hairs are conspicuously absent in large parts of external female spider genitalia. (*Griswoldia acaenata*, from Silva 2003; with permission from AMNH).

in species-specific female traits with which the male structures fit (table 12.1). This coevolutionary interaction may have imposed limits on sexual size dimorphism in some spiders (Ramos et al. 2005).

Summarizing, the external genitalia of female spiders generally have rigid, complex designs that are at least sometimes "selectively cooperative" (see below); they are associated with tortuous ducts through which sperm arrive for storage in the spermatheca, but simple direct ducts through which they leave storage when they fertilize eggs. They fit physically complementary structures of the male genitalia, and are largely devoid of sensation. Why would this unusual set of male and female traits evolve? We will discuss the three most commonly cited hypotheses for genital evolution (for additional reasons to discard two additional hypotheses, see Eberhard & Huber 1998b).

1968, 1973). But the relative rigidity of most external areas of most epigyna seems undeniable. Our tentative conclusion is that female spiders have also evolved an extraordinary absence of mechanical sensitivity in their genitalia (at least on the outer surface) that matches the insensitivity of males! It is not known whether internal portions of the female genitalia such as insemination ducts and spermathecae possess sense organs (such as those described by Foelix & Choms 1979 in walking legs).

Male–Female Fit and Coevolution

A final generalization is that the often complex sculpturing of the external surface of the epigynum fits very precisely with male genitalic structures during copulation. This generalization is based on a much more limited sample of species in which pairs have been killed instantaneously during copulation by freezing or by hot fixatives (van Helsdingen 1965, 1969, 1971; Grasshoff 1968, 1973; Huber 1993a, 1994, 1995a, b; Eberhard & Huber 1998a; Knoflach 1998; Senglet 2004; Uhl et al. 2007). A previous technique, which depended on artificially expanding male palpal bulbs not in contact with the female and attempting to deduce how they fit with females, is likely to lead to erroneous conclusions (Huber 1993a). All available studies document consistent, precise male–female fits: male and female morphology is clearly coevolved, and species-specific male traits are often reflected

WHY THESE STRANGE MALE AND FEMALE GENITALIA?

Lock and Key

At first, the descriptions just given seem to fit perfectly with expectations of male and female morphology that have evolved under selection for species isolation by mechanical "lock and key". This hypothesis supposes that female genital structures evolved to exclude the genitalia of males of other species, to thus enable the female to avoid cross-specific inseminations; males could also benefit (though to a lesser extent because of their cheaper gametes). But there are reasons to doubt the species isolation part of this hypothesis in spiders. Spider species that have evolved in isolation from other close relatives, and that should thus have been free of selection to avoid cross-specific fertilizations, nevertheless have elaborate, species-specific genitalia. Examples include species endemic to particular isolated islands with no other congeners present (figure 12.11; Platnick & Shadab 1980; Peck & Shear 1987; Gertsch & Peck 1992; Hormiga 2002; Hormiga et al. 2003). Multiple congeneric species endemic to different isolated caves that have probably also been isolated and nevertheless have species-specific genitalia are further examples (Gertsch 1974; Deeleman-Reinhold and Deeleman 1980; Hedin 1997), though in these cases strict isolation is less certain. In addition, the genitalic

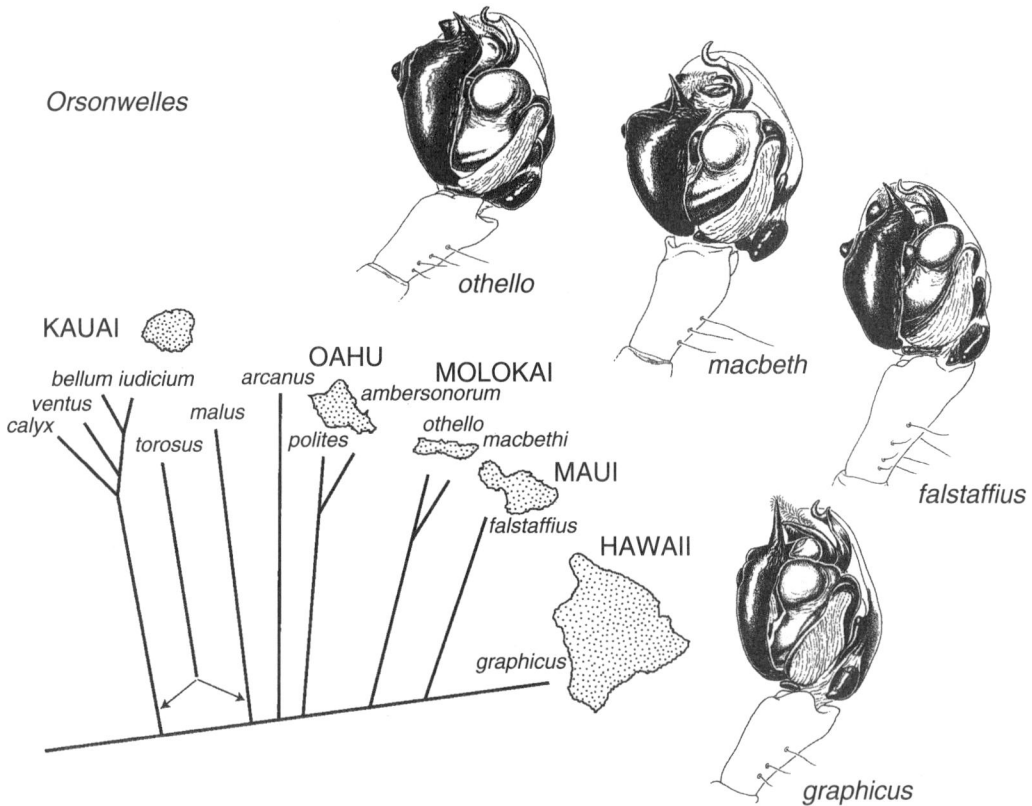

FIGURE 12.11 Male palps (above and at right) of four species of *Orsonwelles* spiders fail to fit the predictions of the hypothesis that species isolation by lock-and-key is responsible for the rapid divergent evolution of these spider genitalia. All of the 13 species are single-island endemics, and most have very small, non-overlapping distributions, usually in high, wet areas and often limited to a single mountain top (Hormiga 2002; Hormiga et al. 2003). A biogeographic pattern of progressive colonization from older to newer islands in the archipelago is consistent with a phylogeny of the spiders based on both morphological and molecular traits (Hormiga et al. 2003). Although there has been substantial intra-island speciation (where strict isolation from congeners is less certain), 4 of 12 cladogenic events occurred between islands (and thus in apparent strict allopatry). Contrary to expectations of lock and key, genitalia are complex and especially useful in distinguishing species throughout the genus (Hormiga et al. 2003), and constituted 53 of 71 phylogenetically informative morphological traits. The islands farther to the right are younger, as are the species endemic to them (phylogeny below). There is only one *Orsonwelles* species on Maui and one on Hawaii; despite this isolation, neither their female nor their male genitalia are simpler. In addition, the female genitalia of the two species sympatric on Molokai (*othello* and *macbethi*) are, contrary to predictions especially similar rather than especially different from each other (after Hormiga et al. 2003; Hormiga 2002; with permission).

character displacement in zones of overlap that is predicted by selection for species isolation did not occur in one pair of species that was carefully chosen to maximize the likelihood that it would occur (Ware & Opell 1989). Character displacement, which should be widespread, seems in fact to be quite rare; we know of only one case (the genitalia of *Argiope trifasciata* are smaller in areas of sympatry with *A. florida*; Levi 1968) (and of course random variation is expected to produce a certain number of apparent confirmations). Detailed study of morphology has showed that cross-specific pairing is not precluded by the female's genitalic design in some spiders (Gering 1953).

TABLE 12.1 Examples of "selectively cooperative" female genital structures and the corresponding male structures

Taxa	Female structures	Male structures	References
Haplogynae			
Pholcidae, various genera	Epigynal and abdominal pits	Cheliceral and bulbal apophyses	Huber 2002, 2003b, 2005b
RTA clade			
Many families	Various folds and pits	RTA	Bristowe 1958; Loerbroks 1983; 1984, Huber 1995a, b
Agelenidae, *Agelenopsis*	Coupling cavity	Conductor	Gering 1953
Oxyopidae, *Peucetia*	Epigynal depression	Ventral paracymbium prong	Exline & Whitcomb 1965
Lycosidae	Epigynal pockets	Bulbal apophyses	Osterloh 1922; Sadana 1972
Miturgidae, *Cheiracanthium*	Epigynal pit	Processes of tibia and paracymbium	Gering 1953
Orbiculariae			
Uloboridae, *Hyptiotes cavatus*	Vaginal invagination	Median apophysis spur	Opell 1983
Linyphiidae, Araneidae	Pits, grooves, and bulges near tip of scape	Projecting point of male suprategular apophysis (= median apophysis), or paracymbium	van Helsdingen 1965, 1969; Blest & Pomeroy 1978; Grasshoff 1968, 1973; Uhl et al. 2007
Linyphiidae, *Neriene* and *Linyphia*	Spiral-shaped atrium	Spiral-shaped bulbal terminal apophysis	Osterloh 1922; van Helsdingen 1969

Sexually Antagonistic Coevolution ("SAC")

One currently popular explanation for rapid divergent evolution in sexual traits like genitalia is sexually antagonistic coevolution ("SAC") of males and females. Briefly (see Chapter 4 of this book for a more detailed discussion), SAC supposes that because male and female interests are not synonymous, conflict between the sexes over control of copulation will lead to coevolutionary races between "aggressive" male traits that enhance the male's control over copulation, and "defensive" female traits that enhance the female's control and thus reduce the damage done to her reproductive output by the male.

One SAC prediction is that female morphology should tend to coevolve with male morphology. As noted above, this prediction is clearly supported in spiders. A second aspect of this predicted coevolution, however, is clearly not confirmed in spiders. If genitalic diversification were due to an arms race between males and females for control of copulation, female genitalia should often have recognizably "defensive" designs, appropriate for excluding male genitalia. We know of no case, however, in the huge array of female spider genitalia illustrated in taxonomic descriptions, in which the female has an erectable spine, or a hood that can be pulled down over the epigynum, and that would thus represent a facultatively imposed female barrier to which males might then be expected to evolve countermeasures under the SAC hypothesis. Such optional barriers (as opposed to fixed barriers which could also be used to filter males under cryptic female choice) would be expected under SAC to defend non-selectively against all male attempts to copulate; if they existed, they would constitute strong evidence in favor of SAC. Instead, many of the traits of female spider genitalia are most easily understood as being "selectively cooperative" structures, such as pits or grooves whose only apparent function is to receive and provide purchase for male structures that have particular forms, aiding the male whose structures fit adequately to perform functions such as to physically couple genitalia together. Examples of selectively cooperative female structures abound in spider genitalia (table 12.1; figures 12.6 and 12.12).

A third prediction of SAC is that rapid divergent genitalic evolution should be associated with certain types of male–female pre-copulatory interactions

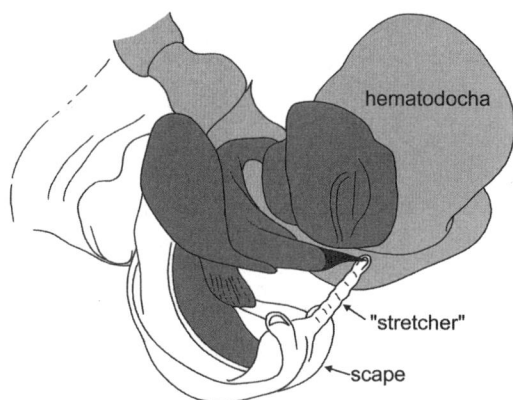

FIGURE 12.12 Female genitalia of *Lepthyphantes leprosus* (white) with male palp coupled (palp and hematodocha light gray, bulb dark gray). The tip of the male median apophysis (black) sits in a small "selectively cooperative" depression near the tip of the "stretcher" (the depression has no other known function); pressure from the median apophysis extends the scape, and allows intromission when the hematodocha is expanded (after van Helsdingen 1965).

but not others (Alexander et al. 1997). Coevolutionary races are most likely in groups in which males are more able to physically coerce or sexually harass unreceptive females ("coercive" interactions—Alexander et al. 1997). In contrast, male–female conflict and coevolutionary races are less likely in groups in which males are, for one reason or another, not able to physically coerce females into copulating, and only interact with females that are receptive (for instance, females which have been lured into the male's vicinity and are thus presumably receptive) ("luring" interactions) (Alexander et al. 1997). A major review of spider mating behavior in more than 150 species (Huber 1998) showed that interactions preceding copulation are typically of the luring type; nevertheless, in contradiction to the SAC prediction, spider genitalia very typically show sustained rapid divergent evolution.

This contradiction of SAC predictions extends to the fine details of the physical coupling between male and female genitalia. It is clear in a number of groups that tiny movements of the female can easily disrupt the difficult process of alignment of the male, arguing against the likelihood that adjustments

of the morphology of the female genitalia are needed as defenses against males, and thus against the idea that such morphological differences in females function in this context. For instance, Gering (1953) noted that "Even relatively slight movements of the female … could effectively preclude the possibility of mating" (p. 53), and concluded that "The cataleptic state of the female is an essential feature in copulation in the genus *Agelenopsis*" (p. 76). The females of *Faiditus* (= *Argyrodes*) *antipodiana*, *Leucauge mariana* and *Nesticus* must flex their abdomens ventrally for the males to be able to couple; the angle of flexion varies, and sometimes it is insufficient for the male to achieve coupling (Whitehouse & Jackson 1994; Eberhard & Huber 1998a, unpublished). In a number of species other female movements are crucial to permit coupling, and sometimes are not executed fully: protrusion of the epigynal area in *Tenuiphantes* (= *Lithyphantes*) (van Helsdingen 1965), the nephilid *Herennia* (Robinson & Robinson 1980), and the theridiosomatid *Wendilgarda* (Coddington 1986); lateral inclination of the abdomen to facilitate intromission in the agelenid *Agelenopsis* (Gering 1953), several lycosids (Rovner 1971; Costa 1979; Stratton et al. 1996) and the ctenid *Cupiennius salei* (Melchers 1963); inflation of the genital area in the mecicobothriid *Mecicobothrium* (Costa & Pérez-Miles 1998); and erection of the scape in araneids (Grasshoff 1968, 1973). Similar examples of female cooperative behavior patterns abound in the papers of U. Gerhardt (Huber 1998). In sum, the idea that female spiders are generally physically coerced via male genitalic structures into copulation is simply not correct.

Cryptic Female Choice

We have proposed (Eberhard & Huber 1998b) an hypothesis that depends on a lock-and-key type of mechanical fit between the male and the female, but in which rapid evolutionary divergence is due to sexual selection by cryptic female choice ("CFC"), rather than natural selection to avoid cross-specific fertilization. Seen from the male's evolutionary perspective, variations in genital morphology that enable him to solve the difficult mechanical challenges of copulation (e.g., more rapid, more reliable, deeper intromission) could confer advantages over other males. Seen from the evolutionary perspective of females, the mechanical problems

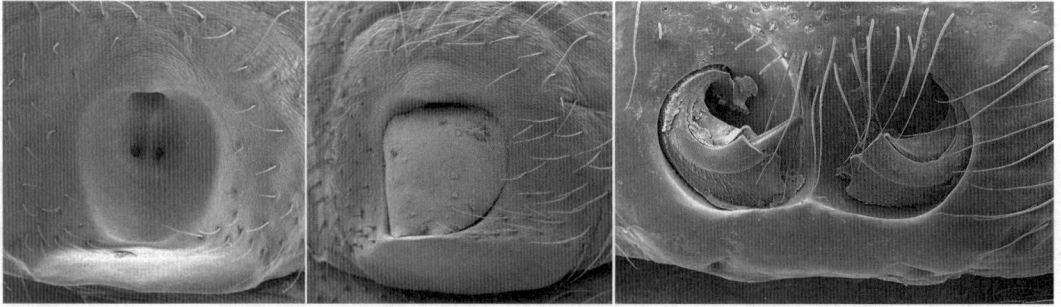

FIGURE 12.13 Mating plugs are common in spiders and vary in many respects. On the left and center unplugged and plugged female specimens of *Theridion varians* (from Knoflach 2004; with permission from Oberösterreichisches Landesmuseum), showing a secretory mating plug. On the right broken portions of male genitalia plug both openings to insemination ducts on the epigynum of a female *Herennia multipuncta* (from Kuntner 2005; with permission from CSIRO).

experienced by males that lack sense organs in their genitalia could lead to selection on females to discriminate against those males least able to achieve effective genitalic alignment, either through the stimuli received or via changes in morphology that bias male abilities to fit mechanically. The female could gain via the production of sons with superior genitalic designs. Such selection to discriminate among male designs could favor changes in female morphology that would make her genitalia more selective, facilitating a male's chances of getting his sperm into her spermathecae only if his genitalia have certain mechanical properties. Selection of this sort could favor rigid female genitalic structures with complex forms that would act as filtering devices (Huber 1993b). The female would thus be exercising sexual selection by cryptic female choice with respect to the male's ability to adjust mechanically to her complex genitalic morphology.

CFC could explain the prevalence of "selective cooperative" female designs that was mentioned above as evidence against the SAC hypothesis. But CFC might seem unable to explain why either male or female genitalia would change, much less change rapidly. Once the males of a species evolved a genitalic design that fits with the corresponding structures of conspecific females, further changes in either males or females would seem to be disadvantageous. A male with variant genitalia should be at a disadvantage because he would couple more poorly with females. And a female with variant morphology that favored non-standard male designs would also stand to lose: she might run

greater risks of not receiving adequate numbers of sperm, and her male offspring might be more likely to have deviant genital morphology because their fathers were atypical.

This description of the disadvantages of changes is based, however, on typological oversimplifications. In the first place, despite the impression given from the usual descriptions in taxonomic papers, neither the genital form of the male nor that of the female is invariant in spiders (figure 12.13; Gering 1953; Lucas & Bücherl 1965; Levi 1968, 1971, 1974, 1977a, b, 1981; Grasshoff 1968; Coyle 1968, 1971, 1974, 1981, 1984, 1986, 1988; Hippa & Oksala 1983; Kraus & Kraus 1988; Ware & Opell 1989; Pérez-Miles 1989; Milasowszky et al. 1999; Azarkina & Logunov 2006). There is also a certain degree of mechanical flexibility in some male genital structures (and perhaps in those of the females of some species) so that morphological variation does not necessarily imply loss of function (Grasshoff 1974, 1975; Loerbroks 1984). In addition, the absolute sizes of male and female genitalia in most if not all species also vary. In six different species measured in five families, the coefficients of variation in the size of male genitalia was of approximately the same order as that of other body parts (Coyle 1985; Eberhard et al. 1998). In sum, there is generally no single genital morphology for a given species. If the pattern of geographic variation in spider genitalia resembles that of some other traits (Mayr 1963), intra-specific differences in genital form could be especially great in small, geographically peripheral populations, where speciation is likely to occur.

An empirical indication that there is indeed a certain amount of imprecision or flexibility in male–female fits (and thus "room" for functional male innovations) is that the males of several groups have changed the side of the female epigynum that they inseminate. A tetragnathid and two distantly related theridiid groups have changed from inserting each palp into the ipsi-lateral insemination duct opening on the female epigynum, and now insert into the contra-lateral side (Huber & Senglet 1997; Agnarsson 2004, 2006). The early stages of such a change must have involved less than perfect male–female fits.

Intraspecific variations in male and female morphology and behavior may often influence the possibility of successful coupling, but their effects are almost completely unstudied, due to the unfortunate typological emphases in studies of the functional morphology of spider genitalia to date (including our own). The problems a male faces are surely not uniform, and a male variant that improves his ability to solve these problems could be favored. These problems could include the need to fit mechanically with the female, to stimulate her effectively, or both. Changes in males could in turn favor changes in females that further bias paternity in favor of certain males, perhaps including

morphological adjustments of females that guide these males' sensorily deprived palps. The combination of male variations and compensatory changes in females could result in rapid evolution under sexual selection by cryptic female choice.

OTHER UNUSUAL TRAITS IN SPIDERS

Lack of a Forceful Grasp on the Female

In insects, the female's reproductive opening is near the tip of her abdomen, and male genitalia often include powerful clasping structures that are capable of largely restraining the movements of the female's abdomen (Snodgrass 1935; Robson and Richards 1936; Tuxen 1970; Wood 1991). In spiders, probably due to the position of the female's epigynum on the anterior portion of her abdomen and the lack of muscles in the palpal bulb, male genitalia are only seldom (Uhl et al., in press) powerful clasping devices (except in Pholcidae—Huber 1999). More delicate clasps, which serve more to hold the palp in contact with the female, rather than restrain her abdomen, are common, however.

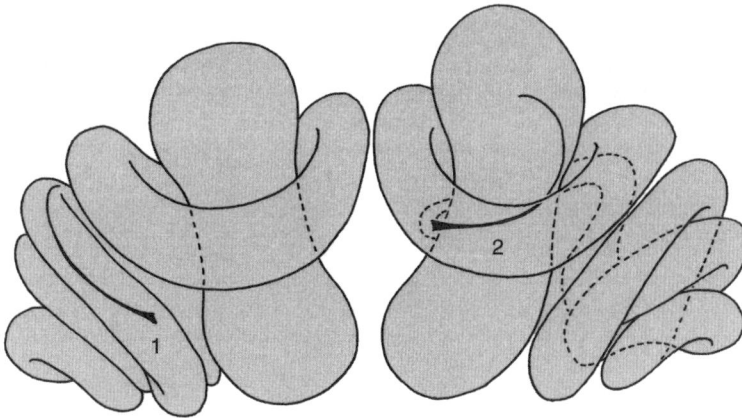

FIGURE 12.14 Broken tips of the male embolus (black) of the redback spider *Latrodectus hasselti* lodged in the female's insemination ducts and spermatheca. When placed at the entrance to the spermatheca (2), the thin, hair-like embolus tip effectively blocks the access of subsequent males to the spermatheca; but when the embolus tip is positioned elsewhere (1), it does not constitute an effective block. The poor morphological design of the tip for blocking is probably due to the tortuous coiling of the female's insemination ducts, which makes it necessary for the embolus to be thin and flexible if it is to arrive at the entrance of the spermatheca (after Snow et al. 2006; with permission from Blackwell).

TABLE **12.2** Genital plugs in spiders. Note that several of the statements about origin and function are not based on detailed observation and need reexamination. Mating plugs consisting of ectomized male body parts are covered in Table 12.3 (largely from Suhm et al. 1995 and Huber 2005a)

	Origin	Barrier for further males	References
Agelenidae: *Agelena labyrinthica*	Female secretions	Possibly	Chyzer & Kulczynski 1897 and Strand 1906 in Suhm et al. 1995; Engelhardt 1910
Agelena limbata	Male palpal glands	Yes (when complete)	Masumoto 1993
Amaurobiidae: *Amaurobius*	Male bulbal gland	Yes (when complete)	Gerhardt 1923; Wiehle 1953; Suhm et al. 1995
Tasmarubrius	?	?	Davies 1998
Anyphaenidae	?	?	Ramírez 1999, 2003
Araneidae: *Metazygia*	?	Possibly [*1]	Levi 1995a
Ctenidae	?	?	Silva 2007
Desidae (sub Toxopidae)	?	?	Forster 1967
Dictynidae	Male?	?	Bertkau 1889
Gnaphosidae	?	?	Grimm 1985; Suhm et al. 1995
Linyphiidae	?	?	Millidge 1991; Stumpf 1990 in Suhm et al. 1995 and Eberhard 1996: 153
Oedothorax	Male?	Yes (depends on copulation duration)	Uhl & Busch personal communication
Lycosidae	?	?	Suhm et al. 1995
Nesticidae: *Nesticus*	?	?	Weiss 1981; B. Huber unpublished data
Oxyopidae: *Peucetia*	?	?	Brady 1964; Exline & Whitcomb 1965; Whitcomb & Eason 1965 in Jackson 1980
Philodromidae: *Philodromus*	Sperm and (male?) secretions	?	Huber 1995a
Pholcidae: *Belisana*	?	?	Huber 2005b (Figures 292–294, 394); common also in other genera, B. A. Huber, unpublished data
Salticidae			
Heliophanus	Sperm plug[*2]	?	Harm 1971; K. Thaler in Harm 1971
Phidippus	Male secretions?	At least in 30%	Jackson 1980
Portia	Sperm plug[*2]	?	Jackson & Hallas 1986
Tetragnathidae: *Leucauge*	Male and female secretions	Depending on plug composition	Eberhard & Huber 1998a; Mendez 2002
Theridiidae			
Argyrodes	?	?	Exline & Levi 1962
Argyrodes argyrodes	Male bulbal secretions	Not necessarily	Knoflach 2004
Argyrodes antipodiana	Sperm plug[*2]	?	Whitehouse & Jackson 1994
Argyrodes and Rhomphaea	?	?	Gertsch 1979: 88
Steatoda bipunctata	Male bulbal secretions	Not necessarily	Knoflach 2004
Steatoda castanea	Male bulbal or oral secretions?	No	Gerhardt 1926
Steatoda triangulosa	Male oral secretions	Possibly	Braun 1956; Knoflach 2004
Theridion	Male genital tract and female vulval secretions	Yes	Gerhardt 1924; Levi 1959; Knoflach 1997, 1998, 2004
Thomisidae: *Misumenops*	?	Probably	Muniappan & Chada 1970
Uloboridae: *Uloborus*	Male palpal (and oral?) secretions	Possibly	Patel & Bradoo 1986
Zodariidae	Male cymbial glands?	?	Jocqué 1991

[*1] The amorphous black secretion was difficult to remove for the observer (Levi 1995a)

[*2] No evidence presented.

Blest and Pomeroy (1978) describe a calliper-like clasp of the female genitalia by two male structures; Grasshoff's (1968, 1973) schemes show male structures clasping female structures; Knoflach and van Harten (2006) describe the *Echinotheridion* palp as functioning like a forceps; and Stratton et al. (1996) describe a palpal process of some lycosids that pinches the sides of the epigynum (further examples in Huber 1993a, 1994, 1995a, b; Uhl et al. 2007).

A second type of forceful activity common in the genitalia of male insects, pulling portions of the female's reproductive tract apart to allow male entry or deeper penetration (Sakai et al. 1991 on a cricket; Byers 1961 on a tipulid fly; Whitman & Loher 1984 on a grasshopper), seems to be absent in spiders. In this case the apparent reason lies with the female, not the male; the genitalia of most female spiders form a single, rigid unit, with few or no moving parts so prying apart female sclerites is often not an option for the male. This description of female rigidity must be tempered, however, by the recent description of muscles that move portions of interrnal female genitalia in one species (figure 12.8) (Burger et al. 2006b), and our current ignorance of internal female musculature.

Mating Plugs are Common

Solid material is often deposited on the genital openings of female spiders (and sometimes the entire epigynum) (figure 12.14). This material or "mating plug" varies with respect to its composition, origin, hardness, and the degree to which it covers the epigynum. The material is generally amorphous. Mating plugs have been described in many spiders (reviewed in Suhm et al. 1995; Huber 2005a; table 12.2), and the taxonomists' practice of 'cleaning' the female genitalia in order to study their morphology almost certainly results in an underestimate of their frequency in the literature. Few studies have gone beyond the traditional assumption that these plugs are produced by the male to impede access of rival males to the female, and other potential functions like preventing sperm leakage, backflow, desiccation, or genitalic infection generally remain to be tested. Alternative explanations are surely important, because some "plugs", such as the sparse blobs of waxy substance in the salticid *Phidippus johnsoni* (Jackson 1980) and the thin and easily broken plugs in some females of the tetragnathid *Leucauge mariana* (Méndez & Eberhard, unpublished data), surely do not impede the access of subsequent males. Plugs constituted by broken pieces of the male's own genitalia inside the female are also common in spiders (figures 12.13 and 12.14).

Several studies suggest that sperm competition is a major factor driving the evolution of spider mating plugs. The clearest evidence comes from a combination of several types of observations: that males more often fail in attempts to insert their palps when a female bears a mating plug; that when a male succeeds in removing a plug he is then able to achieve intromission more frequently; and that males sometimes fail in attempts to remove plugs (Masumoto 1993; Méndez 2002) (both of these studies were incomplete, however, in that they did not demonstrate that subsequent offspring were sired by the male that had removed the plug). The fact that eggs in entelegynes exit via a different opening from the opening used for intromission means that especially tenacious, durable plugs are less damaging to the female than they would be in other groups in which such a plug might interfere with oviposition, and thus may help explain the commonness of plugs in spiders. Plugs utilizing portions of the male's own genitalia may be advantageous in some species due to the possibly low probability that the male will live to encounter another female (Snow et al. 2006). On a more mechanical level, the rigid sclerotized nature of most female external genitalia and copulatory ducts (see above) probably makes physical plugging more feasible than it would be if the female tracts were highly flexible.

Some male plug secretions originate in glands, including bulbal glands, glands in the mouth area, and glands in the genital tract (table 12.2). Other plugs apparently consist mainly of sperm (Huber 1995a; Whitehouse & Jackson 1994). Female production of plugs or at least of components of plugs, has also been known for a long time (e.g., Strand 1906 in Suhm et al. 1995; Engelhardt 1910; Gerhardt 1924), and recent observations have confirmed important female roles in plugging their own genitalia. Females of several species of theridiids and the tetragnathid *Leucauge mariana* contribute a liquid that combines with male products and is crucial if the plug is to form a barrier against further intromissions (Knoflach 1998; Méndez 2002; Méndez & Eberhard, unpublished data). Females of the latter species are more likely to add liquid when the male performs more of certain copulatory courtship behavior patterns (Aisenberg & Eberhard, 2009). Evidence for a less direct female role in plug

TABLE 12.3 Male ectomized genital structures in spiders. Only those cases are listed in which a male structure commonly or obligatorily breaks during or at the end of mating. Occasional breaking is probably much more widespread (e.g., Wiehle 1961, 1967b; Harm 1981)

	Structure	Barrier for further males	Males sterile after mating	References
Araneidae				
Argiope	Embolus tip	Yes, if placed properly	Males die during or shortly after mating	Abalos & Baez 1963; Levi 1965, 1968; Foellmer & Fairbairn 2003; Nessler et al. 2007a, b; Uhl et al. 2007
Larinia (incl. *Drexelia*), *Aculepeira*, *Araneus*, *Metepeira*	Embolus cap	?	Yes	Abalos & Baez 1963; Grasshoff 1968: 43, 1971; Levi 1970, 1973, 1975a, b, 1977b, 1991; Scharff & Coddington 1997; Piel 2001
Acacesia, *Hyposinga*	Embolus scale	?	?	Levi 1972a, b, 1976
Cyclosa	Tooth of conductor	?	?	Levi 1999
Metazygia	Part of embolus	?	?	Levi 1995a
Madrepeira	Appendage of embolus	?	?	Levi 1995b
Singafrotypa	Embolus	?	?	Kuntner & Hormiga 2002
Cybaeidae				
Cybaeus	Conductor	?	?	Ihara 2006, 2007
Nephilidae	Palp broken or only disfigured	Variable[*1]	Yes	Wiehle 1967b; Robinson & Robinson 1978; Schult & Sellenschlo 1983; Fromhage & Schneider 2006; Schneider et al. 2001, 2005a, b; Kuntner 2005, 2007
Oxyopidae				
Peucetia	Paracymbium	?	No	Brady 1964; Exline & Whitcomb 1965; Santos & Brescovit 2003; Ramirez et al. in press
Theridiidae				
Achaearanea	Embolus tip	No	No	Abalos & Baez 1963; Locket & Luczak 1974 in Miller 2007b; Knoflach 2004
Latrodectus	Embolus (part or entire), flagelliform	In some species probably yes [*2]	Variable [*3]	Bhatnager & Rempel 1962; Abalos & Baez 1963; Kaston 1970; Wiehle 1967b; Breene & Sweet 1985; Müller 1985; Berendonck & Greven 2002, 2005; Knoflach & van Harten 2002; Andrade & Banta 2002; Knoflach 2004; Snow et al. 2006
Tidarren/ Echinotheridion	Part of palp, (some species no mutilation)	No (maybe short-term)	Males die during mating	Knoflach & van Harten 2000, 2001, 2006; Knoflach 2002, 2004

[*1] Effective barrier in *Nephila fenestrata* (Fromhage & Schneider 2006), no barrier in *N. plumipes* (Schneider et al. 2001).

[*2] Probably effective barrier in *L. renivulvatus* (Knoflach 2004), *L. revivensis* (Berendonck & Greven 2002, 2005), *L. hasselti* (Snow et al. 2006).

[*3] Male sterility after mating (Abalos & Baez 1963; Andrade & Banta 2002; but: Breene & Sweet 1985) may be due to sperm depletion rather than organ breakage (Snow et al. 2006).

production comes from the behavioral cooperation of the female with the male. For instance, the male of the theridiid *Argyrodes argyrodes* interrupts copulation after sperm transfer and leaves the female, then returns to deposit the plug, with the female continuing to cooperate (Knoflach 2004). Direct female participation in producing a plug is apparently very unusual in other animal groups; the only example that we know of in which females may play a similar role is *Drosophila* (the so-called "insemination reaction", whose significance seems not to have been established) (Markow & Ankney 1988).

One hypothesis that could explain why females sometimes play active roles in forming plugs is related to the fact that most entelegyne spider females have sclerotized external genitalia, and cannot close the openings of their insemination ducts. This might result in possible problems of sperm leakage, backflow during oviposition, and microbial infections (Simmons 2001). This explanation would suggest, however, that some sort of flimsy, self-made plug would also be advantageous *before* copulation, and such plugs are not known (though if they were internal, they could be difficult to discover).

Male genital structures that break off (are "ectomized") and remain in the female can also function as mating plugs (figures 12.13 and 12.14). In entelegyne spiders, routine or obligatory genital ectomization has evolved independently in several groups (Miller 2007b; table 12.3). In some species, males invariably die during copulation and the pedipalp or even the entire male body remains attached to the female for at least a short while, and may function as a short-term mating plug (Knoflach & van Harten 2001; Foellmer & Fairbairn 2003; Knoflach & Benjamin 2003). Genital breakage that leaves pieces inside the female occurs in few other groups of animals (Eberhard 1985).

In some spider species there is a line of weakness at the point where the male genital structure breaks (Bhatnager & Rempel 1962), leaving no doubt that breakage is not accidental, and is advantageous for males. One species has a process that apparently functions only as a plug, and is not involved in insemination (Nessler et al. 2007b). In one and perhaps two species of the theridiid genus *Latrodectus*, genital breakage does not prevent the mutilated male from inseminating subsequent females (Breene & Sweet 1985; Snow et al. 2006), but in others such as the araneid *Argiope bruennichi*,

male breakage leaves the male unable to inseminate additional females (Nessler et al. 2007a). The alternative possibility that "break-away" sclerites function to facilitate male escape from female attempts to cannibalize the male has been ruled out in two species (Snow et al. 2006; Nessler et al. 2007a).

Some of these pieces of male genitalia seem to seal the external opening of the female insemination ducts, as with the plugs just discussed (Levi 1972a; Kuntner 2005) (figure 12.13), while other ectomized structures obstruct internal portions of this duct, permitting intromission by subsequent males but preventing them from reaching more internal portions of the female (Nessler et al. 2007a). In other species, however, there are sometimes pieces from several males inside a single female spermatheca (Abalos & Baez 1963; Müller 1985), suggesting that male ectomized structures are not always effective as plugs (Schneider et al. 2001; Snow et al. 2006). Of course, such plugs could be favored as paternity assurance mechanisms even if they only partially reduce the success of subsequent males.

Recent data indicate even more dynamic, exciting possibilities. In the orb weaver *Argiope bruennichi*, there is variation in whether or not the male's palpal sclerites break off (in 15% of copulations they failed to break), in whether the fragments that broke off remained lodged in the female (3% failure), and in which of two predetermined breakage lines is used (Nessler et al. 2007a; Uhl et al. 2007). Different sized pieces break off in different populations, with the more drastic type of mutilation in only one population (Uhl et al. 2007). Both plugs and ectomized male processes were more common in the epigyna of females of the oxyopid *Peucetia viridans* at drier sites in California (Ramirez et al. 2007), leading the authors to speculate that they serve to resist dessication; this function seems more likely for the plug (which may come from the female) than for the male process. The form of the process that breaks off in the distantly related *Cybaeus* varies among species (Ihara 2006, 2007), and even varies over the geographic range of *C. kuramotoi* in western Japan (Ihara 2007), again suggesting rapid divergence.

In a still another family, broken fragments of the male block access of subsequent males of *Latrodectus hasselti* (Theridiidae) when they lie at the entrance to the female spermatheca (where the insemination duct is narrow and heavily sclerotized (Berendonck & Greven 2005), but for unknown reasons they are sometimes found instead more proximally in the insemination duct, where the lumen is wide; here they do not impede the access of subsequent males (Snow et al. 2006) (figure 12.14). Some male ectomized structures left deep in the female are thin and hairlike, and poorly designed to function as physical plugs (figure 12.14). Probably the reason for this sub-optimal design is the little discussed fact that the female morphology constitutes the "playing field" on which the males must compete to deposit or remove plugs; her morphology imposes limitations on the functional designs that are available to males when they attempt to plug females. Possible coevolutionary male-female interactions remain to be explored.

Perhaps spider males are prone to use such drastic techniques to prevent female remating because males have relatively small expectations of finding and inseminating additional females (Andrade & Banta 2002; Andrade 2003; Fromhage et al. 2005; Kasumovic et al. 2006). A reduced ability to find and inseminate a second female could increase the net advantage of self-sacrifice, which could in turn lead to further reduction in the ability to inseminate other females. Snow and colleagues (2006) speculated that ectomized plugs originated with more costly "accidental" organ breakage, for instance when females attempted to interrupt copulations that males were attempting to prolong. Such "accidental" breakage may be widespread (Wiehle 1961 1967b; Harm 1971).

If plugs do in fact often serve to impede the access of rival males, then males should be under selection to remove plugs. The most obvious male structures for plug removal are sclerites of the male's palp. A partial confirmation of this hypothesis comes from a recent study of *Leucauge mariana*: a hook-shaped process of the conductor is used to snag and remove plugs but does not seem to be crucial for the insertion of the embolus (Eberhard & Méndez, unpublished data). Males of *Agelena limbata* and *Dubiaranea* sp. also remove plugs with their palps (Masumoto 1993; Eberhard 1996), but the particular structures that they use remain to be determined.

The durability of plugs has also been little studied. Lifelong plugs are feasible in entelegyne spiders, because they do not occlude the duct for oviposition (above). Durable plugs may occur in *Amaurobius* (Suhm et al. 1995), and also in *Nesticus cellulanus*, in which a male apophysis ruptures the cuticular cover of the female's vulval pocket and is lodged in this pocket during copulation (Huber 1993a). When a second male copulated immediately after the first copulation, he was able to insert his apophysis, but if two days elapsed before the second copulation, the second male was unable to insert his bulb in the mated side of a half-virgin female, presumably as a result of the hardening of substances in the ruptured female vulval pocket (Huber 1993a). A more extreme case of female mutilation occurs in *Metazygia* orb-weavers, in which the male apparently tears off a portion of the female's epigynum (the scape) during or after mating (Levi 1995a). This mutilation may prevent subsequent males from inseminating the female, because the female scape is crucial in araneid genital mechanics (Grasshoff 1968, 1973a; Uhl et al. 2007).

First Male Sperm Precedence and the "Suitor" Phenomenon

Direct measurements of sperm precedence in doubly-mated female spiders are not common, and have given mixed results (summary, Elgar 1997). Indirect evidence suggests, however, that strong first male sperm precedence is common. Many male spiders associate with immature, penultimate instar females rather than with mature females (the "suitor" phenomenon) (Jackson 1986; Robinson 1982; see also Eberhard et al. 1993; Bukowski & Christenson 1997b). Males associated with penultimate females typically mate with the female soon after she moults to maturity, and then leave. Thus the likely reason for the suitor phenomenon is that the first male achieves appreciable sperm precedence.

Variation and Exaggeration in Female Genitalia

The attention paid by taxonomists to female genitalia in spiders allows a more detailed look at female genital evolution than is possible in many other groups of animals. It may be that female genitalia are more variable intra-specifically than those of the males (Kraus & Kraus 1988 on *Stegodyphus*, especially *S. dufouri*; Baehr & Baehr 1993 on Hersiliidae; Heimer 1989 on Filistatidae; Pérez-Miles 1989 on Theraphosidae; Sierwald 1983 on *Thalassius*; Levi 1997 on *Mecynogea*; Crews & Hedin 2006 and Crews in preparation on Homalonychidae; Bennett 2006 on Amaurobiidae

Aelurillusluctuosus

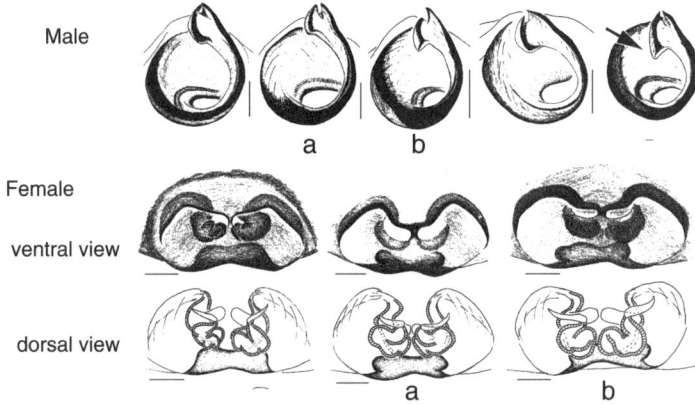

FIGURE 12.15 Intraspecific variation in the genitalia of both males (above) and females (below) of the salticid spider *Aelurillus luctuosus*; the female genitalia are shown in both ventral (external) and dorsal (internal) view. Pronounced intraspecific variation in genitalia is common in this family (from Azarkina & Logunov 2006; with permission).

and Cybaeidae). Another possible intra-specific trend is that both male and female genitalia are especially variable intra-specifically in some groups, such as certain genera of salticids (Azarkina & Logunov 2006) (figure 12.15) (also Crews, unpublished data). The reasons for greater variation in some taxonomic groups than others, or in one sex as opposed to the other are not clear. Further data to evaluate these trends would be welcome.

Not only is it clear that intraspecific variation in genitalia exists in spiders, there is also evidence that such variation has been selectively important. The genitalia of both male and female spiders resemble those of insects in showing negative static allometry (relatively large genitalia in smaller individuals, and relatively small genitalia in large individuals of the same species) (Eberhard et al. 1998; Eberhard, 2008). These low allometric values probably represent special evolutionary adjustments to reduce the amount of difference in genital size between males and females (Eberhard et al. 1998), allowing the male to fit effectively with the most common (intermediate) size of female. The negative allometric pattern in females is surely not just a pleiotropic effect of the male pattern, because completely different structures are involved. Relatively invariant genital size in females could enable them to evaluate more precisely the male's degree of allometric adjustment,

or the genitalic form of the most common (intermediate) sized males (Eberhard, 2009).

In the context of female choice by mechanical fit, the need to evaluate male exaggerations may select for other types of exaggeration (Huber 2006). Females of some species of the pholcid genus *Mesabolivar* have exaggerated external genitalia, and these exaggerations are functionally correlated with extravagant male cheliceral morphology (Huber et al. 2005). In *Mesabolivar* (originally *Kaliana*) *yuruani*, males have unique genitalia, with one specific structure (the 'procursus') about six times as long as usual in the family, and this exaggeration is paralleled in the female internal genitalia (Huber 2006; cf. Jäger 2005 on delenine sparassids). Similar coevolutionary pressures may have obliged the males of some groups with extreme sexual size dimorphisms to evolve such disproportionately large genitalia that they seriously reduce the male's agility, and favor self mutilation behavior in which the male tears off one of his palps shortly after the penultimate molt (Ramos et al. 2004).

CONCLUSION

As we have argued elsewhere (Eberhard 2004b), spiders have several traits that make them well-designed for studies of genitalic function. Despite their unique attributes, they seem to conform to the general evolutionary patterns of genital evolution seen in other groups. They should play an important role in the next generation of studies of genital evolution and function.

Acknowledgements We thank I. Agnarsson, A. Aisenberg, M. Burger, B. Knoflach, M. Kuntner, J. Miller, D. Silva, and G. Uhl for access to unpublished material and use of figures, and G. Hormiga, M. Andrade and G.Uhl for commenting on the entire chapter. Maribelle Vargas helped with figure 12.4, and STRI and the Universidad de Costa Rica (WGE), and ZFMK (BAH) provided financial support.

REFERENCES

Abalos, J. W. & Baez, E. C. 1963. On spermatic transmission in spiders. Psyche 70, 197–207.

Agnarsson, I. 2004. Morphological phylogeny of cobweb spiders and their relatives (Araneae, Araneoidea, Theridiidae). Zoological Journal of the Linnean Society 141, 447–626.

Agnarsson, I. 2006. Asymmetric female genitalia and other remarkable morphology in a new genus of cobweb spiders (Theridiidae, Araneae) from Madagascar. Biological Journal of the Linnean Society 87, 211–232.

Agnarsson, I., Coddington, J. A. & Knoflach, B. 2007. Morphology and evolution of cobweb spider male genitalia (Araneae, Theridiidae). Journal of Arachnology 35, 334–395.

Alberti, G. 1990. Comparative spermatology of Araneae. Acta Zoologica Fennica 190, 17–34.

Alberti, G. & Michalik, P. 2004. Feinstrukturelle Aspekte der Fortpflanzungssysteme von Spinnentieren (Arachnida). Denisia 12, 1–62.

Alexander, R. D., Marshall, D. C. & Cooley, J. R. 1997. Evolutionary perspectives on insect mating. In: Choe, J. C, Crespi, B. J. (eds) The Evolution of Mating Systems in Insects and Arachnids. Cambridge University Press, pp. 4–31.

Andrade, M. C. B. 2003. Risky mate search and male self-sacrifice in redback spiders. Behavioral Ecology 14, 531–538.

Andrade, M.C.B. & Banta, E. M. 2002. Value of male remating and functional sterility in redback spiders. Animal Behaviour 63, 857–870.

Austad, S. 1984. Evolution of sperm priority patterns in spiders. In: Sperm Competition and Animal Mating Systems (R. Smith, ed.). pp. 223–250. Academic Press, New York.

Azarkina, G. N. & Logunov, D. V. 2006. Taxonomic notes on nine *Aelurillus* species of the western Mediterranean (Araneae: Salticidae). Bulletin of the British Arachnological Society 13, 233–248.

Baehr, M. & Baehr, B. 1993. The Hersiliidae of the Oriental Region including New Guinea. Taxonomy, phylogeny, zoogeography (Arachnida, Araneae). Spixiana, Supplement 19, 1–95.

Baum, S. 1972. Zum "Cribellaten-Problem": Die Genitalstrukturen der Oecobiinae und Urocteinae (Arach.: Aran.: Oecobiidae). Abhandlungen und Verhandlungen des naturwissenschaftlichen Vereins Hamburg (NF) 16, 101–153.

Bennett, R. 2006. Ontogeny, variation and synonymy in North American *Cybaeus* spiders (Araneae: Cybaeidae). Canadian Entomologist 138, 473–492.

Berendonck, B. & Greven, H. 2002. Morphology of female and male genitalia of *Latrodectus revivensis* Shulov, 1948 (Araneae, Theridiidae) with regard to sperm priority patterns. In: European Arachnology 2000 (S. Toft & N. Scharff, eds), pp. 157–167. Aarhus University Press, Aarhus.

Berendonck, B. & Greven, H. 2005. Genital structures in the entelegyne widow spider *Latrodectus revivensis* (Arachnida; Araneae; Theridiidae) indicate a low ability for cryptic female choice by sperm manipulation. Journal of Morphology 263, 118–132.

Bertkau, P. 1889. Über ein "Begattungszeichen" bei Spinnen. Zoologischer Anzeiger 12, 450–454.

Bhatnagar, R. D. S. & Rempel, J. G. 1962. The structure, function, and postembryonic development of the male and female copulatory organs of the black widow spider *Latrodectus curacaviensis* (Müller). Canadian Journal of Zoology 40, 465–510.

Blest, A. D. & Pomeroy, G. 1978. The sexual behaviour and genital mechanics of three species of *Mynoglenes* (Araneae: Linyphiidae). Journal of Zoology, London 185, 319–340.

Bond, J. E. & Platnick, N. I, 2007. A taxonomic revision of the trapdoor spider genus *Myrmekiaphila* (Araneae, Mygalomorphae,

Cyrtaucheniidae). American Museum Novitates 3596, 1–30.

Brady, A. R. 1964. The lynx spiders of North America, north of Mexico (Araneae: Oxyopidae). Bulletin of the Museum of Comparative Zoology 131, 429–518.

Braun, R. 1956. Zur Biologie von *Teutana triangulosa* (Walck.) (Araneae; Theridiidae, Asageneae). Zeitschrift für wissenschaftliche Zoologie 159, 255–318.

Breene, R. G. & Sweet, M. H. 1985. Evidence of insemination of multiple females by the male black widow spider, *Latrodectus mactans* (Araneae, Theridiidae). Journal of Arachnology. 13, 331–335.

Bristowe, W. S. 1926. The mating habits of British thomisid and sparassid spiders. Annals and Magazine of Natural History 18, ser 9, nr 103, 114–131.

Bristowe, W. S. 1929. The mating habits of spiders, with special reference to the problems surrounding sex dimorphism. Proceedings of the Zoological Society 21, 309–358.

Bristowe, W. S. 1958. The World of Spiders. Publ. Collins, London.

Bukowski, T. C. & Christenson, T. E. 1997a. Determinants of sperm release and storage in a spiny orbweaving spider. Animal Behaviour 53, 381–395.

Bukowski, T. C. & Christenson, T. E. 1997b. Natural history and copulatory behavior of the orbweaving spider *Micrathena gracilis* (Araneae, Araneidae). Journal of Arachnology 25, 307–320.

Burger, M. 2007. Sperm dumping in the haplogyne spider *Silhouettela loricatula* (Arachnida: Araneae: Oonopidae). Journal of Zoology 273, 74–81.

Burger, M. 2008. Functional genital morphology of armored spiders (Arachnida: Araneae: Tetrablemmidae). Journal of Morphology 296, 1073–1094.

Burger, M. & Kropf, C. 2007. Genital morphology of the haplogyne spider *Harpactea lepida* (Arachnida, Araneae, Dysderidae). Zoomorphology 126, 45–52.

Burger, M., Nentwig, W. & Kropf, C. 2003. Complex genital structures indicate cryptic female choice in a haplogyne spider (Arachnida, Araneae, Oonopidae, Gamasomorphina). Journal of Morphology 255, 80–93.

Burger, M., Michalik, P., Graber, W., Jacob, A., Nentwig, W. & Kropf C. 2006a. Complex genital system of a haplogyne spider (Arachnida, Araneae, Tetrablemmidae) indicates internal fertilization and full female control over transferred sperm. Journal of Morphology 267, 166–186.

Burger, M., Graber, W., Michalik, P. & Kropf C. 2006b. *Silhouetella loricatula* (Arachnida, Araneae, Oonopidae): a haplogyne spider with complex female genitalia. Journal of Morphology 267, 663–677.

Byers, G. 1961. The crane fly genus *Dolichopeza* in North America. University of Kansas Science Bulletin 42, 666–924.

Carico, J. E. & Holt, P. C. 1964. A comparative study of the female copulatory apparatus of certain species in the spider genus *Dolomedes* (Pisauridae : Araneae). Technical Bulletin of the Agricultural Experiment Station Blacksburg, Virginia 172, 1–27.

Chapman, R. F. 1998. The Insects: Structure and Function. 4th Edition. Cambridge, Cambridge University Press. 770 pp.

Coddington, J. A. 1986. The genera of the spider family Theridiosomatidae. Smithsonian Contributions to Zoology 422, 1–96.

Coddington, J. A. 1990. Ontogeny and homology in the male palpus of orb-weaving spiders and their relatives, with comments on phylogeny (Araneoclada: Araneoidea, Deinopoidea). Smithsonian Contributions to Zoology 496, 1–52.

Coddington, J. A. & Levi, H. W. 1991. Systematics and evolution of spiders (Araneae). Annual Review of Ecology and Systematics 22, 565–592.

Comstock, J. 1967. The Spider Book. Cornell University Press, Ithaca, NY.

Cooke, J. A. L. 1966. Synopsis of the structure and function of the genitalia of *Dysdera crocata* (Araneae, Dysderidae). Senckenbergiana Biologica 47, 35–43.

Cooke, J. A. L. 1970. Spider genitalia and phylogeny. Bulletin du Muséum National d'Histoire Naturelle (2ᵉ serie) 41, 142–146.

Costa, F. G. 1979. Analisis de la copula y de la actividad postcopulatoria de *Lycosa malitiosa* (Tullgren) (Araneae, Lycosidae). Revista Brasileira do Biologia 30, 361–376.

Costa, F. G., Pérez-Miles, F. 1998. Behavior, life cycle and webs of *Mecicobothrium thorelli* (Araneae, Mygalomorphae, Mecicobothriidae). Journal of Arachnology 26, 317–329.

Costa, F. G., Pérez-Miles, F. & Corte, S. 2000. Which spermatheca is inseminated by each palp in Theraphosidae spiders?: a study of *Oligoxystre argentinensis* (Ischnocolinae). Journal of Arachnology 28, 131–132.

Coyle, F. A. 1968. The mygalomorph spider genus *Atypoides* (Araneae: Antrodiaetidae). Psyche 75, 157–194.

Coyle, F. A. 1971. Systematics and natural history of the mygalomorph spider genus *Antrodiaetus* and related genera (Araneae: Antrodiaetidae). Bulletin of the Museum of Comparative Zoology 141, 269–402.

Coyle, F. A. 1974. Systematics of the trapdoor spider genus *Aliatypus* (Araneae: Antrodiaetidae). Psyche 81, 431–500.

Coyle, F. A. 1981. The mygalomorph spider genus *Microhexura* (Araneae, Dipluridae). Bulletin of the American Museum of Natural History 170, 64–75.

Coyle, F. A. 1984. A revision of the African myga-lomorph spider genus *Allothele* (Araneae, Dipluridae). American Museum Novitates 2794, 1–20.

Coyle, F. A. 1985. Two-year life cycle and low palpal character variance in a great smoky mountain population of the lamp-shade spider (Araneae, Hypochilidae, *Hypochilus*). Journal of Arachnology 13, 211–218.

Coyle, F. A. 1986. *Chilehexops*, a new funnelweb mygalomorph spider genus from Chile (Araneae, Dipluridae). American Museum Novitates 2860, 1–10.

Coyle, F. A. 1988. A revision of the American funnel-web mygalomorph spider genus *Euagrus* (Araneae, Dipluridae). Bulletin of the American Museum of Natural History 187, 203–292.

Coyle, F. A., Harrison, F. W., McGimsey, W. C. & Palmer, J. M. 1983. Observations of the structure and function of spermathecae in haplogyne spiders. Transactions of the American Microscopical Society 102, 272–280.

Crews, S. C. & Hedin, M. 2006. Studies of morphological and molecular phylogenetic divergence in spiders (Araneae: *Homalonychus*) from the American southwest, including divergence along the Baja California Peninsula. Molecular Phylogenetics and Evolution 38, 470–487.

Davies, V. T. 1998. A redescription and renaming of the Tasmanian spider *Amphinecta milvina* (Simon, 1903), with descriptions of four new species (Araneae: Amaurobioidea: Amaurobiidae). In: Proceedings of the 17th European Colloquium of Arachnology, Edinburgh 1997 (P. A. Selden, ed.), pp. 67–82.

Deeleman-Reinhold, C. & Deeleman, P. R. 1980. Remarks on troglobitism in spiders. Proceedings of the 8th International Arachnological Congress (Vienna). 433–438.

Dimitrov, D., Alvarez-Padilla, F. & Hormiga, G. 2007. The female genital morphology of the orb-weaving spider genus *Agriognatha* (Araneae, Tetragnathidae). Journal of Morphology 268, 758–770.

Eberhard, W. G. 1985. Sexual Selection and Animal Genitalia. Cambridge, Massachusetts: Harvard University Press. 244 p.

Eberhard, W. G. 1996. Female Control: Sexual Selection by Cryptic Female Choice. Princeton, NJ: Princeton University Press. 501 p.

Eberhard, W. G. 2004a. Male-female conflicts and genitalia: failure to confirm predictions in insects and spiders. Biological Reviews 79, 121–186.

Eberhard, W. G. 2004b. Why study spider sex: special traits of spiders facilitate studies of sperm competition and cryptic female choice. Journal of Arachnology 32, 545–556.

Eberhard, W. G. 2004c. Rapid divergent evolution of sexual morphology: comparative tests of antagonistic coevolution and traditional female choice. Evolution 58, 1947–1970.

Eberhard, W. G. in press. Genital evolution: theo-ries and data. In J. Leonard & A. Córdoba-Aguilar (eds) Evolution of Primary Sexual Characters in Animals. Oxford University Press, Oxford, U.K.

Eberhard, W. G. 2009. Static allometry and animal genitalia. Evolution 63, 48–66.

Eberhard, W. G. & Huber B. A. 1998a. Courtship, copulation, and sperm transfer in *Leucauge mariana* (Araneae, Tetragnathidae) with implications for higher classification. Journal of Arachnology 26, 342–368.

Eberhard, W. G. & Huber B. A. 1998b. Possible links between embryology, lack of innerva-tion, and the evolution of male genitalia in spiders. Bulletin of the British Arachnological Society 11, 73–80.

Eberhard, W. G., Guzman-Gomez, S. & Catley, K. 1993. Correlation between genitalic morphol-ogy and mating systems in spiders. Zoological Journal of the Linnean Society 50, 197–209.

Eberhard, W. G., Huber, B. A., Rodriguez, R. L., Briceño, D., Salas, I. & Rodriguez, V. 1998. One size fits all? Relationships between the size of genitalia and other body parts in 20 species of insects and spiders. Evolution 52, 415–431.

Elgar, M. A. 1997. Sperm competition and sexual selection in spiders and other arachnids. Pp. 307–340 In T. R. Birkhead & A. P. Moller (eds.) Sperm Competition and Sexual Selection. Academic Press, New York, pp. 32–56.

Engelhardt, V. von 1910. Beiträge zur Kenntnis der weiblichen Copulationsorgane einiger Spinnen. Zeitschrift für wissenschaftliche Zoologie 96, 32–117.

Exline, H. & Levi, H. W. 1962. American spiders of the genus *Argyrodes* (Araneae, Theridiidae). Bulletin of the Museum of Comparative Zoology 127, 73–204, pl. 1–15.

Exline, H. & Whitcomb, W. H. 1965. Clarification of the mating procedure of *Peucetia viridans* (Araneida: Oxyopidae) by a microscopic examination of the epigyneal plug. Florida Entomologist 48, 169–171.

Foelix, R. F. 1985. Mechano- and chemoreceptive sensilla. In: Barth, F. (ed.) Neurobiology of Arachnids. Springer, New York, pp 118–137.

Foelix, R. F. 1996. Biology of Spiders. 2nd edn. Oxford Univ. Press, G. Thieme Verlag: New York, Oxford.

Foelix, R. F. & Choms, A. 1979. Fine structure of a spider joint receptor and associated synapses. European Journal of Cell Biology 19, 149–159.

Foellmer, M. W. & Fairbairn, D. J. 2003. Spontaneous male death during copulation in an orb-weaving spider. Proceedings of the Royal Society of London B 270, S183–S185.

Forster, R. R. 1967. The spiders of New Zealand. Otago Museum Bulletin 1, 1–124.

Forster, R. R. 1980. Evolution of the tarsal organ, the respiratory system and the female genitalia in spiders. Proceedings of the 8th International Arachnologen-Kongreß, Wien 1980, 269–284.

Forster, R. R. & Platnick, N. I. 1984. A review of the archaeid spiders and their relatives, with notes on the limits of the superfamily Palpimanoidea (Arachnida, Araneae). Bulletin of the American Museum of Natural History 178, 1–106.

Forster, R. R., Platnick, N. I. & Gray, M. R. 1987. A review of the spider superfamilies Hypochiloidea and Austrochiloidea (Araneae, Araneomorphae). Bulletin of the American Museum of Natural History 185, 1–116.

Fromhage, L. & Schneider J. M. 2006. Emasculation to plug up females: the significance of pedipalp damage in *Nephila fenestrata*. Behavioral Ecology 17, 353–357.

Fromhage, L., Elgar, M. A. & Schneider, J. M. 2005. Faithful without care: the evolution of monogyny. Evolution 59, 1400–1405.

Gerhardt, U. 1923. Weitere sexualbiologische Untersuchung an Spinnen. Archiv für Naturgeschichte 89 (A,10), 1–225, pl 1–3.

Gerhardt, U. 1924. Weitere Studien über die Biologie der Spinnen. Archiv für Naturgeschichte 90 (A,5), 85–192.

Gerhardt, U. 1926. Weitere Untersuchungen zur Biologie der Spinnen. Zeitschrift für Morphologie und Ökologie der Tiere 6, 1–77.

Gering, R. L. 1953. Structure and function of the genitalia in some American agelenid spiders. Smithsonian Miscellaneous Collections 121, 1–84.

Gertsch, W. J. 1974. The spider family Leptonetidae in North America. Journal of Arachnology 1, 145–203.

Gertsch, W. J. 1979. American Spiders. 2nd edn. Van Nostrand Reinhold Company: New York.

Gertsch, W. J. & Peck, S. B. 1992. The pholcid spiders of the Galápagos Islands, Ecuador (Araneae: Pholcidae). Canadian Journal of Zoology 70, 1185–1199.

Grasshoff, M. 1968. Morphologische Kriterien als Ausdruck von Artgrenzen bei Radnetzspinnen der Subfamilie Araneinae (Arachnida: Araneae: Araneidae). Abhandlungen der senckenbergischen naturforschenden Gesellschaft 516, 1–100.

Grasshoff, M. 1971. Die Tribus Mangorini, III. Die Gattung *Drexelia* MacCook (Arachnida: Araneae: Araneidae-Araneinae). Senckenbergiana biologica 52, 81–95.

Grasshoff, M. 1973. Bau und Mechanik der Kopulationsorgane der Radnetzspinne *Mangora acalypha* (Arachnida, Araneae). Zeitschrift für Morphologie der Tiere 74, 241–252.

Grasshoff, M. 1974. Transformierungsreihen in der Stammesgeschichte—mechanischer Wandel an Kopulationsorganen von

Radnetzspinnen. Natur und Museum 104, 321–330.

Grasshoff, M. 1975. Reconstruction of an evolutionary transformation - the copulatory organs of *Mangora* (Arachnida, Araneae, Araneidae). Proceedings of the 6th International Arachnology Congress (Amsterdam, 1974): 12–16.

Grimm, U. 1985. Die Gnaphosidae Mitteleuropas (Arachnida, Araneae). Abhandlungen des naturwissenschaftlichen Vereins in Hamburg (NF) 26, 1–316.

Grimm, U. 1986. Die Clubionidae Mitteleuropas: Corinninae und Liocraninae (Arachnida, Araneae). Abhandlungen des naturwissenschaftlichen Vereins in Hamburg (NF) 27, 1–91.

Griswold, C. E. 1987. A review of the sourthern African spiders of the family Cyatholipidae Simon, 1894 (Araneae: Areneomorphae). Annals of the Natal Museum 28, 499–542.

Griswold, C. E. 1990. A revision and phylogenetic analysis of the spider subfamily Phyxelidinae (Araneae, Amaurobiidae). Bulletin of the American Museum of Natural History 196, 1–206.

Griswold, C. E. 1991. A revision and phylogenetic analysis of the spider genus *Machadonia* Lehtinen (Araneae, Lycosidae). Entomologica Scandinavica 22, 305–351.

Griswold, C. E. 1993. Investigations into the phylogeny of the lycosoid spiders and their kin (Arachnida: Araneae: Lycosoidea). Smithsonian Contributions to Zoology 539, 1–39.

Griswold, C. E. 1994. A revision and phylogenetic analysis of the spider genus *Phanotea* Simon (Araneae, Lycosoidea). Annales du Musée Royal de L'Afrique Centrale Tervuren, Belgique, Sciences Zoologiques 273, 1–83.

Griswold, C. E. 1997. The spider family Cyatholipidae in Madagascar (Araneae, Araneoidea). Journal of Arachnology 25, 53–83.

Griswold, C. E., Coddington, J. A., Hormiga, G. & Scharff, N. 1998. Phylogeny of the orb-web building spiders (Araneae, Orbiculariae: Deinopoidea, Araneoidea). Zoological Journal of the Linnean Society 122, 1–99.

Griswold, C. E., Ramirez, M. J., Coddington, J. A. & Platnick, N. I. 2005. Atlas of phylogenetic data for entelegyne spiders (Araneae: Araneomorphae: Entelegynae) with comments on their phylogeny. Proceedings of the

California Academy of Sciences 56, suppl. 2, 1–324.

Harm, M. 1931. Beiträge zur Kenntnis des Baues, der Funktion und der Entwicklung des akzessorischen Kopulationsorgans von *Segestria bavarica* C. L. Koch. Zeitschrift für Morphologie und Ökologie der Tiere 22, 629–670.

Harm, M. 1971. Revision der Gattung *Heliophanus* C. L. Koch (Arachnida: Araneae: Salticidae). Senckenbergiana biologica 52, 53–79.

Harm, M. 1981. Revision der mitteleuropäischen Arten der Gattung *Marpissa* C. L. Koch 1846 (Arachnida: Araneae: Salticidae). Senckenbergiana biologica 61, 277–291.

Haupt, J. 1979. Lebensweise und Sexualverhalten der mesothelen Spinne *Heptathela nishihirai* n. sp. (Araneae, Liphistiidae). Zoologischer Anzeiger 202, 348–374.

Haupt, J. 2003. The Mesothelae—a monograph of an exceptional group of spiders (Araneae: Mesothelae). Zoologica 154, 1–102.

Haupt, J. 2004. The palpal organ of male spiders (Arachnida, Araneae). In: European Arachnology 2002. Proceedings of the 20th European Colloquium of Arachnology, Szombathely 2002 (F. Samu & Cs. Szinetár, eds), pp. 65–71.

Hedin, M. 1997. Speciational history in a diverse clade of habitat-specialized spiders (Araneae: Nesticidae: *Nesticus*): inferences from geographic-based sampling. Evolution 51, 1929–1945.

Heimer, S. 1989. Untersuchungen zur Evolution der Kopulationsorgane bei Spinnen (I) (Arachnida, Araneae). Entomologische Abhandlungen, Staatliches Museum für Tierkunde Dresden 53, 1–25.

Hippa, H. & Oksala, I. 1983. Epigynal variation in *Enoplognatha latimana* Hippa & Oksala (Araneae, Theridiidae) in Europe. Bulletin of the British arachnological Society 6, 99–102.

Hormiga, G. 1994. A revision and cladistic analysis of the spider family Pimoidae (Araneoidea: Araneae). Smithsonian Contributions to Zoology 549, 1–104.

Hormiga, G. 2002. *Orsonwelles*, a new genus of giant linyphiid spiders (Araneae) from the Hawaiian Islands. Invertebrate Systematics 16, 369–448.

Hormiga, G. & Scharff, N. 2005. Monophyly and phylogenetic placement of the spider genus

Labulla Simon, 1884 (Araneae, Linyphiidae) and description of the new genus *Pecado*. Zoological Journal of the Linnean Society 143, 359–404.

Hormiga, G., Arnedo, J. & Gillespie, R. G. 2003. Speciation on a conveyer belt: sequential colonization of the Hawaiian Islands by *Orsonwelles* spiders (Araneae, Linyphiidae). Systematic Biology 52, 70–88.

Huber, B. A. 1993a. Genital mechanics and sexual selection in the spider *Nesticus cellulanus* (Araneae: Nesticidae). Canadian Journal of Zoology 71, 2437–2447.

Huber, B. A. 1993b. Female choice and spider genitalia. Bolletino dell' Accademia Gioenia di Scienze Naturali (Catania) 26, 209–214.

Huber, B. A. 1994. Copulatory mechanics in the funnel-web spiders *Histopona torpida* and *Textrix denticulata* (Agelenidae, Araneae). Acta Zoologica, Stockholm 75, 379–384.

Huber, B. A. 1995a. The retrolateral tibial apophysis in spiders—shaped by sexual selection? Zoological Journal of the Linnean Society 113, 151–163.

Huber, B. A. 1995b. Genital morphology and copulatory mechanics in *Anyphaena accentuata* (Anyphaenidae) and *Clubiona pallidula* (Clubionidae: Araneae). Journal of Zoology, London 235, 689–702.

Huber, B. A. 1998. Spider reproductive behaviour: a review of Gerhardt's work from 1911–1933, with implications for sexual selection. Bulletin of the British Arachnological Society 11, 81–91.

Huber, B. A. 1999. Sexual selection in pholcid spiders (Araneae, Pholcidae): artful chelicerae and forceful genitalia. Journal of Arachnology 27, 135–141.

Huber B. A. 2002. Functional morphology of the genitalia in the spider *Spermophora senoculata* (Pholcidae, Araneae). Zoologischer Anzeiger 241, 105–116.

Huber, B. A. 2003a. Rapid evolution and species-specificity of arthropod genitalia: fact or artifact? Organisms Diversity & Evolution 3, 63–71.

Huber, B. A. 2003b. Southern African pholcid spiders: revision and cladistic analysis of *Quamtana* n. gen. and *Spermophora* Hentz (Araneae: Pholcidae), with notes on male-female covariation. Zoological Journal of the Linnean Society 139, 477–527.

Huber, B. A. 2004a. Evidence for functional segregation in the directionally asymmetric

male genitalia of the spider *Metagonia mariguitarensis* (González-Sponga) (Pholcidae: Araneae). Journal of Zoology, London 262, 317–326.

Huber, B. A. 2004b. The evolutionary transformation from muscular to hydraulic movements in spider genitalia: a study based on histological serial sections. Journal of Morphology 261, 364–376.

Huber, B. A. 2005a. Sexual selection research on spiders: progress and biases. Biological Reviews 80, 363–385.

Huber, B. A. 2005b. High species diversity, male-female coevolution, and metaphyly in Southeast Asian pholcid spiders: the case of *Belisana* Thorell, 1898 (Araneae, Pholcidae). Zoologica 155, 1–126.

Huber, B. A. 2006. Cryptic female exaggeration: the asymmetric female internal genitalia of *Kaliana yuruani* (Araneae: Pholcidae). Journal of Morphology 267, 705–712.

Huber, B. A. & Senglet, A. 1997. Copulation with contralateral insertion in entelegyne spiders (Araneae: Entelegynae: Tetragnathidae). Netherlands Journal of Zoology 47, 99–102.

Huber, B. A., Brescovit A. D & Rheims C. A. 2005. Exaggerated female genitalia in two new spider species (Araneae: Pholcidae), with comments on genital evolution by female choice versus antagonistic coevolution. Insect Systematics and Evolution 36, 285–292.

Ihara, Y. 2006. *Cybaeus jinsekiensis* n. sp., a spider species with protogynous maturation and mating plugs (Araneae, Cybaeidae). Acta Arachnologica 55, 5–13.

Ihara, Y. 2007. Geographic variation and body size differentiation in the medium-sized species of the genus *Cybaeus* (Araneae: Cybaeidae) in northern Kyushu, Japan, with descriptions of two new species. Acta Arachnologica 56, 1–14.

Jackson, R. R. 1980. The mating strategy of *Phidippus johnsoni* (Araneae, Salticidae): II. Sperm competition and the function of copulation. Journal of Arachnology 8, 217–240.

Jackson, R. R. 1986. Cohabitation of males and juvenile females: a prevalent mating tactic of spiders. Journal of Natural History 20, 1193–1210.

Jackson, R. R. & Hallas, S. E. A. 1986. Comparative biology of *Portia africana*, *P. albimana*, *P. fimbriata*, *P. labiata*, and *P. shultzi*, araneophagic, web-building jumping spiders (Araneae: Salticidae): utilisation of webs, predatory versatility, and intraspecific interactions. New Zealand Journal of Zoology 13, 423–489.

Jäger, P. 2005. Lengthening of embolus and copulatory duct: a review of an evolutionary trend in the spider family Sparassidae (Araneae). Acta Zoologica Bulgarica, Suppl. 1, 49–62.

Jocqué, R. 1991. A generic revision of the spider family Zodariidae (Araneae). Bulletin of the American Museum of Natural History 201, 1–160.

Jones, T. M. & Elgar, M. 2008. Male insemination decisions and sperm quality influence paternity in the golden orb-weaving spider. Behavioral Ecology. 19, 285–291.

Juberthie-Jupeau, L. & Lopez, A. 1981. Ultrastructure du tube séminifère chez Leptoneta microphthalma Simon, 1872 (Araneae, Leptonetidae). Revue Arachnologique 3, 65–73.

Kaestner, A. 1968. Invertebrate Zoology Volume II (translated and adapted by H. W. Levi and L. Levi). New York. John Wiley & Sons. 472 pp.

Kaston, B. J. 1970. Comparative biology of American black widow spiders. Transactions of the San Diego Society of Natural History 16, 33–82.

Kasumovic, M. M., Bruce, M. J., Herberstein, M. E. & Andrade, M. C. B. 2006. Risky mate search and mate preference in the golden orb-web spider (Nephila plumipes). Behavioral Ecology 18, 189–195.

Knoflach, B. 1997. Zur Taxonomie und Sexualbiologie von Theridion adrianopoli Drensky (Arachnida: Araneae, Theridiidae). Berichte des naturkundlich-medizinischen Vereins Innsbruck 84, 133–148.

Knoflach, B. 1998. Mating in Theridion varians Hahn and related species (Araneae: Theridiidae). Journal of Natural History 32, 545–604.

Knoflach, B. 2002. Copulation and emasculation in Echinotheridion gibberosum (Kulczynski, 1899) (Araneae, Theridiidae). In European Arachnology 2000 (eds. S. Toft and N. Scharff), pp. 139–144. Aarhus University Press, Aarhus.

Knoflach, B. 2004. Diversity in the copulatory behaviour of comb-footed spiders (Araneae, Theridiidae). Denisia 12, 161–256.

Knoflach, B. & Benjamin, S. P. 2003. Mating without sexual cannibalism in Tidarren sisyphoides (Araneae, Theridiidae). Journal of Arachnology 31, 445–448.

Knoflach, B. & Pfaller, K. 2004. Kugelspinnen— eine Einführung (Araneae, Theridiidae). Denisia 12, 111–160.

Knoflach, B. & van Harten, A. 2000. Palpal loss, single palp copulation and obligatory mate consumption in Tidarren cuneolatum (Tullgren, 1910) (Araneae, Theridiidae). Journal of Natural History 34, 1639–1659.

Knoflach, B. & van Harten, A. 2001. Tidarren argo sp. nov. (Araneae: Theridiidae) and its exceptional copulatory behaviour: emasculation, male palpal organ as a mating plug and sexual cannibalism. Journal of Zoology, London 254, 449–459.

Knoflach, B. & van Harten, A. 2002. The genus Latrodectus (Araneae: Theridiidae) from mainland Yemen, the Socotra Archipelago and adjacent countries. Fauna of Arabia 19, 321–361.

Knoflach, B. & van Harten, A. 2006. The one-palped spider genera Tidarren and Echinotheridion in the Old World (Araneae, Theridiidae), with comparative remarks on Tidarren from America. Journal of Natural History 40, 1483–1616.

Kraus, O. 1978. Liphistius and the evolution of spider genitalia. Symposium of the Zoological Society of London 42, 235–254.

Kraus, O. 1984. Male spider genitalia: evolutionary changes in structure and function. Verhandlungen des naturwissenschaftlichen Vereins Hamburg (NF) 27, 373–382.

Kraus, O. & Kraus, M. 1988. The genus Stegodyphus (Arachnida, Araneae). Sibling species, species groups, and parallel origin of social living. Verhandlungen des naturwissenschaftlichen Vereins Hamburg (NF) 30, 151–254.

Kuntner, M. 2005. A revision of Herennia (Araneae: Nephilidae: Nephilinae), the Australian 'coin spiders'. Invertebrate Systematics 19, 391–436.

Kuntner, M. 2007. A monograph of Nephilengys, the pantropical 'hermit spiders' (Araneae, Nephilidae, Nephilinae). Systematic Entomology 32, 95–135.

Kuntner, M. & Hormiga, G. 2002. The African spider genus Singafrotypa (Araneae, Araneidae). Journal of Arachnology 30, 129–139.

Kuntner, M., Coddington, J. A. & Hormiga, G. 2008. Phylogeny of extant nephilid orb-weaving spiders (Araneae, Nephilidae): testing morphological and ethological homologies. Cladistics 24, 147–217.

Lamoral, B. H. 1973. On the morphology, anatomy, histology and function of the tarsal organ on the pedipalpi of *Palystes castaneus* (Sparassidae, Araneida). Annals of the Natal Museum 21, 609–648.

Levi, H. W. 1959. The spider genera *Achaearanea*, *Theridion* and *Sphyrotinus* from Mexico, Central America and the West Indies (Araneae, Theridiidae). Bulletin of the Museum of Comparative Zoology 121, 55–163, pl. 1–26.

Levi, H. W. 1961. Evolutionary trends in the development of the palpal sclerites in the spider family Theridiidae. Journal of Morphology 108, 1–9.

Levi, H. W. 1965. Techniques for the study of spider genitalia. Psyche 72, 152–158.

Levi, H. W. 1968. The spider genera *Gea* and *Argiope* in America (Araneae: Araneidae). Bulletin of the Museum of Comparative Zoology 136, 319–352.

Levi, H. W. 1970. Problems in the reproductive physiology of the spider palpus. Bulletin du Musée national d'Histoire naturelle, 2ᵉ ser, 41, suppl 1, 108–111.

Levi, H. W. 1971. The *diadematus* group of the orb-weaver genus *Araneus* north of Mexico (Araneae: Araneidae). Bulletin of the Museum of Comparative Zoology 141, 131–179.

Levi, H. W. 1972a. Observations of the reproductive physiology of the spider *Singa* (Araneidae). Arachnologorum Congressus Internationalis V. Loksa, I. Araneae II. Fauna Hungarica 109, 189–192.

Levi, H. W. 1972b. The orb-weaver genera *Singa* and *Hyposinga* in America (Araneae: Araneidae). Psyche 78, 229–256.

Levi, H. W. 1973. Small orb-weavers of the genus *Araneus* north of Mexico (Araneae: Araneidae). Bulletin of the Museum of Comparative Zoology 145, 473–552.

Levi, H. W. 1974. The orb-weaver genera *Araniella* and *Nuctenea* (Araneae: Araneidae). Bulletin of the Museum of Comparative Zoology 146, 291–316.

Levi, H. W. 1975a. Mating behavior and presence of embolus cap in male Araneidae. Proceedings of the 6th International Arachnology Congress, Amsterdam 1974, 49–50.

Levi, H. W. 1975b. The American orb-weaver genera *Larinia*, *Cercidia* and *Mangora* north of Mexico (Araneae, Araneidae). Bulletin of the Museum of Comparative Zoology 147, 101–135.

Levi, H. W. 1976. The orb-weaver genera *Verrucosa*, *Acanthepeira*, *Wagneriana*, *Acacesia*, *Wixia*, *Scoloderus* and *Alpaida* north of Mexico (Araneae: Araneidae). Bulletin of the Museum of Comparative Zoology 147, 351–391.

Levi, H. W. 1977a. The American orb-weaver genera *Cyclosa*, *Metazygia* and *Eustala* north of Mexico (Araneae, Araneidae). Bulletin of the Museum of Comparative Zoology 148, 61–127.

Levi, H. W. 1977b. The orb-weaver genera *Metepeira*, *Kaira* and *Aculepeira* in America north of Mexico (Araneae: Araneidae). Bulletin of the Museum of Comparative Zoology 148, 185–238.

Levi, H. W. 1981. The American orb-weaver genera *Dolichognatha* and *Tetragnatha* north of Mexico (Araneae: Araneidae, Tetragnathinae). Bulletin of the Museum of Comparative Zoology 149, 271–318.

Levi, H. W. 1991. The neotropical and Mexican species of the orb-weaver genera *Araneus*, *Dubiepeira*, and *Aculepeira* (Araneae: Araneidae). Bulletin of the Museum of Comparative Zoology 152, 167–315.

Levi, H. W. 1995a. The neotropical orb-weaver genus *Metazygia* (Araneae: Araneidae). Bulletin of the Museum of Comparative Zoology 154, 63–151.

Levi, H. W. 1995b. Orb-weaving spiders *Actinosoma*, *Spilasma*, *Micrepeira*, *Pronous*, and four new genera (Araneae: Araneidae). Bulletin of the Museum of Comparative Zoology 154, 153–213.

Levi, H. W. 1997. The American orb weavers of the genera *Mecynogea*, *Manogea*, *Kapogea* and *Cyrtophora* (Araneae: Araneidae). Bulletin of the Museum of Comparative Zoology 155, 215–255.

Levi, H. W. 1999. The neotropical and Mexican orb weavers of the genera *Cyclosa* and *Allopecosa* (Araneae: Araneidae). Bulletin of the Museum of Comparative Zoology 155, 299–379.

Loerbroks, A. 1983. Revision der Krabbenspinnen-Gattung *Heriaeus* Simon (Arachnida: Araneae: Thomisidae). Verhandlungen des naturwissenschaftlichen Vereins Hamburg (NF) 26, 85–139.

Loerbroks, A. 1984. Mechanik der Kopulationsorgane von *Misumena vatia* (Clerck, 1757) (Arachnida: Araneae: Thomisidae). Verhandlungen des naturwissenschaftlichen Vereins Hamburg (NF) 27, 383–403.

Lopez, A. 1987. Glandular aspects of sexual biology. In: Nentwig W (ed): Ecophysiology of Spiders. Springer Verlag, Berlin, pp. 121–132.

Lopez, A. & Juberthie-Jupeau, L. 1982. Structure et ultrastructure du bulbe copulateur chez la mygale *Nemesia caementaria* (Latreille, 1798) (Araneae, Ctenizidae). Bulletin de la Société d'Etude des Sciences Naturelles de Béziers NS 8(49), 12–19.

Lopez, A. & Juberthie-Jupeau, L. 1985. Ultrastructure comparée du tube séminifère chez les males d'araignées. Mémoires de biospéléologie 12, 97–109.

Lucas, S. & Bücherl, W. 1965. Importância dos órgãos sexuais na sistemática das aranhas. I. Variação interpopulacional dos receptáculos seminais em *Actinopus crassipes* (Keyserling) 1891, Actinopodidae Sul-Americanas. Memorias do Instituto Butantan 32, 89–94.

Markow, T. A. & Ankney, P. F. 1988. Insemination reaction in *Drosophila*: found in species whose males contribute material to oocytes before fertilization. Evolution 42, 1097–1101.

Masumoto, T. 1993. The effect of the copulatory plug in the funnel-web spider, *Agelena limbata* (Araneae: Agelenidae). Journal of Arachnology 21, 55–59.

Mayr, E. 1963. Animal Species and Evolution. Cambridge, MA: Harvard University Press.

Melchers, M. 1963. Zur Biologie und zum Verhalten von *Cupiennius salei* (Keyserling), einer amerikanischen Ctenide. Zoologisches Jahrbuch für Systematik 91, 1–90.

Mendez, V. 2002. Comportamiento sexual y dinámica de población en *Leucauge mariana* (Araneae: Tetragnathidae). Thesis, MSc, Universidad de Costa Rica, San José.

Merret, P. 1963. The palpus of male spiders of the family Linyphiidae. Proceedings of the Zoological Society, London 140, 347–467.

Michalik, P., Haupt, J. & Alberti, G. 2004. On the occurrence of coenospermia in mesothelid spiders (Araneae: Heptathelidae). Arthropod Structure and Development 33, 173–181.

Michalik, P., Reiher, W., Tintelnot-Suhm, M., Coyle, F. A. & Alberti, G. 2005. Female genital system of the folding-trapdoor spider *Antrodiaetus unicolor* (Hentz, 1842) (Antrodiaetidae, Araneae): ultrastructural study of form and function with notes on reproductive biology of spiders. Journal of Morphology 263, 284–309.

Milasowszky, N., Buchar, J. & Zulka, K. P. 1999. Morphological variation in *Pardosa maisa* Hippa & Mannila 1982 (Araneae, Lycosidae). Senckenbergiana biologica 79, 11–18.

Miller, J. A. 2007a. Review of erigonine spider genera in the Neotropics (Araneae: Linyphiidae, Erigoninae). Zoological Journal of the Linnean Society 149, suppl. 1, 1–263.

Miller, J. A. 2007b. Repeated evolution of male sacrifice behavior in spiders correlated with genital mutilation. Evolution 61, 1301–1315.

Millidge, A. F. 1991. Further linyphiid spiders (Araneae) from South America. Bulletin of the American Museum of Natural History 205, 1–199.

Montgomery, T. H. 1903. Studies on the habits of spiders, particularly those of the mating period. Proceedings of the Academy of Natural Sciences Philadelphia 55, 59–151, plates 4, 5.

Müller, H.-G. 1985. Abgebrochene Emboli in der Vulva der "Schwarzen Witwe" *Latrodectus geometricus* C. L. Koch 1841 (Arachnida: Araneae: Theridiidae). Entomologische Zeitschrift 95, 27–30.

Muniappan, R. & Chada, H. L. 1970. Biology of the crab spider, *Misumenops celer*. Annals of the Entomological Society of America 63, 1718–1722.

Nessler, S. H., Uhl, G. & Schneider, J. M. 2007a. Genital damage in the orb-web spider *Argiope bruennichi* (Araneae: Araneidae) increases paternity success. Behavioral Ecology 18, 174–181.

Nessler, S. H., Uhl, G. & Schneider, J. M. 2007b. A non-sperm transferring genital trait under sexual selection: an experimental approach. Proceedings of the Royal Society B 274, 2337–2342.

Opell, B. D. 1979. Revision of the genera and tropical American species of the spider family Uloboridae. Bulletin of the Museum of Comparative Zoology 148, 443–549.

Opell, B. D. 1983. The female genitalia of *Hyptiotes cavatus* (Araneae: Uloboridae). Transactions of the American Microscopical Society 102, 97–104.

Osterloh, A. 1922. Beiträge zur Kenntnis des Kopulationsapparates einiger Spinnen. Zeitschrift für wissenschaftliche Zoologie 119, 326–421.

Patel, B. H. & Bradoo, B. L. 1986. Observations on sperm induction, courtship, and mating behavior of *Uloborus ferokus* Bradoo (Araneae: Uloboridae). International Congress of Arachnology 9 (Panama), 181–191.

Peck, S. B. & Shear, W. A. 1987. A new blind cavernicolous *Lygromma* (Araneae, Gnaphosidae) from the Galapagos Islands. Canadian Entomologist 119, 105–108.

Pérez-Miles, F. 1989. Variación relativa de caracteres somaticos y genitales en *Grammostola mollicoma* (Araneae, Theraphosidae). Journal of Arachnology 17, 263–274.

Peschke, K. 1978. Funktionsmorphologische Untersuchungen zur Kopulation von *Aleochara curtula* Goeze (Coleoptera, Staphylinidae). Zoomorphologie 89, 157–184.

Peschke, K. 1979. Tactile orientation by mating males of the staphylinid beetle, *Aleochara curtula*, relative to the setal fields of the female. Physiological Entomology 4, 155–159.

Piel, W. H. 2001. The systematics of neotropical orb-weaving spiders in the genus *Metepeira* (Araneae: Araneidae). Bulletin of the Museum of Comparative Zoology 157, 1–92.

Platnick, N. I. 1975. A revision of the palpimanid spiders of the new subfamily Otiothopinae (Araneae, Palpimanidae). American Museum Novitates 2562, 1–32.

Platnick, N. I. 1989. A revision of the spider genus *Segestrioides* (Araneae, Diguetidae). American Museum Novitates 2940, 1–9.

Platnick, N. I. 2007. The world spider catalog, version 8.0. American Museum of Natural History, online at http://research.amnh.org/entomology/spiders/catalog/index.html

Platnick, N. I. & Shadab, M. 1980. A revision of the spider genus *Cesonia* (Araneae: Gnaphosidae). Bulletin of the American Museum of Natural History 165, 335–386.

Ramirez, M. G., Achekian, A. C., Coverley, C. R., Pierece, R. M., Eiman, S. S. & Wetkowski, M.

M. in press. The success of copulations with multiple females by males and the commonness of the post-mating epignal plug in the lynx spider *Peucetia viridans* (Araneae, Oxyopidae). Journal of Arachnology.

Ramirez, M. G., Wight E. C., Chirikian V. A., Escobedo, E. S., Quezada, L. K., Antu Schamberger, A., Kagihara, J. A. & Hoey, C. L. 2009. Evidence for multiple paternity in broods of the green lynx spider Peucetia viridans (Araneae: Oxyopidae). Journal of Arachnology 37, 375–378.

Ramirez, M. J. 1999. New species and cladistic reanalysis of the spider genus *Monapia* (Araneae, Anyphaenidae, Amaurobioidinae). Journal of Arachnology 27, 415–431.

Ramirez, M. J. 2003. The spider subfamily Amaurobioidinae (Araneae, Anyphaenidae): A phylogenetic revision at the generic level. Bulletin of the American Museum of Natural History 277, 1–262.

Ramos, M., Irschick, D. J. & Christenson, T. E. 2004. Overcoming an evolutionary conflict: Removal of a reproductive organ greatly increases locomotor performance. Proceedings of the National Academy of Sciences USA 101, 4883–4887.

Ramos, M., Coddington, J. A., Christenson, T. E. & Irschick, D. J. 2005. Have male and female genitalia coevolved? A phylogenetic analysis of genitalic morphology and sexual size dimorphism in web-building spiders (Araneae: Araneoidea). Evolution 59, 1989–1999.

Rezác, M. 2007. The first record of traumatic insemination in Chelicerata. 17th International Congress of Arachnology, Brazil 2007, Abstracts, p. 139.

Robinson, M. H. 1982. Courtship and mating behavior in spiders. Annual Review of Entomology 27, 1–20.

Robinson, M. H. & Robinson, B. 1973 Ecology and behavior of the giant wood spider *Nephila maculata* (Fabricius) in New Guinea. Smithsonian Contributions to Zoology 149, 1–76.

Robinson, M. H. & Robinson, B. 1978. The evolution of courtship systems in tropical araneid spiders. Symposium of the zoological Society of London 42, 17–29.

Robinson, M. H & Robinson, B. 1980. Comparative studies on the courtship and mating behaviour of tropical araneid spiders. Pacific Insect Monographs 36, 1–218.

Robson, G. C. & Richards, O. W. 1936. *The Variation of Animals in Nature*. London Longmans, Green & Co.

Rovner, J. S. 1966. Courtship in spiders without prior sperm induction. Science 152, 543–544.

Rovner, J. S. 1967. Copulation and sperm induction by normal and palpless male linyphiid spiders. Science 157, 835.

Rovner, J. S. 1971. Mechanisms controlling copulatory behavior in wolf spiders (Araneae: Lycosidae). Psyche 78, 150–165.

Sadana, G. L. 1972. Mechanics of copulation in *Lycosa chaperi* Simon (Araneida: Lycosidae). Bulletin of the British Arachnological Society 2, 87–89.

Sakai, M., Taoda, Y., Mori, K., Fujino, M. & Ohta, C. 1991. Copulation sequence and mating termination in the male cricket *Gryllus bimaculatus* Degeer. Journal of Insect Physiology 37, 599–615.

Santos, A. J. & Brescovit, A. D. 2003. A revision of the Neotropical species of the lynx spider genus *Peucetia* Thorell 1869 (Araneae: Oxyopidae). Insect Systematics and Evolution 34, 95–116.

Scharff, N. & Coddington, J. 1997. A phylogenetic analysis of the orb-weaving spider family Araneidae (Arachnida, Araneae). Zoological Journal of the Linnean Society 120, 355–434.

Schneider, J. M., Thomas, M. L. & Elgar, M. A. 2001. Ectomised conductors in the golden orb-web spider, *Nephila plumipes* (Araneoidea): a male adaptation to sexual conflict? Behavioral Ecology and Sociobiology 49, 410–415.

Schneider, J. M., Fromhage, L. & Uhl, G. 2005a. Copulation patterns in the golden orb-web spider *Nephila madagascariensis*. Journal of Ethology 23, 51–55.

Schneider, J. M., Fromhage, L. & Uhl, G. 2005b. Extremely short copulations do not affect hatching success in *Argiope bruennichi* (Araneae, Araneidae). Journal of Arachnology 33, 163–169.

Schult, J. & Sellenschlo, U. 1983. Morphologie und Funktion der Genitalstrukturen bei *Nephila* (Arach., Aran., Araneidae). Mitteilungen aus dem Hamburger zoologischen Museum und Institut 80, 221–230.

Schulmeister, S. 2001. Functional morphology of the male genitalia and copulation in lower Hymenoptera, with special emphasis on the Tenthredinoidea s. str. (Insecta, Hymenoptera, 'Symphyta'). Acta Zoologica (Stockholm) 82, 331–349.

Scudder, G. G. E. 1971. Comparative morphology of insect genitalia. Annual Review of Entomology 16, 379–406.

Senglet, A. 2004. Copulatory mechanisms in *Zelotes*, *Drassyllus* and *Trachyzelotes* (Araneae, Gnaphosidae) with additional faunistic and taxonomic data on species from Southwest Europe. Mitteilungen der Schweizerischen Entomologischen Gesellschaft 77, 87–119.

Sierwald, P. 1983. Morphological criteria and the discrimination of species of the genus *Thalassius* Simon, 1885 (Arachnida: Araneae: Pisauridae). Verhandlungen des naturwissenschaftlichen Vereins in Hamburg (NF) 26, 201–209.

Sierwald, P. 1990. Morphology and homologous features in the male palpal organ in Pisauridae and other spider families, with notes on the taxonomy of Pisauridae (Arachnida: Araneae). Nemouria 35, 1–59.

Silva D., D. 2003. Higher-level relationships of the spider family Ctenidae (Araneae: Ctenoidea). Bulletin of the American Museum of Natural History 274, 1–86.

Silva-D., D. 2007. *Mahafalytenus*, a new spider genus from Madagascar (Araneae, Ctenidae). Proceedings of the California Academy of Sciences 58, 59–98.

Simmons, L. W. 2001. Sperm Competition and its Evolutionary Consequences. Princeton NJ: Princeton University Press.

Snodgrass, R. E. 1935. Principles of Insect Morphology. McGraw-Hill, New York.

Snow, L. S. E. & Andrade, M. C. B. 2004. Patterns of sperm transfer in redback spiders: implications for sperm competition and male sacrifice. Behavioral Ecology 15, 785–792.

Snow, L.S.E & Andrade, M.C.B. 2005. Multiple sperm storage organs facilitate female control of paternity. Proceedings of the Royal Society London B 272, 1139–1144.

Snow, L.S.E., Abdel-Mesih, A. & Andrade, M.C.B. 2006. Broken copulatory organs are low-cost adaptations to sperm competition in redback spiders. Ethology 112, 279–389.

Song, H. 2006. Systematics of Cyrtacanthacridinae (Orthoptera: Acrididae) with a focus on the genus *Schistocerca* Stål 1873: Evolution of locust phase polyphenism and study of insect genitalia. PhD Thesis, Ohio State University.

Stratton, G. E., Hebets, E. A., Miller, P. R. & Miller, G. L. 1996. Pattern and duration of copulation in wolf spiders (Araneae, Lycosidae). Journal of Arachnology 24, 186–200.

Suhm, M., Thaler, K. & Alberti, G. 1995. Glands in the male palpal organ and the origin of the mating plug in *Amaurobius* species (Araneae: Amaurobiidae). Zoologischer Anzeiger 234, 191–199.

Suzuki, H. 1995. Fertilization occurs internally in the spider *Achaearanea tepidariorum* (C. Koch). Invertebrate Reproduction and Development 28, 211–214.

Thomas, R. H. & Zeh, D. W. 1984. Sperm transfer and utilization strategies in arachnids: ecological and morphological constraints. Pp 179–221 in: R. L. Smith (ed.) Sperm Competition and the Evolution of Animal Mating Systems. Acad. Press, Orlando etc.

Tuxen, S. L. (ed.) 1970. Taxonomist's Glossary of Genitalia in Insects. 2nd edition. Munksgaard, Copenhagen.

Uhl, G. 1994. Genital morphology and sperm storage in *Pholcus phalangioides* (Fuesslin) (Pholcidae; Araneae). Acta Zoologica, Stockholm 75, 13–25.

Uhl, G. 2000. Two distinctly different sperm storage organs in female *Dysdera erythrina* (Araneae: Dysderidae). Arthropod Structure and Development 29, 163–169.

Uhl, G. 2002. Female genital morphology and sperm priority patterns in spiders (Araneae). In: Toft, S., Scharff, N. (eds.) European Arachnology 2000, Aarhus Univ. Press, Aarhus, pp. 145–156.

Uhl, G. & Vollrath, F. 1998. Genital morphology of *Nephila edulis*: implications for sperm competition in spiders. Canadian Journal of Zoology 76, 39–47.

Uhl, G., Nessler, S. & Schneider, J. 2007. Copulatory mechanism in a sexually cannibalistic spider with genital mutilation (Araneae: Araneidae: *Argiope bruennichi*). Zoology 110, 398–408.

Useta, G., Huber, B. A. & Costa, F. G. 2007. Preliminary data on spermathecal morphology and sperm dynamics in the female *Schizocosa maliciosa* (Araneae: Lycosidae).

European Journal of Entomology 104, 777–785.

van Helsdingen, P. J. 1965. Sexual behaviour of *Lepthyphantes leprosus* (Ohlert) (Araneida, Linyphiidae), with notes on the function of the genital organs. Zoologische Mededelingen 41, 15–42.

van Helsdingen, P. J. 1969. A reclassification of the species of *Linyphia* Latreille based on the functioning of the genitalia (Araneida, Linyphiidae). Part I. *Linyphia* Latreille and *Neriene* Blackwell. Zoologische Verhandelingen 105, 1–303.

van Helsdingen, P. J. 1971. The function of genitalia as a useful taxonomic character. Arachnologorum Congressus Internationalis V. Brno 123–128.

von Helversen, O. 1976. Gedanken zur Evolution der Paarungsstellung bei den Spinnen (Arachnida: Araneae). Entomologica Germanica 3, 13–28.

Ware, A. D. & Opell, B. D. 1989. A test of the mechanical isolation hypothesis in two similar spider species. Journal of Arachnology 17, 149–162.

Watson, P. J. 1991. Multiple paternity as genetic bet-hedging in female sierra dome spiders, *Linyphia litigiosa* (Linyphiidae). Animal Behaviour 41, 343–360.

Weiss, I. 1981. Der Kopulationsmechanismus bei *Nesticus cibiensis* n. sp., einer neuen Höhlenspinne aus Rumänien (Arachnida: Araneae: Nesticidae). Reichenbachia 19, 143–152.

Whitehouse, M. E. A. & Jackson, R. R. 1994. Intraspecific interactions of *Argyrodes antipodiana*, a kleptoparasitic spider from New Zealand. New Zealand Journal of Zoology 21, 253–268.

Whitman, D. W. & Loher, W. 1984. Morphology of male sex organs and insemination in the grasshopper *Taeniopoda eques* (Burmeister). Journal of Morphology 179, 1–12.

Wiehle, H. 1953. Spinnentiere oder Arachnoidea (Araneae) IX: Orthognatha-Cribellatae-Haplogynae-Entelegynae (Pholcidae, Zodariidae, Oxyopidae, Mimetidae, Nesticidae). In: Dahl F (ed) Tierwelt Deutschlands 42, viii + 150pp. Jena: G. Fischer.

Wiehle, H. 1961. Der Embolus des männlichen Spinnentasters. Verhandlungen der deutschen zoologischen Gesellschaft (1960), 457–480.

Wiehle, H. 1967a. *Meta*—eine semientelegyne Gattung der Araneae (Arach.). Senckenbergiana biologica 48, 183–196.

Wiehle, H. 1967b. Steckengebliebene Emboli in den Vulven von Spinnen (Arach., Araneae). Senckenbergiana biologica 48,197–202.

Wood, D. M. 1991. Homology and phylogenetic implications of male genitalia in Diptera. The ground plan. Proceedings of the Second International Congress of Dipterology (eds. Weismann, N. Orszagh, G. & Pont, A.), Pp. 255–284. The Hague.

13

Genitalic Evolution in Opiliones

ROGELIO MACÍAS-ORDÓÑEZ, GLAUCO MACHADO,
ABEL PÉREZ-GONZÁLEZ, AND JEFFREY W. SHULTZ

MORPHOLOGY AND SYSTEMATICS OF OPILIONES

The Opiliones, usually known in English as harvestmen or daddy longlegs, are the third largest group in the class Arachnida, with nearly 6,000 described species (Machado et al. 2007). Harvestmen are a common and nearly ubiquitous component of terrestrial environments, being found in all continents, except Antarctica, from the equator to subpolar regions. They occur in a great variety of habitats in most terrestrial ecosystems, including soil, moss, leaf litter, under rocks, stones, and debris, on vertical surfaces from tree trunks to stone walls, among grassy clumps, and high vegetation. Although some species are widely distributed and can be found in a wide range of habitats, many are much more limited in geographic distribution and habitat use, especially in tropical areas. Some species are restricted to caves, and others occur in very specific microhabitats, such as nests of leaf-cutter ants, (see Curtis & Machado 2007).

The harvestman body is compact and has two main parts, an anterior prosoma (or cephalothorax) and a posterior opisthosoma (or abdomen), with a broad and sometimes poorly defined juncture. This body plan contrasts with that of spiders, whip spiders, whip scorpions, and certain other arachnids in which the prosoma and

opisthosoma are separated by a distinct constriction. The prosoma bears the chelicerae, pedipalps, and four pairs of legs, with the second typically elongated and used as a sensory appendage (Shultz & Pinto-da-Rocha 2007). The dorsal plate of the prosoma, the carapace, generally has a pair of median eyes, but visual acuity is likely poor in most groups (see Acosta & Machado 2007). A pair of defensive glands opens laterally on the carapace, a feature unique to the order (Gnaspini & Hara 2007). The genital opening is located ventrally on the second opisthosomal somite, which has shifted anteriorly relative to the dorsal parts, and lies between the last pair of legs (Shultz & Pinto-da-Rocha 2007).

The basic architecture of the male and female reproductive tract has been described by Shultz & Pinto-da-Rocha (2007). The testis and ovary are both U-shaped with a mesodermal gonoduct (sperm duct, oviduct) emerging from each side (figure 13.1). These gonoducts fuse and continue anteriorly as a single duct, which eventually merges with a cuticle-lined tube that travels through either an eversible penis or ovipositor (figure 13.1). The posterior end of the penis and ovipositor attach to the posterior end of a pregenital chamber formed by flexible cuticular walls that often bear accessory glands. The walls of the pregenital chamber have muscles that seem to expand the pregenital lumen and initiate

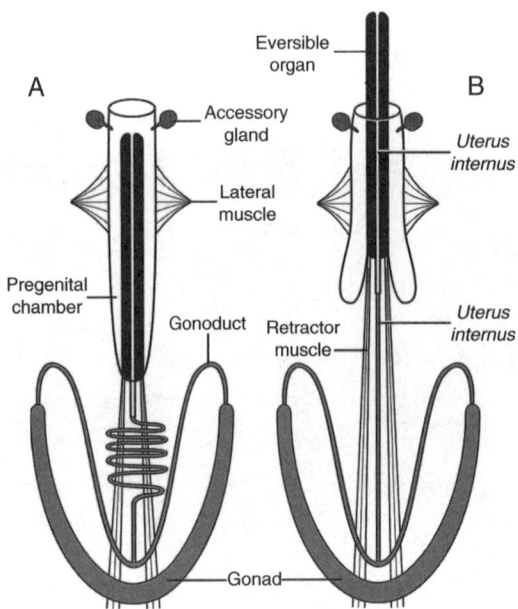

FIGURE 13.1 Schematic representation of the reproductive system of both male and female harvestmen in ventral view showing the eversible organ (A) inverted and (B) everted. Gonad = testis (♂) and ovary (♀); gonoduct = sperm duct (♂) and oviduct (♀); *uterus internus = vas deferens* (♂) and *uterus* (♀); *uterus externus* = propulsive organ + ejaculatory duct (♂) and propulsive organ + vagina (♀); eversible organ = spermatopositor or penis (♂) and ovipositor (♀). Based on de Graaf (1882) and Kästner (1935).

called Dyspnolaniatores (Giribet et al. 1999, 2002, but see Giribet et al. in press). Regardless of the system of higher classification, we recognize 45 families and about 1,500 genera of Opiliones (Giribet & Kury 2007). This diversity is not evenly distributed among the suborders, with Eupnoi and Laniatores comprising together nearly 90% of the species in the order.

Representatives of the suborder Cyphophthalmi are generally small (1 to 3 mm in body length), short-legged, heavily sclerotized inhabitants of soil and caves (Giribet 2007). The suborder Eupnoi comprises two superfamilies: the Phalangioidea, which includes the long-legged forms widely known in the Northern Hemisphere, and the Caddoidea, a small group easily recognized by their huge eyes and spiny pedipalps (Cokendolpher et al. 2007). The suborder Dyspnoi also comprises two super-families, Ischyropsalidoidea and Troguloidea, which are mainly distributed in the Northern Hemisphere (Gruber 2007). The suborder Lania-tores is a diverse lineage of armored harvestmen, typically with large and spiny pedipalps (Kury 2007). This suborder is divided into two infraorders: Insidiatores, comprising the super-families Travunioidea and Triaenonychoidea, and Grassatores, encompassing the superfamilies Zalmoxoidea, Biantoidea, Phalangodoidea, Epedanoidea, and Gonyleptoidea (Giribet & Kury 2007). However, because our goal is to focus on genitalic diversity across suborders, we will tend to avoid details on the relationships of the superfamilies.

eversion (figure 13.1), which is most likely completed by pressure of the hemolymph. Retraction, on the other hand, is accomplished by a pair of muscles that originate at a posterior tergite and attach to the proximal end of the eversible organ (figure 13.1).

Recent phylogenetic studies place Opiliones near the orders Scorpiones, Pseudoscorpiones, and Solifugae, forming a clade called Dromopoda united by synapomorphies associated with the appendages and mouthparts (Shultz 1990, 2007). Currently, there are two hypotheses for the relationship among the suborders of Opiliones (figure 13.2). One proposes that the suborders Eupnoi and Dyspnoi form a monophyletic group called Palpatores (Shultz 1998; Shultz & Regier 2001). The other unites the suborders Dyspnoi and Laniatores in a monophyletic group

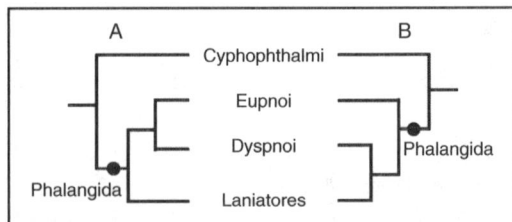

FIGURE 13.2 The two current hypotheses for the relationship among the suborders of Opiliones. One of them (A) proposes that the suborders Eupnoi + Dyspnoi form a monophyletic group called Palpatores (Shultz 1998; Shultz & Regier 2001), whereas the other (B) proposes that Dyspnoi + Laniatores form a monophyletic group called Dyspnolaniatores (Giribet et al. 1999, 2002).

MORPHOLOGY AND FUNCTION OF GENITALIA IN OPILIONES: A MACRO-EVOLUTIONARY PLAY IN FOUR ACTS

Reproductive Strategies: Where We Lay Our Scene

Our knowledge of reproductive strategies in harvestmen has recently been reviewed by Machado & Macías-Ordóñez (2007), based on information available for two species of Cyphophthalmi, five of Eupnoi, four of Dyspnoi, and nine of Laniatores. A few more recent studies have also been considered in this chapter (Willemart et al. 2006; Buzatto & Machado 2008; Nazareth & Machado 2009). In this section, we summarize the reproductive strategies recorded for each suborder, which will provide the context for addressing the evolution of genitalia in Opiliones.

The great majority of harvestmen reproduce sexually, although some species reproduce asexually by parthenogenesis (e.g., Phillipson 1959; Tsurusaki 1986). All sexually reproducing species studied so far are polygynandrous, i.e., both males and females engage in copulation with multiple mates frequently within a single day, throughout their reproductive lives (e.g., Edgar 1971; Mora 1990; Macías-Ordóñez 1997; Buzatto & Machado 2008). Fertilization is internal, and mature spermatozoa are immobile (Morrow 2004). Courtship before intromission is often quick and tactile, but in some cases males may offer a glandular nuptial gift before copulation (Martens 1969). Many studies also mention intense courtship during intromission and mate guarding after copulation (reviewed in Machado & Macías-Ordóñez 2007; see also Nazareth & Machado 2009). Additionally, males of many species defend territories, which are used by females as oviposition sites (Mora 1990; Machado & Oliveira 1998; Macías-Ordóñez 1997; Buzatto & Machado 2008). Given the complexity of the male genitalia and the enormous diversity of sexual dimorphism in the order, sexual selection (be it intra- or intersexual) has most likely played a major role in the evolution of harvestmen, as has been proposed for many other groups (e.g., Hosken & Stockley 2004; Cueva del Castillo & Núñez-Farfán 2008).

Males have been shown to produce spermatophores in at least two genera of Cyphophthalmi, *Cyphophthalmus* and *Stylocellus* (Karaman 2005;

Novak 2005; Schwendinger & Giribet 2005), and this may be the rule throughout the suborder. Roughly half the gametes are allocated to protective layers around the viable sperm, and covered by a mucous secretion from accessory glands of the reproductive tract (Juberthie & Manier 1978). When compared with other arachnids (reviewed in Proctor 1998), spermatophore production in Cyphophthalmi seems to represent a relatively large investment (around 3% of the body volume in *Cyphophthalmus*, based on Fig. 1 from Karaman 2005).

All records on mating in the suborder Eupnoi include copulation by means of fully intromittent male genitalia (reviewed in Machado & Macías-Ordóñez 2007). Copula duration is highly variable and in some cases consists of repeated genital intromissions while the male keeps his long, sexually dimorphic pedipalps clasped to the base of female legs II (occasionally I, III or IV), near the trochanter (e.g., Bishop 1949, 1950; Edgar 1971; Macías-Ordóñez 1997). This form of male grasping has been described for three species of *Leiobunum* (see Machado & Macías-Ordóñez 2007), but it seems to be ubiquitous in Eupnoi (but see Willemart et al. 2006). When copulation occurs, female cooperation seems to be evident in many cases since females are able to reject intromissions by lowering the prosoma against the substrate thus obstructing the entrance to their genital opening (Immel 1955; Edgar 1971; Macías-Ordóñez 1997). Interactions between male genitalia or chelicerae and the female's mouthparts, as well as grasping of the male genitalia by the female, suggest that some kind of nuptial gift is also obtained by the female (Willemart et al. 2006; see also below).

Only five descriptions of copulation in the suborder Dyspnoi exist, but most of them are fairly detailed and show similarities and differences with Eupnoi. In at least one species of *Paranemastoma* (Nemastomatidae, Troguloidea) and two of *Ischyropsalis* (Ischyropsalididae, Ischyropsalidoidea) the bases of the male chelicerae are either offered or somewhat forced into the female's mouth, after which the female obtains a secretion from cheliceral glands (Martens 1969; Meijer 1972). Precopulatory interactions seem to be intense in some species, including male tapping on the female's back, and copulation occurs in a face-to-face position, much as in Eupnoi (Immel 1954; Martens 1969). On the other hand, full intromission in Trogulidae occurs in a belly-to-belly position, and females may be able to reject the male by lowering the anterior end of the body (Pabst 1953). Dyspnoi males seem to rely less on powerful grasping to negotiate with the females and more on precopulatory courtship, including nuptial gifts. Copulatory courtship such as that reported in Eupnoi has been described in all species of Dyspnoi that have been studied, raising the possibility that females may also exert cryptic choice (Pabst 1953; Immel 1954; Martens 1969). The amount and quality of the secretion offered may be the subject of female evaluation and may influence paternity (Machado & Macías-Ordóñez 2007).

In contrast to the Dyspnoi, courtship before intromission is generally quick and tactile in the Laniatores. Courtship during intromission, on the other hand, involve leg tapping and rubbing and may be intense (reviewed in Machado & Macías-Ordóñez 2007; see also Nazareth & Machado 2009). Like the Eupnoi, copulation occurs in a face-to-face position, but in Laniatores males grasp the females' pedipalps, and not their legs. So far, there is no evidence that the male offers any glandular secretion as a nuptial gift (e.g., Miyosi 1941; Matthiesen 1983; Mora 1990; Buzatto & Machado 2008; Willemart et al. 2008; Nazareth & Machado 2009). Oviposition generally occurs immediately after copulation, with the male remaining close to the female, waving his second pair of legs over her and occasionally tapping her legs and dorsum. Mate guarding may last more than 24 hours, during which the male often tries to copulate, and occasionally succeeds. This pattern is especially common in species with maternal care belonging to the subfamily Goniosomatinae (Gonyleptidae), in which a single male can monopolize a harem containing up to six egg-guarding females (e.g., Machado & Oliveira 1998; Buzatto & Machado 2008). In species with paternal care, such as *Zygopachylus albomarginis* (Manaosbiidae), females court egg-guarding males, which sometimes reject females without copulating. As might be expected, *Z. albomarginis* males display behavioral paternity-assurance strategies, including repeated copulations, postcopulatory female guarding, and coercion to lay eggs soon after copulation (Mora 1990).

The (sexual) Characters:
An Ovipositor and a Penis

Most of the research on the reproductive morphology of Opiliones has focused on the structure of the penis and, to a lesser extent, the ovipositor, and inspired primarily by the search for characters of taxonomic or phylogenetic significance. The eversible genitalia of male Cyphophthalmi is often called a penis, but is much shorter than that of other harvestmen (figure 13.3a) and appears to be used in the transfer of spermatophores rather than direct copulation (see below). Consequently, we follow Van der Hammen (1985) in calling this structure a spermatopositor. Unlike the other groups, the spermatopositor has an extensive array of internal muscles and a ring of projections resembling setae (Martens 1986).

The penis of the non-Cyphophthalmi harvestmen, the Phalangida, is typically divided into two main parts: *pars basalis* and *pars distalis*. The *pars basalis* corresponds to most of the long shaft called *truncus*; the *pars distalis* contains the distal end of the *truncus* and the terminal or subterminal glans (figure 13.4). The *pars distalis* is the part that interacts with the ovipositor and is often equipped with spines, sensilla, and other projections, some highly asymmetrical. The glans is the most variable structure of the penis and contains the opening of the *ductus ejaculatorius*, located at the end of the *stylus* (figure 13.4). Typically in Eupnoi and Dyspnoi, the *pars distalis* is composed almost exclusively of a relatively simple glans with an apical *stylus*, with the glans being only slightly differentiated from the *truncus* (figures 13.3B–E; 13.4). The plesiomorphic glans in Laniatores (e.g., Travunioidea) is a little more complex and differentiated from the *truncus* (figure 13.3F). In the remaining Laniatores, the glans is much more complex, with sclerites that vary widely among the families (figures 13.3G,H, 13.4). The sclerites associated with the distal end of the *ductus ejaculatorius* are called *capsula interna* (figures 13.4B–H). Generally the *capsula interna* is surrounded totally or partially by the *capsula externa* formed simply by a soft sac called *follis* (figures 13.4F–H) or by highly modified structures such as the titillators (figure 13.4E) and the *stragulum* (see details in Act IV).

In muscular penes, the movement of the glans relative to the *truncus* is provided by one or two intrinsic muscles that originate from the shaft and insert on a cuticular tendon that ends at the base of the glans. This muscular type of penis occurs in the Eupnoi and Dyspnoi, as well as in the superfamilies Travunioidea and Triaenonychoidea within Laniatores (figures 13.3b–g). Grassatores, on the other hand, have a hydraulic penis, i.e., the intrinsic

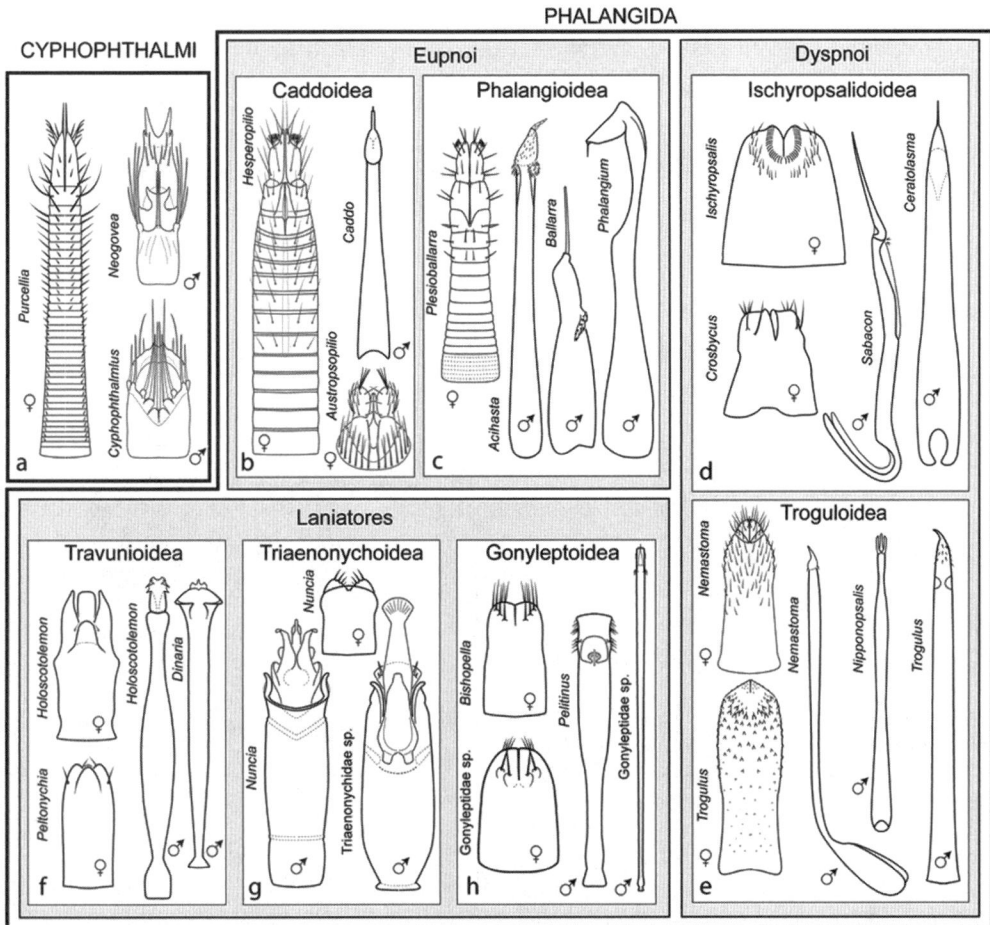

FIGURE 13.3 Male and female genitalia from representatives of the order Opiliones. (a) Cyphophthalmi: *Purcellia*, ovipositor (Hansen & Soerensen 1904); *Neogovea*, *Cyphophthalmus*, spermatopositors (Martens 1986). (b) Caddoidea: *Hesperopilio*, ovipositor (Shultz & Cekalovic 2006); *Austropsopilio*, ovipositor (Shultz & Cekalovic 2003); *Caddo*, penis (Gruber 1975). (c) Phalangioidea: *Pseudoballarra*, ovipositor; *Acihasta*, *Ballarra*, penes (Hunt & Cokendolpher 1991); *Phalangium*, penis (de Graaf 1882). (d) Ischyropsalidoidea: *Ischyropsalis*, ovipositor (Hansen & Soerensen 1904); *Crosbycus*, ovipositor (Shear 1986); *Sabacon*, penis (Martens 1986); *Ceratolasma*, penis (Gruber 1978). (e) Troguloidea: *Nemastoma*, ovipositor (Hansen & Soerensen 1904); *Trogulus*, ovipositor (Martens et al. 1981); *Nemastoma*, *Trogulus*, penes (Hansen & Soerensen 1904), *Nipponopsalis*, penis (Martens & Suzuki 1966). (f) Travunioidea: *Peltonychia*, *Holoscotolemon*, ovipositors (Martens 1986); *Holoscotolemon*, *Dinaria*, penes (Martens 1976). (g) Triaenonychidae: *Nuncia*, ovipositor, penis (Muñoz Cuevas 1972); Triaenonychidae sp., penis (Martens 1976). (h) Gonyleptoidea (representative of the infraorder Grassatores): Gonyleptidae sp., *Bishopella*, ovipositors (Martens et al. 1981); *Peltinus*, Gonyleptidae sp., penes (Martens 1976). All figures redrawn.

muscles are absent and the glans is apparently operated by internal hemolymph pressure (e.g., Gonyleptoidea; figure 13.3h). It is important to note, however, that the terms "muscular" and "hydraulic" refer to the operation of the glans only;

eversion and inversion of the entire penis in all harvestmen are apparently achieved by a combination of muscular and hydraulic mechanisms (Martens 1986; Shultz & Pinto-da-Rocha 2007). Seminal products are apparently pushed through the long

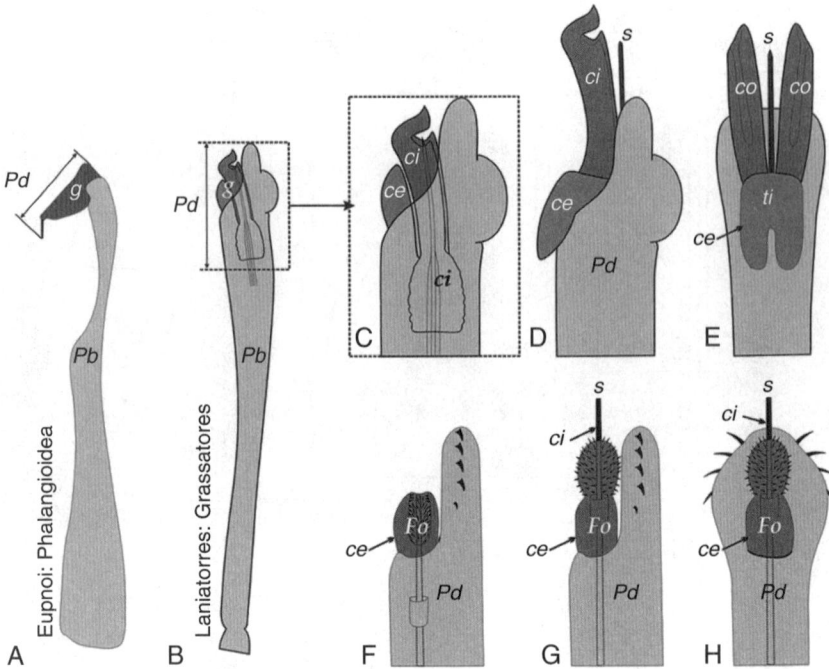

FIGURE 13.4 Schemes contrasting the general morphology of male genitalia in representatives of the suborders (A) Eupnoi and (B) Laniatores. Although the *pars basalis* (*Pb*) are a long shaft in both suborders, the *par distalis* (*Pd*), especially the glans (*g*), is more complex in Laniatores than in Eupnoi. (C) Detail of the *pars distalis* in Laniatores (Biantidae) showing the *capsula interna* (*ci*) and *capsula externa* (*ce*). Pushed by hydraulic pressure, the *capsula interna* is everted exposing the conductors (*co*) and the *stylus* (*s*): (D) lateral view; (E) frontal view. (F) Lateral view of the *pars distalis* of another Laniatores (Assamiidae) showing the *stylus* retracted inside the *capsula externa*. With the eversion of the *follis* (Fo), also promoted by hydraulic pressure, the *stylus* is exposed: (G) lateral view; (H) frontal view.

ejaculatory duct by a muscular propulsive organ located at the base of the penis; this organ is absent in Cyphophthalmi (Kästner 1935).

The ovipositor in Cyphophthalmi and most Eupnoi has a shaft (sometimes as long as or longer than the female body) composed of cuticular rings connected by segmentally arranged muscles (figures 13.5A–C). It ends in paired bilateral processes derived from one or more rings, and the genital opening is located basally between the these processes (figures 13.3a–c, 13.5A,C). Each process generally bears a tuft of sense organs on the latero-subdistal surface (figures 13.3a–c, 13.5A,C). A pair of seminal receptacles (sclerotized vaginal diverticula) is found inside the genital opening, sometimes associated with glands (de Graaf 1882; Martens 1986) (figure 13.5C). Lack of seminal receptacles is often associated with parthenogenesis (Shultz & Pinto-da-Rocha 2007). The ovipositors

of Dyspnoi and Laniatores differ markedly from those of the other suborders (figures 13.3d–h). Their ovipositors are always much shorter than the female body. Additionally, their segmentation is absent or vestigial, a system of circular muscles surrounds the vagina (except in Troguloidea), and vaginal symmetry is biradial (rather than simply bilateral); only in the Laniatores the lumen is X-shaped in cross section (Martens 1986). In contrast to the Eupnoi and Cyphophthalmi, Dyspnoi and Laniatores have one or more sperm receptacles in each of their four lobes, and these receptacles are much shorter (figures 13.5D–F). The end of the ovipositor in Dyspnoi has a pair of valves (figures 13.3d,e), but ends in four lobes in the majority of Laniatores (figures 13.3f–h, 13.5E).

As mentioned in chapter 1, Darwin's original distinction between primary and secondary sexual characters has been blurred by findings in the last

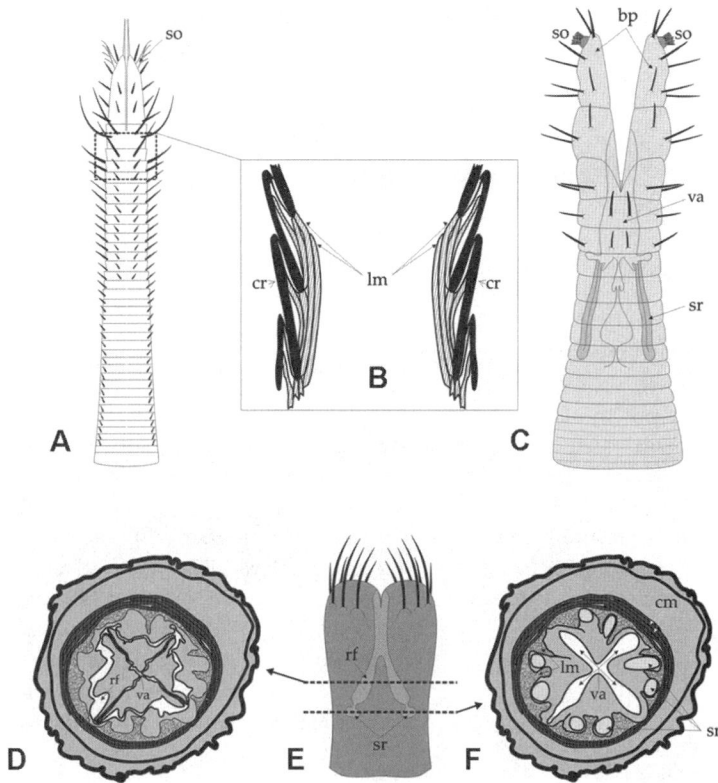

FIGURE 13.5 Schemes contrasting the morphology of ovipositors in representatives of the suborders (A) Cyphophthalmi (*Purcellia* from Hansen & Soerensen 1904), (C) Eupnoi (*Pseudoballarra* from Hunt & Cokendolpher 1991), and (E) Laniatores (*Bishoppella laciniosa* from Martens et al. 1981). Note that the ovipositor of Cyphophthalmi and Eupnoi bears bilateral processes (bp) and tufts of sense organs (so). (B) Longitudinal cut of the distal part of the *truncus* showing the highly sclerotized cuticular rings (cr), the ring folder (rf), and the segmentally arranged longitudinal muscles (lm) (Martens et al. 1981). (D, F) Transversal cuts showing the X-shaped vagina (va), the radial disposition of the seminal receptacles (sr), and circular muscles (cm). All figures redrawn.

few decades on the action of sexual selection (be it by male–male competition or female choice) on the morphology, physiology, and behavior of male and female genitalia. Thus, it is impossible to discuss the evolution of "primary" sexual characters without reference to sexual selection, "secondary" characters, and mating strategies. In order to follow the scope of this book, however, the *characters* in this *play* will be the male and female genitalia, and even their somatic complements (the male and female bodies) will be somewhat seen as their evolutionary context, *the stage*.

Harvestmen are a particularly interesting system to study evolution of sexual characters because they have evolved an intromittent organ independently of other better-studied taxa. Moreover, the penis

and ovipositor in harvestmen are among the largest and most accessible genitalic structures known among terrestrial arthropods and offer exceptional opportunities for exploring the evolutionary factors driving their diversification. The taxonomy of harvestmen at all levels is based largely on the configuration of the penis. The male genitalia of virtually all families has been described at least for one species, and in many cases for several. Nevertheless, no evolutionary, functional or behavioral hypotheses have been put forward to explain the great morphological diversity of male genitalia, and females have been largely ignored due to a widespread assumption that they show little useful diversity. From the standpoint of an evolutionary biologist, this situation is lamentable, but typical for many groups of terrestrial arthropods: genitalic diversity is tapped for its taxonomic information and presented only as illustrations in various publications, where they rest like fossils buried in vast strata of paper. We will attempt to establish the relationship between this vast material and the little behavioral data available. Given the few but important hard facts we do know, the following sections present our hypotheses on the macro-evolutionary pathways of genitalia in Opiliones. Since eversible genitalia is an autapomorphic trait of Opiliones, and since Cyphophthalmi presents a spermatopositor instead of a penis, it is difficult to reconstruct the plesiomorphic states both at the basal node of the order and at the basal node of Phalangida, the clade formed by Eupnoi, Dyspnoi, and Laniatores (figure 13.2). However, the hypotheses presented here make the underlining assumption that the genitalic traits observed in Cyphophthalmi represent the plesiomorphic condition for the order.

Act I: An Almost Intromittent Spermatopositor and a Sensitive Ovipositor

Sperm transfer by means of a spermatophore is probably a basal trait within Opiliones, and within Arachnida (Proctor 1998). Although only a few spermatophores have actually been recorded, the shape of the male genitalia in all Cyphophthalmi suggests it is the rule in this suborder. Spermatophores have been found attached to the female in a way that suggests direct participation of the male. The spermatophore duct is glued to the tip of the ovipositor near the opening of the sperm receptacles (Karaman 2005). The female lacks a genital operculum and the ovipositor is covered with numerous setae, most likely sensitive to micro-conditions of potential oviposition sites (Machado & Macías-Ordóñez 2007). The spermatophore, on the other hand, is a complex structure in which functional sperm is covered by a layer of modified spermatogonia (Alberti 2005).

It is likely that the male produces the spermatophore while interacting with the female, attaches it to her ovipositor by the spermatophore duct using the spermatopositor, and remains there until it solidifies and/or sperm is transferred. Schwendinger & Giribet (2005) reported a male "copulating" with a female in *Fangensis lecrerci* (Stylocellidae), "belly to belly", facing in opposite directions, and taping her anal region with his pedipalps, although no intromission was reported. It is likely that this was a form of pre- or post-spermatoposition courtship, and or mate guarding. Furthermore, this position may also give the female access to the glandular organs in the anal crown and hind legs of the male. As will be detailed below, genital nuptial feeding has been very recently described in Eupnoi, and it may be common in harvestmen. Its origin in male courtship in Cyphophthalmi is a hypothesis worthy of further exploration.

Act II: A Daring Penis Appears on the Scene

The ovipositor of most of the Eupnoi is similar to that of Cyphophthalmi (figures 13.3a–c). Although the ovipositor of most Acropsopilioninae (Caddidae) is short and has only a few segments, its architecture is basically similar to that of other Eupnoi. The male genitalia, however, is strikingly different from the Cyphophthalmi. Without any known intermediate states, an intromittent penis appears in the Phalangida. All known species have it and the oldest fossil record of any intromittent organ is that of the Devonian harvestman *Eophalangium sheari* (Eupnoi) from the Rhynie Chert, Scotland (Dunlop et al. 2003). If intromittent genitalia evolved from a Cyphophthalmi-like state, the evolutionary leap from such a spermatopositor to a penis is not hard to imagine. Direct sperm transfer by means of a penis may have enabled males to increase sperm production by eliminating the material investment in infertile sperm and protective layers of the spermatophore (Machado & Macías-Ordóñez 2007). Genitalic intromission is the rule in the Phalangida, but based on the morphological and behavioral evidence available, there is a great diversity of ways in which male and female genitalia interact.

There are no detailed descriptions of genitalic interaction published for Eupnoi (or any other suborder). Unpublished studies by J. W. Shultz & R. Macías-Ordóñez, in which intense transmitted light was used to reveal internal copulatory events in *Leiobunum verrucosum* (Sclerosomatidae), show that the penis' *stylus* does not go beyond the first third of the ovipositor, where the openings of the sperm receptacles are located, even when the female is immobilized (see Act IV). Thus, although most of the penis shaft enters the female body in *L. verrucosum* and other Eupnoi, it may not penetrate far inside the female reproductive tract. Since the ovipositor is retracted and may bend inside the female, the length of the penis inserted beyond the female's genital operculum does not indicate how far it goes into the ovipositor. This is true for all Opiliones since, unlike other groups with internal fertilization, such as insects or mammals, the entrance to the female reproductive tract is not continuous with the apparent external genital opening. Thus, once inside the females' pregenital chamber, the penis must reach the tip of the ovipositor to enter the female reproductive tract.

Once inside the ovipositor, the shape of the *stylus* seems appropriate to enter the sperm receptacles after going through the ovipositor atrium. Furthermore, the *stylus* (and sometimes the glans) of most Eupnoi has an angle so that it would find the opening to the sperm receptacles and bend when entering them, where it could simply deposit the aflagellate sperm. This reasoning predicts that *stylus* length should be positively correlated to the depth of the sperm receptacles. In figure 13.6 we present original data from nine species of the family Neopilionidae showing such a correlation. Although we controlled for body size, we did not control for possible phylogenetic effects because no generally accepted phylogeny of the family is available.

Another remarkable feature of the Eupnoi penis is a morphologically diverse set of structures (sacs, bulbs, and alae) found at the distal end of the trunk. If the *stylus* was inserted in the sperm receptacles, these structures would fall just outside of the ovipositor, near its tip, potentially in contact with the abundant sensilla found on it. A likely possibility is

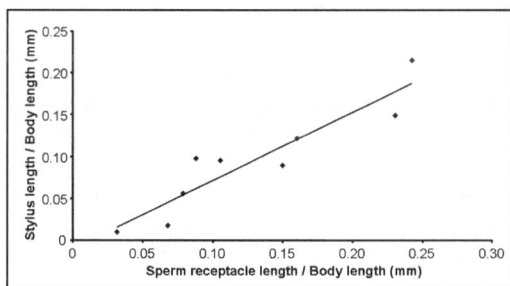

FIGURE 13.6 Positive and significant relation between *stylus* length and sperm receptacle length in nine species of the family Neopilionidae ($y = 0.813x - 0.010$; $R^2 = 0.861$; $p < 0.001$). The length of the structures was divided by body length in order to control for the effect to the size. Data taken from drawings presented in Hunt & Cokendolpher (1991) (figures 5d, f; 8b, d; 12d, e; 14b, d; 17b, d; 18b, d; 20f, g; 22c, f; 24a, c).

that males stimulate these sensilla while accessing the sperm receptacle as a form of copulatory courtship subject to cryptic female choice (Eberhard 1985, 1996). An alternative view, further discussed below, is that males may be exploiting a female sensory bias, by seductively stimulating the ovipositor's sensilla, used by the female to probe optimal sites for egg laying (Machado & Macías-Ordóñez 2007).

Other features of the male genitalia may be relevant to copulation. Several groups within the superfamily Phalangioidea have glands that open at the base of the non-sclerotized section of the everted genitalia. The sacs, bulbs, and alae of the shaft may also serve as "buckets" that convey secretions from the glands to the female. Suggestive evidence of nuptial feeding prior to copulation by means of penis intromission into the female's mouth has recently been reported for the sclerosomatid *L. verrucosum* (Shultz 2005) and the phalangiid *Phalangium opilio* (Willemart et al. 2006). Moreover, unpublished observations by R. Macías-Ordóñez & J. W. Shultz of the sclerosomatid *Leiobunum vittatum* indicate that the female apparently feeds from the base of the male penis during male grasping (figure 13.7A).

Act III: A Demanding Ovipositor Appears on the Scene

The penis of Dyspnoi has the same general ground plan of that found in Eupnoi. The ovipositor, however, is strikingly different from that of Cyphophthalmi and Eupnoi. It is shorter and cuticular segmentation has been lost, although segmental muscles have been retained in some Troguloidea (Martens 1986). Sperm receptacles are smaller and highly variable in number, ranging from one to multiple sacs on each side. In Ischyropsalididae, there are four to ten tube-shaped sperm receptacles, and the number of sensilla at the apex of the ovipositor is greatly reduced (Gruber 2007).

There are detailed records of copulation for two species of the genus *Ischyropsalis* (Martens 1969), showing a few short intromissions (a couple of seconds long) while the female grasps and presumably feeds on the male's cheliceral glands. Although penis microstructures in Dyspnoi have been reported for many species of the seven families that comprise the suborder (see Gruber 2007), detailed genitalic interaction during copulation has never been described. It is evident, however, that the

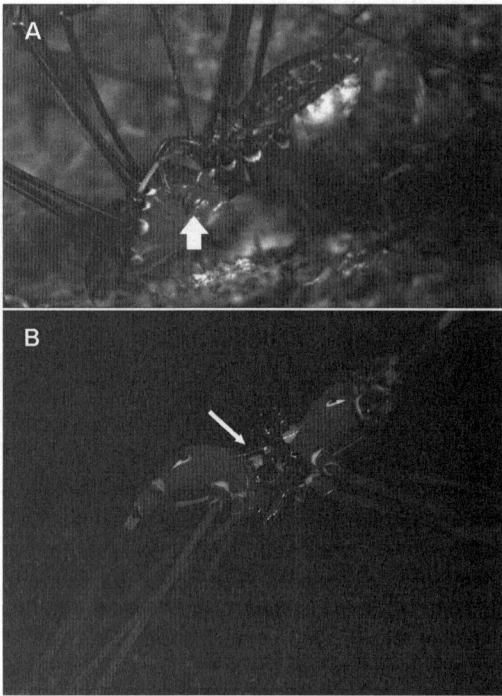

FIGURE 13.7 (A) Female of *Leiobunum vitattum* (Sclerosomatidae, Eupnoi) manipulating the inflated hematodocha of the male using her chelicerae (arrow), probably obtaining a nuptial secretion (photo by Joseph Warfel). (B) Face-to-face copulation in *Acutisoma proximum* (Gonyleptidae, Laniatores). Note that the male grasps the pedipalps of the female using his own pedipalps (photo by Bruno A. Buzatto). The arrow indicates the *truncus* of the penis.

Eupnoi-like penis with its *stylus* does not seem to fit this kind of sperm receptacle. The penis likely leaves the sperm in the lumen of the ovipositor, where the female may have more control of its fate than in Eupnoi. Furthermore, the sensitivity of the ovipositor to the stimulation of the penis may be reduced due to fewer sensilla. However, given the available information, we do not have even a plausible hypothesis on the way the penis and the ovipositor interact in this suborder.

Act IV: A Demanding Ovipositor, a Resourceful Penis

The ovipositor of Laniatores is similar to that of Dyspnoi, but with a narrower atrium and a cross like lumen (figures 13.5D, F). The lumen of the

vagina is X-shaped in cross section; the sperm receptacles are radially oriented, varying widely in position and number (in multipliers of four) (Martens 1986) (figures 13.5D, F). The internal morphology of the ovipositor needs to be better studied in the Travunioidea and in Triaenonychoidea (Giribet & Kury 2007). The design of the penis in Laniatores is clearly different from that of previous groups. It is shorter, not heavily sclerotized, and the morphology of the *pars distalis* is highly variable, particularly the glans, where a set of new sclerites results in many different sizes and forms of the *capsula interna* and *externa*. In the incredibly complex male genitalia of the Fissiphalliidae, for instance, the *capsula externa* bears a ventral plate modified in two tagmata: a rounded *pergula* and a spade-like *rutrum* (Pinto-da-Rocha 2007) (figures 13.8A, B). The *capsula externa* in this family also exhibits modifications, such as a rigid *stragulum*, which is articulated to the *truncus* like a jackknife (Pinto-da-Rocha 2007) (figures 13.8A, B). In Escadabiidae, the *capsula interna* is very wide and can bear modified structures called conductors (Kury & Pérez-González 2007), which are generally sclerotized, blade-like plates located dorsolaterally at the *stylus* (figures 13.8C, D; see also figures 13.4C–E).

Even though the modified structures of the *capsula interna* and *externa* show very different arrangements among the families of Laniatores (see Martens 1976, 1986), they seem to play three main roles. First, they attach the *par distalis* of the penis at the distal end of the ovipositor where the seminal receptacles are located (figure 13.5F). Probably some of the strong setae and sensillae perform this function. Moreover, the morphology of the *pergula* suggests that this structure hampers a deep intromission inside the ovipositor (figure 13.8A, B); the same may be the case of the ventral keel in the *pars distalis* of escadabiids (figures 13.8C, D). Second, the modified structures of the *capsula interna* and *externa* seem to promote penetration of the penis in the ovipositor, as in the acute, spade-like profile of the *stragulum* and *rutrum* of the fissiphalliids (figure 13.8A), also observed in the profile of the *pars distalis* in escadabiids (figure 13.8C). Third, they may open the narrow X-shaped vagina, thus allowing sperm deposition inside the lumen. Some examples are the *stragulum* in fissiphalliids (figure 13.8B), the blade-like conductors in some biantids (figure 13.4E), or even the spiny *follis* in assamiids (figures 13.4F–H). Once inside the

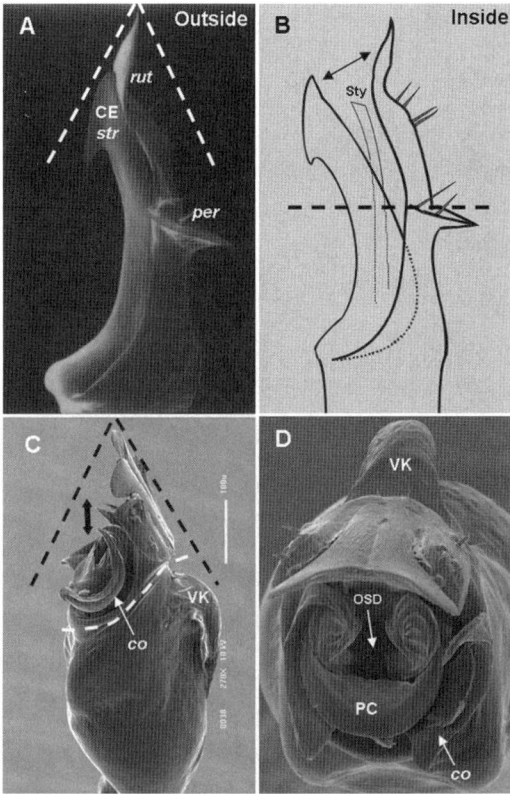

FIGURE 13.8 (A) Lateral view of the *pars distalis* in the male genitalia of the Fissiphaliidae *Fissiphallius martensi*. The *stragalum* (*str*) and the blade-like *rutrum* (*rut*) probably facilitate the penetration in the female reproductive tract when they are closed. The *pergula* (*per*), on the other hand, is likely to hamper a deep intromission inside the ovipositor. (B) Once inside the narrow lumen of the ovipositor (vagina), the *stragalum* opens by hydraulic pressure and expose the *stylus*, which will release sperm in the ovipositor lumen. The dark dashed line shows the putative limit of the penis intromission. (C) Lateral view of the *pars distalis* in the male genitalia of the Escadabiidae *Baculigerus* sp. showing its acuminated profile. In this species, a deep intromission inside the ovipositor is probably prevented by a ventral keel (VK), and the white dashed line shows the putative limit of the penis intromission. The dark arrow indicates that the *capsula interna* bearing the opening of the sperm duct (OSD) is everted by hydraulic pressure. (D) The parastylar collar (PC) and probably a pair of conductors (*co*) help to open space inside the lumen of the ovipositor.

narrow lumen of the ovipositor (vagina), these structures may open or be everted by hydraulic pressure and expose the *stylus*, which will release sperm in the lumen. Like in the Dyspnoi, the vagina of the Laniatores has a set of ringed muscles that allow constriction of the lumen. Thus, the sperm deposited in the lumen may fill up the multiple sperm receptacles when it closes, probably pressed by the circular muscles, when the penis retracts. Such a muscular system may also enable females to reject sperm if the entrance to the sperm receptacles may be obstructed. This hypothesis, however, has yet to be tested in future studies. Although the ovipositors of Laniatores show fewer sensilla as compared to the other suborders, the male genitalia possess a set of more proximal and highly variable and ornamented structures (unlike anything else in the other suborders) that seem in an ideal position to stimulate these sensilla. The females' sensilla invaginate when the penis enters, thus maintaining contact with the ornaments of the penis. In contrast to the Eupnoi, in which the male constantly moves the body and penis during intromission, in the gonyleptid *Acutisoma proximum*, for instance, males are motionless during copulation (Buzatto & Machado 2008) (figure 13.7B). We speculate that whereas Eupnoi males find the opening of the seminal receptacle in order to release sperm, Laniatores males leave their sperm in the lumen, and it enters passively into multiple sperm receptacles.

No detailed description of genitalic interaction between male and female in Laniatores exists, but morphological evidence suggests that the penis may not venture far inside the female reproductive tract. In some families, such as Escadabiidae, the sclerotized base of the *pars distalis* is wider than the ovipositor, thus making the penetration below the penis tip impossible, in which case sperm deposition would be restricted to the apical section of the ovipositor. This may have resulted in male genitalic structures capable of removing sperm, as has been described in many other arthropod groups with intromittent genitalia (see references in Kamimura 2000). A handful of species in six distantly related subfamilies of Gonyleptidae have evolved a structure that could serve that function, the ventral process (figure 13.9). This structure has probably evolved independently at least in some of these subfamilies, based on their phylogenetic relationships (Pinto-da-Rocha 2002; Kury & Pinto-da-Rocha 2007). Although no information is available on the actual role of the ventral process

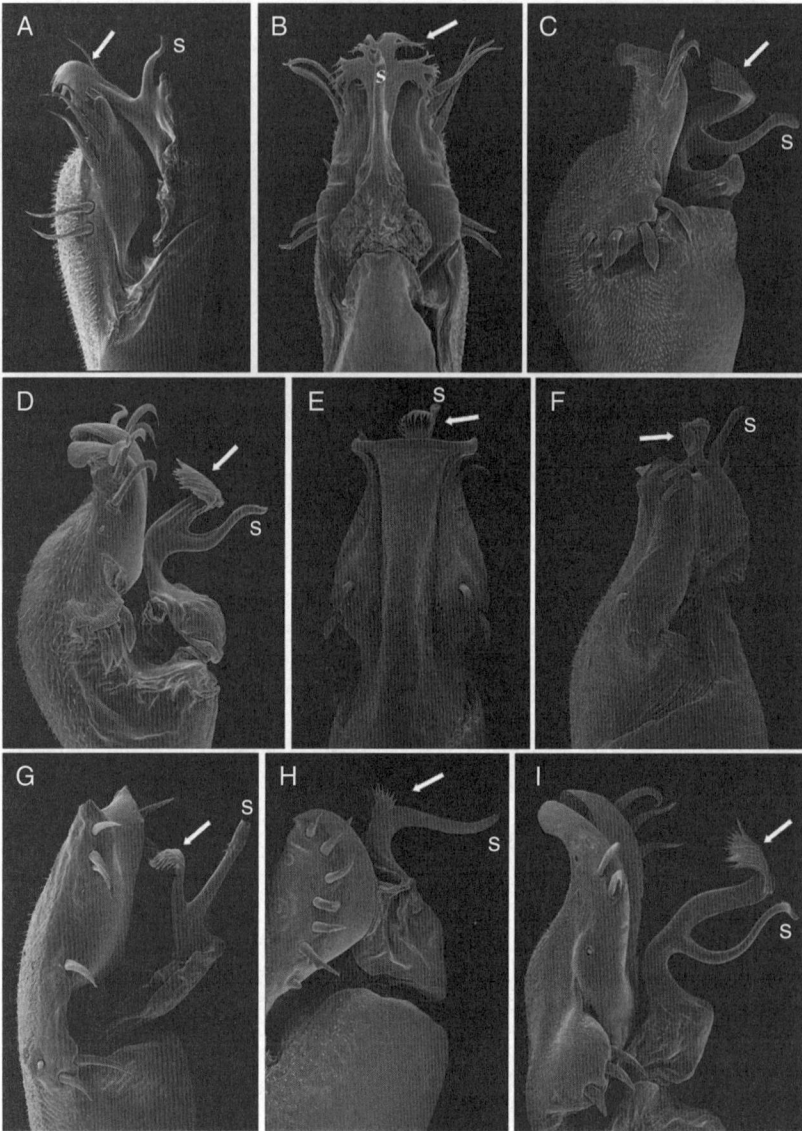

FIGURE 13.9 Male genitalia showing the ventral process (arrow) in representatives of six non-closely related subfamilies: (A) *Bourguyia* sp. and (B) *Asarcus ingenuus* (Bourguyiinae); (C) *Acrogonyleptes unus* (Hernandariinae); (D) *Geraecormobius nanus* (Gonyleptinae); (E) *Promitobates hatschbachi* and (F) *P. ornatus* (Mitobatinae); (G) *Discocyrtus testudineus* and (H) *Metagyndes pulchella* (Pachylinae); and (I) *Stygnobates barbiellinii* (Sodreaninae). Note that the shape and relative position of the ventral process in relation to the *stylus* (s) is remarkably similar among these subfamilies. The ventral part of the penis is always at the left side, except in (E), which is a ventral view of the genitalia.

during intromission, its shape and relative position is remarkably similar among these subfamilies and suggests that it may penetrate the lumen of the ovipositor along with the *stylus*, smoothly "brushing" the inner walls on the way in, but scraping off the same surface on the way out. Both the *stylus* and the ventral process are hardened by hydraulic pressure, thus potentially giving the male some extra control to maneuver these structures inside the female ovipositor (vagina). The process of sperm removal may explain why copulation in gonyleptids may last up to 20 min in some species (B. A. Buzatto & G. Machado, unpublished data).

A MICRO-EVOLUTIONARY SCENARIO: EVOLUTION OF MALE GENITALIC DIVERSITY IN *LEIOBUNUM*

Male Genitalic Diversity

The subfamily Leiobuninae (Sclerosomatidae, Eupnoi) has an essentially Holarctic distribution with major centers of known diversity in North America (especially the Appalachian Region and southeastern U.S.A.), Mesoamerica, East Asia, and the Europe-Mediterranean Region (e.g., Suzuki 1976; Martens 1978; Cokendolpher & Lee 1993). In this section male genitalic diversity in American leiobunine harvestmen is briefly summarized, some details of mating behavior are reported for the first time, and testable hypotheses—based on observations of genitalia in action—are proposed. We focus on male genitalic diversity within leiobunines of eastern North America (*Leiobunum*, *Nelima*, *Hadrobunus*, *Eumesosoma*), with about 30 species. The male genital apparatus comprises a tubular,

sclerotized penis and basal, membranous hematodocha (figures 13.10 to 13.12). The penis has a long shaft that articulates with a short terminal glans, with the tendon of an intrinsic penial muscle operating the shaft–glans joint. The hematodocha is a bag of flexible cuticle that is inflated under hemolymph pressure during mating. The hematodocha acts as a flexible turret for the everted penis and forms the walls of the pregenital chamber when the genital apparatus is withdrawn (figures 13.10 and 13.11).

Penial diversity has been categorized into two broad and probably non-monophyletic groups—sacculate and lanceolate (McGhee 1970, 1977)—based on subterminal modifications of the shaft (figures 13.10 and 13.11). The sacculate condition is plesiomorphic based on outgroup comparisons, with the subterminal apparatus consisting of bilaterally paired, distally opened, chitinous sacs (figures 13.10A and 13.11A). In the inverted state, a pair of glandular papillae at the base of the hematodocha enters the openings to the chitinous sacs (figure 13.11D). The lanceolate condition encompasses a diverse assemblage of penis types united by the absence of sacs. Four North American lanceolate groups have been recognized thus far, the *L. calcar* (figure 13.10C), *L. vittatum* (figure 13.10D), *L. formosum* and *Hadrobunus* species groups, as well as *L. holtae* (J. W. Shultz, unpublished data).

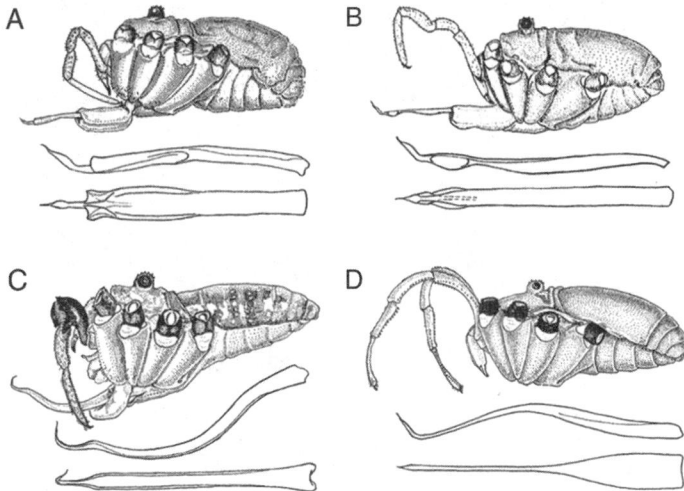

FIGURE 13.10 Representative males from four *Leiobunum* species illustrating genitalic diversity. Each figure depicts a lateral view of a male and enlarged lateral and dorsal views of the penis. (A) *L. aldrichi* with sacculate penis. (B) *L. politum* with bulbate penis. (C) *L. calcar* with lan-ceolate penis and robust, clasping pedipalps. (D) *L. speciosum* with lanceolate penis and elongate, spiny pedipalps. All figures after Bishop (1949).

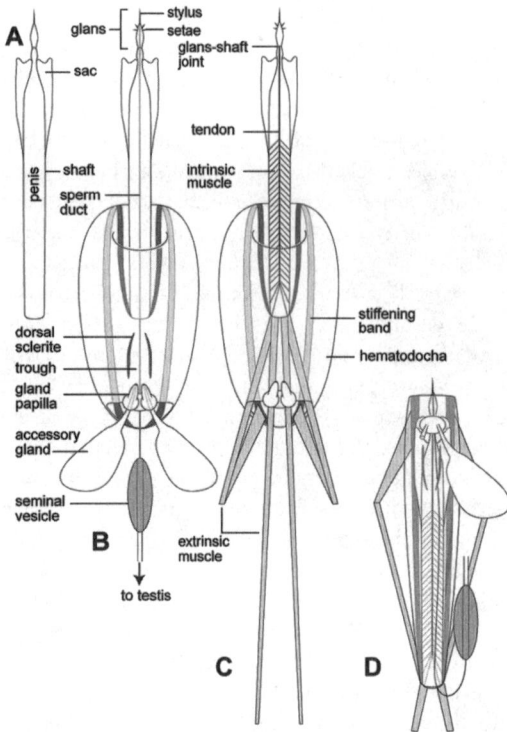

FIGURE 13.11 Diagrammatic dorsal views of the genital apparatus of the primitively sacculate species, *Leiobunum verrucosum*. (A) Penis. The sclerotized portion of the genital apparatus and the only part traditionally described by systematists. (B) Everted genital apparatus emphasizing the cuticular elements, genital tract and accessory glands. (C) Everted genital apparatus emphasizing cuticular elements and muscles. (D) Inverted genital apparatus. Note that the gland papillae are inserted into the penial sacs.

Mating Behavior in a Primitively Sacculate Species

In an attempt to understand the broader significance of penial diversity in leiobunines, Shultz (2005) examined mating behavior in several *Leiobunum* species but focused on a primitively sacculate form, *L. verrucosum*—formerly *L. nigripes* (Shultz 2008). Mating in ten virgin pairs was recorded. In five cases, the female was immobilized by a stick glued to her dorsum to allow detailed inspection of male–female interactions. The latter treatment did not appear to affect the order of mating events, although the male was more likely to mate multiple times as the female could not adopt the "face down" rejection posture (Macías-Ordóñez 2000).

The sequence of events is summarized in figure 13.12 and the following numbers correspond to those in the figure. Upon contact with a male, (1) a receptive female oriented to the male and (2) opened her stomotheca ("mouth") by inflating the membranous portion of the coxapophyses ("lips") of the pedipalps and leg I. Unreceptive females faced the substratum and/or turned away from the male. (3) The male moved rapidly toward the female, face to face, and (4) everted his penis. (5) The male used his pedipalps to clasp the female behind the coxae of her second pair of legs and (6) inserted his penis rapidly into the female's stomotheca, removed it and placed it at the opening of the female pregenital chamber. Flexible walls likely ensure that penial sac contents are deposited in the female's stomotheca upon withdrawal. Some males did not insert the penis into the female's stomotheca. (7) A variable interval was spent with the penis tip positioned just within or at the opening of the

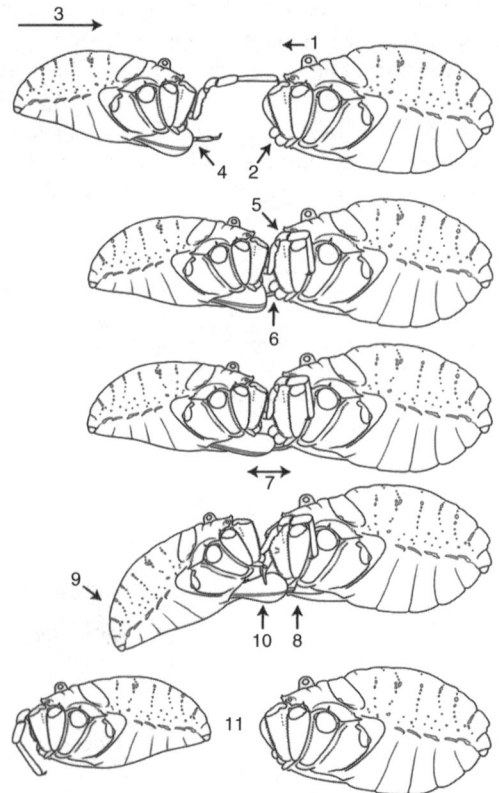

FIGURE 13.12 Summary of events during mating in the sacculate species, *Leiobunum verrucosum*. The events indicated by numbers are described in the text.

pregenital chamber, with the dorsal surface of the hematodocha forming a conduit from basal nuptial glands to the female's mouth. The female manipulated the dorsal surface of male's hematodocha with her chelicerae. (8) After a variable interval (several minutes), the penis gained entry into the female pregenital chamber and (9) the male assumed a "face up" position, exposing the gland papillae at the base of the hematodocha. (10) A variable interval was spent with the female feeding on nuptial secretions and with the male making deep penetrations into the female pregenital chamber, apparently copulating. (11) The male withdrew his penis and departed. No postcopulatory mate guarding was observed nor has been reported in this species. The female sometimes groomed her ovipositor following copulation.

These observations debunk the widespread supposition that *Leiobunum* mate indiscriminately without precopulatory courtship (e.g., Bishop 1949; Edgar 1971), and traces the original error to a remarkable similarity between copulation and the preceding close-contact "courtship" involving the male genitalia. The primitive sacculate genital apparatus apparently serves a dual role as a genital organ and as a delivery mechanism for a male-generated nuptial gift, with the gift being offered in three phases: an initial delivery by the subterminal sacs directly to the female's open stomotheca, a second delivery via a stream of secretion traveling along the dorsum of the hematodocha, and, finally, full exposure of the gland papillae to the female that coincides with the female granting the male access to her pregenital chamber.

Mating Behavior in a Derived Lanceolate Species

A long-term field study of *Leiobunum vittatum* in the eastern Pennsylvania (U.S.A.) has revealed that suitable substrate for oviposition is limited to cracks in rocks and fallen logs, which males actively patrol and fight for with other males (Macías-Ordóñez 1997, 2000). Mating pairs and ovipositing females may be found from late August to early November, when the first frosts kill all the adults in the population. When a female encounters a rock, she slowly goes over the whole surface, inserting the ovipositor inside cracks, and probing potential sites before laying one or many eggs. This behavior is impossible when they encounter a male, since on contact the male eagerly attempts to grasp her using his pedipalps as described above. The female may reject intromission, but grasping seems harder to avoid. If the female escapes grasping, however, she usually must abandon the rock to avoid the male, thus abandoning also the opportunity to find a suitable oviposition substrate. If copulation proceeds, however, a series of short repeated intromissions will take place for a period of a few minutes up to about one hour.

As described in other Eupnoi, between intromissions, the female seems to obtain some sort of "nuptial" secretion from the base of the male genitalia using her chelicera (R. Macías-Ordóñez & J. W. Shultz, unpublished data, figure 13.7A). In fact, the female actively strokes the male around his genital operculum with her pedipalps while his genitalia is not everted seemingly to stimulate genitalic eversion. After close examination, it is clear that most of the times the male everts his genitalia, no intromission takes place but the female always seems to obtain something from the glands at the base of the *truncus*. However, in contrast to *L. verrucosum*, insertion of the penis into the female's stomotheca has never been observed. Once male grasping is over, the male will guard the female by wrapping one or two female legs at the femur or tibia with the terminal tarsi of his own first pair of legs and following her while she walks around. The female seems free to probe the rock at will, undisturbed not only by this male, but by any other male, since the guarding male will aggressively expel any other approaching male. A male will stop guarding in this fashion only when the female abandons the territory. Thus, by accepting copulation a female gains the opportunity to have full and harassment-free access to the scarce rock cracks where the eggs may safely spend the winter. When the end of the short reproductive season is near, females may encounter and copulate with two or three males within a few hours in their search for oviposition substrates. The protection offered by a resident male may be worth taking and may imply yet stronger selection for cryptic mechanisms to influence the paternity of her eggs (Machado & Macías-Ordóñez 2007).

What Factors Could be Driving Male Genital Diversity in *Leiobunum*?

The loss of penial sacs in several leiobunine groups may represent elimination of the first phase of a primitive gift-delivery sequence. Because sac loss is typically associated with novel clasping mechanisms in the male pedipalp (see figures 13.10C, D), the appearance of lanceolate penes may reflect an evolutionary movement along the strategic spectrum from female enticement toward coercion. Significantly, in species in which male anatomy has departed most from the primitive condition (e.g., L. calcar and L. speciosum), females show modifications of the pregenital opening that appear to either exclude or entrap the penis, suggesting that the lanceolate male strategy has promoted evolutionary responses in female morphology. These emerging trends are consistent with a scenario in which male genitalic innovation is driven by sexual conflict, resulting in a possible sexual arms race. In this race, females would have an advantage if they are able to exert cryptic choice after receiving nutritional gifts and or harassment protection by their mates.

It is important to note that the lanceolate penis in Leiobunum appears to be a phenomenon limited primarily to temperate regions; virtually all tropical species examined thus far are sacculate (J. W. Shultz, unpublished data). Furthermore, within the temperate region sacculate species occur in the milder southern regions or, farther north, overwinter as subadults and gain sexual maturity in late spring. There appears to be only one exception (i.e., L. aldrichi) (figure 13.10A), which matures in summer. In contrast, lanceolate species attain sexual maturity in mid to late summer or even later. The theme emerging from these observations is that sacculate species prosper in regions with long reproductive seasons, during which males and females can gather resources for gifts and eggs, respectively. In climate/life-history combinations where young adults have a high probability of abundant future resources, the fitness value of the nuptial gift to the female may be minor compared to that of resources she can gather herself, and the cost of losing a small gift to an unreceptive female may be relatively inexpensive to the male. However, as the duration of resource availability for egg and gift production decreases, the value of the nuptial gift to the female and its potential costs to the male may increase. Under these conditions, male gifts may have a greater impact on the total number of eggs the female can produce. For the male, the cost of giving limited gifts to an unreceptive female may be high, thus leading selection to favor the elimination of the sac-borne initial gift in favor of mechanisms that increase the probability of a *quid pro quo*

exchange of gift and copulation, as in the case of L. *vittatum*. If this scenario is correct, then it would appear to represent a case where male genital diversification has been driven to a large extent by natural selection for efficient use of nutritional resources. Research devoted to testing these proposals is ongoing.

HYPOTHESES OF GENITALIC EVOLUTION

Origin Versus Current Function of Sexual Characters

Any trait may appear and then continue to perform its original function, or it may perform functions completely different from those that provided its initial selective advantage. It is hard to talk about origin without a robust phylogeny in which characters are mapped and ancestral states are identified. Inferring the relative phylogenetic timing of origin, however, does not necessarily reveal original, function although correlation with other traits may help to infer such ancestral functions. Function, however, may also be simply defined as "what the trait does" without any assessment of its contribution to fitness (proximate function). Alternatively, function may imply "how a trait contributes to fitness", i.e., the ultimate function (see discussion in Coddington 1988). From an evolutionary standpoint, maintenance of the trait must be discussed based on current function, but only if "function" is defined as contribution to fitness.

Evolutionary biologists frequently address the potential fitness advantage of a trait by making several assumptions about actual functional morphology, which are rarely corroborated. This has often been the case when debating "function" of genitalic traits (e.g., Eberhard 2004b), most likely because the taxonomic literature provides abundant information on genital morphology, but none on how it works. Such is the case for genitalia in Opiliones. Furthermore, the most useful structures in taxonomy are sclerotized, but soft parts are frequently not depicted. When they are depicted, they are frequently collapsed by preservation or in a resting position, at best. Given that the base of the genitalia in Opiliones usually has a strong hydraulic component, we lack even the morphological information to infer proximate function of these structures. Thus we have been forced, as most everyone else, to speculate on function based on a small data set of behavioral observations, and a larger base of morphological information. It is our hope that such speculation may stimulate empirical research both on proximate function and fitness consequences of genital morphology in this group.

Our main hypothesis for the origin and initial function of the male genitalia is that it appeared as a spermatophore placement structure (a spermatopositor) in the Cyphophthalmi, and then evolved into an intromittent organ (a penis), a function retained in the rest of the suborders. Insemination seems to range from dynamic in Eupnoi to passive in Laniatores. While insemination in Eupnoi seems to demand intense movement of the penis due to a long, highly mobile and flexible ovipositor with a single entrance to a single pair of sperm receptacles, a shorter and less flexible ovipositor with multiple small receptacles in Dyspnoi and Laniatores suggests a more passive copula, as insemination requires simply opening the ovipositor and leaving sperm in the lumen.

Besides taking male gametes closer to female gametes, it seems reasonable that, as it is the case in other animal groups (examples in Eberhard 1985), the penis has a stimulatory and/or coercive function. In the case of the female ovipositor as a terminal organ of the reproductive tract, it may seem obvious that it appeared as an egg-laying structure, capable of inserting eggs deep in the substrate. The female ovipositor may have incorporated a secondary function as a discriminatory organ in charge of screening male stimulatory performance. Such secondary function is puzzling in the suborders Dyspnoi and Laniatores; ovipositor evolution may have been driven by natural selection acting on egg-laying strategies or by sexual selection as a consequence of an arms race between males and females. The implications of ovipositor reduction in size and sensibility due to sexual selection would be profound. Since short and less sensitive ovipositors would not be able to explore deep crevices and assess proper conditions for egg development, females would lay eggs on the substrate and in some cases remain nearby and brood them. Therefore, cryptic female choice and male sperm competition would have been the forces behind changes in egg-laying structures and, consequently, oviposition and brooding strategies (see discussion in Machado & Macías-Ordóñez 2007).

Sperm Competition, Female Choice and Sexual Conflict

Genitalic function involves many and diverse morphological, physiological, and behavioral traits. Penes and ovipositors are trait arrays subject simultaneously to selection on the whole array and on single traits or sub-arrays somewhat independently. In Opiliones, as in many other groups, one of the main selective forces on these arrays is the morphology and behavior of the opposite sex during copulation. The relative importance of sperm competition versus cryptic female choice driving the evolution of primary sexual characters has been the focus of some debate (see chapter 1). Sperm competition (intrasexual selection among sperm or other male products from different males inside the female reproductive tract) seems ubiquitous among polyandrous species, and Opiliones are unlikely to be an exception. The selective power of sperm competition on male genitalia is undisputed and may be the source of much evolutionary change in Opiliones. However, sperm competition has never been studied in any harvestman species.

Antagonistic coevolution (*chase-away*) and positive-feedback coevolution (Fisherian *run-away* and good-genes) have also been contrasted in the last decade, as competing models in the evolution of sexual characters (Holland & Rice 1998; Hosken & Stockley 2004). This debate has been fueled by mixing arguments of origin and function. Many male genitalic traits may have imposed fitness costs to females when they appeared, but may currently be under Fisherian selection. Kokko et al. (2003) have suggested that these and other models of sexual selection by female choice are not mutually exclusive and may represent two sides of the same coin, or two stages during the evolution of sexually selected traits. Once the male and the female are interacting inside the female reproductive tract, the female has more control over the fate of the male's gametes, even more so once the male has withdrawn his genitalia (Eberhard 1996). The elaborate and highly diverse male genitalia of the Opiliones could be expected under a process of cryptic female choice in which females select males based on their stimulatory abilities (Machado & Macías-Ordóñez 2007). A sexual conflict chase-away scenario may also be imagined in which males may "seduce" females by stimulating their ovipositors and inducing the female to mate suboptimally

(Arnqvist & Nilsson 2000). These scenarios are impossible to tell apart without accounting for all costs and benefits involved for both sexes, which are likely to be context dependent. Furthermore, the mating systems of some harvestman species are known to differ sharply between populations, probably due to context dependent cost–benefit relations. Male *Leiobunum vittatum*, for instance, defend mating territories and guard females after copulation in eastern Pennsylvania, U.S.A. (Macías-Ordóñez 1997), but not in Michigan (Edgar 1971), 1,000 kilometers away. Different selective pressure by sperm competition, harassment of females by males, and cryptic female choice are most likely operating in each of these populations. Such inter-population differences in the mating system are well documented in other animal groups (Emlen & Oring 1977; Kokko & Rankin 2006).

Genitalic evolution in harvestmen apparently involved an initial phase of enlargement, then change into an intromittent organ, a general tendency to reduction in ovipositor size and sensitivity, followed by an increase in penis complexity. Given that the entrance to the female reproductive tract is not attached to the genital opening, the penis probably does not go beyond the sperm receptacles, located very near the tip of the ovipositor retracted inside the female body. Male strategies for direct insemination and enhanced female stimulation may have coevolved with female strategies that restrict accessibility to the seminal receptacles by means of ovipositor reduction, loss of sensibility, and even promotion of sperm competition, observed now in the form of male structures that probably remove sperm from previous males from the female reproductive tract (Eberhard 1996; Birkhead & Møller 1998).

The juxtaposition of the genital opening and the mouth, so that the female mouth is close to the base of the male penis, has probably promoted the evolution of "genital nuptial feeding" in Eupnoi, sometimes involving the penis. In such cases, the "primary" sexual organ of Opiliones may be under sexual selection via female choice. Females may also select for size of penial receptacles containing nuptial gifts and or stimulation on sensilla at the tip of their ovipositor. Nuptial gifts in Dyspnoi, however, do not come from the male genitalia, but from different glands located on the chelicerae.

As suggested by Hosken and Stockley (2004), exploring the mechanisms of genitalic evolution is a fertile ground to test predictions derived from different sexual selection models. The Eupnoi penis may have originated as a way to impose insemination on the female by getting the gametes closer to the sperm receptacles, and even "seducing" the ovipositor by exploiting a sensory bias on its sensilla. In time, females seem to respond by restricting access to receptacles by means of shorter, constraining and less sensitive ovipositors, thus turning them into penis-screening devices, which now must not only court the ovipositor but even feed the female. It may be common in other taxa that a male trait originates in a sexual conflict context (as suggested in the section of male genitalic diversity in *Leiobunum*), falling later in a female-screening evolutionary process. Given their abundance, diversity and relatively large genitalia, Opiliones seem to be great candidates to shed light on this and other evolutionary mechanisms. Overall, we hope to motivate colleagues to challenge our hypotheses and pursue research in this fascinating group.

Acknowledgments We are very grateful to Janet Leonard and Alex Córdoba for the invitation to write this chapter, to Jonathan Coddington for a critical review of the manuscript, to Joe Warfel and Bruno Buzatto for providing the photos presented here as figures 13.7A and 13.7B, respectively, to Ricardo Pinto da Rocha for providing the photos presented here as figures 13.8A and 13.9A–I, and to Adriano Kury for providing the photos presented here as figures 13.8C–D and also for valuable discussions about the morphology and evolution of male genitalia in harvestmen. RMO is supported by Instituto de Ecologia, A.C., GM has a research grant from Fundação de Amparo à Pesquisa do Estado de São Paulo (FAPESP, no. 02/00381-0), APG is supported by a post-doctoral fellowship of the Conselho Nacional de Desenvolvimento Científico e Tecnológico (CNPq, no. 155524/2006-2), and JWS is supported by NSF grant DEB-0640179 and the Maryland Agricultural Experiment Station.

REFERENCES

Acosta, L. E. & Machado, G. 2007. Diet and foraging. In: Harvestmen: The Biology of Opiliones (Ed. by R. Pinto-da-Rocha;

G. Machado & G. Giribet), pp. 309–338. Massachusetts: Harvard University Press.

Alberti, G. 2005. Double spermatogenesis in Chelicerata. Journal of Morphology, 266, 281–297.

Arnqvist, G. & Nilsson, T. 2000. The evolution of polyandry: multiple mating and female fitness in insects. Animal Behaviour, 60, 145–164.

Birkhead, T. R. & Møller, A. P. 1998. Sperm Competition and Sexual Selection. San Diego: Academic Press.

Bishop, S. C. 1949. The Phalangida (Opiliones) of New York. Proceedings of the Rochester Academy of Science, 9, 159–235.

Bishop, S. C. 1950. The life of a harvestman. Nature Magazine, 43, 264–267, 276.

Buzatto, B. A. & Machado, G. 2008. Resource defense polygyny shifts to female defense polygyny over the course of the reproductive season of a neotropical harvestman. Behavioral Ecology and Sociobiology, 63, 85–94.

Coddington, J. A. 1988. Cladistic tests of adaptational hypotheses. Cladistics, 4, 3–22.

Cokendolpher, J. C. & Lee, V. F. 1993. Catalogue of the Cyphopalpatores and Bibliography of the Harvestmen (Arachnida, Opiliones) of Greenland, Canada, USA, and Mexico. Privately published.

Cokendolpher, J. C., Tsurusaki, N., Tourinho, A. L., Taylor, C. K., Gruber, J. & Pinto-da-Rocha, R. 2007. Taxonomy: Eupnoi. In: Harvestmen: The Biology of Opiliones (Ed. by R. Pinto-da-Rocha, G. Machado & G. Giribet), pp. 108–131. Massachusetts: Harvard University Press.

Cueva del Castillo, R. & Núñez-Farfán, J. The evolution of sexual size dimorphism: the interplay between natural and sexual selection. Journal of Orthoptera Research, 2008 17: 197–200.

Curtis, D.J. & Machado, G. 2007. Ecology. In: Harvestmen: The Biology of Opiliones (Ed. by R. Pinto-da-Rocha, G. Machado & G. Giribet), pp. 280–308. Massachusetts: Harvard University Press.

de Graaf, H. W. 1882. Sur la Construction des Organes Genitaux des Phalangiens. Leiden: E. J. Brill.

Dunlop, J. A., Anderson, L. I., Kerp, H. & Hass, H. 2003. Preserved organs of Devonian harvestmen. Nature, 425, 916–916.

Eberhard, W. G. 1985. Sexual Selection and Animal Genitalia. Massachusetts: Harvard University Press.

Eberhard, W. G. 1996. Female Control: Sexual Selection by Cryptic Female Choice. Princeton: Princeton University Press.

Eberhard, W. G. 2004a. Male–female conflict and genitalia: failure to confirm predictions in insects and spiders. Biological Review, 79, 121–186.

Eberhard, W. G. 2004b Rapid divergent evolution of sexual morphology: Comparative tests of antagonistic coevolution and traditional female choice. Evolution, 58, 1947–1970.

Edgar, A. L. 1971. Studies on the biology and ecology of Michigan Phalangida (Opiliones). Miscellaneous Publications Museum of Zoology, University of Michigan, 144, 1–64.

Emlen, S. T. & Oring, L. W. 1977. Ecology, sexual selection, and the evolution of mating systems. Science, 197, 215–223.

Giribet, G. 2007. Taxonomy: Cyphophthalmi. In: Harvestmen: The Biology of Opiliones (Ed. by R. Pinto-da-Rocha, G. Machado & G. Giribet), pp. 89–108. Massachusetts: Harvard University Press.

Giribet, G. & Kury, A. B. 2007. Phylogeny and biogeography. In: Harvestmen: The Biology of Opiliones (Ed. by R. Pinto-da-Rocha, G. Machado & G. Giribet), pp. 62-87. Massachusetts: Harvard University Press.

Giribet, G., Rambla, M., Carranza, S., Riutort, M., Baguñà, J. & Ribera, C. 1999. Phylogeny of the arachnid order Opiliones (Arthropoda) inferred from a combined approach of complete 18S, partial 28S ribosomal DNA sequences and morphology. Molecular Phylogenetics and Evolution, 11, 296–307.

Giribet, G., Edgecombe, G. D., Wheeler, W. C. & Babbitt, C. 2002. Phylogeny and systematic position of Opiliones: a combined analysis of chelicerate relationships using morphological and molecular data. Cladistics, 18, 5–70.

Giribet, G., Vogt, L., Perez-Gonzalez, A., Sharma, P. & Kury, A. B. In press. A multilocus approach to harvestman (Arachnida: Opiliones) phylogeny with emphasis on biogeography and the systematics of Laniatores. Cladistics. Published on line: 10.1111/j.1096-0031.2009.0029.x

Gnaspni, P. & Hara, M. R. 2007. Defense mechanisms. In: Harvestmen: The Biology of Opiliones (Ed. by R. Pinto-da-Rocha, G. Machado & G. Giribet), pp. 374–399. Massachusetts: Harvard University Press.

Gruber, J. 1975. Bemerkungen zur Morphologie und systematischen Stellung von *Caddo*,

Acropsopilio und verwandter Formen (Opiliones, Arachnida). Annalen des Naturhistorischen Museums in Wien, 78, 237–259.

Gruber, J. 1978. Redescription of *Ceratolasma tricantha* Goodnight and Goodnight, with notes on the family Ischyropsalidae (Opiliones, Palpatores). Journal of Arachnology, 6, 105–124.

Gruber, G. 2007. Taxonomy: Dyspnoi. In: Harvestmen: The Biology of Opiliones (Ed. by R. Pinto-da-Rocha, G. Machado & G. Giribet), pp. 131–159. Massachusetts: Harvard University Press.

Hansen, H. J. & Sørensen, W. 1904. On Two Orders of Arachnida: Opiliones, Especially the Suborder Cyphophthalmi, and Ricinulei, Namely the Family Cryptostemmatoidae. Cambridge: Cambridge University Press.

Holland, B. & Rice, W. R. 1998. Chase-away sexual selection: antagonistic seduction versus resistance. Evolution, 52, 1–7.

Hosken, D. J. & Stockley, P. 2004. Sexual selection and genital evolution. Trends in Ecology and Evolution, 19, 87–93.

Hunt, G. S. & Cokendolpher, J. C. 1991. Ballarrinae, a new subfamily of harvestmen from the Southern Hemisphere (Arachnida, Opiliones, Neopilionidae). Records of the Australian Museum, 43, 131–169.

Immel, V. 1954. Zur Biologie und Physiologie von Nemastoma quadripunctatum (Opiliones, Dyspnoi). Zoologische Jahrbücher, Abteilung für Systematik, 83, 129–184.

Immel, V. 1955. Einige Bemerkungen zur Biologie von Platybunus bucephalus (Opiliones, Eupnoi). Zoologische Jahrbücher, Abteilung für Systematik, 83, 475–484.

Juberthie, C. & Manier, J. F. 1978. Étude ultrastructurale comparée de la spermiogènese des opilions et son intérêt phylétique. Symposium of the Zoological Society of London, 42, 407–416.

Kamimura, Y. 2000. Possible removal of rival sperm by the elongated genitalia of the earwig, *Euborellia plebeja*. Zoological Science, 17, 667–672.

Karaman, I. M. 2005. Evidence of spermatophores in Cyphophthalmi (Arachnida, Opiliones). Revue suisse de Zoologie, 112, 3–11.

Kästner, A. 1935. Opiliones Sundevall. Weberknechte. In: Handbuch der Zoologie.

Eine Naturgeschichte der Stamme des Tierreiches III, 2 Hälfte, 1 Teil (Ed. by W. Kukenthal), pp. 300–393. Berlin: Walter de Gruyter & Co.

Kokko, H. & Rankin, D. J. 2006. Lonely hearts or sex in the city? Density-dependent effects in mating systems. Proceedings of the Royal Society of London B, 361, 319–334.

Kokko, H., Brooks, R., Jennions, M. D. & Morley, J. 2003. The evolution of mate choice and mating biases. Proceedings of the Royal Society of London B, 270, 653–664.

Kury, A. B. 2007. Taxonomy: Laniatores. In: Harvestmen: The Biology of Opiliones (Ed. by R. Pinto-da-Rocha, G. Machado & G. Giribet), pp. 159–246. Massachusetts: Harvard University Press.

Kury, A. B. & Pérez González, A. 2007. Taxonomy: Laniatores, Escadabiidae. In: Harvestmen: The Biology of Opiliones (Ed. by R. Pinto-da-Rocha, G. Machado & G. Giribet), pp. 191–194. Massachusetts: Harvard University Press.

Kury, A. B. & Pinto-da-Rocha, R. 2007. Taxonomy: Laniatores, Gonyleptidae. In: Harvestmen: The Biology of Opiliones (Ed. by R. Pinto-da-Rocha, G. Machado & G. Giribet), pp. 196–203. Massachusetts: Harvard University Press.

Machado, G. & Macías-Ordóñez, R. 2007. Reproduction. In: Harvestmen: The Biology of Opiliones (Ed. by R. Pinto-da-Rocha, G. Machado & G. Giribet), pp. 414–454. Massachusetts: Harvard University Press.

Machado, G. & Oliveira, P. S. 1998. Reproductive biology of the neotropical harvestman Goniosoma longipes (Arachnida, Opiliones: Gonyleptidae): mating and oviposition behaviour, brood mortality, and parental care. Journal of Zoology, 246, 359–367.

Machado, G., Pinto-da-Rocha, R. & Giribet, G. 2007. What are harvestmen? In: Harvestmen: The Biology of Opiliones (Ed. by R. Pinto-da-Rocha, G. Machado & G. Giribet), pp. 1–13. Massachusetts: Harvard University Press.

Macías-Ordóñez, R. 1997. The mating system of *Leiobunum vittatum* Say, 1821 (Arachnida: Opiliones: Palpatores): resource defense polygyny in the striped harvestman. Ph.D. Thesis, Lehigh University, Bethlehem, PA.

Macías-Ordóñez, R. 2000. Touchy harvestmen. Natural History, 109, 58–61.

Martens, J. 1969. Die Sekretdarbietung während des Paarungsverhaltens von *Ischyropsalis*

C. L. Koch (Opiliones). Zeitschrift für Tierpsychologie, 26, 513–523.

Martens, J. 1976. Genitalmorphologie, System und Phylogenie der Weberknechte (Arachnida: Opiliones). Entomologica Germanica, 3, 51–68.

Martens, J. 1978. Spinnentiere, Arachnida: Weberknechte, Opiliones. In: Die Tierwelt Deutschlands, Vol. 64. (Ed. by K. Senglaub, H.-J. Hannenmann & H. Schumann), pp. 1–464. Jena: Gustav Fischer.

Martens, J. 1986. Die Grossgliederung der Opiliones und die Evolution der Ordnung (Arachnida). In: Proceedings of the 10th International Congress of Arachnology (Ed. by J. A. Barrientos), pp. 289–310. Barcelona: Instituto Pirenaico de Ecología & Grupo de Aracnología.

Martens, J. & Suzuki, S. 1966. Zur systematischen Stellung ostasiatischer Ischyropsalididen-Arten (Arachnoidea, Opiliones, Ischyropsalididae). Annotationes Zoologicae Japonenses, 39, 215–221.

Martens, J., Hoheisel, U. & Götze, M. 1981. Vergleichende Anatomie der Legeröhren der Opiliones als Beitrag zur Phylogenie der Ordnung (Arachnida). Zoologische Jahrbücher, Abteilung für Anatomie und Ontogenie der Tiere, 105, 13–76.

Matthiesen, F. A. 1983. Comportamento sexual de um opilião brasileiro Discocyrtus pectinifemur Mello-Leitão, 1937 (Opiliones: Gonyleptidae). Ciência & Cultura, 35, 1339–1341.

McGhee, C. R. 1970. The sacculate and lanceolate groups of the genus Leiobunum (Arachnida: Phalangida, Phalangiidae) in the eastern United States. Unpublished Doctoral Dissertation, Department of Zoology, Virginia Polytechnic Institute and State University.

McGhee, C. R. 1977. The politum group (bulbate species) of Leiobunum (Arachnida: Phalangida: Phalangiidae) of North America. Journal of Arachnology, 3, 151–163.

Meijer, J. 1972. Some data on the phenology and the activity patterns of Nemastoma lugubre (Müller) and Mitostoma chrysomelas (Herman) (Nemastomatidae: Opilionida: Arachnida). Netherlands Journal of Zoology, 22, 105–118.

Miyosi, Y. 1941. Reproduction and post-embryonic development in the Japanese laniatorid Pseudobiantes japonicus. Acta Arachnologica, 6, 98–107.

Mora, G. 1990. Parental care in a neotropical harvestman, Zygopachylus albomarginis (Arachnida, Opiliones: Gonyleptidae). Animal Behaviour, 39, 582–593.

Morrow, E. H. 2004. How the sperm lost its tail: the evolution of a flagellate sperm. Biological Review, 79, 795–814.

Muñoz-Cuevas, A. 1972. Presencia de la tribo Triaenobunini en Chile. Descripción del nuevo genero y de la nueva especie Americobunus ringueleti (Arachn: Opil: Triaenonychidae). Physis, 31, 1–7.

Nazareth, T. M. & Machado, G. 2009. Reproductive behavior of Chavesincola inexpectabilis (Opiliones: Gonyleptidae), with the description of a new and independently evolved case of paternal care in harvestman. Journal of Arachnology, 37, 127–134.

Novak, T. 2005. Notes on spermatophores in Cyphophthalmus duricorius Joseph (Arachnida: Opiliones: Sironidae). Annales, Series Historia Naturalis, 15, 277–280.

Pabst, W. 1953. Zur Biologie der mitteleuropäischen Troguliden. Zoologische Jahrbücher, Abteilung für Systematik, Ökologie und Geographie der Tiere, 82, 1–156.

Phillipson, J. 1959. The seasonal occurrence, life histories and fecundity of harvest-spiders (Phalangida, Arachnida) in neighborhood of Durham City. Entomologist's Monthly Magazine, 95, 134–138.

Pinto-da-Rocha, R. 2002. Systematic review and cladistic analysis of the Caelopyginae (Opiliones, Gonyleptidae). Arquivos de Zoologia, 36, 357–464.

Pinto-da-Rocha, R. 2007. Taxonomy: Laniatores, Fissiphalidae. In: Harvestmen: The Biology of Opiliones (Ed. by R. Pinto-da-Rocha, G. Machado & G. Giribet), pp. 194–196. Massachusetts: Harvard University Press.

Proctor, H. C. 1998. Indirect sperm transfer in arthropods: behavioral and evolutionary trends. Annual Review of Entomology, 43, 153–174.

Schwendinger, P. J. & Giribet, G. 2005. The systematics of the south-east Asian genus Fangensis Rambla, 1994 (Opiliones: Cyphophthalmi: Stylocellidae). Invertebrate Systematics, 19, 297–323.

Shear, W. A. 1986. A cladistic analysis of the opilionid superfamily Ischyropsalidoidea, with descriptions of the new family Ceratolasmatidae, the new genus *Acuclavella*, and four new species. American Museum of Novitates, 2844, 1–29.

Shultz, J. W. 1990. Evolutionary morphology and phylogeny of Arachnida. Cladistics, 6, 1–38.

Shultz, J. W. 1998. Phylogeny of Opiliones (Arachnida): an assessment of the "Cyphopalpatores" concept. Journal of Arachnology, 26, 257–272.

Shultz, J. W. 2005. Preliminary analysis of mating in *Leiobunum nigripes* (Opiliones) and diversification of male reproductive structures in *Leiobunum*. American Arachnology, 72, 11.

Shultz, J. W. 2007. A phylogenetic analysis of the arachnid orders based on morphological characters. Zoological Journal of the Linnean Society, 150, 221–266.

Shultz, J. W. 2008. *Leiobunum nigripes* is a junior synonym of *Leiobunum verrucosum* (Opiliones, Sclerosomatidae). Journal of Arachnology, 36, 184–186.

Shultz, J. W. & Cekalovic, T. 2003. First species of *Austropsopilio* (Opiliones, Caddoidea, Caddidae) from South America. Journal of Arachnology, 31, 20–27.

Shultz, J. W. & Cekalovic, T. 2006. First species of *Hesperopilio* (Opiliones, Caddoidea, Caddidae) from South America. Journal of Arachnology, 34, 46–50.

Shultz, J. W. & Pinto-da-Rocha, R. 2007. Morphology and functional anatomy. In: Harvestmen: The Biology of Opiliones (Ed. by R. Pinto-da-Rocha, G. Machado & G. Giribet), pp. 14–61. Massachusetts: Harvard University Press.

Shultz, J. W. & Regier, J. C. 2001. Phylogenetic analysis of Phalangida (Arachnida, Opiliones) using two nuclear protein-encoding genes supports monophyly of Palpatores. Journal of Arachnology, 29, 189–200.

Suzuki, S. 1976. The genus *Leiobunum* C.L. Koch of Japan and adjacent countries (Leiobunidae, Opiliones, Arachnida). Journal of Science, Hiroshima University, Series B, Div. 1, 26, 187–260.

Tsurusaki, N. 1986. Parthenogenesis and geographic variation of sex ratio in two species of *Leiobunum* (Arachnida, Opiliones). Zoological Science, 3, 517–532.

van der Hammen, L. 1985. Comparative studies in Chelicerata. III. Opilionida. Zoologische Verhandelingen, 220, 1–60.

Willemart, R. H.; Farine, J. P., Peretti, A. V. & Gnaspini, P. 2006. Behavioral roles of the sexually dimorphic structures in the male harvestman *Phalangium opilio* (Opiliones, Phalangiidae). Canadian Journal of Zoology, 84, 1763–1774.

Willemart, R. H.; Osses, F., Chelini, M. C.; Macías-Ordóñez, R. & Machado, G. 2008. Sexually dimorphic legs in a neotropical harvestman (Arachnida, Opiliones): ornament or weapon? Behavioral Processes, 2009, 80: 51–59.

14

The Evolution of Male and Female Internal Reproductive Organs in Insects

NINA WEDELL AND DAVID J HOSKEN

INTRODUCTION

Darwin's emphasized the fact that exaggerated secondary sexual characters evolved through sexual selection (1874). He paid less attention to the role of sexual selection in the evolution of primary sexual characters. However, since the publication of Eberhard's (1985) monumental book on genital evolution, a large body of work has been published on the evolution of male genitalia (Arnqvist 1998; Bertin & Fairbairn 2005; Cordoba-Aguilar 1999; Eberhard 1993, 2004; Fairbairn et al. 2003; Gwynne 2005; Hosken et al. 2001, 2005; Hosken & Stockley 2004; House & Simmons 2003; Sirot 2003; Stockley 2002; reviewed in chapter 3), and it has become increasingly clear that primary sexual characters can also be sexually selected (Eberhard 1985, 1996, 2010, this volume). Part of the interest in male genital-evolution stems from the fact that they frequently evolve so rapidly and divergently relative to other morphological traits (see chapter 3 this volume), which means they are often species defining characters. In spite of this, evolutionary biologists have, for the most part, continued to pay less attention to other primary reproductive characters, even though they are likely to be subject to similar selection (Darwin 1874; Eberhard 1985). Testis size evolution is the one exception to this generality (e.g., Harcourt et al. 1981; Hosken 1997; Stockley et al. 1997). Nevertheless, like male genital form, it appears that

many other primary reproductive characters also evolve rapidly and divergently (figure 14.1). Here we review patterns of variation in a range of these characters, and potential agents responsible for their evolution. We suggest that post-copulatory sexual selection is likely to be heavily involved in the evolution of these traits, but in almost all instances net sexual selection has not been assessed and alternative explanations have not been ruled out.

Before proceeding further, we need to clarify what it is we are going to restrict our discussion to. We will not consider sperm form and only superficially discuss accessory gland products (Acps): these have both been the subjects of numerous recent reviews (e.g., Birkhead et al. 2009; Chapman 2001; Snook 2005; Ram & Wolfner 2007; Wolfner 1997). We will also not review variation in ovaries and eggs other than to note here that considerable variation exists in these characters across species (e.g., Pont & Meier 2002; Starmer et al. 2003), nor will we comment greatly on the bursa copulatrix (vagina) as relatively little attention has been paid to this aspect of female morphology (but see, e.g., Eberhard et al. 1998; Gage 1998; Morrow & Gage 2000; Siva-Jothy 1987). Instead we will restrict our discussion to testes, spermatophores, mating plugs, male and female accessory reproductive glands, female sperm stores (spermathecae) and the ducts associated with some of these structures. Furthermore, for the purposes of this review,

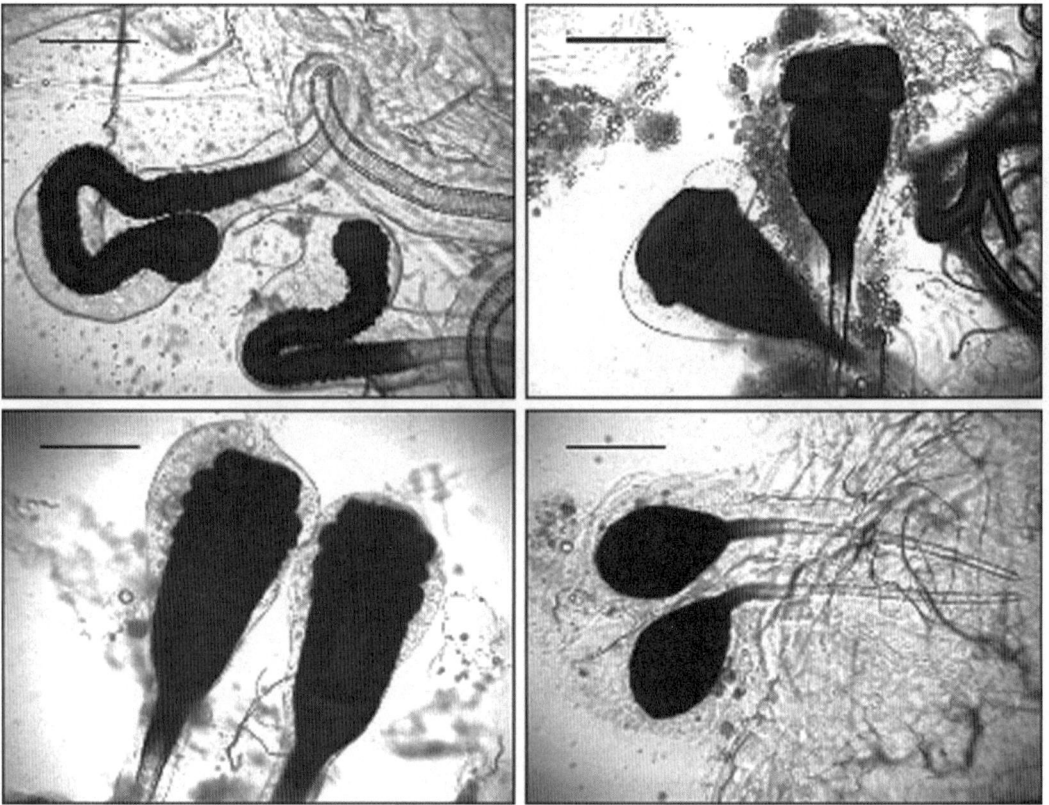

FIGURE 14.1 An example of the variation that exists in spermathecal form and size across the Scathophagidae. Clockwise from top-left: *Norellia spinimana*; *Spaziphora hydromyzina*; *Norellia striolata*; *Megophthalmides unilineatum*. Scale bar = 100 um (images from Minder 2002).

we consider any non-random process that generates variance in reproductive success to be sexual selection, even when this is generated by sexual conflict (e.g., Holland & Rice 1998). We will begin with the male characters before moving onto female characters and then finally, male-female co-evolution. We note here that although we differentiate between the two forms of post-copulatory sexual selection (cryptic female choice and sperm competition), at an operational level this division is often difficult to sustain. Females are the arenas in which post-copulatory sexual selection occurs, so at some fundamental level they are always involved passively or otherwise (Eberhard 1996). We also note that although we discuss male–female co-evolution as a separate section, this is merely a stylistic device and represents nothing more sinister.

MALE CHARACTERS

The basic design of the male internal reproductive tract is quite simple. Males typically have paired testis that are composed of a series of tubules held together by a mesodermal sheath. Tubules open into a mesodermally derived sperm duct (vas deferens) and these unite and connect to an ectodermally derived ejaculatory duct that leads to the gonopore (Davey 1985a; de Wilde & de Loof 1973). Accessory glands are often formed as out-pockets from the vas deferens, and additionally, the ducts themselves are often glandular (e.g., Hosken et al. 1999). The vas deferens can also be expanded over part of its length to act as a seminal vesicle, an area of sperm storage, and there can also be diverticula of the ejaculatory duct itself that are also glandular (Davey 1985a). We reiterate however, that in spite of this rather simple ground plan, there is tremendous variation on this theme across taxa.

Testis Size

The testes are the sperm producing organs in most metazoans (White-Cooper et al. 2007). It has been suggested that because most mutations arise during spermatogenesis, the testis are fundamentally important in generating the variation on which selection acts (Short 1997), and of the male internal reproductive characters we consider, testis size has been most widely studied from an evolutionary perspective.

Inter-Specific Variation

Testis size varies enormously across species (e.g., Demary & Lewis 2007; Gage 1994; Minder et al. 2005; Pitnick & Markow 1994). Investment in testis can be >10% of body mass (Pitnick 1996), and the balance of evidence suggests post-copulatory male–male competition (sperm competition: Parker 1970) has played a major role in testis evolution. Bigger testes have larger sperm producing capacities (e.g., Schärer et al. 2004; Simmons 2001), indicating variation in testis size is likely to be directly related to variation in sperm production. Parker (1982, 1984, 1998) has primarily provided the theoretical foundation for our understanding of testis size evolution. He reasoned that if sperm competition precedes by the raffle principle, where sperm are the tickets and fertilizations the prize, and if testis size was positively associated with sperm production capacity, then, as the risk of sperm competition—the likelihood females mate with more than one male per reproductive window—increased, males should invest more into spermatogenesis (testis size). While these ideas were primarily developed with mammals in mind, the main prediction has been tested across a number of insect taxa. One of the first studies was across butterflies, and using the number of spermatophores found in the copulatory bursa to estimate the average degree of polyandry, testis size was found to be strongly associated with female mating frequency (Gage 1994). Subsequently, many other insect investigations have also upheld the sperm competition risk–testis size association predicted by theory (e.g., Baer & Boomsma 2004; Demary & Lewis 2007; Morrow & Gage 2000; Pitnick & Markow 1994), indicating this is a robust relationship.

In fact the association between testis size and sperm competition risk is so well supported across so many taxa, that testis size is now used as a surrogate measure of sperm competition risk (Gage & Freckleton 2003). Nevertheless, because of the diversity of sperm competition mechanisms in insects, testis size is not always predicted to track sperm competition risk (Simmons 2001), and other explanations for testis size evolution have been proposed. Pitnick (1996) suggested testis size variation across *Drosophila* has more to do with energetics and selection on sperm size than sperm competition per se. He found that testis size was associated with sperm size rather than sperm number, even though testis size is associated with female mating frequency in a subset of the species he investigated: the *nannoptera* group (Pitnick & Markow 1994). Furthermore, it appears that large sperm size is favored by cryptic female choice in at least some *Drosophila* (Miller & Pitnick 2002; Pitnick et al. 2003), so testis size may be subject to indirect selection through female choice on sperm length rather than sperm competition *sensu stricto*. Pitnick (1996) also found that testis size was positively associated with body size across species and suggested this was because larger species are better able to bear the costs of the larger testes that are needed to produce the larger sperm females prefer. While all this seems plausible across *Drosophila*, across other insect groups sperm and testis size are often uncorrelated, and associations between body and testis size can also be lacking (e.g., Minder et al. 2005; Presgraves et al. 1999). Therefore, Pitnick's (1996) explanation is unlikely to be universal, which is unsurprising since *Drosophila* frequently have such unusually large sperm.

Intra-Specific Variation

Testis size also varies considerably intra-specifically. Variation has been documented within and between years (Demary & Lewis 2007), between populations (e.g., Hosken et al. 2003), between morphs (e.g., Simmons et al. 1999), and testis size often scales with body size (e.g., Blanckenhorn et al. 2004; Demary & Lewis 2007) and increases with male age (Wilkinson et al. 2005). Some of this intra-specific variation is probably due to environmental factors (noise), but like testis size variation across species, within species variation can also be adaptive and reflect sperm competition risk. For example, Gage (1995) showed that the numbers of potential competitors influence moth (*Plodia inter-punctella*) investment in testis size. In high-density populations with many potential competitors, males invest more in sperm production than in low-density populations (Gage 1995). Similar work on another moth species found essentially the same result (He & Tsubaki 1992), and higher larval density also resulted in larger testes in yellow dung flies (Stockley & Seal 2001; but also see Hellreigel & Blanckenhorn 2002). These results are paralleled by intra-specific findings in the dung beetle *Onthophagus binodus* (Simmons et al. 1999). Here there are two morphs, sneaking minor males and larger, dominant major males. Major males guard females and only face sperm competition in some matings, and have relatively smaller testes. Minor males on the other hand always face some risk of sperm competition, and have relatively larger testes (Simmons et al. 1999). In another beetle however, there are so many minor males that essentially every mating results in sperm competition and relative testis size does not vary between the two morphs (Simmons et al. 1999). Interestingly, recent work both intra- and inter-specifically in these beetles suggests complicated trade-offs in allocation to testis size and another sexually selected trait, horn size. Testis size seems to have primacy over secondary sexual traits in terms of resource allocation in these beetles (Simmons & Emlen, 2006), indicating that post-copulatory sexual selection is of particular importance in this group.

Experimental evolution in two fly species, *Scathophaga stercoraria* and *D. melanogaster*, also indicates sperm competition selects on testis size intra-specifically (Hosken & Ward 2001; Pitnick et al. 2001). In these studies monogamy (no sperm competition) or polyandry (sperm competition) were experimentally enforced in replicated populations and after several generation of evolution, testis size was found to have evolved in accordance with the predictions of sperm competition theory: polyandrous lines had larger testis (Hosken & Ward 2001; Pitnick et al. 2001) (Figure 14.2). Interestingly, evolving larger testes resulted in reduced immune function in yellow dung flies (Hosken 2001). Both these fly studies also indicate that variation in testis size is heritable, and a significant narrow-sense heritability of testis mass has been directly measured in a dung beetle ($h^2 = 0.97\pm0.45$: Simmons & Kotiaho 2002). This high estimate is suggestive of Y-linkage and taking this into account gives an estimate of ca. 0.5

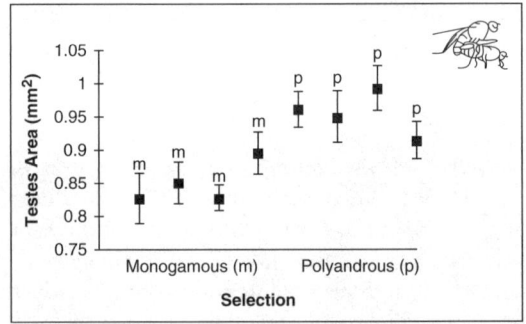

FIGURE 14.2 Testis size evolution in yellow dung fly populations evolving under experimentally enforced monogamy or polyandry. Males evolving under polyandry had larger testis than males from: monogamous populations. Each data point represents a line after 10 generations of selection (redrawn from Hosken & Ward 2001).

(Simmons & Kotiaho 2002). This estimate is similar to those generated from bidirectional selection on testis length in *D. hydei*. Selection resulted in rapid upward and downward divergence and realized h^2 between 0.45 and 0.72 (Pitnick & Miller 2000). Similarly in the cockroach *Nauphoeta cineria*, the heritability of testis mass was 0.50 ± 0.20 (Moore et al. 2004), and for both beetles and roaches, the CV_A for testis mass was very high. Testis mass also had significant broad sense heritability in *Gryllus* crickets in both lab and field environments (H^2 lab $= 0.84\pm0.13$; H^2 field $= 0.56\pm0.20$; Simons & Roff 1994). Together these results suggest there is ample scope for testis size evolution via sperm competition (although the cockroaches are monogamous), or any other mechanism of sexual selection, although in *Teleogryllus oceanicus*, testis mass was not significantly heritable ($h^2 = 0.04\pm0.18$) (Simmons 2003).

In sum, it seems clear that sexual selection acts on testis size indirectly either through sperm number (as predicted by sperm competition theory) or through sperm size.

Male Accessory Glands

The term accessory reproductive glands lumps together a variety of structures. Glands can have one of two developmental origins, they may be paired or unpaired, and cell types and structures in the gland also vary greatly (Adiyodi & Adiyodi 1975; Davey 1985a). Nevertheless, although these structures are not homologous across taxa, for the purposes of this chapter we group them together.

Inter-Specific Variation

In addition to variation in ontogenetic origin, the number of accessory glands varies enormously across groups, indicative of rapid and divergent evolution (Simmons 2001). The glands also produce a wide variety of products, from lipids to prostaglandins to uric acid to sugars to amino acids and proteins, all of which are transferred to females during copulation. These products potentially influence male fitness by facilitating sperm transfer or storage, altering female reproductive behavior, including stimulation of oviposition and suppressing receptivity, removal of rival sperm (Gillott 2003), and they may also activate sperm (Davey 1985a). Each of these functions is potentially subject to sexual selection, and consistent with this, the effects of these various products are often dose-dependent (Eberhard 1996; Simmons 2001; Gillott 2003), so that males able to deliver more products to females will have a reproductive advantage. Eberhard (1996) provided further argument for sexual selection acting on gland products. Evidence includes, the rapid evolution of products and the genes that encode them, product redundancy—more than one gland component can act on the same pathway, which is typical of sexually selected characters (West-Eberhard 1984)—and products often ape the females own hormones (Eberhard 1996). If gland products are subject to sexual selection, as seems likely, an association between the size of the male glands and risk of sperm competition (or intensity of polyandry) is predicted across species (Simmons & Siva-Jothy 1998), assuming larger glands produce more product. Across moths, the size of the male accessory gland was found to be positively associated with sperm competition risk as measured by testis size (Morrow & Gage 2000). In contrast with the Simmons & Siva-Jothy (1998) prediction, Morrow & Gage (2000) predicted a trade-off between these two organs if the glands were involved in nutrient provisioning to females—males should provide less investment as certainty of paternity decreased. Instead, they suggested the pattern documented was more consistent with gland products functioning in paternity protection (Morrow & Gage 2000). Finally, across a few species of firefly, species with greater degrees of polyandry invested more in gland size than

a monogamous species and the difference in investment was huge (0.07% of mass vs. 7.7%) (Demary & Lewis 2007).

As stated, across species gland products have evolved rapidly (Haerty et al. 2007; Swanson & Vacquier 2002), and consistent with this, the effects various products elicit are lower in across-species matings (Simmons 2001). However, the effects of some ejaculatory substances are conserved across large evolutionary distances (Wedell 2005), which seems slightly paradoxical. Additional work on this issue is required.

Intra-Specific Variation

There is also a paucity of data on within-species variation in male accessory gland size. However, gland size has been shown to increase with male age, male morphology can influence the rate at which glands increase in size (Socha & Hodkova 2006), and size has also been found to vary with dietary stress (e.g., Demary & Lewis 2007). Of the few studies that have investigated intra-specific gland size-variation from an evolutionary perspective, all find evidence consistent with glands evolving through sexual selection. The size of the accessory glands has probably been most thoroughly investigated in the stalk-eyed fly, *Cyrtodiopsis dalmanni*. In this species, depletion of the gland, rather than sperm depletion, is a primary constraint on male mating frequency, so that males with larger glands have higher mating frequency (Baker et al. 2003). Furthermore, experimental evolution provides additional evidence for this, and for the heritability of gland size (Rogers et al. 2005). In stalk-eyed flies, glands are apparently costly to build, as males with larger glands have longer development times (Baker et al. 2003). Similar findings have been reported for whirligig beetles (Fairn et al. 2007). However, while it seems gland size correlates with mating frequency in these species, it is unclear if gland size is influenced by post-copulatory sexual selection. If the latter was the case, we predict gland size to be positively correlated with the risk of sperm competition, rather than number of mating opportunities. There is also a positive correlation between gland size and mating rate in *D. melanogaster*, while testis size is not associated with number of matings (Bangham et al. 2002), and similar results have been reported in firebugs where the size of the accessory gland is positively correlated with mating success in both large- and small-winged morphs (Socha 2006). There is also evidence of genetic variation in gland size. In crickets, for example, gland size is significantly heritable (h^2 = 0.42±0.18), has high CV_A, and is strongly genetically correlated with ovary size (Simmons 2003). Similarly, QTLs for male accessory reproductive gland size have also recently been reported for *Bombus terrestris* indicating there is genetic variation for this trait (Wilfert et al. 2007).

The effects of various gland products on females and sperm storage have been extensively investigated (Adams & Wolfner 2007; Chapman 2001; Ram & Wolfner 2007; Wolfner 1997), but other functions have been subject to less investigation. For example, a role for gland secretion in sperm activation has been reported for several insects (e.g., Shepherd 1975), but it is unclear how this affects male fertilization success, especially because in many cases it seems female involvement is required for sperm transport to storage (e.g., Arthur et al. 1998; Hellriegel & Bernasconi 2000; Linley & Simmons 1981). Furthermore, to the best of our knowledge, there have been no investigations of inter-male variation in sperm activation, so it is unclear if this secretion function is subject to sexual selection, or natural selection for that matter. Secretions from the accessory glands of male *D. melanogaster* were also implicated in damaging rival sperm (Harshman & Prout 1994). However, more recent investigation found no evidence for this (Snook & Hosken 2004).

In sum, it appears that gland size can be an important determinant of male mating-success, although the potential for sperm competition and female post-copulatory choice to further affect gland size or shape is not well understood.

Spermatophores

In many insects sperm are packaged together with semen and other ejaculate components to form a spermatophore that can, for example, be deposited on the ground in the absence of the female and later picked up, or transferred directly to the female at mating (Mann 1984). Spermatophores range in complexity, and are useful species diagnostic characters in butterflies and moths (Petersen 1907). They also vary dramatically in size. For example, in *Ephippiger ephippiger* bushcrickets the spermatophore represents as much as 30% of male body weight (Wedell 1993), while in other species the spermatophore is only a small, loosely

mucus-coated, sperm aggregation (Mann 1984). Spermatophores can be produced and stored by males prior to mating, formed inside females during copulation, or constructed during copulation, but retained by males once the ejaculate has been transferred to the female. In general, spermatophores primarily function as storage containers and transport vehicles for the spermatozoa, probably to minimize sperm loss during insemination, and enhance sperm survival both before and after insemination. Furthermore, the spermatophore also contains nutrients critical to female reproductive success in some species. In the bushcricket *Requena verticalis*, for example, females utilize male derived nutrients passed in the spermatophore to increase their fecundity (Gwynne 1984). In other species, spermatophores act as an aphrodisiac when placed on a substrate in the absence of a female, emitting pheromones to attract receptive females (Wertheim et al. 2005). Finally, spermatophores can also directly influence female post-copulatory behavior by transferring compounds that directly affect female reproductive physiology. There is a vast literature dealing with the function and evolution of spermatophores, a summary of which is far beyond the scope of this chapter. Instead we present a few examples to highlight the range and complexity of these reproductive structures, and the potential mechanisms of selection acting on them.

Inter-Specific Variation

As noted above, there is substantial variation in both the size and complexity of the spermatophore between closely related species. To illustrate this variability we will primarily focus on the bushcricket spermatophore (Orthoptera: Tettigoniidae), with only brief discussion of other taxa. The spermatophore of these insects generally consists of two parts: the sperm-containing ampulla and a gelatinous sperm-free spermatophylax that surrounds the ampulla (Boldyrev 1915; Gwynne 1997). Both are attached externally to the females' genital opening during copulation. Following mating, females remove and consume the spermatophylax, during which time the ejaculate moves from the ampulla to the females' reproductive tract. Following spermatophylax consumption, females remove and consume the ampulla. The ampulla itself can be composed of one or more 'compartments' and contains the sperm and seminal fluids. The sperm-free spermatophylax ranges in size from barely covering the ampulla (e.g., *Coptaspis sp. 2*; Wedell 1998), to a large dense structure completely covering it (Wedell 1993). The function of the spermatophylax is two-fold; it protects the ejaculate during insemination (e.g., Wedell 1991), but may also contain nutrients that enhance female fecundity (e.g., Gwynne 1984). In species where the spermatophylax is nutritious, the spermatophylax is larger than necessary to ensure complete ejaculate transfer from the ampulla; it takes substantially longer to consume than is necessary for the ampulla to empty of sperm (e.g., Gwynne 1986). The non-spermatophylax component of the ejaculate itself appear to contain compounds that directly reduce female receptivity as there is a positive relationship between the amount of ejaculate received and the duration of the females' non-receptivity periods (Gwynne 1986; Wedell & Arak 1989). The suggestion that the primary function of the spermatophylax is to protect the ejaculate during insemination is supported by the finding that across taxa there is a positive relationship between spermatophylax size and the duration of the females' non-receptivity period (Vahed 2006; Vahed & Gilbert 1996; Wedell 1993), and between spermatophylax size and ejaculate volume (Vahed 2006; Vahed & Gilbert 1996; Wedell 1993). This implies a direct relationship between the amount of ejaculate transferred and the size of the protective spermatophylax. In addition, and as predicted by the hypothesis that males may additionally invest in offspring production by providing nutrients once high paternity is achieved, species producing spermatophylaces that are larger than required to protect the ejaculate during insemination, pack them with more protein (Wedell 1994). Their production also results in longer periods of male recuperation (Vahed 2007; Wedell 1994) than in taxa whose spermatophylaces primarily function as ejaculate protectors. The elaboration of overall spermatophore size is therefore probably driven by sexual selection. A larger ejaculate volume suppresses females' receptivity reducing the risk of sperm competition for the male, which may impose counter-selection on females to overcome male manipulation, promoting further increases in ejaculate size. As a consequence of this, male bushcrickets need to invest more resources in a protective spermatophylax resulting in overall increased spermatophore size. Once high paternity is achieved, additional modification of the spermatophylax may take place, resulting in the dual function of ejaculate protection and paternal investment

in some species (e.g., Simmons 1995a). This illustrates the impact of sexual selection arising from sperm competition for the elaboration of the spermatophore in these insects. However, female preference for high investing males (Gwynne 2001), and natural selection for increased sperm survival are also likely processes shaping spermatophore size and elaboration in bushcrickets.

Similarly in butterflies, a spermatophore is deposited inside the female at mating, and these can contain male-derived nutrients that enhance female fecundity and longevity, and appear to influence female mating frequency. Across taxa there is a positive relationship between spermatophore size and nutritional value and degree of polyandry (Svärd & Wiklund 1989; Bissoondath & Wiklund 1995). This suggests females may 'forage' for large nutritious spermatophores (Kaitala & Wiklund 1995). However, it is not clear whether the nutritional value of the spermatophores in paternally investing species is directly driving higher female mating frequency. Males attempt to achieve high paternity by suppressing female receptivity, which in part is affected by spermatophore size (e.g., Cook & Wedell 1996; Kaitala & Wiklund 1995). Thus high-investing males may ensure high paternity in spite of the high average female remating rate. These finding indicate that spermatophores in butterflies are also subject to sexual selection.

This is also true more generally and there seem to be good evidence across taxa that spermatophores have evolved through sexual selection, even though their initial evolution may have been via natural selection. Furthermore, while we have largely emphasized sperm competition, and to a lesser degree sexual conflict, as the agents of sexual selection this does not preclude the operation of other forms of sexual selection, such as cryptic female choice.

Intra-Specific Variation

There is also considerable intra-specific variation in spermatophore size and complexity. Part of this variability can be explained by the cost of spermatophore production as evident by the negative association between spermatophore size and copulation number seen in many species. As a consequence of these production costs, males must decide how to allocate finite resources to the various spermatophore components. Again, we refer to bushcrickets to illustrate this point. Male bushcrickets frequently tailor their spermatophores in a manner consistent with fitness maximization. In *R. verticalis*, for example, males adjust their investment in components of the spermatophore in direct relation to their remating interval. Both the size of the sperm-containing ampulla and the nutritious spermatophylax increase when remating interval increases. However, males disproportionally increase their investment in the non-spermatophylax part of the spermatophore (Simmons 1995b). This was interpreted as increasing paternity assurance when paternal investment is high. Similarly, male *Kawanaphila nartee* bushcrickets are also sensitive to variation in the degree of sperm competition they face and alter their spermatophores accordingly (Simmons & Kvarnemo 1997). However, sperm competition is not the only form of sexual selection that affects spermatophore production intra-specifically.

Since spermatophores are costly to produce and are wholly synthesized by males, this means they are potentially indicators of males' quality, and females could use this information in choosing sires, or in their reproductive allocation to offspring (e.g., Wedell 1996). For example, male *Utetheisa ornatrix* moths vary in their ability to produce spermatophores containing toxic compounds that females use to render their offspring unpalatable to predators (González et al. 1999). Males advertise their donating ability by emitting pheromones and females preferentially mate with males providing high quality gifts (Dussourd et al. 1991), and 'reward' high-donating males with increased paternity (LaMunyon & Eisner 1993). These results indicate elaboration of spermatophores can be affected by both pre- and post-copulatory female choice. In addition, there is genetic variation in males' ability to produce a spermatophore. For example, in *P. napi* butterflies, spermatophore size and nutrient provisioning are heritable with an h^2 of 0.48 ± 0.24 (Wedell 2006), and in cockroaches, spermatophore and ampulla mass are also heritable ($h^2 = 0.38 \pm 0.19$ and $h^2 = 0.52 \pm 0.20$, respectively) (Moore et al. 2004). This suggests selection on spermatophores will produce an evolutionary response, all else being equal. Environmental factors also generate variation in males' ability to produce spermatophores, with juvenile and adult diet directly affecting males' spermatophore production capacity in some species (e.g., Simmons 1995b), and male age can affect both the size and quality of the spermatophore (e.g., Wedell & Ritchie 2004).

Overall, there is considerable intra-specific evidence of sexual selection affecting spermatophore size and composition, and the evidence for several mechanisms is quite good. Taken together with the inter-specific data, it would seem that sexual selection is primarily responsible for spermatophore evolution; however, alternative explanations have typically not been investigated.

Mating Plugs

In some insects, males transfer substances to the female at copulation that hardens to form a physical barrier, either internally or externally—the mating plug. Plugs may be an attempt to prevent further mating by females—they represent a form of mate guarding—or they may facilitate sperm transfer and/or prevent sperm loss from the female reproductive tract. Mating plugs (termed sphraga in butterflies and moths), are widespread in dipterans, hymenopterans, and lepidopterans, for example, where they can vary in both size and complexity. Mating plugs are formed by substances produced by males' accessory glands, or by specialized glands, as is the case in *Parnassius* butterflies, for example (Scott 1986). There are many different types of plugs, both internal and external, that vary in degree of specialization. Selection can potentially lead to plug elaboration to increase their effectiveness, as both rival males and females may attempt to remove plugs.

Inter-Specific Variation

The best-studied mating plugs are probably the sphraga of the Lepidoptera. This may be due to the conspicuousness of the external mating plug in some butterflies, where it can surround the female's entire abdomen, potentially interfering with egg laying (Orr 1995). Mating plugs are formed in the Lepidoptera by substances from the males' accessory glands, and can be remarkably large and elaborate. They appear to function as a means of reducing the likelihood of female remating: the number of matings by females decreases with increasing elaboration or size of the mating plug across species (Simmons 2001). In the butterfly *Cressida cressida*, the large sphragis covers the female genital opening providing a life-long chastity belt resulting in monogamy (Orr & Rutowski 1991). Sphraga seem to be costly for male Lepidoptera to produce, but costs to females have not been explored. Some species are only able to produce a few plugs, and the size is often negatively correlated with number of previous copulations (e.g., Matsumoto & Suzuki 1992; Orr 1995, 2002). Elaboration of the sphragis appears to have occurred at the expense of spermatophore size because there is an inverse relationship between spermatophore size and sphragis elaboration (Matsumoto & Suzuki 1995; Orr 1995). This suggests when plugs reduce female mating rate, males need less ejaculate due to relaxed risk of sperm competition, and highlights the importance of mating plugs in regulating female remating. However, other functions, such as plugs preventing sperm loss (i.e., sperm corralling) and enhancing female fertility, remain possible.

Mating plugs are also found in social Hymenoptera and are thought to be effective in preventing female remating (e.g., Sauter et al. 2001). Polyandry is not common in this insect group, although in some bees females mate >100 times (Page 1986), but it is not clear whether mating plugs are the cause of the generally low levels of polyandry seen in social hymenoptera. Mating plugs can represent a substantial investment by males (Boomsma et al. 2005). In attine fungus-growing ants, for example, there appears to be a negative relationship between investment in the mating plug and the number of sperm transferred, as with butterflies (see above). Males of singly-mated-queen species have relatively bigger accessory glands and smaller sperm-containing accessory testes than males of multiply-mated-queen species (Baer & Boomsma 2004; Mikheyev 2004). Since glands probably produce the mating plugs, large male-accessory-gland size indicates male control over female mating frequency could be common in these ants (Baer & Boomsma 2004). Additionally, excessive investment in mating plugs may result in males losing the ability to mate multiply, potentially leading to the evolution of suicidal copulations (Boomsma et al. 2005). In contrast, in highly eusocial bees, polyandry is instead associated with large male accessory gland size (producing the mating plug) such as in *Apis* honeybees, for example, and monandrous stingless bees have reduced gland size (or even complete absence) (Colonello & Hartfelder 2005). The causal relationship between plug size and degree of polyandry in social hymenoptera therefore remains unresolved, and it is also unclear why males are apparently able to determine female mating rates so readily in some members of this taxon.

Intra-Specific Variation

There is also considerable intra-specific variation is size and function of mating plugs. In the butterflies, the sphragis, apart from physically preventing the female from remating, also appears to function as a visual deterrent to rival males, as the larger the plug the less likely males are to attempt to mate with the female in some species (Orr & Rutowski 1991). This suggests pre-copulatory sexual selection operating on plug size. Plugs do not completely prevent female remating however, as females can remate despite the presence of a plug (Matsumoto & Suzuki 1992; Orr & Rutowski 1991). Additionally, in the chalcedon checkerspot butterfly *Euphydryas chalcedona*, females with the plug experimentally removed are just as likely to reject courting males as females with an intact plug (Dickinson & Rutowksi 1989), indicating that other factors affect female receptivity in this species. It is therefore possible that female post-copulatory sexual selection also operates on mating plugs in butterflies. There is also evidence of adaptations by males to circumvent mating plugs. In *Heteronympha penelope* butterflies, males have specialized genitalia capable of removing rival males' sphraga (Orr 2002). However, it is not clear if there is variation in the ability to remove plugs or to prevent plug removal.

In *D. melanogaster* semen coagulates to form a mating plug after copulation. The posterior region of the mating plug is formed by a male ejaculatory-bulb protein, that show sequence similarities with structural proteins from spider silk and byssal threads of mussels (Lung & Wolfner 2001). The distal end seems to contain accessory glands proteins, which affect female receptivity and oviposition (Chapman 2001). Similarly, in *D. hibisci*, a mating plug is formed within the female uterus during copulation and this reduces females' tendency to remate. Experimental reduction of plug size directly increases the speed with which females are courted and remate (Polak et al. 2001). In addition, the plug appears to be important in ensuring successful sperm storage, as reduced plugs fails to prevent back-flow of sperm from females' sperm storage (Polak et al. 1998). Therefore both sexual selection via sperm competition and male reproductive success generally, but also natural selection increasing female fertility levels, probably operate on mating plugs in *Drosophila*.

Stingless bee males often engage in suicidal matings, leaving their genital capsule firmly attached as a mating plug to the female providing a barrier to further matings (Colonello & Hartfelder 2005). The mating plug of the bumblebee *Bombus terrestris*, is formed by the males' accessory glands and apparently permanently switches off female receptivity (Baer et al. 2001), despite documented benefits of polyandry to females (Baer & Schmid-Hempel 1999). The plug is made up of four non-specific fatty acids, and lineolic acid is the suppressive substance (Baer et al. 2001). Interestingly, the same fatty acids are also present in the fire ant *Solenopsis invicta*, for example (Mikheyev 2003). It is surprising that the active chemical component in the mating plug that manipulates female reproductive behavior in several hymenopteran species is a non-specific fatty acid. The non-specificity of this compound may be explained by the long delay between mating and egg laying at the start of colony initiation, which may select against production of factors stimulating immediate increased oviposition rate (Colonello & Hartfelder 2005). In sum, the available evidence suggests that the mating plug in stingless bees is shaped by sexual selection and the same is true of other taxa too. However, in many cases alternative explanations for plug use have not been fully explored.

As is the case for spermatophores, there is considerable intra- and inter-specific evidence of sexual selection affecting both the size and composition of mating plugs. While in some cases the function of mating plugs may be to ensure efficient sperm storage and prevent leakage, there is overwhelming data indicating the primary function is to prevent female remating.

FEMALE CHARACTERS

Female reproductive tract morphology is more complicated and arguably less studied than male morphology. The basic composition of the female genital tract is quite variable but includes paired ovaries, and associated oviducts (all mesodermal in origin). The oviducts meet at an ectodermal medial oviduct (or vestibule) that then forms a vagina (bursa copulatrix) that opens externally (de Wilde & de Loof 1973). The medial oviduct usually carries one to several spermathecae, which act as the sperm stores after copulation and accessory glands with various functions may either enter the medial oviduct or the vagina. There are numerous variations on this theme, with lepidopterans and some beetles having two genital orifices for example and many *Drosophila* and other Diptera having an additional non-spermathecal sperm store called the ventral receptacle which have become the primary, and in some cases the only, sperm store (Pitnick et al. 1999). Furthermore, spermathecal number is also extremely variable across species (Eberhard 1985), and at least some of this variation has been attributed to sexual selection. We note that in what follows we refer to direct sexual selection on female characters when selection results from female-female reproductive competition (Clutton-Brock 2009) and indirect when selection results from associations with male characters (i.e. because of sexual conflict or female choice).

Sperm stores—Spermathecae and Receptacles

The primary function of the various sperm-storage organs is in the name, they store fertile sperm for future female use, and in some cases the duration of storage can amount to years (Davey 1985b). Various glands can be associated with the stores and these secrete a range of products that have been implicated in prolonging sperm lifespan, either through nutrient provisioning or maintaining an appropriate chemical milieu for the sperm (Davey 1985b). The role of spermathecal glands in sperm storage has been clearly demonstrated in a handful of cases where gland ablation or removal results in gradual loss of sperm motility within the female sperm-store (e.g., Villavaso 1975a). Additionally, various components of the spermatheca have been implicated in facilitating sperm movement to or from storage (e.g., Eady 1994; Hosken & Ward 2000; Rodríguez 1994; Villavaso 1975b), but it is unclear if any of these responses depend on male phenotype, and hence if there is any selection imposed by these contrivances or by spermathecal gland secretion. Here we largely restrict ourselves to discussion of the gross morphology of female sperm-stores as there have been few evolutionary investigations of other aspects of the stores. However, we also note that other aspect of the stores, their ducts and the frequent tortuous nature of the female reproductive tract are also likely to facilitate female choice, prevent male monopolization of reproduction, and hence be subject to direct or indirect sexual selection (Eberhard 1996).

Interspecific Variation

As with all the characters discussed so far, there is enormous interspecific variation in spermathecal shape, size and number (figure 14.1 and see, e.g., Eberhard 1985; Minder et al. 2005; Morrow & Gage 2000; Pitkin 1988; Pitnick et al. 1999) (here, unless specifically stated otherwise, when we say spermatheca we include all sperm storage organs). For example, linearized measures of spermathecal area (cube root) range from around 6% to more than 14% of the hind tibia length across dung flies (Minder et al. 2005) and volume varies by two orders of magnitude across moths (Morrow & Gage 2000). The large variation in spermathecal form across groups has been used by taxonomists in defining species (e.g., Evans & Adler 2000; Ilango 2004; Pitkin 1988; Throckmorton 1962), but the reasons for this variation are not well understood. Nevertheless, rapid divergent evolution of reproductive characters has all the hallmarks of evolution through sexual selection, and sexual selection is thought to be involved in the evolution of sperm stores directly or indirectly (Eberhard 1985; Siva-Jothy 1987; Walker 1980; Ward 1993).

As stated, spermathecal shape varies considerably across taxa (figure 14.1). Walker (1980) hypothesized that this variation determined the degree to which males could monopolize paternity by displacing rival sperm, and he found a trend for more displacement in elongated rather than spherical spermathecae. This result suggested variation in spermathecal shape is the result of either cryptic female choice (for better sperm displacers), or of sexual conflict over paternity. Walker's (1980) hypothesis was more rigorously tested by Ridley (1989), but he found no association between the paternity of second males to mate (P2) and spermathecal shape. However, Ridley's (1989) test was also limited because so little was (and is) known about the mechanisms underlying the P2 variation he analyzed, and subsequent warnings about overinterpreting P2-values are justified (Simmons & Siva-Jothy 1998). For example, high P2 can result from sperm death/leakage from storage and have nothing to do with spermathecal shape, and P2 variation within species is typically very large. The validity of Walker's (1980) argument also depends on the elasticity of the sperm store, probably only being applicable in taxa with fixed volume stores (Simmons 2001). In a modification of Walker's

(1980) hypothesis, Ward (1993: and also see Eberhard 1985) suggested that variation in spermathecal *number* could be the result of selection for female control of paternity. More than one sperm store could allow females to spatially separate the sperm of different males, and providing they could then choose which sperm store to use during fertilization, they could cryptically influence paternity. This idea is broadly supported by modeling (Hellriegel & Ward 1998), and if generally correct, predicts an association between spermstore numbers and measures of female multiple mating. To the best of our knowledge this has not been explicitly tested, even though variation in the numbers of storage organs has been documented, as has variation in levels of polyandry. However, measures of sperm storage capacity are often associated with measures of polyandry such as testis size. For example, spermathecal volume is positively associated with testis size across moths and dung flies (Minder et al. 2005; Morrow & Gage 2000). These associations are broadly consistent with Ward's (1993) predictions. In addition to spermathecal size and number, the placement of sperm stores on long ducts away from the insemination site may also facilitate female choice (Eberhard 1985) or be due to conflict over paternity, but again we are not aware of any comparative study directly testing this idea. However, across odonates, variation in spermathecal ducts and how they connect to the sperm stores is thought to represent an attempt to thwart male sperm removal (Siva-Jothy 1987). Direct removal of rival sperm by males occurs in some odonates because males can directly access the female sperm stores (Waage 1979, 1984), but by making sperm stores inaccessible, females could retain more control over paternity (Siva-Jothy 1987), and there is good intraspecific evidence for this (Siva-Jothy & Hooper 1996; and see below). However, Eberhard (1985) has argued against conflicts of interest being the primary determinant of sperm store evolution because if a female evolved to prevent sperm displacement, this suggests she would produce sons that were worse than average at displacing sperm. Instead he suggest cryptic female choice is more likely to be responsible for general tract morphology, although his rejection of the conflict argument seems to hinge on how much sperm displacement females prevent (Siva-Jothy 1987), and on the relative costs and benefits of preventing displacement. Although not directly bearing on this issue, it is interesting to note that males can indirectly remove rival sperm by stimulating females in some odonates (Córdoba-Aguilar 1999).

Possession of multiple kinds of sperm storage organs has also been discussed in the context of female control of paternity. For example, *Drosophila* have two types of storage organ, the spermathecae and the ventral receptacle (VR) and females can have either or both types of store (Pitnick et al. 1999). Pitnick et al. (1999) estimate there have been 13 independent losses of spermathecae across *Drosophila* and one loss of the VR, and in this latter instance all descendent taxa have evolved very elaborate spermathecae. These data additionally indicate how evolutionarily labile these structures are, but in cases of spermathecal loss, only rarely is the structure "refound" (Pitnick et al. 1999). As discussed above, broadly analogous changes in other taxa have been hypothesized to be due to selection favoring females that prevent sperm removal by males (Siva-Jothy 1987), but Pitnick et al. (1999) suggest this is not the case for *Drosophila*. Instead they suggest that the evolution of sperm stores in *Drosophila* is due to a combination of selection for a "better" store and for organs that can specialize in short and long term storage (also see Twig & Yuval 2005). It is not clear what "better" really means in this context, but this could encompass being both better at thwarting male attempts to monopolise paternity, or better at allowing females to cryptically choose sires. Co-evolution of VR and sperm length across *Drosophila* provides good support for both these hypotheses (Pitnick et al. 1999).

While it has been suggested that sexual selection is involved in the evolution of size, shape and/or number of spermathecae, it is clear that storage of fertile sperm is also strongly naturally selected and some of the variation in storage volume may be accounted for by variation in clutch size, copulation frequency, female longevity and/or sperm utilization efficiency (Pitnick et al. 1999). To the best of our knowledge no study has investigate these possibilities. However, clutch size frequently increases with body size in insects (Honek 1993; Starmer et al. 2003), so if fertility selection alone was responsible for variation in sperm store size across species, then associations between body and store size measures would be expected. However, across species this association is often lacking (e.g., Minder et al. 2005; Morrow & Gage 2000; Presgraves et al. 1999). Furthermore, it is unclear if any of these non-sexual selection hypotheses require more than a single sperm store, nor do they account for variation in the shape of the sperm stores, although one could argue that this variation is neutral and only total storage capacity is seen by selection. This is an area that requires more investigation.

Intraspecific Variation

There is considerable variation in spermathecal size, and sometimes number, within species (figure 14.3) (and see, e.g., Cordero Rivera et al. 2005; Fedina & Lewis 2004; Fitz & Turner 2002; Hosken et al. 1999; Parker et al. 1999). Some of this variation is due to allometry (e.g., Bernasconi et al. 2006; Parker et al. 1999), but for some species there is no scaling relationship (e.g., figure 14.3). Additionally, there can be intra-female variation in sperm-store size and sometimes the asymmetries in store size can be very large (e.g., Minder 2002; Pitkin 1988). Females can also have more than one type of storage organ (e.g., Cordero Rivera et al. 2005; Pitnick et al. 1999). Data from *Anastrepha suspensa* fruit flies also indicate that different sperm storage organs receive sperm independently from one another, possibly indicating there is scope for females to sort sperm from different males (Fitz 2004). Across populations, there can also be extreme levels of differentiation in spermathecal length (Cordero Rivera et al. 2005; Pitnick et al. 2003).

As above, sexual selection conflict has been invoked to explain within species variation in sperm-store size/number/complexity. However, only a handful of studies have investigated this possibility and most have done so by looking for associations between spermathecal form or number and P2, the proportion of offspring sired by the second of two males to mate with a female. Two studies have been conducted on red flour beetles (*Tribolium castaneum*) and these have produced contrasting results. Fedina and Lewis (2004) found that P2 declined as spermathecal volume increased, while Bernasconi et al. (2006) found no association between P2 and spermathecal volume, complexity or width. House and Simmons (2005) also found no influence of spermathecal width or outline on P2 in the dung beetle *O. taurus*. Later studies have revealed significant additive genetic variation in spermathecal size and that larger spermathecae favor shorter sperm resulting in genetic covariance between spermatheca size and sperm length (García-González & Simmons 2007; Simmons &

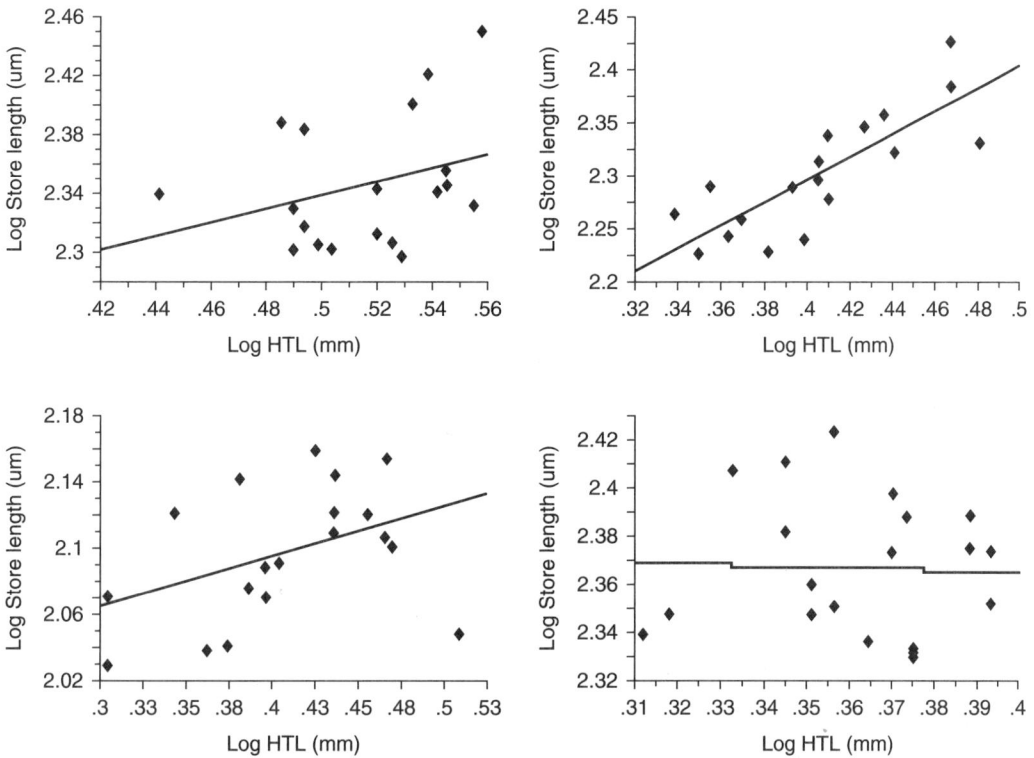

FIGURE 14.3 Intra-specific variation in mean spermathecal length (um log10) as a function of body size (Hind Tibia Length (HTL) mm log$_{10}$) in four species of dung fly. Clockwise from the top left: *Cordilura ciliata*, *Scathophaga taeniopa*, *Norellia spinimana*, *Scathophaga stercoraria*. Least squares regression lines included (data from Minder 2002).

Kotiaho 2007). This indicates female post-copulatory choice may directly influence sperm size evolution in this dung beetle species. The influence of spermathecal number on paternity has been investigated in yellow dung flies (Ward 2000). In this fly most females have three spermathecae (see Hosken 1999), but some females have four (Parker et al. 1999; Ward 2000). There is an additive genetic basis to this variation because artificial selection can alter the proportion of females with four spermathecae (Ward 2000). Spermathecal number was found to influence P2, but the effect was only significant in an interaction with "female quality"—females were dichotomized as high or low quality depending on the number of eggs they laid (Ward 2000). While there is some disagreement about how spermathecal number influences total storage volume—some data suggesting more stores equals greater volume (Ward 2000) while other data suggests not (Parker et al. 1999)—volume did

not influence the P2 result in one study (Ward 2000), but appeared to have an effect in another (Parker et al. 1999). Interestingly, there seems to be some sperm mortality during storage in these flies, and mortality varies across males (Bernasconi et al. 2002). One clear example of how spermathecal size-variation generates selection is that of the damselfly *Calopteryx haemorrhoidalis* (Cordero Rivera et al. 2005). Here spermathecal length determines how much sperm males can displace and only a small proportion of males (4%) have penis extensions that are able to fully access the extremely elongated sperm stores of a large proportion of females in the population (38%). It therefore seems that the sperm stores of many females are beyond the reach of all but a few males (Cordero Rivera et al. 2005). In all the cases cited it should be noted that the failure to detect significant female effects does not mean there are none because if, for example, all females had the same preference, variance attributable to males would be impossible to differentiate from that due to females (Pitnick & Brown 2000). There is also evidence that different types of sperm store differentially store sperm. Siva-Jothy and Hooper (1995) found more genetically diverse sperm in the spermatheca than in the bursal store in another damselfly (*C. splendens*), and that females can differentially use these stores to provide sperm for fertilization (Siva-Jothy & Hooper 1996). However, it is not clear if this results in selection.

In addition to the selection on spermathecal number in dung-flies discussed above, laboratory selection has also been conducted on one of the *Drosophila* sperm stores, the ventral (seminal) receptacle (VR) (Miller et al. 2001). Females were subject to bidirectional selection on VR length and rapid divergence was achieved, with responses tending to be greater in the downward direction. Realized heritabilities for VR length were between 0.18 and 0.27, and there were strong positive genetic correlations between body and VR size. This work is consistent with the pattern of rapid divergence in VR size across *Drosophila* (Pitnick et al. 1999), and there is evidence that variation in VR length is a strong determinant of male fertilization success, although this depends on an interaction between VR and sperm length (Miller & Pitnick 2002). Evidence from mosquitoes and dung flies also indicates that sperm stores may selectively store sperm based on sperm length (Otronen et al. 1997; Klowden & Chambers 2004), but how this influences fertilization success is not clear, although in the dung fly, sexual selection does not seem to act on sperm length (Hosken et al. 2001). Other factors, including diet and age, have also been implicated in within-species variation in spermathecal size. For example, female mosquitoes reared on a low quality diet had smaller stores than females reared on a higher-quality diet (Klowden & Chambers 2004), and juvenile dung flies have incompletely sclerotized spermathecae (Minder 2002).

The available data to date indicate the size, shape, and number of female sperm storage structures can be important generators of sexual selection and may evolve as a result of conflict over fertilization. It is not clear how important female-female reproductive competition is, but sexual conflict could enhance competition between females. It is clear that sperm storage is also strongly naturally selected...' etc.

Accessory Glands

There is enormous variation in the form, function, and location of female accessory reproductive glands across insects (Adiyodi & Adiyodi 1975; Davey 1985b). Functions range from egg coating and cocoon production by collateral glands to embryo nourishment by the milk glands of Tsetse flies. Anti-bacterial peptides have also been isolated from the female accessory reproductive glands (e.g., Manetti et al. 1997; Rosetto et al. 1996). These products probably protect eggs from bacterial attack and are only found in sexually mature flies (Rosetto et al. 1996). It is clear that many of the substances produced by female glands are likely to be naturally selected, but there has been very little work on gland size and function from an evolutionary perspective, and even less that has considered potential sexual selection roles for these glands. For example, differential coating of eggs with antibacterial peptides is a potential mechanism of cryptic female choice, albeit an expensive (and unlikely) one for females. This has never been investigated and because of a general paucity of data, we will keep our discussion of female glands short and will not divide our discussion into inter- and intra-specific sections as we have elsewhere.

While medfly gland products are probably naturally selected, female flies also produce enzymes in the accessory reproductive glands that aid chitin metabolism and appear to facilitate fertilization (Marchini et al. 1989). Similarly, in female *Musca domestica* substances involved in fertilization and sperm degradation are produced in the accessory reproductive glands (Degrugillier 1985). It seems likely that these functions could be subject to sexual selection directly or indirectly, but there have been no investigations to test this. Gland size has been subject to a degree of investigation in yellow dung flies (*Scathophaga stercoraria*). While decreases in gland size volume have been noted following oviposition in several insects (e.g., Sareen et al. 1989; Hosken & Ward 1999), in *S. stercoraria*, the amount of secretion in the female gland—and hence gland volume—is negatively associated with copulation duration (Hosken & Ward 1999). While the precise role of gland secretion in dung flies is unknown, this association suggest it protects females in some way during copulation, and females that copulate more often die faster, which is consistent with this idea (Hosken et al. 2002). Additionally, experimental evolution suggests these glands are influenced by female mating patterns and conflict over fertilization. Here flies where forced to evolve either with or without sexual selection, and in lines evolving with sexual selection, female gland size was greater (Hosken et al. 2001). P2 was also lower in these females, and while preliminary work suggested the glands produce a spermicidal substance, detailed investigation refuted this (Bernasconi et al. 2002). In the experimental evolution lines there were no other detectable differences in female anatomy, which suggests larger glands help females thwart last male paternity monopolization (Hosken et al. 2001). Interestingly, sperm are frequently found in the glands of these (Hosken & Ward 1999), and other flies (Pitnick et al. 1999), which could explain the paternity results. Pitnick et al. (1999) suggested that sperm in *Drosophila* accessory glands could represent the initial stage in the evolution of a new sperm store. Gland size evolution in the dung fly lines also implies there is additive genetic variation for this character (Hosken et al. 2001).

It is clear that many of the substances produced by female glands are likely to be naturally selected to aid survival of eggs and offspring after oviposition. The very limited evidence available suggests female glands can be involved in sexual selection, but precisely how is still unclear. This is an area warranting further investigation.

MALE–FEMALE CO-EVOLUTION

Many mechanisms of sexual selection are predicted to generate co-evolution between male and female reproductive characters. For example, evolutionary change in female morphology may occur to enhance females' ability to choose particular males, leading to co-evolution between male and female form (Eberhard 1996): females prefer particular phenotypes, and because of this, male phenotypes evolve to fit the preference. Additionally, sexual conflict over paternity could also generate male–female co-evolution (Knowlton & Greenwell 1984; Parker 1979), and the potential for such sexually antagonistic selection is ubiquitous, although this potential may not always translate into selection or evolution (Parker 2006). Unfortunately, under some circumstances natural selection could also generate male–female co-evolution as could some neutral processes—pleiotropy for example.

However, many male primary reproductive traits are Y-linked (e.g., Carvalho et al. 2001; Morgan 1910; Simmons & Kotiaho 2002), making pleiotropy unlikely as a general means of generating male–female coevolution in the characters we have focused on. Furthermore, while many sexual selection models directly predict co-evolution of male and female characters (e.g., Lande 1980; Parker 1979), the same is not necessarily true with natural selection. Nevertheless, we are not aware of any studies that have tested for evidence of natural selection generating patterns of male–female co-evolution in the characters we discuss. For example, is there evidence of more or less spermathecal-testis size co-evolution in zones of secondary contact or in isolated island populations? Distinguishing between mechanisms of selection is difficult, and determining precisely what is generating co-evolution even more so. However, co-evolution of internal reproductive organs seems unlikely to be primarily due to natural selection for many of the same reasons genital evolution is unlikely to be due to natural selection (see Eberhard 1985), and as discussed above, there is evidence that many of these characters are involved in sexual selection in one way or another.

Across Species Patterns of Male-Female Coevolution

One of the most pervasive patterns of co-evolution is that between sperm form and aspects of the female sperm stores. For example, many comparative studies find associations between sperm length and the length of the female sperm stores (e.g., Dybas & Dybas 1981; Pitnick et al. 1995) or the length of the duct leading to the sperm stores (e.g., Minder et al. 2005; Morrow & Gage 2000). For readers interested in sperm-female co-evolution we recommend Birkhead et al. (2009). In addition to these patterns, several other associations have also been found. For example, associations between testis and female sperm–store size have been reported across moths and dung flies (Minder et al. 2005; Morrow & Gage 2000). This association could be fertility driven—females with bigger stores need more sperm—but the balance of evidence suggests either sexual conflict over paternity or cryptic female choice has generated these associations (Minder et al. 2005; Morrow & Gage 2000). For example, if cryptic female choice was based on the number of sperm males transferred during copulation, this could generate co-evolution between the size of the female sperm store and testis size: males with bigger testes are better able to fill the store and sire more offspring. Alternatively, selection could act on males to circumvent female control of paternity by completely filling sperm-stores, with counter selection on sperm store size to prevent this, and so on, which could also generate across species correlations between the two reproductive organs. In flies the testes and spermathecae are assumed to have separate genetic control (Sànchez & Guerrero 2001), so pleiotropy can probably be ruled out. Additionally, associations between spermathecal duct length and testis size are also found across dung flies (Minder et al. 2005), and the same forms of selection discussed above—cryptic female choice or sexual conflict—are probably responsible for this association too. Morrow and Gage (2000) also reported significant associations between the size of the female copulatory bursa and male accessory gland and testis size across moths. The suggestion was that this association is probably the result of selection for paternity protection (Morrow & Gage 2000). Unfortunately, there have not been many studies of this ilk and as a result, there is clearly scope for additional comparative work in this area.

Within Species Patterns of Male-Female Co-Evolution

As above, much of this work has focussed on sperm–female co-evolution (e.g., Pitnick et al. 2003), and we were surprised to find there are fewer intra-specific investigations of association between other male–female primary reproductive characters. The main approach used in these intra-specific investigations is experimental evolution (selection studies). One study forced yellow dung flies to evolve either with or without post-copulatory sexual selection and found male testis and female gland size evolved in response to this (Hosken & Ward 2001; Hosken et al. 2001; and see above). However, in another fly (*D. hydei*), direct selection on testis size generated no correlated response in the size of the female sperm store (or any other female trait examined) (Pitnick & Miller 2000). Finally, Simmons (2003) found genetic correlations between ovary and male accessory gland size in crickets, as did Fischer et al. (2008) for egg and spermatophore sizes in butterflies. Again, because of the paucity of data it is not really possible to make any general comments about the types of associations found or the likely causative agents responsible. However, patterns of intra-population associations would be helpful in this context, as seen with sperm–female associations (e.g., Pitnick et al. 2003).

There are plenty of examples indicating sexual selection is responsible for co-evolution of male and female internal reproductive traits. However, under some circumstances natural selection could also generate male–female co-evolution, as could some neutral processes such as pleiotropy. The difficult task that lies ahead will be to separate the relative importance of natural from sexual selection responsible for male–female co-evolution.

CONCLUSIONS AND FUTURE PROSPECTS

When we began writing this chapter we envisaged the co-evolution section as forming the bulk of our discussion. However, as we progressed we soon realized that there are in fact very few studies of male–female co-evolution in insects, especially those that focus on characters other than sperm and intromittent organs. This is clearly an area that would benefit from additional work. We also note that while there is ample evidence for sexual selection on various characters, almost invariably, net sexual selection has not been measured. This may seem trivial, but there is no a priori reason to expect pre and post-copulatory sexual selection to align, and there is evidence that at least sometimes they do not (e.g., Danielsson 2001). Similarly, in most investigations of the characters we discuss, investigators rarely consider natural selection as a potential explanation for character evolution. Nevertheless, the available evidence suggests that the reproductive characters we consider here are frequently subject to sexual selection directly or indirectly. This is most clearly seen for testes and spermatophores, the two characters subject to most intensive investigation. All potential mechanisms of sexual selection are implicated in the evolution of these characters in one way or another, and we are largely left with the difficult task of deciding the relative importance of each, although, as we stressed at the start of the chapter, the distinction between sperm competition and cryptic female choice is in some ways artificial. Having said that, these two mechanisms of selection are potentially discernible using a quantitative genetics approach, as only the latter explicitly predicts genetic and phenotypic correlations between male and female traits. If there are no correlations, then female choice can be ruled out, although the converse is not true. As a final caveat, ruling out all means of cryptic female choice—that is knowing which traits to include in the search for male-female correlations—is a Sisyphean task because of the vast array of potential ways to exercise such choice (see Eberhard 1996).

There is also some evidence that sperm-store characters are subject to or are generators of sexual selection, even though Walker's (1980) hypothesis has not been up-held with any conviction. However, as cautioned (Simmons 2001), this idea is probably limited to taxa with fixed-volume sperm-stores. Additionally, the limited data on the influence of spermathecal number on paternity are also not compelling, and this seems like an area that would benefit from comparative exploration. As recently noted (Clutton-Brock 2009) sexual selection on females via female-female reproductive competition is a neglected area of research, but we think this is possibly important in the evolution of some female characters we discuss. This is an area worthy of more work.' Finally, we encourage researchers to engage in more artificial selection and quantitative genetic exploration of the characters we discuss, particularly focusing on inter-sexual genetic correlations. These are both time and labor intensive undertakings, but have the potential to shed at least some light on the mechanisms of evolution and evolutionary constraints of these characters (Simmons 2003).

Acknowledgments We would like to thank the editors for kindly inviting us to contribute this chapter, BBSRC, NERC, The Leverhulme Trust, The European Social Fund, and The Royal Society for current funding, Darryl Gwynne and the two anonymous referees for many helpful comments on the manuscript, and wish to thank numerous colleagues for stimulating discussion of these matters over the last decades.

REFERENCES

Adams, E. M. & Wolfner, M. F. 2007. Seminal proteins but not sperm induce morphological changes in the *Drosophila melanogaster* female reproductive tract during sperm storage. Journal of Insect Physiology, 53, 319–331.

Adiyodi, K. G. & Adiyodi, R. G. 1975. Morphology and cytology of the accessory sex glands in invertebrates. International Review of Cytology, 43, 353–398.

Arnqvist, G. 1998. Comparative evidence for the evolution of genitalia by sexual selection. Nature, 393, 784–786.

Arthur, B. I., Hauschteck-Jungen, E., Nothiger, R. & Ward, P. I. 1998. A female nervous system is necessary for normal sperm storage in *Drosophila melanogaster*: a masculinized nervous system is as good as none. Proceedings of the Royal Society of London, Series B, 265, 1749–1753.

Baer, B. & Schmid-Hempel, P. 1999. Experimental variation in polyandry affects parasite load and fitness in a bumble-bee. Nature, 397, 151–154.

Baer, B., Morgan, E. D. & Schmid-Hempel, P. 2001. A non-specific fatty acid within the bumblebee mating plug prevents females from remating. Proceedings of the National Academy of Sciences USA, 98, 3926–3928.

Baer, B. & Boomsma, J. J. 2004. Male reproductive investment and queen mating frequency in fungus growing ants. *Behavioral Ecology*, 15, 426–432.

Baker R. H., Denniff, M., Futerman, P., Folwer, K., Pomiankowski, A. & Chapman, T. 2003. Accessory gland size influences time to sexual maturity and mating frequency in the stalk-eyed fly, *Cyrtodiopsis dalmanni*. *Behavioral Ecology*, 14, 607–611.

Bangham, J., Chapman, T. & Partridge, L. 2002. Effects of body size, accessory gland and testis size on pre- and post-copulatory success in *Drosophila melanogaster*. Animal Behaviour, 64, 915–921.

Bernasconi, G., Hellriegel, B, Heyland, A. & Ward, P. I. 2002. Sperm survival in the female repoductive tract in the fly *Scathophaga stercoraria* (L.). Journal of Insect Physiology, 48, 57–63.

Bernasconi, G., Brostaux, Y., Meyer, E. P. & Arnaud, L. 2006. Do spermathecal morphology and inter-mating interval influence paternity in the polyandrous beetle *Tribolium castaneum*? Behaviour, 143, 643–658.

Bertin, A. & Fairbairn, D. J. 2005. One tool, many uses: precopulatory sexual selection on genital morphology in *Aquarius remigis*. J. Evol. Biol. 18:949–961.

Bertin, A. & Fairbairn, D. J. 2007. The form of sexual selection on male genitalia cannot be inferred from within-population variance and allometry, a case study in *Aquarius regemis*. Evolution 61, 825–837.

Birkhead, T. R. & Hosken, D. J. & Pitnick, S. 2009. Sperm Biology: an Evolutionary Perspective. London: Academic Press.

Bissoondath C. J. & Wiklund, C. 1995. Protein content of spermatophores in relation to monandry/polyandry in butterflies. Behavioral Ecology & Sociobiology, 37, 365–371.

Blanckenhorn, W. U., Hellriegel, B,, Hosken, D. J., Jann, P., Altweg, R. & Ward, P. I. 2004. Does testis size track expected mating success in yellow dung flies. Functional Ecology, 18, 414–418.

Boldyrev, B. T. 1915. Contributions à l'étude de la structures des spermatophores et des particularities de la copulation chez Locustodea et Gryllidea. Horae Societatis Entomologicae Rossicae, 41, 1–245.

Boomsma, J. J., Baer, B. & Heinze, J. 2005. The evolution of male traits in social insects. Annual Review of Entomology, 50, 395–420.

Carvalho, A. B., Dobo, B. A., Vibranovski, M. D. & Clark, A. G. 2001. Identification of five new genes on the Y chromosome of *Drosophila melanogaster*. Proceedings of the National Academy of Sciences USA, 98, 13225–13230.

Chapman, T. 2001. Seminal fluid-mediated fitness traits in *Drosophila*. Heredity, 87, 511–521.

Clutton-Brock, T. Sexual selection in females. Animal Behaviour 77, 3–11.

Colonello, N. A. & Hartfelder, K. 2005. She's my girl—male accessory gland products and their function in the reproductive biology of social bees. Apidologie, 36, 231–244.

Cook, P. A. & Wedell, N. 1996. Ejaculate dynamics in butterflies: a strategy for maximizing fertilization success? Proceedings of the Royal Society of London, Series B, 263, 1047–1051.

Cordera Rivera, A., Andrés, J. A., Córdoba-Aguilar, A. & Utzeri, C. 2004. Postmating sexual selection: allopatric evolution of sperm competition mechanisms and genital morphology in Calopterygid damselflies (Insecta: Odonata). Evolution, 58, 349–359.

Córdoba-Aguilar, A. 1999. Male copulatory sensory stimulation induces female rejection of rival sperm in a damselfly. Proceedings of the Royal Society of London B, 266, 779–784.

Danielsson, I. 2001. Antagonistic pre and postcopulatory sexual selection on male body size in a water strider (*Gerris lacustris*). Proceedings of the Royal Society of London B, 268, 77–81.

Darwin, C. 1874. The Descent of Man and Selection in Relation to Sex. Reprinted in 1998 by Prometheus Books, Amherst, New York.

Davey, K. G. 1960. The evolution of spermatophores in insects. Proceeding of the Royal Entomological Society of London Series A, 35, 107–113.

Davey, K. G. 1985a. The male reproductive tract. In: Comprehensive Insect Physiology Biochemistry and Pharmacology. Volume 1. Embryogenesis and Reproduction (Ed. by G. A. Kerkut & L. I. Gilbert), pp. 1–14. Oxford: Pergamon Press.

Davey, K. G. 1985b. The female reproductive tract. In: Comprehensive Insect Physiology Biochemistry and Pharmacology. Volume 1. Embryogenesis and Reproduction (Ed. by G. A. Kerkut & L. I. Gilbert), pp. 15–36. Oxford: Pergamon Press.

Degrugillier, M. E. 1985. In vitro release of house fly, *Musca domestica* L. (Diptera: Muscidae), acrosomal material after treatments with secretion of female accessory gland and micropyle cap substance. International Journal of Insect Morphology & Embryology, 14, 381–391.

Demary, K. C. & Lewis, S. M. 2007. Male reproductive allocation in fireflies (Photinus spp.). Invertebrate Biology, 126, 74–80.

de Wilde, J. & de Loof, A. 1973. Reproduction. In: The Physiology of Insecta (Ed. by M. Rockstein), pp. 12–95. London: Academic Press.

Dickinson, J. L. & Rutowski, R. L. 1989. The function of the mating plug in the chalcedon checkerspot butterfly. Animal Behaviour, 38, 154–162.

Dybas, L. K. & Dybas, H. S. 1981. Coadaptation and taxonomic differentiation of sperm and spermathecae in featherwing beetles. Evolution, 35, 168–174.

Eady, P. 1994. Sperm transfer and storage in relation to sperm competition in

Callosbruchus maculatus. Behavioural Ecology & Sciobiology, 35, 123–129.

Eberhard, W. G. 1985. Sexual Selection and Animal Genitalia. Cambridge: Harvard University Press.

Eberhard, W.G. 1993. Evaluating models of sexual selection: genitalia as a test case. The American Naturalist, 142, 564–571.

Eberhard, W. G. 1996 Female Control: Sexual Selection by Cryptic Female Choice. Princeton: Princeton University Press.

Eberhard, W.G. 2004. Male–female conflict and genitalia: failure to confirm predictions in insects and spiders. Biological Reviews, 79, 121–186.

Eberhard, W. G., Huber, B. A., Rodriguez, R. L, Briceño, R. D., Salas, I. & Rodriguez, V. 1998. One size fits all? Relationship between the size and degree of variation in genitalia and other body parts in twenty species of insects and spiders. Evolution, 52, 415–431.

Evans, C. L. & Adler, P. H. 2000. Microsculpture and phylogenetic significance of the spermatheca of black flies (Diptera: Simuliidae). Canadian Journal of Zoology 78, 1468–1482.

Fairbairn, D. J., Vermette, R., Kapoor N. N. & Zahiri, N. 2003. Functional morphology of sexually selected genitalia in the water strider *Aquarius remigis*. Canadian Journal of Zoology, 81, 400–413.

Fairn, E. R., Schulte-Hostedde, A. I. & Alarie, Y. 2007. Sexual selection on accessory glands, genitalia and protarsal pads in the whirligig beetle *Dineutus nigrior* Roberts (Coleoptera: Gyrinidae). Ethology, 113, 257–266.

Falconer, D. S. 1981. Introduction to Quantitative Genetics, 2nd edition. London: Longman.

Fedina, T. Y. & Lewis, S. M. 2004. Female influence over offspring paternity in the red flour beetle *Tribolium castaneum*. Proceedings of the Royal Society of London B, 271, 1393–1399.

Fischer, K., Zimmer, K. & Wedell, N. 2008. Correlated responses to selection on female egg size in male reproductive traits in a butterfly. Evolutionary Ecology 23: 389–402.

Fitz, A. H. 2004. Sperm storage patterns in singly mated females of the Carribean fruit fly, *Anastrepha suspensa* (Diptera: Tehphritidae). Annals of the Entomological Society of America, 97, 1328–1335.

Fitz, A. H. & Turner, F. R. 2002. A light and electron microscopical study of the spermathecae and ventral receptacle of *Anastrepha suspense* (Diptera: Tephretidae) and implications in female influence of sperm storage. Arthropod Structure & Development, 30, 293–313.

Gage, M. J. G. 1994. Associations between body size, mating pattern, testis size and sperm lengths across butterflies. Proceedings of the Royal Society of London B, 258, 247–254.

Gage, M. J. G. 1995. Continuous variation in reproductive strategy as an adaptive response to population density in the moth *Plodia interpunctella*. Proceedings of the Royal Society of London B, 261, 25–30.

Gage, M. J. G. 1998 Influences of sex, size, and asymmetry on ejaculate expenditure in a moth. Behavioral Ecology 9, 592–597.

Gage, M. J. G. & Freckleton, R. P. 2003. Relative testis size and sperm morphometry across mammals: no evidence for an association between sperm competition and sperm length. Proceedings of the Royal Society of London B, 270, 625–632.

García-González, F. & Simmons, L. W. 2007. Shorter sperm confer higher competitive fertilization success. Evolution, 61, 816–824.

Gillott, C. 2003. Male accessory gland secretions: modulators of female reproductive physiology and behaviour. Annual Reviews of Entomology, 48, 163–184.

González, A., Rossini, C., Eisner, M. & Eisner, T. 1999. Sexually transmitted chemical defense in a moth (*Utetheisa ornatrix*). Proceedings of the National Academy of Sciences USA, 96, 5570–5574.

Gwynne, D. T. 1984. Courtship feeding increases female reproductive success in bushcrickets. Nature, 307, 361–363.

Gwynne, D. T. 1986. Courtship feeing in katydids (Orthoptera: Tettigoniidae): Investment in offspring or in obtaining fertilizations? American Naturalist, 128, 342–352.

Gwynne, D. T. 1997. The evolution of edible 'sperm sacs' and other forms of courtship feeding in crickets, katydids and their kin (Orthoptera: Ensifera). In: The Evolution of Mating Systems in Insects and Arachnids (Ed. By J. C. Choe & B. J. Crespi), pp. 110–129. Cambridge: Cambridge University Press.

Gwynne, D. T. 2001. Katydids and Bush-Crickets: Reproductive Behavior and Evolution of the Tettigoniidae. Ithaca and London: Cornell University Press.

Gwynne, D. T. 2005. The secondary copulatory organ in female ground weta (Hemiandrus pallitaris, Orthoptera: Anostostomatidae): a sexually selected device in females? Biological Journal of the Linnean Society, 85, 463–469.

Haerty, W., Jagadeeshan, S., Kulathinal, R. J., Wong, A., Ram, K. R., Sirot, L. K., Levesque, L., Artieri, C. G., Wolfner, M. F., Civetta, A. & Singh, R. S. 2007. Evolution in the fast lane: Rapidly evolving sex-related genes in Drosophila. Genetics, 177, 1321–1335.

Harcourt, A. H., Harvey, P. H., Larson, S. G. & Short, R. V. 1981. Testis weight, body weight and breeding system in primates. Nature, 293, 55–57.

Harshman, L. G. & Prout, T. 1994. Sperm displacement without sperm transfer in Drosophila melanogaster. Evolution, 48, 758–766.

He, Y. & Tsubaki, Y. 1992. Variation in spermatophore size in the armyworm, Pseudaletia separata (Lepidoptera: Noctuidae) in relation to rearing density. Applied Entomology & Zoology, 27, 39–45.

Hellriegel, B. & Bernasconi, G. 2000. Female-mediated differential sperm storage in a fly with complex spermathecae, Scatophaga stercoraria. Animal Behaviour, 59, 311–317.

Hellriegel, B., & Blanckenhorn, W.U. 2002. Environmental influences on the gametic investment of yellow dung fly males. Evolutionary Ecology. 16: 505–522.

Hellriegel, B. & Ward, P. I. 1998. Complex female reproductive tract morphology: its possible use in postcopulatory female choice. Journal of Theoretical Biology, 190, 179–186.

Holland, B. & Rice, W. R. 1998. Chase-away sexual selection: antagonistic seduction versus resistance. Evolution, 52, 1–7.

Honek, A. 1993. Intraspecific variation in body size and fecundity in insects—a general relationship. Oikos, 66, 483–492.

Hosken, D. J. 1997. Sperm competition in bats. Proceedings of the Royal Society of London B, 264, 385–392.

Hosken, D. J. 1999. Sperm displacement in yellow dung flies: a role for females. Trends in Ecology & Evolution, 14, 251–252

Hosken, D. J. 2001 Sex and death: microevolutionary trade-offs between reproductive and immune investment in dung flies. Current Biology, 11, R379–R380.

Hosken, D. J. & Stockley, P. 2004. Sexual selection and genital evolution. Trends in Ecology and Evolution, 19, 87–93.

Hosken, D. J. & Ward, P.I. 1999. Female accessory reproductive gland activity in the yellow dung fly Scathophaga stercoraria (L.). Journal of Insect Physiology, 45, 809–814.

Hosken, D. J. & Ward, P.I. 2000. Copula in yellow dung flies (Scathophaga stercoraria): investigating sperm competition models by histological observation. Journal of Insect Physiology, 46, 1355–1363.

Hosken, D. J. & Ward, P.I. 2001. Experimental evidence for testis size evolution via sperm competition. Ecology Letters, 4, 10–13.

Hosken, D. J., Meyer, E. P. & Ward, P. I. 1999. Internal female reproductive anatomy and genital interactions during copula in the yellow dung fly, Scathophaga stercoraria (Diptera: Scathophagidae). Canadian Journal of Zoology, 77, 1975–1983.

Hosken, D. J., Jones, K. E., Chipperfield, K. & Dixson, A. 2001. Is the bat penis sexually selected? Behavioural Ecology & Sociobiology, 50, 450–460.

Hosken, D. J., Garner, T. W. J. & Ward, P. I. 2001. Sexual conflict selects for male and female reproductive characters. Current Biology, 11, 489–493.

Hosken, D. J., Uhia, E. & Ward, P. I. 2002. The function of female accessory reproductive gland secretion and a cost to polyandry in the yellow dung fly. Physiological Entomology, 27, 87–91.

Hosken, D. J., Garner, T. W. J. & Blanckenhorn, W. U. 2003. Asymmetry, testis and sperm size in yellow dung flies. Functional Ecology, 17, 231–236.

Hosken, D. J., Minder, A. M. & Ward, P. I. 2005. Male genital allometry in Scathophagidae (Diptera). Evolutionary Ecology, 19, 501–515.

House, C. M. & Simmons, L. W. 2003. Genital morphology and fertilization success in the dung beetle Onthophagus taurus: an example of sexually selected male genitalia. Proceedings of the Royal Society of London B, 270, 447–455.

House, C. M. & Simmons, L. W. 2005. Relative influence of male and female genital morphology on paternity in the dung beetle Onthophagus taurus. Behavioural Ecology, 16, 889–897

Ilango, K. 2004. Phylogeny of the old world phlebotimine sand flies (Diptera: Psychodidae) with special reference to structural diversity of female spermathecae. Oriental Insects, 38, 419–462.

Kaitala, A. & Wiklund, C. 1995. Female mate choice and mating costs in the polyandrous butterfly *Pieris napi* (Lepidoptera: Pieridae). Journal of Insect Behavior, 8, 355–363.

Klowden, M. J. & Chambers, G. M. 2004. Production of polymorphic sperm by anopheline mosquitoes and their fate within the female genital tract. Journal of Insect Physiology, 50, 1163–1170.

Knowlton, N. & Greenwall, S. R. 1984. Male sperm competition avoidance mechanisms: the influence of female interests. In: Sperm Competition and the Evolution of Animal Mating Systems (Ed. by R. L. Smith), pp. 62–85. London: Academic Press.

Lande, R. 1980. Sexual dimorphism, sexual selection, and adaptation in polygenic characters. Evolution, 34, 292–305.

LaMunyon, C. W. & Eisner, T. 1993. Postcopulatory sexual selection in an arctiid moth (*Utetheisa ornatrix*). Proceedings of the National Academy of Sciences USA, 90, 4689–4692.

Linley, J. R. & Simmons, K. R. 1981. Sperm motility and spermathecal filling in lower Diptera. International Journal of Invertebrate Reproduction, 4, 137–145.

Lung, O. & Wolfner, M. F. 2001. Identification and characterization of the major *Drosophila melanogaster* mating plug protein. Insects Biochemistry and Molecular Biology, 31, 543–551.

Manetti, A. G. O., Roseto, M., de Filippis, T., Marchini, D., Baldari, C. T. & Dallai, R. 1997. Juvenile hormone regulates the expression of the gene encoding Ceratotoxin A, an antibacterial peptide from the female reproductive accessory glands of the medfly *Ceratitis capitata*. Journal of Insect Physiology, 43, 1161–1167.

Mann, T. 1984. Spermatophores. Zoophysiology Vol 15. Springer-Verlag.

Marchini, D., Bernini, L. F. & Dallai, R. 1989. B-N-Acetylhexosaminidase in the secretion of the female reproductive accessory glands of Ceratitis capitata (Diptera). Insect Biochemistry 19, 549–555.

Matsumoto, K. & Suzuki, N. 1992. Effectiveness of the mating plug in *Atrophaneura alcinous* (Lepidoptera: Papilionidae). Behavioral Ecology & Sociobiology, 30, 157–163.

Matsumoto, K. & Suzuki, N. 1995. The nature of mating plugs and the probability of reinsemination in Japanese Papilionidae. In: *Swallowtail Butterflies: Their Ecology and Evolutionary Biology* (eds. J. M. Scriber, Y. Tsubaki, Y. and R. C. Lederhauser), pp. 145–155. Gainesville: Scientific Publishers.

Mikheyev, A. S. 2003. Evidence for mating plugs in the fire ant *Solenopsis invicta*. Insectes Sociaux, 50, 401–402.

Mikheyev, A. S. 2004. Male accessory gland size and the evolutionary transition from single to multiple mating in the fungus-gardening ants. Journal of Insect Science, 4, article number 37.

Miller, G. T. & Pitnick, S. 2002. Female–sperm coevolution in *Drosophila*. Science, 298, 1230–1233.

Miller, G. T., Starmer W. T. & Pitnick, S. 2001. Quantitative genetics of seminal receptacle length in *Drosophila melanogaster*. Heredity, 87, 25–32.

Miller, G. T., Starmer W. T. & Pitnick, S. 2003. Quantitative genetic analysis of among population variation in sperm and female sperm-storage organ length in *Drosophila mojavensis*. Genetical Research, 81, 213–220.

Minder, A. M. 2002. Co-evolution of sperm, testis and female reproductive tract morphology in Scathophagidae. Diploma Thesis, University of Zürich.

Minder, A. M., Hosken, D. J. & Ward, P. I. 2005. Co-evolution of male and female reproductive characters across the Scathophagidae. Journal of Evolutionary Biology, 18, 60–69.

Morgan, T. H. 1910. Science, 32, 120–122.

Moore, P. J., Harris, W. E., Montrose, V. T., Levin, D. & Moore A. J. 2004. Constraints on evolution and postcopulatory sexual selection: trade-offs among ejaculate characteristics. Evolution, 58, 1773–1780.

Morrow, E. H. & Gage, M. J. G. 2000. The evolution of sperm length in moths. Proceedings of the Royal Society of London B, 267, 307–313.

Orr, A. G. 1995. The evolution of sphragis in the Papilionidae and other butterflies. In: Swallowtail Butterflies: Their Ecology and Evolutionary Biology (eds. J. M. Scriber, Y. Tsubaki & R. C. Lederhauser), pp. 155–164. Gainesville: Scientific Publishers.

Orr, A. G. 2002. The sphragis of Heteronympha penelope Waterhouse (Lepidoptera: Satyridae): its structure, formation and role in sperm guarding. Journal of Natural History, 36, 185–196.

Orr, A. G. & Rutowski, R. L. 1991. The function of the sphragis in Cressida cressida (Fab) (Lepidoptera, Papilionidae): A visual deterrent to copulation attempts. Journal of Natural History, 25, 703–710.

Otronen M., Reguera P. & Ward P. I. 1997. Sperm storage in the yellow dung fly Scathophaga stercoraria: Identifying the sperm of competing males in separate female spermathecae. Ethology 103, 844–854.

Page, R. E. 1986. Sperm utilization in social insects. Annual Reviews in Entomology, 31, 297–320.

Parker, G. A. 1970. Sperm competition and its evolutionary consequences in the insects. Biological Reviews, 45, 525–567.

Parker, G. A. 1979. Sexual selection and sexual conflict. In: Sexual Selection and Reproductive Competition in Insects (Ed. by M. S. Blum & N. A. Blum), pp. 123–166. New York: Academic Press.

Parker, G. A. 1982. Why are there so many tiny sperm? Sperm competition and the mainte-nance of two sexes. Journal of theoretical Biology, 96, 281–294.

Parker, G. A. 1984. Sperm competition and the evolution of animal mating strategies. In: Sperm Competition and the Evolution of Animal Mating Systems (Ed. by R. L. Smith), pp. 1–60. London: Academic Press.

Parker, G. A. 1998. Sperm competition and the evolution of ejaculates: towards a theory base. In: Sperm Competition and Sexual Selection (Ed. by T. R. Birkhead & A. P. Møller), pp. 1–54. Academic Press: London.

Parker, G. A. 2006. Sexual conflict over mating and fertilization: an overview. Philosophical Transactions of the Royal Society B, 361, 235–260.

Parker, G. A., Ball, M. A., Stockley, P. & Gage, M. J. G. 1996. Sperm competition games: individual assessment of sperm competition intensity by groups spawners. Proceedings of the Royal Society of London B, 263, 1291–1297.

Parker, G. A., Simmons, L. W., Stockley, P., McChristie, D. M. & Charnov, E. L. 1999. Optimal copula duration in yellow dung flies: effects of female size and egg content. Animal Behaviour, 57, 795–805.

Petersen, W. 1907. Über die Spermatophoren der Schmetterlinge. Zeitschrift für Wissenschaften Zoologie, 88, 117–130.

Pitkin, B. R. 1988. Lesser Dung Flies. Diptera:Sphaeroceridae. Handbook for the Identification of British Insects, Vol. 10, Part 5E. Royal Entomological Society of London.

Pitnick, S. 1996. Investment in testes and the costs of making long sperm in Drosophila. The American Naturalist, 148, 57–80.

Pitnick, S. & Brown, W. D. 2000. Criteria for demonstrating female sperm choice. Evolution, 54, 1052–1056.

Pitnick, S. & Markow, T. A. 1994. Male gametic strategies: sperm size, testis size, and alloca-tion of ejaculates among successive mates by the sperm limited fly Drosophila pachea and its relatives. The American Naturalist, 143, 785–819.

Pitnick, S. & Miller, G. T. 2000. Correlated response in reproductive and life history traits to selection on testis length in Drosophila hydei. Heredity, 84, 416–426.

Pitnick, S., Markow, T.A. & Spicer, G.S. 1995. Delayed male maturity is a cost of producing large sperm in Drosophila. Proceedings of the National Academy of Sciences USA, 92, 10614–10618.

Pitnick, S., Markow, T.A. & Spicer, G.S. 1999. Evolution of multiple kinds of female sperm storage organs in Drosophila. Evolution, 53, 1804–1822.

Pitnick, S., Miller, G.T., Reagan, J., & Holland, B. 2001. Evolutionary responses by males to experimental removal of sexual selection. Proceedings of the Royal Society of London B, 268, 1071–1080.

Pitnick, S., Miller, G.T., Schneider, K. & Markow, T.A. 2003. Ejaculate-female coevolution in Drosophila mojavensis. Proceedings of the Royal Society of London B, 270, 1507–1512.

Polak, M., Starmer, W. T. & Barker, J. S. F. 1998. A mating plug and male choice in Drosophila hibisci Bock. Animal Behaviour, 56, 919–926.

Polak, M., Wolf, L. L., Starmer, W. T & Barker, J. S. F. 2001. Function of the mating plug in Drosophila hibisci Bock. Behavioral Ecology & Sociobiology, 49, 196–205.

Pont, A. C. & Meier, R. 2002. The Sepsidae (Diptera) of Europe. Fauna Entomologica Scandinavica Vol. 37. Leiden: Brill.

Presgraves, D.C., Baker, R.H. & Wilkinson, G.S. 1999. Coevolution of sperm and female reproductive tract morphology in stalk-eyed flies. Proceedings of the Royal Society of London B, 266, 1041–1047.

Ram, K. R. & Wolfner, M. F. 2007. Seminal influences: *Drosophila* Acps and the molecular interplay between males and females during reproduction. Integrative and Comparative Biology, 47, 427–445.

Ridley, M. 1989. The incidence of sperm displacement in insects: four conjectures, one corroboration. Biological Journal of the Linnean Society, 38, 349–367.

Rodriguez, V. 1994. Function of the spermathecal muscle in *Chelymorpha alternans* Boheman (Coleoptera: Chrysomelidae: Cassidinae). Physiological Entomology, 19, 198–202.

Rosetto, M., Manetti, A. G., Giordano, P. C., Marri, L., Amons, R., Baldri, C. T. Marchini, D. & Dallai, R. 1996. Molecular characterisation of ceratotoxin C, a novel antibacterial female-specific peptide of the ceratoxin family from the medfly *Ceratitis capitata*. European Journal of Biochemistry, 241, 330–337.

Sànchez, L. & Guerrero, I. 2001. The development of the *Drosophila* genital disc. BioEssays, 23, 698–707.

Sareen ML, Sood P & Thukrad D. 1989. The female accessory glands in Musca *domestica nebulo* F. Diptera Muscidae. Research Bulletin of the Punjab University of Science 40, 213–216.

Sauter, A., Brown, M. J. F., Baer, B. & Schmid-Hempel, P. 2001. Males of social insects can prevent queens from multiple mating. Proceedings of the Royal Society of London B, 268, 1449–1454.

Schärer, L., Ladurner, P. & Rieger, R.M. 2004. Bigger testes do more work: experimental evidence that testis size reflects testicular cell proliferation activity in the marine invertebrate, the free-living flatworm Macrostorum sp. Behavioral Ecology and Sociobiology, 65, 420–425.

Scott, J. A. 1986. The Butterflies of North America. Stanford: Stanford University Press.

Shepherd, J. G. 1975. Polypeptide sperm activator from male saturnid moths. Journal of Insect Physiology, 21, 9–22.

Short, R. V. 1997. The testis: the witness of the mating system, the site of mutation and the engine of desire. Acta Paediatrica Supplement, 422, 3–7.

Simmons, L.W. 1995a. Courtship feeding in katydids (Orthoptera: Tettigoniidae): investment in offspring and in obtaining fertilizations. American Naturalist, 146, 307–315.

Simmons, L.W. 1995b. Male bushcrickets tailor their spermatophores in relation to their remating intervals. Behavioral Ecology, 9, 881–886.

Simmons, L. W. 2001. Sperm Competition and its Evolutionary Consequences in the Insects. Princeton: Princeton University Press.

Simmons, L. W. 2003. The evolution of polyandry: patterns of genotypic variation in female mating frequency, male fertilization success and a test of the sexy sperm hypothesis. Journal of Evolutionary Biology, 16, 624–634.

Simmons, L. W. & Emlen, D. J. 2006. Evolutionary trade-off between weapons and testes. Proceedings of the National Academy of Science USA, 103, 16346–16351.

Simmons, L. W. & Kotiaho, J. S. 2002. Evolution of ejaculates: patterns of phenotypic and genotypic variation and condition dependence in sperm competition traits. Evolution, 56, 1622–1631.

Simmons, L. W. & Kotiaho, J. S. 2007. Quantitative genetic correlation between trait and preference support a sexually selected sperm process. Proceedings of the National Academy of Sciences, 104, 16604–16608.

Simmons, L. W. & Kvarnemo, C. 1997. Ejaculate expenditure by male bushcrickets decreases with sperm competition intensity. Proceedings of the Royal Society of London B, 264, 1203–1208.

Simmons, L. W. & Siva-Jothy, M. T. 1998. Sperm competition in insects: mechanisms and the potential for selection. In: Sperm Competition and Sexual Selection (Ed. by T. R. Birkhead & A. P. Møller), pp. 341–434. London: Academic Press.

Simmons, L. W., Tomkins, J. L. & Hunt, J. 1999. Sperm competition games played by dimorphic beetles. Proceedings of the Royal Society of London B, 266, 145–150.

Simons, A. M. & Roff, D. A. 1994. The effect of environmental variability on the heritabilities of traits of a field cricket. Evolution, 48, 1637–1649.

Sirot, L. K. 2003. The evolution of insect mating structures through sexual selection. Florida Entomologist, 86, 124–133.

Siva-Jothy, M. T. 1987. The structure and function of the female sperm-storage organs in libellulid dragonflies. Journal of Insect Physiology, 33, 559–567.

Siva-Jothy, M. T. & Hooper, R. E. 1995. The disposition and genetic diversity of stored sperm in females of the damselfly *Calopteryx splendens xanthostoma*. Proceedings of the Royal Society of London B, 259, 313–318.

Siva-Jothy, M. T. & Hooper, R. E. 1996. Differential use of stored sperm during oviposition in the damselfly *Calopteryx splendens xanthostoma* (Charpentier). Behavioural Ecology & Sociobiology, 39, 389–394.

Snook, R. R. 2005. Sperm in competition: not playing by the numbers. Trends in Ecology & Evolution, 20, 46–53.

Snook, R. R. & Hosken, D. J. 2004. Sperm death and dumping in *Drosophila*. Nature, 428, 939–941.

Socha, R. 2006. Endocrine control of wing morph-related differences in mating success and accessory gland size in male firebugs. Animal Behaviour, 71, 1273–1281.

Socha, R. & Hodkova, M. 2006. Corpus allatum volume-dependent differences in accessory gland maturation in long- and short-winged males of *Pyrrhocoris apterus* (Heteroptera: Pyrrhocoridae). European Journal of Entomology, 103, 27–32.

Starmer, W. T, Polak, M., Pitnick, S., McEvey, S. F., Barker, J. S. F. & Wolf, L. L. 2003. Phylogenetic, geographical ad temporal analysis of female reproductive tradeoff in *Drosophilidae*. Evolutionary Biology, 33, 139–171.

Stockley, P. 2002 Sperm competition risk and male genital anatomy: comparative evidence for reduced duration of female sexual receptivity in primates with penile spines. Evolutionary Ecology, 16, 123–137.

Stockley, P & Seal, N. J. 2001. Plasticity in reproductive effort of male dung flies (*Scatophaga stercoraria*) as a response to larval density. Functional Ecology, 15, 96–102.

Stockley, P., Gage, M. J. G., Parker, G. A. and Møller, A. P. 1997. Sperm competition in fishes: the evolution of relative testis size and ejaculate characteristics. The American Naturalist, 149, 933–954.

Svärd, L. & Wiklund, C. 1989. Mass and production rate of ejaculates in relation to monandry/polyandry in butterflies. Behavioral Ecology & Sociobiology, 24, 395–402.

Swanson, W. J. and Vacquier, V. D. 2002. The rapid evolution of reproductive proteins. Nature Genetics, 3, 137–144.

Throckmorton, L.H. 1962. The problem of phylogeny in the genus *Drosophila*. University of Texas Publications, 6205, 207–343.

Twig, E. & Yuval, B. 2005. Function of multiple sperm storage organs in female Mediterranean fruit flies (*Ceratitis capitata*, Diptera: Tephritidae). Journal of Insect Physiology, 51, 67–74.

Vahed, K. 2006. Larger ejaculate volumes are associated with a lower degree of polyandry across bushcricket taxa. Proceedings of the Royal Society of London B, 273, 2387–2394.

Vahed, K. 2007. Comparative evidence for a cost to males of manipulating females in bush-crickets. Behavioral Ecology, 18, 499–506.

Vahed, K. & Gilbert, F. S. 1996. Differences across taxa in nuptial gift size correlate with differences in sperm number and ejaculate volume in bushcrickets. Proceedings of the Royal Society of London B, 263, 1257–1265.

Villvaso, E. J. 1975a. The role of the spermathecal gland of the boll weevil, *Anthonomus grandis*. Journal of Insect Physiology, 21, 1457–1462.

Villvaso, E. J. 1975b. Functions of the spermathecal muscle of the boll weevil, *Anthonomus grandis*. Journal of Insect Physiology, 21, 1275–1278.

Waage, J. K. 1979. Dual function of the damselfly penis: sperm removal and transfer. Science, 203, 916–918.

Waage, J. K. 1984. Sperm competition and the evolution of Odonate mating systems. In: Sperm Competition and the Evolution of Animal Mating Systems. (Ed. by R.L. Smith) pp. 251–290. London: Academic Press.

Walker, W. F.1980. Sperm utilization strategies in nonsocial insects. The American Naturalist, 115, 780–799.

Ward, P.I. 1993. Females influence sperm storage and use in the yellow dung fly *Scathophaga stercoraria* (L.). Behavioral Ecology & Sociobiology, 32, 313–319.

Ward, P. I. 2000. Cryptic female choice in the yellow dung fly *Scathophaga stercoraria* (L.). Evolution, 54, 1680–1686.

Wedell, N. 1991. Sperm competition selects for nuptial feeding in a bushcricket. Evolution, 45, 1975–1978.

Wedell, N. 1993. Spermatophore size in bushcrickets: comparative evidence for nuptial gifts as a sperm protection device. Evolution, 47, 1203–1212.

Wedell, N. 1994. Dual function of the bush cricket spermatophore. Proceedings of the Royal Society of London Series B, 258, 181–185.

Wedell, N. 1996. Mate quality affects reproductive effort in a paternally investing species. The American Naturalist, 148, 1075–1088.

Wedell, N. 1998. Sperm protection and mate assessment in the bushcricket *Coptaspis sp. 2*. Animal Behaviour, 56, 357–363.

Wedell, N. 2005. Female receptivity in butterflies and moths. Journal of Experimental Biology 208, 3433–3440.

Wedell, N. 2006. Male genotype affects female fitness in a paternally investing species. Evolution, 60, 1638–1645.

Wedell, N. & Arak, A. 1989. The wartbiter spermatophore and its effect on female reproductive output (Orthoptera: Tettigoniidae, *Decticus verrucivorus*). Behavioural Ecology & Sociobiology, 24, 117–125.

Wedell, N. & Ritchie, M. G. 2004. Male age, mating status and nuptial gift quality in a bushcricket. Animal Behaviour, 67, 1059–1065.

Wertheim, B., van Baalen, E.-J. A., Dicke, M. & Vet, L. E. M. 2005. Pheromone-mediated aggregation in non-social arthropods: An evolutionary ecological perspective. Annual Reviews in Entomology, 50, 321–346.

West-Eberhard, M. J. 1984. Sexual selection, competitive communication and species-specific signals in insects. In: Insect Communication (Ed. by T. Lewis), pp. 283–324. New York: Academic Press.

White-Cooper, H., Dogett, K. & Ellis, R. 2009. Spermatogenesis. In: Sperm Biology: an Evolutionary Perspective (Ed. by T.R. Birkhead, D.J. Hosken & S. Pitnick) pp. 151–183. London: Academic Press.

Wilfert, L., Gadau, J. & Schmid-Hempel, P. 2007. The genetic architecture of immune defence and reproduction in male *Bombus terrestris* bumblebees. Evolution, 61, 8040–815.

Wilkinson, G. S., Amitin, E. G. & John, P. M. 2005. Sex-linked correlated responses in female reproductive traits to selection on male eye span in stalk-eyed flies. Integrative and Comparative Biology, 45, 500–510.

Wolfner, M. F. 1997. Tokens of love: function and regulation of *Drosophila* male accessory gland products. Insect Biochemistry & Molecular Biology, 27, 179–192.

15

Selective Forces Propelling Genitalic Evolution in Odonata

ADOLFO CORDERO-RIVERA AND ALEX CÓRDOBA-AGUILAR

INTRODUCTION

Odonates are considered to be among the oldest insects (Silsby 2001), with fossil odonatoid insects known from the upper Carboniferous (about 300 million years ago). Although they have changed very little in morphology since the Jurassic (65 million years), their behavior is by no means simple, showing pre-copulatory courtship, intense male–male contests, post-copulatory associations between males and females, and other elaborate behaviors (figure 15.1). Odonates are the only insect group whose males do not have the penis directly connected to the testis, but use a seminal vesicle (figure 15.2) for temporary sperm storage. This fact explains the need for intra-male sperm transfer (figure 15.1c), before each mating, when the male translocates his sperm from the testis, opening at the end of the abdomen, to the seminal vesicle, situated under the second and third abdominal segments (figure 15.2). In some species, males perform elaborate precopulatory courtship, slowly flying around the female, and simultaneously exposing wing and body coloration (figure 15.1a,b), for instance in Calopterygidae (Heymer 1973). In other cases, males simply capture mature females with their anal appendages forming the precopulatory tandem, and then perform an "invitation" to copulate (Robertson and Tennessen 1984), by elevating the abdomen and vibrating their wings. Only if the female touches the male's secondary

genitalia, does the male proceed to sperm translocation and copulation.

When a mating couple finishes copulation, the male may or may not guard the female during oviposition (either in tandem or by remaining close to her). Reproductive behavior in odonates is "classical" in the sense that males compete for females, who are the limiting resource for reproduction, and sexual selection is intense, especially on males (Banks & Thompson 1985; Conrad and Pritchard 1992; Córdoba-Aguilar 2002b; Fincke 1986; Fincke and Hadrys 2001).

Odonates are popular for research perhaps because their reproductive behavior is a typical text-book example for postcopulatory sexual selection: males use their penis for a dual function, removal of rivals' sperm during the first part of copulation, and transfer of their own sperm, during the final part (Córdoba-Aguilar et al. 2003b). The description of this fact by Waage (1979) opened a new era in sexual selection studies, clearly under the influence of the *seminal* work on sperm competition by Parker (1970). Removing or repositioning rivals' sperm is obviously advantageous for males, and there is no doubt that selective forces (sperm competition, see below) have contributed to the evolution of sperm removal behavior, not only in odonates but other insects as well (Haubruge et al. 1999; Kamimura 2000; Ono et al. 1989; Yokoi 1990; reviewed by Simmons 2001). However, research on genitalic evolution and sexual biology

332

FIGURE 15.1 Reproductive behavior of a typical territorial odonate (*Calopteryx haemorrhoidalis*, Calopterygidae). Males perform precopulatory courtship, sometimes landing over water or appropriate oviposition substrates (a), and slowly flying around the female (b). If the female remains motionless, the male is able to grasp her, and perform the intra-male sperm translocation (c), transferring the sperm from the testes, whose opening is at the end of the abdomen, to the secondary genitalia, situated under the segments 2 and 3. Copulation follows (d) and is usually divided into a first stage, where rivals' sperm removal takes place, and a stage II, when insemination occurs. At the end of copulation the male flies directly to the territory, and the female sometimes remains in a "postcopulatory rest" (e). In some cases females release a drop of sperm (f) after mating. Photos by A. Cordero Rivera.

in general (see for example explanations for male postcopulatory behavior by Waage 1984) has been biased or shaped by sperm competition theories, and sexual selection in general (Fincke et al. 1997), and tests for other selective forces have been rare. For example, what about female interests during sperm competition in odonates? Fincke (1997) provides a comprehensive review of the potential for female choice and benefits to females of mating multiply, and offers alternative (or complementary) explanations for mating patterns in odonates. If the last male to copulate with a female always removes the sperm, females would certainly lose when re-mating with a low quality male after having mated with a good male. Therefore we should expect females to retain control over fertilization, and exercise cryptic female choice (Eberhard 1996). The fact that female odonates mate multiply (e.g., Córdoba-Aguilar et al. 2003a) is the key for the existence of a conflict of interests between the sexes (Arnqvist and Rowe 2005).

This chapter reviews the available evidence for sexual selection forces shaping genitalic evolution. We also review briefly other hypotheses that have not been tested in odonates and that are not related to sexual selection. These are the lock and key (Shapiro & Porter 1989) and pleiotropy hypotheses (Mayr 1963). Odonate genitalia have been mainly studied from the point of view of sexual selection, as secondary sexual characters, and little has been done from the standpoint of natural selection, which is bizarre given that genitalia are traditionally considered primary sexual characters. This is probably due to the existence of secondary genitalia in this order, which is a unique character of odonates.

GENITALIC MORPHOLOGY AND FUNCTION

The copulatory apparatus of male odonates is not homologous with any organ in the Animal Kingdom (Tillyard 1917). It is developed from the second sternite, and consists of a penis, a seminal vesicle (*vesica spermalis*), and a series of accessory structures (hamuli anteriores, hamuli posteriores) that protect the penis, and presumably help to achieve genital connection during copulation (figure 15.2). Little is known about the evolution of these structures, but they seem to be already present in the Mesozoic fossil *Tarsophlebia eximia*, and can be

observed in some Protozygoptera (Fleck et al. 2004). In the odonate literature, these structures are known as the "secondary genitalia", to distinguish them from the "primary genitalia" found at abdominal segment 9 (in males, reduced to two scales closing the genital pore). This distinction is not related with the primary and secondary sexual characters which are the focus of this book. For further morphological details, the reader can consult Tillyard's (1917) monograph, which remains a rich source of information for the anatomy of this order, and the detailed functional morphology work of Pfau (1971, 1991, 2005). Fleck et al. (2004) discuss the possible origin of the secondary genital apparatus in the Odonata, and suggest that the stem species of Odonata did not remove rival's sperm, but rather transferred a spermatophore to the female. The oldest dragonflies known from the fossil record (Odonata-like insects from the Upper Carboniferous) seem to have had a paired penis with a pair of lateral parameres and a pair of segmented, leaf-like gonopods at the end of the abdomen, and therefore were unlikely to form a copulatory wheel like modern odonates (Bechly et al. 2001). This makes it possible for several interpretations of the origin of the secondary copulatory apparatus of modern odonates to be constructed (Bechly et al. 2001). Unfortunately, the fossil record cannot say much about precopulatory behavior.

Three different structures act as intromittent organ (hereafter penis) in the Odonata. Given their use for sperm removal, the most parsimonious explanation implies that this specialized behavior has evolved three times independently in modern odonates (Bechly et al. 2001). In damselflies (Zygoptera), the intromittent organ is the *ligula* or aedeagus (figure 15.2). It is a chitinized arc, with an inflatable membrane, and variable morphology at the distal part (spoon-like, a variable number of stout appendages, "horns", flagella, and so on). It has no direct connection with the *vesica spermalis*. The sperm is conducted in a furrow of the *ligula* during insemination (figure 15.2) (Pfau 1991). In contrast with the penis of the Anisoptera, which is supplied with internal muscles and tracheae, the penes of the Zygoptera seem not to have such structures (Tillyard 1917), but in both suborders some nerves are present inside the penis (Uhía & Cordero Rivera 2005).

In dragonflies (Anisoptera), the distal part of the *vesica spermalis* acts as a penis (figure 15.2), and

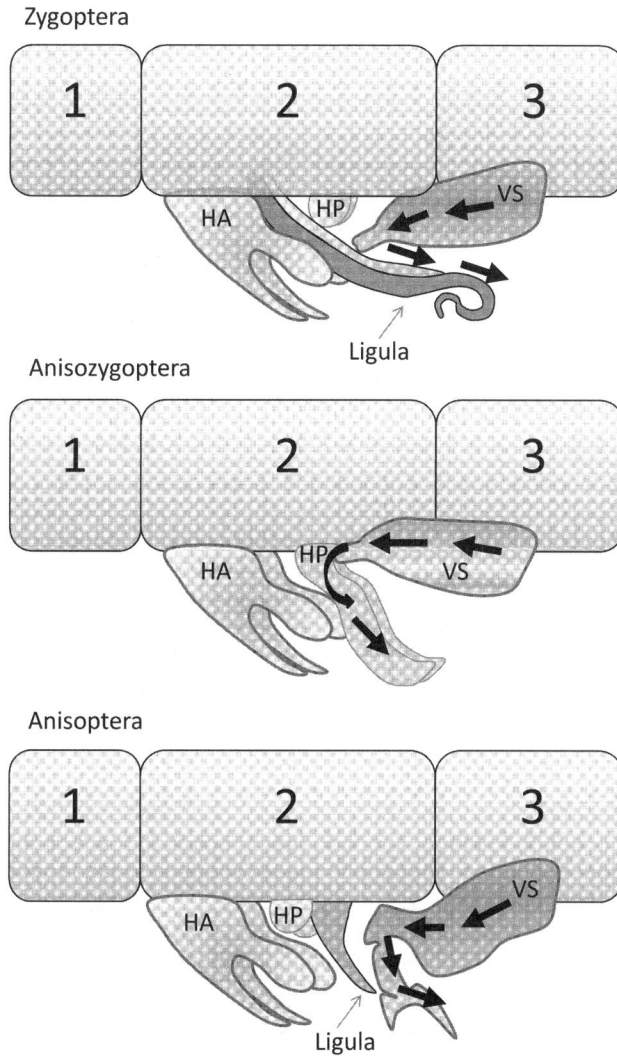

FIGURE 15.2 External genitalic structures of male odonates, in schematic representation. Note that the intromittent organ (the functional penis) is a different structure in each taxon. Numbers indicate abdominal segments, and the arrows show the way of sperm during insemination. HA: *hamuli anteriores*, HP: *hamuli posteriores*, VS: *vesica spermalis*.

has therefore been modified into three parts, a basal joint, which is strongly chitinized, a second element, usually curved, and with the orifice at the end, and a third element, very variable in form, and sometimes with flagella (figure 15.3c, e) or short lateral flaps, only visible when the penis is erected. The penis projects forwards from the seminal vesicle, situated ventrally on the third abdominal segment (figure 15.2), to which it is directly connected. The sperm, which in dragonflies is usually

transferred in groups or "spermatodems" (see below), is temporally stored in the vesicle, and transferred to the female at the end of copulation.

In the Anisozygoptera (previously considered a separate suborder but now included among the Anisoptera; Bybee et al. 2008), a group which only has two extant species, the penis consists of the paired *hamuli posteriores* (figure 15.3), which are pressed against each other and form a tube of two halves (Pfau 1991).

During copulation, nerve cells in the penis are likely used to detect the presence of sperm inside the female, probably by means of chemical sensilla (Andrés & Cordero Rivera 2000; Uhía & Cordero Rivera 2005). The first part of copulation (named "stage I") is characterized by rhythmic movements of the penis, which remove sperm from the bursa and spermatheca (figure 15.1d) (Miller & Miller 1981). This stage takes up most of the copulation time in odonates (Córdoba-Aguilar et al. 2003b; Córdoba-Aguilar and Cordero Rivera 2008). As we discuss below, this phase could serve not only to remove sperm, but might also be prolonged due to copulatory courtship. A few minutes (or seconds) before the end of copulation, males change their behavior, flex their abdomen, and inseminate. This is stage II, which, in some species, slowly progresses into a motionless phase known as stage III, when insemination ends (Miller & Miller 1981). In some species, females show a "postcopulatory rest", occasionally associated with sperm ejection (figure 15.1e, f). The copulatory process has been studied in detail in a number of Zygoptera (Córdoba-Aguilar 2003a; Miller 1987a, b). Figure 15.1 shows these phases in a model species, *Calopteryx haemorrhoidalis*.

Male genitalic diversity and function in the Anisoptera have been recently reviewed in detail (Pfau 2005). No such study is available for the Zygoptera (but see Pfau 1971). Some genitalic structures, like the hamuli (figure 15.2), are considered to function in guidance of the ovipositor (Tillyard 1917), and as such, might be examples of naturally selected traits. Here we focus on structures likely to have arisen as sexually selected traits, because a comprehensive analysis of genitalic diversity of the order is premature.

Anisoptera (Dragonflies)

Pfau (2005) has shown that the distal part of the penis in Anisoptera functions like a pressure pump or a two-way tap, that allows males to wash out rivals' sperm and simultaneously inject their own sperm into the female tract. This organ is inflated during copulation, and therefore its three-dimensional configuration is not easily deduced from dried specimens. The diversity of this structure does not seem related to the taxonomic position of the species (Miller 1991). For instance, *Gomphus pulchellus* (figure 15.3a, b) has a penis with two distal tubes (perhaps a two-way tap), but *Onychogomphus uncatus*, from the same family

(Gomphidae), has a complex three-dimensional penis, with no clear tubes (figure 15.3c, d). Many Libellulidae have a well-developed distal segment of the penis, with inflatable parts and a variable number of flagella (Garrison et al. 2006; Siva-Jothy 1984). Some species have one distal tube, which is accompanied by a long thin flagellum, presumably used to remove sperm from the spermatheca (Miller 1991); *Oxygastra curtisii* (Corduliidae) is a typical example (figure 15.3e, f), but this is also found in other cordulids like *Macromia splendens* (Córdoba-Aguilar & Cordero Rivera 2008).Two flagella are found in the penis of some of the Libellulidae (Córdoba-Aguilar et al. 2003b), Austropetaliidae, Aeshnidae, Gomphidae and Corduliidae (Garrison et al. 2006).

Female anisopterans have a large bursa copulatrix and from none to two spermathecae, variable in size and shape, but with uniformity in histology, musculature and sensory structures, at least in the Libellulidae (Siva-Jothy 1987). There are even examples of species without a bursa copulatrix (Miller 1991) which is very interesting as this structure seems less derived than the spermathecae (the bursa is more widespread across species, and the spermatheca usually emerges from the bursa). In *Anax*, the cuticular intima of the spermathecae is thin and lightly corrugated while that of the bursa copulatrix is thick and heavily folded (Andrew and Tembhare 1997). Many representatives of this suborder ejaculate sperm in groups ("spermatodesms"), particularly those larger-bodied taxa that utilize non-defendable resources (Siva-Jothy 1997), and these spermatodesms are dissociated inside the bursa copulatrix, probably under the action of compounds produced by the female. If that is the case, female control over this process is highly likely. The variability of genitalic structures in female anisopterans is far from being well studied, and the possible functions of accessory glands (Andrew and Tembhare 1997) need to be established. Siva-Jothy (1997) discusses several hypotheses as to why some species of Anisoptera use spermatodesms, including their possible function as a nutritional gift to the female and their evolution as sexually-selected traits in the context of postcopulatory sexual selection.

Zygoptera (Damselflies)

The intromittent organ in the Zygoptera, the *ligula* (figure 15.2 and 15.4), has arisen from a sternal

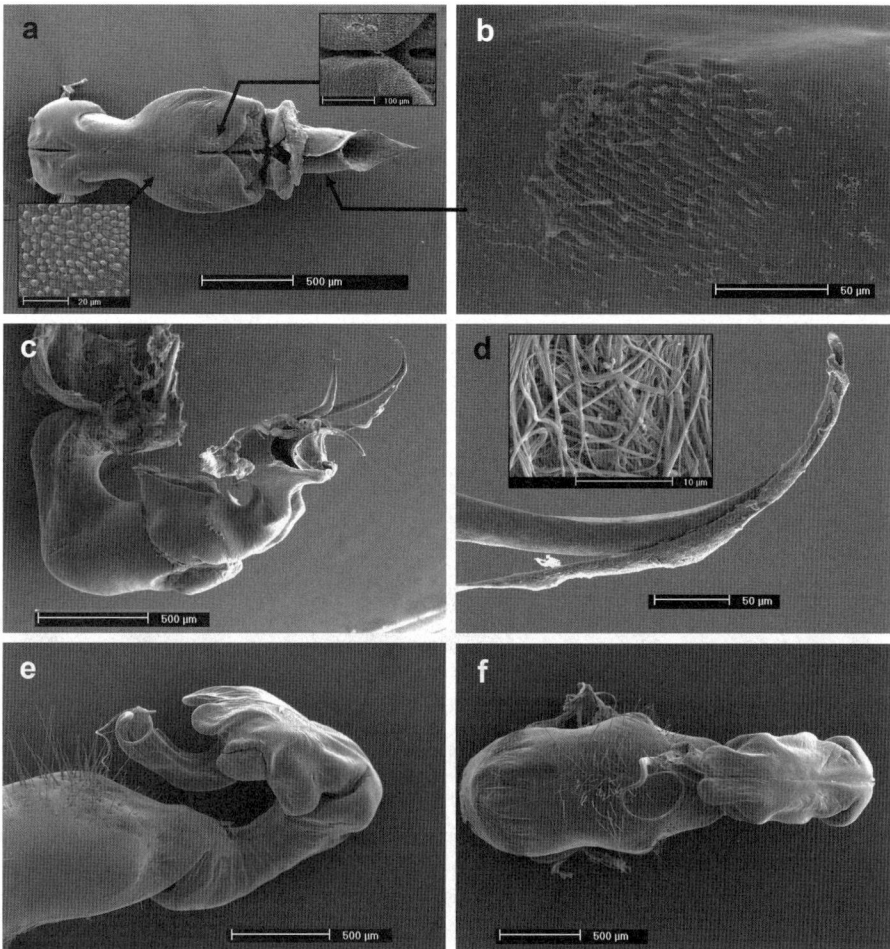

FIGURE 15.3 Typical examples of male genitalia in Anisoptera. (a) Ventral view of penis in *Gomphus pulchellus* (Gomphidae), with details of the microstructure. In this species, the head of the penis ends in two tubes, only one visible in this picture. (b) A field of short spines found in the sides of the penis in *G. pulchellus*. The arrow indicates the approximate position. (c) Lateral view of the penis in *Onychogomphus uncatus* (Gomphidae), showing details of the sperm (d) trapped in one of the horns. (e) Lateral and (f) ventral view of the genitalia in male *Oxygastra curtisii* (Corduliidae), an example of genitalia with just one flagelum.

abdominal appendix (Pfau 1971), and is used to transfer sperm to the female, and to remove sperm stored from previous matings (Waage 1979). In general, zygopterans have a long *ligula* whose head is a mobile element, like a hinge, allowing a flexible joint between both parts. The lateral distal parts of the *ligula*, before the junction of the head, are usually covered with microspines (figure 15.4b, d, f), which presumably help in removing sperm from the bursa copulatrix. Some species lack spines on the penis head, but have, nevertheless, an extensive covering of micro-spines on the *ligula*. One example is the Cuban endemic *Protoneura capillaris* (A. Cordero, personal observation). In *Coenagrion scitulum* this morphology is associated with limited sperm removal ability (Cordero et al. 1995).

A recent paper has experimentally demonstrated that, in two species of Calopterygidae, the head of the *ligula* is used to remove sperm from the bursa copulatrix, and the lateral processes are used to remove sperm from the spermatheca (Tsuchiya & Hayashi 2008). Surgically removing the lateral processes of the *ligula* produced a reduction of movements during copulation in *Calopteryx cornelia*, and no sperm removal from the spermatheca, while in *Mnais pruinosa*, a species whose males are unable to physically remove sperm from the spermatheca, cutting these processes had no effect on copulation or sperm removal from the bursa (Tsuchiya and Hayashi 2008). This is the best direct evidence we have for the function of male genitalia as a device to remove sperm in an odonate.

The head of the *ligula* can be classified into four main types (Waage 1986) (pending a description of genitalia of some tropical families). Kennedy (1920) also describes four groups of damselflies, based on

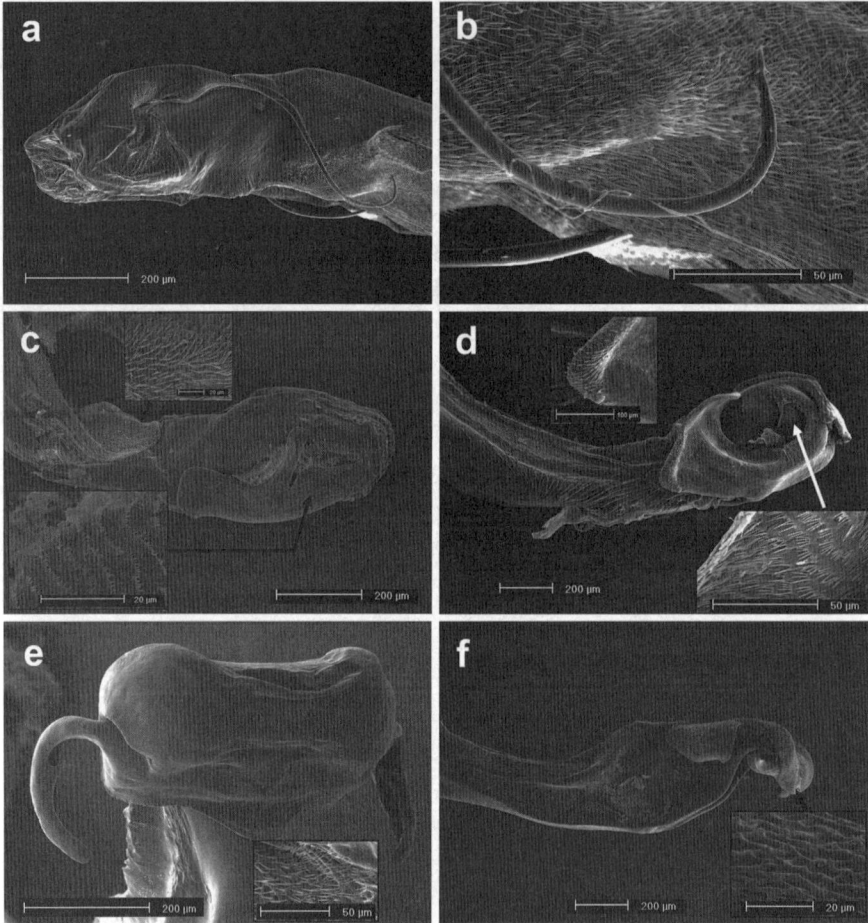

FIGURE 15.4 Typical examples of male genitalia in Zygoptera. (a) Lateral view of the penis in *Coenagrion mercuriale* (Coenagrionidae), showing two long thin flagella. The tip ends in a hook (b) presumably used to trap and remove sperm. (c) Lateral view of the penis in *Telebasis dominicana* (Coenagrionidae), a typical example of "spoon-shaped" aedeagus, with back-orientated microspination. (d) The aedeagus of *Hetaerina vulnerata* (Calopterygidae), representative of morphologies with short prolongations, also covered by fine spination. (e) The penis head of *Chalcolestes viridis* (Lestidae), with no flagella, but with a complex tridimensional structure. The insert shows a zone with some structures that seem chemical sensilla, and could be used to detect the presence of sperm inside the female. (f) Detail of the penis head of *Platycnemis pennipes* (Platycnemididae), a species with two short flagella, with a detail of the microspination found at the basis of the aedeagus.

the morphology of the ligula, and illustrates many representatives of each group, although his grouping is not exactly the same as the proposed here. Many of the Coenagrionidae (e.g., *Ischnura elegans, I. graellsii, I. ramburii, I. hastata, Coenagrion mercuriale*, personal observation; *Argia translata*, Von Ellenrieder & Lozano 2008), have two long thin flagella (figure 15.4a, b), corresponding with the presence of a long thin spermathecal duct. This morphology is also found in representatives of other families (e.g., Platycnemididae; Gassmann 2005, Protoneuridae; Pessacq 2008). In *Ischnura graellsii* and *I. senegalensis* there is experimental evidence for males being able to introduce these flagella into the spermatheca, and removing sperm in this way (Cordero & Miller 1992; Sawada 1995), but in *Ischnura elegans* males are apparently unable to introduce their flagella into the spermatheca (Miller 1987b). There is also at least one species which has only one long thin flagellum (*Podopteryx selysi*; see figures 105 and 106 in Kennedy 1920). A second group of species has a wide head, with a well developed flexible joint, and back-oriented spines (figure 15.4c). This morphology is common in the Coenagrionidae, like *Ceriagrion tenellum* (Andrés & Cordero Rivera 2000), *Enallagma cyathigerum, Telebasis dominicana* (figure 15.4), and species of the genera *Acanthagrion, Aceratobasis, Argentagrion, Cyanallagma, Enallagma, Homeoura,*

Hylaeonympha, Oxyagrion, Phoenicagrion, Schistolobos and *Telagrion* (Von Ellenrieder 2008; Von Ellenrieder & Garrison 2008a,c; Von Ellenrieder & Lozano 2008), but is also found in the protoneurid *Epipleoneura venezuelensis* (Pessacq 2008), and some Southeast Asian Platycnemididae (Gassmann 2005; Gassmann & Hämäläinen 2002). These species probably remove sperm by using the head of the *ligula* like a spoon. The third morphology shows a wide head, which ends into a variable number (2 or 4) of short appendages, finely covered with spines. This is typical of the Calopterygidae (Adams & Herman 1991; Cordero Rivera et al. 2004; Garrison 2006; Orr & Hämäläinen 2007; Waage 1984) (figure15.4d) and Platycnemididae (Dijkstra et al. 2007; Gassmann 1999, 2000, 2005; Gassmann and Hämäläinen 2002; Uhía and Cordero Rivera 2005) (figure 15.4e). Some of the Protoneuridae also show this morphology, in agreement with their phylogenetic affinity with the Platycnemididae (Pessacq 2008; Von Ellenrieder & Garrison 2008b). Finally, some species have no flexible joint, and a penis head with a variable number of lobules, and very little spination, like the Lestidae (Uhía & Cordero Rivera 2005; Waage 1982) (figure 15.4f).

The genitalia of female zygopterans consist of a weakly chitinized vagina, that has two chitinized plates with embedded sensilla, where the oviducts

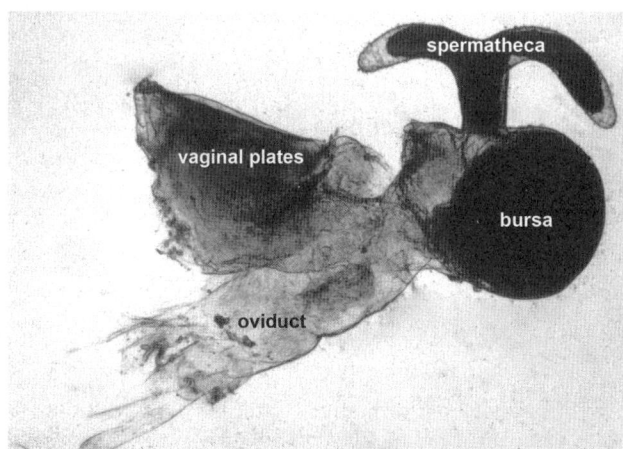

FIGURE 15.5 Female genitalia in the zygopteran *Calopteryx haemorrhoidalis*, showing the typical configuration for odonates. Variations on this pattern include the presence of one spheroid spermatheca, its complete absence, and the presence of accessory glands, whose function is poorly known.

open, together with the openings of the bursa copulatrix and spermatheca (figure 15.5). Histological evidence has shown that sperm maintenance is the primary function for these organs (e.g., Córdoba-Aguilar 2003a). During the fertilization of eggs, the vaginal plates are deformed by the egg that is about to be laid, and this elicits the release of sperm (Miller 1987a). This fact allows males to exploit this sensory channel, by stimulating the sensilla, and thereby eliciting the ejection of sperm during copulation, even if no eggs are laid at that moment (Córdoba-Aguilar 2002a). As with Anisoptera, there is substantial interspecific variation in the size and number of spermathecae (Córdoba-Aguilar et al. 2003b), and part of this variation might be due to an arms race between sexes to control the fertilization process (see below).

HYPOTHESES OF GENITALIC EVOLUTION

Sperm Competition: Tests, Predictions and Results

Sperm competition may occur when two or more males mate with a female during a single reproductive event. This competition is particularly important in insects because females store sperm in special organs, and fertilize eggs only at the moment of oviposition (Parker 1970). Sexual selection theory predicts that males should either (1) reduce the likelihood of their sperm competing with rivals' sperm, by minimizing female mating rate, or (2) maximize their probability of fertilization by removing, displacing, or incapacitating rivals' sperm, when sperm competition is unavoidable.

The first hypothesis predicts the evolution of claspers or other structures to maintain a secure hold of the female, as in water striders (Arnqvist and Rowe 2002) or aquatic beetles (Bergsten et al. 2001). Odonates have two anal appendages, which are used to grasp the prothorax (Zygoptera) or head (Anisoptera) of females. These structures are species specific, and are therefore good taxonomic characters. In some Zygoptera, male anal appendages stimulate particular areas of the prothorax of the female, and may contribute to species recognition (Robertson & Paterson 1982). As far as we know, odonate genitalia do not have internal claspers used to secure females during copulation, but the *hamuli* (figure 15.2) could be externally

used for this. Some stout spines found at the basis of the *ligula*, like in *Ischnura* (see figure 3 in Córdoba-Aguilar & Cordero Rivera 2008), might be an example of such structures.

The second hypothesis has stimulated a fruitful field of research in odonates. The pioneering work of Jonathan Waage and Peter Miller (among others) in the 1980s and 1990s, showed that the penis is covered by spines, oriented backwards, which trap sperm stored in the female genitalia (figures 15.3 and 15.4) that can therefore be removed during copulation (Miller 1987b, 1991, 1995; Miller & Miller 1981; Waage 1979, 1984, 1986). In some cases, the penis lack spines, but is used to reposition rivals' sperm far from the fertilization sites (Siva-Jothy 1988). In other cases, spines are present but the penis cannot physically remove sperm from the spermatheca (Cordero et al. 1995).

Predictions derived from sperm competition theory have been very successful at explaining odonate genital diversity. In general, there is a good concordance between male genitalic structure and female sperm storage organs (Waage 1984). Two possibilities have been recognized. First, in many species, notably in the Calopterygidae and Coenagrionidae, the penis has a form and a size that enables males to situate the *ligula* well inside the females' bursa copulatrix, and sometimes, spermatheca. In these cases, sperm competition theory predicts the evolution of spines and other structures that trap sperm and eject it to the outside during genitalic movements of stage I (figure 15.3c, d and 15.4c). Second, in some cases males have genitalia that cannot be inserted inside the spermatheca, thereby making the presence of spines useless for sperm removal. Nevertheless, if male and female genitalia are in an evolutionary arms race, the evolution of sperm storage organs in females that cannot be accessed by males, may explain cases of "useless" spines in males, as primitive characters that have not been lost possibly because selection against them is weak.

Cryptic Female Choice: Tests, Predictions and Results

Females can exert postcopulatory choice only if males are unable to remove sperm from the bursa and spermatheca completely. The fact that many studies on odonate reproductive biology have reported that P_2 values, the proportion of sperm fertilized by the second of two males mated to the same female, is nearly 100% (for a review see Córdoba-Aguilar et al. 2003b) is strong evidence in favor of male control of fertilization, that is sperm competition. Nevertheless, the last male advantage is only clear when eggs are laid shortly after mating. The pattern is less clear for eggs fertilized some days after copulation (Cordero & Miller 1992; McVey & Smittle 1984; Sawada 1998; Siva-Jothy & Tsubaki 1989). Behavioral studies have shown that females sometimes lay eggs without re-mating, even in the presence of territorial males (Siva-Jothy & Hooper 1996), and this is certainly common in species that lay eggs unguarded, like many *Ischnura* (Cordero 1994) or *Calopteryx* (Cordero Rivera & Andrés 2002; Hooper & Siva-Jothy 1997; Waage 1987). Therefore, the potential for cryptic female choice (CFC) in odonates is high. Cryptic female choice mechanisms are possible when some of the sperm stored inside the genitalic trait of females remain unreachable for males. The evolution of two sperm storage organs, the bursa and the spermatheca, is consistent with predictions of CFC theories (but certainly not the only possibility for females retaining some control of sperm stores).

Variation in copulation duration (figure 15.6) has been used to infer mechanisms of genital evolution (Córdoba-Aguilar & Cordero Rivera 2008). Cordero (1990) experimentally showed that copulation duration in some damselflies varies with female mating history. Apparently males control copulation duration (Miller 1987a), and perform short copulations with virgin females and long copulations with previously mated females (Andrés & Cordero Rivera 2000; Cordero 1990). Males can physically remove sperm from both the bursa and the spermatheca(e), in only a fraction of odonate species. In these cases, females seem to have lost their control over fertilization, and male and female genitalia diversity is well explained by sperm competition predictions.

In many cases, nevertheless, males have a spoon-shaped *ligula* (figure 15.4c), as in the Lestidae (Uhía & Cordero Rivera 2005; Waage 1982), and some of the Coenagrionidae (Andrés & Cordero Rivera 2000; Uhía & Cordero Rivera 2005), or short genitalic processes, that clearly cannot remove sperm from the spermatheca, as in the Platycnemididae (Uhía & Cordero Rivera 2005). Some species have no spermatheca (Uhía & Cordero Rivera 2005). This diversity is well suited for controlled experiments to test predictions of CFC and sperm competition hypotheses. It is easy to see that if

females do not have spermatheca, males can remove sperm from the bursa, and if sperm removal is fast, then males should not prolong copulation with mated females compared to virgins (figure 15.6; case of *Lestes barbarus* and *L. virens*). This assumes that sperm removal from the bursa is fast. This seems reasonable, because removing sperm from the bursa needs only 5 minutes in *Ceriagrion tenellum* (Andrés & Cordero Rivera 2000), and *Calopteryx* males are able to completely empty the bursa in less than two minutes (Cordero Rivera et al. 2004). The behavior of species without a spermatheca and with a bursa accessible to male genitalia could be explained by sperm competition alone, which suggests that females have little control over the process. If sperm removal were slow, then males should mate for longer periods with mated females, to maximize the amount of sperm removed, irrespective of the presence/absence of spermathecae. We are unaware of any example of this in Odonata. On the contrary, if males cannot remove sperm from the spermatheca, but they remove it from the bursa, then sperm competition predicts the same copulation duration with virgin and mated females, but cryptic female choice predicts longer matings with mated females, because males should perform "copulatory courtship" (Eberhard 1994) to increase their paternity success, that is, mated females can cryptically choose, but virgins do not (although if the virgin is going to store sperm before egg-laying she may choose later). In this case, the sperm stored in the spermatheca, which is inaccessible to male genitalia, is the key that allows females to exercise cryptic choice. There is experimental evidence in Coenagrionidae, Lestidae, and Platycnemididae in agreement with CFC predictions (figure 15.6; Uhía & Cordero Rivera 2005).

Even after copulation females might exert cryptic choice, for instance by selectively ejecting sperm from a particular male (figure 15.1f). The ejection of sperm after copulation has been overlooked in studies of odonate behavior, until Eberhard (1996) highlighted observations on *Paraphlebia quinta* females, that were seen to expel a drop of sperm after copulation (González Soriano & Córdoba-Aguilar 2003). This led to further experimental and observational work that suggested this is a case of cryptic female choice of sperm (Córdoba-Aguilar 2006). Even females of *Ischnura graellsii* have been observed to expel sperm after their first mating, before oviposition, in a laboratory environment

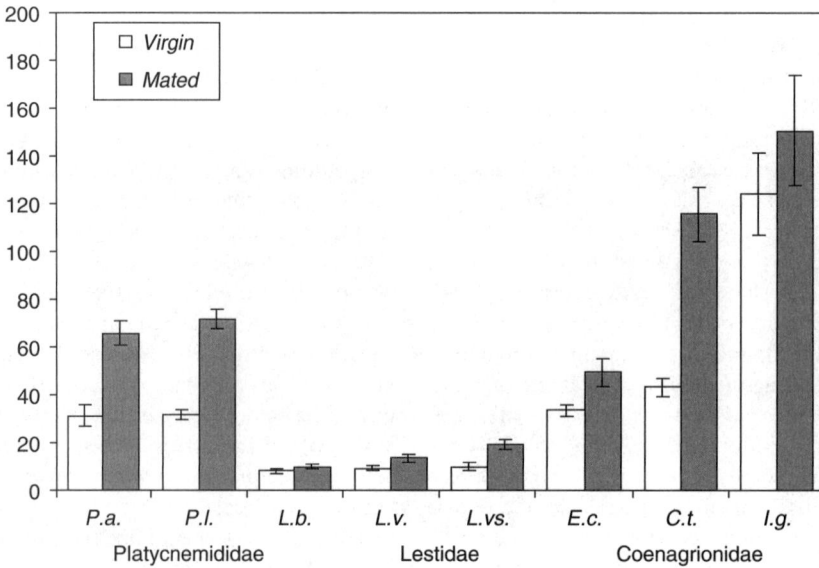

FIGURE 15.6 Copulation duration (mean±SE) in laboratory matings of virgin and mated females of three families of Zygoptera: *Platycnemis acutipennis (P.a.), P. latipes (P.l.)* (Platycnemididae); *Lestes barbarus (L.b.), L. virens (L.v.), Lestes viridis (L.vs.)* (Lestidae), and *Enallagma cyathigerum (E.c.), Ceriagrion tenellum (C.t.), Ischnura graellsii (I.g.)* (low density) (Coenagrionidae). Note that copulation lasts longer when females have stored sperm, but this effect is not significant for species without a spermatheca (*L.b.* and *L.v.*). data from Andrés & Cordero Rivera 2000; Cordero 1990; Uhía & Cordero Rivera 2005).

(A. Cordero, personal observation). The control of sperm release from the spermatheca is so good that females have apparently evolved a mechanism by which they can control from which spermathecal duct they can eject more sperm, in *C. haemorrhoidalis*, a species whose females have two spermathecal ducts (figure 15.5) (Córdoba-Aguilar & Siva-Jothy 2004). Nevertheless, this sperm ejection behavior could be a by-product of sperm competition, if females were simply ejecting the sperm removed by their mate during copulation (Lindeboom 1998). Future studies should use molecular markers to determine the identity of the sperm, by comparing DNA fingerprints of the last male and the ejected sperm. If a CFC mechanism is at work, then in many cases the sperm ejected should belong to the copulating male.

Sexual Conflict: Tests, Predictions and Results

The interpretation of sexual selection as a conflict of interests is not an alternative to the sperm competition and CFC. Rather, conflict will be ubiquitous given that males and females have different interests. In some cases, males seem to be ahead in the interaction, and a sperm competition approach is then the best to explain and predict patterns (for instance in *Ischnura graellsii*; Cordero & Miller 1992). In other cases, although sperm competition, in the form of sperm removal, occurs, CFC mechanisms prevail (a good example is *Ceriagrion tenellum*; Andrés & Cordero Rivera 2000). Nevertheless, there are some novel predictions about genitalic evolution that cannot be derived from the male or female standpoint alone. From the previous discussion, the sperm stores in the spermatheca seem to be the focus of sexual conflict. Long term maintenance of sperm in this organ allows females to lay fertile eggs over their whole life, even after just one mating (Cordero 1990; Fincke 1987; Grieve 1937). Females can control which organ releases sperm during fertilization, and it has been shown that they use both organs in different contexts (Nakahara & Tsubaki 2007, Siva-Jothy & Hooper 1996). An elegant and recent paper on *Ischnura senegalensis* has revealed that the spermatheca is a safer place (in terms of mortality) for sperm than the bursa, perhaps because sperm in the bursa are more likely to be removed (Nakahara & Tsubaki 2007). This may explain patterns of sperm use in this species in which females use bursal sperm immediately after copulation while spermathecal sperm are used over the long-term (Nakahara & Tsubaki 2007). Furthermore, patterns of sperm survival are associated with whether sperm are removed or not. In *Mnais pruinosa*, where spermathecal sperm is not removed, sperm survival in the spermatheca is reduced as compared to *Calopteryx cornelia* where spermathecal sperm are removed (Hayashi & Tsuchiya 2005). If males are able to reach the spermatheca to remove sperm, one possible evolutionary response by females is a reduction of the size of this organ, as it is no longer useful for sperm storage. This is what Hayashi and Tsuchiya (2005) found for *Mnais pruinosa*, where the spermatheca is almost vestigial. Nevertheless, other alternatives do exist that allow females to regain control: longer spermathecal ducts, which impede males from reaching the tip of the spermatheca (a likely case is *Calopteryx splendens*; A. Cordero, personal observations), the evolution of mechanical barriers in the ducts, larger spermathecae internally convoluted, and so on. In any case, such an interspecific difference in the function of sperm storage organs suggests antagonistic coevolution between the sexes (Holland & Rice 1998) as females may derive benefits from storing sperm and such benefits are lost once males are able to have access to the spermatheca. In fact, a reduction in female mating rate has been associated with increased male ability to displace sperm (Córdoba-Aguilar 2009). This means that when females accrue no more benefits of storing sperm, there is no need to mate multiply.

In relation to genitalic evolution, a straightforward prediction is that male traits that manipulate females in ways that reduce the female's direct fitness (Eberhard 2006), like reducing re-mating frequency, or increasing egg-laying even if this has survival costs for females, are to be expected. Examples are hooks and other structures that damage the internal female genital tract (Crudgington & Siva-Jothy 2000), or seminal products that are toxic to females (Rice 1996). Nevertheless, damaging females would be a bizarre male adaptation. Morrow et al. (2003) experimentally showed in two beetles and *Drosophila melanogaster* that females do not delay re-mating or increase their reproductive rate after being harmed, but re-mate sooner and lay fewer eggs in some cases. This suggests that male harmful "adaptations" in the context of sexual conflict are more parsimoniously explained as byproducts or pleiotropic side effects of other male adaptations.

Nothing is known about the composition of odonate ejaculates, but hooks and spines are common in the penis (figure 15.3 and 15.4). Nevertheless, these structures seem better explained as sperm removal devices rather than "sexual weapons" in odonates. There is little evidence for negative effects on females of multiple mating, except in some polymorphic *Ischnura* (Cordero et al. 1998;Gosden & Svensson 2007; Sirot & Brockmann 2001), and no evidence for genital damage during mating in odonates. Dunkle (1991) found dragonfly females whose heads were damaged, presumably by the male abdominal appendages during mating attempts prior to secure tandem formation, but this would be a "weaponry" case for a non-intromittent genitalic trait. This topic merits further study.

Other Hypotheses: Lock and Key and Pleiotropy

Rooted in evolution textbooks and traditional evolutionary thinking, is the idea that animals may continuously face the risk of mating with members of a different species. This should promote the evolution of physiological, morphological and behavioral traits aimed to reduce such risk and complex genitalia may be a such a set of traits (Dufour 1844 in Mayr 1963). Complex genitalia, being species-specific, prevent males from mating with females of different species. This lock and key hypothesis "purports to explain species-specific genitalic morphology in terms of mechanical reproductive isolation" (Shapiro & Porter 1989), and although defended by a few people (e.g., Nagata et al. 2007; Takami et al. 2007), has not been supported either by recent tests (Arnqvist et al. 1997; Arnqvist 1998; Arnqvist & Thornhill 1998), or by comparative evidence (Eberhard 1985). The lock and key hypothesis, however, has not been tested in odonates. Watson (1966) in a study of the size of secondary genitalia in five species of *Tramea* (Libellulidae), concluded that there is a clear correlation between the size of male hamuli and female vulvar scales, and suggested that this is an example of a lock and key mechanism in the Odonata. This example is suggestive but manipulative experiments are needed for a formal test of the hypothesis.

On the other hand, Paulson (1974) performed experiments with five species of male Coenagrionidae and ten species of females, and only in one of the possible combinations was the male able to achieve the tandem position easily, suggesting that the primary genitalia in this family may act as a mechanical barrier to interspecific matings.

In fact, there is evidence for a match between male anal appendages and female mesostigmal plates in the Zygoptera, acting as a mechanical barrier to interspecific tandems (Robertson & Paterson 1982), although it never prevents all interspecific tandems (Corbet 1999, reviewed in Utzeri & Belfiore 1990). The fact that interspecific matings and hybrids (Leong & Hafernik 1992; Monetti et al. 2002; Tynkkynen et al. 2008) may be more common than usually thought, supports the assumption that there is a risk of heterospecific mating.

Genitalia may have a role in avoiding heterospecific matings, at least in those species that do not show pre-copulatory courtship. Whether the lock and key hypothesis can be the prime explanation for genitalic evolution in these animals is still open to discussion. It may also be that a lock and key process is incidentally reinforced by a sexual selection mechanism. It may actually be that the two mechanisms are not mutually exclusive. Take the case of two sister, sometimes sympatric, species for which interspecific matings and hybrids have been documented: *Ischnura graellsii* and *I. elegans*. It has been shown that these species produce hybrids and are undergoing an incipient isolation (Monetti et al. 2002). Thus, one would expect that there must be selection to avoid mating mistakes. There is evidence which suggests that secondary male genitalia in both species are under strong sexual selection (Cordero & Miller 1992; Miller 1987a, b). But also, a lock and key mechanism may apply for the primary genitalia of these species: the prothoracic tubercle of female *I. elegans* impedes males of *I. graellsii* from achieving a firm grasp for a precopulatory tandem. As a consequence, these hybrid matings are almost never observed (Monetti et al. 2002). The opposite is nevertheless not true: male *I. elegans* have no physical impediments in grasping female *I. graellsii*, and matings between the two species are frequent in the laboratory and the field, resulting in viable hybrids (R.A. Guillén & A. Cordero, unpublished).

The pleiotropy hypothesis supposes a nonfunctional basis for genitalic traits. It implies that the same genes that control other adaptive traits, incidentally control genitalic morphology. Thus, genitalic evolution is driven by the evolution of other adaptive traits (Arnold 1973; Mayr 1963). Although this hypothesis is tremendously difficult to test, the few tests performed with insects have not rejected it (Arnqvist et al. 1997; Arnqvist & Thornhill 1998). In fact, in one group of Jamaican millipedes, this hypothesis seems to match the gradual evolution of male genitalia (Bond et al. 2003). Whether the pleiotropy hypothesis operates on odonate genitalic traits also needs to be checked although we cannot foresee an easy experimental test using these animals. (Evidence that specific genital traits are strongly correlated with fitness would argue against a pleiotropy hypothesis).

SEASONAL EFFECTS: AN OVERLOOKED SOURCE OF GENITAL VARIATION?

Body size in insects is affected by seasonality: adults that emerge early in the reproductive season are larger than those emerging late in the season (Roff 1980, 1986). This pattern is mainly due to the time, accrued food, and developmental strategies, that larvae use depending on when they were laid as eggs (for a review of this in odonates, see Stoks et al. 2008). In those temperate places in which animals have a restricted season (e.g., a few months), this effect is particularly strong as early emerging individuals have spent nearly a year as larvae. This has given them more time to acquire more food, unlike late emerging individuals which may have started and completed their development in the same season that they emerge. Little is known about the effect of this change on genital size and, if this is the case, how genital functions are affected. Some evidence in a dung fly indicates that the seasonal effect may have an evolutionary impact on sperm competition: late emerging individuals have smaller testes which produce less sperm (Ward & Simmons 1991). These late emerging males are less successful as in this species the more successful males are those that transfer relatively more sperm (Simmons & Parker 1992). A recent study in two calopterygids, *Calopteryx haemorrhoidalis asturica* and *Hetaerina americana*, has uncovered more details (Córdoba-Aguilar 2009). These species vary considerably in the extent of their reproductive seasons: from three to four months for *C. h. asturica* and nearly the

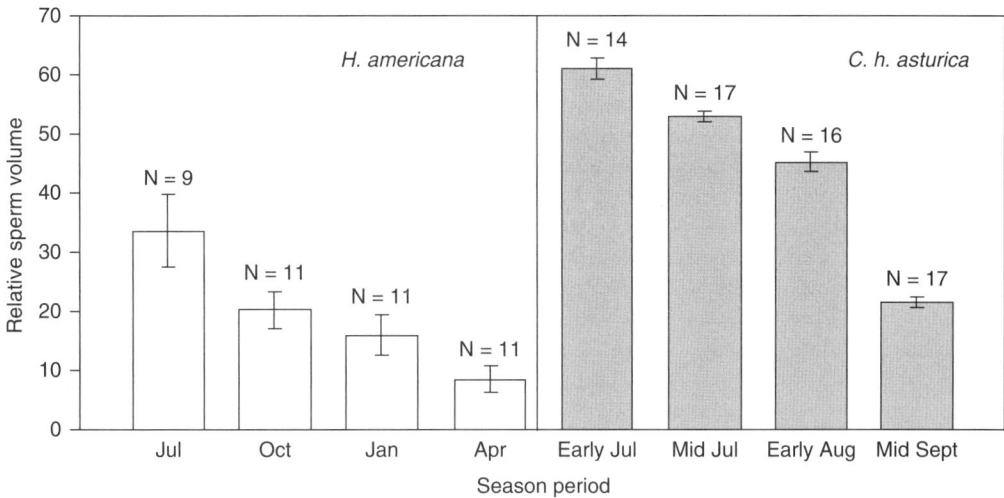

FIGURE 15.7 Variation (mean ± SE) in relative sperm remaining in the spermatheca (the sperm that males were unable to displace after aedeagal stimulation within the female vagina) according to representative months during the reproductive season in *H. americana* and *C. h. asturica*.

whole year for *H. americana*. Since the width of the aedeagus is important in displacing sperm in both species (the wider the aedeagus, the more intense the stimulation, which induces females to eject previously stored sperm) and this trait correlates with body size, seasonality has extensive effects. Interestingly, larger males tend to obtain the territories in which females arrive to mate (for *C. h. asturica* see Córdoba-Aguilar 2009; for *H. americana* see Serrano-Meneses et al. 2007) so that, in general, large males are more successful. However, early emerging males that gain a territory tend to mate with females that match their size, which is a situation different to late emerging males which mate with females that vary considerably in size (as different cohorts overlap) but in general are relatively smaller. In fact, if one measures the sexual size dimorphism of mating couples over the season, skew toward larger males becomes more pronounced at the end of the season. In terms of copulation, this means that late in the season males with larger aedeagi with respect to the female zone that the aedeagus stimulates, become more successful in eliciting female sperm ejection than males early in the season (see figure 15.7). Thus, late in the season females are less able to keep away the sperm they stored. The fact that this pattern is consistent in two species that differ in the extent of their

reproductive season, implies that this phenomenon may apply to other species including non-odonate species. In terms of sperm competition and/or sexual conflict, then there will be varying regimes of selection intensity along the season and the female benefits of storing sperm and keeping it unreachable during male displacement will vary depending on when females emerge. In fact, late emerging females tend to mate less frequently and male harassment increases both as possible consequences of the reduction in mating frequency (Córdoba-Aguilar 2009).

EVOLUTION OF GENITALIA AS AN ENGINE FOR SPECIES DIVERGENCE?

Genitalic diversification has long ago been proposed as an important engine for species divergence via sexual selection (Eberhard 1985). In many animal groups there is a general pattern for male genitalia to be more diverse than female genitalia, and this has been interpreted in terms of cryptic female choice (Eberhard 1996). In odonates, there is evidence for allopatric divergence of male and female genitalia in the Calopterygidae (Cordero Rivera et al. 2004). The genus *Calopteryx* has been enlightening in this respect. Research in different species has shown that males have evolved different genitalic morphologies aimed to displace spermathecal sperm (but never seem able to remove all) and that, due to sexual co-evolution, there may be representatives of different "situations". For example, in *C. maculata*, the penis is narrow enough to penetrate the spermatheca and remove the sperm located in this site (Córdoba-Aguilar 2003b; Waage 1979). This does not apply to *C. splendens xanthostoma* where the *ligula* is larger than the spermatheca (Córdoba-Aguilar 2003b) which explains why males cannot remove the sperm present in this organ (Siva-Jothy and Hooper 1996). These differences are not only inter- but also intraspecific. In *C. haemorrhoidalis*, there is variation in sperm displacement mechanisms with males of some populations being able to displace spermathecal sperm while in other populations, males are unable to do so (Cordero Rivera et al. 2004). These interpopulational differences are only present in genitalia and not in other traits which suggests that sexual selection, at the copulatory level, has been key in species divergence. It may be that post-copulatory sexual selection may be stronger than pre-copulatory sexual selection and this is why the evolution of genitalic traits has been the engine of species divergence. This idea can be tested now with the genus *Calopteryx* where information is available as to characters being selected during pre- (i.e., pigmentation; Córdoba-Aguilar & Cordero Rivera 2005) and post-copulatory events.

CONCLUSIONS AND SUGGESTIONS FOR FUTURE RESEARCH

Although it seems that sexual selection, particularly sperm competition, is an important force shaping genital morphology and function, other selective forces cannot be disregarded. Other sexual selection forces are cryptic female choice and sexual conflict. A similar argument can be made for natural selection hypotheses, especially the lock and key hypothesis. Further investigations should test hypotheses from both sexual and natural selection.

Our knowledge of genital functional morphology is still rather poor for many families of Anisoptera (but see Pfau 2005; for a comprehensive work see Siva-Jothy 1997), and this is especially

true for females. Another research priority is tropical families, and also species-poor and primitive taxa, like the Hemiphlebiidae or Petaluridae. Furthermore, study of the genital morphology of highly diverse and localized taxa, like *Megalagrion* in Hawaii (Polhemus and Asquith 1996) or *Nesobasis* in Fiji (Donnelly 1990), both with more than 20 species, would be appropriate tests of hypotheses of genital evolution and speciation on islands.

As we have mentioned above, there is limited evidence for mating frequency having negative effects on females, and we lack direct evidence for genital damage, two predictions derived from sexual conflict hypotheses, and therefore open to future studies. Finally, the lock-and-key and pleiotropy hypotheses are still not formally tested with odonates, a group that offers high rewards for future studies of genital diversity.

Acknowledgments   This work was directly related to part of the goals of the research grants CGL2005-00122 and CGL2008-02799 (Dirección General de Proyectos de Investigación, Ministerio de Ciencia e Innovación de España including FEDER funding). We are very grateful to the reviewers for their clever suggestions to improve this manuscript.

REFERENCES

Adams JA, Herman TH (1991). A comparison of the male genitalia of three *Calopteryx* species (Odonata: Calopterygidae). Canadian Journal of Zoology, 69, 1164–1170.

Andrés JA, Cordero Rivera A (2000). Copulation duration and fertilization success in a damselfly: An example of cryptic female choice? Animal Behaviour, 59, 695–703.

Andrew RJ, Tembhare DB (1997). The post-ovarian genital complex in *Anax guttatus* (Burmeister) (Anisoptera: Aeshnidae). Odonatologica, 26, 385–394.

Arnold EN (1973). Relationships of the palaearctic lizards assigned to the genera *Lacerta*, *Algyroids* and *Psammordromus* (Reptilia: Lacertidae). Bulletin of the British Museum of Natural History (Zoology), 25, 291–366.

Arnqvist G (1998). Comparative evidence for the evolution of genitalia by sexual selection. Nature, 393, 784–785.

Arnqvist G, Rowe L (2002). Correlated evolution of male and female morphologles in water striders. Evolution, 56, 936–947.

Arnqvist G, Rowe L (2005). Sexual Conflict. Princeton University Press, Princeton, NJ.

Arnqvist G, Thornhill R (1998). Evolution of animal genitalia: Patterns of phenotypic and genotypic variation and condition dependence of genital and non-genital morphology in water strider (Heteroptera: Gerridae: Insecta). Genetical Research, 71, 193–212.

Arnqvist G, Thornhill R, Rowe L (1997). Evolution of animal genitalia: Morphological correlates of fitness components in a water strider. Journal of Evolutionary Biology, 10, 613–640.

Banks MJ, Thompson DJ (1985). Lifetime mating success in the damselfly *Coenagrion puella*. Animal Behaviour, 33, 1175–1183.

Bechly G, Brauckmann C, Zessin W, Groning E (2001). New results concerning the morphology of the most ancient dragonflies (Insecta: Odonatoptera) from the Namurian of Hagen-Vorhalle (Germany). Journal of Zoological Systematics and Evolutionary Research, 39, 209–226.

Bergsten J, Toyra A, Nilsson AN (2001). Intraspecific variation and intersexual correlation in secondary sexual characters of three diving beetles (Coleoptera: Dytiscidae). Biological Journal of the Linnean Society, 73, 221–232.

Bond JE, Beamer DA, Hedin MC, Sierwald P (2003). Gradual evolution of male genitalia in a sibling species complex of millipedes (Diplopoda: Spirobolida: Rhinocricidae: *Anadenobolus*). Invertebrate Systematics, 17, 711–717.

Bybee SM, Ogden TH, Branham MA, Whiting MF (2008). Molecules, morphology and fossils: a comprehensive approach to odonate phylogeny and the evolution of the odonate wing. Cladistics, 24, 477–514.

Conrad KF, Pritchard G (1992). An ecological classification of odonate mating systems: the relative influence of natural, inter- and intra-sexual selection on males. Biological Journal of the Linnean Society, 45, 255–269.

Corbet PS (1999). Dragonflies. Behaviour and Ecology of Odonata. Harley Books, Essex, UK.

Cordero A (1990). The adaptive significance of the prolonged copulations of the damselfly, *Ischnura graellsii* (Odonata: Coenagrionidae). Animal Behaviour, 40, 43–48.

Cordero A (1994). Inter-clutch interval and number of ovipositions in females of the damselfly *Ischnura graellsii* (Odonata: Coenagrionidae). Etología, 4, 103–106.

Cordero A, Miller PL (1992). Sperm transfer, displacement and precedence in *Ischnura graellsii* (Odonata: Coenagrionidae). Behavioral Ecology and Sociobiology, 30, 261–267.

Cordero A, Santolamazza Carbone S, Utzeri C (1995). Male disturbance, repeated insemination and sperm competition in the damselfly *Coenagrion scitulum* (Zygoptera: Coenagrionidae). Animal Behaviour, 49, 437–449.

Cordero A, Santolamazza Carbone S, Utzeri C (1998). Mating opportunities and mating costs are reduced in androchrome female damselflies, *Ischnura elegans* (Odonata). Animal Behaviour, 55, 185–197.

Cordero Rivera A, Andrés JA (2002). Male coercion and convenience polyandry in a Calopterygid damselfly (Odonata). Journal of Insect Science, 2, 14—Available online: insect-science.org/2.14.

Cordero Rivera A, Andrés JA, Córdoba-Aguilar A, Utzeri C (2004). Postmating sexual selection: allopatric evolution of sperm competition mechanisms and genital morphology in calopterygid damselflies (Insecta: Odonata). Evolution, 58, 349–359.

Córdoba-Aguilar A (2002a). Sensory trap as the mechanism of sexual selection in a damselfly genitalic trait (Insecta: Calopterygidae). American Naturalist, 160, 594–601.

Córdoba-Aguilar A (2002b). Wing pigmentation in territorial male damselflies, *Calopteryx haemorrhoidalis*: a possible relation to sexual selection. Animal Behaviour, 63, 759–766.

Córdoba-Aguilar A (2003a). A description of male and female genitalia and a reconstruction of copulatory and fertilisation events in *Calopteryx haemorrhoidalis* (Vander Linden) (Zygoptera : Calopterygidae). Odonatologica, 32, 205–214.

Córdoba-Aguilar A (2003b). Predicting mechanisms of sperm displacement based on genitalic morphometrics in the Calopterygidae (Odonata). Journal of Insect Behavior, 16, 153–167.

Córdoba-Aguilar A (2006). Sperm ejection as a possible cryptic female choice mechanism in

Odonata (Insecta). Physiological Entomology, 31, 146–153.

Córdoba-Aguilar A (2010). Seasonal variation in genital and body size, sperm displacement ability, female mating rate and male harassment in two calopterygid damselflies (Odonata: Calopterygidae). Biological Journal of the Linnean Society, 96, 815–829.

Córdoba-Aguilar A, Cordero Rivera A (2005). Evolution and ecology of Calopterygidae (Zygoptera: Odonata): Status of knowledge and future research perspectives. Neotropical Entomology, 34, 861–879.

Córdoba-Aguilar A, Cordero Rivera A (2008). Cryptic female choice and sexual conflict. In A Córdoba-Aguilar, ed. Dragonflies. Model Organisms for Ecological and Evolutionary Research, pp 189–202. Oxford University Press, Oxford.

Córdoba-Aguilar A, Siva-Jothy MT (2004). Sperm displacement ability in *Calopteryx haemorrhoidalis* (Vander Linden): Male and female roles, male limits in performance, and female neural control (Zygoptera: Calopterygidae). Odonatologica, 33, 245–252.

Córdoba-Aguilar A, Salamanca-Ocaña JC, Lopezaraiza M (2003a). Female reproductive decisions and parasite burden in a calopterygid damselfly (Insecta: Odonata). Animal Behaviour, 66, 81–87.

Córdoba-Aguilar A, Uhía E, Cordero Rivera A (2003b). Sperm competition in Odonata (Insecta): the evolution of female sperm storage and rivals' sperm displacement. Journal of Zoology, 261, 381–98.

Crudgington HS, Siva-Jothy MT (2000). Genital damage, kicking and early death. Nature, 407, 855–856.

Dijkstra KDB, Clausnitzer V, Martens A (2007). Tropical African *Platycnemis* damselflies (Odonata: Platycnemididae) and the biogeographical significance of a new species from Pemba Island, Tanzania. Systematics and Biodiversity, 5, 187–198.

Donnelly TW (1990). The Fijian genus *Nesobasis* Part 1: Species of Viti Levu, Ovalau, and Kadavu (Odonata: Coenagrionidae). New Zealand Journal of Zoology, 17, 87–117.

Dunkle SW (1991). Head damage from mating attempts in dragonflies (Odonata: Anisoptera). Entomological News, 102, 37–41.

Eberhard W (2006). Sexually antagonistic coevolution in insects is associated with only limited morphological diversity. Journal of Evolutionary Biology, 19, 657–681.

Eberhard WG (1985). Sexual Selection and Animal Genitalia. Harvard University Press, Cambridge, Massachusetts.

Eberhard WG (1994). Evidence for widespread courtship during copulation in 131 species of insects and spiders, and implications for cryptic female choice. Evolution, 48, 711–733.

Eberhard WG (1996). Female Control: Sexual Selection by Cryptic Female Choice. Princeton University Press, Princeton.

Fincke OM (1986). Lifetime reproductive success and the opportunity for selection in a nonterritorial damselfly (Odonata: Coenagrionidae). Evolution, 40, 791–803.

Fincke OM (1987). Female monogamy in the damselfly *Ischnura verticalis* Say (Zygoptera: Coenagrionidae). Odonatologica, 16, 129–143.

Fincke OM (1997). Conflict resolution in the Odonata: implications for understanding female mating patterns and female choice. Biological Journal of the Linnean Society, 60, 201–220.

Fincke OM, Hadrys H (2001). Unpredictable offspring survivorship in the damselfly, *Megaloprepus coerulatus*, shapes parental behavior, constrains sexual selection, and challenges traditional fitness estimates. Evolution, 55, 762–772.

Fincke OM, Waage JK and Koenig WD (1997). Natural and sexual selection components of odonate mating patterns. In JC Choe, BJ Crespi, eds. The Evolution of Mating Systems in Insects and Arachnids, pp 58–74. Cambridge University Press, Cambridge.

Fleck G, Bechly G, Martínez-Delclòs X, Jarzembowski EA, Nel A (2004). A revision of the Upper Jurassic–Lower Cretaceous dragonfly family Tarsophlebiidae, with a discussion on the phylogenetic positions of the Tarsophlebiidae and Sieblosiidae (Insecta, Odonatoptera, Panodonata). Geodiversitas, 26, 33–60.

Garrison RW (2006). A synopsis of the genera *Mnesarete* Cowley, *Bryoplathanon* gen. nov., and *Ormenophlebia* gen. nov. (Odonata: Calopterygidae). Contributions in Science, 506, 1–84.

Garrison RW, Von Ellenrieder N and Louton J (2006). Dragonfly Genera of the New World.

The Johns Hopkins University Press, Baltimore.

Gassmann D (1999). Taxonomy and distribution of the inornata species-group of the Papuan genus *Idiocnemis* Selys (Odonata: Zygoptera: Platycnemididae). Invertebrate Taxonomy, 13, 977–1005.

Gassmann D (2000). Revision of the Papuan *Idiocnemis bidentata*-group (Odonata: Platycnemididae). Zoologische Mededelingen, 74, 375–402.

Gassmann D (2005). The phylogeny of Southeast Asian and Indo-Pacific Calicnemiinae (Odonata, Platycnemididae). Bonner zoologische Beiträge, 53, 37–80.

Gassmann D, Hämäläinen M (2002). A revision of the Philippine subgenus *Risiocnemis* (*Igneocnemis*) Hämäläinen (Odonata: Platycnemididae). Tijdschrift voor Entomologie, 145, 213–266.

González Soriano E, Córdoba-Aguilar A (2003). Sexual behaviour in *Paraphlebia quinta* calvert: Male diamorphism and a possible example of female control (Zygoptera: Megapodagrionidae). Odonatologica, 32, 345–353.

Gosden TP, Svensson EI (2007). Female sexual polymorphism and fecundity consequences of male mating harassment in the wild. PLoS ONE, 2, e580.

Grieve EG (1937). Studies on the biology of the damselfly *Ischnura verticalis* Say, with notes on certain parasites. Entomologica Americana, 17, 121–153.

Haubruge E, Arnaud L, Mignon J, Gage MJG (1999). Fertilization by proxy: Rival sperm removal and translocation in a beetle. Proceedings of the Royal Society of London Series B Biological Sciences, 266, 1183–1187.

Hayashi F, Tsuchiya K (2005). Functional association between female sperm storage organs and male sperm removal organs in calopterygid damselflies. Entomological Science, 8, 245–252.

Heymer A (1973). Étude du comportement reproducteur et analyse des mécanismes déclencheurs innés (MDI) optiques chez les Calopterygidae (Odon. Zygoptera). Annales de la Societé entomologique de France (Nouvelle Série), 9, 219–255.

Holland B, Rice WR (1998). Perspective: Chase-away sexual selection: Antagonistic seduction versus resistance. Evolution, 52, 1–7.

Hooper RE, Siva-Jothy MT (1997). "Flybys": A prereproductive remote assessment behavior of female *Calopteryx splendens xanthostoma* (Odonata: Calopterygidae). Journal of Insect Behavior, 10, 165–175.

Kamimura Y (2000). Possible removal of rival sperm by the elongated genitalia of the earwig, *Euborellia plebeja*. Zoological Science, 17, 667–672.

Kennedy CH (1920). The phylogeny of the zygopterous dragonflies as based on the evidence of the penes. Ohio Journal of Science, 21, 19–32.

Leong JM, Hafernik JEJ (1992). Hybridization between two damselfly species (Odonata: Coenagrionidae): Morphometric and genitalic differentiation. Annals of the Entomological Society of America, 85, 662–670.

Lindeboom M (1998). Post-copulatory behaviour in *Calopteryx* females (Insecta, Odonata, Calopterygidae). International Journal of Odonatology, 1, 175–184.

Mayr E (1963). Animal Species and Evolution. Harvard University Press, Harvard.

McVey ME, Smittle BJ (1984). Sperm precedence in the dragonfly *Erythemis simplicicollis*. Journal of Insect Physiology, 30, 619–628.

Miller PL (1987a). An examination of the prolonged copulations of Ischnura elegans (Vander Linden) (Zygoptera: Coenagrionidae). Odonatologica, 16, 37–56.

Miller PL (1987b). Sperm competition in *Ischnura elegans* (Vander Linden) (Zygoptera: Coenagrionidae). Odonatologica, 16, 201–207.

Miller PL (1991). The structure and function of the genitalia in the Libellulidae (Odonata). Zoological Journal of the Linnean Society, 102, 43–73.

Miller PL (1995). Sperm competition and penis structure in some libellulid dragonflies (Anisoptera). Odonatologica, 24, 63–72.

Miller PL, Miller CA (1981). Field observations on copulatory behaviour in Zygoptera, with an examination of the structure and activity of male genitalia. Odonatologica, 10, 201–218.

Monetti L, Sánchez-Guillén RA, Cordero Rivera A (2002). Hybridization between *Ischnura graellsii* (Vander Linder) and *I. elegans* (Rambur) (Odonata: Coenagrionidae): are they different species? Biological Journal of the Linnean Society, 76, 225–235.

Morrow EH, Arnqvist G, Pitnick S (2003). Adaptation versus pleiotropy: why do males harm their mates? Behavioral Ecology, 14, 802–806.

Nagata N, Kubota K, Yahiro K, Sota T (2007). Mechanical barriers to introgressive hybridization revealed by mitochondrial introgressive patterns in *Ohomopterus* ground beetles assemblages. Molecular Ecology, 16, 4822–4836.

Nakahara M, Tsubaki Y (2007). Function of multiple sperm-storage organs in female damselflies (*Ischnura senegalensis*): Difference in amount of ejaculate stored, sperm loss, and priority in fertilization. Journal of Insect Physiology, 53, 1046–1054.

Ono T, Siva-Jothy MT, Kato A (1989). Removal and subsequent ingestion of rivals' semen during copulation in a tree cricket. Physiological Entomology, 14, 195–202.

Orr AG, Hämäläinen M (2007). *The Metalwing Demoiselles (Neurobasis and Matronoides) of the Eastern Tropics*. Natural History Publications, Borneo.

Parker GA (1970). Sperm competition and its evolutionary consequences in the insects. Biological Reviews, 45, 525–567.

Paulson DR (1974). Reproductive isolation in damselfies. Systematic Zoology, 23, 40–49.

Pessacq P (2008). Phylogeny of Neotropical Protoneuridae (Odonata: Zygoptera) and a preliminary study of their relationship with related families. Systematic Entomology, 33, 511–528.

Pfau HK (1971). Struktur und Funktion des sekundären Kopulationsapparates der Odonaten (Insecta, Palaeoptera), ihre Wandlung in der Stammesgeschichte und Bedeutung für die adaptive Entfaltung. Zeitschrift fuer Morphologie und Oekologie der Tiere, 70, 281–371.

Pfau HK (1991). Contributions of functional morphology to the phylogenetic systematics of the Odonata. Advances in Odonatology, 5, 109–141.

Pfau HK (2005). Structure, function and evolution of the 'glans' of the anisopteran vesica spermalis (Odonata). International Journal of Odonatology, 8, 259–310.

Polhemus D, Asquith A (1996). Hawaiian Damselflies. A Field Identification Guide. Bishop Museum Press, Hawaii.

Rice WR (1996). Sexually antagonistic male adaptation triggered by experimental arrest of female evolution. Nature, 381, 232–234.

Robertson HM, Paterson HEH (1982). Mate recognition and mechanical isolation in *Enallagma* damselflies (Odonata: Coenagrionidae). Evolution, 36, 243–250.

Robertson HM, Tennessen KJ (1984). Precopulatory genital contact in some zygoptera. Odonatologica, 13, 591–595.

Roff DA (1980). Optimizing development time in a seasonal environment: The 'ups and downs' of clinal variation. Oecologia, 45, 202–208.

Roff DA (1986). Predicting body size with life history models. Bioscience, 36, 316–323.

Sawada K (1995). Male's ability of sperm displacement during prolonged copulations in *Ischnura senegalensis* (Rambur) (Zygoptera: Coenagrionidae). Odonatologica, 24, 237–244.

Sawada K (1998). Sperm precedence in the damselfly *Ischnura senegalensis* (Rambur): Is prolonged copulation advantageous to sperm precedence? (Zygoptera: Coenagrionidae). Odonatologica, 27, 425–431.

Serrano-Meneses MA, Córdoba-Aguilar A, Méndez V, Layen SJ, Székely T (2007). Sexual size dimorphism in the American rubyspot: male body size predicts male competition and mating success. Animal Behaviour, 73, 987–997.

Shapiro A, Porter A (1989). The lock-and-key hypothesis: evolutionary and biosystematic interpretation of insect genitalia. Annual Review of Entomology, 34, 231–245.

Silsby J (2001). Dragonflies of the Wold.The Natural History Museum, London.

Simmons LW (2001). Sperm Competition and Its Evolutionary Consequences in the Insects. Princeton University Press, Princeton.

Simmons LW, Parker GA (1992). Individual variation in sperm competition success of yellow dung flies, *Scatophaga stercoraria*. Evolution, 46, 366–375.

Sirot LK, Brockmann HJ (2001). Costs of sexual interactions to females in Rambur's forktail damselfly, *Ischnura ramburi* (Zygoptera: Coenagrionidae). Animal Behaviour, 61, 415–424.

Siva-Jothy MT (1984). Sperm competition in the family Libellulidae (Anisoptera) with special reference to *Crocothemis erythraea* (Brullé) and *Orthetrum cancellatum* (L.). Advances in Odonatology, 2, 195–207.

Siva-Jothy MT (1987). The structure and function of the female sperm-storage organs in libellulid dragonflies. Journal of Insect Physiology, 33, 559–567.

Siva-Jothy MT (1988). Sperm "repositioning" in *Crocothemis erythraea*, a libellulid with a brief copulation. Journal of Insect Behavior, 1, 235–245.

Siva-Jothy MT (1997). Odonate ejaculate structure and mating systems. Odonatologica, 26, 415–437.

Siva-Jothy MT, Hooper RE (1996). Differential use of stored sperm during oviposition in the damselfly *Calopteryx splendens xanthostoma* (Charpentier). Behavioral Ecology and Sociobiology, 39, 389–393.

Siva-Jothy MT, Tsubaki Y (1989). Variation in copulation duration in *Mnais pruinosa pruinosa* Selys (Odonata: Calopterygidae) 2. Causal factors. Behavioral Ecology and Sociobiology, 25, 261–267.

Stoks R, Johansson F and De Block M. (2008). Life-history plasticity under time stress in damselfly larvae. In A Córdoba-Aguilar, ed. Dragonflies and Damselflies: Model Organisms for Ecological and Evolutionary Research, pp 39–50.Oxford University Press, Oxford.

Takami M, Nagata N, Sasabe M, Sota T (2007). Asymmetric in reproductive isolation and its effects on directional mitochondrial introgression in the parapatric ground beetles *Carabus yamato* and *C. albrechti*. Population Ecology, 49, 337–346.

Tillyard RJ (1917). The Biology of Dragonflies. Cambridge University Press, Cambridge.

Tsuchiya K, Hayashi F (2008). Surgical examination of male genital function of calopterygid damselflies (Odonata). Behavioral Ecology and Sociobiology, 62, 1417–1425.

Tynkkynen K, Grapputo A, Kotiaho JS, Rantala MJ, Vaananen S, Suhonen J (2008). Hybridization in *Calopteryx* damselflies: the role of males. Animal Behaviour, 75, 1431–1439.

Uhía E, Cordero Rivera A (2005). Male damselflies detect female mating status: importance for postcopulatory sexual selection. Animal Behaviour, 69, 797–804.

Utzeri C, Belfiore C (1990). Tandem anomali fra Odonati. Fragmenta Entomologica, 22, 271–287.

Von Ellenrieder N (2008). *Phoenicagrion* gen. nov. for *Leptagrion flammeum*, with description of a new species, *P. paulsoni*, from Peru (Odonata: Coenagrionidae). International Journal of Odonatology, 11, 81–93.

Von Ellenrieder N, Garrison RW (2008a). A redefinition of *Telagrion* Selys and *Aceratobasis* Kennedy stat. rev. and the description of *Schistolobos* gen. nov. for *Telagrion boliviense* Daigle (Odonata: Coenagrionidae). Transactions of the American Entomological Society, 134, 1–22.

Von Ellenrieder N, Garrison RW (2008b). *Drepanoneura* gen. nov. for *Epipleoneura letitia* and *Protoneura peruviensis*, with descriptions of eight new Protoneuridae from South America (Odonata: Protoneuridae). Zootaxa, 1842, 1–34.

Von Ellenrieder N, Garrison RW (2008c). Oreiallagma gen. nov. with a redefinition of *Cyanallagma* Kennedy 1920 and *Mesamphiagrion* Kennedy 1920, and the description of *M. dunklei* sp. nov. and *M. ecuatoriale* sp. nov. from Ecuador (Odonata: Coenagrionidae). Zootaxa, 1805, 1–51.

Von Ellenrieder N, Lozano F (2008). Blues for the red *Oxyagrion*: a redefinition of the genera *Acanthagrion* and *Oxyagrion* (Odonata: Coenagrionidae). International Journal of Odonatology, 11, 95–113.

Waage JK (1979). Dual function of the damselfly penis: sperm removal and transfer. Science, 203, 916–918.

Waage JK (1982). Sperm displacement by male *Lestes vigilax* Hagen (Zygoptera: Lestidae). Odonatologica, 11, 201–209.

Waage JK (1984). Sperm competition and the evolution of odonate mating systems. In RL Smith, ed. Sperm Competition and the Evolution of Animal Mating Systems, pp 251–209. Ademic Press, Orlando.

Waage JK (1986). Evidence for widespread sperm displacement ability among Zygoptera (Odonata) and the means for predicting its presence. Biological Journal of the Linnean Society, 28, 285–300.

352 Primary Sexual Characters in Selected Taxa

Waage JK (1987). Choice and utilization of oviposition sites by female *Calopteryx maculata* (Odonata: Calopterygidae). I. Influence of site size and the presence of other females. Behavioral Ecology and Sociobiology, 20, 439–446.

Ward PI, Simmons LW (1991). Copula duration and testes size in the yellow dung fly, *Scathophaga stercoraria* (L)—The effects of diet, body size, and mating history. Behavioral Ecology and Sociobiology, 29, 77–85.

Watson JAL (1966). Genital structure as an isolating mechanism in Odonata. Proceedings of the Royal Entomological Society of London Series A—General Entomology, 41, 171–174.

Yokoi N (1990). The sperm removal behavior of the Yellow Spotted Longicorn Beetle *Psacothea hilaris* (Coleoptera: Cerambycidae). Applied Entomology and Zoology, 25, 383–388.

16

Postcopulatory Sexual Selection in the Coleoptera

Mechanisms and Consequences

PAUL EADY

INTRODUCTION

With over 350,000 named species, beetles are the most varied group of insects, representing the largest order in the Animal Kingdom. Roughly one in every four species of animal is a beetle and they range in size from minute featherwing beetles (Ptiliidae), about 0.25 mm in length, to the titan beetle (*Titanus giganteus*) which can grow up to 20 cm in length. They have colonized the land, air, and water, and occupy niches as diverse as soil, dung, carrion, rotting vegetation, fungi, herbaceous plants, bushes, trees, lakes, and streams. Many live in close association with humans, being found on agricultural crops, house timbers, furs, hides, museums, and dry food stores. Their success appears to be connected to the evolution of heavily sclerotized forewings (elytra) that protect the abdomen and hind wings, enabling them to bore into wood, tunnel into dung and soil and squeeze into tight galleries under rocks and bark. Flight is possible via foldable, membranous hind wings that can be packed neatly and safely beneath the protective elytra.

The order is divided into four sub-orders (Hunt et al. 2007); the Archostemata (40 species), Myxophaga (94 species), Adephaga (~10% of species) and the Polyphaga (~90% of species) and, as with many species, their external diversity is matched, if not surpassed, by the diversity of their reproductive organs. The value of primary reproductive characters (especially those of males)

in beetle taxonomy is testament to the rapid and divergent evolution of these traits (Sharp & Muir 1912; Jeannel 1955), although like many other groups, the selection pressures that underpin this divergent evolution remain largely speculative. Despite this, a number of recent studies have begun to elucidate the underlying mechanisms responsible for the bewildering array of primary reproductive traits found in Coleoptera, and it is likely that beetle research will play a key role in identifying the selective agents that bring about this evolutionary diversity.

Here I provide a brief description of the reproductive biology of beetles before reviewing the evidence that post-copulatory sexual selection is a central agent driving the evolutionary divergence of primary reproductive traits in the Coleoptera.

GENERAL REPRODUCTIVE BIOLOGY

Given the vast diversity of Coleoptera it is difficult to present typical male and female reproductive systems. However, some general patterns are apparent. The male reproductive system consists of the testes, vasa deferentia, one or more pairs of accessory glands, and a median ejaculatory duct. Vesiculae seminalis are often present as swellings of the vasa deferentia. Within the Coleoptera, two types of reproductive system are recognized. In the Adephaga, the testes are simple and tubular and

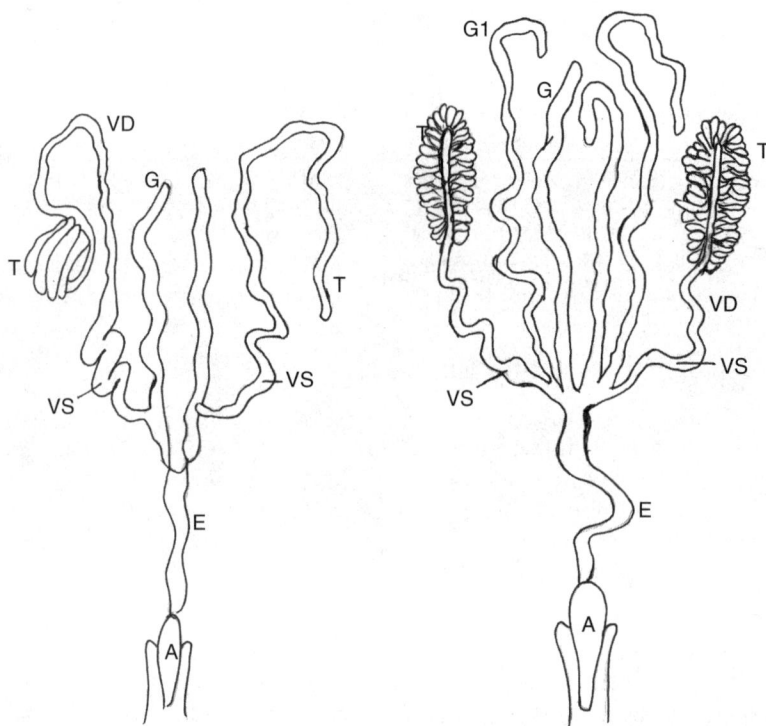

FIGURE **16.1** The male reproductive organs of: left Adephaga; right, Polyphaga. The right testis in the Adephaga is represented as uncoiled. A, aedeagus; E, ejaculatory duct; G, accessory gland (ectadenes); G₁, accessory gland (mesadenes); T, testis; VD, vad deferene; VS, vesicular seminalis. Redrawn from Richards & Davies 1977.

more or less closely coiled. In the Polyphaga, the testes are compound and divided into a number of separate follicules (figure 16.1) (Richards & Davies 1977). Accessory glands differ greatly with respect to number, position and mode of origin.

The male reproductive system is divided into two parts: the zygotic portion describes the paired seminal ducts leading from the testes to where they fuse into a common duct. The azygos comprises the entire unpaired portion of the tube from the divergence of the seminal ducts to the body wall. The azygotic section is further sub-divided into the stenazygotic portion, the long slender duct that originates from the divergence of the seminal ducts to the eurazygotic portion, an enlargement of the azygotic section. This enlarged section of the azygos is eventually reflected outwards to join the body wall. This reflected portion is sometimes referred to as the phallic portion (Sharp & Muir 1912). Part of the eurazygotic portion is invaginated upon itself, and this invaginated section is known as the

internal sac (endophallus) which is usually more or less evaginated during copulation. The phallic portion is further divided into two sections: the median lobe and the tegmen. The median lobe is a complex of individual sclerites that are situated at the distal end of the phallic portion, whilst the tegmen, which consists of the lateral lobes (parameres) and basal piece, sits at the base of the phallic section. Within the Coleoptera, the tegmen, median lobe, and internal sac vary widely in size, shape, and armature, thus it is difficult to present a typical beetle form. For a comprehensive review of the male genitalia see Sharp and Muir (1912) and Jeannel (1955).

Studies of coleopteran female genitalia are relatively scarce in comparison to studies of the male aedeagus (Dupuis 2005; Lopez-Guerro & Halffter 2002) and there are often inconsistencies in the terminology used (see Dupuis 2005 for a list of synonyms). The female reproductive system of the Adephaga and the Polyphaga can (like the male reproductive system) be divided into two types.

In the Polyphaga the ovarioles are acrotrophic (sometimes referred to as teleotrophic), meaning the trophocytes (nurse cells that contribute nutrition to the developing oocytes) are confined to the germarium (the site at which the primary oocytes are produced) and remain attached to the oocytes by cytoplasmic strands, as the oocyte moves down the ovariole (Gullan & Cranston 2005). In the Adephaga, the ovarioles are polytrophic in character, meaning a number of trophocytes are connected to each oocyte and move down the ovariole with the oocyte. In general, the vagina receives the male intromittent organ. A bursa copulatrix is often present as a diverticulum of the vaginal wall and this is often the site at which sperm are received. A spermatheca (generally the primary site of sperm storage) opens into the bursa or the vagina via a long, slender spermathecal duct. An accessory gland is generally found in connection with the spermatheca. In some Coleoptera, a second passage or 'canal of fecundation' leads from the spermatheca to the point of union of the two oviducts.

The external genitalia (figure 16.2) consist of dorsal, ventral and lateral sclerites (tergites, sternites and pleurites, respectively), that are derived from abdominal segments XIII to X (Dupuis 2005). However, there is considerable variation on this theme with the loss, reduction, enlargement or fusion of genital sclerites.

Some features of the external genitalia are taxonomically useful. For example, in the genus Heterogomphus (Dynastidae: Oryctinae) the coxosubcoxites (a fusion of subcoxite IX and coxite IX) are species specific, whilst in the sub-family Trodinae (Trodidae) there are clear species specific differences in the style, coxosubcoxite, epipleurite, and sympleurite (figure 16.3). Within this sub-family, there are also species-specific differences in the internal genital structures (Dupuis 2005); (figure 16.4), see also Lopez-Guerrero and Halffter (2002) for species-specific differences in spermathecal structure.

COPULATION AND SPERM TRANSFER

In insects, sperm transfer is via a spermatophore or free sperm transfer into the female reproductive tract. According to Gerber (1970) there are four basic methods of spermatophore formation;

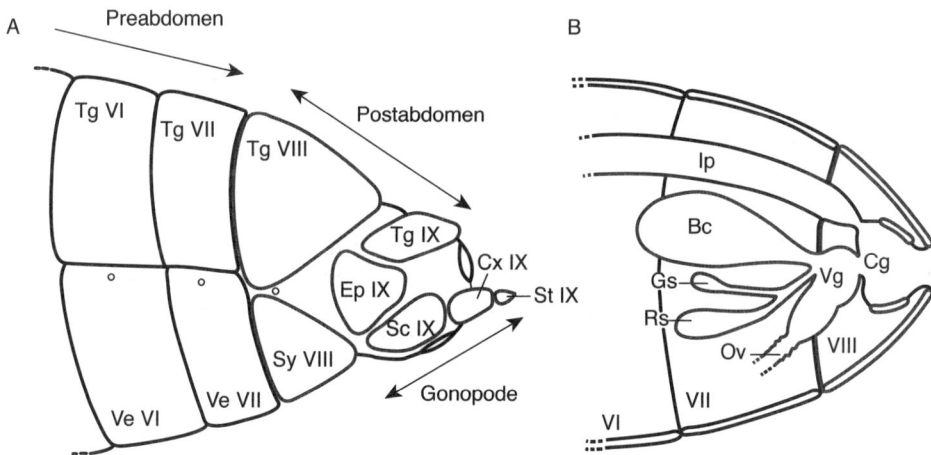

FIGURE **16.2** Simplified representation of the abdomen and genitalia of the Scarabaeoidae. A. external genitalia: Tg (tergite), Ve (ventrite), Sy (sympleurite), Ep (epipleurite), Sc (subcoxite), Cx (coxite), St (style). B. internal genitalia: Ip (posterior intestine), Bc (bursa copulatrix), Cg (anogenital chamber), Vg (vagina), Gs (spermathecal gland), Rs (seminal receptacle), Ov (Oviduct). Reproduced with permission fom Dupuis (2005) L'abdomen et les genitalia des femelles de coléoptères Scarabaeoidea (Insecta, Coleoptera). Zoosystema, 27(4), 733–823. © Publications Scientifiques du Muséum national d'Histoire naturelle, Paris.

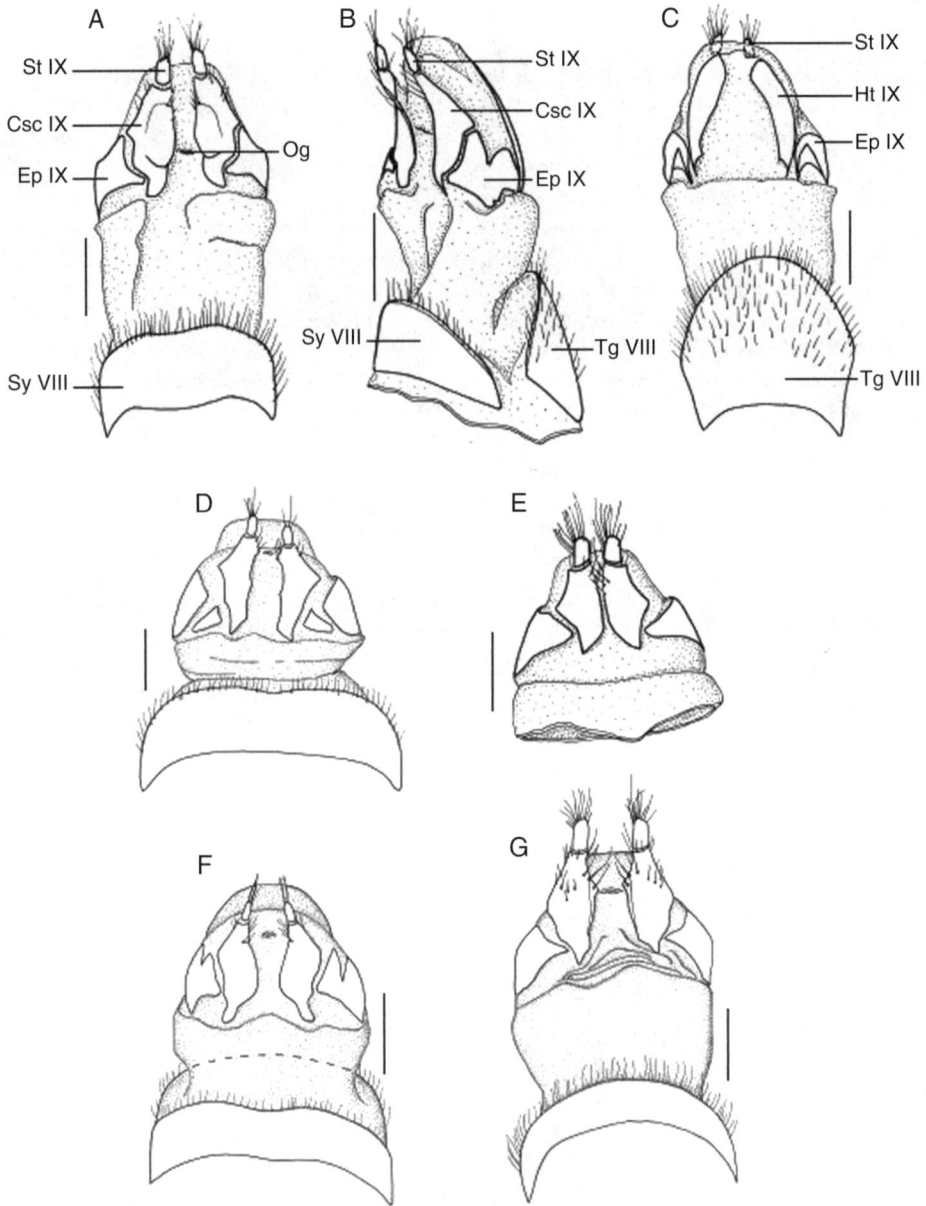

FIGURE 16.3 External genitalia of female Trogidae. A–C *Trox perlatus* (A ventral, B lateral, and C dorsal view. D *Trox scaber* (ventral view), E *Trox sabulosus* (ventral view), F *Trox perrisi* (ventral view), G *Trox hispidus* (ventral view). Csc (coxosubcoxite), Ep (epipleurite), St (style), Sy (sympleurite), Tg (tergite). Scale bar 0.5 mm. Reproduced with permission from Dupuis (2005) L'abdomen et les genitalia des femelles de coléoptères Scarabaeoidea (Insecta, Coleoptera). Zoosystema, 27(4), 733–823. © Publications Scientifiques du Muséum national d'Histoire naturelle, Paris.

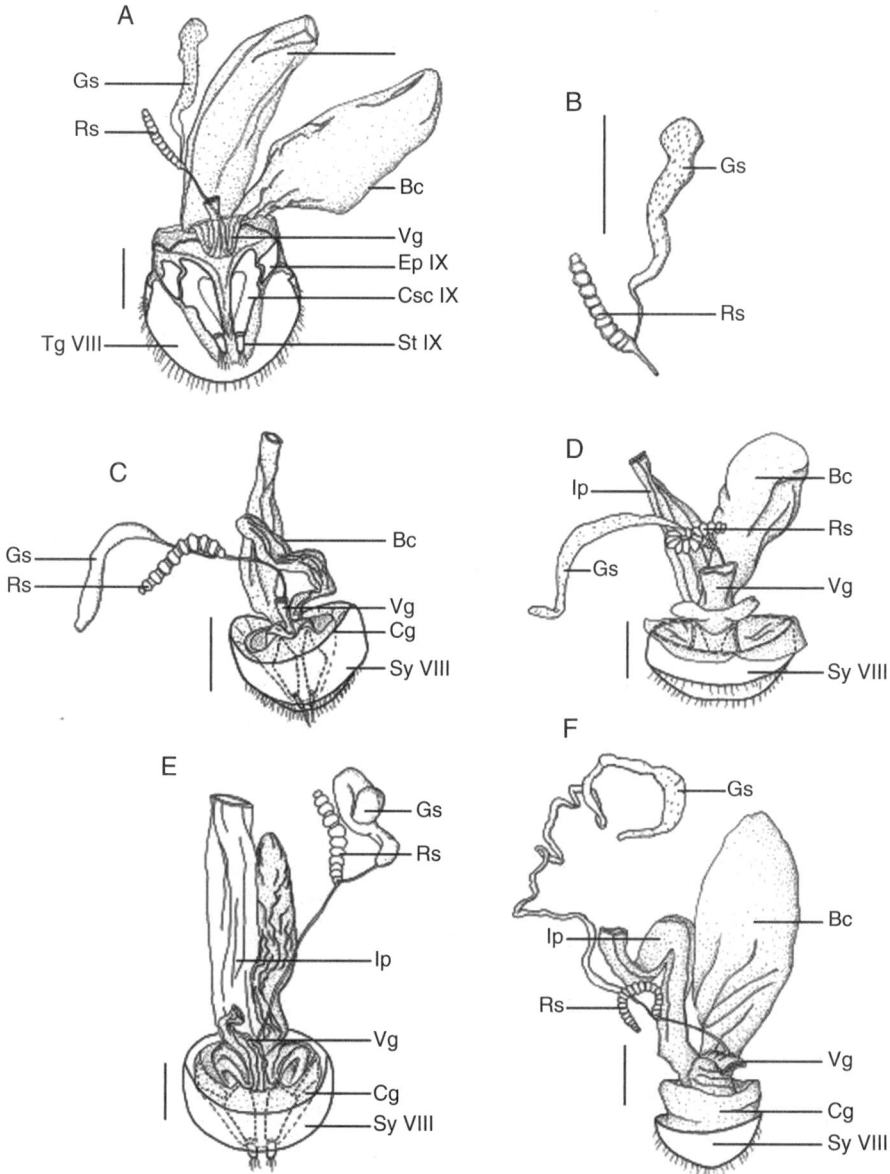

FIGURE 16.4 Internal genitalia of female Trogidae (all ventral view). A *Trox perlatus*, B detail of spermatheca of *Trox perlatus*, C *Trox perrisi*, D *Trox scaber*, E *Trox hispidus*, F *Trox sabulosus*. Bc (bursa copulatrix), Csc (coxosubcoxite), Ep (epipleurite), Gs (spermathecal gland), Ip (posterior intestine), Rs (seminal receptacle), St (style), Tg (tergite), Vg (vagina). Scale bar = 0.5 mm. Reproduced with permission from Dupuis (2005) L'abdomen et les genitalia des femelles de coléoptères Scarabaeoidea (Insecta, Coleoptera). Zoosystema, 27(4), 733–823. © Publications Scientifiques du Muséum national d'Histoire naturelle, Paris.

1. the spermatophore is moulded in the anterior end of the male reproductive tract prior to being transferred to the female,
2. the spermatophore is moulded in the copulatory sac of the male copulatory organ, while this sac is everted into the bursa copulatrix of the females,
3. the spermatophore is moulded by the female bursa copulatrix, following the ejaculation of sperm and seminal fluid by the male, and
4. male sperm and accessory gland material are ejected into the female reproductive tract, but the accessory gland products do not encapsulate the spermatozoa.

All four methods have been observed in Coleoptera (Gerber 1970).

Unsurprisingly, given the large number of species, copulation in the Coleoptera is very diverse. In *Aleochara curtula* (Coleoptera: Staphylinidae), the spermatophore consists of a number of amorphous secretions, whose final shape is determined inside the female genital chamber (Forster et al. 1998). A capsular sperm sac is distinguishable within the spermatophore and, shortly before the termination of copulation, a tube begins to grow out of the sperm sac into the spermathecal duct. Following copulation, the tube continues to grow up through the spermathecal duct and into the spermatheca. This tubule growth appears to be a two-stage process, in which a primary tube extends to a certain length, before bursting and releasing the secondary tube. The secondary tube also elongates, reaching the spermatheca, at which point the tip of the tube inflates to form a balloon and then ruptures, releasing sperm into the female sperm storage organ (Forster et al. 1998). Interestingly, the aedeagus of male *Aleochara* is equipped with a flagellum, a cuticular rod that arises from the soft endophallus [similar flagella/pseudoflagella are found in some Chrysomelidae (Flowers & Eberhard 2006) and in the Sisyphina (Scarabaeidae) Lopez-Guerrero & Halffter (2002)]. In *Aleochara*, the flagellum does not serve as a duct through which the spermatophore is passed because the ejaculatory duct is at the base of the flagellum. Rather, the flagellum is introduced into the spermathecal duct during copulation where it is thought to function as a guiding sturucure for the tube growing out of the spermatophore. The tip of the flagellum reaches a site very close to the opening of the spermatheca. The flagellum is approximately 16 mm long (more than twice the length of the beetle) and whilst retracted in the aedeagus, is coiled like a spring. The female spermathecal duct is also a coiled structure that approximates 16 mm in length. The extraordinary length and very narrow width of the male's intromittent organ poses an interesting physical challenge to the male. During the retraction of the flagellum from the female, the male secures the flagellum between the wing shoulder and the pronotum, holding it taut for about one half of its length, allowing the gradual retraction of the flagellum in an orderly fashion (figure 16.5), presumably to prevent entanglement and damage (Gack & Peschke 2005).

In the bruchid beetle, *Callosobruchus maculatus* little pre-copulatory courtship is evident. Males chase after females antennating on their elytra. Should a female stop, the male mounts the female and continues to antennate, whilst probing the female genital opening with his aedeagus. If successful, the male's intromittent organ, which is tipped with spines, is everted within the female reproductive tract. At this point the male stops antennating on the elytra of the female. The male's internal sac is then everted into the female's bursa copulatrix, where a spermatophore is formed (personal observation). Approximately two-thirds of the way through copulation, the female begins to kick at the male with her hind legs, and continues until copulation is terminated. Following copulation sperm migrate from the bursa copulatrix to the spermatheca via the spermathecal duct (Eady 1994a). In *Macrodactylus* (Scrabaeidae: Melolonthinae), copulation involves several stages including rubbing and tapping movements of the parameres prior to the male gaining access to the female genital chamber, the 'wedging' of the parameres (and inflatable sacs near the tip of the parameres) into the female genital opening, eversion of the internal sac through the vulva and into the vagina, the formation and transfer of a spermatophore, and finally the positioning of the spermatophore (Eberhard 1993). Observations of pairs frozen in copula revealed that in some cases the entrance of the male's internal sac into the vagina was prevented by the vulva being closed and/or the vaginal walls being contracted (Eberhard 1993). Eberhard (1993) interprets these observations within the framework of the sustained energetic copulatory courtship in these beetles, in which males are required to court during copulation in order to induce females to accept full

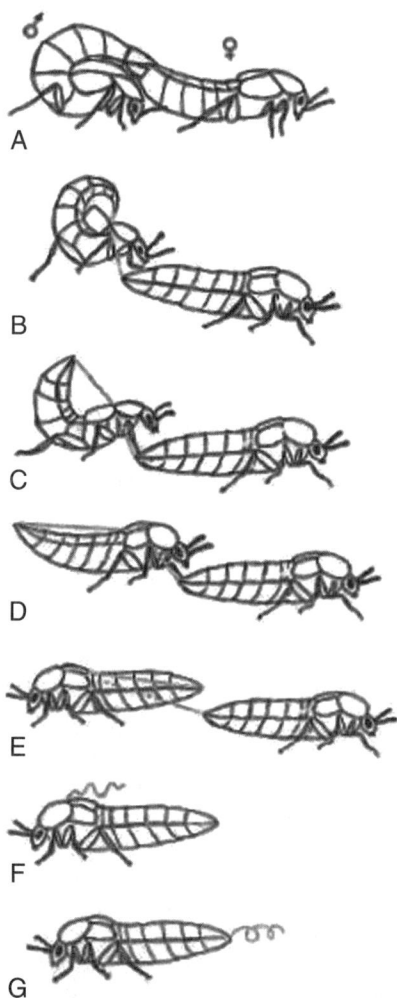

FIGURE 16.5 *Aleochara tristis*. A, position of mating; B, male pulls away from female exposing a small portion of the flagellum which he secures between the pronotum and the mesothorax ("shouldering"); C and D, he then moves his abdomen backwards, extracting the flagellum from the female to about half its overall length; E, he turns 180° and continues to feed the flagellum back into his aedeagus; F and G, male and female separate, and the male continues to feed the flagellum into his aedeagus. The sequence of retraction lasts about 90 seconds. Reproduced with permission from Gack & Peschke (2005) 'Shouldering' exaggerated genitalia: a unique behavioural adaptation for the retraction of the elongate intromittent organ by the male rove beetle (*Aleochara tristis* Gravenhorst). *Biological Journal of the Linnean Society*. Wiley-Blackwell.

copulatory penetration and successful spermatophore deposition. The sheer diversity of copulatory linking is exemplified by the study of Flowers and Eberhard (2006) on Neotropical Chrysomelidae. For example, in the sub-family Galerucinae, tribe Alticini, the endophallus of *Alagoasa gemmata* and *Walterianella* sp. are very short and appear not to enter the female reproductive tract. Instead, the apical parts of the female bursae are everted and enter the tip of the male median lobe, where sclerites of the median lobe clamp around the vaginal membrane. By contrast in the tribe Galerucini, the male median lobe is long and narrow and penetrates deep into the female bursa, where eversion of the endophallus results in the opening of a fan-like arrangement of needle-like sclerites that pierce the wall of the bursa. Further variation in form and function can be found in the sub-family Eumolpinae, in which the male endophalli are long, generally filling the female vagina, being secured in place by sclerotized lateral and basal appendages in *Colaspis sanjoseana* and *Brachyphoea irazuensis*, a pair of membranous swellings in *Metaxyonycha amasia*, and apical microspicules in *Xanthonia*. In *B. irazuensis* the endophallus is tipped with a flagellum that has been observed to penetrate the female spermathecal duct. In summary, the way male and female genitalia link during copulation is very diverse even between closely related species.

POST-COPULATORY SEXUAL SELECTION

The primary role of traits associated with reproduction is the formation of the zygote, and ultimately offspring production via the fusion of eggs and sperm. However, because females of many insects mate with more than one male and have the ability to store sperm for relatively long periods of time, it is likely that post-copulatory sexual selection has played a significant role in the evolution of these traits (Birkhead & Moller 1998; Eberhard 1985; Eberhard 1996; Simmons 2001; Arnqvist & Rowe 2005). Essentially, any male trait that confers an advantage at fertilization (behavior, morphology, physiology) can become subject to post-copulatory sexual selection, and the genes responsible for these traits will become established in the population.

Tightly associated with the notion of post-copulatory sexual selection is sperm competition, the competition between the sperm of two or more males for the fertilization of a given set of ova (Parker 1998). In its strictest sense, sperm competition occurs only when there is a temporal and spatial overlap of ejaculates from competing males (Simmons 2001). Under these conditions selection is likely to act on individual sperm to out compete their rivals. However, the selection pressures generated by sperm competition have led to the evolution of male traits that avoid sperm competition (i.e., prevent the overlap of ejaculates), such as sperm removal and/or mate guarding (Simmons 2001).

Male success at sperm competition is typically recorded as P_2 (the proportion of eggs fertilized by the second of two males to mate, in controlled mating trials; Boorman & Parker 1976). That P_2 is a good estimate of last male paternity when the female has copulated with more than two males has been confirmed in the beetles *Tribolium castaneum*; (Lewis & Jutkiewicz 1998) and *Callosobruchus maculatus* (Eady & Tubman 1996). In beetles, species specific P_2 values range between 0.11 and 0.94, but there is considerable intraspecific variation in P_2, ranging between 0 and 1 (e.g., Eady 1991; Jones et al. 2006). Following Simmons

(2001) a histogram of Coleopteran species specific P_2 values reveals a number of cases of sperm mixing (P_2 between 0.4 and 0.6), but then a skew towards higher values of P_2, indicative of sperm displacement or sperm removal (figure 16.6). Overall, the mean (+ se) P_2 from 36 species of beetle = 0.65 + 0.02.

The mechanisms that underlie species specific P_2 values are largely unknown. High P_2 may be indicative of sperm removal by the last male to mate, sperm stratification within female sperm storage organs (last in first out principle), or simply sperm loss from spermathecae during the interval between copulations. Intraspecific variation in P_2 is generally more informative with respect to mechanisms of sperm competition. For example, if sperm stratification resulted in high P_2, one might expect P_2 to decline over successive bouts of oviposition as the last sperm to enter the sperm stores either become depleted or have had sufficient time to become mixed with sperm inseminated by previous males. In *Tenebrio molitor*, P_2 is initially high after the second male of a pair has copulated but declines over time, indicating sperm stratification (Siva-Jothy et al. 1996). By contrast, in *Callosobruchus maculatus*, P_2 does not decline over

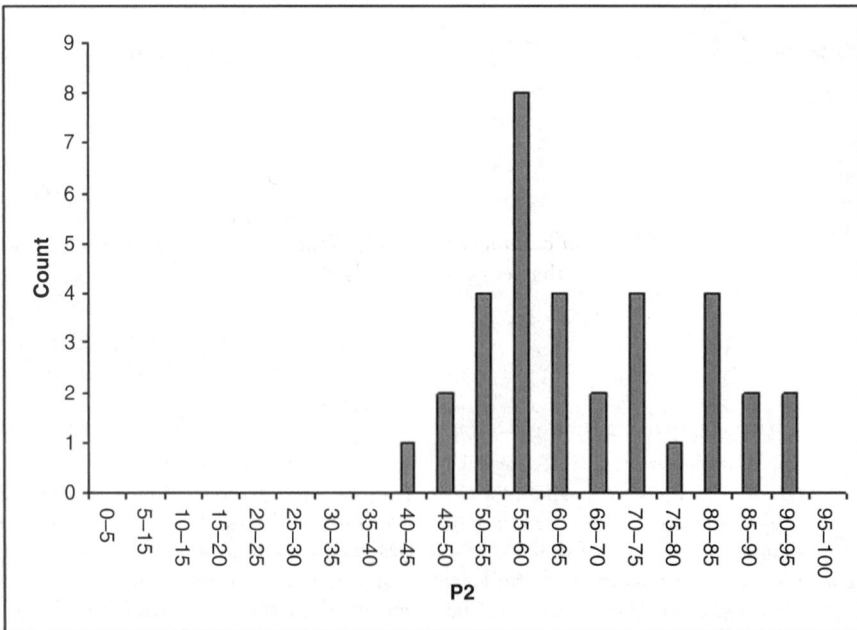

FIGURE **16.6** Frequency histogram of Coleopteran species specific mean P_2 values.

time indicating that sperm stratification is unlikely to be responsible for high P_2 in this species (Eady 1994b). The mechanism responsible for high P_2 in *C. maculatus* is also unlikely to be passive sperm loss because at the time of the second copulation, the fertilization set (primary site of sperm storage, which in *C. maculatus* is the spermatheca) contained approximately 60% of its capacity. Thus, by simply topping up the female's sperm stores the second mating male would expect to sire—40% of subsequent eggs, substantially less than the observed 83% (Eady 1995). The most likely mechanism affecting the high P_2 in *C. maculatus* is a combination of passive sperm loss and indirect sperm removal (sperm flushing). However, this conculsion was derived from empirical data matching the predictions of a model of constant random sperm displacement (Parker & Simmons 1991), when the parameter p (the proportion of inseminated sperm that enter the spermatheca) was estimated. Thus, without sophisticated techniques of following the fate of inseminated sperm, it is often difficult to elucidate the underlying mechanisms of sperm competition.

Sperm removal—direct sperm removal has been reported in the longicorn beetle *Psacothea hilaris* (Yokio 1990). In this species there are two stages of copulation. During stage 1 the male's penis is repeatedly inserted into and withdrawn from the female reproductive tract, after which the pair enter stage 2, in which they remain in copula. The number of sperm in the female sperm storage organs is reduced by up to 98% during stage 1 of copulation. To facilitate this sperm removal the male intromittent organ is equipped with a comb of microbristles and a scoop-like structure at the distal end.

The intromittent organ of *Triboluim castaneum* also has an array of chitinous spines that entrap rival sperm from the bursa copulatrix of the female. In *T. castaneum*, P_2 is approximately 60%, thus it appears that sperm are not removed from the primary sperm storage site, the spermatheca. However, removal of rival sperm from the bursa may increase the fertilization success of the copulating male, as this behavior reduces the number of rival sperm entering the fertilization set from the bursa. Interestingly, adherence of rival sperm to the aedeagus of the copulating male may lead to fertilization by proxy, as non-self sperm may be transferred during subsequent copulations with different females (Haubruge et al. 1999), although

Tigreros et al. (2009) found no evidence to support fertilization by proxy in this species.

The presence of genitalic spines is not always associated with direct sperm removal. In *Callosobruchus maculatus*, Eady (1994) found no evidence of sperm adherence to the spines of the male intromittent organ. By contrast, in *Tenebrio molitor*, sperm were found on the intromittent organs of males following copulation. However, subsequent analysis by Siva-Jothy et al. (1996) revealed that the sperm most likely belonged to the copulating male, indicating that care should be taken in the interpretation of sperm competition mechanisms based only on direct observations of sperm attached to intromittent organs.

Strategic ejaculation—Despite the underlying assumption that male reproductive success is limited by access females (Bateman 1948), Dewsbury (1982) highlighted that ejaculates are often costly and males sperm limited. For example, in *Callosobruchus maculatus*, the initial ejaculate may represent >5% of male body weight (Savali & Fox 1999), and this expenditure declines markedly during subsequent ejaculations (Savali & Fox 1999; Eady 1995). Such apparent costs have prompted a theoretical treatment of ejaculate strategy based on the trade-off between investment of resources in current copulation versus those of additional copulations (Parker 1998). Across species, ejaculate expenditure is predicted to increase with both sperm competition risk (probability of female mating with a second male; Parker et al. 1997) and intensity (number of males in competition for a given batch of ova; Parker et al. 1996) and positive associations between testes size and the degree of polyandry have been found in butterflies (Gage 1994) and *Drosophila* (Pitnick & Markow 1994). Within species, the outcomes of theoretical models depend critically on the roles assigned to males (i.e., whether one or other male in competition is favored at fertilization). When roles are random, the evolutionary stable strategy (ESS) is for ejaculate expenditure to be equal for both males, but when one male is consistently in the disfavored role, the ESS is for an increase in ejaculate expenditure. Such an outcome may be apparent in the dung beetle *Onthophagus bimodis*. In this species males are dimorphic in relation to head horns. Horned males monopolise females by guarding the entrances to tunnels beneath the dung, in which males and females cooperate to provision young, whereas hornless males sneak into guarded tunnels to copulate with females (see Zunino & Halffter (in press) for a wider discussion of cephalic ornamentation in Onthophagus). Thus, hornless males always face sperm competition, whereas horned males face a lower probability of sperm competition. As predicted, the hornless males invested more in testes and ejaculates (Simmons et al. 1999). Similar alternative mate securing tactics are also evident in the congeneric O. taurus. However, in this species relative testes mass was found to be equivalent between horned and hornless males. Simmons et al. (1999) suggest the high frequency of sneaks in the O. taurus population means that both morphs face a high probability of sperm competition which, in accord with theory, should reduce the discrepancy in ejaculate expenditure between the two morphs.

Males may also adjust ejaculate expenditure in relation to current information on sperm competition risk (i.e., exhibit phenotypic plasticity in ejaculate expenditure; Parker et al. 1997). For example, in Tenebrio molitor males ejaculate more sperm per copulation when in the presence of other males than when alone (Gage & Baker 1991). However, in the two-spot ladybird (Adalia bipunctata; Ransford 1997) and the Eucalyptus snout-beetle (Gonipterus scutellatus; Carbone & Rivera 2003) sperm competition risk had no effect on ejaculate expenditure. These contrasting findings highlight the importance of elucidating the mechanisms of sperm competition. For example, if male success at sperm competition is independent of ejaculate size, as would be the case with direct sperm removal, then one would not predict ejaculate size to vary with sperm competition risk. However, if the outcome of sperm competition was determined by indirect sperm displacement or the numerical representation of sperm in the fertilization set, then ejaculate size (or more specifically the number of sperm inseminated) would be predicted to increase with increasing sperm competition risk.

SPERMATOZOA DIVERSITY AND EVOLUTION

As with genitalia (Eberhard 1985), insect sperm exhibit remarkably rapid and divergent evolution (Jamieson 1987). In general, such patterns of evolutionary divergence are indicative of sexual selection, and given that post-copulatory sexual selection is widespread in insects (Eberhard 1996; Simmons 2001) it is likely that traits that function to enhance fertilization success will become established in the population. Long sperm have been associated with increased sperm competition risk in birds (Briskie et al. 1997), mammals (Gomendio & Roldan 1991; but see Gage & Freckleton 2003), butterflies (Gage 1994), moths (Morrow & Gage 2000), and nematodes (LaMunyon & Ward 1999), whilst in fish, Stockley et al. (1997) found a negative relationship between sperm length and sperm competition risk, and in bats Hosken (1997) found no relationship. In a comparative analysis of 15 species of bruchid beetle Rugman-Jones and Eady (2008) found no relationship between sperm length (range 71.85 um in Kytorhinus sharpianus to 721.07 um in Zabrotes subfasciatus) and relative testes size (an indirect measure of sperm competition risk; Morrow & Gage 2000). By way of contrast, in the dimorphic dung beetle Onthophagus bimodis, Simmons et al. (1999) found hornless sneak males (which experience high levels of sperm competition) to have longer sperm than the horned guards, which generally experience lower levels of sperm competition. In the congener, O. taurus, horned and hornless males had equivalent sized sperm, although in this population both male morphs face high levels of sperm competition, thus similar selection pressures (Simmons et al. 1999). Thus, at present there is little evidence that sperm length is a reliable indicator of sperm competition in beetles.

A number of studies have reported correlated evolution between sperm length and areas of the female reproductive tract, indicating that the female reproductive environment exerts considerable selection on sperm morphology. In moths (Morrow & Gage 2000), scathophagid flies (Minder et al. 2005), featherwing beetles (Dybas & Dybas 1981), and bruchid beetles (Rugman-Jones & Eady 2008), sperm length has been shown to be positively related to the length of the ducts leading to the sperm storage organs. Sperm length has also been shown to correlate positively with the size of the female sperm storage organ in featherwing beetles (Dybas & Dybas 1981), stalk-eyed flies (Presgraves et al. 1999), and Drosophila (Pitnick et al. 1999), but negatively in the Bruchidae (Rugman-Jones & Eady 2008). Several explanations might be invoked to explain such cases of correlated evolution. For example, variation in female spermathecal duct length may arise due to genetic drift, pleiotropy and/or selection. If evolved differences in duct length affected the efficiency of sperm function, as has been shown in *Drosophila* (Miller & Pitnick 2002), then it is likely adaptations that optimize sperm function within the 'new' female reproductive environment would become fixed within the population. Indeed, Werner et al. (2007) have shown that in the rove beetle (*Drusilla canaliculata*) sperm motility in vivo is much greater than in vitro and suggest individual spermatozoa gain purchase against the narrow spermathecal duct, pushing themselves forward towards the spermatheca. In the dung beetle *O. taurus*, Kotiaho et al. (2001) found inverse genetic covariance between male condition and sperm length, such that males in good condition produced shorter sperm. Thus, theoretically, by selectively fertilizing their ova using short sperm, females could produce offspring of greater condition (good-genes process) and sons with short sperm, who might be more successful in sperm competition (sexy-sperm process). Indeed, short sperm have been shown to have a fertilization advantage during sperm competition, and that this advantage depends on the size of the female's spermatheca; short sperm have a greater advantage when competing in females with large spermathecae (Garcia-Gonzalez & Simmons 2007). Simmons and Kotiaho (2007) extended this analysis by arguing that because small sperm are 'preferred' by large spermathecae, then according to the sexy-sperm hypothesis, a negative genetic correlation between these traits is expected. Using a quantitative genetic approach in conjunction with geometric morphometrics to quantify variation in the size and shape of spermathecae, Simmons and Kotiaho (2007) found a significant negative genetic correlation between spermathecal size and sperm length such that fathers that sired sons with small sperm also sired daughters with large spermathecae, supporting the notion that postcopulatory preferences are an important selective agent driving the evolutionary divergence of sperm morphology. In many respects, evolved differences in female reproductive traits appear analogous to the sensory biases (sensory drives; Boughman 2002) observed in precopulatory displays (Ryan et al. 1990). As such they may represent the starting point for rapid evolutionary diversification based on Fisherian, good-genes or sexual conflict mechanisms of sexual selection, and thus a potentially fertile field for future research on female preferences.

MALE GENITALIA

In ground beetles of the subgenus *Ohomopterus*, there is a close species-specific match between the sclerotized, hook-like copulatory piece on the endophallus of the male and the counterpart membranous pocket of the vagina (vaginal appendix; Sota & Kubota 1998). Upon genital contact the endophallus is everted and inserted into the female reproductive tract (Takami 2002, 2003) (figure 16.7).

At this point the spermatophore is deposited at the innermost part of the vagina. Through microdissection Takami (2003) shortened the length of the copulatory piece in *Carabus insulicola*, and observed that 53% of manipulated males either failed to produce a spermatophore or failed to deposit a spermatophore at the correct site. Amongst the manipulated males, spermatophore deposition and sperm transfer to the spermatheca was higher in those males with a relatively long copulatory piece. This suggests the copulatory piece locks the male genitalia within the female reproductive tract, permitting correct spermatophore transfer. In the congeneric *C. maiyasanus* and *C. iwawakianus*, that share a hybrid zone, heterospecific matings are known to occur (Sota & Kubota 1998). However, a mis-match between the copulatory piece and the vaginal appendix frequently resulted in elevated levels of female mortality, probably as a result of damage to the vaginal wall. In addition 50% of male *C. maiyasanus* that engaged in copulation with *C. iwawakianus* females had a fracture to their copulatory piece, almost certainly rendering them infertile in future mating attempts. The mis-match between the genitalia of these two species almost certainly reduce the likelihood of hybridizations, which leads to the conclusion that the intricate match between male and female genitalia represents an important mechanism of reproductive isolation (the lock and key hypothesis). However, the expected advantage driving the lock and key mechanism, the avoidance of heteropsecific fertilizations by females, seems unlikely, as females matched with heterospecific males die. Thus, it appears that the mis-match and subsequent reproductive isolation may have arisen as an incidental by-product of other selective agents acting on genitalic form.

House and Simmons (2003) have shown that variation in male genitalia influences fertilization success during sperm competition in the dung beetle, *Onthophagus taurus*. The genitalia of

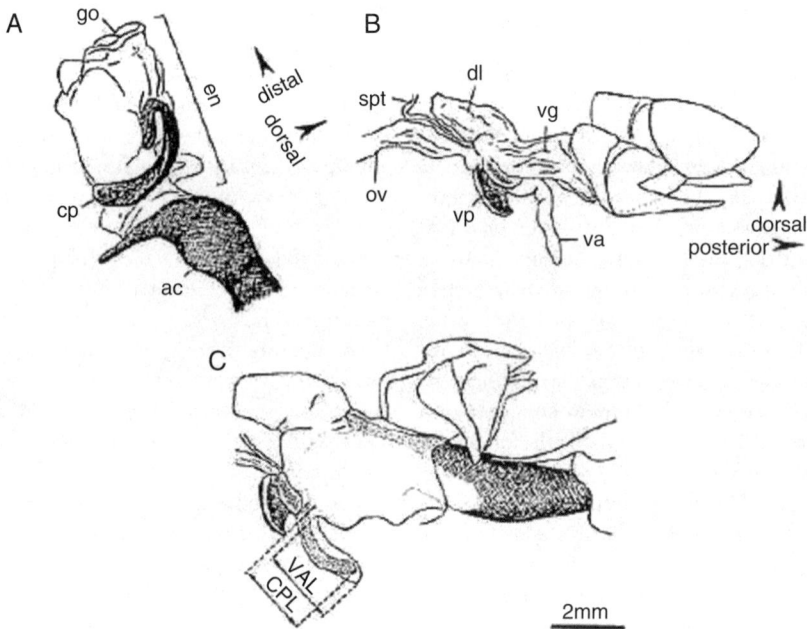

FIGURE **16.7** Genital morphology of *Carabus insulicola*. (A) Male genitalia: ae (aedeagus); cp (copulatory piece); en (endophallus); go (gonopore). (B) Female genitalia: dl (dorsal lobe); ov (oviduct); spt (spermatheca); va (vaginal appendix); vg (vagina); vp (vaginal apophysis). (C) Male and female genitalia in copula: male parts inserted into the vagina are dotted; the copulatory piece is inserted into the vaginal appendix; CPL and VAL refer to the lengths of the copulatory piece and vaginal appendix, respectively. Reproduced with permission from Takami (2003) Experimental analysis of the effect of genital morphology on insemination success in the ground beetle *Carabus insulicola* (Coleoptera Carabidae). *Ethology, Ecology & Evolution*, 15, 51–61.

O. taurus consists of several chitinous sclerites, embedded within an inflatable endophallus, which extends from the phallotheca. Copulation in this species is very complex. Initially the parameres engage with pits under the female pygidium. At this stage the force of body fluids causes the endophallus to extrude through the apex of the phallotheca into the bursa (Palestrini et al. 2000). The horn-like protrusion of the endophallus is inserted into the rectum of the female, presumably to anchor the endophallus in the female reproductive tract, whilst sclerite #4 also appears to function as a mechanical holdfast on the margin of the female gonoporus. At this point the endophallus is fully extruded and sclerites #2, #3, and #5 are flipped by 90° so that their tips are brought into the opening of the spermathecal duct (Werner & Simmons 2008). Variation in the dimensions of four out of the five measured sclerites influenced male fertilization success, with high P_1 being associated with a small sclerite #4 and

a large sclerite #5, while high P_2 was associated with a small sclerite #1 and a large sclerite #2 (House & Simmons 2003). When males were mated in either a defensive role (first male to mate, P_1) or an offensive role (second male; P_2) their success was positively correlated (House & Simmons 2006), suggesting that the genital sclerites do not function independently. Thus positive genetic correlations are predicted between sclerites #1 & #4 and #2 & #5 and negative genetic correlations between sclerites #2 & #4 and #1 & #5, predictions that were largely matched in terms of direction of genetic correlation rather than statistical significance (House & Simmons 2005a). Further support for the notion that male genitalic form affects success in sperm competition comes from the study of Wenninger & Averill (2006) on the oriental beetle *Anomala orientalis* (Scarabaeidae). They found the size of the genital spicule of the first mating male was negatively related to P_2 (i.e., first male success was enhanced if in possession of a large spicule). In typically reductionist terms, how variation in sclerite or spicule size/shape influences fertilization success remains to be seen.

The quantitative genetic study of House and Simmons (2005a) highlights the importance of single species studies on the genetic architecture of genital morphology. For example, the lock and key hypothesis predicts low coefficients of variation in genital traits and absent or weak genetic correlations with general morphological traits (Arnqvist 1997). By contrast, the pleiotropy model of genitalic evolution predicts strong genetic correlations between genital and general morphological traits and consequently high coefficients of variation in these traits (Arnqvist 1997), although which of the many traits are, or are not, anticipated to correlate with genitalia has yet to be formally discussed.

In *Phyllophaga hirticula* (Scarabaeidae), male and female genital characters exhibited greater phenotypic variability than non-sexual traits (Polihrouakis 2006), whilst in O. *taurus*, genital sclerites were found to be as genetically and phenotypically variable as general morphological traits (House & Simmons 2005b). Within O. *taurus* there was a large environmental effect on genitalic sclerite expression which is inconsistent with the lock and key hypothesis, whilst the low additive genetic variation (V_A) of the genital sclerites of O. *taurus* is inconsistent with the pleiotropy hypothesis, although a negative genetic correlation between the principle components describing sclerite shape and body size, suggests a trade-off between investment in genitalia and general morphological traits and thus a role for pleiotropy (House & Simmons 2005a). Population genetic theory predicts strong directional selection will reduce additive genetic variation in traits because alleles conferring highest fitness will be driven to fixation (Kimura 1958). Thus the low V_A in the genital sclerites of O. *taurus* may reflect a history of sexual selection. However, there appears to be considerable genetic variation in sexually selected traits (Andersson 1994) and a number of mechanisms may operate to increase the additive genetic variation of such fitness traits. For example, fitness traits are likely to harbor greater levels of genetic variation because they are influenced by a greater number of loci than non-fitness traits and these are more likely to capture genetic variation and accumulate mutations (Merila & Sheldon 1999). Therefore, with respect to

genitalia, it is difficult to interpret patterns of V_A given the theoretical uncertainty as to how V_A should respond to sexual selection, and thus it is premature to conclude whether the low V_A reported in male genitalia support or otherwise, predictions of the sexual selection hypothesis.

Using geometric morphometrics Pizzo et al. (2008) analyzed shape variation in the head, pronotum and genitalia (sclerotized regions of the female vagina and parameres of the male) of O. *taurus* from its native Italian distribution and two recently introduced (exotic) populations in Western Australia and Eastern United States. They found that shape differences in the head that distinguished major and minor morphs varied across populations and that the degree and nature of the covariation between the head shape and the pronotum shape also differed between populations. Interestingly, their analysis revealed little variation in the shape of the sclerotized regions of the female vagina, but did reveal substantial divergence in the shape of the male parameres, indicating that shape divergence between populations occurs in the absence of sympatry (and thus reinforcement) and can occur extremely rapidly, given the exotic populations were established less than 50 years ago. Given that investment in cephalic horns and genitalia appear to be negatively related (see below), Parzer and Moczek (2008) have suggested that the genitalic divergence between O. *taurus* populations may be a byproduct of sexual selection on horn length.

Developmental trade-offs are likely to arise when two or more structures compete for a shared and limited pool of resources, necessary for their development. Observations of horn-dimorphic *Chalcosoma* species are consistent with a trade-off between the development and evolution of beetle horns and genitalia. *C. caucasus* males that occurred in sympatry with *C. atlas*, expressed the minor (hornless) morph over a much wider range of body sizes and had relatively larger genitalia than when they occurred in allopatry. By contrast, *C. atlas* exhibited the opposite pattern for both traits (Kawano 2002). In O. *taurus* the genitalia and head horns develop from separate imaginal disk-like tissues. The genital disks grow early in the last larval instar, whereas the horn disks undergo brief, but explosive growth approximately 8–10 days later, when larvae enter the pre-pupal stage (Moczek & Nijhout 2004). Males that had their genital disks ablated developed disproportionately large horns indicating a substantial resource allocation trade-off between head-horns and genitalia. In a similar vein, O. *nigriventris* engineered not to invest in horns (through their ablation during development) invested in relatively larger testes (Simmons & Emlen 2006). Such intraspecific trade-offs lead to the prediction that across species relative horn and testes size should be negatively related. However, across 25 species of Onthophagus beetles Simmons and Emlen (2006) found no relationship between horn and testes size, although they did find a significant negative relationship between the steepness of horn allometry and testes allometry, indicating that species with the strongest condition-dependent investment in horn size have correspondingly evolved the most protected (canalized) patterns of testes growth. The opposite is also true; species in which horn size is largely nutrition-insensitive (canalized) exhibit condition dependent patterns of investment in testes. The analysis of the developmental trajectories of 'primary' and secondary sexual traits within a phylogenetic framework, looks to be a very promising method of understanding the complex nature of genitalic evolution. Indeed, a comparison of relative investment into copulatory organs (length of parameres and phallobase combined) with that into head horns across four allopatric populations of O. *taurus* revealed a strong negative association between the two traits; males from populations that invested most in cephalic horns invested least in genitalia, a pattern that held across *Onthophagus* species (Parzer & Moczek 2008). One interpretation of this pattern (see also Pizzo et al. 2008) is that copulatory organ divergence is a byproduct of evolved changes in horn length investment. Between the O. *taurus* populations studied by Parzer and Moczek (2008) there is considerable variation in the extent of male–male competition, such that differences in local densities have selected for differential investment in horn expression, resulting in the diversification of the genitalia. This would indicate that evolved differences in the male copulatory organ are (at least partly) pleiotropic side-effect of selection acting on a secondary sexual trait. However, the opposite is also possible: population density may have selected for differential investment in aedeagus or testes size (Gage 1995) bringing about correlated changes in horn investment across populations (Parzer & Moczek 2008). Whatever the mechanism turns out to be, the results of Parzer and Moczek (2008) and Pizzo et al. (2008) indicate

that the genitalia of O. *taurus* populations have diverged very rapidly (<50 years) in the absence of sympatry.

ALLOMETRY

A number of studies have examined the allometry (or scaling) of genitalia. In essence traits subject to directional selection are typically thought to be associated with positive allometry (i.e., traits increase in size faster than increases in body size), whereas those subject to stabilizing selection are though to exhibit negative allometry or isometry. For example, the head horns of sexually dimorphic Scarabiidae and the enlarged mandibles of sexually dimorphic Cerambycidae and Lucanidae (generally used as weapons during male–male combat) exhibit strong, positive allometry (mean allometric exponent of horn length in 12 species of Scarabiidae = 4.48, range 2.57–8.95; of mandible length in 17 species of Cerambycidae and Lucanidae = 2.15, range 1.11–3.43; Kawano 2006). These data illustrate the general trend of positive allometry in male characters subject to pre-copulatory sexual selection. Such trends in genitalic traits might thus be taken as evidence of sexual selection operating on genitalia. In an analysis of insects and spiders Eberhard et al. (1998) found the allometric exponents of genitalia to be consistently lower than those of other body parts. Across the eight species of beetle included in Eberhard et al. (1998), all of the allometric exponents pertaining to genitalic traits were smaller than those generated from non-genitalic traits. A similar pattern was found by Kawano (2006). Of 38 species of Coleoptera, genitalic allometry was smaller than hind wing allometry in 37 cases and smaller than pronotum allometry in all cases, indicating that genitalic structures tend to have relatively low allometric values (see also Bernstein & Bernstein 2002; Schulte-Hostedde & Alarie 2006; Tatsuta et al. 2001; Tatsuta et al. 2007). At first glance this may be seen as evidence against sexual selection shaping the evolution of genitalia, although Eberhard et al. (1998) have argued that the selection pressures on traits used as weapons and/or threat signals (e.g., the cephalic horns of beetles) and those used internally during copulatory courtship are likely to be quite different, and consequently so will their allometry. For example, if visual perception was important in pre-copulatory mate competition, one might anticipate the evolution of exaggerated visual traits in males. However, within a post-copulatory arena, selection is likely to favor males that provide courtship during copulation that is appropriate for the typical female phenotype. Hence, the negative allometry of genitalic traits would tend to suggest that 'one size fits all' (Eberhard et al. 1998).

Unfortunately, the negative allometry of genitalia does not allow us to differentiate between models of sexual selection (Hosken & Stockley 2004). For example, Fisherian selection, good genes and sexual conflict would favor genitalic characters that were appropriate for the majority female phenotype. A conflict analogy would be a burglar in possession of a skeleton key that can open several locks. Clearly there is a conflict of interest between the burglar and the owner of the lock, but the skeleton key does not take on an exaggerated form, rather it remains the same size as a regular key.

MALE-INDUCED GENITAL DAMAGE

A number of recent studies have documented male traits that appear to harm their mates during copulation. For example, in *Drosophila melanogaster* male seminal products reduce female longevity (Chapman et al. 1995; Rice 1996) and in the dung fly *Sepsis cynipsea*, the male intromittent organ is equipped with spines that scrape the female reproductive tract (Blanckenhorn et al. 2002). Genital wounding has also been observed in the Chrysomelid beetle, *Metrioidea elongate*, in which needle-like endophallic spines penetrate the wall of the female bursa (Flowers & Eberhard 2006) and a similar situation occurs in the bruchid beetle *Callosobruchus maculatus* (Crudgington & Siva-Jothy 2000). In this species the male intromittent organ is tipped with sclerotized spines that puncture the female reproductive tract. Approximately 16 h following copulation, these sites of damage can be seen as melanized scarring within the female reproductive tract (Crudgington & Siva-Jothy 2000). Copulation in this species follows two phases. During the first phase the male and female achieve genital contact and adopt a fairly motionless stance, except for rhythmic rocking on the part of the male. About two-thirds of the way through copulation, females use their hind legs to kick at the male until copulation is terminated (Eady 1994b). Through the ablation of the hind legs of females, Crudgington and Siva-Jothy (2000) were able to demonstrate that female kicking reduced both the duration of copulation and the level of scarring to the female tract. They further demonstrated that twice mated females died sooner than single mated females, suggesting that double mated females suffered more damage and consequently had reduced longevity.

Two hypotheses have been proposed for the evolution of harmful male traits: the pleiotropic harm hypothesis and the adaptive harm hypothesis. The pleiotropic harm hypothesis states that male harm is a side-effect of some other male traits that is adaptive within another context. For example, the fitness benefits to a male of avoiding take-over during competition for mates, via the secure anchoring of genitalia within the female reproductive tract, might outweigh the costs of reducing the longevity of his mate. The adaptive harm hypothesis is centered on the notion that males may benefit directly from harming their mates, if such harm alters female reproductive behavior to their benefit. For example, if harm was to prevent or delay female remating with rival males and/or increase female investment in current reproduction (sometimes referred to as the terminal investment effect due to the reallocation of resources from maintenance to reproduction), then harm would be directly beneficial to males.

Morrow et al. (2003) tested the adaptive harm hypothesis in *C. maculatus*, *Tribolium castaneum*, and *Drosophila melanogaster*, by inflicting artificial damage on females (e.g., leg, antennae or wing ablation, or thorax or abdomen puncture) and recording the female response. In all cases there was no evidence that harm reduced the propensity of females to remate, nor increase current investment in reproduction (i.e., upgrade egg laying following damage). A similar conclusion was drawn by Edvardsson and Tregenza (2005) studying *C. maculatus*. They found females who had their legs ablated prior to copulation (and thus received more genital damage; Crudgington & Siva-Jothy 2000) did not respond by delaying remating or by increasing their reproductive rate. Natural variation in levels of genital damage were also found to be unrelated to female propensity remate and female oviposition rate (Eady et al., unpublished), which in conjunction with the results of Morrow et al. (2003) and Edvardsson and Tregenza (2005) suggest that male harm is most likely a pleiotropic side-effect of some other male adaptation in *C. maculatus*.

There is even some question as to just how costly genital scarring is to females. Although Crudgington and Siva-Jothy (2000) found that double mated females died sooner than single mated females, Fox (1993) found double mated females to have greater longevity and lay more eggs than single mated females, while Eady et al. (2007) found no effect of mating frequency on longevity, after differences in fecundity were incorporated into their analytical model. Furthermore, by reducing the ejaculate contribution of males by sequentially mating them to a series of females in quick succession, Eady et al. (2007) were able to show that females mated to ejaculate limited males suffered slightly elevated levels of genital damage, yet lived longer than females mated to non-ejaculate limited males, suggesting compounds in the ejaculate reduce longevity to a greater extent than the physical damage received. Indeed, through the direct measurement of genital scarring, Eady et al. (unpublished) found no relationship between damage and longevity, indicating that genital scarring carries little or no cost in this species.

This is interesting at two levels. First, if damage is not costly to females then there should be no 'terminal investment' effect, which appears to be the case in C. maculatus (see above). Second, if genital scarring carries no obvious cost, then there is no obvious sexual conflict over harm per se, a conclusion that is counter-intuitive: puncturing the female reproductive tract to the extent that melanized scars form, should be costly. Indeed, Rönn et al. (2007) provide evidence that damage is costly. Across seven species of bruchid beetle the degree of harmfulness of male genitalia is positively related to the extent to which the female reproductive tract is reinforced with connective tissue, suggesting that females invest resources into reducing the harmfulness of male damage. When analyzed in a univariate regression model Rönn et al. (2007) show that across species, the level of harm was not related to the cost of mating (measured as the reduction in female longevity following multiple as opposed to single mating). However, when analyzed as a multivariate model that included female genitalic robustness, the level of damage was positively related to the reduction in female longevity. In effect, when male harm is high and female resistance low, the costs of mating become apparent. Of interest, the amount of scarring suffered by females did not correlate with the cost of mating, suggesting that scarring per se is a poor measure of mating costs within the

Bruchidae, a result which supports the findings of Eady et al. (unpublished) who found no relationship between scarring and longevity in C. maculatus. Rönn et al. (2007) suggest that across species, a lack of association between scarring and longevity may be explained by females evolving other resistance adaptations, such as increased investment in immunocapacity, whilst the study of Eady et al. (unpublished) may reflect the coevolutionary nature of male harm and female resistance, such that a particular snap-shot in evolutionary time may reveal females to be adequately defended against male harm. Clearly, further studies of male harm are required in order to fully understand the evolutionary implications of male-induced genital damage. For example, the positive association between male harmfulness and female investment in reproductive tract reinforcement (Rönn et al. 2007) could be interpreted as cryptic female choice, with female investment in connective tissue a mechanism to filter males with regard to their ability to gain direct access of seminal products to target organs within the female body (Eberhard, personal communication.)

FEMALE GENITALIA

Copulation and ultimately fertilization is the union of male/females and sperm/ova, thus it should come as no surprise that female behavior, morphology and physiology are likely to influence male success at fertilization. This was demonstrated in *Callosobruchus maculatus*, in which male fertilization success during sperm competition was largely dependent on female genotype (or more precisely, the interaction between male and female genotypes; Wilson et al. 1997). Similar male/female compatibility was found by Nilsson et al. (2003) in *Tribolium castaneum*, although House and Simmons (2005b) found no such effect in *Onthophagus taurus*.

In *T. castaneum* males transfer a spermatophore as an invaginated tube that everts inside the female bursa and which is filled with sperm during copulation. Approximately 4% of inseminated sperm are transferred to the primary site of sperm storage, the tubular spermatheca (Bloch Quazi et al. 1998; Fedina 2007), which is generally filled to capacity after two copulations. When females mate with two or more males, sperm appear to become stratified within the tubular spermatheca, as P_2 is initially high during the first 48 h after copulation, but declines over subsequent days, such that after one week P_2 does not differ from that expected under a model of random mixing of the sperm from both males (Lewis & Jutkiewicz 1998). Stratification of this nature appears to offer some protection against the displacement of the first male's sperm, should the female mate for a third time (Lewis et al. 2005). Similar temporal patterns of high P_2 soon after the second copulation, followed by a gradual decline over time has been reported in another tenebrionid beetle, *Tenebrio molitor* (Siva-Jothy et al. 1996), which also has a tubular spermatheca. By contrast, *Tribolium confusum* has a chitinised, U-shaped spermatheca with no temporal decline in P_2, suggesting that spermathecal morphology may influence the pattern of sperm utilization within this taxa and possibly more widely (see Walker 1980; Ridley 1989).

Within *T. castaneum* there is considerable variation in spermathecal morphology, which differ with respect to total volume, the position, number, and length of spermathecal tubules, and the position, number, and length of smaller secondary tubules (Fedina & Lewis 2004). In this species the volume of the spermatheca has been shown to be negatively related to P_2, indicating that the female reproductive environment in part sets the rules by which sperm competition is played out (Fedina & Lewis 2004). The reduced level of P_2 found in females with large spermathecae may arise as a result of lower levels of first male sperm displacement and/or the acceptance of more sperm from the first mating male. Interestingly there is good evidence that females influence the movement of sperm from the bursa to the spermatheca. Bloch Quazi et al. (1998) found an 11-fold reduction in the number of sperm entering the spermatheca in females anaesthetised with CO_2 for 30 minutes following copulation. The administration of CO_2 immediately after copulation with the first male led to a reduction in first male fertilization success during sperm competition, indicating that females can potentially exert considerable influence over male fertilization success.

Few studies have directly compared the extent of variation in male and female genitalia. In a study of *Phyllophagahirticula* (Scarabaeidae:Melolonthinae) Polihronakis (2006) found greater levels of shape diversity in the female pubic process (a chitinous structure located at the entrance of the female bursa

copulatrix) in comparison to shape diversity of the left genital paramere of the male. Both male and female genitalia exhibited greater shape variability than elytra (a non-sexually selected trait), possibly reflecting a history of sexual selection, although as discussed above, it is difficult to attribute modes of selection to patterns of genitalic diversity.

Historically, the general consensus has been that across species female genitalia are less variable than male genitalia (Eberhard 1996). This may reflect the fact that female reproductive tracts are often made of soft connective tissues with few easily quantifiable, chitinous structures which may have biased attention towards male genitalic diversity. However, those studies that have examined female genitalic diversity appear to show considerable interspecific variability. For example, the comparative studies of Dybas and Dybas (1981) on featherwing beetles and Rugman-Jones and Eady (2008) on bruchid beetles are founded on interspecific variability in female genital morphology. Indeed, across species Rugman-Jones and Eady (2008) found an eight-fold difference in spermathecal duct length in comparison to only a two-fold difference in elytra length. Across neotropical tortoise beetles (Chrysomelidae) spermathecal duct length varies by approximately two orders of magnitude (1.03–101.8 mm). This variation is unrelated to female size, but there is strong correlated evolution with the length of the male genitalic flagellum (an open-ended, tubular, lower ejaculatory duct; Rodriguez et al. 2004). During copulation in *Chelymorpha alternans*, a gelatinous spermatophore is deposited in the bursa near the entrance to the spermathecal duct. At about this time, the flagellum is threaded up the spermathecal duct. Dissections of pairs in copula revealed the flagellum to be looped within an ampulla located along the spermathecal duct, near the spermatheca, but on occasion, the flagellum was observed to pass through the ampulla and enter the spermatheca. Although some sperm were seen in the lumen of the flagellum, the relative contribution of sperm passing along the flagellum and those moving along the spermathecal duct to the spermatheca remains unclear. However, flagellum length does appear to be important in determining male fertilization success during sperm competition. Male flagellum length was positively associated with fertilization success during sperm competition. This appears to result from sperm dumping, the emission of sperm droplets by the female during copulation. Females mated to males with either naturally short flagella or males with a surgically reduced flagella, were more likely to emit sperm droplets and consequently stored fewer sperm (Rodriguez et al. 2004).

Thus, in a number of beetle taxa, female genital traits exhibit much variation, an observation that has not escaped the attention of taxonomists (Figure 16.8).

In species of Macrodactylus (Scarabaeidae: Melolonthinae) the hemisterna of females, sclerotized structures located at the vulva (entrance to the

vagina), are quite clearly species specific (Eberhard 1993). (Figure 16.9). Differences in hemisterna shape are echoed in the form of the male parameres, the distal end of which press against the dorsal portion of the hemisterna during copulation, creating a "mechanical mesh" between male and female reproductive characters during copulation (Eberhard 1993), although the mesh appears not to be sufficiently tight to preclude coupling between males and females with different forms. Similar examples of correlated evolution between male and

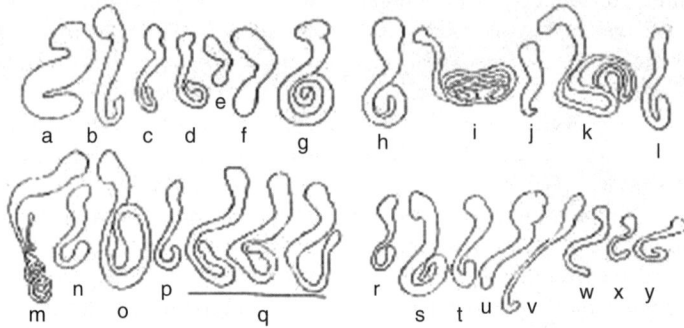

FIGURE **16.8** A selection of female spermathecae from 25 species of beetle belonging to the Atheta group. Reproduced with permission from Skidmore (1991). *Insects of the British Cow-dung Community.* Shrewsbury: Field Studies Council.

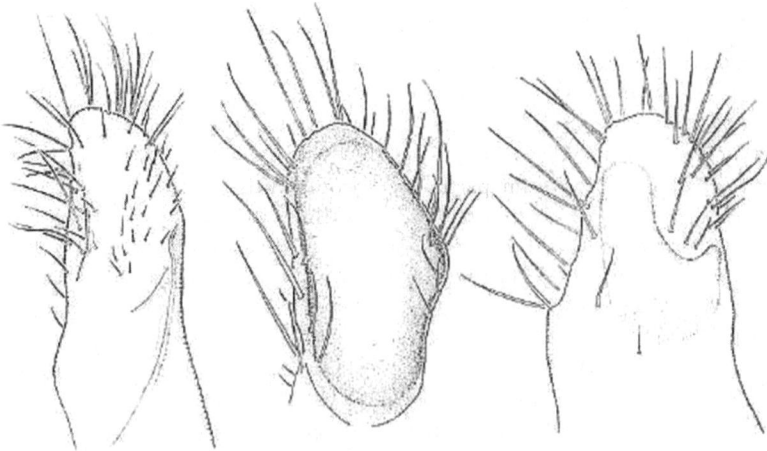

FIGURE **16.9** Hemisterna of female *Macrodactylus costulatus* (left), *M. sericinus* (middle) and *M. sylphis* (right). The hemisternites of *M. sericinus* differend in being deeply concave while the others were nearly flat. Reproduced with permission from Eberhard (1993) Copulatory courtship and genital mechanics of three species of *Macrodactylus* (Coleoptera Scarabaeidae Melolonthinae). *Ethology, Ecology & Evolution,* 5, 19–63.

female genitalia have been found in bruchid beetles (Rönn et al. 2007) and in ground beetles of the subgenus Ohomopterus, where there is a close species specific match between the sclerotized, hook-like copulatory piece on the endophallus of the male and the counterpart membranous pocket of the vagina (vaginal appendix; Sota & Kubota 1998).

EVOLUTIONARY CONSEQUENCES

The biological species concept (Dobzhansky 1951; Mayr 1942) considers species to be groups of populations reproductively isolated from other such groups by genetically based traits that prevent gene exchange (isolating mechanisms; Coyne & Orr 1998). Isolating mechanisms fall into two broad categories: those acting before fertilization (prezygotic mechanisms) and those acting after fertilization (postzygotic mechanisms) (Coyne and Orr 1998). Prezygotic reproductive isolation has traditionally been studied from a behavioral perspective because of the intuitive notion that two populations undergoing sexual selection diverge rapidly in both female preferences and male traits, resulting in sexual isolation (Lande 1981). However, the recognition that sexual selection continues beyond copulation up to the point of fertilization implicates postcopulatory sexual selection in the rapid and divergent evolution of primary reproductive traits (Eberhard 1996; Arnqvist 1998; Howard 1999; Gavrilets 2000; Eady 2001). Should populations embark on different evolutionary trajectories, then it is likely that the reproductive systems of the populations will become incompatible at one or more functional levels, resulting in partial or complete postcopulatory, prezygotic reproductive isolation. In essence, this argument considers reproductive isolation a by product of sexual selection (see also the specific-mate recognition system of Paterson (1985)).

Conspecific sperm precedence (CSP), the favored use of conspecific sperm at fertilization when both conspecifics and heterospecifics have inseminated a female (Howard 1999), points to the evolution of functional incompatibility between the fertilization systems of closely related species. CSP has been demonstrated in a number of beetle species, including ladybirds (Nakano 1985), flour beetles (Wade et al. 1994; Fricke & Arnqvist 2004) and

bruchid beetles (Rugman-Jones & Eady 2007). However, its demonstration tells us little about the mechanisms that bring about CSP nor the selection pressures that drive the evolutionary divergence of fertilization systems. Theoretically, CSP could arise after precopulatory or postzygotic isolation and thus have little to do with the speciation process.

Callosobruchus subinnotatus females readily mate with *C. maculatus* males (sometimes found in sympatry) although the opposite is not true; *C. maculatus* females do not mate with *C. subinnotatus* males. The fecundity and fertility of *C. subinnotatus* females mated to con- or heterospecific males were equivalent, indicating that the sperm of *C. maculatus* males function perfectly well within the reproductive environment of *C. subinnotatus* females. Indeed, the number of con- and heterospecific sperm that entered spermathecae, and the rate at which these sperm were lost from spermathecae were equivalent (Rugman-Jones & Eady 2007). However, when *C. subinnotatus* females mated with both con- and heterospecific males the conspecific male fertilized the majority of the female's eggs regardless of mating order. By observing that the sperm lengths of the two species differed, Rugman-Jones and Eady (2007) were able to follow the fate of heterospecific and conspecific sperm in the female's spermatheca, demonstrating that under conditions of sperm competition, heterospecific sperm were more likely to be lost from the spermatheca than conspecific sperm, accounting for some of the fertilization advantage of conspecific males. An additional advantage to conspecific sperm at fertilization was revealed by quantifying the relative numbers of the two sperm types in the spermathecae of females from which P_2 had been derived earlier. Using this method, Rugman-Jones and Eady (2007) were able to show that conspecific sperm fertilized more eggs, and heterospecific sperm fewer eggs, than predicted on the basis of their numerical representation in spermathecae. How conspecific sperm gain an advantage from the fertilization set (spermatheca) remains unknown, although it might be related to the correlated evolution of sperm and spermathecal duct length in this family (Rugman-Jones & Eady 2008). If several sperm are expelled from the spermatheca through the contraction of the spermathecal muscle, perhaps in response to an egg moving down the oviduct (Lopez-Guerro & Halffter 2002), and sperm motility is partly determined by the interaction between the sperm and the walls of the spermathecal duct (Werner et al. 2007), then conspecific sperm might win the 'race' from the spermatheca to the site of fertilization.

In the flour beetle, *Tribolium castaneum*, the utilization of sperm at fertilization decreased with increasing phylogenetic distance between the female and the focal male, suggesting that postmating incompatibilities accumulate rapidly (Fricke & Arnqvist 2004). How these incompatibilities arise is open to conjecture. One recent avenue of investigation involves studying the results of cross-population matings. Under sexually antagonistic coevolution females should respond more weakly than average to the signals of males with which they are coevolved (Parker & Partridge 1998; Andrés & Arnqvist 2001; Arnqvist & Rowe 2005), whereas under conventional Fisherian/good genes models of sexual selection, females should respond more strongly than average to males with which they are coevolved (Arnqvist & Rowe 2005; Brown & Eady 2001). Using different populations of *C. maculatus* Brown and Eady (2001) demonstrated that males derived from the same population as females outcompeted allopatric rival males with respect to sperm precedence, sperm protection (female refractiveness), and ability to stimulate oviposition, supporting the predictions of conventional models of sexual selection. By contrast, in *T. castaneum* the interaction between male and female population origin had inconsistent effects on P_2 (Nilsson et al. 2003) and female reproductive rate (Nilsson et al. 2002) whilst allopatric males were better able to delay female receptivity than same population males (Nilsson et al. 2003), lending support to sexually antagonistic coevolution. These contrasting results highlight a subtle problem with the original predictions; namely the extent to which populations have diverged before being used in cross-population mating experiments. Should divergence have continued for sufficient time, then both sexual conflict and conventional models of sexual selection predict reproductive incompatibility (i.e., females responding more strongly than average to males with which they are coevolved (Arnqvist & Rowe 2005)).

In conclusion, the results of hetero-population and heterospecific matings add considerable weight to the idea that females play a key role in determining male postcopulatory success. However, patterns of male–female interaction appear to be mixed, thus it is difficult to interpret the evolutionary mechanisms underlying these postcopulatory processes (see also Rowe et al. 2003). As more cross-population studies accumulate, more general conclusions might be possible.

FINAL THOUGHTS

Research into the reproductive behavior, morphology, and physiology of beetles has added greatly to the literature on the evolution of primary reproductive traits. There is convincing evidence that these traits are subject to post-copulatory sexual selection, although, just as with pre-copulatory mate choice, we are still debating the exact mechanisms that operate. To fully appreciate the nature of post-copulatory sexual selection will require very clever experimentation and it is unlikely that the results of a single experiment will resolve this debate. Combining experimental and comparative methods (Rönn et al. 2007), the embracing of modern geometric morphometric techniques of shape analyses in conjunction with quantitative genetics (Simmons & Kotiaho 2007) and the use of model organisms that can be exposed to artificial conditions (Rice 1996) are, in combination, likely to play a key role in understanding the evolution of primary reproductive traits. In the opinion of the author, the key questions that remain to be answered are: what are the mechanisms of post-copulatory sexual selection? Why are female genitalia so diverse? What role does post-copulatory sexual selection play in the process of speciation? The intuitive link between micro evolutionary change at the level of sperm, genitalia and seminal fluid components and the macro evolutionary processes of reproductive isolation and speciation, makes the study of primary reproductive traits especially relevant and exciting.

Acknowledgments I am most grateful to Alex Córdoba-Aguilar, William Eberhard, and Mario Zunino for commenting on earlier drafts of the MS, pointing out gaps in my knowledge and directing me to references I should really have unearthed my-self. I am also grateful to Janet Leonard for showing such patience and to Severine Ligout and Anais Racca for helping translate articles.

REFERENCES

Andersson, M. 1994. Sexual Selection. Princeton, Princeton University Press.

Andrés, J. A. & Arnqvist, G. 2001. Genetic divergence of the seminal signal-receptor system in houseflies: the footprints of sexually antagonistic coevolution. Proceedings of the Royal Society of London, Series B. 268, 399–405.

Arnqvist, G. 1997. The evolution of animal genitalia: distinguishing between hypotheses by single species studies. Biological Journal of the Linnean Society. 60, 365–379.

Arnqvist, G. 1998. Comparative evidence for the evolution of genitalia by sexual slection. Nature, 393, 784–786.

Arnqvist, G. & Rowe, L. 2005. Sexual Conflict. Princeton University Press.

Bateman, A. J. 1948. Intra-sexual selection in *Drosophila*. Heredity, 2, 349–368.

Bernstein, S. & Bernstein, R. 2002. Allometry of male genitalia in a species of soldier beetle: support for the one-size-fits-all hypothesis. Evolution, 56, 1707–1710.

Birkhead, T. R. & Møller, A. P. 1998. Sperm Competition and Sexual Selection. London: Academic Press.

Blanckenhorn, W. U., Hosken, D. J., Martin, O. Y., Reim, C., Teuschl, Y. & Ward, P. I. 2002. The costs of copulating in the dung fly *Sepsis cynipsea*. Behavioural Ecology, 13, 353–358.

Bloch Quazi, M. C., Aprille, J. R. & Lewis, S. M. 1998. Female role in sperm storage in the red flour beetle, *Tribolium castaneum*. Comparative Biochemistry and Physiology A, 120, 641–647.

Bloch Quazi, M. C., Herbeck, J. T. & Lewis, S. M. 2006. Mechanisms of sperm transfer and storage in the red flour beetle (Coleoptera: Tenebrionidae). Annals of the Entomological Society of America, 89, 892–897.

Boorman, E. & Parker, G. A. 1976. Sperm (ejaculate) competition in *Drosophila melanogaster* and the reproductive value of females to males in relation to female age and mating status. Ecological Entomology, 1, 145–155.

Boughman, J.W. 2002. How sensory drive can promote speciation. Trends in Ecology & Evolution, 17, 571–577.

Brown, D.V. & Eady, P. E. 2001. Functional incompatibility in the fertilization systems of two allopatric populations of *Callosobruchus maculatus* (Coleoptera: Bruchidae). Evolution, 55(11), 2257–2262.

Briskie, J. V., Montgomerie, R. & Birkhead, T. R. 1997. The evolution of sperm size in birds. Evolution, 51, 937–945.

Carbone, S. S. & Rivera, A. C. 2003. Fertility and paternity in the Eucalyptus snout-beetle *Gonipterus scutellatus*: females might benefit from sperm mixing. Ethology, Ecology & Evolution, 15, 283–294.

Chapman, T., Liddle, L. F., Kalb, J. M., Wolfner, M. F. & Partridge, L. 1995. Cost of mating in *Drosophila melanogaster* females is mediated by male accessory gland products. Nature, 373, 241–244.

Coyne, J. A. & Orr, H. A. 1998. The evolutionary genetics of speciation. Philosophical Transactions of the Royal Society of London, B. 353, 287–305.

Crudgington, H. S. & Siva-Jothy, M. T. 2000. Genital damage, kicking and early death. Nature, 407, 655–656.

Dewsbury, D. A. 1982. Ejaculate cost and male choice. American Naturalist, 119, 601–610.

Dobzhansky, T. 1951. Genetics and the Origin of the Species. New York: Columbia University Press.

Dupuis, F. 2005. L'abdomen et les genitalia des femelles de coleopteres Scarabaeoidea (Insecta, Coleoptera). Zoosystema, 27 (4), 733–823.

Dybas, L. K. & Dybas, H. S. 1981. Coadaptation and taxonomic differentiation of sperm and spermathecae in featherwing beetles. Evolution. 35, 168–174.

Eady, P. E. 1994a. Sperm transfer and storage in relation to sperm competition in *Callosobruchus maculatus*. Behavioural Ecology and Sociobiology, 35, 123–129.

Eady, P. E. 1994b. Intraspecific variation in sperm precedence in the bruchid beetle *Callosobruchus maculatus*. Ecological Entomology, 19, 11–16.

Eady, P. E. 1995) Why do male *Callosobruchus maculatus* males inseminate so many sperm? Behavioural Ecology and Sociobiology, 36, 25–32.

Eady, P.E. 2001. Postcopulatory, prezygotic reproductive isolation. Journal of Zoology. 253, 47–52.

Eady, P. E. & Tubman, S. 1996. Last-male sperm precedence does not break down when females mate with three males. Ecological Entomology. 21, 303–304.

Eady, P. E., Hamilton, L. & Lyons, R. E. 2007. Copulation, genital damage and early death in *Callosobruchus maculatus*. Proceedings of the Royal Society of London, Series B. 274, 247–252.

Eberhard, W. G. 1985. Sexual Selection and Animal Genitalia. Cambridge: Harvard University Press.

Eberhard, W. G. 1993. Copulatory courtship and genital mechanics of three species of *Macrodactylus* (Coleoptera Scarabaeidae Melolonthinae). Ethology, Ecology & Evolution, 5, 19–63.

Eberhard, W. G. 1996. Female Control: Sexual Selection by Cryptic Female Choice. Princeton University Press.

Eberhard, W. G., Huber, B. A., Rodriguez, R. L., Briceño, R. D., Salas, I. & Rodriguez, V. 1998. One size fits all? Relationships between the size and degree of variation in genitalia and other body parts in twenty species of insects and spiders. Evolution, 52 (2), 415–431.

Edvardsson, M. & Tregenza, T. 2005. Why do male *Callosobruchus maculatus* harm their mates? Behavioural Ecology, 16, 788–793.

Fedina, T. Y. 2007. Cryptic female choice during spermatophore transfer in *Tribolium castaneum* (Coleoptera: Tenebrionidae). Journal of Insect Physiology, 53, 93–98.

Fedina, T. Y. & Lewis, S. M. 2004. Female influence over offspring paternity in the red flour beetle *Tribolium castaneum*. Proceedings of the Royal Society of London, Series B. 271, 1393–1399.

Flowers, R. W. & Eberhard, W. G. 2006. Fitting together: copulatory linking in some Neotropical Chrysomeloidea. Revista de Biologia Tropical, 54 (3), 829–842.

Forster, M., Gack, C. & Peschke, K. 1998. Morphology and function of the spermatophore in the rove beetle, *Aleochara curtula* (Coleoptera: Staphylinidae). Zoology, 101, 34–44.

Fox, C. W. 1993. Multiple mating, lifetime fecundity and female mortality of the bruchid beetle, *Callosobruchus maculatus* (Coleoptera: Bruchidae). Functional Ecology, 7, 203–208.

Fricke, C. & Arnqvist, G. 2004. Conspecific sperm precedence in flour beetles. Animal Behaviour, 67, 729–732.

Gack, C. & Peschke, K. 2005. 'Shouldering' exaggerated genitalia: a unique behavioural adaptation for the retraction of the elongate intromittent organ by the male rove beetle (*Aleochara tristis* Gravenhorst). Biological Journal of the Linnean Society, 84, 307–312.

Gage, M. J. G. 1994. Associations between body size, mating pattern, testis size and sperm lengths across butterflies. Proceedings of the Royal Society of London, Series B, Biological Sciences, 258, 247–254.

Gage, M. J. G. 1995. Continuous variation in reproductive strategy as an adaptive response to population density in the moth *Plodia interpunctella*. Proceedings of the Royal Society of London, Series B. 261, 25–30.

Gage, M. J. G. & Baker, R. R. 1991. Ejaculate size varies with socio-sexual situation in an insect. Ecological Entomology, 16, 331–337.

Gage, M. J. G. & Freckleton, R. P. 2003. Relative testis size and sperm morphometry across mammals: no evidence for an association between sperm competition and sperm length. Proceedings of the Royal Society of London, Series B, Biological Sciences 270, 625–632.

Garcia-Gonzales, F. & Simmons, L. W. 2007. Shorter sperm confer higher competitive fertilization success. Evolution. 61 (4), 816–824.

Gavrilets, S. 2000. Rapid evolution of reproductive barriers driven by sexual conflict. Nature, 403, 886–889.

Gerber, G. H. 1970. Evolution of the methods of spermatophore formation in pterygotan insects. Canadian Entomologist, 102, 358–362.

Gomendio, M. & Roldán, E. R. S. 1991. Sperm competition influences sperm size in mammals. Proceedings of the Royal Society of London, Series B, Biological Sciences, 243, 181–185.

Gullan, P. J. & Cranston, P. S. 2005. The Insects: an Outline of Entomology. Blackwell Publishing.

Haubruge, E., Arnaud, L., Mignon, J. & Gage, M. J. G. 1999. Fertilization by proxy: rival sperm removal and translocation in a beetle. Proceedings of the Royal Society of London, Series B, 266, 1183–1187.

Hosken, D. J. 1997. Sperm competition in bats. Proceedings of the Royal Society of London, Series B. 264, 385–392.

Hosken, D. J. & Stockley, P. 2004. Sexual selection and genital evolution. Trends in Ecology and Evolution. 19 (2), 87–93.

House, C. M. & Simmons, L. W. 2003. Genital morphology and fertilization success in the dung beetle *Onthophagus taurus*: an example of sexually selected male genitalia. Proceedings of the Royal Society of London, Series B, 270, 447–445.

House, C. M. & Simmons, L. W. 2005a. The evolution of male genitalia: patterns of genetic variation and covariation in the genital sclerites of the dung beetle *Onthophagus taurus*. Journal of Evolutionary Biology, 18, 1281–1292.

House, C. M. & Simmons, L. W. 2005b. Relative influence of male and female genital morphology on paternity in the dung beetle *Onthophagus taurus*. Behavioural Ecology, 16, 889–897.

House, C. M. & Simmons, L. W. 2006. Offensive and defensive sperm competition roles in the dung beetle *Onthophagus taurus* (Coleoptera: Scarabaeidae). Behavioural Ecology & Sociobiology, 60, 131–136.

Howard, D. J. 1999. Conspecific sperm and pollen precedence and speciation. Annual Review of Ecology & Systematics, 30, 109–132.

Hunt, T., Bergsten, J., Levkanicova, Z., Papadopoulou, A., St. John, O., Wild, R. M. Hammond, P. M., Ahrens, D., Balke, M., Caterino, M. S., Gómez-Zurita, J., Ribera, I., Barraclough, T. G., Bocakova, M., Bocak, L. & Vogler, A. P. 2007. A comprehensive phylogeny of beetles reveals the evolutionary origins of a superradiation. Science, 318, 1913–1916.

Jamieson, B. G. M. 1987. The Ultrastructure and Phylogeny of Insect Spermatozoa. Cambridge, Cambridge University Press.

Jeannel, R. 1955. L'édéage. Initiation aux recherches sur la systématique des coléoptères. Paris: Publications diverses du Muséum national d'Histoire naturelle

Kawano, K. 2002. Character displacement in giant rhinoceros beetles. American Naruralist, 159, 255–271.

Kawano, K. 2006. Sexual dimorphism and the making of oversized male characters in beetles (Coleoptera). Annals of the Entomological Society of America. 99 (2), 327–341.

Kimura, M. 1958. On the change of population fitness by natural selection. Heredity, 12, 145–167.

Kotiaho, J. S., Simmons, L. W. & Tomkins, J. L. 2001. Towards a resolution of the lek paradox. Nature, 410, 684–686.

LaMunyon, C. W. & Ward, S. 1999. Evolution of sperm size in nematodes: sperm competition favours larger sperm. Proceedings of the Royal Society of London, Series B, Biological Sciences, 266, 263–267.

Lande, R. 1981. Models of speciation by sexual selection on polygenic traits. Proceedings of the National Academy of Science, 78, 3721–3725.

Lewis, S. M. & Jutkewicz, E. 1998. Sperm precedence and sperm storage in multiply mated red flour beetles. Behavioural Ecology and Sociobiology. 43, 365–369.

Lewis, S. M., Kobel, A., Fedina, T. & Beeman, R. W. 2005. Sperm stratification and paternity success in red flour beetles. Physiological Entomology, 30, 303–307.

Lopez-Guerro, Y. & Halffter, G. 2002. Evolution of the spermatheca in the Scarabaeoidea. Fragmenta Entomologica, 32 (2), 225–285.

Mayr, E. 1942. Systematics and the Origin of Species. New York: Columbia University Press.

Merilä, J. & Sheldon, B. C. 1999. Genetic architecture of fitness and nonfitness traits: empirical patterns and development of ideas. Heredity, 83, 103–109.

Minder, A. M., Hosken, D. J. & Ward, P. I. 2005. Co-evolution of male and female reproductive characters across the Scathophagidae (Diptera). Journal of Evolutionary Biology. 18, 60–69.

Miller, G. T. & Pitnick, S. 2002. Sperm–female coevolution in Drosophila. Science. 298, 1230–1233.

Moczek, A. P. & Nijhout, H. F. 2004. Trade-offs during the development of primary and secondary sexual traits in a horned beetle. American Naturalist, 163, 184–191.

Morrow, E. H. & Gage, M. J. G. 2000. The evolution of sperm length in moths. Proceedings of the Royal Society of London, Series B, Biological Sciences, 267, 307–313.

Morrow, E. H., Arnqvist, G. & Pitnick, S. 2003. Adaptation versus pleiotropy: why do males harm their mates? Behavioural Ecology, 14 (6), 802–806.

Nakano, S. 1985. Effect of interspecific mating on female fitness in two closely related ladybirds (*Henosepilachna*). Kontyû, 53, 112–119.

Nilsson, T., Fricke, C. & Arnqvist, G. 2002. Patterns of divergence in the effects of mating on female reproductive performance in flour beetles. Evolution, 56, 111–120.

Nilsson, T., Fricke, C. & Arnqvist, G. 2003. The effects of male and female genotype on variance in postcopulatory fertilization success in the red flour beetle (*Tribolium castaneum*). Behavioural Ecology & Sociobiology, 53, 227–233.

Palestrini, C., Rolando, A. & Laiolo, P. 2000. Allometric relationships and character evolution in *Onthophagus taurus* (Coleoptera: Scarabaeidae). Canadian Journal of Zoology, 78, 1199–1206.

Parker, G. A. 1998. Sperm competition and the evolution of ejaculates: towards a theory base. In: Sperm Competition and Sexual Selection. (Ed. T. R. Birkhead & A. P. Møller) pp 3–54 London: Academic Press.

Parker, G. A. & Partridge, L. 1998. Sexual conflict and speciation. Philosophical Transactions of the Royal Society of London, Series B. 353, 261–274.

Parker, G. A. & Simmons, L. W. 1991. A model of constant random sperm displacement during mating: evidence from Scatophaga. Proceedings of the Royal Society of London, Series B. 246, 107–115.

Parker, G. A., Ball, M. A., Stockley P. & Gage, M. J. G. 1996. Sperm competition games: individual assessment of sperm competition intensity by group spawners. Proceedings of the Royal Society of London, Series B. 263, 1291–1297.

Parker, G. A., Ball, M. A., Stockley, P. & Gage, M. J. G. 1997. Sperm competition games: a prospective analysis of risk assessment. Proceedings of the Royal Society of London, Series B. 264, 1793–1802.

Parzer, H. & Moczek, A. P. 2008. Rapid antagonistic coevolution between primary and secondary sexual characters in horned beetles. Evolution, 62, 2423–2428.

Paterson, H. E. H. 1985. The recognition concept of species. In: Species and Speciation (Ed. By E. S. Vrba), pp 21–29. Pretoria: Transvaal Museum Monograph No. 4, Transvaal Museum.

Pitnick, S., & T. A. Markow. 1994. Male gametic strategies: Sperm size, testes size, and the allocation of ejaculate among successive mates by the sperm-limited fly *Drosophila pachea* and its relatives. American Naturalist, 143, 785–819.

Pitnick, S., Markow, T. A. & Spicer, G. S. 1999. Evolution of multiple kinds of female sperm-storage organs in *Drosophila*. Evolution. 53, 1804–1822.

Pizzo, A., Roggero, A., Palestrini, C., Moczek, A. P. & Rolando, A. 2008. Rapid shape divergence between natural and introduced populations of a horned beetle partly mirror divergences between species. Evolution and Development, 10 (2), 166–175.

Polihronakis, M. 2006. Morphometric analysis of intraspecific shape variation in male and female genitalia of *Phyllophaga hirticula* (Coleoptera: Scarabaeidae: Melolonthinae). Annals of the Entomological Society of America. 99 (1), 144–150.

Presgraves, D. C., Baker, R. H. & Wilkinson, G. S. 1999. Coevolution of sperm and female reproductive tract morphology in stalk-eyed flies. Proceedings of the Royal Society of London, Series B, Biological Sciences. 266, 1041–1047.

Ransford, M. O. 1997. Sperm competition in the 2-spot ladybird *Adalia bipunctata*. Unpublished PhD Thesis, University of Cambridge.

Rice, W. R. 1996. Sexually antagonistic male adaptation triggered by experimental arrest of female evolution. Nature. 381, 232–234.

Richards, O. W. & Davies, R. G. 1977. Imm's General Textbook of Entomology. Vol. Two. 10th edn. Chapman & Hall.

Ridley, M. 1989. The incidence of sperm displacement in insects: four conjectures, one corroboration. Biological Journal of the Linnean Society, 38, 349–367.

Rodriguez, V., Windsor, D. M. & Eberhard, W. G. 2004. Tortoise beetle genitalia and demonstrations of a sexually selected advantage for flagellum length in *Chelymorpha alternans* (Chrysomelidae, Cassidini, Stolaini). In: New Developments in the Biology of Chrysomelidae (Ed. P. Jolivet, J. A. Santiago-Blay & M. Schmitt), pp 739–748. The Hague: SPB Academic Publishing.

Rönn, J., Katvala, M. & Arnqvist, G. 2007. Coevolution between harmful male genitalia and female resistance in seed beetles. Proceedings of the National Academy of Science, 104, 10921–10925.

Rowe L., Cameron, E. & Day, T. 2003. Detecting sexually antagonistic coevolution with population crosses. Proceedings of the Royal Society of London, Series B. 270, 2009–2016.

Rugman-Jones, P. F. & Eady, P. E. 2007. Conspecific sperm precedence in Callosobruchus subinnotatus (Coleoptera: Bruchidae): mechanisms and consequences. Proceedings of the Royal Society of London, Series B. 274, 983–988.

Rugman-Jones, P. F. & Eady, P. E. 2008. Coevolution of male and female reproductive traits across the Bruchidae (Coleoptera). Functional Ecology. 22, 880–886.

Ryan, M. J., Fox, J. H., Wikzynski, W. & Rand, A. S. 1990. Sexual selection for sensory exploitation in the frog Physaemus postulosus. Nature. 343, 66–68.

Savalli, U. M. & Fox, C. W. 1999. The effect of male mating history on paternal investment, fecundity and female remating in the seed beetle Callosobruchus maculatus. Functional Ecology. 13, 169–177.

Schulte-Hostedde, A. & Alarie, Y. 2006. Morphological patterns of sexual selection in the diving beetle Graphoderus liberus. Evolutionary Ecology Research. 8, 891–901.

Sharp, D. & Muir, F. 1912. The comparative anatomy of the male genital tube in Coleoptera. Transactions of the Entomological Society of London, 477–642.

Simmons, L. W. 2001. Sperm Competition and its Evolutionary Consequences in the Insects. Princeton University Press.

Simmons, L. W. & Emlen, D. J. 2006. Evolutionary trade-off between weapons and testes. Proceedings of the National Academy of Sciences. 103 (44), 16346–16351.

Simmons, L. W. & Kotiaho, J. S. 2007. Quantitative genetic correlation between trait and preference supports sexually selected sperm process. Proceedings of the National Academy of Sciences. 104 (42), 16604–16608.

Simmons, L. W., Tomkins, J. L. & Hunt, J. 1999. Sperm competition games played by dimorphic male beetles. Proceedings of the Royal Society of London, Series B. 266, 145–150.

Siva-Jothy, M. T., Blake, D. E., Thompson, J. & Ryder, J. J. 1996. Short- and long-term sperm precedence in the beetle Tenebrio molitor: a test of the 'adaptive sperm removal' hypothesis. Physiological Entomology, 21, 313–316.

Skidmore, P. 1991. Insects of the British Cow-dung Community. Shrewsbury: Field Studies Council.

Sota, T. & Kubota, K. 1998. Genital lock-and-key as a selective agent against hybridization. Evolution, 52 (5), 1507–1513.

Stockley, P., Gage, M. J. G., Parker, G. A. & Møller, A. P. 1997. Female reproductive biology and the coevolution of ejaculate charcteristics in fish. Proceedings of the Royal Society of London, Series B, Biological Sciences, 263, 451–458.

Takami, Y. 2002. Mating behaviour, insemination and sperm transfer in the ground beetle Carabus insulicola. Zoological Science, 19, 1067–1073.

Takami, Y. 2003. Experimental analysis of the effect of genital morphology on insemination success in the ground beetle Carabus insulicola (Coleoptera Carabidae). Ethology, Ecology & Evolution, 15, 51–61.

Tatsuta, H., Mizota, K. & Akimoto, S-I. 2001. Allometric patterns of heads and genitalia in the stag beetle Lucanus maculifemoratus (Coleoptera: Lucanidae). Annals of the Entomological Society of America. 94 (3), 462–466.

Tatsuta, H., Fujimoto, K., Mizota, K., Reinhardt, K. & Akimoto, S-I. 2007. Distinctive developmental variability of genital parts in the sexually dimorphic beetle, Prosopocoilus inclinatus (Coleoptera: Lucanidae). Biological Journal of the Linnean Society. 90, 573–581.

Tigreros, N., South, A., Fedina, T. & Lewis, S. 2009. Does fertilization by proxy occur in Tribolium beetles? A replicated study of a novel mechanism of sperm transfer. Animal Behaviour, 77(2), 555–557.

Wade, M. J., Patterson, H., Chang, N. W. & Johnson, N. A. 1994. Postcopulatory, prezygotic isolation in flour beetles. Heredity, 72, 163–167.

Walker, W. F. 1980. Sperm utilization strategies in non-social insects. American Naturalist, 115, 780–799.

Wenninger, E. J. & Averill, A. L. 2005. Influence of body and genital morphology on relative male fertilization success in oriental beetle. Behavioral Ecology, 17, 656–663.

Werner, M. & Simmons, L. W. 2008. The evolution of male genitalia: functional integration of genital sclerites in the dung beetle *Onthophagus taurus*. Biological Journal of the Linnean Society, 93, 257–266.

Werner, M., Gack, C., Speck, T. & Peschke, K. 2007. Queue up please! Spermathecal filling in the rove beetle *Drusilla canaliculata* (Coleoptera, Staphylinidae). Naturwissenschaften. 94, 837–841.

Wilson, N., Tubman, S. C., Eady, P. E. & Robertson, G. W. 1997. Female genotype affects male success in sperm competition. Proceedings of the Royal Society of London Series B. 264, 1491–1495.

Yokio, N. 1990. The sperm removal behaviour of the yellow spotted longicorn beetle *Psacothea hilaris* (Coleoptera: Cerambycidae). Applied Entomology & Zoology, 25, 383–388.

17

Fertilization Mode, Sperm Competition, and Cryptic Female Choice Shape Primary Sexual Characters in Fish

MICHAEL TABORSKY AND FRANCIS C. NEAT

GENERAL INTRODUCTION

While the principle of fertilization may be simple, the means by which a male ensures its sperm is that which forms the zygote is a subtle and complex process subject to sexual selection. Likewise the means by which a female ensures a particular male's sperm (that of her choice) will fertilize her eggs also involves reproductive adaptations shaped by evolutionary mechanisms. More often than not the genetic interests of the male and the female are not in harmony, which leads to sexual conflict and manipulative tactics on the part of both sexes. As a consequence of these evolutionary processes there is a bewildering variety of behavioral, morphological, and physiological adaptations to reproduction.

Nowhere in the animal kingdom is this more apparent than in fishes, which renders them an ideal model group to study adaptations to reproduction (Stockley et al. 1996; Kunz 2004; Taborsky 2008). Fishes reproduce by both external and internal fertilization, some even have buccal or gastric fertilization tactics. Many species spawn in guarded nests, vegetation, under stones or in holes, while others simply broadcast their gametes into the open ocean. Most fishes are gonochoristic (sex determined at zygote formation) but a number of species show simultaneous or sequential hermaphroditism. Fishes exhibit every class of mating system so far described and the spectrum of parental investment ranges through no care to advanced paternal, maternal, biparental, and even alloparental brood care (Breder & Rosen 1966; Wisenden 1999). The coexistence of alternative male tactics within a species has been demonstrated in over 170 species of fishes belonging to 32 families and may be considered a rule in fishes rather than an exception (Taborsky 1994, 1998, 2008). Interestingly, even within species gonochorism and successive hermaphroditism may coexist (Reinboth 1967), sex change may work in both directions (Kuwamura et al. 1994, 2002, 2007), and alternative tactics are exhibited systematically either by different, specialized individuals or by the same individuals in different contexts (Taborsky 1994).

All these factors have important ramifications for the concomitant variety of primary and secondary sexual characters, both between and within species. Here we illustrate this variety at the level of primary and secondary reproductive organs in teleost fishes. We shall put our primary focus on the male sex because of their prevalence of alternative reproductive tactics that apparently results in much greater trait variation than in females. In our endeavor to illustrate the wealth of reproductive traits exhibited by fishes we shall aim to highlight the role sexual selection has played in their evolution.

DIVERSITY OF THE MALE REPRODUCTIVE APPARATUS

The male gonad is often thought of simply as a sperm-producing organ. Of course this is its primary function, but there is great variability in the

organization of the testis and reproductive apparatus, the process of spermatogenesis, the composition of ejaculates, and the morphology of
spermatozoa themselves (Birkhead & Moller 1998;
Alavi et al. 2008). Grier et al. (1980), Grier (1981),
and later revised by Parenti and Grier (2004) proposed two basal types of testicular organization in
teleosts that largely reflect phylogeny; (1) an anas-
tomizing and tubular testis (the germinal compartments are highly bifurcated and looping with no
clear terminus) characteristic of the primitive bony
fishes, and (2) a lobular testis (the germinal compartments extend to the periphery of the testis and
terminate blindly) characteristic of the neo-teleosts.
The lobular testis can be further categorized as being
of: (a) the unrestricted type (most neo-teleosts)
which is characterized by spermatogonia occurring
along the greater part of the testicular tubules, or b)
the restricted type (exemplified by members of
the Atherinomorpha) in which the spermatogonia
are confined to the distal end of the lobules,
and spermatogenesis proceeds as the germ cells
approach the efferent ducts (Parenti & Grier 2004).

Until recently, efferent ducts were generally considered to be absent in unrestricted spermatogonial
testes, so that germ cysts form along the entire
length of the testicular lobules (Lahnsteiner et al.
1994). However, since Grier's classification it has
become apparent that there is a further variant of
testicular organization and spermatogenesis in
which well-developed networks of ducts collect the
spermatids produced by the germinal epithelium
and transfer them into the main sperm ducts where
they mature into spermatozoa. This has been
termed the 'semi-cystic' mode of spermatogenesis
(Manni & Rasotto 1997; Giacomello et al. 2008).
In addition to the basic seminiferous tissue, the
male gonad is frequently accompanied by a variety
of accessory structures such as seminal vesicles and
glands (Eggert 1931; Setchel & Brooks 1988;
Fishelson 1991; Rasotto 1995; Neat et al. 2003;
Mazzoldi et al. 2007). In internally fertilizing species, the mechanism by which sperm is delivered
into the female reproductive tract has undergone
extensive diversification in the form of intromittent
organs (figure 17.1).

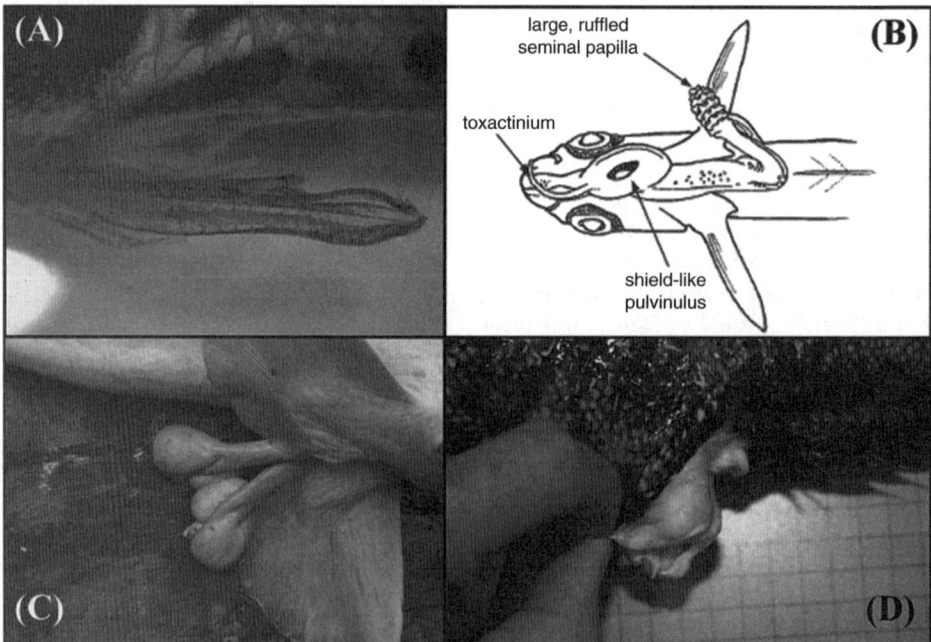

FIGURE 17.1 Examples of variation in external intromittent organs of males of internally fertilizing species.
(A) *Poecilia reticulata* (guppy), (B) ventral view of a member of the Phallostithidae, (C) the shortnose
chimaera *Hydrolagus pallidus* (Chimaeridae), (D) the deep-sea fish *Cataetyx laticeps* (Bythitidae). Picture
(A) by Miguel Aires Tinoco Andrade; (B) by Carpenter & Niem 1999 (The Living Marine Resources of the
Western Central Pacific, Vol. 4, part 2, pp. 2069–2790; Rome, Italy: FAO); (C) and (D) Francis Neat.

A FUNCTIONAL FRAMEWORK FOR MALE REPRODUCTIVE DIVERSITY

This extraordinary variety in male reproductive traits has two important implications for the study of reproductive traits because it suggests that: (1) successful fertilization often requires more than just sperm, and (2) the male reproductive apparatus has secondary reproductive functions enhancing the fertilization probability, economizing the process of spermatogenesis or more derived functions related to sexual conflict. We shall attempt to ask how this diversity in reproductive morphology is related to factors such as the mechanism of sex determination, the mode of fertilization, the occurrence of alternative reproductive tactics, sperm competition, mating systems, and sexual conflict.

Sperm are tiny relative to eggs, can be produced by the billion and released by the hundreds of thousand if not million. Exceptions do occur, for example in seahorses (Sygnathidae), but generally fertilization is a probabilistic process from a male perspective and those males that increase their probability of successfully fertilizing eggs will be favored in the evolutionary process (Parker 1984). The male reproductive apparatus varies across fishes, and that is reflected also in the mass of the testis relative to the overall body mass. In most species the testis represents less than 2% of the body mass (Stoltz et al. 2005), but increases to 4% in the anadromous Atlantic salmon, *Salmo salar* (Vladic & Jarvi 2001) and to 18% in large broadcast spawning species such as Atlantic cod, *Gadus morhua*. It is important to appreciate that increasing the probability of fertilizing eggs is more than just releasing as many sperm as possible. Furthermore the cost of producing sperm is not trivial (Nakatsuru & Kramer 1982) and natural and sexual selection will act to tailor sperm production to that which is optimal for the particular circumstances and associated probability of fertilization in a species (Wedell et al. 2002).

There are multiple factors affecting sperm economy that interact to determine the overall mechanisms of delivery, quantity of sperm, and composition of an optimal ejaculate. Sperm expenditure during a reproductive event will relate to the life-time reproductive pattern of the species, that is whether the species is annual, semelparous, or iteroparous. Semelparous species face no trade-off between current and future reproduction and therefore high investment in reproduction and sperm production is expected. Iteroparous species on the other hand must weigh their current investment in reproduction against the probability that they will live to reproduce again in the future and thus allocate accordingly across lifetime. Equally important is the degree to which reproduction is seasonally determined. Reproduction is often strongly coupled to seasonality and the opportunity for fertilization may range from less than a day to many months, requiring greater or lesser sperm reserves, respectively. Indeed it has been hypothesized that adaptation to seasonality may be the main reason underlying divergence between the three basal types of testis structure described earlier (Parenti & Grier, 2004). On top of this, the absolute fecundity of the species clearly shapes sperm investment patterns; the more fecund the females of a species are the more sperm will be needed (Stockley et al. 1996), even if this is determined to a large extent by body size (Stockley et al. 1997; Stoltz et al. 2005).

A second set of factors concern the mode of fertilization and the environmental conditions into which sperm are released, which are crucial for the performance and survival of sperm. Sperm are quickly diluted in water and those devoid of protective accessory substances cannot resist osmotic stress for more than a minute. Motility is at best in the range of minutes, although some capacity can be retained up to a few hours in salt water or brackish water (Trippel & Morgan 1994; Elofsson et al. 2003; Cosson et al. 2008). Broadcast external fertilizers face radically different conditions than internal fertilizers with sperm quantity being more important than quality and vice versa. Likewise marine species must cope with very different conditions from freshwater species and whether spawning takes place in fast running rivers or oceanic up-wellings or in areas without significant water motion will have a significant influence for an optimal ejaculation tactic.

Third, the mating system of the species and the intensity of sexual selection will significantly shape the evolution of ejaculate tactics. Generally, as the degree of polygyny increases, so the frequency of mating and intensity of sexual selection increases and consequently greater amounts of sperm will be needed. Intrinsically related to this is the existence and degree of sperm competition and the presence of alternative reproductive tactics that will positively select for investment in sperm (Parker 1984). The frequency with which females spawn may also be important for short-term allocation of sperm. However, sperm production is not limitless or without cost and the degree to which spermatogenesis needs to be economized may depend upon whether costly epigamic traits or behaviors are borne by the male or whether prezygotic (e.g., nest building) and postzygotic investment (e.g., guarding and caring of eggs and fry) by males entails significant costs. In internally fertilizing species there is a potential for sperm storage and selective utilization by the female, which introduces further complexities via the process of sexual conflict. Thus the suite of factors that shape the evolution of testes morphology and ejaculate characteristics is subtle and complex arising from both natural and sexual selection. Before we focus on the of role sexual selection, first it is necessary to consider sex determination in fish as it is critical for understanding the prevailing lconstraints and opportunities for reproduction.

SEX DETERMINATION, HERMAPHRODITISM AND SEX CHANGE

In most vertebrates sex determination is controlled genetically, and sex is fixed for life. Exceptions are some reptiles and a large number of fishes, where social or abiotic factors take effect (Crews et al. 1994; Godwin et al. 2003). The mechanisms of sex determination and differentiation in fish exemplify a fantastic playground of evolution (Yamamoto 1969; Mank et al. 2006). Phenotypic sex in teleosts may depend on external and internal factors, with a potential role for endocrine (e.g., steroids), environmental (e.g., temperature) or social (e.g., dominance status) factors in sex differentiation (Fishelson 1970; Francis 1992; Baroiller et al. 1999; Devlin & Nahagama 2002; Godwin et al. 2003). The majority of fish species are gonochoristic. In these species, a simple heterogametic sex determination mechanism (XX/XY or ZW/ZZ) may be at work, with male or female heterogamy. Males *and* females may be heterogametic even within a species (Kallman 1984). In contrast to species where sex differentiation is genetically controlled without relying on endogenous sex steroid production, in many species endogenous sex steroids are instrumental for sex differentiation, which is often also associated with environmental influences (Strussmann & Nakamura 2002). Temperature is frequently an important environmental determinant of sex in gonochoristic

fishes. In contrast, sex inversion (the terms "change" and "inversion" have both been used in regard to sex change in fishes, but there is continuing debate as to which term is more appropriate) in sequential hermaphrodites is often triggered by social factors, notwithstanding that sex steroids play a principal role in sex differentiation and the regulation of sex change (Frisch 2004; Nakamura & Kobayashi 2005; Munday et al. 2006). Interactions between environmental factors and genotype may affect sex determination in both gonochoristic and hermaphroditic fish (Baroiller et al. 1999). When viewed on a broad, comparative basis, changes in the mode of sex determination seem to have happened frequently in evolution and they apparently involve a variety of distinct ancestral-descendant pathways in fishes (Mank et al. 2006).

Functional hermaphroditism has been hitherto described in 94 genera belonging to 27 teleost families from seven orders, with the greatest diversity found among tropical, marine perciforms (de Mitcheson & Liu 2008). Hermaphroditism has been proposed or inferred, but not yet confirmed (de Mitcheson & Liu 2008) in 31 additional genera belonging to 21 families from six orders. This apparently great hermaphroditic potential might indicate a proto-hermaphroditic condition in teleosts, but the highly patchy distribution of different sex-determination mechanisms implies numerous transitions between alternative modes, which is yet to be tested by a rigorous analysis of the phyletic pattern. Hermaphroditism comes in every possible form: it may be simultaneous, with individuals of a population producing eggs and sperm concurrently, or male and female functions may be expressed successively, with either the female (protogyny) or male (protandry) function preceding the other. Bidirectional sex change is also possible (Kuwamura et al. 1994, 2002, 2007; Ohta et al. 2003). In some species with sequential hermaphroditism only part of the population changes sex, whereas the other fraction shows the final sex right from the start (Reinboth 1967), that is gonochoristic and hermaphroditic individuals coexist within a population. This shows up in the gonadal structure, which in those individuals that have changed sex (i.e., "secondary" males or females) reveals traces of the previously expressed sex (Reinboth 1962). In general, in fishes changing sex the gonadal tissues of their first sex are degenerated while the tissues of the final sex proliferate during a sexually transitional phase.

With this broad range of possibilities, can we recognize any patterns that might hint at ecological or evolutionary mechanisms responsible for one or the other form of sex determination in a species? Numerous attempts to this end have indeed revealed some striking patterns (e.g., Warner 1975, 1984; Charnov 1982; Munoz & Warner 2003; Munday et al. 2006; Molloy et al. 2007). Simultaneous hermaphroditism is particularly widespread in the deep sea where population densities are exceedingly low, so doubling the probability that individuals will encounter a mate may be particularly advantageous (Ghiselin 1969). The benefit of bidirectional sex change may as well be that individuals maximize their lifetime reproductive fitness by searching as little as possible for a new mate (Nakashima et al. 1995; Munday 2002). On the other hand, functional hermaphroditism of any kind is exceedingly rare in freshwater fishes; for example, in the most speciose freshwater families, cyprinids, characids, and cichlids, not a single case of functional hermaphroditism is known (de Mitcheson & Liu 2008), despite examples with a juvenile bisexual phase and social determination of sex at the juvenile stage (Oldfield 2005), which demonstrates the inherent hermaphroditism potential. In contrast, sequential hermaphroditism, and especially protogyny is very common in tropical marine reef habitats, especially among perciforms. The prevalence of gonochorism in freshwater species might have to do with the cost/benefit ratios involved in hermaphroditism, but it is presently unclear why these ratios might differ between limnetic and marine environments. Pelagic dispersal of marine fishes in contrast to freshwater species might be one important factor (see below), and phylogenetic inertia certainly seems to be important but explains only part of the pattern.

One model that has been used extensively to explain the occurrence of sequential hermaphroditism and the timing of sex change is the size advantage hypothesis (SAH; Ghiselin 1969; Warner 1975, 1988; Munoz & Warner 2003). In short, it predicts that if the size-dependent fertility functions differ between males and females of a population and if they cross at some specific body size, selection should favor sex change at this intersection (figure 17.2). If the model is framed in terms of reproductive value instead of fertility, sex-specific differences in growth and mortality can also be accounted for when predicting the optimal point of time for sex change (Warner 1988; Munday et al. 2006). A comparative analysis of 52 species showed that most species change sex when the individuals

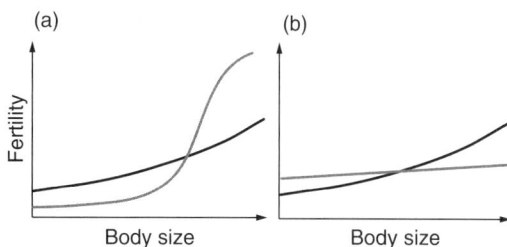

FIGURE 17.2 The size advantage hypothesis of sex change predicts that sequential hermaphroditism is favored if the size-dependence of the reproductive potential differs significantly between the sexes, so that their fitness functions cross. Female and male fertility will usually increase with body size, as a larger female can produce more or larger eggs and a larger male may have better access to partners due to his greater competition potential (intrasexual selection) and attractiveness (intersexual selection). However, the slopes and exponents of male and female fitness functions may diverge considerably, for instance when (a) large males can monopolize several females or resources required by females, whereas small males cannot compete successfully; or when (b) males fall short to reach the size fecundity advantage of females, because their body size does not strongly relate to the number of reproductive partners they can obtain. Blue line: offspring production of females; red line: offspring production of males. Condition (a) selects for protogyny, condition (b) for protandry. From Munday et al. (2006).

have obtained between 80 and 90% of their maximum body size, that is relatively late in reproductive life (Munday et al. 2006). Interestingly, despite such general patterns the timing of sex change may vary adaptively among populations (Gust 2004) or even among social groups within a population (Munoz & Warner 2004). This illustrates the enormous potential of such flexible patterns of reproduction towards maximizing lifetime genetic fitness, and the importance of social environment and behavior. Why this potential appears to be commonly used in marine fish in the tropics, but not by any primary freshwater fish of any region or habitat (except synbranchiform eels; Liem 1963; 1968) remains yet to be explained. Of course a sex-specific size advantage may not favor sex change if it is offset by other life-history trade-offs (Charnov 1986), but it seems unlikely that this fact alone would account for the highly divergent pattern among different fish taxa and environments. Selection for large size of offspring (and hence, eggs) in many freshwater habitats, which might be related to the absence of a pelagic dispersal stage of young, may increase physical differences in male and female gonad anatomy, possibly causing a physical constraint to hermaphroditism (Warner 1978).

One post-mating sexual selection mechanism, sperm competition (see below), may be of particular importance for the fitness accrued by male and female reproductive functions (Munoz & Warner 2003; Molloy et al. 2007) and hence for the evolution of sex allocation. The SAH predicts protogyny if males benefit from large size by an increased potential to monopolize mates (harem potential). However, sperm competition elicited by smaller males can strongly lower the expectation of paternity obtained as a sex-changed, bourgeois male (Taborsky 1998), which should influence the payoff from sex change and consequently the optimal switch size (Munoz & Warner 2003, 2004). In accordance with this, protogynous fishes should be characterized by lower levels of sperm competition than comparable gonochoristic taxa. A comparative analyses of 116 species using testis size as a quantitative measure of sperm competition confirmed this prediction (Molloy et al. 2007). Also, we should expect that in protogynous species sex change should be deferred with higher average levels of sperm competition (which may coincide with habitat characteristics; e.g., seagrass beds), which seems consistent with data (Munoz & Warner 2003).

MODE OF FERTILIZATION

Internal Fertilization

In this review we have restricted ourselves to the study of teleosts, but it should be noted that the Chondrichthyans are all internal fertilizers and that they represent some of the most derived and peculiar of all adaptations to reproduction (e.g., chimaeras, figure 17.1c; see reviews in Wourms 1977; Pratt & Tanaka 1994; Pratt & Carrier 2001). It is estimated that 2–3% of teleosts are internal fertilizers (Helfman et al. 1997; Jobling 1995) being represented in the following families: Adrianichthidae; Anablepidae (Four-eyes); Auchenipteridae (Driftwood catfishes); Brotulidae (Brotula); Characidae (Tetras); Clinidae (Blennies); Coelocanthidae (Coelocanth); Comephoridae (Baikal oilfish); Embitocidae (Surfperches);

Goodeidae (Splitfins); Hemiramphidae (Half-beaks); Phallostethidae (Priapum fishes); Poecelidae (Guppies); Rivulidae (Rivulines); Scorpaenidae (Scorpionfish); and Zoarcidae (Eelpouts). Roughly half of these families belong to the order Atheriniformes.

Gonad and Genital Morphology and Function

The male reproductive apparatus differs between teleost species with external fertilization (shown by the vast majority of species), and those with internal fertilization. This can be illustrated particularly well in taxa where both modes of fertilization are displayed, such as certain catfish families (Mazzoldi et al. 2007). This difference concerns the testes including mucin secreting seminal vesicles, the formation of sperm packages or bundles, the location of the genital opening, and often the presence of intromittent organs consisting of modified fins that help to transmit sperm into the female genital tract (figure 17.1).

Perhaps the best-studied testes of fish with internal fertilization are those of the livebearing Poeciliidae (Grier et al. 1978; Grier 1981; Billard 1990). In their "restricted spermatogonial type" of testis, numerous lobules radiate from the central cavity towards the periphery of the testis. Spermatogonia are located in the blind end of the lobule, where they are associated with Sertoli cells, which reorganize to form cysts when the spermatogonia transform into primary spermatocytes. As spermatogenesis proceeds, the cysts migrate along the lobule to the efferent ducts in the center of the testis. The secondary spermatocytes in the cysts transform into spermatids, which differentiate into spermatozoa. With the flagella pointing inwards, sperm nuclei become associated with the surrounding Sertoli cells to form spermatozeugmata, which are unencapsulated sperm bundles. At the time of spermiation, the cysts open and the spermatozeugmata are drained into the efferent duct system, which ends in a central cavity. The total duration of spermatogenesis in guppy lasts 36 days (at 25°C; Billard 1969). Many variations of this type of testis do exist within the Atherinomorpha, but the functional significance of each of these types has yet to be determined. Nevertheless there is a disproportionate number of internal fertilizers within this clade and it may be that the "restricted" testicular organization somehow predisposes the evolution of internal fertilization.

Gametes

In guppies spermatozeugmata contain about 20,000 sperm each. But sperm packages that are formed in various ways are frequently found also in other species with internal fertilization (figure 17.3; e.g., in Goodeidae: Grier et al. 1978; Hemiramphidae: Grier & Collette 1987; Downing & Burns 1995; Characidae: Pecio & Rafinski 1994; Burns et al. 1995; Pecio et al. 2001; Javonillo et al. 2007; Phallostethidae: Grier & Parenti 1994; Auchenipteridae: Meisner et al. 2000; Burns et al. 2002; Mazzoldi et al. 2007; Anablepidae: Martinez & Monasterio de Gonzo 2002; Clinidae: Fishelson et al. 2007). Apparently, the transfer or storage of bundled sperm seems advantageous in fish with internal fertilization, which is reminiscent of sperm cooperation as found in terrestrial taxa with internal fertilization (Mackie & Walker 1974; Moore & Taggart 1995; Hayashi 1998; Moore et al. 2002; Immler et al. 2008; Johnston et al. 2007). In rodents, sperm cooperation correlates positively with testis size and hence with the degree of sperm competition (Immler et al. 2008). Grouping sperm may help to accelerate swimming speed of sperm or serve to overcome resistance, caused for example by accessory substances released by the female or by other males (see Immler 2008; Pizzari & Foster 2008 for review). On the other hand, sperm bundles may also block or hamper the way towards fertilizable eggs for subsequent inseminations from competing males. Interestingly, colloidal sperm packaging occurs also in mouthbrooders with oral fertilization, which may be viewed as an intermediate form between external and internal fertilization (Wickler 1965a; Grier & Fishelson 1995). We know of no study that has looked at the function of spermatozeugmata or potential sperm cooperation in fishes yet.

With internal fertilization, ejaculates are released into a protective environment, so sperm of internal fertilizers are selected to survive much longer than sperm released into water by external fertilizers. Sperm of embiotocids and poecilids, for example, may remain motile for at least 20 h in isotonic media containing glucose (Billard 1978; Gardiner 1978). Guppy sperm contain glycogen that is used during transfer and residence in the female reproductive tract (Billard & Jalabert 1973). However, exogenous glucose can also be taken up by spermatozoa of viviparous fishes and sperm metabolism is likely supported by ovarian sugars (Gardiner 1978), that will prolong the motility of spermatozoa in the female genital tract and thereby increase the sperm competition potential. Sperm may be stored and kept alive in the female genital tract for months, but within the ovaries there are usually no specialized, differentiated storage structures (Vila et al. 2007). In the Cottidae, spermatozoa may be attached to the ooplasimc membrane within the female genital tract after copulation, but these sculpins spawn unfertilized eggs, that is fertilization occurs externally after egg deposition (Munehara et al. 1989, 1991). This pattern may lead to allopaternal care of eggs that were fertilized by another male (which had copulated with the female before; Munehara et al. 1994).

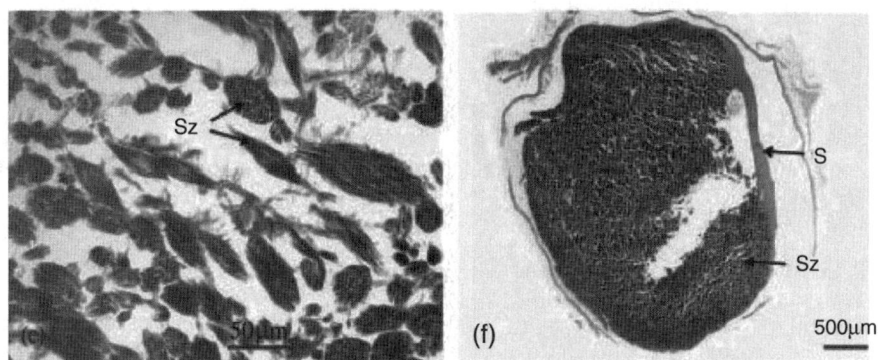

FIGURE 17.3 Spermatozeugmata (Sz) in the sperm duct of the catfish *Auchenipterus nuchalis* (left), and a cross section of the anal fin showing spermatozeugmata in the sperm duct embedded by secretion, S (right). From Mazzoldi et al. (2007).

The structure of spermatozoa of fishes with internal fertilization differs from those with external fertilization. The sperm head and midpiece are on average twice as big as those of the "aquasperm" of teleosts with external fertilization, whereas the flagellum length does not differ systematically (Lahnsteiner & Patzner 2008). Sperm of internal fertilizers are usually more complex, with a well developed midpiece reminiscent of mammalian spermatozoa and containing more mitochondria (figure 17.4), which allows them to survive much longer. For example, spermatozoa of the elkhorn sculpin *Alcichthys alcicornis* (Cottidae) stayed motile for 7–14 days in artificial ovarian fluid (i.e., solutions isotonic to body fluids; Koya et al. 2002), which is orders of magnitude longer than the aquasperm of external fertilizers that usually survive only up to a few minutes (in freshwater), or up to 1 or 2 hours (in sea water). In contrast, the initial sperm swimming velocity of these sperm is very low in comparison to other fishes (Lahnsteiner & Patzner 2008), which illustrates the general trade-off between swimming speed and survival of spermatozoa (Stockley et al. 1997). The prolonged

motility of spermatozoa in species with internal fertilization is probably an adaptation to sperm competition in the female genital tract, because most species show multiple copulations. Females might select for sperm with a high fertilization potential by choosing particular males as suggested by recent studies on guppies, where male coloration predicted sperm swimming speed and viability (Locatello et al. 2006; Pitcher et al. 2007) and where body size correlated positively with sperm length (Skinner & Watt 2007). This suggests that pre-copulatory and post-copulatory sexual selection mechanisms may be functionally linked (but see Evans & Rutstein 2008), which may also hold for the maternal mouthbrooder *Ophthalmotilapia ventralis*. In this cichlid the eggs are fertilized in the maternal buccal cavity and females collect sperm from several males before, during and after spawning (Haesler 2007; Immler & Taborsky 2009; Haesler et al. 2008).

Accessory Structures

In many species with internal fertilization, accessory substances are produced in seminal vesicles that support the formation and transfer of spermatozoa or spermatozeugmata. Seminal vesicles are glandular outgrowths of the common sperm duct producing seminal fluid. For example, the mucins produced there are involved in the formation of spermatozeugmata (Downing & Burns 1995; Meisner et al. 2000). Seminal vesicles are typically paired, multi-chambered, and connected with the sperm duct, and play both, glandular and storage functions (Sneed & Clemens 1963; Lahnsteiner et al. 1990; van den Hurk & Resink 1992; Fishelson et al. 1994; Chowdhury & Joy 2007). In catfish, for example, seminal vesicles may secrete steroids and steroid glucuronides with hormonal and pheromonal functions (Resink et al. 1989; Singh & Joy 1998), but primarily they seem to produce mucoproteins, acid mucopolysaccharides, and phospholipids (Nayyar & Sundararaj 1970; Joy & Singh 1998; Singh & Joy 1998; Santos et al. 2001), providing substrates for a variety of metabolic pathways. Seminal vesicle secretions are regarded as important for the nutrition and protection of spermatozoa, and the presence of glycosidases and catabolic enzymes demonstrates a lytic potency that implies regulatory functions including the removal of necrotic spermatozoa (Chowdhury & Joy 2007). Seminal vesicles also store sperm (Meisner et al.

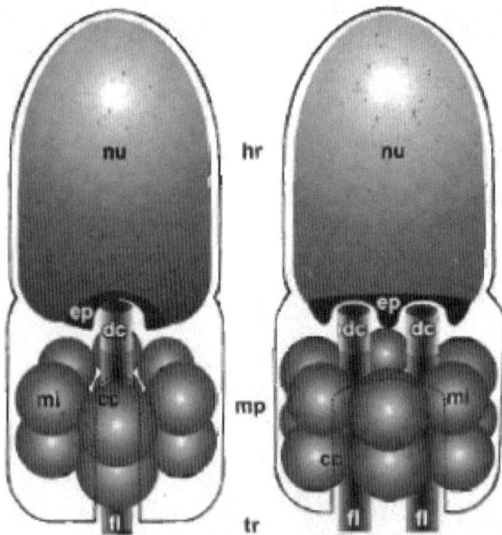

FIGURE 17.4 In *Apogon imberbis*, a species with internal fertilization, two rows of mitochondria surround the central cytoplasmic channel. Twenty percent of the spermatozoa are biflagellate (right picture), but the functional significance of the coexistence of monoflagellate and biflagellate sperm within one ejaculate is not yet known (Lahnsteiner 2003). From Lahnsteiner & Patzner (2008).

2000; Abe & Munehara 2007), where they remain immobile or completely quiescent with a very low level of metabolism until being released.

Mazzoldi et al. (2007) compared the male reproductive apparatus in relation to fertilization modalities in two catfish families. As expected, seminal vesicles were well developed, producing mucins in the form of syalo- and sulpho-glycoproteins, serving to embed sperm and in spermatozeugmata formation in the species with internal and the "sperm drinking type" of fertilization, whereas in other species of the same family that reproduce by external fertilization seminal vesicles are lacking (figures 17.5 and 17.6). It is noteworthy that in all the analyzed species, regardless of fertilization type, the female reproductive apparatus showed a similar organization; that is apparently the fertilization mode only affected the organization of the male reproductive apparatus but not that of females.

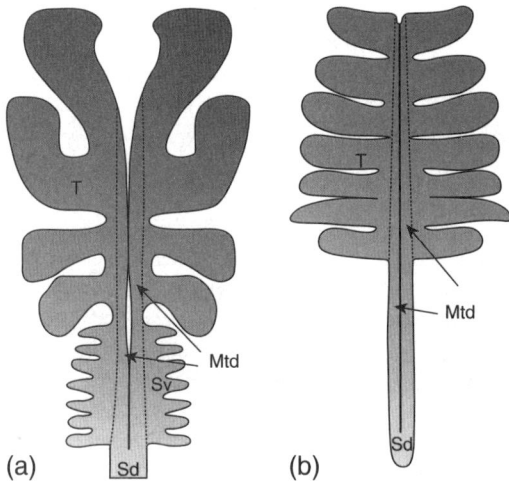

FIGURE 17.5 A comparison of the male reproductive apparatus of two species of the catfish family Auchenipteridae, illustrating differences in complexity between species with internal and external fertilization. In *Auchenipterus nuchalis* with internal fertilization (a), a pair of seminal vesicles secretes mucus serving to release sperm in discrete bundles (spermatozeugmata), and the sperm duct extends to the tip of the anal fin, which is modified as an intromittent organ (not shown). In *Tatia intermedia* with external fertilization (b), seminal vesicles are lacking, sperm are not bundled and the sperm duct ends in the urogenital opening at the base of the anal fin. Mtd, main testicular duct; Sd, sperm duct; Sv, seminal vesicle; T, testis. From Mazzoldi et al. (2007).

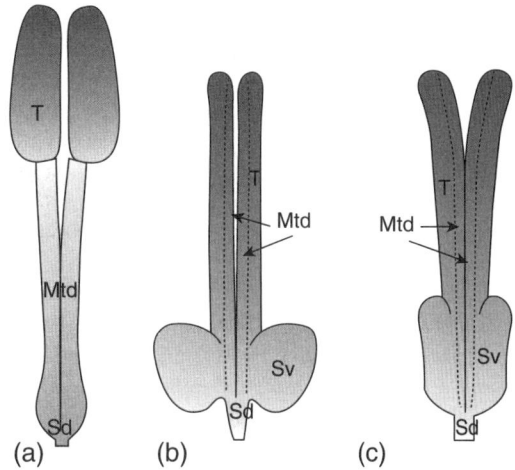

FIGURE 17.6 Several species of the catfish family Callichthyidae show a peculiar spawning pattern involving "sperm drinking" (Burgess 1989): the female takes the sperm up directly from the male's genital opening; then the sperm pass through her intestine and are discharged together with eggs into a protective pouch formed by her pelvic fins, where fertilization takes place (Kohda et al. 1995). The complexity of the male reproductive apparatus in this family seems to coincide with the spawning pattern. *Callichthys callichthys* (a), which shows a conventional spawning pattern, lacks seminal vesicles, whereas in species showing sperm drinking such as *Corydoras spp.* (b) and *Hoplosternum littorale* (c), seminal vesicles serve to produce a mucous secretion embedding sperm before release. The mucus might serve to keep the sperm packed and protected during the passage through the female gut. Mtd, main testicular duct; Sd, sperm duct; Sv, seminal vesicle; T, testis. From Mazzoldi et al. (2007).

Internal fertilization in teleosts is often obtained with the help of specific intromittent organs, which usually involve modifications of the anal fin. For example, of two auchenipterid catfish species, one with external, the other one with internal fertilization, the genital opening of the former is at the base of the anal fin, whereas in the latter it is located at the terminal end of the first ray of the anal fin (Mazzoldi et al. 2007). Anal fins modified as intromittent organs have been found in all auchenipterid species with internal fertilization (Loir et al. 1989) In poeciliids, gonopodia are modified anal fins that are used for both solicited and forced copulations (Bisazza 1993). Specialization in one or the other copulation tactic relates to male body size, with small males using surreptitious gonopodial thrusts instead of courting for mates, which may be genetically determined (Zimmerer & Kallmann 1989; Ryan et al. 1990). Gonopodia vary substantially not only among species, but also among geographical regions within a species (Kelly et al. 2000; Jennions & Kelly 2002). In two live-bearing *Gambusia* species, opposing forces of sexual and natural selection mechanisms have been shown to affect gonopodial length (Langerhans et al. 2005). Females choose males with longer gonopodia, whereas predation risk selects in the opposite direction, presumably because longer gonopodia impede burst-swimming performance (cf. Basolo & Alcaraz 2003). Interestingly, not post-mating but pre-mating sexual selection seems to mainly affect the size of this intromittent organ, demonstrating its additional role as secondary sexual trait.

Gestation

After fertilization, fertilized eggs develop within either the follicle or ovarian cavity (Schindler & Hamlett 1993) and hatching either precedes or coincides with parturition. The exchange of gas, electrolytes, nutrients, metabolic waste, and possibly effector molecules and immunoglobulins between mother and embryos is mediated by embryotrophe, because usually there is no attachment between embryonic and maternal tissues (Schindler & Hamlett 1993). In ovoviparous species, eggs usually develop within a modified section of the oviduct and are retained until hatching. Developing embryos obtain nutrients from the egg yolk and oxygen from the female via highly vascularized oviduct walls. In contrast, viviparous teleosts show intraluminal gestation within the ovarian cavity. Nutrient requirements are fulfilled by both egg yolk and additional maternal secretions that are absorbed through the yolk sac; sometimes a primitive placenta may develop, which allows more efficient nutrient transfer (Amoroso 1960; Wourms 1981). Viviparity is accompanied by a decrease in the number of eggs and a gestation period lasting until a large proportion of embryonic development is completed (Wourms 1981).

An alternative form of viviparity in teleosts exists in the sex-role reversed seahorses and pipefishes. In the ca. 230 species of the family Syngnathidae males provide the brood care, which bears a striking resemblance to that of taxa with maternal care, particularly mammals (Stolting & Wilson 2007). Depending on temperature, male pregnancy lasts for a period of 9–69 days (Foster & Vincent 2004). Tending males not only protect the developing embryos and supply them with oxygen, but in a number of species they apparently provide also nutrients to them via placenta-like epithelia (Drozdov et al. 1997; Carcupino et al. 2002; Laksanawimol et al. 2006; Ripley & Foran 2006), and they may secrete proteins serving an immuno-protective function (Melamed et al. 2005). In contrast to the apparently high investment of syngnathid males in brood care, they seem to economize on pre-spawning reproductive investment. Due to prevailing internal fertilization in the pouch and an apparent lack of sperm competition (Kvarnemo & Simmons 2004; Ah-King et al. 2008) the number of spermatozoa produced by seahorses and pipefishes is greatly reduced (as few as 150 per testis in seahorses; Van Look et al. 2007). In connection with the sex-role reversal in syngnathids, "egg competition" and "cryptic male choice" may occur in male brood pouches instead of sperm competition and cryptic female choice as frequently found in mating systems with conventional sex roles (Ahnesjš 1996).

External Fertilization

The majority of fishes fertilize their gametes externally and there is a huge variety in the environmental and social circumstances under which fertilization takes place. At one extreme are the broadcast spawners that release eggs and sperm into the currents of the open ocean, at the other extreme are those species that release their gametes in highly confined spaces such as constructed nests, holes, crevices, or even into the mouth of the female as with some cichlids (Cichlidae) or into the gastro-intestinal tract as with some catfish (Callichthyidae).

Testis Morphology and Accessory Structures

In externally fertilizing fishes, the spermatocrit (percentage of a given volume of milt that is sperm cells) is often a minor component of the entire ejaculate, for example in arctic charr (*Salvelinus alpinus*) spermatocrit is less than 20% of the ejaculate (Rudolfsen et al. 2006) and in rainbow trout (*Onchrynchus mykiss*) it is less than 40% (Hoysack & Liley 2001). Spermatocrit can also be highly variable between individuals and across the spawning period, ranging from 20 to 90% in haddock, *Melanogrammus aeglefinnus* (Rideout et al. 2004). The remainder of the ejaculate is a solution of seminal fluid composites such as polysaccharide mucins, sugars, acids, proteins, hormones, and other molecules (Mann & Lutwak-Mann 1981; Setchel & Brooks 1988). These substances originate from the interstitial cells of the testis and/or from reproductive accessory structures such as the seminal vesicles. The male reproductive apparatus varies from a simpler testicular organization in species that freely broadcast sperm into the water to rather more elaborate organizations with accessory structures and unusual tactics involving a carefully placed and prolonged release of sperm.

An example of a simply organized male reproductive apparatus is the Atlantic cod, *Gadus morhua*, a highly fecund broadcast spawner that congregates in huge numbers to spawn. The species is almost certainly promiscuous and sperm competition is consequently expected to be high (Hutchings et al. 1999). The testis is of the unrestricted lobular type. It is paired and suspended from the dorsal wall of the body cavity by a peritoneal membrane. The part of the testes nearest the membrane is an elongate tubular structure comprising of several sperm ducts and from it arises a frill-shaped extension composed of numerous lobules that become greatly enlarged as the spawning period approaches. The testes join before opening into the genital pore via the sperm duct and sperm is released in close proximity to the female during a ventral mount behavior. There are no obvious accessory structures and this pattern generally fits with the prediction that this species would be selected to produce large quantities of sperm per ejaculate. We were unable to find any reference to accessory structures in broadcast spawning species; however, we should not over-generalize because in tuna (Scombridae) the process of spermatogenesis differs with evidence of an elaborated efferent duct system (Abascal et al. 2004) and there is a report of a fat body lying adjacent to the testis (Ratty et al. 1990).

The male reproductive apparatus is often more complex in species that have more specialized fertilization and spawning strategies such as nesting tactics, demersal spawning, buccal fertilization, and gastric fertilization. The testis is usually paired and simply lobed in form although it can be fringed and irregular (Lopes et al. 2004). In many species there are accessory structures that are anatomically distinct from the testis, although they eventually release their products into the common urogenital tract where they mix with sperm prior to ejaculation.

Seminal vesicles or sperm-duct glands as they are also referred to, have been described from demersal spawning fish including gobies (Gobiidae) (Miller 1984; Fishelson 1991; Cinquetti 1997; Mazzoldi & Rasotto 2002) and catfish, Siluridae (Nayyar & Sundararaj 1970) and Callichthyidae (Franchescini-Vincentini et al. 2007). Seminal vesicles arise on either side of the reproductive tract and are primarily sites of mucus synthesis, but they may also serve as sperm storage sites (Mazzoldi & Rasotto 2002). Mucins have a wide range of properties including a high affinity for binding water (preventing desiccation), providing a matrix for prolonged release of sperm, mediating osmotic shock as sperm are released, and resisting microbiotic infection (Perez-Villar & Hill 1999). Several possible functions of these seminal vesicles have been suggested, including better nest defense by allowing sperm release and fertilization to become temporally and spatially uncoupled to economize the process of spermatogenesis, regulating the quantity of sperm needed and protecting eggs. Structures resembling seminal vesicles are also found in male cardinal fish (Apogonidae; Fishelsen et al. 2006). Cardinal fish are not demersal spawners, but fertilization dynamics may be similar as males use their pelvic fins to collect small batches of eggs before fertilizing them and orally incubating them (Kuwamura 1983).

There can also be glandular structures associated with the testis. The testicular gland found in the male gonads of blennies (Blenniidae) can occupy up to 70% of the male gonad (figure 17.7) and shows large variability across species (Lahnsteiner et al. 1990; Rasotto 1995; Richtarski & Patzner 2000; Giacomello et al. 2008; Patzner & Lahnsteiner 2009). The gland consists of a network of irregularly arranged tubules through which spermatids pass during metamorphosis, before eventually passing into the spermatic ducts and becoming mature spermatozoa (Lahnsteiner & Patzner 1990a). Thus the gland clearly plays a role in the semi-cystic type of spermatogenesis exhibited by blennies. It is also a site of steroid synthesis (Lahnsteiner et al. 1990), although not a major site of mucus secretion (Neat et al. 2003). In the black goby (*Gobius niger*) the testis is accompanied by a mesorchial gland, although the function of this structure is thought to

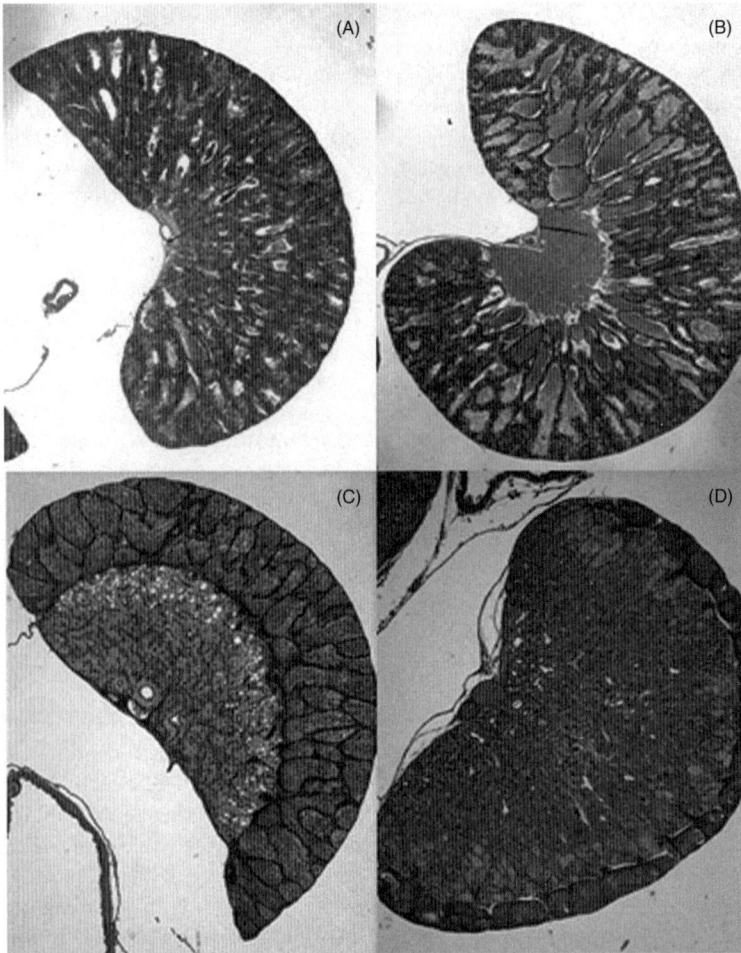

FIGURE 17.7 Transverse histological sections through the testes of three blenny species showing the variation in the ratio of accessory glandular structure (stained pink) to seminiferous tissue (stained blue) within and between species; (A) *Salaria pavo* (parasitic male), (B) *Salaria pavo* (territorial male), (C) *Coryphoblennius galerita* (territorial male), and (D) *Lipophrys canavae* (territorial male). Variable magnification scales. All pictures by Francis Neat.

be pheromone production as it is a site of steroid conjugates that have been shown to attract females into nests (Colombo et al. 1977). Jawfish (Opistognathidae) inhabit, spawn and care for eggs in small holes and are reported to have an accessory gland associated with the male gonad (Rasotto et al. 1992), but no functional interpretation is given.

The sperm ducts that carry the sperm from the testis to the urogenital papilla are sometimes modified and have been shown to secrete mucins and glycogen in blennies (Lahnsteiner & Patzner 1990b) and buccally fertilizing cichlids (Immler & Taborsky 2009). It is thought they may play further

roles in the nourishment and transfer of spermatozoa. In blennies the distal portion of the sperm ducts have evaginated into paired structures termed 'blind pouches' (Lahnsteiner & Patzner 2008). The blind pouches of belnnies have been shown to be sites of steroid synthesis, but their functionality remains unknown. Comparable accessory structures have also been described in the demersal spawning toadfish (Batrichoididae) in which they have been shown to produce mucins and contain sperm (Barni et al. 2001).

Many blennies show specialized accessory glands on the anal fin situated immediately posterior to the urogenital opening (Northcutt & Bullock 1991). The highly convoluted anal fin glands of parental male *Scartella cristata* are mucus secreting (Neat et al. 2003) and Giacomello et al. (2006) showed that the functional role of anal fin glands in *Ophioblennius atlanticus* and *Salaria pavo* is to protect the fertilized eggs from microbial attack, as well as being sites of pheromone synthesis (Laumen et al. 1974; Goncalves et al. 2002; Barata et al. 2008). Similar glandular structures are also present on the other unpaired fins of other blenniids (Northcutt & Bullock 1991) and triplefin blennies (Northcutt & James 1996).

Interestingly the presence of accessory structures is not ubiquitous in gobies and blennies. At least one blenniid species (*Escensius bicolor*) reportedly lacks a testicular gland (Patzner 1989). Observations suggest that this species has a semi-pelagic lifestyle, something altogether different from other blennies (Wickler 1965b). In at least two species of goby, there is no development of seminal vesicles and these both happen to exhibit a monogamous mating system (Mazzoldi 2002). This suggests either that the seminal vesicles play a role in sperm competition (present in the promiscuous gobies, but absent in monogamous ones), or that monogamy somehow emancipates the male from the need to produce pheromones or secretions to protect eggs. Overall it appears that accessory structures play roles in androgen and pheromone synthesis (Colombo et al. 1977), sperm storage (Scaggiante et al. 1999), economizing the process of sperm production and mediating sperm output (Neat et al. 2003; Giacomello et al. 2008). Their biochemical products act to nourish developing sperm (Lahnsteiner et al. 1990) and prolong sperm longevity (Scaggiante et al. 1999).

Alternative Reproductive Tactics (ARTs): How Intraspecific Polymorphisms Can Help Interpret Interspecific Diversity

The occurrence of intraspecific alternative reproductive tactics (ARTs) is very widespread in fishes (Taborsky 2008) and provides a particular insight into the functional significance of reproductive morphology, because they represent contrasting reproductive tactics of individuals that are otherwise comparatively similar (Taborsky et al. 2008). The typical "bourgeois" male (Taborsky 1997) invests in primary access to females by competitively

excluding other males and by attracting females. Parasitic males cuckold the bourgeois males and tend to be small and inconspicuous, often mimicking the appearance of females (Taborsky1997, 2001).

These contrasting tactics result in very different fertilization opportunities and dynamics. ARTs experience asymmetric conditions of sperm competition (Parker 1990; Taborsky 2008). Bourgeois males only face occasional sperm competition from parasitic males. Parasitic males, on the other hand, will always face sperm competition with the bourgeois males and possibly also other parasitic males (Gage et al. 1995; Leach & Montgomerie 2000; Burness et al. 2004). Sperm competition theory predicts a greater investment in sperm will occur with increasing levels of sperm competition (Parker 1998). Thus we should expect to find that the reproductive apparatus of parasitic males facilitates the production and storage of relatively more sperm than their bourgeois counterparts. ARTs also face fundamentally different fertilization opportunities. Bourgeois males have close, regular, and controlled access to eggs or females whereas parasitic males have few and irregular fertilization opportunities (Sato et al. 2004). Parasitic males do not care for offspring and overall will not bear costs of reproduction to the same degree as bourgeois males (Taborsky 2008). Costs to the bourgeois males may include maintaining costly ornaments, displaying to females, defending a territory, nest building, and caring for eggs and young (Taborsky 1994, 1999). Consequently bourgeois males may have lowered feeding opportunities (Taborsky et al. 1987; Schütz et al. 2009) and we might expect bourgeois males to be more economic in their reproduction than parasitic males. One way in which males may offset such costs is by being conservative with their sperm expenditure (Shapiro et al. 1994; Rasotto & Shapiro 1998). Thus we might expect some morphological adaptation and diversity on the reproductive apparatus in bourgeois males but not parasitic males, which often benefit in sperm competition mainly from delivering high sperm numbers. The relationship between ARTs and reproductive morphology has been particularly well studied in blennies and gobies and we now draw upon examples from each to illustrate this.

Several species of goby have a promiscuous mating system with typical bourgeois males defending a territory and attracting females. As discussed earlier, several species also possess accessory structures such as seminal vesicles or sperm-duct glands. In the grass goby, *Zosterisessor ophiocephalus*, a pair of large seminal vesicles are present in the reproductive apparatus of bourgeois males. The seminal vesicles mainly produce mucus that facilitates sperm to be deposited in a 'sperm trail' on the inside of the brood chamber, allowing prolonged and asynchronous fertilization (Marconato et al. 1996; Ota et al. 1996). Experimental removal of the seminal vesicles resulted in a significant reduction of mucus production and longevity of the ejaculate of bourgeois males (Scaggiante et al. 1999). Parasitic males, on the other hand, have relatively smaller seminal vesicles and instead use them to store sperm rather than produce mucus. This results in the release of a short-lived, but sperm-rich ejaculate that is of advantage to an opportunistic fertilization tactic. The testis morphology therefore closely mirrors the fertilization dynamics of the alternative tactics.

Blennies are small, demersally spawning fishes mainly found in shallow seas. Their mating system is promiscuous and they also show paternal care of fertilized eggs (Neat & Lengkeek 2009). The male reproductive tract is comprised of a set of highly variable accessory structures, including a testicular gland, a pair of 'blind pouches', and anal fin glands. In *Scartella cristata* which shows a male polymorphism with typical bourgeois and parasitic tactics the accessory structures are consistently well developed and actively secretory in bourgeois males, whereas the testicular gland is functionally absent in parasitic males (Neat et al. 2003). In contrast the testes of parasitic males are relatively larger and are comprised almost exclusively of seminiferous tissue, with the sperm ducts packed with concentrated sperm. As such parasitic males are ready for the rare opportunity of flooding a nest with sperm, which is what they try to do. The bourgeois tactic is much less compromised by sperm competition, although in blenniids that show a particularly high frequency of reproductive parasitism, for example, in *Salaria pavo*, the testicular gland is reduced in bourgeois males and the volume of sperm released is notably high (Giacomello et al. 2008). Similar patterns have been found in other blennioids (DeJonge et al. 1989; Ruchon et al. 1995; Oliveira et al. 1999; Neat 2001). This in many ways parallels the observations in Gobiids and reaffirms the correspondence between fertilization tactic and testis morphology.

Simple Versus Complex Reproductive Apparatus

Generally parasitic males invest relatively more in sperm than bourgeois males and their testis tends to be simply organized without accessory structures. The simple organization of the parasitic male testis is reminiscent of the testis structure of broadcast spawning species referred to earlier. The complex reproductive apparatus of bourgeois males is unlikely to be an adaptation to sperm competition because these males are exposed to relatively low levels of sperm competition, and a number of species that do not exhibit ARTs still exhibit accessory structures, for example *A. sphinx* (Neat & Locatello 2002) and the Mediterranean *Microlipophrys* sp. (Richtarski & Patzner 2000; Giacomello et al. 2008). Instead the complex reproductive apparatus and accessory structures appear to be adaptations to a particular set of bourgeois reproductive traits that include mate attraction through pheromone secretion, paternal care via egg protection, the need for relatively few sperm for effective fertilization due to spawning being confined and relatively low levels of sperm competition. Sperm economy is therefore at premium to bourgeois males and, at least in blennies, this is in part facilitated by the semi-cystic mode of spermatogenesis. It is interesting to contrast these conclusions with a pelagic spawner that exhibits ARTs, the Hawaiian wrasse, *Thalassoma duperrey*. In this species accessory structures are not observed, but as predicted, the testis is significantly larger in parasitic males (Hourigan et al. 1991). This fits precisely with a species that sheds gametes into the open water and has no parental care; no specialization of the reproductive tract is necessary and sperm is all that is needed, particularly for parasitic males.

Sperm Traits

Sperm morphology varies broadly across teleosts (Jamieson 1991; Alavi et al. 2008); however, few studies have attempted to relate this directly to reproductive strategies. In Tanganyikan cichlids, Balshine et al. (2001) found that sperm traits were related to the mating system and to the site of fertilization. The sperm of monogamous species tended to be shorter than that of their polygynous counterparts, and species that orally fertilize tended to have shorter sperm than externally fertilizing species. However, there is a report of sperm with extraordinarily long sperm in the socially monogamous cichlid species *Pelivicachromis taeniatus* (Thunken et al. 2007). Clearly it would be premature to draw any general conclusions at this stage. Nevertheless it is instructive to look at the significance of ARTs in this respect. No studies to date have found significant differences in sperm morphology between males adopting ARTs (i.e., bourgeois and parasitic males; Gage et al. 1995; Leach & Montgomerie 2000; Neff et al. 2003). There is, however, some evidence that parasitic male sperm are more motile and contain more ATP (Vladic & Jarvi 2001) leading to them having greater swimming speed, but reduced longevity (Burness et al. 2004; see Taborsky 2008 for review). This is clearly an area where further study is needed because not many species have yet been studied in this respect.

CONFLICT BETWEEN THE SEXES

Apart from intrasexual conflict, which mainly affects reproductive traits of males (except in sex-role reversed Syngnathids; Berglund & Rosenqvist 2003), the diverging fitness interests of males and females are responsible for a range of adaptations at the level of primary sexual characters. Sexual conflict occurs when traits enhancing the fitness of one sex lead to a reduction in fitness of the other (Stockley 1997; Maklakov et al. 2005; Parker 2006; Ojanguren & Magurran 2007). With conventional sex roles, where adaptations for increasing male fertilization success negatively affect female fitness, females are selected to avoid or reduce such costs imposed by males (Parker 1984; Chapman et al. 2003). Sexual conflict is widespread and obvious at the behavioral level (Alonzo & Warner 2000; Arnqvist & Rowe 2005; Reichard et al. 2007; Alonzo 2008; Maan & Taborsky 2008), particularly about multiple mating (Andersson 1994; Stockley 1997; Fitze et al. 2005; Hardling & Kaitala 2005; Rice et al. 2006; see also Cameron et al. 2003; Cordero & Eberhard 2003). Conflict is not confined to the behavioral level, however, which is demonstrated by numerous female traits resulting from post-mating sexual selection mechanisms that are usually referred to as cryptic female choice (Thornhill 1983; Eberhard & Cordero 1995; Eberhard 1996; Reyer et al. 1999; Jennions & Petrie 2000), and by male counterstrategies (Cameron et al. 2007; Holman & Snook 2008).

Sperm production is costly (Nakatsuru & Kramer 1982; Wedell et al. 2002) and therefore multiply mating males, which is the rule in fishes, are selected to optimize ejaculate expenditure (Shapiro et al. 1994; Shapiro & Giraldeau 1996). This may reduce the probability that eggs are fertilized if females spawn with the most attractive males in a population, which have the highest spawning rates (Warner et al. 1995). Males adjust sperm production to the risk of sperm competition, which depends on the number of rivals expected to participate in spawning, their size and behavior (Shapiro et al. 1994; Fuller 1998; Alonzo & Warner 2000; Evans et al. 2003a; Zbinden et al. 2003, 2004). Especially in species with external fertilization, this may diminish female fertility levels at low levels of male–male competition (Ball & Parker 1996). In response, females may be selected to attract more males or wait for more males to arrive to induce simultaneous spawning by several partners (Smith & Reichard 2005). Males, on the other hand, should not only respond to intrasexual competition but also to female fecundity when deciding about ejaculate expenditure. In fish with external fertilization mode this seems indeed to be the case, as revealed by field observations (Shapiro et al. 1994) and an interspecific comparison of stripped ejaculate sizes (Stockley et al. 1996).

Even less conspicuous adaptations to intersexual conflict may occur at the level of gonads, gametes, and accessory products. At the gametic level, sperm traits evolve in response to competition among ejaculates of different males (Parker 1970; Gage et al. 1998, 2004; Vladic et al. 2002; Schulte-Hostedde & Burness 2005). In this fertilization race, sperm quality and hence fertilization potential may be congruent with or antagonistic to male physical quality (e.g., coloration, size), as revealed for example in guppies, *Poecilia reticulata* and Arctic charr, *Salvelinus alpinus* (Evans et al. 2003b; Locatello et al. 2006; Pitcher et al. 2007; Skinner & Watt 2007; Liljedal et al. 2008; but see Evans & Rutstein 2008; Pilastro et al. 2008). In addition, sperm traits should also respond to the morphological and biochemical mechanisms of post-mating (i.e., cryptic) female choice. The potential for such selection mechanisms differs between species with internal and external fertilization. In the former, ejaculate or sperm selection might occur in the female genital tract (e.g., in sperm storage organs; McMillan 2007) and by egg traits. Female guppies are able to manipulate the number of sperm transferred or retained at copulation in favor of attractive males (Pilastro et al. 2004). However, the main or perhaps only mechanism by which this is achieved may be simply the duration of copulation, which varies in accordance with male attractiveness

With external fertilization, only the eggs themselves and substances released with or adhering to them can influence sperm selection. Egg discharge is usually accompanied by a simultaneous release of ovarian fluid in which the eggs are stored in the coelomic cavity. In salmonids, for example, the quantity of ovarian fluid released with the eggs amounts to ~10–30% of total egg volume (Lahnsteiner et al. 1999). Ovarian fluid reduces osmotic stress of sperm in water and influences sperm velocity and swimming trajectories, sperm longevity, and the proportion of spermatozoa being motile (e.g., Litvak & Trippel 1998; Turner & Montgomerie 2002; Elofsson et al. 2003; Dietrich et al. 2008). The ovarian fluid of Arctic charr, for example, contains a variety of compounds that sperm could metabolize (Lahnsteiner et al. 1995) and different ions and peptides that are known to influence sperm motility (Morisawa and Suzuki 1980; Lahnsteiner et al. 1995). In the same species, variation in sperm velocity was found to depend on individual female–male interactions when ejaculates were exposed to water-diluted ovarian fluid (Urbach et al. 2005), suggesting that females vary in the effect of their ovarian fluid by stimulating sperm velocity according to individual characteristics of sperm. In addition, competitive in vitro fertilization trials showed that sperm of males sharing more alleles with females fertilized more eggs, which suggests assortative cryptic female choice (Liljedal et al. 2008), possibly by the specific action of ovarian fluid. A motility-enhancing function of ovarian fluid that differed substantially between individual females was found also in rainbow trout (Wojtczak et al. 2007; Dietrich et al. 2008).

Numerous adaptations of sperm to polygamous mating patterns have been attributed to the selective force of sperm competition. This includes sperm morphology (Jamieson 1991; Stockley et al. 1997; Gage et al. 1998, 2002; Balshine et al. 2001; Tuset et al. 2008a, b) and performance, especially swimming speed or velocity (Kazakov 1981; Leach & Montgomerie 2000; Gage et al. 2004; Burness et al. 2004; Casselman et al. 2006; Locatello et al. 2006), sperm longevity (Gage et al. 1995, 2002; Neff et al. 2003) and the percentage of motile sperm

(de Fraipont et al. 1993; Lahnsteiner et al. 1998; Uglem et al. 2001; Linhart et al. 2005). Also, as we have outlined ejaculates are sometimes provided with accessory substances (Piironen & Hyvärinen 1983; Marconato et al. 1996; Scaggiante et al. 1999) that are produced in special accessory glands (Lahnsteiner et al. 1990; Rasotto & Mazzoldi 2002; Immler et al. 2004; Mazzoldi et al. 2005). It has not been studied yet whether and to what extent these male traits are influenced also by intersexual conflict. For example, it is unclear to which extent sperm morphology might generate variance in sperm competence to penetrate the egg cytoplasm (cf. Gage et al. 2002). Results from in vitro fertilization experiments with sperm competition do not necessarily conform to the fair raffle principle, with fertilization biases between males that might well result from interactions with eggs or ovarian fluid (Hoysak et al. 2004). In species with alternative reproductive tactics, parasitic males may be more successful in sperm competition, but this may result either from components of sperm quality other than speed or length (Stoltz & Neff 2006), or from cryptic female choice. Time delays to fertilization may reflect sexual conflict when females are selected to produce gametic traits facilitating complete fertilization and/or multiple paternity (Bakker et al. 2006). In marine invertebrates, females can improve the chances of complete fertilization also by varying egg size (Levitan 1996) and accessory substances (Podolsky 2004). However, this may influence not only the probability of sperm meeting and penetrating the egg (Levitan 1993), but also the effects of sperm competition (Bode & Marshall 2007). How egg size affects sperm options even after they contact an egg may be illustrated by the large eggs of salmonids, where the short time sperm keep alive and moving does not suffice to swim round even only half the circumference of the egg in search of the micropyle (Kime et al. 2001).

A little explored field of research is the chemical communication between ova and sperm of fishes (Nordeide 2007). Already in 1998, Al-Anzi and Chandler found that the jelly surrounding *Xenopus laevis* (an amphibian) eggs releases a diffusible protein into the water that serves as sperm chemoattractant (see also Burnett et al. 2008). A concentration gradient of egg jelly extracts elicits chemotactic behavior of spermatozoa, which probably sample the concentration field along circular and helical swimming paths (Friedrich & Julicher 2007). The mechanisms underlying the guidance of sperm to eggs has been mainly studied in marine invertebrates (Hirohashi et al. 2008; Kaupp et al. 2008) and mammals (Eisenbach & Giojalas 2006). As different chemoattractants are involved and sperm might react specifically, this could hint at a sperm selection mechanism (Eisenbach & Giojalas 2006). In most teleost fishes, the chorion surrounding the eggs is multilayered and the eggs cannot penetrate it, but swim through a hole called a micropyle. It is thought that sperm might be guided to the micropyle after striking the egg by guidance channels on the egg surface (Riehl & Kokoscha 1993), but also by chorionic (ZP) glycoproteins in a broad diluted mucous area on the chorion surface, which have an affinity for spermatozoa (Iwamatsu et al. 1997; Mengerink & Vacquier 2001). There is ample opportunity for chemically mediated sperm selection on this path, both in the surrounding medium (ovarian fluid) and on the chorion, probably also in the micropyle (Yanagimachi et al. 1992). It is of particular interest to unravel how sperm performance (velocity, swimming path, thrust) may interact with the chemical attraction and resistance of eggs and accessory substances, which is an urgent research topic to understand the specificity of sperm performance and fertilization potential under female influence such as demonstrated in Arctic charr (Urbach et al. 2005; Liljedal et al. 2008). Egg activation by sperm and sperm activation by egg substances are other potential but little understood mechanisms where specificity and selection might take effect (Morisawa et al. 1992; Litvak & Trippel 1998; Coward et al. 2002).

As so little is known about the interaction between male and female gametes in fishes it might seem pointless to ponder about "who wins" the conflict. It should be expected, though, that males have more to win because for them fitness benefits from increasing offspring numbers are at stake, whereas in females merely an increase in offspring quality can be gained. Therefore, male traits should be under stronger selection. However, it may be much easier for female gametes to keep control, and male countermeasures may entail much higher costs, especially as they might compromise adaptations to sperm competition. The outcome of sexual conflict will therefore depend on the relative costs and benefits to both sexes (Stockley 1997) and on physical or physiological constraints. One should keep in mind also that many traits will reflect joint fitness interests, in other words reproductive

characters are not necessarily antagonistic between the sexes. This may be exemplified by positive correlations of reproductive traits between the sexes as found in comparative analyses, such as the positive correlation between egg numbers and sperm length, and sperm longevity with ovum diameter, both in fish with external fertilization (Stockley et al. 1996).

Acknowledgments We thank Janet Leonard for inviting us to write this review and for her patience and encouragement. We are very grateful for comments and suggestions of Chris Petersen and David Noakes. Michael Taborsky was supported by grant 3100A0–122511 of the Swiss National Science Foundation (SNF).

REFERENCES

Abascal, F.J., Megina, C. & Medina, A. 2004. Testicular development in migrant and spawning bluefin tuna (*Thunnus thynnus* (L.)) from the eastern Atlantic and Mediterranean. *Fisheries Bulletin* 102, 407–417.

Abe, T. & Munehara, H. 2007. Histological structure of the male reproductive organs and spermatogenesis in a copulating sculpin, *Radulinopsis taranetzi* (Scorpaeniformes: Cottidae). *Ichthyological Research* 54, 137–144.

Ah-King, M., Elofsson, H., Kvarnemo, C., Rosenqvist, G. & Berglund, A. 2008. Why is there no sperm competition in a pipefish with externally brooding males? Insights from sperm activation and morphology. Journal of Fish Biology 68, 9 58–962.

Ahnesjš, I. 1996. Apparent resource competition among embryos in the brood pouch of a male pipefish. Behavioral Ecology and Sociobiology 38, 167–172.

Al Anzi, B. & Chandler, D.E. 1998. A sperm chemoattractant is released from Xenopus egg jelly during spawning. Developmental Biology 198, 366–375.

Alavi, S.M.H., Cosson, J.J., Coward, K. & Raffiee, G. 2008. Fish Spermatology. Oxford, UK: Alpha Science International.

Alonzo, S.H. 2008. Conflict between the sexes and alternative reproductive tactics within a sex. In: Alternative Reproductive Tactics: an

Integrative Approach (Ed. by R.F. Oliveira, M. Taborsky & H.J. Brockmann), pp. 435–450. Cambridge, UK, Cambridge University Press.

Alonzo, S.H. & Warner, R.R. 2000. Allocation to mate guarding or increased sperm production in a Mediterranean wrasse. American Naturalist 156, 266–275.

Amoroso, E.C. 1960. Viviparity in fishes. Symposium of the Zoological Society London 1, 153–181.

Andersson, M. 1994. Sexual Selection. Princeton, N.J.: Princeton Univ. Press.

Arnqvist, G. & Rowe, L. 2005. Sexual Conflict. Princeton: Princeton Univ. Press.

Bakker, T.C.M., Zbinden, M., Frommen, J.G., Weiss, A. & Largiader, C.R. 2006. Slow fertilization of stickleback eggs: the result of sexual conflict? BMC Ecology 6: 7.

Ball, M.A. & Parker, G.A. 1996. Sperm competition games: external fertilization and "adaptive" infertility. Journal of Theoretical Biology 180,141–150.

Balshine, S., Leach, B.J., Neat, F., Werner, N.Y. & Montgomerie, R. 2001. Sperm size of African cichlids in relation to sperm competition. Behavioral Ecology 12, 726–731.

Barata, E.N., Serrano, R.M., Miranda, A., Nogueira, R., Hubbard, P.C. & Canario, A.V.M. 2008. Putative pheromones from the anal glands of male blennies attract females and enhance male reproductive success. Animal Behavior 75, 379–389.

Barni, A., Mazzoldi, C. & Rasotto, M.B. 2001. Reproductive apparatus and male accessory structures in two batrachoid species (Teleostei, Batrachoididae). Journal of Fish Biology, 58 1557–1569.

Baroiller, J.F., Guiguen, Y. & Fostier, A. 1999. Endocrine and environmental aspects of sex differentiation in fish. Cellular and Molecular Life Sciences 55, 910–931.

Basolo, A.L. & Alcaraz, G. 2003. The turn of the sword: length increases male swimming costs in swordtails. Proceedings of the Royal Society of London Series B-Biological Sciences 270, 1631–1636.

Berglund, A. & Rosenqvist, G. 2003. Sex Role Reversal in Pipefish. San Diego, CA, USA: Academic Press.

Billard, R. 1969. La spermatogenenese de *Poecilia reticulata* I.—Estimation du nombre de generations goniales et rendement de la spermatogenese. Annales de Biologie Animale Biochimie Biophysique 9, 251–271.

Billard, R. 1978. Changes in structure and fertilizing ability of marine and freshwater fish spermatozoa diluted in media of various salinities. Aquaculture 14, 187–198.

Billard, R. 1990. Spermatogenesis in teleost fish. In: Reproduction in the Male (Ed. by G.E. Lamming), pp. 183–213. Edinburgh, UK: Churchill Livingstone.

Billard, R. & Jalabert, B. 1973. Le glycogène au cours de la formation des spermatozo" des et de leur transit dans les tractus gènitaux mâle et femelle chez le guppy (poisson poecilidè). Annales de Biologie Animale Biochimie Biophysique 13, 313–320.

Birkhead, T.R. & Moller, A.P. 1998. Sperm Competition and Sexual Selection. San Diego: Academic Press.

Bisazza, A. 1993. Male competition, female mate choice and sexual size dimorphism in poeciliid fishes. Marine Behavior and Physiology 23, 257–286.

Bode, M. & Marshall, D.J. 2007. The quick and the dead? Sperm competition and sexual conflict in sea. Evolution 61, 2693–2700.

Breder, C.M. & Rosen, D.E. 1966. Modes of Reproduction in Fishes. Garden City, NY, USA: Natural History Press.

Burgess, W.E. 1989. An Atlas of Freshwater and Marine Catfishes. A Preliminary Survey of the Siluriformes. Neptune City, NJ, USA: T.F.H. Publications.

Burness, G., Casselman, S.J., Schulte-Hostedde, A.I., Moyes, C.D. & Montgomerie, R. 2004. Sperm swimming speed and energetics vary with sperm competition risk in bluegill (*Lepomis macrochirus*). Behavioral Ecology and Sociobiology 56, 65–70.

Burnett, L.A., Boyles, S., Spencer, C., Bieber, A.L. & Chandler, D.E. 2008. *Xenopus tropicalis allurin*: Expression, purification, and characterization of a sperm chemoattractant that exhibits cross-species activity. Developmental Biology 316, 408–416.

Burns, J.R., Weitzman, S.H., Grier, H.J. & Menezes, N.A. 1995. Internal fertilization, testis and sperm morphology in glandulocaudine fishes (Teleostei, Characidae, Glandulocaudinae). Journal of Morphology 224, 131–145.

Burns, J.R., Meisner, A.D., Weitzman, S.H. & Malabarba, L.R. 2002. Sperm and spermatozeugma ultrastructure in the inseminating catfish, *Trachelyopterus lucenai* (Ostariophysi: Siluriformes: Auchenipteridae). Copeia 2002, 173–179.

Cameron, E.T., Day, T. & Rowe, L. 2003. Sexual conflict and indirect benefits. Journal of Evolutionary Biology 16, 1055–1060.

Cameron, E., Day, T. & Rowe, L. 2007. Sperm competition and the evolution of ejaculate composition. American Naturalist 169, E158–E172.

Carcupino, M., Baldacci, A., Mazzini, M. & Franzoi, P. 2002. Functional significance of the male brood pouch in the reproductive strategies of pipefishes and seahorses: a morphological and ultrastructural comparative study on three anatomically different pouches. Journal of Fish Biology 61, 1465–1480.

Casselman, S.J., Schulte-Hostedde, A.I. & Montgomerie, R. 2006. Sperm quality influences male fertilization success in walleye (*Sander vitreus*). Canadian Journal of Fisheries and Aquatic Sciences 63, 2119–2125.

Chapman, T., Arnqvist, G., Bangham, J. & Rowe, L. 2003. Sexual conflict. Trends in Ecology and Evolution 18, 41–47.

Charnov, E.L. 1982. The Theory of Sex Allocation. Princeton, NJ, USA: Princeton University Press.

Charnov, E.L. 1986. Size advantage may not always favor sex-change. Journal of Theoretical Biology 119, 283–285.

Chowdhury, I. & Joy, K.P. 2007. Seminal vesicle and its role in the reproduction of teleosts. Fish Physiology and Biochemistry 33, 383–398.

Cinquetti, R. 1997. Histochemical, enzyme histochemical and ultrastructural investigation on the sperm-duct gland of *Padogobius martensi* (Pisces, Gobiidae) Journal of Fish Biology 50, 978–991.

Colombo, L., Colombo-Belvedere, P. & Pilati, A. 1977. Biosynthesis of free and conjugated androgens by the testis of the black goby *Gobius jozo* L. Bollettino di Zoologia 44, 131–134.

Cordero, C. & Eberhard, W.G. 2008. Female choice of sexually antagonistic male adaptations: a crucial review of some current research. Journal of Evolutionary Biology 16, 1–6.

Cosson, J., Groison, A.L., Suquet, M., Fauvel, C., Dreanno, C. & Billard, R. 2008. Studying sperm motility in marine fish: an overview on the state of the art. Journal of Applied Ichthyology 24, 460–486.

Coward, K., Bromage, N.R., Hibbitt, O. & Parrington, J. 2002. Gamete physiology, fertilization and egg activation in teleost fish. Reviews in Fish Biology and Fisheries 12, 33–58.

Crews, D., Bergeron, J.M., Bull, J.J., Flores, D., Tousignant, A., Skipper, J.K. & Wibbels, T. 1994. Temperature-dependent sex determination in reptiles: proximate mechanisms, ultimate outcomes, and practical applications. Developmental Genetics 15, 297–312.

DeFraipont, M., FitzGerald, G.J. & Guderley, H. 1993. Age-related differences in reproductive tactics in the three-spined stickleback, *Gasterosteus aculeatus*. Animal Behavior 46, 961–968.

DeJonge, J., De Ruiter A.J.H. & Van Den Hurk, R. 1989. Testis–testicular gland complex of two *Tripterygion* species (Blennioidei, Teleostei): differences between territorial and non-territorial males. Journal of Fish Biology 35, 497–508.

De Mitcheson, Y.S. & Liu, M. 2008. Functional hermaphroditism in teleosts. Fish and Fisheries 9, 1–43.

Devlin, R.H. & Nagahama, Y. 2002. Sex determination and sex differentiation in fish: an overview of genetic, physiological, and environmental influences. Aquaculture 208, 191–364.

Dietrich, G.J., Wojtczak, M., Slowinska, M., Dobosz, S., Kuzminski, H. & Ciereszko, A. 2008. Effects of ovarian fluid on motility characteristics of rainbow trout (*Oncorhynchus mykiss* Walbaum) spermatozoa. Journal of Applied Ichthyology 24, 503–507.

Downing, A. L. & Burns, J. R. 1995. Testis morphology and spermatozeugma formation in three genera of viviparous halfbeaks, *Nomorhamphus, Dermogenys*, and *Hemirhamphodon* (Teleostei, Hemiramphidae). Journal of Morphology 225, 329–343.

Drozdov, A.L., Kornienko, E.S. & Krasnolutsky, A.V. 1997. Reproduction and development of *Syngnathus acusimils*. Russian Journal of Marine Biology 23, 265–268.

Eberhard, W.G. 1996. Female Control: Sexual Selection by Cryptic Female Choice. Princeton, NJ, USA: Princeton University Press.

Eberhard, W.G. & Cordero, C. 1995. Sexual selection by cryptic female choice on male seminal products: a new bridge between sexual selection and reproductive physiology. Trends in Ecology & Evolution 10, 493–496.

Eggert, B. 1931. Die Geschlechtsorgane der Gobiiformes und Blenniiformes. Zeitschrift für wissenschaftliche Zoologie 139, 249–558.

Eisenbach, M. & Giojalas, L.C. 2006. Sperm guidance in mammals - an unpaved road to the egg. Nature Reviews Molecular Cell Biology 7, 276–285.

Elofsson, H., Mcallister, B.G., Kime, D.E., Mayer, I. & Borg, B. 2003. Long lasting stickleback sperm; is ovarian fluid a key to success in fresh water? Journal of Fish Biology 63, 240–253.

Evans, J.P., Pierotti, M. & Pilastro, A. 2003 (a). Male mating behavior and ejaculate expenditure under sperm competition risk in the eastern mosquitofish. Behavioral Ecology 14, 268–273.

Evans, J.P., Zane, L., Francescato, S. & Pilastro, A. 2003 (b). Directional postcopulatory sexual selection revealed by artificial insemination. Nature 421, 360–363.

Evans, J.P. & Rutstein, A.N. 2008. Postcopulatory sexual selection favours intrinsically good sperm competitors. Behavioral Ecology and Sociobiology 62, 1167–1173.

Fishelson, L. 1970. Protogynous sex reversal in the fish Anthias squamipinnis (Teleostei, Anthiidae) regulated by the presence or absence of a male fish. Nature 227, 90–91.

Fishelson, L. 1991. Comparative cytology and morphology of seminal vesicles in male Gobiid fishes. Japanese Journal of Ichthyology 38, 17–30.

Fishelson, L., Van Vuren, J.H.J. & Tyran, A. 1994. Ontogenesis and ultrastucture of seminal vesicles of the catfish, Clarias gariepinus. Journal of Morphology 219, 59–71.

Fishelson, L., Delarea, Y., Gon, O. 2006. Testis structure, spermatogenesis, spermatocytogenesis, and sperm structure in cardinal fish (Apogonidae, Perciformes). Anatomy and Embryology 211, 31–46

Fishelson, L., Gon, O., Holdengreber, V. & Delarea, Y. 2007. Comparative spermatogenesis, spermatocytogenesis, and spermatozeugmata formation in males of viviparous species of clinid fishes (Teleostei: Clinidae, Blennioidei). Anatomical Record-Advances in Integrative Anatomy and Evolutionary Biology 290, 311–323.

Fitze, P.F., Le Galliard, J. - F., Federici, P., Richard, M. & Clobert, J. 2005. Conflict over multiple partner mating among males and females of polygynandrous common lizards. Evolution 59, 2451–2459.

Foster, S.J. & Vincent, A. 2004. Life history and ecology of seahorses: implications for conservation and management. Journal of Fish Biology 65, 1–61.

Francis, R.C. 1992. Sexual lability in teleosts: developmental factors. The Quarterly Review of Biology 67, 1–18.

Friedrich, B.M. & Julicher, F. 2007. Chemotaxis of sperm cells. Proceedings of the National Academy of Sciences of the United States of America 104, 13256–13261.

Frisch, A. 2004. Sex-change and gonadal steroids in sequentially-hermaphroditic teleost fish. Reviews in Fish Biology and Fisheries 14, 481–499.

Fuller, R.C. 1998. Sperm competition affects male behavior and sperm output in the rainbow darter. Proceedings of the Royal Society of London Series B-Biological Sciences 265, 2365–2371.

Gage, M.J.G., Stockley, P. & Parker, G.A. 1995. Effects of alternative male mating strategies on characteristics of sperm production in the Atlantic salmon (Salmo salar): Theoretical and empirical investigations. Philosophical Transactions of the Royal Society of London Series B—Biological Sciences 350, 391–399.

Gage, M.J.G., Stockley, P. & Parker, G.A. 1998. Sperm morphometry in the Atlantic salmon. Journal of Fish Biology 53, 835–840.

Gage, M.J.G., MacFarlane, C., Yeates, S., Shackleton, R. & Parker, G.A. 2002. Relationships between sperm morphometry and sperm motility in the Atlantic salmon. Journal of Fish Biology 61, 1528–1539.

Gage, M.J.G., Macfarlane, C.P., Yeates, S., Ward, R.G., Searle, J.B. & Parker, G.A. 2004. Spermatozoal traits and sperm competition in Atlantic salmon: Relative sperm velocity is the primary determinant of fertilization success. Current Biology 14, 44–47.

Gardiner, D. M. 1978. Utilization of extracellular glucose by spermatozoa of two viviparous fishes. Comparative Biochemistry and Physiology A-Physiology 59, 165–168.

Ghiselin, M. T. 1969. The evolution of hermaphroditism among animals. The Quarterly Review of Biology 44, 189–208.

Giacomello, E., Marchini D. & Rasotto, M.B. 2006. A male sexually dimorphic trait provides antimicrobials to eggs in blenny fish. Biology Letters 2, 330–333.

Giacomello, E., Neat, F.C. & Rasotto, M.B. 2008. Mechanisms enabling sperm economy in blenniid fishes. Behavioral Ecology and Sociobiology 62, 671–680.

Godwin, J., Luckenbach, J.A. & Borski, R.J. 2003. Ecology meets endocrinology: environmental sex determination in fishes. Evolution & Development 5, 40–49.

Goncalves, D. Barata, E.N, Oliveira, R.F. & Canario, A.V.M. 2002. The role of male visual and chemical cues on the activation of female courtship in the sex-role reversed peacock blenny. Journal of Fish Biology, 61, 96–105.

Grier, H.J. 1981. Cellular organization of the testis and spermatogenesis in fishes. American Zoologist 21, 345–357.

Grier, H. J. & Collette, B. B. 1987. Unique spermatozeugmata in tests of halfbeaks of the genus *Zenarchopterus* (Teleostei: Hemiramphidae). Copeia 1987, 300–311.

Grier, H. J. & Fishelson, L. 1995. Colloidal sperm-packaging in mouthbrooding tilapiine fishes. Copeia 1995, 966–970.

Grier, H. J. & Parenti, L. R. 1994. Reproductive biology and systematics of phallostethid fishes as revealed by gonad structure. Environmental Biology of Fishes 41, 287–299.

Grier, H. J., Fitzsimons, J. M. & Linton, J. R. 1978. Structure and ultrastructure of the testis and sperm formation in goodeid teleosts. Journal of Morphology 156, 419–435.

Grier, H. J., Linton, J. R., Leatherland, J. F. & DeVlaming, V. L. 1980. Structural evidence for two different testicular types in teleost fishes. American Journal of Anatomy 159, 331–345.

Gust, N. 2004. Variation in the population biology of protogynous coral reef fishes over tens of kilometres. Canadian Journal of Fisheries and Aquatic Sciences 61, 205–218.

Haesler, M. P. 2007. Sequential Mate Choice Decisions and Sperm Competition in Mouthbrooding Cichlids. PhD thesis, University of Bern, Switzerland.

Haesler, M. P. Lindeyer, C. M., Otti, O., Bonfils, D. & Taborsky, M. submitted. Pre-and post-mating sexual selection and sperm competition in thr mouth of cichlid fish.

Hardling, R. & Kaitala, A. 2005. The evolution of repeated mating under sexual conflict. Journal of Evolutionary Biology 18, 106–115.

Hayashi, F. 1998. Sperm co-operation in the Fishfly, *Parachauliodes japonicus*. Functional Ecology 12, 347–350.

Helfman, G.S., Collette, B.B. & Facey, D.E. 1997. The diversity of fishes. Malden, MA, USA: Blackwell Science.

Hirohashi, N., Kamei, N., Kubo, H., Sawada, H., Matsumoto, M. & Hoshi, M. 2008. Egg and sperm recognition systems during fertilization. Development Growth & Differentiation 50, S221–S238.

Holman, L. & Snook, R. R. 2008. A sterile sperm caste protects brother fertile sperm from female-mediated death in *Drosophila pseudoobscura*. Current Biology 18, 292–296.

Hourigan, T.F., Nakamura, M., Nagahama, Y., Yamauchi, K. & Grau, E.G. 1991. Histology, ultrastructure, and in vitro steroidogenesis of the testes of two male phenotypes of the protogynous fish, *Thalassoma duperrey* (Labridae). General and Comparative Endocrinology, 83, 193–217.

Hoysack, D.J. & Liley, N.R. 2001. Fertilization dynamics in sockeye salmon and a comparison of sperm from alternative male phenotypes. Journal of Fish Biology 58, 1286–1300.

Hoysak, D.J., Liley, N.R. & Taylor, E.B. 2004. Raffles, roles, and the outcome of sperm competition in sockeye salmon. Canadian Journal of Zoology 82, 1017–1026.

Hutchings, J. A., Bishop, T. D. & McGregor-Shaw, C. R. 1999. Spawning behavior of Atlantic cod, *Gadus morhua*: evidence of mate competition and mate choice in a broadcast spawner. Canadian Journal of Fisheries and Aquatic Sciences 56, 97–104.

Immler, S. 2008. Sperm competition and sperm cooperation: the potential role of diploid and haploid expression. Reproduction 135, 275–283.

Immler, S. & Taborsky, M. 2009. Sequential polyandry affords post-mating sexual selection in the mouths of cichlid females. Behavioral Ecology and Sociobiology (in press).

Immler, S., Mazzoldi, C. & Rasotto, M. B. 2004.
    From sneaker to parental male: Change of
    reproductive traits in the black goby, *Gobius
    niger* (Teleostei, Gobiidae). Journal of
    Experimental Zoology Part A-Comparative
    Experimental Biology 301A, 177–185.

Immler, S., Moore, H. D. M., Breed, W. G. &
    Birkhead, T. R. 2008. By hook or by crook?
    Morphometry, competition and cooperation
    in rodent sperm. PLoS ONE 2 e170.

Iwamatsu, T., Yoshizaki, N. & Shibata, Y. 1997.
    Changes in the chorion and sperm entry into
    the micropyle during fertilization in the tele-
    ostean fish, *Oryzias latipes*. Development
    Growth & Differentiation 39, 33–41.

Jamieson, B. G. M. 1991. Fish Evolution and
    Systematics: Evidence From Spermatozoa.
    Cambridge: Cambridge University Press.

Javonillo, R., Burns, J. R. & Weitzman, S. H.
    2007. Reproductive morphology of
    Brittanichthys axelrodi (Teleostei:
    Characidae), a miniature inseminating fish
    from South America. Journal of Morphology
    268, 23–32.

Jennions, M. D. & Kelly, C. D. 2002.
    Geographical variation in male genitalia
    in *Brachyphaphis episcopi* (Poeciliidae): is it
    sexually or naturally selected? Oikos 97,
    79–86.

Jennions, M.D. & Petrie, M. 2000. Why do
    females mate multiply? A review of the
    genetic benefits. Biological Reviews 75,
    21–64.

Jobling, M. 1995. Environmental Biology of
    Fishes. London: Chapman and Hall.

Johnston, S.D., Smith, B., Pyne, M., Stenzel, D. &
    Holt, W.V. 2007. One-sided ejaculation of
    echidna sperm bundles. American Naturalist
    170, e162–e164.

Joy, K.P. & Singh, M.S. 1998. A comparative
    study on distribution, morphology, and some
    biochemical constituents of seminal vesicles in
    catfishes. Journal of Endocrinology and
    Reproduction 2, 26–33.

Kallman, K.D. 1984. A new look at sex
    determination in poeciliid fishes. In:
    Evolutionary Genetics of Fishes (Ed. by
    B. J. Turner), pp. 95–171. New York, USA:
    Plenum Publishing Corporation.

Kaupp, U.B., Kashikar, N.D. & Weyand, I. 2008.
    Mechanisms of sperm chemotaxis. Annual
    Review of Physiology 70, 93–117.

Kazakov, R.V. 1981. Peculiarities of sperm
    production by anadromous and parr Atlantic
    salmon (*Salmo salar* L.) and fish cultural
    characteristics of such sperm. Journal of
    Fish Biology 18, 1–8.

Kelly, C.D., Godin, J. - G.J. & Abdallah, G. 2000.
    Geographical variation in the male
    intromittent organ of the Trinidadian guppy
    (*Poecilia reticulata*). Canadian Journal of
    Zoology—Revue Canadienne de Zoologie 78,
    1674–1680.

Kime, D.E., Van Look, K.J.W., Mcallister, B.G.,
    Huyskens, G., Rurangwa, E. & Ollevier, F.
    2001. Computer-assisted sperm analysis
    (CASA) as a tool for monitoring sperm qual-
    ity in fish. Comparative Biochemistry and
    Physiology C—Toxicology & Pharmacology
    130, 425–433.

Kohda, M., Tanimura, M., Kikue-Nakamura, M.
    & Yamagishi, S. 1995. Sperm drinking by
    female catfish: a novel mode of insemination.
    Environmental Biology of Fishes 42, 1–16.

Koya, Y., Munehara, H. & Takano, K. 2002.
    Sperm storage and motility in the ovary of the
    marine sculpin *Alcichthys alcilcornis*
    (Teleostei: Scorpaeniformes), with internal
    gametic association. Journal of Experimental
    Zoology 292, 145–155.

Kunz, Y.W. 2004. *Developmental Biology of
    Teleost Fishes*. Dordrecht, Netherlands:
    Springer; pp. 636.

Kuwamara, T. 1983. Spawning behavior and
    timing of fertilization in the mouthbreeding
    cardinal fish, *Apogon notatus*. Japanese
    Journal of Ichthyology 30, 61–71.

Kuwamura, T., Nakashima, Y. & Yogo, Y. 1994.
    Sex change in either direction by growth rate
    advantage in the monogamous coral goby,
    Paragobiodon echinocephalus. Behavioral
    Ecology 5, 434–438.

Kuwamura, T., Tanaka, N., Nakashima, Y.,
    Karino, K. & Sakay, H. 2002. Reversed
    sex-change in the protogynous reef fish
    Labroides dimidiatus. Ethology 108,
    443–450.

Kuwamura, T., Sakai, S., Tanaka, N., Ouchi, E.,
    Karino, K. & Nakashima, Y. 2007. Sex
    change of primary males in a diandric labrid
    Halichoeres trimaculatus: coexistence of
    protandry and protogyny within a species.
    Journal of Fish Biology 70, 1898–1906.

Kvarnemo, C. & Simmons, L.W. 2004. Testes
    investment and spawning mode in
    pipefishes and seahorses (Syngnathidae).
    Biological Journal of the Linnaean Society 83,
    369–376.

Lahnsteiner, F. 2003. The spermatozoa and eggs of the cardinal fish. Journal of Fish Biology 62, 115–128.

Lahnsteiner, F. & Patzner, R.A. 1990a. Spermiogenesis and structure of mature spermatozoa in blenniid fishes (Pisces, Blenniidae). Journal of Submicroscopic Cytology and Pathology 22, 565–576.

Lahnsteiner, F. & Patzner R.A. 1990b. The spermatic duct of blenniid fish (Teleostei, Blenniidae): Fine structure, histochemistry and function. Zoomorphology 110, 63–73.

Lahnsteiner, F. & Patzner, R.A. 2008. Sperm morphology and ultrastructure in fish. In: Fish Spermatology (Ed. by S.M.H. Alavi, J. Cosson, K. Coward & G. Raffiee), pp. 1–61. Oxford, UK, Alpha Science International.

Lahnsteiner, F., Richtarski, U. & Patzner, R.A. 1990. Functions of the testicular gland in two blenniid fishes, *Salaria* (=*Blennius*) *pavo* and *Lipophrys* (=*Blennius*) *dalmatinus* (Blenniidae, Teleostei) as revealed by electron microscopy and enzyme histochemistry. Journal of Fish Biology, 37, 85–97.

Lahnsteiner, F., Nussbaumer, B. & Patzner, R.A. 1993. Unusual testicular accessory organs, the testicular blind pouches of blennies (Teleostei, Blenniidae). Fine structure, (enzyme-) histochemistry and possible functions. Journal of Fish Biology, 42, 227–241.

Lahnsteiner, F., Patzner, R.A. & Weismann, T. 1994. Testicular main ducts and spermatic ducts in some cyprinid fishes: 1. morphology, fine-structure and histochemistry. Journal of Fish Biology 44, 937–951.

Lahnsteiner, F., Weismann, T. & Patzner, R.A. 1995. Composition of the ovarian fluid in 4 salmonid species: *Oncorhynchus mykiss, Salmo trutta lacustris, Salvelinus alpinus* and *Hucho hucho*. Reproduction Nutrition Development 35, 465–474.

Lahnsteiner, F., Berger, B., Weismann, T. & Patzner, R.A. 1998. Determination of semen quality of the rainbow trout, *Oncorhynchus mykiss*, by sperm motility, seminal plasma parameters, and spermatozoal metabolism. Aquaculture 163, 163–181.

Lahnsteiner, F., Weismann, T. & Patzner, R.A. 1999. Physiological and biochemical parameters for egg quality determination in lake trout, *Salmo trutta lacustris*. Fish Physiology and Biochemistry 20, 375–388.

Laksanawimol, P., Damrongphol, P. & Kruatrachue, M. 2006. Alteration of the brood pouch morphology during gestation of male seahorses, *Hippocampus kuda*. Marine and Freshwater Research 57, 497–502.

Langerhans, R.B., Layman, C.A. & DeWitt, T.J. 2005. Male genital size reflects a tradeoff between attracting mates and avoiding predators in two live-bearing fish species. Proceedings of the National Academy of Sciences of the United States of America 102, 7618–7623.

Laumen, J., Pern, U. & Blum, V. 1974. Investigations on the function and hormonal regulation of the anal appendices in *Blennius pavo* (Risso). Journal of Experimental Zoology, 190, 47–56.

Leach, B. & Montgomerie, R. 2000. Sperm characteristics associated with different male reproductive tactics in bluegills (*Lepomis macrochirus*). Behavioral Ecology and Sociobiology 49, 31–37.

Levitan, D.R. 1993. The importance of sperm limitation to the evolution of egg size in marine invertebrates. American Naturalist 141, 517–536.

Levitan, D. R. 1996. Effects of gamete traits on fertilization in the sea and the evolution of sexual dimorphism. Nature 382, 153–155.

Liem, K. F. 1963. Sex reversal as a natural process in the synbranchiform fish *Monopterus albus*. Copeia 1963, 303–312.

Liem, K. F. 1968. Geographic and taxonomic variation in the pattern of natural sex reversal in the teleost fish order Synbranchiformes. Journal of Zoology London 156, 225–283.

Liljedal, S., Rudolfsen, G. & Folstad, I. 2008. Factors predicting male fertilization success in an external fertilizer. Behavioral Ecology and Sociobiology 62, 1805–1811.

Linhart, O., Rodina, M., Gela, D., Kocour, M. & Vandeputte, M. 2005. Spermatozoal competition in common carp (*Cyprinus carpio*): what is the primary determinant of competition success? Reproduction 130, 705–711.

Litvak, M. K. & Trippel, E. A. 1998. Sperm motility patterns of Atlantic cod (*Gadus morhua*) in relation to salinity: effects of ovarian fluid and egg presence. Canadian Journal of Fisheries and Aquatic Sciences 55, 1871–1877.

Locatello L., Mazzoldi C. & Rasotto M.B. (2002) Ejaculate of sneaker males is pheromonally inconspicuous in the black goby, *Gobius niger* (Teleostei, Gobiidae). Journal of Experimental Zoology 293, 601–605.

Locatello, L., Rasotto, M. B., Evans, J. P. & Pilastro, A. 2006. Colourful male guppies produce faster and more viable sperm. Journal of Evolutionary Biology 19, 1595–1602.

Loir, M., Cauty, C., Planquette, P. & Le Bail, P. Y. 1989. Comparative study of the male reproductive tract in seven families of South-American catfish. Aquatic Living Resources 2, 45–56.

Lopes, D.C.J.R., Bazzoli, N., Brito, M.F.G. & Maria, T.A. 2004. Male reproductive system in the South American catfish *Conorhynchus conirostris*. Journal of Fish Biology 64, 1419–1424.

Maan, M. E. & Taborsky, M. 2008. Sexual conflict over breeding substrate causes female expulsion and offspring loss in a cichlid fish. Behavioral Ecology 19, 302–308.

Mackie, J. B. & Walker, M. H. 1974. A study of the conjugate sperm of the Dytiscid water beetles *Dytiscus marginalis* and *Colymbetes fuscus*. Cell Tissue Research 148, 505–519.

Maklakov, A. A., Bilde, T. & Lubin, Y. 2005. Sexual conflict in the wild: elevated mating rate reduces female lifetime reproductive success. American Naturalist 165, S38-S45.

Mank, J. E., Promislow, D. E. L. & Avise, J. C. 2006. Evolution of alternative sex-determining mechanisms in teleost fishes. Biological Journal of the Linnaean Society 87, 83–93.

Mann, T. & Lutwak-Mann, C. 1981. Male Reproductive Function and Semen—Themes and Trends in Physiology, Biochemistry and Investigative Andrology. Heidelberg: Springer-Verlag.

Manni, L. & Rasotto, M.B. 1997. Ultrastructure and histochemistry of the testicular efferent duct system and spermiogenesis in *Opistognathus whitehursti* (Teleostei, Trachinoidei). Zoomorphology, 117, 93–102.

Marconato, A., Rasotto, M. B. & Mazzoldi, C. 1996. On the mechanism of sperm release in three gobiid fishes (Teleostei: Gobiidae). Environmental Biology of Fishes 46, 321–327.

Martinez, V. H. & Monasterio de Gonzo, G. A. M. 2002. Testis morphology and sperma-tozeugma formation in *Jenysia multidentata*.

II International Symposium on Livebearing Fishes Queretaro, Mexico (Poster Abstract).

Mazzoldi, C. 2002. Reproductive apparatus and mating system in two tropical goby species. Journal of Fish Biology 59, 1686–1691.

Mazzoldi, C. & Rasotto, M.B. 2002. Alternative male mating tactics in Gobius niger. Journal of Fish Biology 61, 157–172.

Mazzoldi, C., Scaggiante, M., Ambrosin, E. & Rasotto, M. B. 2000. Mating system and alternative male mating tactics in the grass goby *Zosterisessor ophiocephalus* (Teleostei: Gobiidae). Marine Biology 137, 1041–1048.

Mazzoldi, C., Petersen, C. W. & Rasotto, M. B. 2005. The influence of mating system on seminal vesicle variability among gobies (Teleostei, Gobiidae). Journal of Zoological Systematics and Evolutionary Research 43, 307–314.

Mazzoldi, C., Lorenzi, V. & Rasotto, M. B. 2007. Variation of male reproductive apparatus in relation to fertilization modalities in the catfish families Auchenipteridae and Callichthyidae (Teleostei: Siluriformes). Journal of Fish Biology 70, 243–256.

McMillan, D. B. 2007. Fish histology: Female reproductive systems. Dordrecht, Netherlands: Springer; pp. 598.

Meisner, A. D., Burns, J. R., Weitzman, S. H. & Malabarba, L. R. 2000. Morphology and histology of the male reproductive system in two species of internally inseminating South American catfishes, *Trachelyopterus lucenai* and *T. galeatus* (Teleostei: Auchenipteridae). Journal of Morphology 246, 131–141.

Melamed, P., Xue, Y., Poon, J. F. D., Wu, Q., Xie, H. M., Yeo, J., Foo, T. W. J. & Chua, H. K. 2005. The male seahorse synthesizes and secretes a novel C-type lectin into the brood pouch during early pregnancy. Febs Journal 272, 1221–1235.

Mengerink, K. J. & Vacquier, V. D. 2001. Glycobiology of sperm–egg interactions in deuterostomes. Glycobiology 11, 37R–43R.

Miller, P.J. 1984. The tokology of gobioid fishes. In: Fish Reproduction: Strategies and Tactics. (G.W. Potts & R.J. Wootton Eds). pp. 119–153. London, UK: Academic Press.

Molloy, P. R., Goodwin, N. B., Cote, I. M., Reynolds, J. D. & Gage, M. J. G. 2007. Sperm competition and sex change: A comparative analysis across fishes. Evolution 61, 640–652.

Moore, H. D. M. & Taggart, D. A. 1995. Sperm pairing in the opossum increases the efficiency of sperm movement in a viscous environment. Biology of Reproduction 52, 947–953.

Moore, H. D. M., Dworakova, K., Jenkins, N. & Breed, W. G. 2002. Exceptional sperm cooperation in the wood mouse. Nature 418, 174–177.

Morisawa, M. & Suzuki, K. 1980. Osmolality and potassium-ion: their roles in initiation of sperm motility in teleosts. Science 210, 1145–1147.

Morisawa, M., Tanimoto, S. & Ohtake, H. 1992. Characterization and partial purification of sperm-activating substance from eggs of the herring, *Clupea palasii*. Journal of Experimental Zoology 264, 225–230.

Munday, P. L. 2002. Bi-directional sex change: testing the growth-rate advantage model. Behavioral Ecology and Sociobiology 52, 247–254.

Munday, P. L., Buston, P. M. & Warner, R. R. 2006. Diversity and flexibility of sex-change strategies in animals. Trends in Ecology & Evolution 21, 89–95.

Munehara, H., Takano, K. & Koya, Y. 1989. Internal gametic association and external fertilization in the elkhorn sculpin, *Alcichthys alcicornis*. Copeia 1989, 673–678.

Munehara, H., Takano, K. & Koya, Y. 1991. The little dragon sculpin *Blespias cirrhosus*: Another case of internal gametic association and external fertilization. Japanese Journal of Ichthyology 37, 391–394.

Munehara, H., Takenaka, A. & Takenaka, O. 1994. Alloparental care in the marine sculpin *Alcichthys alcicornis* (Pisces, Cottidae): copulating in conjunction with paternal care. Journal of Ethology 12, 115–120.

Munoz, R. C. & Warner, R. R. 2003. A new version of the size-advantage hypothesis for sex change: Incorporating sperm competition and size-fecundity skew. American Naturalist 161, 749–761.

Munoz, R. C. & Warner, R. R. 2004. Testing a new version of the size-advantage hypothesis for sex change: sperm competition and size-skew effects in the bucktooth parrotfish, *Sparisoma radians*. Behavioral Ecology 15, 129–136.

Nakamura, M. & Kobayashi, Y. 2005. Sex change in coral reef fish. Fish Physiology and Biochemistry 31, 117–122.

Nakashima, Y., Kuwamura, T. & Yogo, Y. 1995. Why be a both-ways sex change? Ethology 101, 301–307.

Nakatsuru, K. & Kramer, D. L. 1982. Is sperm cheap? Limited male fertility and female choice in the Lemon Tetra (Pisces, Characidae). Science 216, 753–755.

Nayyar, S. K. & Sundararaj, B. I. 1970. Seasonal reproductive activity in the testes and seminal vesicles of the catfish, *Heteropneutes fossilis* (Bloch). Journal of Morphology 130, 207–226.

Neat, F.C. 2001. Male parasitic spawning in two species of triplefin blenny (Tripterigiidae): contrasts in demography, behavior and gonadal characteristics. Environmental Biology of Fishes 55, 57–64.

Neat, F.C & Lengkeek, W. 2009. Sexual selection in blennies. In: Biology of Blennies. (eds R.A. Patzner, E.J. Gonçalves, P.A. Hastings, B.G. Kapoor). Science Publishers Inc, New Hampshire.

Neat, F.C. & Locatello, L. 2002. No reason to sneak: why males of all sizes can breed in the hole-nesting blenny, *Aidablennius sphinx*. Behavioral Ecology and Sociobiology 52, 66–73.

Neat F.C., Locatello, L. & Rasotto, M.B. 2003. Reproductive morphology of alternative male reproductive tactics in *Scartella cristata* (Teleostei: Blenniidae). Journal of Fish Biology 62, 1381–1391.

Neff, B.D., Fu, P. and Gross, M.R. 2003. Sperm investment and alternative mating tactics in bluegill sunfish (*Lepomis macrochirus*). Behavioral Ecology 14, 634–641.

Nordeide, J. T. 2007. Is there more in 'gamete quality' than quality of the gametes? A review of effects of female mate choice and genetic compatibility on offspring quality. Aquaculture Research 38, 1–16.

Northcutt, S.J. & Bullock, A.M. 1991, The morphology of the club glands on the dorsal fin of mature male shannies, *Lipophrys pholis* (L.) (Blenniidae: Teleostei). Journal of Fish Biology 39, 795–806.

Northcutt, S. J. & James, M.A. 1995. Ultrastructure of the glandular epidermis of the fins of male estuarine triplefins *Fosterygion nigripenne*. Journal of Fish Biology 49. 95–107.

Ohta, K., Sundaray, J. K., Okida, T., Sakai, M., Kitano, T., Yamaguchi, A., Takeda, T. & Matsuyama, M. 2003. Bi-directional sex change and its steroidogenesis in the wrasse,

Pseudolabrus sieboldi. Fish Physiology and Biochemistry 28, 173–174.

Ojanguren, A. F. & Magurran, A. E. 2007. Male harassment reduces short-term female fitness in guppies. Behavior 144, 503–514.

Oldfield, R. G. 2005. Genetic, abiotic and social influences on sex differentiation in cichlid fishes and the evolution of sequential hermaphroditism. Fish and Fisheries 6, 93–110.

Oliveira, R.F., Almada, V.C., Forsgren, E. & Goncalves, E.J. 1999. Temporal variation in male traits, nesting aggregations and mating success in the peacock blenny. Journal of Fish Biology 54, 499–512.

Oliveira, R.F., Canario, A.V.M., Grober, M.S, Serrao Santos, R. 2001a. Endocrine correlates of male polymorphism and alternative reproductive tactics in the Azorean rock-pool blenny, *Parablennius snaguinolentis parvicornis*. General and Compartative Endocrinology 121, 278–288.

Oliveira, R. F., Goncalves, E. J. & Santos, R. S. 2001b. Gonadal investment of young males in two blenniid species with alternative mating tactics. Journal of Fish Biology, 59, 459–462.

Ota, D., Marchesan, M. & Ferrero E.A. 1996. Sperm release behavior and fertilisation in the grass goby. Journal of Fish Biology, 49, 246–256.

Parenti, L. R. & Grier, H. J. 2004. Evolution and phylogeny of gonad morphology in bony fishes. Integrative and Comparative Biology 44, 333–348.

Parker, G. A. 1970. Sperm competition and its evolutionary consequences in the insects. Biological Reviews 45, 525–567.

Parker, G. A. 1984. Sperm competition and the evolution of animal mating strategies. In: Sperm Competition and the Evolution of Animal Mating Systems (Ed. by R. L. Smith), pp. 1–60. Orlando, USA: Academic Press.

Parker, G.A. 1990. Sperm competition games: sneaks and extra-pair copulations. Proceedings of the Royal Society London Series B - Biological Sciences 242, 127–133.

Parker, G.A. 1998. Sperm competition and the evolution of ejaculates: towards a theory base. In: Sperm Competition and Sexual Selection (Ed. by T.R. Birkhead & A.P. Moller), pp. 3–54. San Diego, USA, Academic Press.

Parker, G. A. 2006. Sexual conflict over mating and fertilization: an overview. Philosophical Transactions of the Royal Society B—Biological Sciences 361, 235–259.

Patzner, R.A. 1989. Morphology of the male reproductive systems of two Indopacific blenniid fishes, *Salarias fasciatus* and *Ecsenius bicolor* (Blenniidae, Teleostei). Zeitschrift für Zoologische Systematik und Evolutionsforschung 27, 135–141.

Patzner, R.A. (1991) Morphology of the male reproductive system in *Coralliozetus angelica* (Pisces, Blennioidei, Chaenopsidae). Journal of Fish Biology 39, 867–872.

Patzner, R.A. & Lahnsteiner, F. 2009 Reproductive organs in Blennies. In: Biology of Blennies. (eds R.A. Patzner, E.J. Gonçalves, P.A. Hastings, B.G. Kapoor). Science Publishers Inc, New Hampshire.

Pecio, A. & Rafinski, J. 1994. Structure of the testes, spermatozoa and spermatozeugmata of *Mimagoniates barberi Regan*, 1907 (Teleostei, Characidae), an internally fertilizing, oviparous fish. Acta Zoologica 75, 179–185.

Pecio, A., Lahnsteiner, F. & Rafinski, J. 2001. Ultrastructure of the epithelial cells in the aspermatogenic part of the testis in Mimagoniates barberi (Teleostei: Characidae: Glandulocaudinae) and the role of their secretions in spermatozeugmata formation. Annals of Anatomy—Anatomischer Anzeiger 183, 427–435.

Perez-Villar, J. & Hill, R.L. (1999) The structure and assembly of secreted mucins. Journal of Biological Chemistry 274, 31751–31754.

Petersen, C.W. & Warner, R.R. (1998). Sperm competition in fishes. In: Sperm Competition and Sexual Selection. (Birkhead, T.R. & Miller A.P. eds). pp. 435–463. Academic Press, London.

Piironen, J. & HyvŠrinen, H. 1983. Composition of the milt of some teleost fishes. Journal of Fish Biology 22, 351–361.

Pilastro, A., Simonato, M., Bisazza, A. & Evans, J. P. 2004. Cryptic female preference for colorful males in guppies. Evolution 58, 665–669.

Pilastro, A., Gasparini, C., Boschetto, C. & Evans, J. P. 2008. Colorful male guppies do not provide females with fecundity benefits. Behavioral Ecology 19, 374–381.

Pitcher, T. E., Rodd, F. H. & Rowe, L. 2007. Sexual colouration and sperm traits in guppies. Journal of Fish Biology 70, 165–177.

Pizzari, T. & Foster, K. R. 2008. Sperm sociality: Cooperation, altruism, and spite. PLoS Biology 6, e130.

Podolsky, R. D. 2004. Life-history consequences of investment in free-spawned eggs and their accessory coats. American Naturalist 163, 735–753.

Pratt, H.L. 1991. The storage of spermatozoa in the oviducal glands of western North-Atlantic sharks. Environmental Biology of Fishes 38, 139–149.

Pratt, H. L. & Carrier, J. C. C. 2001 A review of elasmobranch reproductive behavior with a case study on the nurse shark, *Ginglymostoma cirratum*. Environmental Biology of Fishes 60, 157–188.

Pratt, H. L. & Tanaka, S. 1994. Sperm storage in male elasmobranchs: a description and survey. Journal of Morphology 219, 297–308.

Rasotto, M. 1995. Male reproductive apparatus of some blennioidei (Pisces, Teleostei). Copeia 1995: 907–914.

Rasotto, M. B. & Mazzoldi, C. 2002. Male traits associated with alternative reproductive tactics in *Gobius niger*. Journal of Fish Biology 61, 173–184.

Rasotto, M.B. & Shapiro, D.Y. 1998. Morphology of gonoducts and male genital papilla, in the bluehead wrasse: implications and correlates on the control of gamete release. Journal of Fish Biology 52, 716–725.

Rasotto, M., Marconato, A. & Shapiro, D. Y. 1992. Reproductive apparatus of two jawfish species (Opistothognathidae) with a description of a juxtatesticular body. Copeia 1992, 1046–1053.

Ratty, F. J. Laurs, R.M. & Kelly, R.M. (1990) Gonad morphology, histology and spermatogenesis in south Pacific Albacore Tuna *Thunnus alalunga* (Scombridae). Fisheries Bulletin 88, 207–216.

Reichard, M., Le Comber, S. C. & Smith, C. 2007. Sneaking from a female perspective. Animal Behavior 74, 679–688.

Reinboth, R. 1962. Morphologische und funktionelle Zweigeschlechtigkeit bei marinen Teleostiern (Serranidae, Sparidae, Centracanthidae, Labridae). Zoologische Jahrbücher Physiologie 69, 405–480.

Reinboth, R. 1967. Zum Problem der amphisexuellen Fische. Verhandlung der Deutschen Zoologischen Gesellschaft München 314–325.

Resink, J. W., Voorthuis, P. K., Van den Hurk, R., Peters, R. C. & Van Oordt, P. G. W. J. 1989. Steroid glucuronides of the seminal vesicle as olfactory stimuli in African catfish, *Clarias gariepinus*. Aquaculture 83, 153–166.

Reyer, H. U., Frei, G. & Som, C. 1999. Cryptic female choice: frogs reduce clutch size when amplexed by undesired males. Proceedings of the Royal Society B—Biological Sciences 266, 2101–2107.

Rice, W. R., Stewart, A. D., Morrow, E. H., Linder, J. E., Orteiza, N. & Byrne, P. G. 2006. Assessing sexual conflict in the *Drosophila melanogaster* laboratory model system. Philosophical Transactions of the Royal Society B—Biological Sciences 361, 287–299.

Richtarski, U. & Patzner, R.A.. 2000. Comparative morphology of male reproductive systems in Mediterranean blennies (Blenniidae). Journal of Fish Biology 56, 22–36.

Rideout, R.M., Trippel, E.A., Litvak, M.K. 2004. Relationship between sperm density, spermatocrit, sperm motility and spawning date in wild and cultured haddock. Journal of Fish Biology 65, 319–332.

Riehl, R. & Kokoscha, M. 1993. A unique surface pattern and micropylar apparatus in the eggs of *Luciocephalus* sp. (Perciformes, Luciocephalidae). Journal of Fish Biology 43, 617–620.

Ripley, J. L. & Foran, C. M. 2006. Differential parental nutrient allocation in two congeneric pipefish species (Syngnathidae: Syngnathus spp.). The Journal of Experimental Biology 209, 1112–1121.

Ruchon, F., Laugier, T. & Quignard, J.P. 1995. Alternative male reproductive strategies in the peacock blenny. Journal of Fish Biology 47, 826–840.

Rudolfsen, G., Figenschou, L., Folstad, I., Tveiten, H. & Figenschou, M. 2006. Rapid adjustments of sperm characteristics in relation to social status. Proceedings of the Royal Society B—Biological Sciences 273, 325–332.

Ryan, M. J., Hews, D. K. & Wagner, W. E. 1990. Sexual selection on alleles that determine body size in the swordtail *Xiphophorus nigrensis*. Behavioral Ecology and Sociobiology 26, 231–237.

Santos, J. E., Bazzoli, N. & Santos, G. B. 2001. Morphofunctional organization of the male reproductive system of the catfish

Iheringichthys labrosus (Lütken, 1874). Tissue and Cell 33, 533–540.

Sato, T., Hirose, M., Taborsky, M. & Kimura, S. 2004. Size-dependent male alternative reproductive tactics in the shell-brooding cichlid fish *Lamprologus callipterus* in Lake Tanganyika. Ethology 110, 49–62.

Scaggiante, M., Mazzoldi, C., Petersen, C. W. & Rasotto, M. B. 1999. Sperm competition and mode of fertilization in the grass goby *Zosterisessor ophiocephalus* (Teleostei: Gobiidae). Journal of Experimental Zoology 283, 81–90.

Scaggiante, M., Grober, M. S., Lorenzi, V. & Rasotto, M. B. 2004. Changes along the male reproductive axis in response to social context in a gonochoristic gobiid, *Zosterisessor ophiocephalus* (Teleostei, Gobiidae), with alternative mating tactics. Hormones and Behavior 46, 607–617.

Schindler, J. F. & Hamlett, W. C. 1993. Maternal embryonic relations in viviparous teleosts. Journal of Experimental Zoology 266, 378–393.

Schulte-Hostedde, A. I. & Burness, G. 2005. Fertilization dynamics of sperm from different male mating tactics in bluegill (*Lepomis macrochirus*). Canadian Journal of Zoology—Revue Canadienne de Zoologie 83, 1638–1642.

Schütz, D., Pachler, G., Ripmeester, E., Goffinet, O. & Taborsky, M. 2010. Reproductive investment of giants and dwarfs: Specialized tactics in a cichlid fish with alternative male morphs. Functional Ecology 24, 131–140.

Setchel, B.P. & Brooks, D.E. 1988. Anatomy, vasculature, innervation and fluids of the male reproductive tract. In: The Physiology of Reproduction (Knobil, E & Neill, J. Eds). pp. 753–819. New York, Raven Press.

Shapiro, D. Y. & Giraldeau, L. A. 1996. Mating tactics in external fertilizers when sperm is limited. Behavioral Ecology 7, 19–23.

Shapiro, D. Y., Marconato, A. & Yoshikawa, T. 1994. Sperm economy in a coral reef fish, *Thalassoma bifasciatum*. Ecology 75, 1334–1344.

Simmons, L.W., Tomkins, J. L. & Hunt, J.C. (1999) Sperm competition games played by dimorphic male beetles. Proceedings of the Royal Society B-Biological Sciences 266, 145–150.

Singh, M. S. & Joy, K. P. 1998. A comparative study on histochemical distribution of some enzymes related to steroid and glucuronide synthesis in seminal vesicle and testis of the catfish, Clarias batrachus. Zoological Science 15, 955–961.

Skinner, A. M. J. & Watt, P. J. 2007. Phenotypic correlates of spermatozoon quality in the guppy, *Poecilia reticulata*. Behavioral Ecology 18, 47–52.

Smith, C. & Reichard, M. 2005. Females solicit sneakers to improve fertilization success in the bitterling fish (*Rhodeus sericeus*). Proceedings of the Royal Society B-Biological Sciences 272, 1683–1688.

Sneed, K. E. & Clemens, H. P. 1963. The morphology of the testes and accessory reproductive glands of the catfishes (Ictaluridae). Copeia 1963, 606–611.

Sorenson P.W. 1992. Hormones, pheromones and chemoreception. In. Fish Chemoreception (Hara T. J. ed.). pp. 199–228. Chapman and Hall, London.

Spehr, M., Gisselmann, G., Poplawski, A., Riffell, J. A., Wetzel, C. H., Zimmer, R. K. & Hatt, H. 2003. Identification of a testicular odorant receptor mediating human sperm chemotaxis. Science 299, 2054–2058.

Stockley, P. 1997. Sexual conflict resulting from adaptations to sperm competition. Trends in Ecology & Evolution 12, 154–159.

Stockley, P., Gage, M. J. G., Parker, G. A. & Moller, A. P. 1996. Female reproductive biology and the coevolution of ejaculate characteristics in fish. Proceedings of the Royal Society of London Series B—Biological Sciences 263, 451–458.

Stockley, P., Gage, M. J. G., Parker, G. A. & Moller, A. P. 1997. Sperm competition in fishes: The evolution of testis size and ejaculate characteristics. American Naturalist 149, 933–954.

Stolting, K. K. & Wilson, A. B. 2007. Male pregnancy in seahorses and pipefish: beyond the mammalian model. Bioessays 29, 884–896.

Stoltz, J. A. & Neff, B. D. 2006. Sperm competition in a fish with external fertilization: the contribution of sperm number, speed and length. Journal of Evolutionary Biology 19, 1873–1881.

Stoltz, J. A., Neff, B. D. & Olden, J. D. 2005. Allometric growth and sperm competition in fishes. Journal of Fish Biology 67, 470–480.

Strussmann, C. A. & Nakamura, M. 2002. Morphology, endocrinology, and environmental modulation of gonadal sex differentiation in teleost fishes. Fish Physiology and Biochemistry 26, 13–29.

Taborsky, M. 1994. Sneakers, satellites, and helpers: parasitic and cooperative behavior in fish reproduction. Advances in the Study of Behavior 23, 1–100.

Taborsky, M. 1997. Bourgeois and parasitic tactics: do we need collective, functional terms for alternative reproductive behaviors? Behavioral Ecology and Sociobiology 41, 361–362.

Taborsky, M. 1998. Sperm competition in fish: 'bourgeois' males and parasitic spawning. Trends in Ecology & Evolution 13, 222–227.

Taborsky, M. 1999. Conflict or cooperation: what determines optimal solutions to competition in fish reproduction? In: Behavior and Conservation of Littoral Fishes (Ed. by R. F. Oliveira, V. Almada & E. Goncalves), pp. 301–349. Lisboa, Portugal: ISPA.

Taborsky, M. 2001 The evolution of bourgeois, parasitic and cooperative reproductive behaviors in fishes. Journal of Heredity 92, 100–110.

Taborsky, M. 2008. Alternative reproductive tactics in fish. In: Alternative Reproductive Tactics: an Integrative Approach (Ed. by R. F. Oliveira, M. Taborsky & H. J. Brockmann), pp. 251–299. Cambridge, UK, Cambridge University Press.

Taborsky, M., Hudde, B. & Wirtz, P. 1987. Reproductive behavior and ecology of *Symphodus (Crenilabrus) ocellatus*, a European wrasse with four types of male behavior. Behavior 102, 82–118.

Taborsky, M., Oliveira, R. F. & Brockmann, H. J. 2008. The evolution of alternative reproductive tactics: concepts and questions. In: Alternative Reproductive Tactics: an Integrative Approach (Ed. by R. F. Oliveira, M. Taborsky & H. J. Brockmann), pp. 1–21. Cambridge, UK: Cambridge University Press.

Thornhill, R. 1983. Cryptic female choice and its implications in the scorpionfly *Harpobittacus nigriceps*. American Naturalist 122, 765–788.

Thunken, T., Bakker, T.C.M. & Kullmann. 2007. Extraordinary long sperm in the socially monogamous cichlid fish *Pelvicachromis taeniatus*. Naturwissenschaften 94, 489–491.

Trippel, E. A. & Morgan, M. J. 1994. Sperm longevity in Atlantic cod (*Gadus morhua*). Copeia 1025–1029.

Trivers, R. L. 1972. Parental investment and sexual selection. In: Sexual Selection and the Descent of Man (Ed. by B. Campbell), pp. 1871–1971.

Turner, E. & Montgomerie, R. 2002. Ovarian fluid enhances sperm movement in Arctic charr. Journal of Fish Biology 60, 1570–1579.

Tuset, V. M., Dietrich, G. J., Wojtczak, M., Slowinska, M., de Monserrat, J. & Ciereszko, A. 2008a. Relationships between morphology, motility and fertilization capacity in rainbow trout (*Oncorhynchus mykiss*) spermatozoa. Journal of Applied Ichthyology 24, 393–397.

Tuset, V. M., Trippel, E. A. & de Monserrat, J. 2008b. Sperm morphology and its influence on swimming speed in Atlantic cod. Journal of Applied Ichthyology 24, 398–405.

Uglem, I., Galloway, T. F., Rosenqvist, G. & Folstad, I. 2001. Male dimorphism, sperm traits and immunology in the corkwing wrasse (*Symphodus melops* L.). Behavioral Ecology and Sociobiology 50, 511–518.

Urbach, D., Folstad, I. & Rudolfsen, G. 2005. Effects of ovarian fluid on sperm velocity in Arctic charr (*Salvelinus alpinus*). Behavioral Ecology and Sociobiology 57, 438–444.

Van den Hurk, R. & Resink, J. W. 1992. Male reproductive system as sex pheromone producer in teleost fish. Journal of Experimental Zoology 261, 204–213.

Van Look, K. J. W., Dzyuba, B., Cliffe, A., Koldewey, H. J. & Holt, W. V. 2007. Dimorphic sperm and the unlikely route to fertilisation in the yellow seahorse. Journal of Experimental Biology 210, 432–437.

Vila, S., Munoz, M., Sabat, M. & Casadevall, M. 2007. Annual cycle of stored spermatozoa within the ovaries of *Helicolenus dactylopterus dactylopterus* (Teleostei, Scorpaenidae). Journal of Fish Biology 71, 596–609.

Vladic, T. V. & Jarvi, T. 2001. Sperm quality in the alternative reproductive tactics of Atlantic salmon: the importance of the loaded raffle mechanism. Proceedings of the Royal Society of London Series B—Biological Sciences 268, 2375–2381.

Vladic, T. V., Afzelius, B. A. & Bronnikov, G. E. 2002. Sperm quality as reflected through morphology in salmon alternative life histories. Biology of Reproduction 66, 98–105.

Warner, R. R. 1975. The adaptive significance of sequential hermaphroditism in animals. American Naturalist 109, 61–82.

Warner, R. R. 1978. The evolution of hermaphroditism and unisexuality in aquatic and terrestrial vertebrates. In: Contrasts in Behavior (Ed. by E. S. Reese & F. J. Lighter), pp. 77–101. New York: J. Wiley.

Warner, R. R. 1984. Mating behavior and hermaphroditism in coral-reef fishes. American Scientist 72, 128–136.

Warner, R. R. 1988. Sex change in fishes: hypotheses, evidence, and objections. Environmental Biology of Fishes 22, 81–90.

Warner R.R. & Harlan R.H. 1982. Sperm competition and sperm storage as determinants of sexual dimorphism in the dwarf surf-perch, *Micrometrus minimus*. Evolution 36, 44–55.

Warner, R. R., Shapiro, D. Y., Marconato, A. & Petersen, C. W. 1995. Sexual conflict: males with highest mating success convey the lowest fertilization benefits to females. Proceedings of the Royal Society of London Series B—Biological Sciences 262, 135–139.

Wedell, N., Gage, M. J. G. & Parker, G. A. 2002. Sperm competition, male prudence and sperm-limited females. Trends in Ecology & Evolution 17, 313–320.

Wickler, W. 1965a. Signal value of the genital tassel in the male *Tilapia macrochir* Blgr. (Pisces: Cichlidae). Nature 208, 595–596.

Wickler, W. 1965b. Zur Biologie und Ethologie von *Escenius bicolor* (Pisces, Teleostei, Bl.). Zeitschrift für Tierpsychologie 22, 36–49.

Wisenden, B. D. 1999. Alloparental care in fishes. Reviews in Fish Biology and Fisheries 9, 45–70.

Wojtczak, M., Dietrich, G. J., Slowinska, M., Dobosz, S., Kuzminski, H. & Ciereszko, A. 2007. Ovarian fluid pH enhances motility parameters of rainbow trout (*Oncorhynchus mykiss*) spermatozoa. Aquaculture 270, 259–264.

Wourms, J. P. 1977. Reproduction and development in chondrichthyan fishes. American Zoologist 17, 379–410.

Wourms, J. P. 1981. Viviparity: the maternal-fetal relationship in fishes. American Zoologist 21, 473–515.

Yamamoto, T. 1969. Sex differentiation. In: Fish Physiology III (Ed. by W. S. Hoar & D. J. Randall), pp. 117–175. New York, USA: Academic Press.

Yanagimachi, R., Cherr, G. N., Pillai, M. C. & Baldwin, J. D. 1992. Factors controlling sperm entry into the micropyles of salmonid and herring eggs. Development Growth & Differentiation 34, 447–461.

Zbinden, M., Mazzi, D., Kunzler, R., Largiader, C. R. & Bakker, T. C. M. 2003. Courting virtual rivals increase ejaculate size in sticklebacks (*Gasterosteus aculeatus*). Behavioral Ecology and Sociobiology 54, 205–209.

Zimmerer, E. J. & Kallmann, K. D. 1989. Genetic basis for alternative reproductive tactics in the pygmy swordtail, *Xiphophorus nigrensis*. Evolution 43, 1298–1307.

18

Evolution of Primary Sexual Characters in Amphibians

LYNNE D. HOUCK AND PAUL A. VERRELL

INTRODUCTION

In the first modern text dedicated to the biology of the Amphibia, Noble (1931) devoted an entire chapter to secondary sexual characters. We focus briefly on these secondary characters to distinguish them from our main consideration, the primary reproductive traits. Noble defined secondary traits operationally (p. 108) as "all the differences between the two sexes other than those connected with the gonads and their ducts" (note that Noble makes no mention of either natural or sexual selection). As might be expected for the Anura, Noble devotes a good deal of space to discussing sexually dimorphic traits, such as nuptial pads on thumbs, enlarged larynxes, and vocal sacs of various kinds. For both the Anura and Urodeles, Noble also mentions certain kinds of skin glands that are related to reproduction. For anurans, we now believe these glands are used in mate attraction and, in a sense, are the chemical equivalents of mating calls (e.g., Wabnitz et al. 1999; Pearl et al. 2000; see also Brizzi et al. 2002). In salamanders, specialized skin glands are related to identification of species and sex (Dawley 1984, 1986), and also to influencing female receptivity to a courting male (Houck & Reagan 1990; Houck et al. 1998; Rollmann et al. 1999). These specialized traits all have been promoted by sexual selection.

In contrast to the specializations above, Noble (1931) considered that, for all amphibians, only differences between the sexes that related to the gonads and their ducts would be classified as primary sexual characters. To quote Darwin (1871, pp. 253, 254), "Unless indeed we confine the term 'primary' to the reproductive glands, it is scarcely possible to decide which [reproductive traits] ought to be called primary and which secondary" (also see Grant 1995 for similar comments on plants). Here, following both Darwin and Noble, we restrict our discussion of amphibians primarily to gametes, gonads, oviducts, and associated structures required for ova and sperm to unite. As natural selection also acts on other reproductive traits, however, we also include observations of evolutionary change in basic reproductive characteristics. In particular, we consider mating behaviors associated with sperm transfer, physiological cycles (e.g., annual changes in levels of androgen), and female retention of fertilized ova until larvae (or metamorphosed young) are released. We consider these additional reproductive traits to be governed by natural selection (e.g., survivorship of offspring) rather than sexual selection. Also, in contrast to expectations for sexually selected traits, we expect relatively less variation among conspecific males or females in primary characters shaped by natural selection (although trait expression may vary with age or body size). Presumably, these primary traits are under strong stabilizing selection, as mentioned below for the "tail-straddling walk" that has characterized the mating behavior of plethodontid salamanders for over 15 million years.

In many areas, much more information has been published on anurans and salamanders than is available for the third group of amphibians, caecilians.

However, caecilians are well characterized in terms of the development of offspring within the oviduct (e.g., Wake 1993, 2003; Wake & Dickie 1998), and few other amphibians share this reproductive mode. We focus in particular on primary sexual traits related to the production and survivorship of offspring. In this context, we document significant diversity from a presumed common ancestral phenotype of external fertilization in water.

COMMON REPRODUCTIVE TRAITS IN AMPHIBIANS

Annual Cycles of Gametogenesis in North Temperate Amphibians

Of great importance to reproductive ability is the production of mature gametes in a timely fashion. Amphibians in the North Temperate Zone have been most studied in this respect, and, the typical timing of both the spermatogenic and oogenic cycles is surprisingly similar among diverse amphibian species (Joly 1986; Houck & Woodley 1995). This similarity apparently is related to the physiological capabilities of ectothermic animals. In essence, gamete production is a temperature-dependent process. In North Temperate areas (typically characterized by low spring temperatures), amphibians require multiple months to initiate and conclude the many steps required for the formation of mature spermatozoa and ova. Sperm production, for example, begins with the development of primary spermatogonia, which mature into secondary spermatogonia, and so on until culminating in the final stage of mature spermatozoa, typically occurring when testosterone levels are highest (Houck and Woodley 1995). In this context, the gametogenic cycle reflects dependency on external conditions (especially temperature), as well as sequential physiological changes (gradually increasing levels of sex steroids: androgens and estrogens). Thus, gametogenesis typically is initiated in early spring, culminating in mature spermatozoa or ova is summer or early fall (Joly 1986; Houck & Woodley 1995).

Although the general timing of amphibian gametogenesis follows this pattern in North Temperate areas with relatively low spring temperatures, amphibian species still can differ greatly in the timing of breeding. Some species breed in the late summer/early fall (when testosterone levels typically are high: Woodley 2007), but many other species breed in spring (when testosterone typically is at basal levels: e.g., Norris et al. 1985; Houck & Woodley 1995). These seasonal differences in mating have been categorized as either an Associated or Dissociated Reproductive Pattern (Crews 1987; Crews & Moore 1986). Salamander species that breed in the fall have an Associated Pattern in that levels of sex steroids are maximal and mature gametes are available at the same time that mating occurs. In contrast, male tree frogs that mate in the spring are doing so when testosterone levels are low, and thus these levels are dissociated with the occurrence of mating (cf .Norris et al. 1985; Houck et al. 1996). The interpretation of the Associated and Dissociated patterns is very much linked to natural selection favoring a particular timing for gametogenesis, but also favoring the timing of mating (and the resulting production of young) to occur when survivorship of offspring would be maximized. In short, selective changes in the relative timing of gametogenesis, mating, and production of offspring permit the temporal dissociation of these primary reproductive events to the benefit of increased offspring survivorship.

We consider other adaptations that promote offspring survivorship, but exclude the many examples of parental care of oviposited eggs that are known for anurans and salamanders (clearly summarized by Duellman & Trueb 1986). Instead, we focus on traits that provide protection prior to the female releasing eggs, larvae or metamorphosed juveniles into the environment. This focus deliberately restricts our attention to the primary traits most likely to be influenced by natural selection.

Larviparity and Viviparity

We adopt the term "larviparity" as defined by Greven (2003b) to indicate the female's retention of fertilized eggs until the development of larvae that are capable of living independently. In anurans and salamanders, this development typically occurs using yolk as the only food source (i.e., there is no additional maternal nutrition). In caecilians, a female often supplies additional nutrition to support larval development. In either case, when larvae are fully developed, the female returns to water and the larvae are released.

Viviparity is similar to larviparity, except that larval development is completed within the female's oviducts, and young emerge having the adult body form. Species that are viviparous are more likely to provide supplemental maternal nutrition to the developing embryos.

Most anurans rely on external fertilization that occurs at (or shortly after) oviposition. We focus here on the relatively few anuran examples in which eggs are retained within the oviducts until larvae or juveniles are released externally. In these examples, egg size often is increased while clutch size is decreased. The golden coqui frog (*Eleutherodactylus jasperi*), for example, has a maximum clutch size of about five eggs. Following internal fertilization, these eggs are retained in the oviducts (Wake 1978). While developing, the young apparently rely on egg yolk as their only food source, even throughout metamorphosis. Young that are "born" already have developed into juveniles having the appearance of the adult form. In contrast, toads in the genus *Nectophrynoides* (endemic to Tanzania) hold many fertilized eggs within the oviducts for differing developmental durations: offspring can emerge either as larvae or as fully metamorphosed individuals. In these unusual toad species, non-yolk nutrition often is provided by specialized secretions from the oviducts (Xavier 1977, 1986).

In certain salamanders that are larviparous, the uterine portion of the oviduct in certain species can provide nutrition for embryos until all have developed, either into larvae or metamorphosed young (Greven 2003b). Oviductal maintenance of developing young has been described in detail for certain salamandrids (e.g., *Pleurodeles waltl*, Boisseau 1980; *Mertensiella luschmani*, Polymeni and Greven, 1992; *Salamandra salamandra*, Greven 1977, 1998) and this topic was reviewed in detail by Greven (1998, 2003b). Of particular interest is the observation that growth rates for intrauterine young may vary during the year. Amend and Greven (1996) showed that growth rates were substantially lower in the winter than in spring (when most young complete development and are released from the oviducts). These differential growth rates apparently reflect: (a) physiological processes being negatively affected by colder temperatures, and (b) reduced food availability for the female during the colder months. Overall, of course, the female is investing substantial resources in young that are retained and nourished. This commitment to

provisioning retained larvae presumably results in more surviving offspring, but at a substantial metabolic cost to the female.

Among amphibians, few examples of viviparity occur in anurans, in part because internal fertilization is rare. In salamanders, internal fertilization is relatively common, but viviparity still is rare, occurring in less than 1% of the species. Viviparous species include the alpine salamander, *Salamandra atra*, and the Turkish salamander, *Mertensiella luschani* (reviewed by Greven 1998, 2003b). In these species, viviparity is obligatory in the sense that oviposition and larviparity do not occur. As in other viviparous species, supplementary maternal nutrition typically is provided to the developing young by the oviducts.

Larviparity and viviparity only can occur, of course, when fertilization is internal. Caecilians differ from most anurans and many salamanders in that all examined to date have internal fertilization (Wake & Dickie 1998). In fact, a male caecilian typically has an "intromittent organ" (see figure 18.1) in the sense that a portion of the male's cloaca can be everted and inserted into the female's cloaca, thus facilitating sperm transfer. Caecilians have a variety of reproductive modes, ranging from oviposition on land (with aquatic larvae subsequently wriggling into nearby water) to viviparity. Obligate viviparity is very common in three families of caecilians (Typhlonectidae, Caeciliidae, and Scolecomorphidae; Wake & Dickie 1998). The amount of maternal provisioning (in addition to yolk) varies among species (Wake & Dickie 1998). As with other viviparous amphibians, physiological stimuli (including hormones) are responsible for

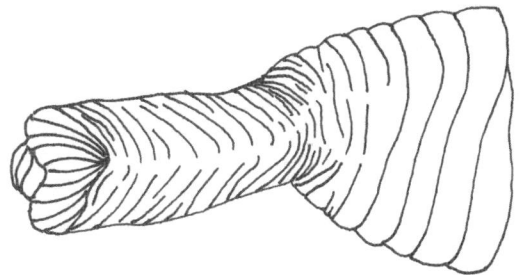

FIGURE **18.1** An intromittent organ of a caecilian, *Geotrypetes seraphini*. This "phallodeum" actually is an eversion of the cloaca. Adapted from Duellman & Trueb (1986).

major changes to the oviducts that result in nutritional provisioning.

Obligate viviparity in salamanders and caecilians is supported by adaptations that include alterations in oviduct structure and function. Another apparent adaptation is the change in egg size (Wake 2003). Eggs in these viviparous species appear to be characteristically smaller (e.g., Wake 1977) than are eggs in larviparous species that do not provide additional maternal nutrition. The difference in egg size emphasizes the necessity of substantial maternal nutrition for developing embryos of viviparous species. In fact, females that fail to acquire sufficient resources prior to the reproductive period may forgo reproducing that year.

Reproductive Changes Related to Habitat

Changes in body shape that permit access to a novel ecological niche occur with consequences for the structure of internal reproductive organs. In comparison with anurans, most salamanders have a relatively elongated torso. Among plethodontid salamanders in particular, however, the elongation of the torso has been taken to extremes in many species. The neotropical genus *Oedipina*, for example, contains salamander species that are surprisingly slender (Ehmcke et al. 2003). Not only are the torso and tail highly elongated, but limbs are reduced to the extent that these animals are referred to as "worm salamanders" (Savage 2003). The slender bodies of these species are related to the fossorial and semi-fossorial habitats of these salamanders. Similarly, the plethodontid genus *Batrachoseps*, found in California, also features worm-like forms.

Overview of Naturally Selected Traits Common to Amphibians

For anurans, salamanders and caecilians, the examples above reveal two main areas in which natural selection can significantly affect primary reproductive traits. First are the physiological changes underlying reproduction. In particular, elevated sex steroid hormone levels (associated with gametogenesis) may not be elevated during the mating season. Second is the protection of offspring by retaining fertilized ova until larval or juvenile development has occurred. We now focus on particular primary traits in the Anura and Urodela.

ANURANS

Male Anurans: Testes and Sperm.

A description of the generalized reproductive system of male anurans has been provided by Tyler (2003) and is briefly summarized here. Each of the paired testes is physically attached to a kidney. Ducts from the testes (typically vasa defferentia; in some species, vasa efferentia) enter the kidney and empty into the Wolffian ducts, which carry (at different times) both urine and sperm to the cloaca. Glands that line the Wolffian ducts are of uncertain function, but likely contribute to the fluid in which sperm are released from the cloaca during mating. As in other vertebrates, the anuran testis produces steroidal hormones (androgens) that are essential for spermatogenesis. Androgens also influence reproductive structures (e.g., the larynx used in vocalization) and sites in the brain that regulate sexual behavior. The testes, of course, also produce spermatozoa.

Sperm Competition vs. Fertilization Success

We consider now the different selective pressures on males and females when a female's clutch may have multiple sires. From the male viewpoint, strong sexual selection can occur when a male cannot monopolize access to the full complement of a female's ova. In these cases, the "parental" male (e.g., the male that is amplexing a female) is vulnerable in that he may not be able to prevent additional males from providing sperm that will fertilize at least some ova in a female's clutch. Perhaps the most thorough test for such sperm competition was provided by Vieites et al. (2004) for the common European frog, *Rana temporaria*. In the population studied by these authors, eggs were abandoned by both parents after fertilization. However, "pirate"males would clasp unattended clutches and fertilize any eggs that were inadvertently unfertilized by the "parental" male (as determined by molecular-genetic analyses of clutches collected from the field; also see Laurila & Seppa 1998; Lode et al. 2005). Furthermore, observations of a wide range of taxa indicate that multiple males may be in the vicinity of a single female when she lays her eggs (reviewed by Halliday 1998), thus providing an arena in which sperm competition may operate (also see Davis & Verrell 2005). In fact, numerous studies have demonstrated that the threat of sperm competition is a potent selective force influencing the evolution of a number of traits (testis size, number of sperm produced) that classically were considered to be primary sexual characters (e.g., see Birkhead & Moller 1998). On the other hand, if the amplexing male failed to fertilize some of the female's ova, she would benefit greatly should other males provide sperm for those ova. From this female viewpoint, the competing males may increase the potential number of offspring she produces.

Given the above example, a reproductive system in which traits are subjected to strong sexual selection may require rigorous analysis in order to reveal effects of natural selection. An excellent example is a study of the influence of various aspects of reproductive strategy on sperm size in the Australian myobatrachid frogs (Byrne et al. 2002; Byrne et al. 2003; and see Scheltinga and Jamieson 2003). Controlling for phylogenetic effects, these researchers found that the risk of sperm competition accounted for much of the variation among taxa in the length of both the sperm head and (especially) the sperm tail. However, an influence of egg size also was detected on sperm size, and the egg size itself was affected by whether eggs were laid on land or in water. Thus, while Byrne et al. (2003) revealed strong effects of sperm competition, the relationship of sperm size to egg size also was significant, and presumably reflects the influence of natural selection operating on myobatrachid sperm.

Both natural and sexual selection also may influence anuran testes size. While several studies have tested for sperm competition effects (e.g., Prado & Haddad 2003; Byrne 2004; Hettyey & Roberts 2007), fewer studies have considered potential influences in the context of natural selection. Emerson (1997) conducted a comparative analysis for frog taxa in which one male typically gained exclusive access to a female during mating, perhaps (but not necessarily) with a lowered risk of sperm competition. She found that relative testes size increased as relative clutch size increased, suggesting that natural selection favors the production of more sperm when there are more eggs to fertilize. A similar analysis was conducted by Byrne et al. (2002) for Australian myobatrachid and hylid frogs. Ranking taxa according to the probability of group spawning, these authors found that oviposition location (land vs. water) and female clutch size explained less variation in relative testes size among taxa than did the risk of sperm competition.

The anuran studies summarized above show that, in terms of sperm (form and amount) and testes (size, as related to sperm production), the considerable diversity that exists within this group begs further investigation from a functional perspective.

Male Anurans: Sperm Transfer

First proposed by Eberhard (1985), the hypothesis that intromittent organs may be subject to sexual selection has been tested and supported for a growing number of arthropod and chordate taxa having internal fertilization. The issue of selection on intromittent organs is almost moot for anurans, however, as fertilization is external in the vast majority of taxa. Presumably, the likelihood of successful fertilization is maximized by the postures adopted by males and females while mating, including (but not limited to) various forms of clasping, or amplexus, of the female by the male (a sampling of such postures is given in figure 3.25 in Duellman & Trueb 1986).

However, internal fertilization in anurans does occur in four genera of African bufonid toads, in two species within the New World genus *Eleutherodactylus* (Leptodactylidae), and in two species of the New World genus *Ascaphus* (Ascaphidae). As reviewed by Sever et al. (2003), sperm transfer in the bufonid genera and in *Eleutherodactylus* species is thought to occur by cloacal apposition: the male positions his cloaca in contact with the female's cloaca, and sperm is directly transferred. In contrast, the two *Ascaphus* species accomplish internal fertilization by transferring sperm via an erectile cloacal protuberance that the male inserts into the female's cloaca while the pair is in amplexus (see figure 18.2). In terms of internal anatomy, this *Ascaphus* structure (once believed to be a tail) bears considerable homoplastic similarity to the hemodynamic penes of amniotes. Some earlier authors reported that the epithelium of the *Ascaphus* intromittent organ bears horny spines, reminiscent of those of the hemipenes of snakes and lizards. However, no such ornamentation was observed by Sever et al. (2003), beyond some areas of thickened stratum corneum.

Stephenson and Verrell (2003) described mating in *Ascaphus montanus* in detail, and noted two points in particular. First, mating between a single male and female often occurs in multiple bouts, separated by intervals when the male releases his

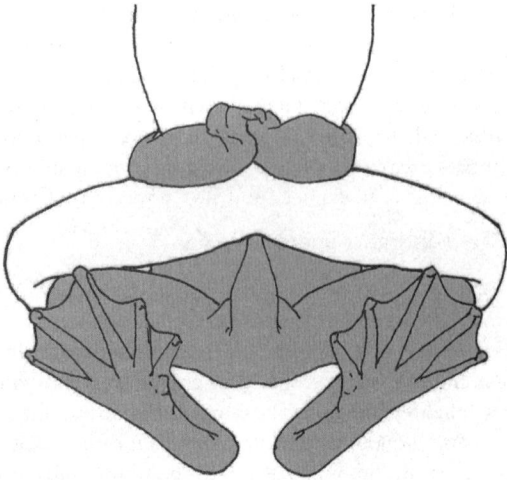

FIGURE **18.2** Ventral view of male *frog (Ascaphus truei)* inserting his everted cloacal protuberance into the cloaca of the female. Adapted from Duellman & Trueb (1986).

hold on the female. Second, a male produces slight thrusts of the pelvic region that press his cloaca closer to his partner's cloaca. Thus, although the intromittent organ of *Ascaphus* does not have much in the way of external ornamentation, the thrusting movements during intromission may increase the success of sperm transfer, making it more likely that all ova will be fertilized. In addition, the physical stimulation provided by this thrusting motion (similar to physical stimulation provided by anurans having external fertilization) may combine with other courtship stimuli to increase the female's readiness to oviposit. Moreover, the intromittent organ also may be subject to sexual selection as a courtship device in the context of cryptic female choice (Eberhard 1985; Hosken & Stockley 2004)

Female Anurans: Ovaries and Ova

Each of the paired ovaries typical of a female anuran lies close to a kidney. Eggs released from an ovary are carried to the cloaca via the oviduct, a convoluted tube that may show substantial regional specialization of form (reviewed by Tyler 2003). Glands within the oviduct secrete substances that form the jelly layer(s) surrounding each ovum (e.g., Greven 2003a). Most commonly, ova are released into an aquatic environment where each

is fertilized. A different reproductive mode, larviparity, has been documented in anurans. Larviparity occurs when (a) sperm are introduced into the female's cloaca, (b) the sperm move into the oviducts and fertilize available ova, and (c) the resulting eggs are retained within the oviduct until embryonic development is completed. At the appropriate time and place, larvae move down the oviduct and through the cloaca, typically into (or very near) a body of water. In these cases, parts of the oviduct may be specialized for retention of developing embryos (see Wake 1993; Wake & Dickie 1998), but no maternal nutrition (beyond the initial yolk) is provided.

Female Anurans: Sperm Storage Tubules

Among anurans that have internal fertilization, only in the western tailed frog (*Ascaphus truei*) are females known to have distinct sperm storage tubules. These tubules are situated in the lower parts of the oviducts, just distal to where jelly coats are applied to descending eggs (Sever et al. 2001). Sperm storage by anuran females certainly may promote sperm competition. The extent of sperm competition may depend on the relative number of sires, the order in which each delivers sperm, and the amount of sperm each sire delivers. Whether sperm competition occurs in *Ascaphus* is not known. In any event, mating with multiple males (if that occurs) may assure the female of sufficient sperm for all ova to be fertilized. Coupled with sperm storage, such multiple mating also may provide females with an opportunity to regulate the paternity of their offspring as a form of cryptic female mate choice.

SALAMANDERS

In terms of reproduction, salamanders are generally similar to anurans in reproductive structures and in their response to endocrine stimulation (Duellman & Trueb 1986; Kikuyama et al. 2003). However, one very basic difference between these two amphibian groups is that anurans produce and retain mature spermatozoa within the testes, while salamanders evacuate mature spermatozoa to the ducts (vasa deferentia) leading from the testes to the cloaca. Thus, the paired vasa are the storage sites in which mature sperm are retained (and maintained) in male salamanders until (and during) the mating season. In some salamanders, mature sperm continue to be evacuated from testicular lobules during the mating season (e.g., in smooth newts, *Triturus vulgaris*, and in long-toed salamanders, *Ambystoma macrodactylum*: Verrell 2004; Verrell et al. 1986). In other species, even when the testes are completely regressed and sperm production terminated, the male still can respond to females by producing viable spermatophores that contain mature spermatozoa transferred from the vasa (e.g., in the marbled salamander, *Ambystoma opacum*; Houck, personal observations).

Salamanders also differ significantly from anurans in two particular aspects of their reproductive patterns. First, most salamander species have internal fertilization, even though the male lacks an intromittent organ. In a typical mating (whether on land or in water), the male delivers spermatozoa to the female via a spermatophore(s) that he deposits near the female. A spermatophore is composed of a gelatinous base that supports an apical cap of sperm. The female positions herself over the spermatophore, lowers her cloaca so that most or all of the spermatophore is within her cloaca, and then lifts up, retaining only the sperm cap in her cloaca (the base remains on the substrate). Second, salamanders differ from anurans in that females of many species have well developed sperm storage organs (spermathecae) in which sperm can be maintained until needed for fertilization of their ova. In some cases, fertilization can occur weeks or even months after insemination. Moreover, fertilization can occur without the male being present, making it possible for the female to locate a secure oviposition site.

Male salamanders: Testes and Sperm

The testes of a male salamander are elongated structures that lie in parallel with the kidneys (mesonephros) (Norris et al. 1985; Üribe 2001). Each testis is composed of lobules within which spermatogenesis occurs. Spermatogenesis is first completed in the most cephalic lobules and, lastly, in the caudal lobules, thus representing a cephalocaudal 'wave' of sperm development along the testis (Houck 1977a). In some species (e.g., the mushroomtongue plethodontid *Bolitoglossa rostrata*),

older males have more lobules than do younger adults (Houck 1977b) and thus are able to produce relatively more sperm. This ability may be reproductively advantageous if more sperm are allocated per spermatophore or if more spermatophores can be produced. This advantage would be age-specific, of course, and not a specialized trait that only some males could exhibit.

Male Salamanders: Spermiation

Mature spermatozoa are released into intratesticular ducts and then into the ducts (vasa deferentia) that each lie in parallel with a testis (cf. Armstrong 1989; Norris 1987). Spermatozoa can remain in these ducts for weeks or months, and later be transferred into the cloaca to be incorporated into a spermatophore.

An important question concerning salamanders that have lengthy mating seasons (~3–9 months) is whether or not a male's supply of spermatozoa could be depleted before the next annual spermatogenic cycle has been completed. In the ocoee salamander (*Desmognathus ocoee*, Plethodontidae), for example, courtship encounters staged in the laboratory revealed that mating behavior and sperm transfer can occur during every month of the year (L. Houck & S. Arnold, personal observations). In the field, a male *D. ocoee* from Macon Co., North Carolina, has a single annual cycle of spermatogenesis, with mature sperm stored in vasa deferentia and available for spermatophores from late summer through the end of the following July. Mating occurs in fall until cold temperatures cause salamanders to retreat underground in late September or October. Mating also may occur underground, but certainly occurs after the animals emerge (usually some time in April or early May), and continues until females oviposit in late July. The cyclical pattern of sperm production, combined with all-year mating presumably reflects the continual availability of sperm stored in the vasa. At the same time, the viable storage of sperm is not likely to exceed one year (Sever 2003). Thus, male *D. ocoee* that mate in the late spring probably retain enough viable sperm from the prior spermatogeneic cycle to provide a female with a typical spermatophore. This supposition could be tested by histological investigation of sperm levels in a male's vasa during June, or by comparing the amount of sperm in the sperm masses produced by males collected in late September versus those collected in June.

Male salamanders: Cloacal Glands and Spermatophores

In salamanders that have internal fertilization, a male may produce one or many spermatophores (e.g., Arnold 1976) to transfer sperm to the female during a mating encounter. The sperm cap is a combination of (a) spermatozoa released from the male's vasa, and (b) cloacal gland secretions that contain (and perhaps maintain) the spermatozoa. The base and sperm cap are produced by secretions from distinct glands within the cloaca (see Sever 1994a, 2003 for details); these glands respond to a seasonal increase in androgen levels by becoming enlarged and secretory (Norris 1987; Norris and Moore 1975, Kikuyama et al. 2000; Sever 1994b).

When a receptive female encounters a spermatophore, she uses tactile cues to position herself, then lowers her cloaca over the spermatophore and lodges the sperm mass within her cloaca. When the female is ready, she lifts up (leaving the base on the substrate). At this point, the female either departs (which is typical of plethodontid females) or she will attempt to obtain additional sperm by repeating this process with new spermatophores (ambystomatid and salamandrid females) (see http://oregonstate.edu/~arnoldst/shermani%20 transter.avi to observe sperm transfer in the red-legged salamander *Plethodon shermani*).

Males of most species are capable of producing multiple spermatophores during a given night of mating. The actual rate of spermatophore production, however, is related to the basic reproductive pattern for each species, as summarized below.

Male Salamanders: Sperm Transfer

Transfer Directly to Ova (External Fertilization)

The three most basal salamander families (Hynobiidae, Cryptobranchidae, and Sirenidae) all have external fertilization, strongly suggesting that this is the ancestral condition for urodeles (Houck & Arnold 2003; but see Selmi et al. 1997). Among these families, sperm transfer is known best for hynobiids. In pond-dwelling species, a breeding male focuses on the two egg sacs released by a female (one sac from each ovary; sacs are joined at the caudal end). A male takes hold of the egg sacs and arches his body to release sperm (Sasaski 1924; Hasumi 1994). Rival males may be attracted to this action, and also may attempt to fertilize ova (Park and Park 2000). This general scenario of external fertilization is similar to that described for many anurans, except that the male salamander only clutches the egg sacs, and does not clasp the female.

Transfer Via Spermatophores

After the female lodges the sperm in her cloaca (described above), the sperm move out of the cap (Zalisko et al. 1984; Sever 2003) and migrate to the female's spermatheca. Sperm are stored for durations varying from minutes to months (discussed below). Spermatophores are used for sperm transfer by salamander species that have aquatic mating, as well as by species that have terrestrial mating.

Transfer Via Cloacal Apposition

Sperm transferred to the female via cloacal apposition is highly unusual in salamanders. In fact, only one salamandrid genus (*Euproctus*) features the three species in which a male uses cloacal apposition. In these species of *Euproctus* (*asper, platycephalus,* and *montanus*), courtship begins when the male captures the female by holding her in his jaws and with his tail (see figure 10.6 in Houck & Arnold 2003). While completely restraining the female, the male transfers sperm to her by pressing his spermatophore(s) into the female's cloaca (Ahrenfeldt 1955; Houck & Arnold 2003). We include this version of sperm transfer simply to present the complete range in salamanders. Field observations (L. Houck & S. Arnold, unpublished data) suggest, however, that the male *Euproctus* behavior is strongly related to male–male competition for a mate (and thus sexual selection): by monopolizing a female via physical restraint, the dominating male would achieve greater reproductive success. This behavior also represents a possible example of sexual conflict to the extent that the female is forced to mate with an undesired male.

Female Salamanders: Ovaries, Oviducts and Ova

We provide here a brief summary of the basic structure and function of the female reproductive tract. In a separate section below, we consider specializations of the oviducts that promote larviparity and viviparity (as defined earlier).

Ovaries

The basic structure of salamander ovaries is very similar to that of anurans (Lofts 1984). The paired ovaries of salamanders are elongated structures situated along the dorsal wall of the body cavity. The sac-like structure of each ovary includes the germinal epithelium, which produces oocytes during the reproductive season (Dodd 1977; Tokarz 1978). As in other vertebrates, the urodele ovary also functions to secrete sex steroids (e.g., estrogens) necessary for (a) the production of mature oocytes, and (b) the release of these oocytes (ovulation) that allows their transfer into the oviducts. Shortly before ovulation, the ova are maximally yolked and the ovaries have expanded to occupy most of the space in the female's abdominal cavity (Lofts 1984).

Oviducts

The oviducts are derived from the Müllerian ducts in response to ovarian estrogens (e.g., Norris et al. 1997). Oviducts are thin strand-like structures in juvenile females. In reproductive females, each oviduct becomes enlarged and convoluted in concert with the endocrine support that underlies enlargement (yolking) of the ova (e.g., Adams 1940; and see figure 5 in Greven 2003a). While transporting the mature oocytes to the cloaca, the oviducts also secrete a gelatinous substance ("egg jelly") that surrounds each ovulated oocyte. Multiple jelly layers come to surround an oocyte as it travels down the oviduct (Greven 2003a). In fact, an ovum cannot be fertilized if it is not surrounded by gelatinous coats that have sperm-binding properties (e.g., Watanabe & Onitake 2003). Generic differences in the amount of gelatinous material surrounding the ova have been described (e.g., Ambystoma has thicker jelly layer; Triturus has thinner layer; Greven 2003a). However, these and other differences are not related to reproductive mode, egg size, or oviposition site (water vs. land) as was proposed by Salthe (1963; and see Salthe & Mecham 1974). Thus, functions of the gelatinous layers include: (a) activation of sperm, (b) preventing egg desiccation (Marco & Blaustein 1998), and (c) reducing predation on eggs (Hardy & Lucas 1991), all directly related to basic reproduction and survival.

Among salamander genera, oviducts vary in the degree of convolution. In mole salamanders (many

Ambystoma spp.), for example, large convolutions of the oviducts span most of the width of the body cavity. In contrast, the oviduct of a smooth newt (Triturus vulgaris) is thinner, more tightly convoluted, and does not span as much of the body width. In another case, adult females of a neotropical worm salamander (Oedipina uniformis) have comparatively straight oviducts, thus reflecting the extreme slenderness of this fossorial species (Ehmcke et al. 2003). Changes in body shape that permit a salamander to access a novel ecological niche therefore have consequences for the structure of internal reproductive organs.

Ova

The development of oogonia into mature ova is a process that can take up to three reproductive seasons (Norris 1997). Therefore the ovary may contain oogonia that are pre-vitellogenic (lacking yolk), as well as oogonia that are vitellogenic and will become pre-ovulatory oocytes (Dodd 1977; Ûribe 2001). For species that lay eggs, only the oocytes that continue to develop into mature ova comprise the female's clutch.

In some cases, clutch size apparently has been shaped by natural selection. Consider, first, certain plethodontid species that are not fossorial or arboreal, and that occur either in the neotropics (e.g., Bolitoglossa rostrata) or North America (e.g., large eastern Plethodon spp.). Neotropical species have average clutch sizes that are significantly larger than are clutches of North American species of comparable body size (Houck 1977a). However, eggs of neotropical species are smaller on average than are those of North American species (ibid.). Differences in clutch volume and egg size presumably are linked to geographic differences in survivorship of young: the "fewer-but-larger" strategy may be effective in North America, as opposed to the derived (apomorphic) condition of "more-but-smaller" eggs found in neotropical species.

Another example within the Plethodontidae also illustrates the likely action of natural selection. In many plethodontid species, clutch size appears not to change once a set of ova are substantially yolked. In other words, after the ovulation and oviposition of yolky ova, any oocytes remaining in the ovary are substantially smaller, even if they appear to have at least some yolk platelets (Houck, personal observations). In contrast, a more flexible clutch size apparently occurs in the California slender salamander (*Batrachoseps attenuatus*). As the common name suggests, these salamanders are relatively worm-like and have tiny limbs. In specimens preserved prior to ovulation, clutch size can be assessed accurately by counting the number of distinctly enlarged yolky ova within the ovaries. However, a careful scrutiny of clutches in this species revealed that—in response to an unpredicted increase in food availability (hence increased food consumption)— an individual female may augment her clutch by secondarily yolking an additional ovum (Maiorana 1976). The female's ability to increase her clutch size in this manner apparently reflects flexibility of reproductive productivity in a fluctuating environment.

Female Salamanders: Sperm Storage Organs

In all salamander species having internal fertilization, glands in the dorsal portion of a female's cloaca form spermathecal tubules that store spermatozoa (Houck and Sever 1994). Among species, these tubules range from simple (e.g., in the one-toed Amphiuma, *Amphiuma pholeter*) to complex (e.g., in the ocoee salamander, *D. ocoee* and other terrestrial plethodontids (see figure 9.17 in Sever 2003). These spermathecae are not found in any other vertebrates (Sever 1994a), and thus confer uncommon benefits to female salamanders. These benefits have been discussed in detail (Sever and Brizzi 1998; Brizzi et al. 1889), and we describe only one example here.

One of the primary advantages to a female storing sperm is that mating can be uncoupled from oviposition. In species of large eastern *Plethodon* (e.g., the red-legged salamander, *P. shermani*) from the Appalachian Mountains in eastern USA, for example, mating typically occurs from the end of July until these salamanders retreat from the cold by moving underground in late September or October. In *P. shermani*, pairs kept in the laboratory at normal mating temperatures (~15–18°C) will court at least through early December (L. Houck, personal observations). The timing of natural oviposition for these salamanders is not known: the females move deep underground to oviposit and the discovery of clutches is extremely rare. In the laboratory, however, we have treated female *P. shermani* with injections of hormones typically associated with ova maturation and ovulation (FSH, HCG; cf. Crespi & Lessig 2004). This treatment usually will induce a gravid female to fertilize and oviposit a complete clutch in early January (Houck, personal observations), thus suggesting that early January may be the usual time that *P. shermani* oviposit. The main point here is that a female has time to locate a sequestered site in which to oviposit and then brood her eggs, with brooding being a multi-month process. This selected site most likely provides significant refuge from egg-eating predators, and so increases the female's chance of producing offspring. In addition, the duration of brooding is directly correlated with the time when young will hatch (each appearing as a miniature adult) and can emerge above ground. We first observe young *P. shermani* at the surface in May, a time when food is typically available. Given these circumstances, we consider the spermatheca to be a primary reproductive trait: sperm storage is essential for fertilization to occur at times that maximize the survivorship of the resulting offspring.

Female Salamanders: Oviposition

The ancestral mode of producing offspring is by oviposition, which is accomplished either by (a) external fertilization: releasing ova that are fertilized only when outside the female's body, or (b) internal fertilization: sperm that a male transferred to the female are used to fertilize ova that have reached the female's cloaca. During oviposition, ova typically are connected with gelatinous strands produced by the oviducts (one exception is the northern roughskin newt, *Taricha granulosa*, in which a female oviposits a single egg at a time). A departure from the typical "strand of eggs" pattern is found in certain hynobiid species (Houck and Arnold 2003). Instead of having "strands" of eggs emerging from the oviducts, the hynobiid female has an oviductal specialization: the caudal area of the oviducts are joined to form a "uterine" section. In this uterine area, ova from both ovaries are present in a two-pronged sac (having the shape of a tuning fork). This sac is extruded from the female (often with "help" as the male amplexes the sac). Keeping the ova together within this sac may increase the chances that all ova will be fertilized (a benefit to the female), and perhaps provides a benefit to the amplexing male in terms of temporary safety from rival males competing for access to egg sacs (Hasumi 1994).

CONCLUSIONS AND FUTURE RESEARCH

For all three amphibian groups, the effects of natural selection on primary reproductive traits is evident. One main theme that emerges is the evolutionary promotion of traits that increase the survivorship of offspring. Three events in particular represent substantial advances over the presumed ancestral reproductive mode in which females produced offspring by oviposition and external fertilization.

The first advance is internal fertilization. Although this trait is rare in anurans, caecilians have developed a cloacal protuberance (phallodeum, figure 18.1) that facilitates internal fertilization. Salamanders also have achieved internal fertilization mainly through sperm transfer via the male production of spermatophores (but, in a few species, by cloacal apposition). These structures permit internal fertilization in the absence of any cloacal modifications related to a phallus. Sperm transferred to the cloaca increases the chances that the male will fertilize at least some ova in a female's clutch (Houck et al. 1985).

Second, the development of a spermatheca (to store sperm until ovulation) is a key innovation that allowed the female to decouple the presence of the male with fertilization of her ova. Consequently, a female could select a more secretive oviposition site that—*sans* male—could provide increased protection for the developing embryos. The female also could adjust the timing of oviposition (even weeks or months after insemination) so that hatchlings emerged at favorable times.

The third advance in reproductive mode is represented by a variety of specializations in the oviducts that permit the female to retain fertilized ova. This retention provides security and nutrition to developing embryos, larvae, and metamorphosing juveniles. The caecilians, in particular, are exceptional in this regard.

By restricting ourselves to primary reproductive traits, however, we focused only on aspects from gametogenesis, to fertilization, to a female releasing ova, eggs, or offspring into the environment. Thus, we have excluded other examples of embryonic protection, such as oviposited eggs that develop into larvae in a pouch on the female's back (e.g., del Pino 1980), or male care of eggs that develop into larvae on his back or hind limbs (Márquez 1993;

and see Duellman & Trueb 1986 for other examples). We also excluded the important evolutionary event of direct development: terrestrial oviposition of eggs, with each embryo completing all development within the egg capsule and emerging as a miniature adult. This mode of development eliminates the phase of free-living aquatic larvae, a stage at which offspring are highly vulnerable to predation (Bruce 2003).

Another excluded trait is the evolution of mating behavior, including long-term stasis in male–female behaviors that promote sperm transfer. In plethodontid salamanders, for example, tail-straddling walk occurs when the female follows the male, straddling his tail and with her chin on the dorsal base of the male's tail. The pair moves forward in tandem, and then halts while the male deposits a spermatophore. When deposition is completed, the pair move forward together until the female's cloaca is over the spermatophore. She then lowers her cloaca over the spermatophore, lodging the apical sperm cap in her cloaca (Arnold 1977). This male–female behavior has existed for millions of years (see Houck & Arnold 2003). Under a broader view of reproduction, one might construe the behavioral component of sperm transfer to be a primary reproductive trait.

We now appreciate that selection acting on primary characters can be complex, and may be masked by effects of sexual selection. Ironically, research on sexual selection may provide more evidence of the action of natural selection (as in Byrne et al. 2003). While natural selection favors sperm able to swim through a fluid medium to reach an egg, more complex pressures may act on sperm morphology, size, and number: the Pitkin et al. (2003) study of *Drosophila* is classic in this regard. In short, some actions of natural selection may only be revealed by thorough studies that also investigate sexual selection.

Two other areas of research affecting primary traits in amphibian reproduction continue to be neglected, despite earlier pleas for additional work. First, effects of endocrine and other physiological measures related to primary reproductive patterns are not typically investigated (but see Houck & Woodley 1995; Greven 2003b; Kikuyama et al. 2003; Wake 2003). We do know that egg size can be related to different modes of reproduction: eggs are larger in species having direct development, relative to species that provide maternal nutrition to the offspring developing in the oviducts (Wake 2003). However, within the group of plethodontid salamanders that have direct development, we do not know why the neotropical species have significantly smaller eggs than do their similarly sized counterparts in North America (Houck 1977a, b). This difference suggests another area where research is needed: tropical species of amphibians, particularly salamanders and anurans. Much work has been devoted to North Temperate species; with the exception of in-depth studies of caecilians by Marvalee Wake and colleagues, comparatively few studies have addressed species that experience very different environmental regimes. The focus on tropical amphibians for investigations of the mechanistic bases for primary reproductive traits is an approach likely to yield surprising information.

Acknowledgments We appreciate the assistance of Sarah L. Eddy in preparing the manuscript. This work was supported in part by grants from the National Science Foundation: IOS-0818554 to L.D. Houck and IOS-0818649 to R.C. Feldhoff.

REFERENCES

Adams, A. E. 1940. Sexual conditions in *Triturus viridescens*. III. The reproductive cycle of the adult aquatic form of both sexes. Amer. J. Anatomy 66, 235–271.

Ahrenfeldt, R. J. 1955. Mating behaviour of *Euproctus asper* in captivity. Br. J. Herpetol. 2, 194–197.

Amend, R & Greven, H. 1996. Zur Ossifikation des Skeletts intra- und extrauteriner Larven des Feuersalamanders *Salamandra salamandra* (Amphibia, Urodela). Abhandlungen und Berichte fur Naturkunde, Magdeburg 19, 31–67.

Armstrong, J. B. 1989. Spermatogenesis. In: Developmental Biology of the Axolotl (Ed. by J. B. Armstrong and G. M. Malacinski), pp. 36–41. New York: Columbia University Press.

Arnold, S. J. 1976. Sexual behavior, sexual interference and sexual defense in the salamanders *Ambystoma maculatum, Ambystoma tigrinum*, and *Plethodon jordani*. Zeitschrift fur Tierpsychologie. 42: 247–300.

Arnold, S. J. 1977. The courtship behavior of North American salamanders with some comments on Old World salamandrids. In: The Reproductive Biology of Amphibians (Ed. by D. H. Taylor & S. I. Guttman), pp. 141–183. New York: Plenum Press.

Birkhead, T. R. & Møller, A. P., Eds. 1998. Sperm Competition and Sexual Selection. New York: Academic Press.

Boisseau, C. 1980. Étude ultrastructurale de l'oviducte du triton *Pleurodels waltlii* Michah. V. Ultrastructure et cytochimie de l'oviducte postérieur et de l' »utérus » de la femelle adulte. Ann. Sci.. Nat. Zool. 2, 67–89.

Brizzi, R., Delfino, G., Calloni, C. 1989. Cloacal anatomy in the spectacled salamander, *Salamandrina terdigitata* (Amphibia: Salamandridae). Herpetologica 45, 310–322.

Brizzi, R., Delfino, G. & Pellegrini, R. 2002. Specialized mucous glands and their possible adaptive roles in males of some species of *Rana* (Amphibia, Anura). J. Morphol. 254, 328–341.

Bruce, R. C. 2003. Life histories. In: Reproductive Biology and Phylogeny of Urodela (Ed. by D. M. Sever), pp. 477–525. Volume 1 of Series: Reproductive Biology and Phylogeny (Series Ed. by B.G.M. Jamieson). New Hampshire: Science Publishers.

Byrne, P. G. 2004. Male sperm expenditure under sperm competition risk and intensity in quacking frogs. Behav. Ecol. 15, 857–863.

Byrne, P. G., Roberts, J. D. & Simmons, L. W. 2002. Sperm competition selects for increased testes mass in Australian frogs. J. Evol. Biol. 15, 347–355.

Byrne, P. G., Simmons, L. W. & Roberts, J. D. 2003. Sperm competition and the evolution of gamete morphology in frogs. Proc. R. Soc. Lond. B 270, 2079–2086.

Crespi, E. J. & Lessig, H. 2004. Mothers influence offspring size through post-oviposition maternal effects in the redbacked salamander, *Plethodon cinereus*. Oecologia 138, 306–311.

Crews, D. 1987. Diversity and evolution of behavioral controlling mechanisms. In: Psychobiology of Reproductive Behavior: An Evolutionary Perspective (Ed. by D. Crews), pp. 88–119. New Jersey: Prentice-Hall.

Crews, D. & Moore, M. C. 1986. Evolution of mechanisms controlling mating behavior. Science 231, 121–125.

Darwin, C. 1871. The Descent of Man, and Selection in Relation to Sex. London: Murray.

Davis, A. B. & Verrell, P. A. 2005. Demography and reproductive ecology of the Columbia spotted frog (*Rana luteiventris*) across the Palouse. Can. J. Zool. 83, 702–711.

Dawley, E. M. 1984. Recognition of individual sex and species odours by salamanders of the *Plethodon glutinosus – Plethodon jordani* complex. Animal Behav. 32, 353–361.

Dawley, E. M. 1986. Evolution of chemical signals as a premating isolating mechanism in a complex of terrestrial salamanders. In: Chemical Signals in Vertebrates, Vol 4 (Ed. by D. Duvall, D. Muller-Schwarze and R. M. Silverstein), pp. 221–224. New York: Plenum Press.

del Pino, E. M. 1980. Morphology of the pouch and incubatory integument in marsupial frogs (Hylidae). Copeia, 1980, 10–17.

Dodd, J. M. 1977. The structure of the ovary of nonmammalian vertebrates. In: The Ovary, Volume 1, Second Edition (Ed. by S. Zuckerman & B. J. Weir), pp. 117–264. New York: Academic Press.

Duellman, W. E. & Trueb, L. 1986. Biology of Amphibians. New York: McGraw-Hill.

Eberhard, W. G. 1985. Sexual Selection and Animal Genitalia. Massachusetts: Harvard University Press.

Ehmcke, J. Clemen, G. & Greven, H. 2003. Structural diversity of secretory products in the glandular parts of the oviduce of two plethodontid salamanders (Amphibia, Urodela). Folia Zool. 52(2), 203–211.

Emerson, S. B. 1997. Testis size variation in frogs: testing the alternatives. Behav. Ecol. Sociobiol. 41, 227–235.

Grant, V. 1995. Sexual selection in plants: pros and cons. Proc. Natl. Acad. Sci., USA 92, 1247–1250.

Greven, H. 1977. Comparative ultrastructural investigations of the uterine epithelium in the viviparous *Salamandra atra* Laur. and the ovoviviparous Salamandra salamandra (L.) (Amphibia, Urodela). Cell Tissue Res. 181, 215–237.

Greven, H. 1998. Survey of the oviduct of salamandrids with special reference to the viviparous species. Jour. Exper. Zool., 282, 507–525.

Greven, H. 2003a. Oviduct and egg-jelly. In: Reproductive Biology and Phylogeny of Urodela (Ed. by D. M. Sever), pp. 151–181. Volume 1 of Series: Reproductive Biology and Phylogeny (Series Ed. by B.G.M. Jamieson). New Hampshire: Science Publishers.

Greven, H. 2003b. Larviparity and pueriparity. In: Reproductive Biology and Phylogeny of Urodela (Ed. by D. M. Sever), pp. 447–475. Volume 1 of Series: Reproductive Biology and Phylogeny (Series Ed. by B.G.M. Jamieson). New Hampshire: Science Publishers.

Halliday, T. 1998. Sperm competition in amphibians. In: Sperm Competition and Sexual Selection (Ed. by T. R. Birkhead & A. P. Möller), pp. 463–502. New York: Academic Press.

Hardy, L. M. & Lucas, M. C. 1991. A crystalline protein is responsible for dimorphic egg jellies in the spotted salamander, *Ambystoma maculatum* (Shaw) (Caudata: Ambystomatidae). Compar. Biochem. and Physiol.100A, 653–660

Hasumi, M. 1994. Reproductive behavior of the salamander *Hynobius nigrescens*: monopoly of egg sacs during scramble competition. J. of Herpetol., 28, 264–267.

Hettyey, A. & Roberts, J. D. 2007. Sperm traits in the quacking frog (*Crinia georgiana*), a species with plastic alternative mating tactics. Behav. Ecol. Sociobiol. 61, 1303–1310.

Hosken, D. J. & Stockley, P. 2004. Sexual selection and genital evolution. Trends Ecol. Evol. 19, 87–93.

Houck, L. D. 1977a. Life history patterns and reproductive biology of neotropical salamanders. In: Reproductive Biology of Amphibians (Ed. by D. H. Taylor & S. I. Guttman), pp. 43–72. New York: Plenum Press.

Houck, L. D. 1977b. The reproductive biology of a neotropical salamander, *Bolitoglossa rostrata* (Urodela: Plethodontidae). Copeia 1977, 70–83.

Houck, L.D., Arnold, S.J. and Tilley, S.G. 1985. Sperm competition in a plethodontid salamander: preliminary results. J. Herpetol. 19: 420–423.

Houck, L. D. & Arnold, S. J. 2003. Courtship and mating behavior. In: Reproductive Biology and Phylogeny of Urodela (Ed. by D. M. Sever), pp. 383–424. Volume 1 of Series: Reproductive Biology and Phylogeny (Series Ed. by B.G.M. Jamieson). New Hampshire: Science Publishers

Houck, L. D. & Reagan, N. L. 1990. Male courtship pheromones increase female receptivity in a plethodontid salamander. Animal Behav. 39, 729–734.

Houck, L. D. & Sever, D.M. 1994. The role of the skin in reproduction and behavior.

In: Amphibian Biology, Vol. 1, The Integument (Ed. by H. Heatwole & G. Barthalamus), pp. 351–381. Australia: Surrey Beatty and Sons.

Houck, L. D. & Woodley, S. K. 1995. Field studeis of steroid hormones and male reproductive behaviour in amphibians. In: Amphibian Biology, Vol. 2, Social Behaviour. (Ed. by H. Heatwole & B. Sullivan), pp. 677–703. Australia: Surrey Beatty and Sons.

Houck, L. D., Mendonça, M. T., Lynch, T. K, and Scott, D. E. 1996. Courtship behavior and plasma levels of androgens and corticosterone in male marbled salamanders, *Ambystoma opacum* (Ambystomatidae). Gen. & Comp. Endocrinol. 104, 243–252.

Houck, L. D., Bell, A. M., Reagan-Wallin, N. L. & Feldhoff, R. C. 1998. Effects of experimental delivery of male courtship pheromones on the timing of courtship in a terrestrial salamander, *Plethodon jordani* (Caudata: Plethodontidae). Copeia 1998(1):214–219.

Joly, J. 1986. La reproduction de la salamandre terrestre (*Salamandra salamandra* L.). In: Traité de Zoologie Amphibiens, Vol. 14 (Ed. by P. P. Grasse & M. Delsol), pp. 471–486. Paris: Masson.

Kikuyama, S., Yazawa, T., Abe, S., Yamamoto, K., Iwata, T., Hoshi, K., Hasunuma, I., Mosconi, G. & Polzonetti-Magni, A. M. 2000. Newt prolactin and its involvement in reproduction. Canad. J. Pharmacol. 78, 984–993.

Kikuyama, S., Tanaka, S. & Moore, F. 2003. Endocrinology of reproduction. In: Reproductive Biology and Phylogeny of Urodela (Ed. by D. M. Sever), pp. 275–321. Volume 1 of Series: Reproductive Biology and Phylogeny. B.G.M. Jamieson (Series Ed.). New Hampshire:Science Publishers.

Laurila, A. & Seppa, P. 1998. Multiple paternity in the common frog (*Rana temporaria*): genetic evidence from tadpole kin groups. Biol. J. Linn. Soc. 63, 221–232.

Lode, T., Holveck, M.-J. & Lesbarreres, D. 2005. Asynchronous arrival pattern, operational sex ratio and occurrence of multiple paternities in a territorial breeding anuran, *Rana dalmatina*. Biol. J. Linn. Soc. 86, 191–200.

Lofts, B. 1984. Amphibians. In: Marshall's Physiology of Reproduction. Vol. 1 Reproductive Cycles of Vertebrates. (Ed. by G. E. Lamming), pp. 127–205. Edinburgh: Churchill Livingstone.

Maiorana, V.C. 1976. Size and environmental predictability for salamanders. Evolution 30, 599–613.

Marco, A. & Blaustein, A. R. 1998. Egg gelatinous matrix protects Ambystoma gracile embryos from prolonged exposure to air. Herpetol. J. 8, 207–211.

Márquez, R. 1993. Male reproductive success in two midwife toads, *Alytes obstetricans* and *A. cisternasii*. Behav. Ecol. and Sociobiol. 32, 283–291.

Noble, G. K. 1931. The Biology of the Amphibia. New York: McGraw-Hill.

Norris, D.O. 1987. Regulation of male gonaducts and sex accessory structures. In: Hormones and Reproduction in Fishes, Amphibians and Reptile (Ed. by D. O. Norris & R.E. Jones), pp. 327–354. New York: Plenum Press.

Norris, D.O. & Moore, F. L. 1975. Antagonism of testosterone-induced cloacal development by estradiol-17 in immature larval tiger salamanders (*Ambystoma tigrinum*). Herpetologica 31, 255–263.

Norris, D.O., Norman, M. F., Pankak, M. K. & Duval, D. 1985. Seasonal variation in spermatogenesis, testicular weights, vasa deferentia and androgen levels in neotenic tiger salamanders, *Ambystoma tigrinum*. Gen. and Compar. Endocrino. 60, 51–57.

Norris, D.O., Carr, J. A., Summers, C. H. & Featherston, R. 1997. Interactions of androgens and estradiol on sex accessory ducts of larval tiger salamanders, *Ambystoma tigrinum*. Gen. and Compar. Endocrinol. 106, 348–355.

Park, D. & Park, S. R. 2000. Multiple insemination and reproductive biology of *Hynobius leechii*. J. Herpetol. 34, 594–598.

Pearl, C. A., Cervantes, M., Chan, M., Ho, U., Shoji, R. & Thomas, E. O. 2000. Evidence for a mate-attracting chemosignal in the dwarf African clawed frog *Hymenochirus*. Horm. Behav. 38: 67–74.

Pitkin, S., Miller, G. T., Schneider, K. & Markow, T. A. 2003. Ejaculate–female coevolution in *Drosophila mojavensis*. Proc. R. Soc. Lond. B 270, 1507–1512.

Polymeni, R.M. & Greven, H. 1992. Histology and fine. structure of the oviduct in

Mertensiella luschani (Stein-. dachner, 1891). J. Exptl. Zoology, Part A, Exper. Compar. Biol. 282, 507–525.

Prado, C. P. de A. & Haddad, C. F. B. 2003. Testes size in leptodactylid frogs and occurrence of multimale spawning in the genus *Leptodactylus* in Brazil. J. Herpetol. 37, 354–362.

Rollmann, S. M., Houck, L. D., and Feldhoff, R. C. 1999. Proteinaceous pheromone affecting female receptivity in a terrestrial salamander. Science 285, 1907–1909.

Salthe, S. N. 1963. The egg capsules in the Amphibia. J .Morphol. 113, 161–171.

Salthe, S.N. & Mecham, J. S. 1974. Reproductive and courtship patterns. In: Physiology of the Amphibia. Vol. II. (Ed by B. Lofts), pp. 309–521. New York: Academic Press.

Sasaki, M. 1924. On a Japanese salamander in Lake Kuttarush which propagates like the axolotl. Journal of the College Agriculture Hokkaido Imperial University 15, 12–23.

Savage, J. M. 2002. The Amphibians and Reptiles of Costa Rica: A Herpetofauna Between Two Continents, Between Two Seas. Chicago: University of Chicago Press.

Scheltinga, D. M. & Jamieson, B. G. M. 2003. Spermatogenesis and the mature spermatozoon: form, function and phylogenetic implications. Reproductive Biology and Phylogeny of Anura (Ed by B.G. M. Jamieson), pp. 120–251. New Hampshire: Science Publishers.

Selmi, M. G., Brizzi, R. & Bigliardi, E. 1997. Sperm morphology of salamandrids (Amphibia, Urodela): implications for phylogeny and fertilization biology. Tissue and Cell 29, 651–664.

Sever, D. M. 1994a. Comparative anatomy and phylogeny of the cloacae of salamanders (Amphibia: Caudata). VII. Plethodontidae. Herpetol. Monogr. 8, 276–337.

Sever, D. M. 1994b. Observations on regionalization of secretory activity in the spermathecae of salamanders and comments on phylogeny of sperm storage in female salamanders. *Herpetologica* 50, 383–397.

Sever, D. M. 2002. Female sperm storage in amphibians. J. Exp. Zool. 292, 165–179.

Sever, D. M. 2003. Courtship and mating glands. In: Reproductive Biology and Phylogeny of Urodela (Ed. by D. M. Sever), pp. 323–381.

Volume 1 of Series: Reproductive Biology and Phylogeny (Series Ed. by B.G.M. Jamieson). New Hampshire: Science Publishers.

Sever, D. M. & Brizzi, R. 1998. Comparative biology of sperm storage in female salamanders. J. Exp. Zool. 282, 460–476.

Sever, D. M., Moriarty, E. C., Rania, L. C. & Hamlett, W. C. 2001. Sperm storage in the oviduct of the internal fertilizing frog Ascaphus truei. J. Morphol. 248, 1–21.

Sever, D. M., Hamlett, W. C., Slabach, R., Stephenson, B. & Verrell, P. A. 2003. Internal fertilization in the Anura with special reference to mating and female sperma storage in *Ascaphus*. In: Reproductive Biology and Phylogeny of Anura (Ed. by B. G. M. Jamieson), pp. 319–341. New Hampshire: Science Publishers.

Stephenson, B. & Verrell, P. 2003. Courtship and mating of the tailed frog (*Ascaphus ruei*). J. Zool., Lond. 259, 15–22.

Tokarz, R. R. 1978. Oogonial proliferation, oogenesis, and folliculogenesis in nonmammalian vertebrates. In: The Vertebrate Ovary (Ed. by R. E. Jones), pp. 145–179. New York: Plenum Press.

Tyler, M. J. 2003. The gross anatomy of the reproductive system. In: Reproductive Biology and Phylogeny of Anura (Ed. by B. G. M. Jamieson), pp. 19–26. New Hampshire: Science Publishers, Inc.

Üribe, M.C.A. 2001. Reproductive systems of caudata amphibians. In: Vertebrate Functional Morphology (Ed. by H. M. Dutta and J. S. Datta Munshi), pp. 267–294. Horizons of Research in the 21st Century. New Hampshire: Science Publishers.

Verrell, P. A. 2004. The male reproductive cycle of the North American salamander *Ambystoma macrodactylum columbianum*. Amphibia–Reptilia 25, 349–356.

Verrell, P. A., Halliday, T. R. & Griffiths, M. L. 1986. The annual reproductive cycle of the smooth newt (*Triturus vulgaris*) in England. J. Zool., Lond. 210, 101–119.

Vieites, D. R., Nieto-Roman, S., Barluenga, M., Palanca, A., Vences, M. & Meyer, A. 2004. Post-mating clutch piracy in an amphibian. Nature 431, 305–308.

Wabnitz, P. A., Bowie, J. H., Tyler, M. J., Wallace, J. C. & Smith, B. P. 1999. Aquatic sex pheromone from a male tree frog. Nature 401, 444–445.

Watanabe, T. & Onitake, K.. 2003. Sperm Activation. In: Reproductive Biology and Phylogeny of Urodela (Ed. by D. Sever); Volume 1 of Series: Reproductive Biology and Phylogeny (Series Ed. by B. G. M. Jamieson), pp. 425–445. New Hampshire: Science Publishers.

Wake, M. H. 1977. The reproductive biology of caecilians: an evolutionary perspective. In: The Reproductive Biology of Amphibians (S .I. Guttman & D. H. Taylor (Eds.), pp. 73–102. New York: Plenum Press.

Wake, M. H. 1978. The reproductive biology of Eleutherodactylus jasperi (Amphibia, Anura, Leptodactylidae), with comments on the evolution of live-bearing systems. J. Herpetol., 12, 121–133.

Wake, M. H. 1993. Evolution of oviductal gestation in amphibians. J. Exp. Zool. 266, 394–413.

Wake, M. H. 2003. Reproductive modes, ontogenies, and the evolution of body form. In Animal Biology, Vol. 53, pp. 209–223. Leiden: Brill NV.

Wake, M. H. & Dickie, R. 1998. Oviduct structure and function and reproductive modes in amphibians. J. Exp. Zool. 282, 477–506.

Watanabe, A. & Onitake K. 2003. Sperm activation. In: Reproductive Biology and Phylogeny of Urodela (Ed. by D. M. Sever), pp. 425–455. Volume 1 of Series: Reproductive Biology and Phylogeny (Series Ed. by B.G.M. Jamieson). New Hampshire: Science Publishers.

Woodley, S. K. 2007. Sex steroid hormones and sexual dimorphism of chemosensory structures in a terrestrial salamander (Plethodon shermani). Brain Research 1138, 95–103.

Xavier, F. 1977. An exceptional reproductive strategy in Anura: Nectophrynoides occidentalis Angel (Bufonidae), an example of adaptation to terrestrial life by viviparity. In: Major Patterns in Vertebrate Evolutions (Ed. by M. K. Hecht, P. C. Goodya & B.M. Hecht), pp. 545–552. New York: Plenum Publishers.

Xavier, F. 1986. La reproduction des Nectophrynoides. In: Traité de Zoologie Amphibiens. Vol. 14 (Ed. by P. P. Grasse & M. Delsol), pp. 497–513. Paris: Masson.

Zalisko, E. K., Brandon, R. A. and Martan, J. 1984. Microstructure and histochemistry of salamander spermatophores (Ambystoma, Salamandridae and Plethodontidae). Copeia 1984, 741–749.

19

Evolution of Primary Sexual Characters in Reptiles

TOBIAS ULLER, DEVI STUART-FOX, AND MATS OLSSON

INTRODUCTION

Variation in sexual traits within and among species arises from past and current biotic and abiotic selective regimes, environmental conditions experienced during ontogeny, and developmental and phylogenetic constraints (Andersson 1994). Our aim with this review is to assess variation in primary sexual characters in relation to abiotic and biotic factors in reptiles. However, as a consequence of modern evolutionary biology, the main focus will be on processes directly targeting the size and function of gametes and their storage from a perspective of ongoing selection, that is mostly postcopulatory sexual selection. In order to do that without replicating previous work in this area, we review the literature with most emphasis on work published since 1997, that is, the year of submission of the last major review in this field (Olsson and Madsen 1998). Pre-1997 information is only considered if crucial for a more complete understanding of conceptual issues. We also deliberately minimize treatment of some areas that we think are still essentially up to date in Olsson and Madsen (1998), including hormone cycles and copulatory organs. A substantial part of the current review is also a phylogenetic analysis of variation in testis size lacking in Olsson and Madsen's (1998) descriptive presentation, now extended with additional data from the published literature since 1997.

To put our treatment into perspective, we start with a brief introduction into general reptilian reproductive biology.

A Brief Overview of Reptilian Biology

Reptilia (excluding birds) is a diverse group of ectotherm animals comprising Rhynchocephalians (tuataras), crocodilians, turtles, and squamates (lizards and snakes) (see Hedges & Poling 1999 and Townsend et al. 2004 for phylogenies). In most species, parental care of hatchlings is absent or rudimentary (Shine 1988) and males do not provide any direct resources to the female before, during, or after mating. Mating systems are generally categorized by intense male–male competition for females and, in many species, female or resource defense polygyny (e.g., Martins 1994; Shine 2003). Female mate choice on male quantitative traits has rarely been documented (Tokarz 1995; Olsson & Madsen 1995; but see, e.g., Lopez et al. 2003) and females frequently mate with more than one male within each ovarian cycle, both in captivity and in the wild (e.g., crocodilians: Davis et al. 2001; turtles: Pearse et al. 2002; Jensen et al. 2006; lizards: Zamudio & Sinervo 2000; Laloi et al. 2004; snakes: Schwartz et al. 1989; Prosser et al. 2002; reviewed in Uller & Olsson 2008). Such polyandrous mating is a prerequisite for the operation of postcopulatory sexual selection (e.g., sperm competition and cryptic female choice).

Reproductive intervals range from days and weeks to years within lizards and turtles (e.g., Cogger 1978; Pearse & Avise 2001), whereas they are more consistently long (≥ 1 year) in snakes and crocodilians (e.g., Seigel & Ford 1987). Breeding normally occurs according to a seasonal pattern even in tropical species (e.g., James & Shine 1985; Seigel & Ford 1987). In most species of vertebrates, sperm production, insemination, and fertilization show close temporal association. In reptiles, however, and in particular in snakes, sperm production, insemination, and fertilization is frequently decoupled in time (reviewed in Duvall et al. 1982; Saint-Girons 1982; Crews 1984; Schuett 1992; Aldridge & Duvall 2002). This may have important implications for the evolution of male and female primary sexual traits for a number of reasons, for example, by generating selection on sperm storage and ejaculates. Clutch size ranges from one to over 50 in squamates (Fitch 1970), whereas turtles and crocodilians can lay more than 100 eggs in a single clutch (Greer 1975; Lutz & Musick 1996).

The weak evidence for wide-spread pre-copulatory mate choice (Tokarz 1995; Olsson & Madsen 1995), high incidence of multiple mating (Olsson & Madsen 1998; Uller & Olsson 2008), and female sperm storage (Sever & Hamlett 2002) suggests that post-copulatory sexual selection on male primary sexual traits, such as testis size and ejaculates should be strong in reptiles (Olsson & Madsen 1998). Similarly, in females, high multiple mating should strongly select for traits that facilitate post-copulatory paternity bias (cryptic female choice, Eberhard 1996) and may influence selective regimes on patterns of ovulation and sperm storage capacity. Many reptiles are thus highly suited for addressing the role of sexual selection for the evolution of primary sexual traits. Here, we provide an overview of our current understanding of primary sexual traits in reptiles and the evidence for sexual selection as a driving force in generating within- and among-species variation.

TESTIS SIZE AND SPERM PRODUCTION—A PHYLOGENETIC ANALYSIS

Sperm is produced in the testis and testis size in reptiles is maximal at the time of spermiogenesis, suggesting that large testes are indicative of a high sperm production at the individual level (Licht 1984). Testis size is also sensitive to food intake and general male health (Olsson & Madsen 1998), which could provide a link between male phenotype and the size or quality of his ejaculate (at least when spermiogenesis and breeding are temporally associated). The link between testis size and sperm production at the intraspecific level could also lead to patterns at the inter-specific (or inter-population) level. More specifically, relative testis size is predicted to be positively related to the strength of sperm competition resulting from female polyandry (Short 1979; Møller & Briskie 1995; Parker et al. 1997; Birkhead & Møller 1998), which has been confirmed in both comparative (reviewed in Parker et al. 1997) and experimental (Hosken & Ward 2001) studies of other taxa. In reptiles, however, there has been no phylogenetically controlled test of this hypothesis. Furthermore, if production of sperm is energetically costly (Olsson et al. 1997), we may predict that species with a long breeding season would show lower peak testis mass compared to more "explosive breeders" where the cost is paid only during a brief period of time (Olsson & Madsen 1998).

Therefore, we conducted an analysis where we assessed the phylogenetically independent effects on relative testis size (Gonado-Somatic Index, GSI) of variables that we have reason to believe are related to sperm competition intensity or costs of sperm production (see Olsson & Madsen 1998): male–female synchrony of reproductive cycles (present vs. absent), and territoriality (present vs. absent), both of which could be related to the strength of sperm competition, for example, by adjusting the operational sex ratio and opportunity for multiple mating (Olsson & Madsen 1998); latitude (tropical vs. subtropical or temperate) and altitude (lowland or generalist vs. montane specialist), both related to the length of breeding season. Finally, we tested the effect of the mode of reproduction (viviparity vs. oviparity) since this is a fundamental aspect of reptilian life history with many potential carry-over effects on the strength of selection on both reproductive and non-reproductive traits (Shine 1983, 2005). The methods and results are summarized in Box 19.1.

Patterns of Testis Size in Reptiles

Our analyses provided only limited support for a relationship between testis size and intensity of

Methods

Because of differences of reporting relative testis size in the literature, we used two different indices as given in Olsson and Madsen (1998; taken directly from the literature when appropriate, or calculated from provided information in text and illustrations): (i) testis mass (g) divided by body mass (g) times 100, or (ii) testis mass (mg) divided by snout-vent length (mm). When testis mass was given as a volume, we converted volume to mass using the volume–mass relationship of ellipsoid testes in Swain and Jones (1994). When testis mass was given for one testis, we multiplied this figure by two. Thus, we ignored the possibility that the left and right testis may differ in size but any error in these estimates should be random across predictors. To ensure that the two indices of relative testis size were comparable, we first confirmed that there was no significant difference between the two (GSI index 1 N = 76, GSI index 2 N=72, $F_1 = 0.39$, $p = 0.54$) then standardized each to range between 0 and 1. We also assessed the ratio between the peak and trough GSI measurements (GSI ratio) when both were available for an annual cycle, to assess the possibility that seasonal variation in intensity of sperm competition selects for changes in testis size. All data are available from the authors upon request.

We compiled a phylogeny (Figure 19.1) for the 148 species in our dataset from recent published molecular and morphological phylogenies (see references) and assumed branches to be equal length for the purposes of the analysis. Comparative methods follow those in Ord and Stuart-Fox (2006) and Stuart-Fox and Moussalli (2007). Briefly, we used a Phylogenetic General Least Squares (PGLS) multiple regression model (Martins & Hansen 1997) with GSI or GSI ratio as the dependent variable with the following independent variables: snout-vent length (SVL), latitude, altitude, male–female synchrony, territoriality and mode of reproduction. Due to different sample sizes for different combinations of variables, we used the model selection procedure described in Purvis et al. (2000) to identify the best models. PGLS estimates a parameter α, which measures the extent of phenotypic variation across taxa that can be explained by phylogeny and subsequently controls for this effect in the model. If α is set to 0, results are identical to Felsenstein's (1985) independent contrasts (FIC), and when α is large (>15.5), it is equivalent to ignoring phylogeny (Tips). The parameter α can be interpreted as a measure of phylogenetic conservatism in the trait data (Martins & Hansen 1997). All analyses were carried out in COMPARE v4.6 (Martins 2004).

Results

The only significant predictor of GSI was latitude. Subtropical and temperate species had greater relative testis mass than tropical species (PGLS: $N = 125$, $r^2 = 0.05$, p = 0.02; Tips: $r^2 = 0.07$, $p < 0.001$). In non-phylogenetic analyses (Tips), territoriality was also a significant predictor of GSI in a multiple regression model with latitude ($N = 82$, model $r^2 = 0.15$, latitude $p < 0.001$ and territoriality $p = 0.02$), with territorial species tending to have greater GSI than non-territorial species. However territoriality was not a significant predictor in the equivalent PGLS model ($N = 82$, model $r^2 = 0.08$, latitude $p = 0.03$ and territoriality $p = 0.1$) because territoriality is highly conserved within genera and families. None of the other independent variables (altitude, SVL, male–female synchrony and mode of reproduction) predicted GSI. Similarly, none of the independent variables predicted the ratio of peak to trough relative testis size over an annual cycle. For most of the models we ran, α values were low to moderate, indicating that the variables are phylogenetically conservative.

(Continued)

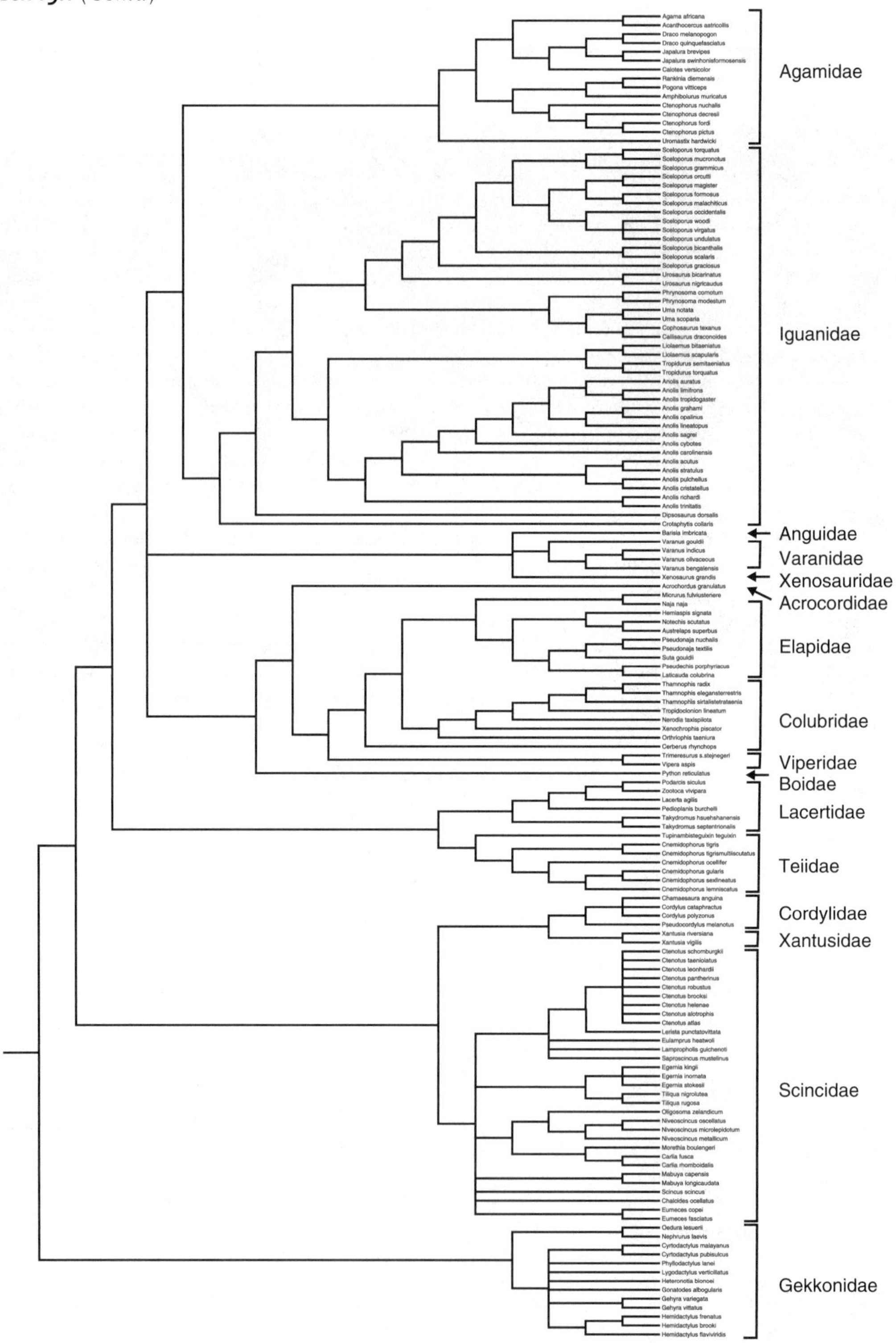

BOX **19.1** (*Contd*)

FIGURE 19.1 Phylogeny used for comparative analysis of Gonado-Somatic Index (GSI). Sources used to construct the composite tree are given in Appendix 1. Polytomies were randomly resolved by inserting branch lengths of 0.00001. All other branch lengths were set to one.

sperm competition or length of breeding season. For example, we failed to identify any effect of territoriality on relative testis size when controlling for phylogeny. In fact, the only statistically significant pattern was larger testes in temperate regions, that is, where the mating season in general is more strictly defined. Because virtually all temperate zone reptiles in our data set have an associate reproductive pattern (i.e., maximum testis size and spermiogenesis co-occur with mating season; Crews 1984), we cannot with confidence separate the two effects at present. Thus, although the observed pattern may relate to a bias in the proportion of reproductively active males and females and therefore high scramble competition for fertilizations in temperate breeders, it could also arise because of a cost of maintaining large gonads through a prolonged mating season in tropical species (see Simmons & Emlen 2006). Cost of sperm production and testis size in reptiles include the direct energetic cost of ejaculate production (Olsson et al. 1997) but also costs associated with increased thermoregulation (Olsson et al. 1997; Herczeg et al. 2007). The latter should be relatively minor for tropical species, however.

Across a wide range of taxa, relative testis size has repeatedly been linked to mating system variation, with larger testis consistently being found in mating systems with more promiscuous females and hence stronger sperm competition (e.g., insects: Gage 1994; amphibians: Byrne et al. 2002; birds: Møller & Briskie 1995; mammals: Hosken 1998; early work summarized in Parker et al. 1997). Why do reptiles deviate from this pattern? One reason could be high phylogenetic conservatism of mating systems, which may be driven by coevolved neurosensory systems and reproductive traits. For instance, virtually all skinks are non-territorial mate guarders that rely on olfactory perception, whereas almost all iguanid and agamid species are visually orientated, territorial species. This reduces the power in a phylogenetically controlled analysis to dissect out the effect of mating system and associated effects. Furthermore, the classification scheme is necessarily coarse due to the limited data available from natural populations. Although both territoriality and breeding system (synchronous vs. asynchronous male and female reproductive cycles) have been suggested to co-vary with the strength of sperm competition, this may simply reflect our ignorance of factual patterns of female polyandry in the wild. For example, in contrast to the prediction that territorial species should have higher GSI

indices (Olsson & Madsen 1998), a negative relationship between territoriality and GSI is also plausible. Both pair-bonding skinks (Australian genera *Egernia* and *Tiliqua*) and highly territorial Australian agamids have lower multiple paternity than non-territorial lizards (Uller & Olsson 2008), suggesting that successful guarding of partners, and hence reduced sperm competition, could arise via multiple routes. More direct estimates may be obtained via studies of multiple mating and multiple paternity. Both are common in turtles and squamates (Olsson & Madsen 1998; Uller & Olsson 2008) but the overlap between species for which we have data on multiple mating, paternity and testis size is too low at present to allow a phylogenetically robust analysis. Furthermore, variation among populations in multiple mating and paternity, possibly arising from variation in operational sex ratios, is almost as high as the total interspecific variation for some species [e.g., multiple paternity ranges from 30 to 92% in Olive ridley sea turtles, *Lepidochelys olivacea* (Jensen et al. 2006) and 17 to 80% in adders, *Vipera berus* (Höggren 1995); Uller & Olsson 2008]. This suggests that species averages may not be informative, in particular for small sample sizes, and that robust tests will be difficult to generate. Similarly, despite our attempts to minimize errors in several ways, large intra-annual variation in testis size and problems with combining data obtained in different ways may introduce errors in the data set to significantly reduce the reliability of the results (Calhim & Birkhead 2007).

Finally, from a selection perspective, it may simply be that testis size and its correlate sperm production are relatively unimportant traits in sperm competition in reptiles. For example, under prolonged sperm storage (see below), scramble competition models of sperm competition do not capture the complexities of sperm survival in the female reproductive tract. Indeed, there is a growing awareness that other ejaculate characteristics than sperm number or sperm concentration may be just as, or even more, important in predicting fertilization success in polyandrous species (e.g., sperm longevity, sperm size, ejaculate composition; Snook 2005).

EJACULATES

Fertilization is internal in all reptiles. Thus, selection acting on sperm and ejaculate composition arises from natural selection on sperm morphology and physiology due to (i) the environment of the female reproductive tract, (ii) temporal differences between ejaculation and fertilization, (iii) spatial differences between the site of insemination and the site of fertilization, and sexual selection arising via (iv) the presence of ejaculates from multiple males competing for fertilization, and (v) sexual conflict over fertilization. The first three categories are undisputed and could explain certain characteristics of both semen (such as the presence of antioxidants; Breque et al. 2003) and sperm (such as swimming ability; Snook 2005). These are undoubtedly of importance also in reptiles. However, given that multiple mating in reptiles is widespread with few exceptions (Olsson & Madsen 1998; Uller & Olsson 2008), post-copulatory sexual selection is also likely to mold ejaculate traits. We provide an overview of ejaculate composition and sperm morphology and discuss to what extent variation among taxa could be explained by natural selection (in particular temporal separation of sperm production, copulation, and fertilization) or post-copulatory sexual selection.

Seminal Fluids and Copulatory Plugs

Ejaculates of vertebrates and invertebrates contain a variety of fluids and substances that can increase sperm longevity, facilitate sperm storage, increase fertilization success under sperm competition, and manipulate female re-mating or reproductive investment (see Gillott 2003; Poiani 2006; Ram & Wolfner 2007 for reviews). In squamates, seminal fluid is produced by the epididymis and the renal sexual segment (RSS), a hypertrophied region of the distal urinary ducts (Fox 1977; Sever & Hopkins 2005). Both hypertrophy of the RSS and number and densities of sexual granules in the cytoplasm show seasonal variation that closely correspond to circulating levels of plasma testosterone and testicular action (Fox 1977; Krohmer 2004; Sever & Hopkins 2005). This implies a role for the RSS during spermiation and mating (at least in species with associated breeding cycles), suggesting that the primary role of the RSS is to provide seminal fluids during copulation. However, the exact functions of the RSS are not well understood. It has been suggested to produce courtship pheromones (Volsøe 1944; Devine 1975), sustain and activate sperm (e.g., Cuellar 1966), and form material for

copulatory plugs (Volsøe 1944; Devine 1975) (see Fox 1977; Sever & Hamlett 2002; Sever & Hopkins 2005 for overviews). Although there is currently little evidence for sperm activation and nutrition, studies of garter snakes (*Thamnophis* spp.) have shown that seminal fluids (although not necessarily of RSS origin) have both pheromonal action and form copulatory plugs (Devine 1975; Ross & Crews 1977, 1978; Shine et al. 2000).

Copulatory plugs are more or less thick gelatinous structures that form in the female cloaca at the end of copulation subsequent to sperm transfer (Volsøe 1944; Devine 1975; Fox 1977; Shine et al. 2000). They are quite common in mammals and insects, but have also been described in several species of snakes and lacertid lizards (Devine 1984; in den Bosch 1994). Copulatory plugs were initially believed to prevent sperm from leaking out of the cloaca (Volsøe 1944; Fox 1977), a role that would benefit both males and females. However, with the advent of studies of intra- and intersexual conflict (Parker 1970; Trivers 1972; Arnqvist & Rowe 2005), copulatory plugs have increasingly been viewed as manipulative tools by which copulating males ensure their own reproductive success while compromising that of competing males and potentially that of the female (enforced chastity; Shine et al. 2000). This outcome could be achieved in three ways: first, by serving as physical plugs preventing successful hemipenis intromission and sperm transfer by other males; second, by producing pheromones that makes females unattractive to other males or suppress female re-mating; or third, by reducing the competitive ability of sperm from other males. Initial studies of garter snakes supported both the physical barrier and pheromone hypotheses (Ross & Crews 1977, 1978; Devine 1984) but subsequent work has suggested that, although pheromones from the seminal fluids reduce female attractiveness, this is largely independent of the plug itself (Shine et al. 2000).

Although logical, there are two problems with plugs as physical barriers. First, subsequent studies on garter snakes have found a high incidence of multiple paternity (Schwartz et al. 1989), suggesting that copulatory plugs are ultimately inefficient as chastity belts. However, this conclusion relies on the assumption that multiple paternity arises primarily from multiple mating within a mating season and not via sperm storage across seasons (see below). Furthermore, in *Thamnophis sirtalis*, plugs were shown to be effective for two days after the initial mating, which may still confer sufficient benefits to be under selection (Shine et al. 2000). Second, studies of the Iberian rock lizard (*Lacerta monticola*) have shown that female re-mating probability and male intromission is not prevented by the presence of a copulatory plug (Moreira & Birkhead 2003). In fact, second mating males have a higher fertilization success when they mate 30 minutes after the first males, when the plug is still intact, than when mating 4 h after the first mating, when the plug has started to disintegrate (Moreira et al. 2007). This suggests that the mating plug contains sperm that are actively dislodged by second males and, consequently, that the plug has evolved as a means of sperm transfer rather than to prevent female remating. Furthermore, copulatory plugs may be used as a means of assessing presence or absence of sperm competition and rival quality (Olsson et al. 2004; Moreira et al. 2006), which can explain highly sophisticated patterns of ejaculate allocation by males that further increase the fertilization success of second mated males. In sand lizards, relatedness to the female predicts paternity under sperm competition (Olsson et al. 1996). Male sand lizards adaptively adjust copula duration (and hence sperm transfer) in relation to the previous male's relatedness with his female, possibly via pheromonal cues obtained from the copulatory plug (Olsson et al. 2004). Finally, using vasectomized males (that only can transfer RSS secretions and no sperm), Olsson et al. (1994b) showed that seminal fluid itself does not reduce the probability of paternity for subsequently mating rivals. Thus, there is little direct evidence that seminal fluids have evolved via postcopulatory sexual selection in reptiles.

Sperm Morphology

Sperm of squamates (Furieri 1970; Jamieson 1995), turtles (Furieri 1970; Hess et al. 1991), crocodilians (Jamieson et al. 1997), and tuatara (Jamieson & Healey 1992) show a typical morphology, with a distinct head, midpiece and tail, similar to that of mammals, birds, and amphibians. There is no evidence for polymorphism in sperm morphology or the presence of non-fertilizing sperm as is commonly found in invertebrates (Swallow & Wilkinson 2002). Differences between major taxonomic groups in sperm ultrastructure mainly arise from variation in the structure of the acrosome and midpiece (Furieri 1970), but minor differences can be found also among closely related species and have

been suggested to provide important phylogenetic information (reviewed in Jamieson 1995; Vieira et al. 2007).

Sperm morphometrics (i.e., the size of different parts) may be selected both via co-evolution with female reproductive traits (e.g., the female reproductive tract; Miller & Pitnick 2002; Anderson et al. 2006), or via intrasexual selection (e.g., via sperm competition, Gage 1994; Briskie et al. 1997; Byrne et al. 2003; see Parker 1998 for theoretical overview). For example, flagellum length may be under positive selection due to selection for increased velocity (Gage 1998; Malo et al. 2006) and midpiece size may be under selection via an increase in mitochondrial number or size, thereby generating greater power output (Cardullo & Baltz 1991; Anderson et al. 2005; Immler et al. 2007). Thus, under strong sperm competition, both flagellum and midpiece length are predicted to increase, which could lead to a positive relationship between sperm length and intensity of sperm competition unless there is a trade-off between sperm size and number (Gage 1994; Byrne et al. 2003; Gomendio et al. 2007 but see Gage & Freckleton 2003; Garcia-Gonzalez & Simmons 2007). However, under prolonged sperm storage, small sperm, and in particular a small midpiece, may be favored if it enhances sperm longevity (Immler & Birkhead 2007; Immler et al. 2007).

Of the total 49 reptilian taxa for which sperm morphology could be obtained, only seven were species for which we also had information on gonadosomatic index, preventing an analysis of the relationship between sperm morphology and this index of the strength of postcopulatory sexual selection. In total, we found information on midpiece and flagellum length for 36 squamate taxa (figure 19.2; Appendix 2). Sperm lengths range from approximately 20 μm in some crododilians (Ferguson 1985) to 170 μm in the blindsnake *Ramphotyphlops waitii* (Harding et al. 1995). Within lizards, the group from which the majority of data derives, total sperm length ranges from 28.5 μm in the bearded dragon, *Pogona barbata* (Olivier et al. 1996), to 98.8 μm in the leopard lizard, *Gambelia wislizenii* (Vieira et al. 2007).

After controlling for phylogenetic relationships among species (figure 19.2; see Box 19.1), there was a significant allometric relationship between midpiece and flagellum length (table 19.1). This suggests co-evolution between kinetic (flagellum length) and energetic (midpiece size) aspects of sperm morphology among reptiles, similarly to

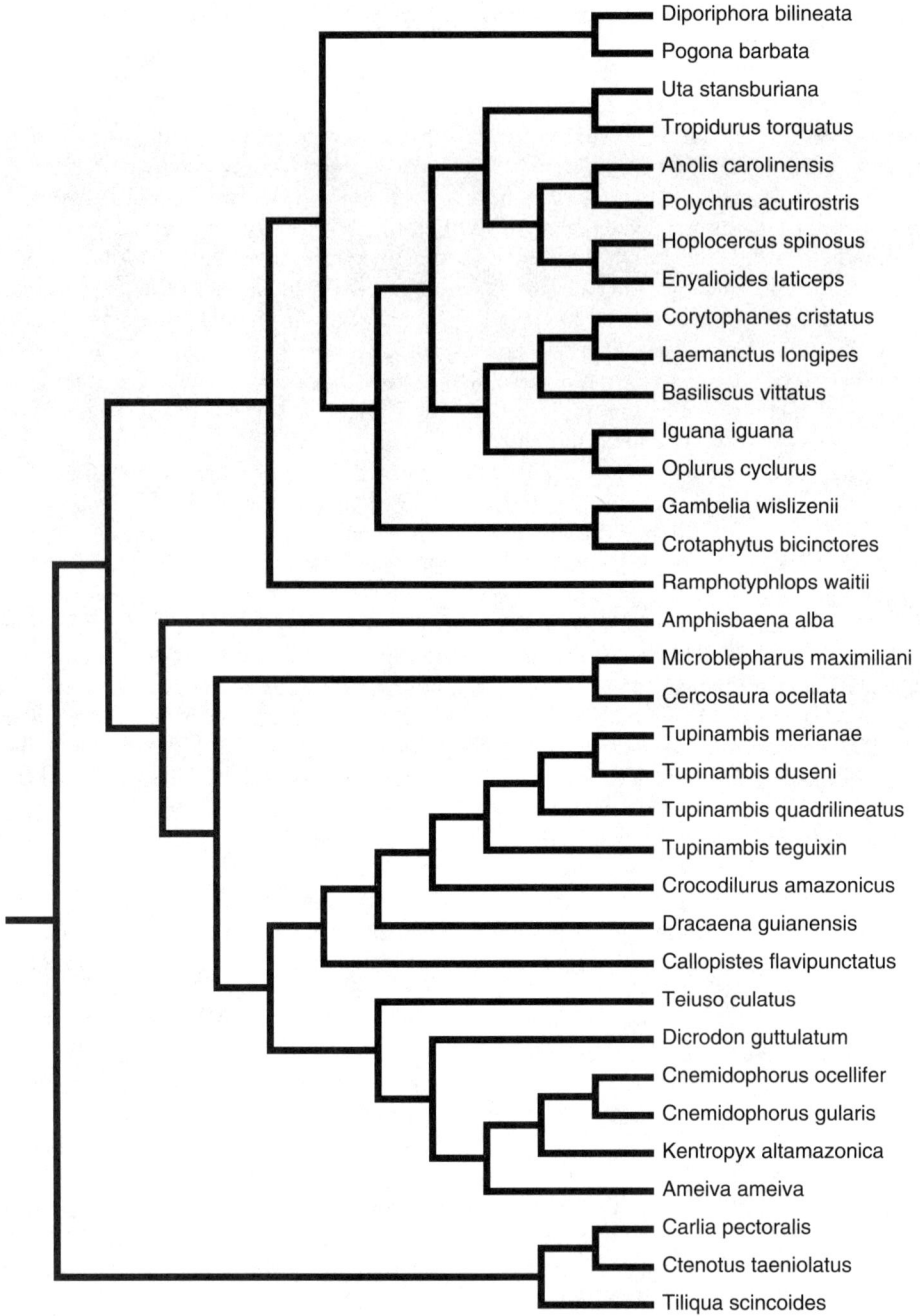

FIGURE 19.2 Phylogeny used for comparative analysis of sperm morphology. Sources used to construct the composite tree are given in Appendix 1. All branch lengths were set to one.

TABLE 19.1 Correlation between lengths of different parts of sperm (head, midpiece and flagellum)

			PGLS		Tips	
Variables	N	α	r	p	r	p
Head and midpiece	30	1.64	−0.18	0.33	−0.1	0.61
Head and flagellum	30	2.41	0.30	0.10	0.42	0.01
Midpiece and flagellum	35	3.82	0.60	<0.0001	0.53	<0.0001

PGLS = Phylogenetic generalized least squares, Tips = non-phylogenetic analysis. Alpha is a measure of phylogenetic conservatism (see methods). The values of alpha are low, indicating that sperm morphometrics are phylogenetically conserved. *Boa constrictor* was removed from analyses because it was an outlier (due to its unusually long midpiece).

results for mammals and birds (Gage 1998; Immler et al. 2007). However, the underlying reason for allometric relationships cannot be inferred from these data and robust conclusions regarding sperm morphology and their relation to the strength of sexual selection should await accumulation of data from more species. In an attempt to provide a starting point for further research, we document here some existing patterns of sperm morphometry in relation to reptilian mating systems.

Boa constrictor has one of the longest reptilian sperm (Tourmente et al. 2006). Furthermore, it has a greatly elongated mid-piece (approx 10 times longer than in lizards). This suggests that sperm competition is relatively intense in this species. Strong sperm competition in snakes is also supported by their mating systems (Duvall et al. 1993; Shine 2003), high levels of multiple paternity (e.g., Schwartz et al. 1989; Uller & Olsson 2008), hemipene morphology (Keough 1999), large relative testis size and long copulation times (Olsson & Madsen 1998). However, comparable data on sperm morphometrics for other snake species and the convergent legless lizards are currently lacking (Appendix 19.2; but see Hamilton & Fawcett 1968; Teixeira et al. 1999a; Tavares-Bastos et al. 2007).

In the turtle, *Chrysemys picta*, mitochondria are unusually laminated, which has been suggested to be an adaptation for prolonged survival in the female oviduct during storage (Hess et al. 1991). Prolonged sperm storage in both the female reproductive tract and in the male epididymis commonly occurs also in snakes with dissociated reproductive cycles (Saint-Girons 1982; Aldridge & Duvall 2002; see below), which makes this group well suited for addressing the role of mitochondrial reorganization in adaptation to prolonged sperm storage.

In addition to the large variation among species, there is also often substantial variation among males

in sperm and ejaculate traits within species (e.g., Harris et al. 2007). Of particular interest is the link between mating strategies and ejaculate composition. For example, small males that are less competitive for access to females or territories may be under stronger post-copulatory sexual selection and, for example, transfer more sperm per ejaculate than larger males (Parker 1990). This was supported by a study of northern watersnakes (*Nerodia sipedon*) in which Schulte-Hostedde and Montgomerie (2006) found that smaller males produced ejaculates with a higher sperm concentration than did larger males, which may compensate for a lower mating frequency in terms of paternity success (Weatherhead et al. 2002; see also Olsson et al. 2009). Most traits (sperm velocity, spermatocrit) showed substantial variation among ejaculates. However, sperm length was largely invariable, suggesting stabilizing postcopulatory sexual selection on sperm length in *N. sipedon*. To what extent these results reflect more general intraspecific variation in ejaculate composition in snakes and other reptiles remains to be investigated (see Calhim et al. 2007 for a comparative study of passerine birds). Importantly, ejaculates can be obtained using non-invasive techniques (Schulte-Hostedde & Montgomerie 2006), which provides exciting scope for combining descriptive studies of ejaculate traits with experimental manipulations of, for example, mating opportunities.

Sperm Motility

Sperm motility could be important in scramble competition for fertilization (Birkhead et al. 1999). In ectotherms, female body temperature should have the potential to adjust sperm motility subsequent to mating, possibly serving as a behavioral mechanism for cryptic female choice via female basking. However, in the turtle *Chrysemys picta*, sperm motility was highest at low temperatures, suggesting local adaptation of sperm motility in relation to the decreasing ambient temperatures during the timing of copulation in this species (Gist et al. 2000).

Although a high sperm motility can be favored under situations of sperm competition, low motility could be favored in species with substantial sperm storage (Gist et al. 2000). This may be supported by the lower motility of turtle sperm compared to lizards (Depeiges & Dacheux 1985; Gist et al. 2000), although the lack of data makes this hypothesis largely suggestive. However, in *C. picta*, 70% of sperm remained viable after 40 days in vitro, suggesting strong selection on sperm viability in species with prolonged storage of sperm in the male or female reproductive systems (Gist et al. 2000, 2002).

FEMALE SPERM STORAGE

Storage of sperm in the female reproductive tract is an important aspect of the reproductive biology of both invertebrates and vertebrates (Howarth 1974; Neubaum & Wolfner 1999). In reptiles, hatching or parturition is restricted to the warmer parts of the year, which means that ovulation and embryonic development normally occurs in spring and summer, respectively. However, sperm production also requires sufficiently high temperatures, which can lead to a decoupling of male and female reproductive cycles when there is insufficient time in spring for sperm production (Licht 1984; Aldridge & Duvall 2002). Despite the frequent decoupling of sperm production, mating and fertilization, long-term female sperm storage is not particularly common in reptiles (Saint-Girons 1982; Duvall et al. 1982; Schuett 1992; Aldridge & Duvall 2002; Murphy et al. 2006). Instead, sperm is commonly stored in the male reproductive tract for long periods of time (e.g., Gist et al. 2002) and mating normally occurs only weeks or a few months before fertilization.

Classification of sperm storage in female reptiles is not straightforward (see Saint-Girons 1982; Schuett 1992; Aldridge & Duvall 2002 for different schemes). For example, sperm storage could be classified according to whether it occurs within ovarian cycles or across cycles (i.e., from matings that occurred before oviposition of the previous clutch). However, for each of these two categories,

sperm may be subject to either short-term sperm storage (STSS; days to weeks) or long-term sperm storage (LTSS; months to years). Furthermore, sperm storage may occur within a female ovarian cycle, but across mating periods (as is the case when there is both an autumn and a spring mating season and sperm is stored from autumn until fertilization in spring–summer; Schuett 1992; Murphy et al. 2006). Finally, very long-term sperm storage (VLTSS; many years; i.e., for a period of time longer than both the female ovarian cycle and across mating seasons) has been documented in some species. Thus, although a classification based on absolute time is artificial since the characterization of "long" and "short" may not be related to the life cycle of the species, characterizations based on reproductive cycles are also problematic as male and female cycles can be decoupled in time. Here, we will therefore be explicit with respect to the absolute time of sperm storage and its relationship to male and female reproductive cycles without attempting further classification or introduction of new terminology (see Schuett 1992; Aldridge & Duvall 2002; for useful discussions).

Short term sperm storage within ovarian cycles is probably common in many reptiles as insemination often occurs over a period of time before completion of vitellogenesis. For example, in the European adder, *Vipera berus*, sperm from spring matings is stored in the female reproductive tract for weeks or even months before fertilization occurs in late spring–early summer (Saint-Girons 1982). To some extent, such STSS should have evolved to be flexible, as the temporal dissociation of mating and egg production will vary as a result of environmental conditions, such as annual variation in temperature and food availability. Thus, even in species that do not normally have to store sperm for more than a few weeks, sperm may remain viable for longer if females are prevented from egg production.

Long-term sperm storage has been documented in many species of snakes (Duvall et al. 1982; Schuett 1992; Almeida-Santos & da Graca Salomão 2002), turtles (Pearse & Avise 2001; Pearse et al. 2001) and some lizards (Smyth & Smith 1968; Wapstra et al. 1999; Murphy et al. 2006) and is necessary when the mating season is separated from egg production for long periods of time, such as in species with autumn matings and spring vitellogenesis and ovulation. However, as pointed out by Saint-Girons (1982), most snakes with autumn matings also mate subsequent to hibernation, suggesting that sperm storage over hibernation is not necessary to ensure fertilization. Nevertheless, at least in some American pitvipers, mating in spring may not occur at all (Schuett 1992; Almeida-Santos & da Graca Salomão 2002). Long-term sperm storage resulting from temporal separation of mating and fertilization seems to be rarer in lizards, maybe as a result of the generally shorter reproductive cycle of lizards compared to snakes (Licht 1984; but see Wapstra et al. 1999; Murphy et al. 2006).

Very long term sperm storage (> 1 year) is known from snakes and turtles. The longest reported storage of viable sperm in the female reproductive tract is seven years in a file snake (*Achrochordus arafurae;* Magnusson 1979) and storage for more than three years has repeatedly been recorded in turtles (Sever & Hamlett 2002). Importantly, sperm storage for such long periods of time will extend across mating seasons and, hence, mating opportunities and, in most cases, also extend across female reproductive cycles. However, although data on female fertility in the absence of males suggests VLTSS, previous anecdotal reports of facultative virgin births in squamates have recently been substantiated using molecular verification of offspring genotype (Groot et al. 2003; Watts et al. 2006). Thus, reports of VLTSS in female reptiles should be treated with caution until facultative parthenogenesis can be ruled out (Schuett 1992). However, VLTSS in snakes does not seem unlikely, given that females of many species reproduce infrequently (e.g., Bull & Shine 1979).

Natural and Sexual Selection on Female Sperm Storage

As noted above, STSS and LTSS within ovarian cycles are common in many reptiles and have probably initially evolved via natural selection to ensure fertilization. Nevertheless, presence of sperm storage will generate strong post-copulatory sexual selection on males as it increases the intensity of sperm competition (provided that females mate multiply; Olsson & Madsen 1998). In some species, sperm competition has led to male strategies to prevent female remating, such as mate guarding (Olsson et al. 1996), whereas in others it has led to changes in ejaculate composition (see above). More importantly from the female perspective, however, is that an increase in the number of ejaculates

present in the reproductive tract at the timing of fertilization would select for mechanisms that ensure fertilization of sperm that is 'optimal' with respect to genetic composition (either because of high genetic quality or complementarity; Madsen et al. 1992; Olsson et al. 1996; Birkhead & Pizzari 2002).

Although sperm storage from mating to fertilization is obligate in species with temporal separation between the two, it is more difficult to explain why females store sperm across ovarian cycles or across mating seasons (table 19.2). Surprisingly, very few studies have addressed the role of sperm storage across ovarian cycles for male and female reproductive success, although evidence from two species (*Ctenophorus pictus* and *Uta stansburiana*), suggest that it can lead to significant changes in selection on male mating strategies (Zamudio & Sinervo 2000; Olsson et al. 2007a, 2009). Early workers suggested that female sperm storage across reproductive events has evolved as a strategy to ensure fertilization when access to new sperm is uncertain (e.g., Connor & Crews 1980). At low population densities, for example, mate encounter rates may not be sufficiently high to ensure mating during each ovarian cycle. Storage of sperm would thereby prevent females from being sperm limited. Furthermore, some species can have a two-phased mating season, one in autumn and one in spring (e.g., Schuett 1992; Wapstra et al. 1999; Murphy et al. 2006). Storage of sperm from autumn matings may ensure that high-quality females can progress rapidly through ovulation in spring, without wasting time and energy on mating, in particular under high male harassment (see Løvlie & Pizzari 2007 for a similar scenario). Thus, it could allow earlier oviposition or parturition, traits that are likely to be favored in temperate-zone reptiles (e.g., Olsson & Shine 1997a; Warner & Shine 2007). Despite these two widely inferred selective pressures, there is virtually no evidence that female sperm storage has evolved because it ensures fertilization of eggs in the absence of males. Such evidence would require showing that (i) mate encounter rates are sufficiently low to cause sperm depletion under natural conditions and (ii) females that store sperm have a higher fitness (or at least fertilization success) than female that do not store sperm. Species with both autumn and spring matings would be ideal candidates for an experimental approach to address this issue. For example, manipulation of the operational sex ratio, perceived and actual

TABLE **19.2** Hypotheses for the evolution of female sperm storage in reptiles

Hypothesis	Comment	References
Consequence of selection for optimal timing of birth or parturition	Likely in many snakes and turtles with selection against mating in spring, but cannot explain sperm storage across ovarian cycles.	Saint Girons 1982; Schuett 1992
Consequence of selection for optimal timing of spermiogenesis	Could lead to female sperm storage if selection favors summer or autumn mating, but is unlikely to explain storage across ovarian cycles.	Saint-Girons 1982
Fertilization insurance under low mating encounter rate	Possible in some turtles and snakes with low population densities. Unlikely in most lizards.	Phillipp 1979; Conner & Crews 1980
Fertilization insurance under risk of male infertility	Unlikely as male infertility should be highest in early emerging males with sperm that is not yet mature.	Olsson & Madsen 1996
Reduction in copulation frequency when copulation is costly	Unlikely in species with annual mating seasons. Little evidence that females reject males when receptive.	Conner & Crews 1980
Allowing cryptic female choice of viable sperm via passive sperm loss	Possible, but no direct evidence.	Olsson & Madsen 1998
Allowing cryptic female choice via manipulation of ejaculates	Possible but no direct evidence. May be unlikely considering that sperm storage organs are relatively undifferentiated compared to, for example, insects.	Olsson & Madsen 1998
Selection on sperm to remain viable across ovarian cycles	Possible. Evidence that stored sperm contribute to male reproductive success in two short-lived species with multiple clutches.	Zamudio & Sinervo 2000; Olsson et al. 2007

mate encounter, and monitoring of female reproductive behavior, sperm storage and paternity could shed light on the extent to which female sperm storage is flexible and contingent upon mate encounter rates.

If sperm storage across reproductive cycles has not evolved in response to sperm depletion, what can explain its presence in so many reptiles? One explanation that has been favored by many authors is that sperm storage facilitates cryptic female choice (Olsson and Madsen 1998). For example, in some insects, females may have control over the entry of sperm into sperm storage organs via muscular contractions (Simmons 2001) and, in redback spiders, sperm storage organs facilitate discrimination among ejaculates (Snow & Andrade 2005). Thus, the morphology and position of sperm storage organs may yield important information regarding the potential for female control.

Morphology of Sperm Storage and its Implications

The reptilian oviduct consists of the vagina, uterus, isthmus, tuba (or uterine tube) and infundibulum (Girling 2002). The main difference between squamates, turtles and crocodilians is that the latter two have a much longer tuba region containing tubular glands, which release albumen during egg formation (Gist & Jones 1987). During copulation, sperm is deposited in the cloaca and subsequently reaches the vagina. Sperm is eventually transferred to the tuba or infundibulum where fertilization takes place (Saint-Girons 1975; Girling 2002). Whether sperm transport occurs via muscular contraction of the oviduct or via sperm movement is unknown, although both hypotheses are plausible (Saint-Girons 1975; Halpert et al. 1982).

Sperm storage occurs in receptacles (tubules) in the vagina and infundibulum in squamates, whereas it is confined to the tuba and uterus in chelonians (Fox 1956, 1963; Gist & Jones 1987, 1989; Gist & Congdon 1998; Girling 2002; Sever & Hamlett 2002). Over-winter sperm storage has also been reported to occur via uterine muscular twisting (Yamanouye et al. 2004). The ancestral state for squamates is infundibular sperm storage tubules, and vaginal sperm storage is less common (Sever & Hamlett 2002; Sever & Hopkins 2004). Regardless of the position of the sperm storage tubules, they seem to be unspecialized structures that show little cytological difference from those that lack sperm (e.g., Gist & Fischer 1993; Sever & Ryan 1999; Gist & Jones 1987; Girling 2002; Sever & Hamlett 2002). Sperm are present in bundles or arranged head-first towards the luminal wall. In squamates, sperm heads are in close contact with the epithelium (Cuellar 1966), whereas this does not seem to be the case in turtles (Gist & Jones 1987). Many authors have suggested that sperm receive nourishment from the oviduct (in particular in squamates; Cuellar 1966; Conner & Crews 1980; Halpert et al. 1982), but there is little direct evidence for this (Gist & Jones 1987; Olsson & Madsen 1998; Girling 2002; Sever & Hamlett 2002).

The available data on morphology of sperm storage tubules in reptiles suggest a much lower level of sophistication than that of insects and birds (Bakst 1987; Pitnick et al. 1999; Simmons 2001). Indeed, Gist & Jones (1987) concluded that the receptacles containing sperm are indistinguishable from normal glands in the area and recent evidence support this conclusion (Sever & Hamlett 2002). A notable exception, however, is that sperm storage tubules in *Anolis sagrei* lack cilia and secretory products at the place of sperm concentration (Sever & Hamlett 2002; see also Conner & Crews 1980), an anatomy similar to that in birds (Bakst 1987), suggesting selection on female sperm storage. Interestingly, in contrast to other lizards but similar to birds, Anoles produce one egg at a time but have a prolonged period of egg laying (Smith et al. 1972), which may select for sperm storage to avoid the need for remating between eggs (Sever & Hamlett 2002). Indeed, sperm storage for up to two months has been shown in *Anolis sagrei* (Calsbeek et al. 2007; see also Conner & Crews 1980) and variation in paternity among eggs is common (Calsbeek et al. 2007).

Given that specialized sperm storage tubules are not found in most reptiles, selection on female sperm storage across reproductive events may be weak in most reptiles. Consequently, it is unknown whether sperm storage can contribute to, or is a result of selection for, cryptic female choice. Furthermore, there is no current evidence for co-evolution of sperm length and size of sperm receptacles (as may be the case in birds; Briskie & Montgomerie 1993) although this may simply be due to lack of attention. Instead, prolonged sperm survival in the female reproductive tract may be a (near) neutral trait for female fitness, suggesting that it may evolve simply because of strong selection on male sperm viability, in particular in species with a rapid turn-over of ovarian cycles. If so, males may differ in their ability to produce sperm that survive in the oviduct of females and, hence, variation in this trait and the timing of mating may strongly affect reproductive success in species with multiple clutches (Oring et al. 1992; Zamudio & Sinervo 2000; Olsson et al. 2009), with potential consequences for offspring development (e.g., offspring sex: Olsson et al. 2007b). Documenting the anatomy and physiology of sperm storage together with patterns of parentage in (related) species that differ in key aspects of life history (such as associated vs. dissociated reproductive cycles and clutch interval) should be a research priority for the future.

CLUTCH SIZE AND OVULATION PATTERNS

Follicular growth in single-clutched reptiles either begins in early spring with ovulation in late spring or starts in summer or autumn (directly subsequent to oviposition or birth) and is completed the following spring (with the exception of a few viviparous species; Licht 1984). The first situation is common in lizards and snakes and the second in turtles. Ovulation is autochronic (i.e., synchronous among follicles) in the vast majority of reptile species (Licht 1984). However, in anoles, ovulation of one egg alternates between the ovaries (allochrony), leading to the presence of multiple eggs at different stages in the female reproductive tract at a given point in time, similar to the situation in birds (Smith et al. 1973). The ultimate reason for this peculiar reproductive system is unknown. Nevertheless, it may increase the scope for female control of sex-specific resource allocation (Uller 2006; Uller et al. 2007) and cryptic female choice of sperm (Calsbeek et al. 2007). Anoles would therefore be a suitable model system in which to address how an evolutionary shift in ovulation patterns may lead to changes in the direction and strength of sexual selection on both primary and secondary sexual traits.

Variation in clutch size is normally addressed from a perspective of natural selection. Indeed, there are a number of studies on reptiles that show that clutch size is under selection to maximize life-time reproductive output via the trade-off between offspring size and number and a negative effect of reproductive effort on female survival (e.g., Schwartzkopf 1994; Bonnet et al. 2002). However, evidence from other taxa suggests that sexual selection may also be important, primarily via two routes: First, males of some insects may be able to increase female reproductive output either via resource provisioning (e.g., nuptial gifts, Vahed 1998) or via manipulation of female oviposition rate by transfer of proteins in seminal fluids (Ram & Wolfner 2007). Second, increased sexual dimorphism can select for reduced clutch size because of increased demands on production of the larger sex (Carranza 1996) or to reduce the temporal and spatial overlap between offspring of different resource demands (Uller 2006, Badyaev et al. 2006; Kuhl et al. 2007). The latter is perhaps particularly likely to occur in viviparous animals (Uller 2003, 2006). In a comparative study of 106 species of mammals, Carranza (1996) found a negative relationship between litter size and sexual size dimorphism. We suggest that similar patterns also exist in squamate reptiles. The large number of independent evolutionary shifts from oviparity to viviparity provides further opportunities to test whether the relationship is stronger in species with a higher degree of developmental overlap between the sexes.

Nuptial gifts do not occur in reptiles and male manipulation of female clutch size has not been documented and may be unlikely considering that females normally oviposit all their eggs simultaneously. Furthermore, in some species (primarily capital breeders), reproductive allocation decisions may be inflexible at the time of mating (e.g., Bonnet et al. 2001). In contrast, a study of the sand lizard, *Lacerta agilis,* suggested that females adjust their reproductive investment in relation to partner quality by increasing their reproductive output when mated to males with large sexual ornaments (Olsson et al. 2005). As this study manipulated male ornamentation independently of male quality, it suggests that females are able to differentially allocate resources in relation to perceived male quality (Sheldon 2000).

CONSTRAINTS AND
CO-EVOLUTION

Current research in sexual selection on primary sexual traits emphasizes the dynamic role of co-evolution between males and females (Arnqvist & Rowe 2005; Andersson & Simmons 2006; Parker 2006). Thus, an evolutionary response to selection in one of the sexes is likely to generate a correlated response in the other (even in the absence of genetic correlations). For example, the evolution of female sperm storage creates an arena for sperm competition and may select for increased sperm transfer, changes in seminal fluid composition, sperm morphology, and sperm longevity. Reptiles with female sperm storage are therefore key candidates for studies of sperm competition and the evolution of male primary reproductive traits. Furthermore, sperm storage has the potential to have a strong impact on selection on mating strategies and secondary sexual traits in males, for example, by increasing the benefits of copulations early in the breeding season in species with multiple clutches (Olsson et al. 2007a).

We argue that an understanding of the evolution of male and female primary sexual characteristics in reptiles is facilitated by an appreciation of the importance of environmental constraints on reproduction and the co-evolution of male and female traits (figure 19.3). Strong selection on matching embryonic development and spermiogenesis to benign environmental conditions has frequently led to a decoupling of male and female reproductive cycles (Licht 1984; Aldridge & Duvall 2002), which dictates the framework in which post-copulatory sexual selection can occur (figure 19.3). For example, it may lead to differences in the timing of emergence from hibernation (Olsson & Madsen 1996; Olsson et al. 1999), with potential impact on selection on male–male competition, female choice, and fitness of both sexes via changes in the operational sex ratio. Risk of mating with infertile males (with immature sperm) may also partly explain high levels of female promiscuity in some species (Olsson & Madsen 1996; Olsson & Shine 1997b; Uller & Olsson 2006). Second, the decoupling of sperm production from mating activities and fertilization reduces or even eliminates the link between peak testosterone levels, maximum testis size, sperm number and quality, and male phenotype at mating (Licht 1984; Duvall et al. 1982; Olsson & Madsen 1998; Murphy et al. 2006). This modifies the cost of reproduction in males and could have important consequences for the application of certain pre- and post-copulatory sexual selection scenarios (Crews 1984; Birkhead & Møller 1993), such as the immunocompetence handicap hypothesis,

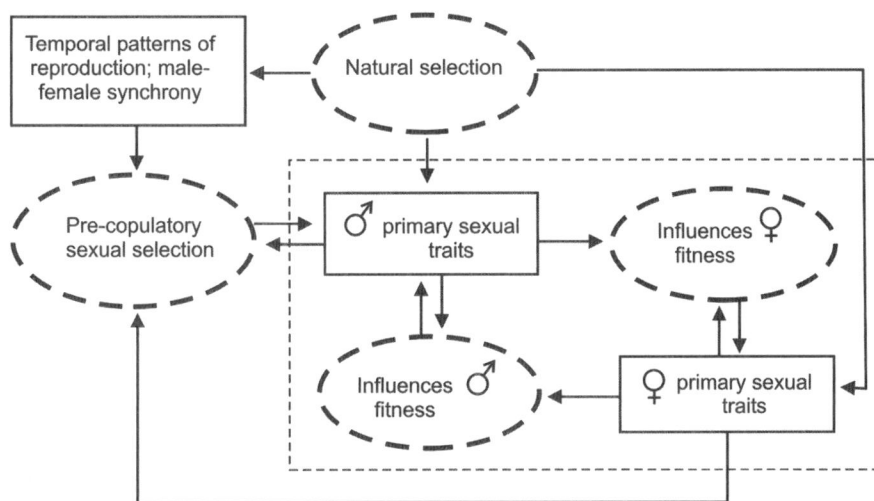

FIGURE 19.3 Evolutionary processes (dashed ovals) influencing male and female primary sexual traits (black rectangles). Male primary sexual traits include testis size, sperm morphology, seminal fluids and copulatory plugs. Female primary sexual traits include characteristics of the female reproductive tract, especially in relation to sperm storage as well as clutch size and ovulation patterns. Thin black dashed rectangle indicates processes of post-copulatory sexual selection and male–female coevolution.

which proposes a trade-off between testosterone-dependent sexual traits and immune function. Furthermore, environmental effects on mating systems may explain why peak testis mass is larger in associate temperate zone reptiles. Specifically, a prolonged or dissociated mating season may change the rules by which ejaculates compete for fertilization, for example, by reducing the importance of rapid and high sperm production in favor of other aspects of ejaculate composition and sperm longevity (Snook 2005; Dean et al. 2007; Pizzari et al. 2008). Finally, sperm is frequently stored for several weeks and storage across ovarian cycles in short-lived species with multiple clutches has been documented (Olsson et al. 2007a,b; Calsbeek et al. 2007). However, the lack of evolution of sophisticated sperm storage organs despite prolonged storage of sperm (Sever & Hamlett 2002) suggests a relatively minor role of the female reproductive tract for sperm survival and viability. This could imply that sperm storage serves as a filter to ensure fertilization of males that are able to produce highly viable and long-lived sperm, perhaps thereby accruing genetic benefits for the offspring. Furthermore, the presence of a simple sperm storage system suggests that cryptic female choice for genetically compatible males may occur primarily via gamete

interactions (Swanson & Vacquier 2002; see Madsen et al. 1992; Olsson et al. 1994a, 1996; Olsson & Madsen 2001 for discussion in reptiles). Ultimately, disentangling the selective causes for the observed variation among species will require studies of the fitness consequences of variation in male and female primary sexual characters in natural populations. Furthermore, as primary sexual traits are unlikely to evolve in isolation of secondary traits, we advocate an integrative whole organism approach, something for which many reptiles are uniquely suited (Lailvaux & Irschick 2006; Irschick et al. 2007).

SUMMARY: SEXUAL SELECTION AND PRIMARY SEXUAL TRAITS IN REPTILES

Evidence for sexual selection on male and female primary reproductive traits is strong in many taxa as evident from the contributions to this volume. In reptiles, however, the evidence is currently more circumstantial than direct. The scarcity of studies directly addressing this issue makes it difficult to evaluate to what extent this reflects a relatively minor effect or simply a lack of attention. However, the ubiquitous occurrence of multiple mating and multiple paternity strongly suggests that postcopulatory sexual selection should be strong in general in reptiles (Olsson & Madsen 1998; Uller & Olsson 2008). Our analyses of relative testis size found a significant relationship with climate, testis size being larger in temperate zone species. This may be an effect of a more intense sperm competition or a reduced cost of sperm production compared to tropical species with a prolonged breeding season.

In snakes, the available data suggest that copulatory plugs prevent intromission by foreign males or, alternatively, reduce female attractiveness via pheromonal cues. In lizards, however, there is no evidence that copulatory plugs prevent remating and plugs are more likely to serve as a source of sperm being taken up by females subsequently to mating. To what extent this allows cryptic female choice via biased uptake or ejection of sperm as has been described in fowl (Pizzari & Birkhead 2000), remains to be addressed.

Postcopulatory sexual selection on sperm should also be strong in reptiles. Comparative data do not yet allow direct tests of theory but the variation among investigated species suggests that this is a promising research approach. Within-species, among-individual variation in ejaculate and sperm traits and their relation to fertilization success under sperm competition have rarely been investigated and likewise deserve greater attention. The documented non-random variation in fertilization success (e.g., in relation to genetic similarity, Olsson et al. 1996) and wide distribution of sperm storage, suggest that sperm–female co-evolution (Miller & Pitnick 2002) should occur.

Sperm storage over periods extending the time from mating to fertilization seems to be common in female reptiles and is unlikely to be explained by selection to ensure fertilization. Surprisingly, however, there are virtually no studies that directly address the extent to which sperm storage facilitates cryptic female choice or its consequences for fertilization success among males (but see Zamudio & Sinervo 2000; Calsbeek & Sinervo 2004; Olsson et al. 2007a, b; 2009). However, evidence from an agamid lizard that stored sperm gives rise to a larger proportion of sons compared to recently inseminated sperm suggests that sperm storage could relate to sex allocation strategies (Olsson et al. 2007b).

TABLE 19.3 Suggested questions that could be addressed in future work on reptiles using a combination of field and laboratory experiments, molecular paternity assignment, and comparative methods

Testis size and sperm production/allocation

- What are the relative importance of sperm number, sperm morphology, and sperm longevity for male reproductive success under sperm competition? Does this differ for associate and dissociate breeders?
- Is sperm allocation facultative in relation to the level of sperm competition and environmental conditions (e.g., mate availability) and are there consistent differences between males (e.g., sneaks versus territorial males)?
- What is the cost of sperm production and are the relatively smaller testes of tropical species due to higher costs in prolonged breeders?

Sperm and ejaculates

- Is sperm competition important for the evolution of sperm morphology and longevity?
- Do ejaculates and female reproductive systems co-evolve, for example, does midpiece morphology and sperm aging relate to presence of sperm storage at the interspecific level?
- To what extent does dissociate mating systems select for changes in ejaculates and their relation to primary sexual characters?

Sperm storage

- Can sperm storage facilitate cryptic female choice via biased use of sperm from different males?
- Does sperm storage lead to differences in the strength and direction of post-copulatory sexual selection, for example by favoring increased sperm longevity over sperm numbers?
- What are the consequences of sperm storage for the strength of pre- and post-copulatory sexual selection and male mating strategies?

Clutch size and ovulation patterns

- Do females adjust their reproductive effort in relation to male traits and to what extent does capital breeding constrain adaptive allocation?
- Does sequential ovulation facilitate cryptic female choice of sperm and differential allocation?

Furthermore, the resulting selective pressures on male sexual traits, including ejaculate composition, sperm number and morphology, suggest that studies of species with sperm storage could generate valuable insights into the evolution of ejaculates and male and female co-evolution.

CONCLUDING REMARKS

Our survey of the literature on primary sexual traits in reptiles shows that there is an exciting scope for addressing key issues in sexual selection using reptilian model systems, both via within- and among-species studies. However, it also reveals our ignorance of the proximate mechanisms and selective forces behind the observed variation. We therefore outline some of the routes that seem to us the most interesting and feasible to take over the next ten years (figure 19.3; table 19.3).

Reptiles generated important insights in the early days of sexual selection studies (Noble & Bradley 1933; Trivers 1976; Stamps 1977) and have continued to generate insights into postcopulatory phenomena (e.g., Madsen et al. 1992; Olsson et al. 1996). However, we would argue that reptilian systems are being underutilized with respect to our understanding of the evolution of primary sexual traits. The field has been invigorated with novel conceptual and experimental approaches regarding both inter- and intraspecific variation over the last two decades (e.g., Arnqvist & Rowe 2005; Snook 2005). Thus, a concerted research agenda using reptilian model systems could add to our understanding of the evolution of gametes and their storage and provide important steps towards an integration of proximate and ultimate levels of explanation in post-copulatory sexual selection.

Acknowledgments Phil Byrne and Rebecca Dean provided valuable comments on the chapter. T. Uller was supported by the Wenner-Gren Foundations and the Australian Research Council. M. Olsson and D. Stuart-Fox are each supported by the Australian Research Council.

REFERENCES

Aldridge, R. D. & Duvall, D. 2002. Evolution of the mating season in the pitvipers of North America. Herpetological Monographs, 16, 1–25.

Almeida-Santos, S. M. & Graça Salomão, M. 2002. Reproduction in Neotropical Pitvipers, with Emphasis on Species of the Genus *Bothrops*. In: Biology of Vipers. (Ed. by G. W. Schuett, M. Höggren, M. E. Douglas & H. W. Greene), pp. 445–462, Eagle Mountain, UT: Eagle Mountain Publishing.

Anderson, M. J., Nyholt, J. & Dixson, A. F. 2005. Sperm competition and the evolution of sperm midpiece volume in mammals. Journal of Zoology, 267, 135–142.

Anderson, M. J., Dixson, A. S. & Dixson, A. F. 2006. Mammalian sperm and oviducts are sexually selected: evidence for co-evolution. Journal of Zoology, 270, 682–686.

Andersson, M. 1994. Sexual Selection. Princeton: Princeton University Press.

Andersson, M. & Simmons, L. W. 2006. Sexual selection and mate choice. Trends in Ecology & Evolution, 21, 296–302.

Arnqvist, G. & Rowe, L. 2005. Sexual Conflict. Princeton: Princeton University Press.

Badyaev, A. V., Seaman, D. A., Navara, K. J., Hill, G. E. & Mendonca, M. T. 2006. Evolution of sex-biased maternal effects in birds: III. Adjustment of ovulation order can enable sex-specific allocation of hormones, carotenoids, and vitamins. Journal of Evolutionary Biology, 19, 1044–1057.

Bakst, M. R. 1987. Anatomical basis of sperm-storage in the avian oviduct. Scanning Microscopy, 1, 1257–1266.

Birkhead, T. R. & Møller, A. P. 1993. Sexual selection and the temporal separation of reproductive events—sperm storage data from reptiles, birds and mammals. Biological Journal of the Linnean Society, 50, 295–311.

Birkhead, T. R. & Møller, A. P. 1998. Sperm Competition and Sexual Selection. New York: Academic Press.

Birkhead, T. & Pizzari, T. 2002. Postcopulatory sexual selection. Nature Reviews Genetics, 3, 262–273.

Birkhead, T. R., Martinez, J. G., Burke, T. & Froman, D. P. 1999. Sperm mobility determines the outcome of sperm competition in the domestic fowl. Proceedings of the Royal Society of London, Series B, 266, 1759–1764.

Bonnet, X., Naulleau, G., Shine, R. & Lourdais, O. 2001. Short-term versus long-term effects of food intake on reproductive output in a viviparous snake (*Vipera aspis*). Oikos, 92, 297–308.

Bonnet, X., Lourdais, O., Shine, R. & Naulleau, G. 2002. Reproduction in a typical capital breeder: Costs, currencies, and complications in the aspic viper. Ecology, 83, 2124–2135.

Breque, C., Surai, P. & Brillard, J. P. 2003. Roles of antioxidants on prolonged storage of avian spermatozoa in vivo and in vitro. Molecular Reproduction and Development, 66, 314–323.

Briskie, J. V. & Montgomerie, R. 1993. Patterns of sperm storage in relation to sperm competition in passerine birds. Condor, 95, 442–454.

Briskie, J. V., Montgomerie, R. & Birkhead, T. R. 1997. The evolution of sperm size in birds. Evolution, 51, 937–945.

Bull, J. J. & Shine, R. 1979. Iteroparous animals that skip opportunities for reproduction. American Naturalist, 114, 296–303.

Byrne, P. G., Roberts, J. D. & Simmons, L. W. 2002. Sperm competition selects for increased testes mass in Australian frogs. Journal of Evolutionary Biology, 15, 347–355.

Byrne P. G., Simmons, L. W. & Roberts, J. D. 2003. Sperm competition and the evolution of gamete morphology in frogs. Proceedings of the Royal Society of London, Series B, 270, 2079–2086.

Calhim, S. & Birkhead, T. R. 2007. Testes size in birds: quality versus quantity—assumptions, errors, and estimates. Behavioral Ecology, 18, 271–275.

Calhim, S., Immler, S. & Birkhead, T. R. 2007. Postcopulatory sexual selection is associated with reduced variation in sperm morphology. PLoS ONE, 5, e413.

Calsbeek, R. & Sinervo, B. 2004. Within-clutch variation in offspring sex determined by differences in sire body size: cryptic mate choice in the wild. Journal of Evolutionary Biology, 17, 464–470.

Calsbeek, R., Bonneaud, C., Prabhu, S., Manoukis, N. & Smith, T. B. 2007. Multiple paternity and sperm storage lead to increased genetic diversity in Anolis lizards. Evolutionary Ecology Research, 9, 495–503.

Cardullo, R. A. & Baltz, J. M. 1991. Metabolic-regulation in mammalian sperm—mitochondrial volume determines sperm length and flagellar beat frequency. Cell Motility and the Cytoskeleton, 19, 180–188.

Carranza, J. 1996. Sexual selection for male body mass and the evolution of litter size in mammals. American Naturalist, 148, 81–100.

Cogger, H. G. 1978. Reproductive-cycles, fat-body cycles and socio sexual behavior in the mallee dragon, Amphibolurus-fordi (Lacertilia-Agamidae). Australian Journal of Zoology, 26, 653–672.

Colli, G. R., Teixeira, R. D., Scheltinga, D. M., Mesquita, D. O., Wiederhecker, H. C. & Bao, S. N. 2007. Comparative study of sperm ultrastructure of five species of teiid lizards (Teiidae, Squamata), and Cercosaura ocellata (Gymnophthalmidae, Squamata). Tissue & Cell, 39, 59–78.

Conner, J. & Crews, D. 1980. Sperm Transfer and Storage in the Lizard, *Anolis-carolinensis*. Journal of Morphology, 163, 331–348.

Crews, D. 1984. Gamete production, sex hormone secretion, and mating behavior uncoupled. Hormones and Behavior, 18, 22–28.

Cuellar, O. 1966. Oviducal anatomy and sperm storage structures in lizards. Journal of Morphology, 119, 7–19.

Davis, L. M., Glenn, T. C., Elsey, R. M., Dessauer, H. C. & Sawyer, R. H. 2001. Multiple paternity and mating patterns in the American alligator, *Alligator mississippiensis*. Molecular Ecology, 10, 1011–1024.

Dean, R., Bonsall, M. B. & Pizzari, T. 2007. Aging and sexual conflict. Science, 316, 383–384.

Depeiges, A. & Dacheux, J. L. 1985. Acquisition of sperm motility and its maintenance during storage in the lizard, Lacerta-vivipara. Journal of Reproduction and Fertility, 74, 23–27.

Devine, M. C. 1975. Copulatory plugs in snakes: Enforced chastity. Science, 187, 844–845.

Devine, M. C. 1977. Copulatory plugs, restricted mating opportunities, and reproductive competition among male garter snakes. Nature, 267, 345–346.

Devine, M. C. 1984. Potential for sperm competition in reptiles: Behavioral and physiological consequences. In: Sperm Competition and the Evolution of Mating Systems (Ed. by R. L. Smith), pp. 509–521. New York: Academic Press.

Duvall, D., Guillette L. J. & Jones, R. E. 1982. Environmental control of reptilian reproductive cycles. In: Physiology and Physiological Ecology: Biology of the Reptilia, (Ed. by

C. Gans & F. H. Pough), pp. 201–231. New York: Wiley.

Duvall, D., Schuett, G. W. & Arnold, S. J. 1993. Ecology and evolution of snake mating systems. In: Snakes: Ecology and Behavior (Ed. by R. A. Seigel & J. T. Collins), pp. 165–200. New York: McGraw-Hill.

Eberhard, W. G. 1996. Femal Control: Sexual Selection by Cryptic Female Choice. Princeton: Princeton University Press.

Felsenstein, J. 1985. Phylogenies and the comparative method. American Naturalist, 125, 1–15.

Ferguson, M. W. J. 1985. Reproductive biology and embryology of the Crocodilians. In: Biology of the Reptilia, Vol 14. Development A (Ed. by A. F. Billett & P. F. A. Madserson), pp. 329–491. New York: John Wiley & Sons.

Fitch, H. S. 1970. Reproductive Cycles in Lizards and Snakes. Miscellaneous Publication 52, University of Kansas Museum of Natural History.

Fox, W. 1956. Seminal receptacles in snakes. The Anatomical Record, 124, 415–450.

Fox, W. 1963. Special tubules for sperm storage in female lizards. Nature, 198, 500–501.

Fox, H. 1977. The urogenital system of reptiles. In: Biology of the Reptilia. Vol. 6, Morphology (Ed. by C. Gans & T. S. Parsons), pp. 1–157. New York: Academic Press.

Furieri, P. 1970. Sperm morphology of some reptiles: Squamata and Chelonia. In: Comparative Spermatology (Ed. by B. Baccetti), pp. 115–131. New York: Academic Press.

Gage, M. J. G. 1994. Associations between body-size, mating pattern, testis size and sperm lengths across butterflies. Proceedings of the Royal Society of London, Series B, 258, 247–254.

Gage, M. J. G. 1998. Influences of sex, size, and symmetry on ejaculate expenditure in a moth. Behavioral Ecology, 9, 592–597.

Gage, M. J. G. & Freckleton, R. P. 2003. Relative testis size and sperm morphometry across mammals: No evidence for an association between sperm competition and sperm length. Proceedings of the Royal Society of London, Series B, 270, 625–632.

Garcia-Gonzalez, F. & Simmons, L. W. 2007. Shorter sperm confer higher competitive fertilization success. Evolution, 61, 816–824.

Gillott, C. 2003. Male accessory gland secretions: Modulators of female reproductive physiology and behavior. Annual Review of Entomology, 48, 163–184.

Girling, J. E. 2002. The reptilian oviduct: A review of structure and function and directions for future research. Journal of Experimental Zoology, 293, 141–170.

Gist, D. H. & Congdon, J. D. 1998. Oviductal sperm storage as a reproductive tactic of turtles. Journal of Experimental Zoology, 282, 526–534.

Gist, D. H. & Fischer, E. N. 1993. Fine-structure of the sperm storage tubules in the box turtle oviduct. Journal of Reproduction and Fertility, 97, 463–468.

Gist, D. H. & Jones, J. M. 1987. Storage of sperm in the reptilian oviduct. Scanning Microscopy, 1, 1839–1849.

Gist, D. H. & Jones, J. M. 1989. Sperm storage within the oviduct of turtles. Journal of Morphology, 199, 379–384.

Gist, D. H., Turner, T. W. & Congdon, J. D. 2000. Chemical and thermal effects on the viability and motility of spermatozoa from the turtle epididymis. Journal of Reproduction and Fertility, 119, 271–277.

Gist, D. H., Dawes, S. M., Turner, T. W., Sheldon, S. & Congdon, J. D. 2002. Sperm storage in turtles: A male perspective. Journal of Experimental Zoology, 292, 180–186.

Giugliano, L. G., Teixeira, R. D., Colli, G. R. & Bao, S. N. 2002. Ultrastructure of spermatozoa of the lizard Ameiva ameiva, with considerations on polymorphism within the family Teiidae (Squamata). Journal of Morphology, 253, 264–271.

Gomendio, M., Malo, A. F., Garde, J. & Roldan, E. R. S. 2007. Sperm traits and male fertility in natural populations. Reproduction, 134, 19–29.

Greer, A. E. 1975. Clutch size in crocodilians. Journal of Herpetology, 9, 319–322.

Groot, T. V. M., Bruins, E. & Breeuwer, J. A. J. 2003. Molecular genetic evidence for parthenogenesis in the Burmese python, Python molurus bivittatus. Heredity, 90, 130–135.

Halpert, A. P., Garstka, W. R. & Crews, D. 1982. Sperm transport and storage and its relation to the annual sexual cycle of the female red-sided garter snake, Thamnophis sirtalis parietalis. Journal of Morphology, 174, 149–159.

Hamilton, D. W. & Fawcett, D. W. 1968. Unusual features of neck and middle-piece of snake spermatozoa. Journal of Ultrastructure Research, 23, 81–97.

Harding, H. R., Aplin, K. P. & Mazur, M. 1995. Ultrastructure of spermatozoa of Australian blindsnakes, Ramphotyphlops spp. (Typhlopidae, Squamata): first observations on the mature spermatozoon of scolecophidian snakes. In: Advances in Spermatozoal Phylogeny and Taxonomy (Ed. by Jamieson, B. G. M., Ausio, J. & Justine, J. L.) Mém. Mus. Natn. Hist. Nat., 166, 385–396. Paris

Harris, W. E., Moore, A. J. & Moore, P. J. 2007. Variation in sperm size within and between ejaculates in a cockroach. Functional Ecology, 21, 598–602.

Hedges, S. B. & Poling, L. L. 1999. A molecular phylogeny of reptiles. Science, 283, 998–1001.

Herczeg, G., Saarikivi, J., Abigel, G., Perälä, J., Toumola, A. & Merilä, J. 2007. Suboptimal thermoregulation in male adders (Vipera berus) after hibernation imposed by spermiogenesis. Biological Journal of the Linnean Society, 92, 19–27.

Hess, R. A., Thurston, R. J. & Gist, D. H. 1991. Ultrastructure of the turtle Spermatozoon. The Anatomical Record, 229, 473–481.

Hoggren, M. 1995. Mating strategies and sperm competition in the adder (Vipera berus). PhD Thesis, Uppsala University.

Hosken, D. J. 1998. Testes mass in megachiropteran bats varies in accordance with sperm competition theory. Behavioral Ecology and Sociobiology, 44, 169–177.

Hosken, D. J. & Ward, P. I. 2001. Experimental evidence for testis size evolution via sperm competition. Ecology Letters, 4, 10–13.

Howarth, B. 1974. Sperm storage: as a function of the female reproductive tract. In: The Oviduct and its Functions (Ed. by A. D. Johnson & C. W. Foley), pp. 237–270. New York: Academic Press.

Immler, S. & Birkhead, T. R. 2007. Sperm competition and sperm midpiece size: no consistent pattern in passerine birds. Proceedings of the Royal Society of London, Series B, 274, 561–568.

Immler, S., Saint-Jalme, M., Lesobre, L., Sorci, G., Roman, Y. & Birkhead, T. R. 2007. The evolution of sperm morphometry in pheasants. Journal of Evolutionary Biology, 20, 1008–1014.

in den Bosch, H. A. J. 1994. First record of mating plugs in lizards. Amphibia-Reptilia, 15, 89–93.

Irschick, D. J., Herrel, A., Vanhooydonck, B. & Van Damme, R. 2007. A functional approach to sexual selection. Functional Ecology, 21, 621–626.

James, C. & Shine, R. 1985. The seasonal timing of reproduction - a tropical-temperate comparison in Australian lizards. Oecologia, 67, 464–474.

Jamieson, B. G. M. 1995. The ultrastructure of spermatozoa of the Squamata (Reptilia) with phylogenetic considerations. In: Advances in Spermatozoal Phylogeny and Taxonomy (Ed. by Jamieson, B. G. M., Ausio, J. & Justine, J. L.). Mem. Mus. Natn. Hist. Nat. 166, pp. 359–383.

Jamieson, B. G. M. & Healy, J. M. 1992. The phylogenetic position of the Tuatara, Sphenodon (Sphenodontida, Amniota), as indicated by cladistic-analysis of the ultrastructure of spermatozoa. Philosophical Transactions of the Royal Society of London, Series B, 335, 207–219.

Jamieson, B. G. M. & Scheltinga, D. M. 1994. The ultrastructure of spermatozoa of the Australian skinks, Ctenotus taeniolatus, Carlia pectoralis and Tiliqua scincoides scincoides (Scincidae, Reptilia). Memoirs of the Queensland Museum, 37, 181–193.

Jamieson, B. G. M., Scheltinga, D. M. & Tucker, A. D. 1997. The ultrastructure of spermatozoa of the Australian freshwater crocodile, Crocodylus johnstoni Krefft, 1873 (Crocodylidae, Reptilia). Journal of Submicroscopic Cytology and Pathology, 29, 265–274.

Jensen, M. P., Abreu-Grobois, F. A., Frydenberg, J. & Loeschcke, V. 2006. Microsatellites provide insight into contrasting mating patterns in arribada vs. non-arribada olive ridley sea turtle rookeries. Molecular Ecology, 15, 2567–2575.

Keogh, J. S. 1999. Evolutionary implications of hemipenial morphology in the terrestrial Australian elapid snakes. Zoological Journal of the Linnean Society, 125, 239–278.

Krohmer, R. W. 2004. Variation in seasonal ultrastructure of sexual granules in the renal sexual segment of the Northern Water Snake, Nerodia sipedon sipedon. Journal of Morphology, 261, 70–80.

Kuhl, A., Mysterud, A., Erdnenov, G. I., Lushchekina, A. A., Grachev, I. A., Bekenov, A. B. & Milner-Gulland, E. J. 2007. The 'big spenders' of the steppe: sex-specific maternal allocation and twinning in the saiga antelope. Proceedings of the Royal Society London, Series B, 274, 1293–1299.

Lailvaux, S. P. & Irschick, D. J. 2006. A functional perspective on sexual selection: insights and future prospects. Animal Behaviour, 72, 263–273.

Laloi, D., Richard, M., Lecomte, J., Massot, M. & Clobert, J. 2004. Multiple paternity in clutches of common lizard Lacerta vivipara: data from microsatellite markers. Molecular Ecology, 13, 719–723.

Licht, P. 1984. Reproductive cycles of vertebrates. Reptiles. In: Marshall's Physiology of Reproduction vol 1 (Ed. by G. E. Lamming), pp 206–282. New York: Churchill Livingstone.

Lopez, P., Aragon, P. & Martin, J. 2003. Responses of female lizards, Lacerta monticola, to males' chemical cues reflect their mating preference for older males. Behavioral Ecology and Sociobiology, 55, 73–79.

Lovlie, H. & Pizzari, T. 2007. Sex in the morning or in the evening? Females adjust daily mating patterns to the intensity of sexual harassment. American Naturalist, 170, E1–E13.

Lutz, P. L. & Musick, J. A. 1996. The Biology of Sea Turtles, vol I. Boca Raton, FL: CRC Press.

Madsen, T., Shine, R., Loman, J. & Håkansson, T. 1992. Why do female adders copulate so frequently? Nature, 355, 440–441.

Magnusson, W. E. 1979. Production of an embryo by an Achrochordus javanicus isolated for seven years. Copiea, 1979, 744–745.

Malo, A. F., Gomendio, M., Garde, J., Lang-Lenton, B., Soler, A. J. & Roldan, E. R. S. 2006. Sperm design and sperm function. Biology Letters, 2, 246–249.

Martins, E. P. 1994. Phylogenetic perspectives on the evolution of lizard territoriality. In: Lizard Ecology—Historical and Experimental Perspectives (Ed. by L. J. Vitt & E. R. Pianka), pp. 117–144. Princeton: Princeton University Press.

Martins, E. P. 2004. COMPARE 4.6: statistical analysis of comparative data, version 4.6.

Department of Biology, Indiana University, Bloomington.

Martins, E. P. & Hansen, T. F. 1997. Phylogenies and the comparative method: a general approach to incorporating phylogenetic information into the analysis of interspecific data. American Naturalist, 149, 646–667.

Miller, G. T. & Pitnick, S. 2002. Sperm–female coevolution in Drosophila. Science, 298, 1230–1233.

Møller, A. P. & Briskie, J. V. 1995. Extra-pair paternity, sperm competition and the evolution of testis size in birds. Behavioral Ecology and Sociobiology, 36, 357–365.

Moreira, P. L. & Birkhead, T. R. 2003. Copulatory plugs in the Iberian Rock Lizard do not prevent insemination by rival males. Functional Ecology, 17, 796–802.

Moreira, P. L., Lopez, P. & Martin, J. 2006. Femoral secretions and copulatory plugs convey chemical information about male identity and dominance status in Iberian rock lizards (Lacerta monticola). Behavioral Ecology and Sociobiology, 60, 166–174.

Moreira, P., Nunes, V., Martin, J. & Paulo, O. S. 2007. Copulatory plugs do not assure high first male fertilisation success: sperm displacement in a lizard. Behavioral Ecology and Sociobiology, 62, 281–288.

Murphy, K., Hudson, S. & Shea, G. 2006. Reproductive seasonality of three cold-temperate viviparous skinks from southeastern Australia. Journal of Herpetology, 40, 454–464.

Neubaum, D. M. & Wolfner, M. F. 1999. Wise, winsome, or weird? Mechanisms of sperm storage in female animals. Current Topics in Developmental Biology, Vol 41, 41, 67–97.

Newton, W. D. & Trauth, S. E. 1992. Ultrastructure of the spermatozoon of the lizard Cnemidophorus sexlineatus (Saura: Teiidae). Herpetologica, 48, 330–343.

Noble, G. K. & Bradley, H. T. 1933. The mating behavior of lizards: Its bearing on the theory of sexual selection. Annals of the New York Academy of Sciences, 35, 25–100.

Olivier, S. C., Jamieson, B. G. M. & Scheltinga, D. M. 1996. The ultrastructure of spermatozoa of squamata. II. Agamidae, varanidae, colubridae, elapidae, and boidae (Reptilia). Herpetologica, 52, 216–241.

Olsson, M. & Madsen, T. 1995. Female choice on male quantitative traits in lizards—why is it so rare? Behavioral Ecology Sociobiology, 36, 179–184.

Olsson, M. & Madsen, T. 1996. Costs of mating with infertile males selects for late emergence in female sand lizards. Copeia, 2, 462–464.

Olsson, M. & Madsen, T. 1998. Sexual selection and sperm competition in reptiles. In: Sperm Competition and Sexual Selection (Ed. by T. R. Birkhead & A. P. Møller), pp. 503–578. Cambridge: Academic Press.

Olsson, M. & Madsen, T. 2001. Promiscuity in sand lizards (Lacerta agilis) and adder snakes (Vipera berus): Causes and consequences. Journal of Heredity, 92, 190–197.

Olsson, M. & Shine, R. 1997a. The seasonal timing of oviposition in sand lizards: why early clutches are better. Journal of Evolutionary Biology, 10, 369–381.

Olsson, M. & Shine, R. 1997b. Advantages of multiple matings to females: a test of the infertility hypothesis using lizards. Evolution, 51, 1684–1688.

Olsson, M., Gullberg, A., Tegelström, H., Madsen, T. & Shine, R. 1994a. Can female adders multiply? Nature 369, 528.

Olsson, M., Gullberg, A. & Tegelström, H. 1994b. Sperm competition in the sand lizard. Animal Behaviour, 48, 193–200.

Olsson, M., Gullberg, A. & Tegelstrom, H. 1996. Mate guarding in male sand lizards (Lacerta agilis). Behaviour, 133, 367–386.

Olsson, M., Shine, R., Madsen, T., Gullberg, A. & Tegelström, H. 1996. Sperm selection by females. Nature, 383, 585–585.

Olsson, M., Madsen, T. & Shine, R. 1997. Is sperm really so cheap? Costs of reproduction in male adders, Vipera berus. Proceedings of the Royal Society of London, Series B, 264, 455–459.

Olsson, M., Birkhead, T. & Shine, R. 1999. Can relaxed time constraints on sperm production eliminate protandry in an ectotherm? Biological Journal of the Linnean Society, 66, 159–170.

Olsson, M., Madsen T., Ujvari, B. & Wapstra, E. 2004. Fecundity and MHC affects ejaculation tactics and paternity bias in sand lizards. Evolution, 58, 906–909.

Olsson, M., Wapstra, E. & Uller, T. 2005. Differential sex allocation in sand lizards: bright males induce daughter production in a species with heteromorphic sex chromosomes. Biology Letters, 1, 378–380.

Olsson, M., Healey, M., Wapstra, E., Schwartz, T., LeBas, N. & Uller, T. 2007a. Mating system variation and morph fluctuations in a polymorphic lizard. Molecular Ecology, 16, 5307–5315.

Olsson, M., Schwartz, T., Uller, T. & Healey, M. 2007b. Sons are made from old stores: sperm storage effects on sex ratio in a lizard. Biology Letters, 3, 491–493.

Olsson, M., Schwartz, T., Uller, T. & Healey, M. 2009. Effects of sperm storage and male colour on probability of paternity in a polychromatic lizard. Animal Behaviour, 77: 419–424.

Ord, T. J. & Stuart-Fox, D. 2006. Ornament evolution in dragon lizards: multiple gains and widespread losses reveal a complex history of evolutionary change. Journal of Evolutionary Biology, 19, 797–808.

Oring, L. W., Fleischer, R. C., Reed, J. M. & Marsden, K. E. 1992. Cuckoldry through stored sperm in the sequentially polyandrous spotted sandpiper. Nature, 359, 631–633.

Parker, G. A. 1970. Sperm competition and its evolutionary consequences in insects. Biological Reviews of the Cambridge Philosophical Society, 45, 525–567.

Parker, G. A. 1990. Sperm competition games— raffles and roles. Proceedings of the Royal Society of London, Series B, 242, 120–126.

Parker, G. A. 1998. Sperm competition and the evolution of ejaculates: towards a theory base. In: Sperm Competition and Sexual Selection (Ed. by T. R. Birkhead & A. P. Møller), pp. 3–55. Cambridge: Academic Press.

Parker, G. A. 2006. Sexual conflict over mating and fertilization: an overview. Philosophical Transactions of the Royal Society of London, Series B, 361, 235–259.

Parker, G. A., Ball, M. A., Stockley, P. & Gage, M. J. G. 1997. Sperm competition games: a prospective analysis of risk assessment. Proceedings of the Royal Society of London, Series B, 264, 1793–1802.

Pearse, D. E. & Avise, J. C. 2001. Turtle mating systems: Behavior, sperm storage, and genetic paternity. Journal of Heredity, 92, 206–211.

Pearse, D. E., Janzen, F. J. & Avise, J. C. 2001. Genetic markers substantiate long–term storage and utilization of sperm by female painted turtles. Heredity, 86, 378–384.

Pearse, D. E., Janzen, F. J. & Avise, J. C. 2002. Multiple paternity, sperm storage, and reproductive success of female and male painted turtles (Chrysemys picta) in nature.

Behavioral Ecology and Sociobiology, 51, 164–171.

Pitnick, S., Markow, T. & Spicer, G. S. 1999. Evolution of multiple kinds of female sperm-storage organs in *Drosophila*. Evolution, 53, 1804–1822.

Pizzari, T. & Birkhead, T. R. 2000. Female feral fowl eject sperm of subdominant males. Nature, 405, 787–789.

Pizzari, T., Dean, R., Pacey, A., Moore, H. & Bonsall, M. B. 2008. The evolutionary ecology of pre- and post-meiotic sperm senescence. Trends in Ecololgy & Evolution, 23, 131–140.

Poiani, A. 2006. Complexity of seminal fluid: A review. Behavioral Ecology and Sociobiology, 60, 289–310.

Prosser, M. R., Weatherhead, P. J., Gibbs, H. L. & Brown, G. P. 2002. Genetic analysis of the mating system and opportunity for sexual selection in northern water snakes (*Nerodia sipedon*). Behavioral Ecology, 13, 800–807.

Purvis, A., Gittleman, J. L., Cowlishaw, G. & Mace, G. M. 2000. Predicting extinction risk in declining species. Proceedings of the Royal Society of London, Series B, 267, 1947–1952.

Ram, K. R. & Wolfner, M. F. 2007. Seminal influences: *Drosophila* Acps and the molecular interplay between males and females during reproduction. Integrative and Comparative Biology, 47, 427–445.

Ross, P. & Crews, D. 1977. Influence of seminal plug on mating-behavior in garter snake. Nature, 267, 344–345.

Ross, P. & Crews, D. 1978. Stimuli influencing mating-behavior in garter snake, *Thamnophis-radix*. Behavioral Ecology and Sociobiology, 4, 133–142.

Saint-Girons, H. 1975. Sperm survival and transport in the female genital tract of reptiles. In: The Biology of the Spermatozoa, (Ed. by E. Hafez & C. Thibault), pp. 105–113. Basel: Karger.

Saint-Girons, H. 1982. Reproductive cycles of male snakes and their relationship with climate and female reproductive cycles. Herpetologica, 38, 5–16.

Scheltinga, D. M., Jamieson, B. G. M., Trauth, S. E. & Mcallister, C. T. 2000. Morphology of the spermatozoa of the iguanian lizards *Uta stansburiana* and *Urosaurus ornatus* (Squamata, Phrynosomatidae). Journal of Submicroscopic Cytology and Pathology, 32, 261–271.

Scheltinga, D. M., Jamieson, B. G. M., Espinoza, R. E. & Orrell, K. S. 2001. Descriptions of the mature spermatozoa of the lizards *Crotaphytus bicinctores*, *Gambelia wislizenii* (Crotaphytidae), and *Anolis carolinensis* (Polychrotidae) (Reptilia, Squamata, Iguania). Journal of Morphology, 247, 160–171.

Schuett, G. W. 1992. Is long-term sperm storage an important component of the reproductive biology of temperate pitvipers? In: Biology of the Pitvipers (Ed. by J. A. Campbell & E. D. Brodie III), pp. 169–184. Tyler, TX: Selva.

Schulte-Hostedde, A. I. & Montgomerie, R. 2006. Intraspecific variation in ejaculate traits of the northern watersnake (*Nerodia sipedon*). Journal of Zoology, 270, 147–152.

Schwartz, J. M., McCracken, G. F. & Burghardt, G. M. 1989. Multiple paternity in wild populations of the garter snake, *Thamnophis sirtalis*. Behavioral Ecology and Sociobiology, 25, 269–273.

Schwarzkopf, L. 1994. Measuring trade-offs: a review of studies of costs of reproduction in lizards. In: Lizard Ecology: Historical and Experimental Perspectives (Ed. by L. J. Vitt & E. R. Pianka), pp. 7–29. Princeton: Princeton University Press.

Seigel, R. A. & Ford, N. B. 1987. Reproductive ecology. In: Snakes: Ecology and Evolutionary Biology (Ed. by R. A. Seigel, J. T. Collins & S. S. Novak), pp. 210-252. New York: Macmillan.

Sever, D. M. & Hamlett, W. C. 2002. Female sperm storage in reptiles. Journal of Experimental Zoology, 292, 187–199.

Sever, D. M. & Hopkins, W. A. 2004. Oviductal sperm storage in the ground skink *Scincella laterale holbrook* (Reptilia : Scincidae). Journal of Experimental Zoology Part a—Comparative Experimental Biology, 301A, 599–611.

Sever, D. M. & Hopkins, W. A. 2005. Renal sexual segment of the ground skink, *Scincella laterale* (Reptilia, Squamata, Scincidae). Journal of Morphology, 266, 46–59.

Sever, D. M. & Ryan, T. J. 1999. Ultrastructure of the reproductive system of the black swamp snake (*Seminatrix pygaea*): Part I. Evidence for oviducal sperm storage. Journal of Morphology, 241, 1–18.

Sheldon, B. 2000. Differential allocation: tests, mechanisms and implications. Trends in Ecology & Evolution, 15, 397–402.

Shine, R. 1983. Reptilian reproductive modes—the oviparity-viviparity continuum. Herpetologica, 39, 1–8.

Shine, R. 1988. Parental care in reptiles. In: Biology of the Reptilian, vol 16 (Ed. by C. Gans & R. B. Huey), pp. 275–330. New York: Alan R. Liss Inc.

Shine, R. 2003. Reproductive strategies in snakes. Proceedings of the Royal Society of London, Series B, 270, 995–1004.

Shine, R. 2005. Life-history evolution in reptiles. Annual Review of Ecology Evolution and Systematics, 36, 23–46.

Shine, R., Olsson, M. M. & Mason, R. T. 2000. Chastity belts in gartersnakes: the functional significance of mating plugs. Biological Journal of the Linnean Society, 70, 377–390.

Short, R. V. 1979. Sexual selection and its component parts, somatic and genital selection, as illustrated by man and the great apes. Advances in the Study of Behavior, 9, 131–158.

Simmons, L. W. 2001. Sperm Competition and its Evolutionary Consequences in the Insects. Princeton: Princeton University Press.

Simmons, L. W. & Emlen, D. J. 2006. Evolutionary trade-off between weapons and testes. Proceedings of the National Academy of Sciences of the United States of America, 103, 16346–16351.

Smith, H. M., Sinelnik, G., Fawcett, J. D. & Jones, R. E. 1972. A survey of the chronology of ovulation in Anoline lizard genera. Transactions of the Kansas Academy of Sciences, 75, 107–121.

Smyth, M. & Smith, M. J. 1968. Obligatory sperm storage in skink Hemiergis peronii. Science, 161, 575–576.

Snook, R. R. 2005. Sperm in competition: not playing by the numbers. Trends in Ecology & Evolution, 20, 46–53.

Snow, L. S. E. & Andrade, M. C. B. 2005. Multiple sperm storage organs facilitate female control of paternity. Proceedings of the Royal Society of London, Series B, 272, 1139–1144.

Stamps, J. A. 1977. Social behavior and spacing patterns in lizards. In: Biology of the Reptilia Vol 7. Ecology and Behavior. (Ed. by C. Gans & D. W. Tinkle), pp. 265–334. New York: Academic Press.

Stuart-Fox, D. & Moussalli, A. 2007. Sex-specific ecomorphological variation and the evolution of sexual dimorphism in dwarf chameleons (Bradypodion spp.). Journal of Evolutionary Biology, 20, 1073–1081.

Swain, R. & Jones, S. M. 1994. Annual cycle of plasma testosterone and other reproductive parameters in the Tasmanian skink, Niveoscincus metallicus. Herpetologica, 50, 502–509.

Swallow, J. G. & Wilkinson, G. S. 2002. The long and short of sperm polymorphisms in insects. Biological Reviews, 77, 153–182.

Swanson, W. J. & Vacquier, V. D. 2002. Reproductive protein evolution. Annual Review of Ecology and Systematics, 33, 161–179.

Tavares-Bastos, L., Teixeira, R. D., Colli, G. R. & Bao, S. N. 2002. Polymorphism in the sperm ultrastructure among four species of lizards in the genus Tupinambis (Squamata: Teiidae). Acta Zoologica, 83, 297–307.

Tavares-Bastos, L., Cunha, L. D., Colli, G. R. & Bao, S. N. 2007. Ultrastructure of spermatozoa of scolecophidian snakes (Lepidosauria, Squamata). Acta Zoologica, 88, 189–197.

Teixeira, R. D., Colli, G. R. & Bao, S. N. 1999a. The ultrastructure of the spermatozoa of the worm lizard Amphisbaena alba (Squamata, Amphisbaenidae) and the phylogenetic relationships of amphisbaenians. Canadian Journal of Zoology, 77, 1254–1264.

Teixeira, R. D., Colli, G. R. & Bao, S. N. 1999b. The ultrastructure of the spermatozoa of the lizard Micrablepharus maximiliani (Squamata, Gymnophthalmidae), with considerations on the use of sperm ultrastructure characters in phylogenetic reconstruction. Acta Zoologica, 80, 47–59.

Teixeira, R. D., Colli, G. R. & Bao, S. N. 1999c. The ultrastructure of the spermatozoa of the lizard Polychrus acutirostris (Squamata, Ploychrotidae). Journal of Submicroscopic Cytology and Pathology, 31, 387–395.

Teixeira, R. D., Vieira, G. H. C., Colli, G. R. & Bao, S. N. 1999d. Ultrastructural study of spermatozoa of the neotropical lizards, Tropidurus semitaeniatus and Tropidurus torquatus (Squamata, Tropiduridae). Tissue & Cell, 31, 308–317.

Teixeira, R. D., Scheltinga, D. M., Trauth, S. E., Colli, G. R. & Bao, S. N. 2002. A comparative ultrastructural study of spermatozoa of the teiid lizards Cnemidophorus gularis gularis, Cnemidophorus ocellifer, and Kentropyx altamazonica (Reptilia,

Squamata, Teiidae). Tissue & Cell, 34, 135–142.

Tokarz, R. 1995. Mate choice in lizards: a review. Herpetological Monographs, 9, 17–40.

Tourmente, M., Cardozo, G., Bertona, M., Guidobaldi, A., Giojalas, L. & Chiaraviglio, M. 2006. The ultrastructure of the spermatozoa of *Boa constrictor occidentalis*, with considerations on its mating system and sperm competition theories. Acta Zoologica, 87, 25–32.

Townsend, T. M., Larson, A., Louis, E. & Macey, J. R. 2004. Molecular phylogenetics of Squamata: The position of snakes, Amphisbaenians, and Dibamids, and the root of the Squamate tree. Systematic Biology, 53, 735–757.

Trivers, R. L. 1972. Mother-offspring conflict. American Zoologist, 12, 648–648.

Trivers, R. L. 1976. Sexual selection and resource-accruing abilities in *Anolis garmani*. Evolution, 30, 253–269.

Uller, T. 2003. Viviparity as a constraint on sex-ratio evolution. Evolution, 57, 927–931.

Uller, T. 2006. Sex-specific sibling interactions and offspring fitness in vertebrates: patterns and implications for maternal sex ratios. Biological Reviews, 81, 207–217.

Uller, T. & Olsson, M. 2006. Multiple copulations in natural populations of lizards: evidence for the fertility assurance hypothesis. Behaviour, 142, 45–56.

Uller, T. & Olsson, M. 2008. Multiple paternity in reptiles: patterns and processes. Molecular Ecology, 17, 2566–2580.

Uller, T., Astheimer, L. & Olsson, M. 2007. Consequences of egg yolk testosterone for offspring development and survival: experimental test in a lizard. Functional Ecology, 21, 544–561.

Vahed, K. 1998. The function of nuptial feeding in insects: review of empirical studies. Biological Reviews, 73, 43–78.

Vieira, G. H. C., Colli, G. R. & Bao, S. N. 2004. The ultrastructure of the spermatozoon of the lizard *Iguana iguana* (Reptilia, Squamata, Iguanidae) and the variability of sperm morphology among iguanian lizards. Journal of Anatomy, 204, 451–464.

Vieira, G. H. C., Colli, G. R. & Bao, S. N. 2005. Phylogenetic relationships of corytophanid lizards (Iguania, Squamata, Reptilia) based on partitioned and total evidence analyses of sperm morphology, gross morphology,

and DNA data. Zoologica Scripta, 34, 605–625.

Vieira, G. H. C., Cunha, L. D., Scheltinga, D. M., Glaw, F., Colli, G. R. & Bao, S. N. 2007. Sperm ultrastructure of hoplocercid and oplurid lizards (Sauropsida, Squamata, Iguania) and the phylogeny of Iguania. Journal of Zoological Systematics and Evolutionary Research, 45, 230–241.

Volsøe, H. 1944. Structure and seasonal variation of the male reproductive organs of *Vipera berus* (L.). Spolia Zoologica Musei Heuni Ensis V, 5, 1–157.

Wapstra, E., Swain, R., Jones, S. M. & O'Reilly, J. 1999. Geographic and annual variation in reproductive cycles in the Tasmanian spotted snow skink, *Niveoscincus ocellatus* (Squamata : Scincidae). Australian Journal of Zoology, 47, 539–550.

Warner, D. & R. Shine. 2007. Fitness of juvenile lizards depends on seasonal timing of hatching, not offspring body size. Oecologia, 154, 65–73.

Watts, P. C., Buley, K. R., Boardman, W., Ciofi, C. & Gibson O. R. 2006. Parthenogenesis in komodo dragons. Nature, 444, 1021–1022.

Weatherhead, P. J., Prosser, M. R., Gibbs, H. L. & Brown, G. P. 2002. Male reproductive success and sexual selection in northern water snakes determined by microsatellite DNA analysis. Behavioral Ecology, 13, 808–815.

Yamanouye, N., Silveira, P., Abdalla, F. M. F., Almeida-Santos, S. M., Breno, M. C. & Salomao, M. G. 2004. Reproductive cycle of the Neotropical *Crotalus durissus terriicus*: II. Establishment and maintenance of the uterine muscular twisting, a strategy for long-term sperm storage. General and Comparative Endocrinology, 139, 151–157.

Zamudio, K. R. & Sinervo, E. 2000. Polygyny, mate-guarding, and posthumous fertilization as alternative male mating strategies. Proceedings of the National Academy of Sciences of the United States of America, 97, 14427–14432.

APPENDIX 1. SOURCES USED TO CONSTRUCT PHYLOGENY

Brandley, M. C., Schmitz, A. & Reeder, T. W. 2005. Partitioned Bayesian analyses, partition choice, and the phylogenetic relationships of scincid lizards. Systematic Biology, 54, 373–390.

Carranza, S. & Arnold, E. N. 2006. Systematics, biogeography, and evolution of *Hemidactylus* geckos (Reptilia: Gekkonidae) elucidated using mitochondrial DNA sequences. Molecular Phylogenetics and Evolution, 38, 531–545.

Castoe, T. A. & Parkinson, C. L. 2006. Bayesian mixed models and the phylogeny of pitvipers (Viperidae: Serpentes). Molecular Phylogenetics and Evolution, 39, 91–110.

de Queiroz, A. Lawson, R. & Lemos-Espinal, J. A. 2002. Phylogenetic relationships of north American garter snakes (*Thamnophis*) based on four mitochondrial genes: how much DNA sequence is enough? Molecular Phylogenetics and Evolution, 22, 315–329.

Fitch, A. J., Goodman, A. E. & Donnellan, S. C. 2006. A molecular phylogeny of the Australian monitor lizards (Squamata:Varanidae) inferred from mitochondrial DNA sequences. Australian Journal of Zoology, 54, 253–269.

Flores-Villela, O., Kjer, K. M., Benabib, M. & Sites Jr, J. W. 2000. Multiple data sets, congruence, and hypothesis testing for the phylogeny of basal groups of the lizard genus *Sceloporus* (Squamata, Phrynosomatidae). Systematic Biology, 49, 713–739.

Giugliano, L. G., Collevatti, R. G. & Colli, G. R. 2007. Molecular dating and phylogenetic relationships among Teiidae (Squamata) inferred by molecular and morphological data. Molecular Phylogenetics and Evolution, 45, 168–179.

Han, D., Zhou, K. & Bauer, A. M. 2004. Phylogenetic relationships among gekkotan lizards inferred from Cmos nuclear DNA sequences and a new classification of the Gekkota. Biological Journal of the Linnean Society, 83, 353–368.

Lawson, R., Slowinski, J. B., Crother, B. I. & Burbrink, F. T. 2005. Phylogeny of the Colubroidea (Serpentes): new evidence from mitochondrial and nuclear genes. Molecular Phylogenetics and Evolution, 37, 581–601.

Macey, J. R., Schulte II, J. A., Larson, A., Ananjeva, N. B., Wang, Y., Pethiyagoda, R., Rastegar-Pouyani, N. & Papenfuss, T. J. 2000. Evaluating trans-Tethys migration: an example using acrodont lizard phylogenetics. Systematic Biology, 49, 233–256.

Nagy, Z. T., Lawson, R., Joger, U. & Wink, M. 2004. Molecular systematics of racers, whipsnakes and relatives (Reptilia: Colubridae) using mitochondrial and nuclear markers. Journal of Zoological Systematics and Evolutionary Research, 42, 223–233.

Odierna, G., Canapa, A., Andreone, F., Aprea, G., Barucca, M., Capriglione, T. & Olmo, E. 2002. A phylogenetic analysis of Cordyliformes (Reptilia: Squamata): comparison of molecular and karyological data. *Molecular Phylogenetics and Evolution*, 23, 37–42.

Poe, S. 2004. Phylogeny of anoles. Herpetological Monographs, 18, 37–89.

Reeder, T. W. 2003. A phylogeny of the Australian Sphenomorphus group (Scincidae: Squamata) and the phylogenetic placement of the crocodile skinks (*Tribolonotus*): Bayesian approaches to assessing congruence and obtaining confidence in maximum likelihood likelihood inferred relationships. Molecular Phylogenetics and Evolution, 27, 384–397.

Reeder, T. W., Cole, C. J. & Dessauer, H. C. 2002. Phylogenetic relationships of whiptail lizards of the genus *Cnemidophorus* (Squamata: Teiidae): a test of monophyly, reevaluation of karyotypic evolution, and review of hybrid origins. American Museum Novitates, 3365, 1–61.

Schulte II, J. A., Melville, J. & Larson, A. 2003a. Molecular phylogenetic evidence for ancient divergence of lizard taxa on either side of Wallace's Line. Proceedings of the Royal Society of London Series B, 270, 597–603.

Schulte II, J. A., Valladares, J. P. & Larson, A. 2003b. Phylogenetic relationships within Iguanidae inferred using molecular and morphological data and a phylogenetic taxonomy of iguanian lizards. Herpetologica, 59, 399–419.

Slowinski, J. B. & Keogh, J. S. 2000. Phylogenetic relationships of elapid snakes based on cytochrome b mtDNA sequences. Molecular Phylogenetics and Evolution, 15, 157–164.

Townsend, T., Larson, A., Louis, E. & Macey, J. R. 2004. Molecular phylogenetics of Squamata: the position of snakes, amphisbaenians, and dibamids, and the root of the squamate tree. Systematic Biology, 53, 735–757.

Vidal, N. & Hedges, S. B. 2005. The phylogeny of squamate reptiles (lizards, snakes, and amphisbaenians) inferred from nine nuclear protein-coding genes. Comptes Rendus Biologies, 328, 1000–1008.

Vidal, N., Delmas, A-S., David, P., Cruaud, C., Couloux, A. & Hedges, S. B. 2007. The phylogeny and classification of caenophidian snakes inferred from seven nuclear protein-coding genes. Comptes Rendus Biologies, 330, 182–187.

Wiens, J. J. 2000. Decoupled evolution of display morphology and display behaviour in phrynosomatid lizards. Biological Journal of the Linnean Society, 70, 597–612.

Wiens, J. J. & Hollingsworth, B. D. 2000. War of the iguanas: conflicting molecular and morphological phylogenies and long-branch attraction in iguanid lizards. Systematic Biology, 49, 143–159.

APPENDIX 2. Data on sperm morpology in reptiles

Family	Genus	Species	Sperm length	Head length	Midpiece length	Flagellum length	Reference
Agamidae	Diporiphora	bilineata	68.27	12.1	4.05	50.36	Vieira et al. 2007
Agamidae	Pogona	barbata	28.5	-	3.7	18.4	Olivier et al. 1996
Amphisbaenidae	Amphisbaena	alba	84.3	14.3	4.3	65.7	Teixeira et al. 1999a
Chamaeleonidae	Bradypodion	karrooicum	-	.	9.13	.	Vieira et al. 2007
Gymnophthalmidae	Microblepharus	maximiliani	60	11	2.5	46.5	Teixeira et al. 1999b
Gymnophthalmidae	Cercosaura	ocellata	67.49	14.75	2.18	50.57	Colli et al. 2007
Iguanidae	Anolis	carolinensis	83.2	16.5	4.6	63.3	Scheltinga et al. 2001
Iguanidae	Anolis	carolinensis	83.22	16.99	4.38	63.24	Vieira et al. 2007
Iguanidae	Basiliscus	vittatus	90.31	18.31	2.91	68.17	Vieira et al. 2005
Iguanidae	Corytophanes	cristatus	96.11	18.43	3.02	68.17	Vieira et al. 2005
Iguanidae	Crotaphytus	bicinctores	95.22	19.88	4.06	72.5	Vieira et al. 2007
Iguanidae	Crotaphytus	bicinctores	85.5	20	4	57.2	Scheltinga et al. 2001
Iguanidae	Enyalioides	laticeps	102.76	19.31	3.73	78.53	Vieira et al. 2007
Iguanidae	Gambelia	wislizenii	99.33	20.15	5.07	77.52	Vieira et al. 2007
Iguanidae	Gambelia	wislizenii	98.8	20.2	5	76.8	Scheltinga et al. 2001
Iguanidae	Hoplocercus	spinosus	110.58	27.61	4.75	74.32	Vieira et al. 2007
Iguanidae	Iguana	iguana	71.7	18.2	3.4	53.5	Viera et al. 2004
Iguanidae	Iguana	iguana	71.69	18.22	3.36	49.5	Vieira et al. 2005
Iguanidae	Laemanctus	longipes	97.65	18.37	3.02	75.7	Vieira et al. 2005
Iguanidae	Oplurus	cyclurus	85.85	22.36	3.83	60.41	Vieira et al. 2007
Iguanidae	Polychrus	acutirostris	-	-	7.5	-	Teixiera et al. 1999c
Iguanidae	Polychrus	acutirostris	83.7	17.15	3.84	62.94	Vieira et al. 2005
Iguanidae	Tropidurus	semitaeniatus	-	-	2.8	-	Teixeira et al. 1999d
Iguanidae	Tropidurus	semitaeniatus	-	.	2.52	.	Vieira et al. 2007
Iguanidae	Tropidurus	torquatus	-		2.8	-	Teixeira et al. 1999
Iguanidae	Tropidurus	torquatus	93.17	19.53	2.63	70.67	Vieira et al. 2005
Iguanidae	Urosaurus	ornatus	-	-	3.9	-	Scheltinga et al. 2000
Iguanidae	Uta	stansburiana	88.2	18.2	3.8	67.5	Scheltinga et al. 2000
Scincidae	Carlia	pectoralis	96.5	-	11.5	79.8	Jamieson et al. 1994
Scincidae	Ctenotus	taeniolatus	84	-	7.1	64.8	Jamieson et al. 1994
Scincidae	Tiliqua	scincoides	-	5.5	5.5	42.8	Jamieson et al. 1994
Teiidae	Ameiva	ameiva	68	15.4	4.6	48	Guigliano et al. 2002
Teiidae	Callopistes	flavipunctatus	104.98	16.81	1.82	85.56	Colli et al. 2007
Teiidae	Cnemidophorus	gularis	54.5	10.8	3.5	40.25	Teixeira et al. 2002
Teiidae	Cnemidophorus	ocellifer	56.4	13.3	3.35	40.1	Teixeira et al. 2002
Teiidae	Cnemidophorus	sexlineatus	-	-	4	-	Newton & Trauth 1992
Teiidae	Crocodilurus	amazonicus	84.98	16.6	3.7	63.85	Colli et al. 2007
Teiidae	Dicrodon	guttulatum	78.34	15.6	3.52	54.01	Colli et al. 2007
Teiidae	Dracaena	guianensis	90.5	18.08	3.82	66.66	Colli et al. 2007
Teiidae	Kentropyx	altamazonica	75.6	14.7	7.55	53.6	Teixeira et al. 2002
Teiidae	Teius	oculatus	74.74	18.91	3.54	53.2	Vieira et al. 2005
Teiidae	Tupinambis	duseni	82.3	23.5	4.2	55.5	Tavarez-Bastos et al. 2002
Teiidae	Tupinambis	merianae	82.3	23.5	4.2	55.5	Tavarez-Bastos et al. 2002
Teiidae	Tupinambis	quadrilineatus	82.3	23.5	3.3	55.5	Tavarez-Bastos et al. 2002
Teiidae	Tupinambis	teguixin	82.3	23.5	3.3	55.5	Tavarez-Bastos et al. 2002
Varanidae	Varanus	gouldii	-	-	3.1	-	Olivier et al. 1996
Boidae	Boa	constrictor	97.1	12.5	40.5	44.1	Tourmente et al. 2006
Colubridae	Nerodia	sipedon	112	-	-	-	Schulte-Hostedde & Montgomerie 2006
Typhlopidae	Ramphotyphlops	waitii	179	-	55	108	Harding et al. 1995

20

Sexual Conflict and the Intromittent Organs of Male Birds

ROBERT MONTGOMERIE

INTRODUCTION

Birds are unique in the animal kingdom in that internal fertilization is the rule but male intromittent organs (IOs) are the exception. The otherwise almost universal association between IOs and internal fertilization in animals (see table 1 in Briskie & Montgomerie 1997) led many early authors to suggest that IOs were necessary to facilitate insemination and were thus typical primary sexual traits shaped by natural selection. Thus interspecific variation in IO structure was often thought to reduce hybridization and its costs. As a result, complex male and female genitalia were presumed to fit together in a lock–and–key fashion specifically to facilitate within-species matings and reduce the efficacy of interspecific coupling. Eberhard (1985, 1990) and others (see Hosken & Stockley 2004), however, suggested that male genitalia might serve other, more cryptic, functions, including male advertisement, internal courtship, sexual coercion, and sperm competition, all of which are mechanisms of sexual selection. These insights blurred the distinction between primary and secondary sexual traits such that that dichotomy is no longer particularly useful, as Darwin (1871, pp. 253–254) had insightfully intimated.

The study of bird IOs has made some useful contributions to our modern understanding of the function of male external genitalia in general, even though both the structure and, especially, the

function of bird IOs are notoriously difficult to study. This difficulty stems from their relative rarity (occurring in only 3% of bird species), their soft tissue, their erectile anatomy, and their internal location in the male except at the moment of insertion into the female (Brennan et al. 2010). Thus although some early anatomists examined the genitalia of a few bird species (Müller 1836; Eckhard 1876; Owen 1879; Müller 1908; see figure 20.1), only recently has there been renewed interest in studying and quantifying phallic structures of non–domestic species (Oliveira & Mahecha 2000; Brennan et al. 2007, 2008), and the position and behavior of bird IOs during copulation has only just been determined (Brennan et al. 2010). From today's perspective, it seems as if lack of interest, or possibly interesting theory to test, is partly responsible for the dearth of knowledge about the male genitalia of birds. Thus, even though we expect, for phylogenetic reasons, that about 300 bird species have IOs, we have convincing evidence for the presence of IOs from only about 25% of those species, and important discoveries are still being made (Brennan et al. 2010).

For a long time, ornithologists believed that the interesting thing about the male phallus in birds was that it is present in so few species. Recent work, however, has pointed out that the absence of IOs in almost 10,000 species of obligate internal fertilizers is really the mystery (Briskie & Montgomerie 1997). Thus it is the loss of IOs in so many lineages

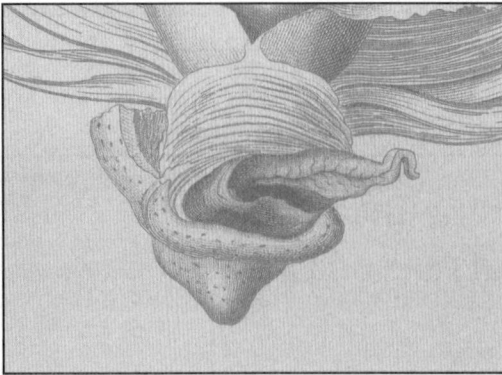

FIGURE 20.1 Intromittent organ of the male brown kiwi (*Apteryx australis*). Detail from drawing by Richard Owen (1879, plate III) showing the phallus (s) and the external sphincter (r).

EVOLUTION OF EXTERNAL GENITALIA IN MALE BIRDS

The 'external' genitalia of male birds are found in the form of either a true phallus, a pseudophallus, or a cloacal protuberance. True phalluses and pseudophalluses may or may not be intromittent, whereas apparently none of the cloacal protuberances are. Examples of each of these external genitalia are shown in figure 20.2, and I refer the reader to previous reviews for more detail (Lake 1981; King 1981a, b; Briskie & Montgomerie 1997; Montgomerie & Briskie 2007) and a summary of anatomical and histological studies dating back into the early nineteenth century. Although we think we have a pretty good idea of the taxonomic distribution of each of these forms of male 'external' genitalia in birds, the presence or absence of these structures has been documented in fewer than 5% of the 10,000 or so extant bird species, and less than half of those have had their external genitalia studied in any detail.

The true phallus of male birds is clearly homologous to that of their closest extant relatives—the reptiles—with which it shares both anatomical and histological characters (King 1981). Thus the most recent ancestors of birds, including presumably the dinosaurs (figure 20.3), appear to have all possessed a phallus that, like the true phallus of birds, is a single elongated structure of erectile tissue emerging from the wall of the cloaca (e.g., figures 20.1, 20.2B, C, and 20.8). The true phallus in birds is comprised of two fused fibroelastic bodies with a median ventral ejaculatory groove (*phallic sulcus*) where these bodies join (figures 20.2B, C and 20.8). As the avian phallus becomes erect, this groove often becomes a closed channel along which semen flows during ejaculation (Brennan et al. 2010). The size, surface structures, shape, and histology of true phalluses varies widely in birds (figures 20.1, 20.2B, C, and 20.8; see also Montgomerie & Briskie 2007), and we are only just beginning to understand some of the adaptive significance of that diversity. I explore some of this variation in the waterfowl in a later section of this chapter.

As far as we know, the true phallus (figures 20.1, 20.2B, C and 20.8) is intromittent in most species of Paleognathae (ostriches, rheas, emu, cassowaries, kiwis, and tinamous; $n = 59$ species) and Anseriformes (ducks, geese, swans; $n = 162$ species), but is rudimentary, non–intromittent, and very different looking in most of the 268 species of

of birds that is the exception that might help us to understand the adaptive significance of bird genitalia in particular, and may provide insights into the function of genitalia in general. Unfortunately the information currently available suggests that IO loss has occurred only a few times in the evolution of birds. For that reason, researchers have recently turned to studying the extensive variation in IO structure among species to gain some insights into IO function, and that is the main focus of this chapter.

Because there have been some recent extensive reviews on the intromittent organs of birds (Briskie & Montgomerie 1997; Montgomerie & Briskie 2007), I refer the reader to those accounts for background information. Instead, I focus here on some new evidence for the adaptive function of bird genitalia in the context of sexual conflict, particularly in the waterfowl (order Anseriformes). I begin with a brief overview of the phylogenetic distribution of genitalia in birds, updating previously published phylogenetic analyses with some new information (Hackett et al. 2008; Brennan et al. 2008). I then outline several hypotheses that have been proposed to explain the general pattern of presence and absence of IOs in different avian lineages. Finally, I use evidence from the waterfowl to test and extend some of those ideas, particularly in the context of sexual conflict. Waterfowl are particularly good subjects for the analysis of genitalia evolution in birds because males of all species probably have an IO, and there is extreme diversity in both size and structure of IOs in this taxon.

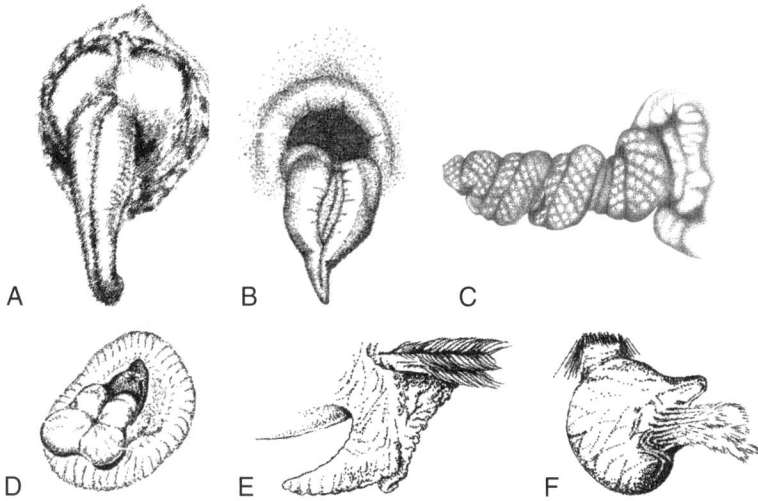

FIGURE 20.2 Some examples of external male genitalia in crocodiles and birds. Intromittent true phalluses of male (A) crocodile, (B) North Island kiwi (*Apteryx australis mantelli*), and (C) spotted tinamou; non-intromittent true phallus of (D) turkey; non-intromittent pseudophallus of (E) red-billed buffalo weaver; and non-intromittent cloacal protuberance and pseudophallus of (F) superb fairy wren (*Malurus cyaneus*). (A), (B), and (D–F) modified from Montgomerie & Briskie (2007); (C) drawn from photograph in Oliveira & Mahecha (2000: figure 18).

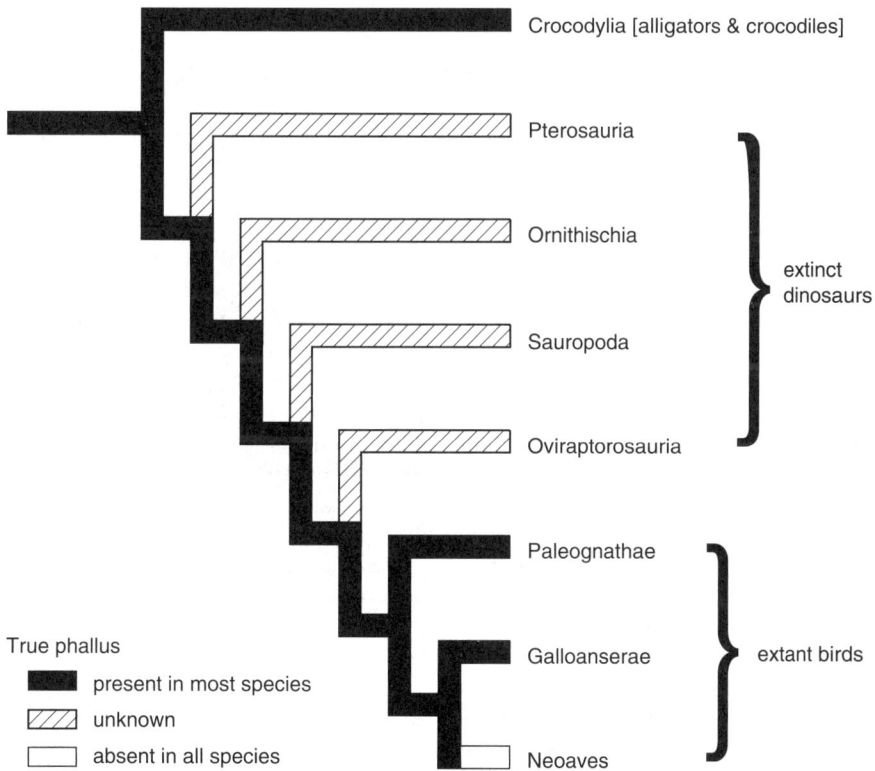

FIGURE 20.3 Phylogeny of birds and their closest extinct and extant relatives showing the distribution of true phalluses in males. See figure 20.4 for details in extant birds. Modified from Montgomerie & Briskie (2007).

455

Galliformes (megapodes, curassows, pheasants, quails, and their relatives). The remaining 9500+ extant bird species do not possess a true phallus and only a handful of those have a phallus–like structure of any form. Thus the true phallus was lost early in the lineage of most extant bird species (figure 20.4), at the base of the huge clade (Neoaves) that is the sister taxon to the Galloansere. As mentioned above, it is the pattern of loss of the true phallus, and particularly of the intromittent organ, that is particularly interesting in birds because all species are internal fertilizers. This pattern, however, does not lend itself well to comparative analysis because there appear to be very few evolutionary transitions, and thus the statistical power of any such analysis is necessarily low.

In addition to the loss of the true phallus (and thus an IO) early in the Neoaves lineage, the true phallus is not intromittent in some tinamous and megapodes, nor in any of the Galliformes in the families Phasianidae (turkeys, grouse, pheasants, and partridges), Numididae (guineafowl), or Odontophoridae (New World quail), to the best of our present knowledge (figure 20.4). The non–intromittent true phallus has not been well-studied in birds, except in the domesticated chicken and turkey. In those species, the phallus is a flattened oval or heart-shaped protrusion on the internal wall of the cloaca (figure 20.2D), but when the phallus is erect the cloaca is everted, exposing the phallus externally for a few seconds during copulation. Like other true phalluses, the non–intromittent phallus has a median groove down which semen flows during ejaculation.

In the tinamous (family Tinamidae), the 20 species in the genus *Crypturellus* appear to have a non–intromittent phallus (Brennan et al. 2008) whereas the remaining 27 tinamou species (in 8 genera) likely have an intromittent phallus. Unfortunately, only a few tinamou species have been studied in any detail, but it seems likely that the evolutionary transition from intromittent to non–intromittent phallus occurred only once in the tinamous, in the lineage leading to *Crypturellus* (figure 20.5A).

Even less research has been done on male genitalia in the megapodes (family Megapodiidae), but a recent study shows that the male Australian brush turkey (*Alectura lathami*) has a non–intromittent true phallus, and that the male malleefowl (*Leipoa ocellata*) has no phallus at all (Brennan et al. 2008). The phallus may also be absent in the male orange-footed scrubfowl (*Megapodius reinwardt*) based on casual observations (Brom & Dekker 1992; Brennan et al. 2008), but some detailed anatomical work is needed to confirm this. The remaining 18 species of megapodes are thought to have an intromittent true phallus, but the evidence for this is slim indeed (Montgomerie & Briskie 2007). Nonetheless, the clear absence of a phallus in the malleefowl (Brennan et al. 2008) means that the phallus has been lost at least twice in the evolutionary history of birds—once in the Paleognathae (in the Megapodes, figure 20.5B) and once at the base of the Neoaves (figure 20.4)—and not just once as had previously been assumed (Briskie & Montgomerie 1997; Montgomerie & Briskie 2007). It will be interesting to discover whether those two losses of the same trait have a similar cause.

To the best of our knowledge, a phallus–like structure, or pseudophallus, has arisen only three times in the evolution of the Neoaves—in the greater and lesser vasa parrots (genus *Coracopsis*), in the red–billed and white–billed buffalo weavers (genus *Bubalornis*; figure 20.2E), and in some species of fairy wren (genus *Malurus*; figure 20.2F). In the vasa parrots this pseudophallus is erectile, blood–engorged, and intromittent, and the copulating pair remains locked together for up to 30 min (Wilkinson & Birkhead 1995). In the buffalo weavers (figure 20.2E) and fairy wrens (figure 20.2F), the pseudophallus is an external phalloid structure composed of cartilage or connective tissue, located beside the cloaca (Birkhead et al. 1993; Mulder & Cockburn 1993; Tuttle et al. 1996). In the fairy wrens its use and function are unknown, but in the buffalo weavers the male rubs his pseudophallus vigorously against the female during copulation (Winterbottom et al. 2001), though the reason for this behavior is unknown.

In all species of passerine birds (order Passeriformes) that have been examined to date, the male develops a swollen cloacal region, called the cloacal protuberance (CP), in the breeding season (figures 20.6 and 20.7). This CP surrounds the vent and usually contains the seminal glomera—the distal portion of the *ductus deferens*—that develop as the breeding season approaches (figure 20.6). These seminal glomera store sperm, and it has been speculated that the CP might enhance sperm storage as sperm in the seminal glomera within the CP are kept cooler than body temperature (Wolfson 1954b), though not all CPs enclose the seminal glomera (figure 20.6B). In the bearded tit (*Panurus biarmicus*), the seminal glomera are relatively small

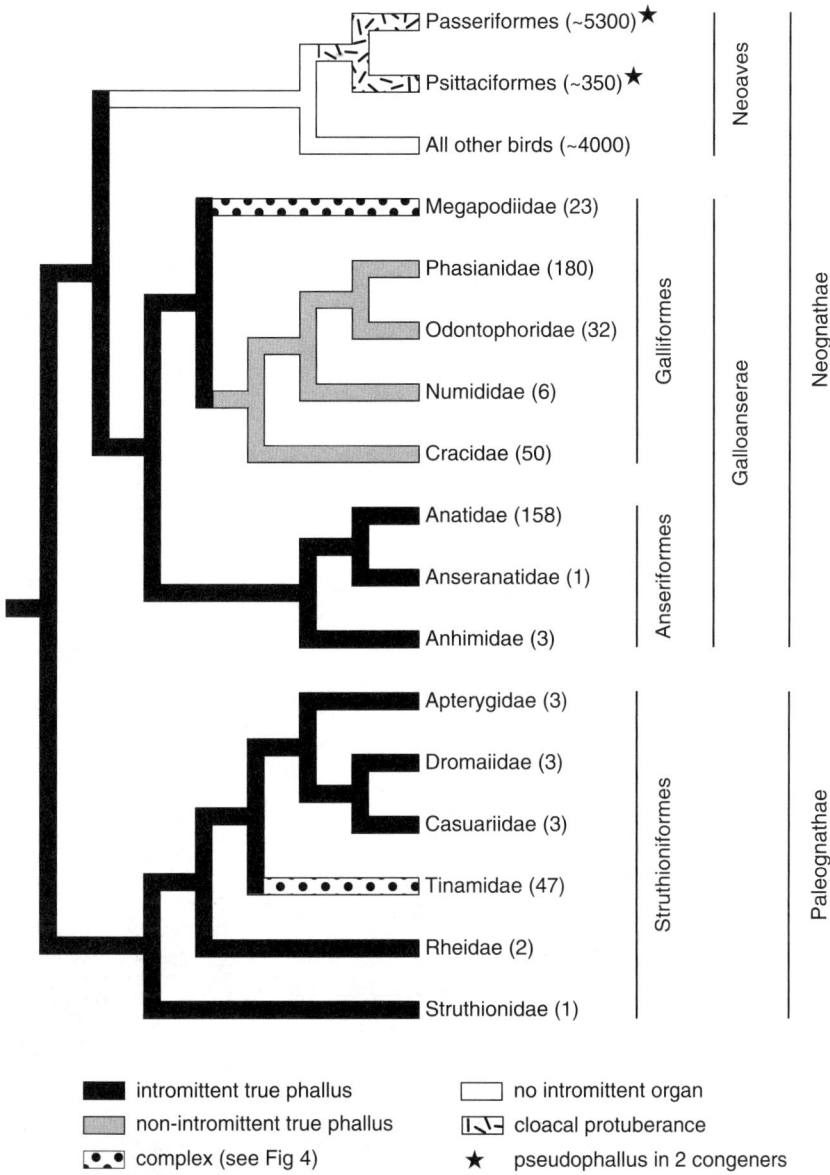

FIGURE 20.4 Distribution of external male genitalia in birds, mapped onto a phylogenomic analysis of avian evolution (Hackett et al. 2008), showing only taxa mentioned in the text. Some taxon names and phylogenetic details from Cracraft et al. (2004). Approximate number of species in each taxon is shown in brackets. Note that the cloacal protuberance has so far been documented in only one species of Psittaciformes (see text). See figure 20.5 for more details about the families Tinamidae and Megapodiidae.

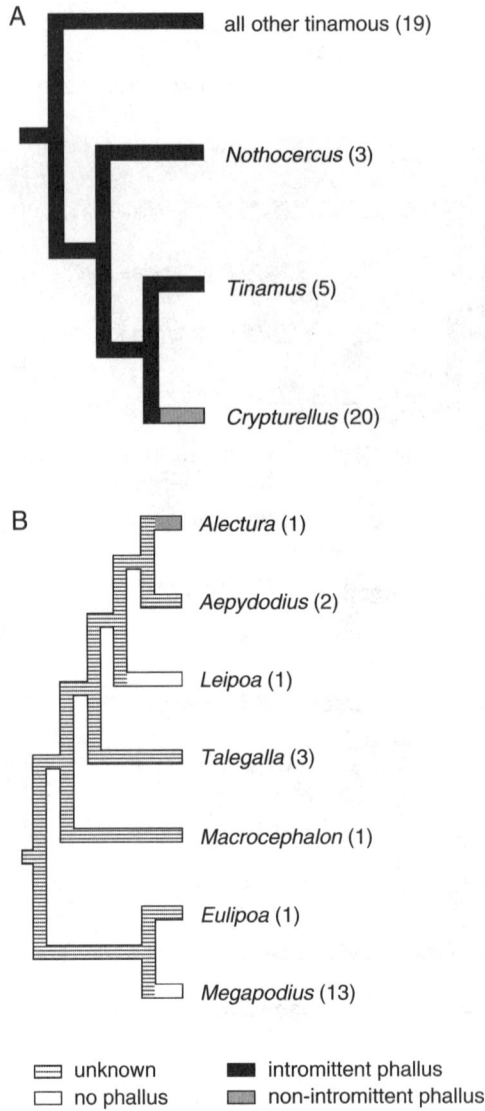

FIGURE 20.5 Taxonomic distribution of external male genitalia in the (A) Tinamidae and (B) Megapodiidae. Information on genitalia from Brennan et al. (2008); phylogenies from (A) Bertelli and Porzecanski (2004) and (B) Birks and Edwards (2002).

and are located at the base rather than inside the CP (Birkhead & Hoi 1994). Moreover, in that species, a large, red, phallus–like structure appears when the CP is everted. This structure terminates in a small papilla from which semen can be extruded, suggesting that it has some copulatory function. Such papilla have been recorded in other passerine birds (e.g., Chiba & Nakamura 2003) but their function and structure have not been studied.

Male budgerigars (*Melopsittacus undulatus*) also have a CP around the seminal glomera (Samour 2002) and this CP is present only during the breeding season. This suggests that the CP may be present in some form in other (or all) parrots (order Psittaciformes) though there are no data available on any other species. It would not be surprising to find the CP present in male parrots as they are the sister taxon to the Passeriformes and the CP may

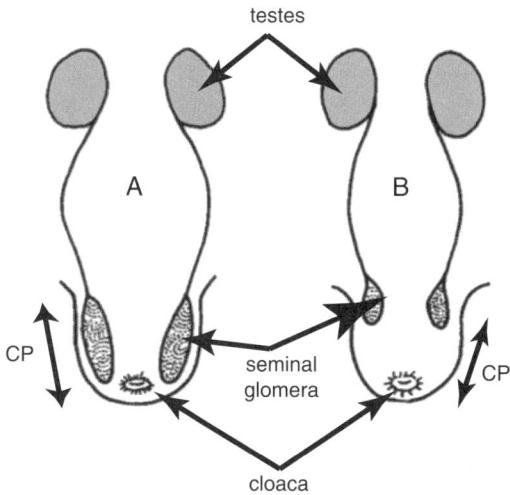

FIGURE **20.6** General structure of the cloacal protuberance (CP) showing location of seminal glomera in (A) typical passerine and (B) bearded tit. Double-headed arrows show the extent of the cloacal protuberance. Modified after Birkhead and Hoi (1994).

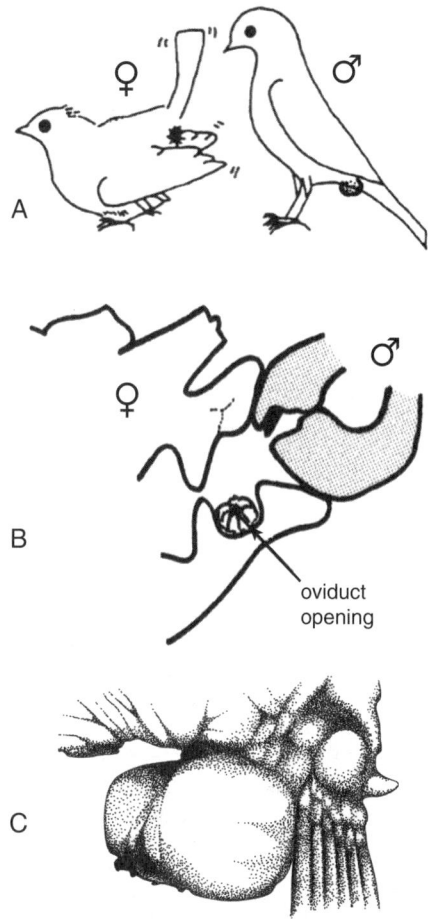

FIGURE **20.7** Cloacal protuberance of the alpine accentor. Location of male and female genitalia (A) prior to and (B) during copulation. (C) External structure of male cloacal protuberance shown in same orientation as in (B). (A) and (B) modified from drawings in figure 7 and (C) redrawn from a photograph in figure 2 of Chiba and Nakamura (2003).

well have evolved in their common ancestor (figure 20.4). Certainly detailed anatomical studies of male parrot genitalia would resolve this issue and potentially provide insights into the function of the CP, especially if it is not found in all species.

Among passerine bird species, the full size of the male's CP varies considerably, even among closely related species, with those species copulating more frequently having larger CPs (Birkhead et al. 1993). There is also some evidence that the CP might enhance a male's efficiency at copulation because it elevates the male's vent (figure 20.6) and may facilitate union with the female's vent (figure 20.7B; Wolfson 1954a; Chiba & Nakamura 2003). The alpine accentor (*Prunella collaris*), for example, has the largest reported CP relative to its body size (figure 20.7C), and averages 32 copulations per female per day (Nakamura 1990), one of the highest rates known in birds (Briskie 1993). The female of this species also has a cloacal protuberance (figure 20.7A; Nakamura 1990), but this has not been reported in other species, though in some species the female's cloacal region is swollen (e.g., Birkhead et al. 1991). In both the alpine accentor and the dunnock (*Prunella modularis*), the female cloacal region is red and is shown to males, apparently to solicit copulations (Nakamura 1990; Birkhead et al. 1991). In the unrelated New Zealand stitchbird

(*Notiomystis cincta*), the large male CP does not contain the seminal glomera, but does become 'erect' during copulation, changing orientation and presumably allowing a male to more easily align his vent with that of the female during the remarkable face-to-face copulations in which this species engages (Low et al. 2005). Thus, while we know a little bit about the CP in a few species that have particularly large or unusual protuberances, this structure has been studied very little and both the extent and causes of interspecific, and indeed intraspecific, variation are not yet known.

Interestingly, the remaining 4000+ species of birds in the Neoaves (figure 20.4) apparently have no CP or any other form of 'external' male genitalia, copulating solely by a usually brief juxtaposition of the external lips of male and female cloacas. If the CP is important for some aspect of copulation, then the complete absence of any copulatory structure in such a large clade within the Neoaves is of particular interest.

THE FUNCTION OF MALE INTROMITTENT ORGANS IN BIRDS

Because the males of most bird species lack an IO, and at least 40% of bird species lack any sort of external genitalia (cloacal protuberances, pseudophalluses, or phalluses), it is clear that these structures are not necessary for sperm transfer in birds. Nonetheless, it has been argued that the pattern of IO presence/absence in birds can be explained as adaptations to sperm transfer in different situations, and seven different hypotheses have been proposed (table 20.1; Montgomerie & Briskie 2007). There is actually some limited empirical evidence in support of most of these hypotheses but there are clear exceptions to all of them, suggesting possibly that the IO (and possibly other forms of external male genitalia) serve different functions in different taxa. In this section I briefly review each of these hypotheses (table 20.1), and suggest some potentially fruitful avenues for further research.

Two of these hypotheses suggest that the male IO has been retained in some birds to enhance the efficacy of sperm transfer during copulation on the water (H1), or when cloacal contact is difficult

(H2)—for example in large-bodied, long-legged, or flightless birds. The many species that are clear exceptions to both of these predictions suggest that these hypotheses cannot provide a general explanation, though they may apply in some circumstances (Montgomerie & Briskie 2007). Neither of these hypotheses addresses the absence of an IO in the majority of species.

Three other hypotheses acknowledge that an IO may be useful for sperm transfer in internally-fertilizing species, but that the costs of an IO in male birds have selected against such a structure in most birds. Thus in combination with H1 and H2, these additional three hypotheses could potentially explain the entire pattern of IO presence/absence in birds, though the combination of correct explanations may vary among taxa. H3 suggests that selection for copulations of exceptionally short duration (typically a few seconds) in birds—due to the dangers of predation, and so on—has resulted in selection against structures like IOs that would increase copulation time (Wesołowski 1999). Current evidence on copulation duration in a wide variety of species does not support this hypothesis (Briskie & Montgomerie 2001), but few reliable quantitative data are available (Wesołowski 2001) and better information is clearly needed. H4 suggests that the IO has been lost in the Neoaves due to the high costs of flight. While many birds with an IO are flightless or relatively poor flyers, the mass of the male IO is unlikely to exert a significant flight cost (Briskie & Montgomerie 1997), and many species that fly well do have an IO. The recently discovered (Brennan et al. 2008) absences of IOs in Crypturellus tinamous (which do not fly more than other tinamous), and in the malleefowl and Australian brush turkey (neither of which spend

TABLE 20.1 Seven hypotheses to explain the phylogenetic pattern of presence/absence of the male intromittent organ in birds. The initial references for each hypothesis are cited, and the expected form of selection is listed

	Hypothesis	Type of selection	Reference
H1	Prevent water damage	Natural	Owen 1866; Lake 1981
H2	Maintain genital contact	Natural	King 1981a, b
H3	Reduce copulation duration	Natural	Wesołowski 1999
H4	Reduce flight costs	Natural	Briskie & Montgomerie 1997
H5	Reduce STDs	Natural	Briskie & Montgomerie 1997
H6	Improve sperm competition	Sexual (intrasexual conflict)	Briskie & Montgomerie 1997
H7	Enhance forced copulations	Sexual (intersexual conflict)	Briskie & Montgomerie 1997

much time flying) also provide evidence against this hypothesis. H5 proposes that copulation exposes male birds to potentially high levels of sexually transmitted diseases (STDs) and that this would favor the disappearance of the IO because such a structure would increase the contact between male and female tissues. Briskie & Montgomerie (1997) argued that STDs might be particularly prevalent in birds because, along with the eponymous Monotremes, they are the only homeothermic animals to copulate via a common urogenital cloaca, which might well be a good environment for STDs to flourish. Unfortunately, there are too few data available on STDs in birds to allow this hypothesis to be evaluated quantitatively.

The remaining two hypotheses to explain IO presence/absence in birds suggest that sexual rather than natural selection is the agent, in the form of sexual conflict, and that the pattern of presence/absence in birds is the result of a trade–off between potential costs (including STDs) and benefits. The first of these (H6) suggests that the IO is maintained by intrasexual conflict resulting from sperm competition (Briskie & Montgomerie 1997). Thus the benefits of IOs exceed the costs when sperm competition is intense and males benefit from an IO that inserts sperm farther into the female reproductive tract. This mechanism should be especially important in species with extensive male parental care where the opportunities for a male to acquire more than one mate are severely restricted. In support of this hypothesis, Briskie & Montgomerie (1997) found that males in species with exclusively male incubation and brood care were significantly more likely to have an IO than males in species with biparental or female-only incubation. While this is an intriguing pattern, there are many exceptions that would need to be explained.

H7 proposes that the functional significance of the male IO in birds (and potentially other animals) is to enable male birds to force copulations on females, and thus that the pattern of presence/absence in birds is the result of intersexual conflict (Briskie & Montgomerie 1997). This hypothesis gains some support from the fact that the males of many birds with IOs, particularly waterfowl, are well known to force copulations (Adler 2010). To test this idea quantitatively, Briskie & Montgomerie (1997) compared the relative egg masses of species with and without male IOs. The rationale for this comparison is that females with relatively large eggs would incur a larger cost of egg or clutch abandonment than females with relatively small eggs. Thus females with relatively small eggs should be more willing to abandon eggs fertilized by forced copulations from unwanted male partners, potentially selecting against males that force copulations, and concomitantly against IOs. The interspecific pattern clearly supports this prediction as species in which males have an IO have significantly larger egg mass (figure 3.15C in Montgomerie & Briskie 2007), but there are many species that are exceptions to this pattern.

Given the difficulty of properly testing any of these hypotheses while controlling for the influence of phylogeny, and the pervasiveness of obvious exceptions to the general patterns of IO presence and absence, a different approach is needed. Certainly, further detailed anatomical work on the tinamous and megapodes might reveal some interesting insights, but even in those taxa the patterns of IO loss may be constrained phylogenetically, making comparative analysis intractable statistically. One alternative approach to assessing IO function is to look at within-taxon patterns of IO size and shape, rather than focusing on presence/absence. This might prove particularly fruitful in the tinamous and megapodes where there appears to be interspecific variation in both IO morphology and the presence/absence of an intromittent phallus (tinamous), as well as the absence of a phallus of any kind (megapodes). In the next section, I apply this approach to an interspecific analysis of IOs in the waterfowl.

SEXUAL CONFLICT AND THE INTROMITTENT ORGANS OF MALE WATERFOWL

Males of all 162 extant waterfowl species appear to have an intromittent organ that is a true phallus. While details on only about 60 species have been published so far, most of the remaining species have been studied in the wild or in captivity and the examination of genitalia is a common method of sexing in this taxon. Thus it seems likely that only a very few waterfowl species have not been examined for male genitalia and, so far, the absence of a male IO has not been reported in any species.

It has been known for a long time that there is considerable variation in IO morphology—at least in relative size—among species of Anseriformes. For example, there have long been at least anecdotal reports of extremely large IOs in some species of stiff-tailed ducks (subfamily Oxyurinae; e.g., Marchant & Higgins 1990; McCracken 2000), as well as relatively small IOs, for their body size, in the geese and swans (subfamily Anserinae; Frank McKinney personal communication 1997). Thus the Anseriformes is an excellent taxon in which to explore the potential causes of variation in IO morphology in relation to some of the hypotheses presented in the preceding section. The ready availability of many species breeding in waterfowl parks and zoos also makes this taxon especially tractable for detailed behavioral observations (e.g., Johnsgard 1965) and experiments (e.g., Burns et al. 1980). Unfortunately, there has been little work done to date on copulation in this taxon (but see Burns et al. 1980), and none with respect to the influence of IO morphology on copulation and fertilization success. Since 2002, two studies have examined in some detail the interspecific variation in IO morphology in the family Anatidae (about 140 species of ducks, geese, and swans), and those studies (Coker et al. 2002; Brennan et al. 2007) are the main focus of this section.

The IO of the male mallard (*Anas platyrhynchos*) has been particularly well-studied so I begin by describing it in some detail to provide an overview of general IO structure in the Anatidae. In its flaccid state, the mallard's phallus sits coiled in a thin-walled sac inside the cloaca, on its ventral wall. The phallus becomes erect via peristaltic contractions of the cloacal sphincter which increase the flow of lymph to the phallus (Guzsal 1974), causing the base of the phallus to fill the male's vent (external opening of the cloaca), and the erect phallus to protrude about 4 cm out of the vent (figure 20.8). Like other true phalluses, the mallard's IO is comprised of two fibrolymphatic bodies fused longitudinally, forming a deep ejaculatory groove on its ventral surface. The left fibrolymphatic body is considerably larger than the right, and the phallus twists in a counterclockwise spiral for 3–4 turns from its base to its tip (figure 20.8). When the phallus is erect, the lips of the ejaculatory groove are pressed together such that semen stays within this channel during ejaculation, exiting only at the tip (see also Brennan et al. 2010). After ejaculation the phallus detumesces—assisted by both muscular contractions and elastic fibres inside the phallus that had been stretched during erection—resulting in the flow of lymph back into the circulatory system. Thus the entire phallus is first retracted back into the cloaca and is then folded, over a

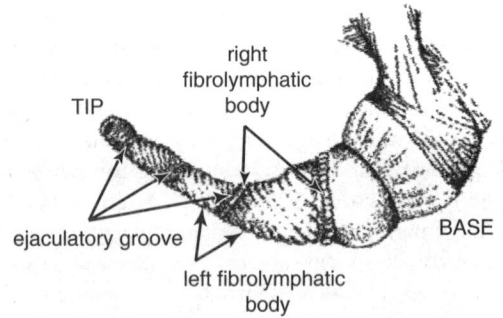

FIGURE 20.8 Erect intromittent organ (a true phallus) of the male mallard (modified after Briskie & Montgomerie 1997).

period of <5 min, into the peritoneal sac in the wall of the cloaca. The surface of the erect phallus is relatively smooth at its base but soon becomes cornified with rough transverse ridges about 2 mm apart, extending all the way to the tip (figure 20.8).

Coker et al. (2002) were the first to examine and quantify interspecific variation in IO structure within any group of birds. To do this, they capitalized on an existing set of scale drawings of male IOs from 54 species of Anatidae prepared earlier (but unpublished) by Helen Hays (one of the coauthors of that study). These drawings were made from genitalia dissected out of museum specimens that had been collected during the breeding season and preserved in formalin. From each drawing, they measured IO length and circumference, and then quantified the numbers and sizes of spikes (knobs) and ridges on the surface of the IO (figure 20.9) in each species. All of these traits vary considerably within this taxon. IO length, for example, ranged

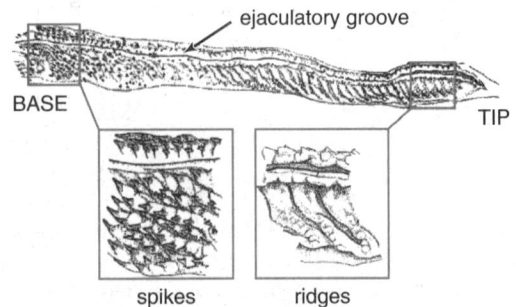

FIGURE 20.9 Drawing of the flaccid intromittent organ (true phallus) of the male ruddy duck (*Oxyura jamaicensis*) showing detail of the spikes and ridges (modified from Montgomerie & Briskie 2007).

from 1.25 cm in the 1.3 kg red-breasted goose (*Branta ruficollis*) to 28.5 cm in the 0.85 kg Australian blue-billed duck (*Oxyura australis*). In the full sample of 54 species there was no relation between IO length and male body length ($r = -0.11$, $P = 0.36$), but species more likely to engage in forced extra-pair copulations (FEPCs) had significantly longer IOs than monogamous species where FEPCs were thought to be rare (figure 20.10A). This relation between IO length and forced copulations provides some support for the intersexual conflict hypothesis (H7) described above. Coker et al. (2002) also found that IO length was significantly correlated with testes mass, controlling both variables for body size (figure 20.10B). Since there is considerable evidence that residual testes mass (controlling for body mass) is a useful index of the intensity of sperm competition (Calhim & Birkhead 2007), this latter finding is consistent with the intrasexual conflict hypothesis (H6). These apparently conflicting patterns and interpretation can be reconciled if selection for increased IO length was the result of sperm competition due to FEPCs.

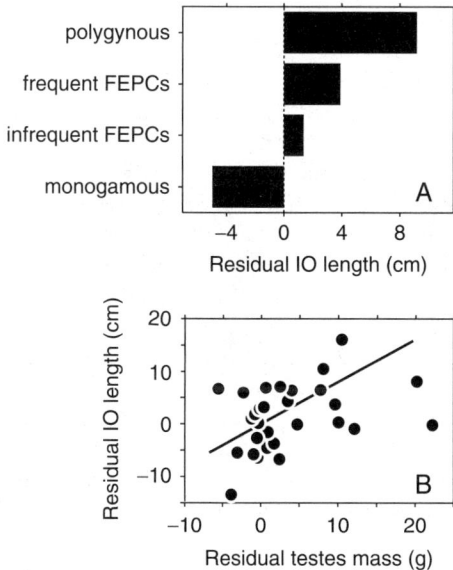

FIGURE 20.10 Interspecific variation in IO size in the waterfowl. (A) Residual IO length (controlling for body mass) varies with mating system, increasing with assumed level of forced extrapair copulations. (B) Relation between residual IO length and residual testes mass (both controlling for body mass). Redrawn from data in Coker et al. (2002) and data supplied by Kim Cheng (personal communication).

Coker et al. (2002: figures 3 and 4) also found that both the extent and sizes of ridges and spikes on the phallus were significantly greater in species with polygynous mating systems, which are assumed to have higher incidence of FEPCs (Adler 2010). Moreover, the ratio of spikes to ridges also increased with IO size, leading Coker et al. (2002) to suggest that species with larger IOs, and thus higher expected levels of sperm competition, might use these surface structures to displace or remove the sperm of rival males from the female's reproductive tract.

While the study by Coker et al. (2002) was ground breaking and their results intriguing, several aspects of that study limit the usefulness of their findings as a test of different hypotheses to explain variation in bird IOs. First, the drawings that they analyzed were prepared by examination of a single male of each species, preserved in formalin, with the IO dissected out, then longitudinally split and laid flat so that the whole surface could be illustrated. As accurate as these drawings might be, it is impossible to know whether each specimen was typical of the species, and long-term preservation in formalin is certain to have distorted some features and resulted in some shrinkage. Moreover, the original drawings are not available for comparison with live or freshly-killed specimens (Kim Cheng, personal communication 2005). In fairness, the measurements of IO length of eight species studied by Coker et al. (2002) are significantly correlated ($r = 0.87$, $P = 0.005$; figure 20.11) with measurements taken from more recently preserved material (Brennan et al. 2007) that was unlikely to have suffered any adverse effects of formalin preservation. That analysis does, however, suggest that some shrinkage might have occurred in some of the Coker et al. (2002) specimens (figure 20.11). Note also that both studies measured a small number ($n = 1$–2) of individuals per species, so some of the unexplained differences between the measurements from these two studies could well be due to intraspecific variation. Second, the expected frequency of forced extrapair copulations for many species studied by Coker et al. (2002) was inferred from their mating system rather than by direct observation. While there are some general relations between mating systems and the incidence of both forced and unforced extra-pair copulations in birds, there are many exceptions (Birkhead and Møller 1995) and reliance on such an inference is fraught with error. Third, while Coker et al. (2002) did examine specimens taken during the breeding season, it is

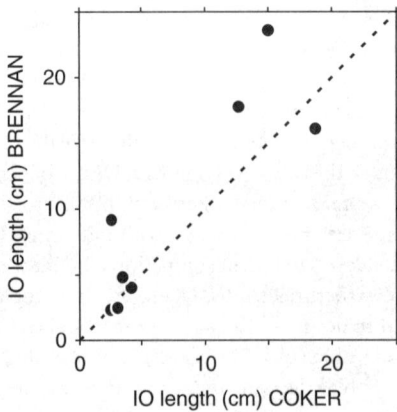

FIGURE 20.11 Relation between size of recently- and long-preserved male IOs from eight species of waterfowl (data from Brennan et al. 2007; Coker et al. 2002, respectively). All specimens were preserved in 10% formalin. Dashed line indicates equality.

critically important to examine males that are known to be breeding and to be in the copulation phase of the reproductive cycle, to ensure that genitalia are at full size (Höhn 1960). Finally, some of the significant patterns described by Coker et al. (2002)—for example, the relation between IO size and mating system—became nonsignificant when controlling for the effects of phylogeny. A reanalysis of these patterns is certainly warranted as both new phylogenetic information and more sophisticated methods for comparative analysis (e.g., Freckleton et al. 2002) become available.

More recently, Brennan et al. (2007) took the study of avian IOs to a new level by examining fresh specimens, by also looking at female reproductive tracts, and by using state-of-the-art comparative methods to examine interspecific variation. To do that study, they removed oviducts, phalluses, and testes from males and females of 16 waterfowl species (family Anatidae) collected during their breeding seasons. Specimens for 13 of the species studied were taken from socially paired individuals, whereas specimens for the remaining three species were taken from commercial farms (two species), and a museum collection (one species, frozen samples). For each male with well-developed testes, the phallus was dissected out and manually everted and fixed in formalin for later measurement. For each female with either eggs in the oviduct or a well-developed oviduct, they preserved both oviduct and

ovaries in formalin for later study. The length of the vagina was measured from the lip of the vent to the uterovaginal junction (UVJ; figure 20.12). This measurement is important because the UVJ is the site of sperm storage tubules (SSTs) in birds (e.g., Briskie & Montgomerie 1993), and it is possible that most fertilizations in birds result from stored sperm (K. Persaud & R. Montgomerie, unpublished data). Thus the vent-to-UVJ distance should be the optimal length for the male IO to maximize the chance of sperm being ejaculated nearest to the site of sperm storage. Brennan et al. (2007) also examined and quantified the internal structure of the vagina (see below).

Interspecific variation in the internal structure of the female vagina turned out to be remarkable. In five species (*Branta canadensis*, *Bucephala clangula*, *Histrionicus histrionicus*, *Somateria mollissima*, and *Anser cygnoides*), the vagina was a simple tube, not unlike that of the few bird species whose vagina had been examined previously (e.g., figure 4.11 in Jacob & Bakst 2007), and widely assumed to be typical of all bird species. In the remaining eight species, however, the vagina was complex, with up to three blind–ending pouches near the cloaca and as many as eight full clockwise spirals ending at the uterus (figure 20.12). These findings are extraordinary both because such variation had not previously been noticed, even in well-studied species like the mallard, and because they suggested, for the first time, that female birds might have some cryptic control over copulation and fertilization.

Indeed, Brennan et al. (2007) suggested that the vaginal pouches might sometimes be used to prevent the male phallus from everting inside the female's vagina, and this is supported by recent experiments (Brennan et al. 2010). In waterfowl, at least, phallus eversion occurs during, rather than prior to the juxtaposition of male and female cloacas during copulation, so those pouches near the female's cloaca can trap the everting phallus and hamper further intromission and insemination (Brennan et al. 2010). Moreover, sperm deposited in these pouches would have a long way to travel to the SSTs and might be more easily ejected by the female (Davies 1983; Pizzari & Birkhead 2000). Brennan et al. (2007) also suggested that the spirals might be able to prevent penetration of the male phallus and this too is supported by recent experiments (Brennan et al. 2010). The phalluses of all waterfowl spiral counterclockwise from their base (e.g., figures 20.8 and 20.12) but the vaginal spirals

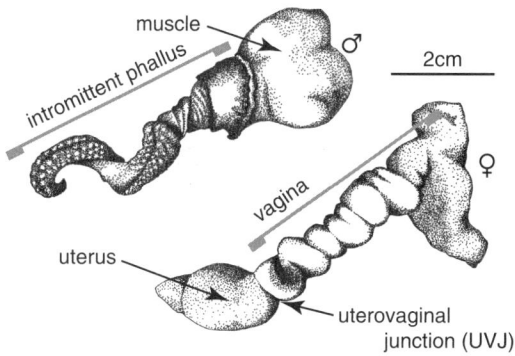

FIGURE 20.12 Male phallus (IO) and female vagina and uterus in the long-tailed duck (*Clangula hyemalis*). Male IO shows counterclockwise spiral and surface ridges. Female vagina shows seven clockwise spirals (pouches are hidden inside the base of the vagina). Orientation shown here is similar to that during copulation, with the base of each structure at the right. Redrawn from photograph in figure 2C of Brennan et al. (2007).

are coiled, unexpectedly, in the opposite direction (figure 20.12). Thus the structures of both phallus and vagina suggest that some intersexual cooperation might be necessary for successful insemination, and thus that the evolution of these genitalia has been influenced by sexual conflict.

To test this sexual conflict idea, Brennan et al. (2007) looked at the relation between phallus length and vaginal structure, arguing from previous work (e.g., Coker et al. 2002) that phallus length is an index of both FEPCs and the intensity of sperm competition. Consistent with the intersexual conflict hypothesis, they found a significant positive relation between phallus length and the number of spirals (partial $r = 0.58$, $P = 0.02$, $n = 13$ species), but not quite significant with the number of pouches (partial $r = 0.07$, $P = 0.8$, $n = 13$ species), in the female vagina (figure 20.13). For this analysis, they controlled for the effects of vagina length, female mass, and phylogeny using generalized linear models, but it is not clear to me that such statistical control is biologically relevant. Vagina length was also significantly correlated with both phallus length ($r = 0.75$, $P = 0.0002$, $n = 13$) and female mass ($r = 0.63$, $P = 0.004$, $n = 13$), again controlling statistically for the effects of phylogeny. These analyses are exemplary for using up-to-date methods to control for phylogeny and for constructing a robust phylogeny based on three mitochondrial and parts of four nuclear gene sequences.

Mapping male phallus length on this phylogeny, Brennan et al. (2007) also showed that large phallus size must have evolved independently at least three times in the Anatidae (figure 20.14). Since this analysis is based on only 10% of the species in this family, a much larger number of independent evolutionary transitions to larger phalluses is to be expected in this taxon. Based on the relations between phallus size and vaginal anatomy described above, the authors concluded that the correlated evolution of male and female genitalia has occurred several times in the waterfowl. This analysis plus recent evidence that female vaginal structures are able to hinder rather than assist male intromission (Brennan et al. 2010) provides the clearest argument to date in support of the idea that IO evolution is influenced by intersexual conflict.

CONCLUSIONS

It should be clear from this and previous reviews that we do not really know very much about the external genitalia of male birds, compared to the rich literature on other taxa reviewed in this volume. This is particularly surprising because the behavior, anatomy, physiology, and evolution of birds has been so well studied (Birkhead 2008), and the general absence of IOs in internally-fertilizing species is such an anomaly. There are several reasons for this dearth of research on the external genitalia of birds, and particularly on the intromittent organ. First, and probably foremost, many bird species with male IOs are large, long-lived birds that do not lend themselves well to experimental study in captivity—including the current, and quite understandable, necessity for often restrictive permits for capture, collecting, and experimental manipulation of live birds. Second, the phallus of male birds is flaccid and internal except for a few seconds during copulation, and so is not easy to study in its operational (erect) state (but see Brennan et al. 2010). Techniques need to be developed to induce phallus erection (e.g., Cary et al. 2004, Brennan et al. 2010), to quantify IO size and structure, and to study copulation experimentally (e.g., Pizzari et al. 2003) so that we have a better idea how the bird IO functions in relation to both its own and the female's anatomy. Finally, we simply need more and better anatomical data on a wider variety of species. Of the 300 bird species known to have an IO, the size and structure of the male phallus is well-known

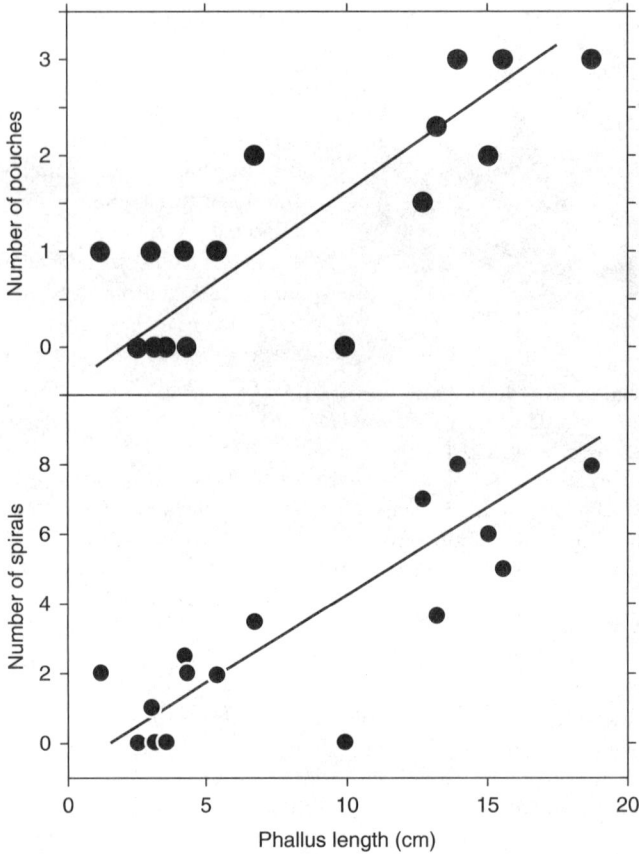

FIGURE 20.13 Relations between female vaginal structure and male IO length in the waterfowl. (A) Number of pouches increases with IO length, but this relation is not significant after controlling for female vagina length. (B) Number of spirals increases with IO length, and this relation remains significant after controlling for vagina length. Redrawn from figure 3 in Brennan et al. (2007).

for only about 20, and even for all of those more details would be useful.

Despite the potential handicaps facing researchers interested in the avian IO, there are some attractive prospects for further study. Based on the recent studies of waterfowl IOs reported above (Coker et al. 2002; Brennan et al. 2007, 2010), it seems clear that this is an excellent taxon for even more detailed analysis. For example, there remain about 130 waterfowl species for which we lack any useful data on female vaginal morphology, and at least 100 species for which even the barest details of IO anatomy are unknown. Most waterfowl breed readily in captivity (e.g., Burns et al. 1980) and their genital anatomy is large enough to be easily studied and quantified. Moreover, the phylogeny of

the Anseriformes is now well enough understood (e.g., Brennan et al. 2007) that we can be reasonably confident about controlling for the influence of phylogeny on any analysis.

The tinamous and megapodes should also be studied in more detail as the male IO appears to have been lost in both taxa. At least some species in each of these taxa breed readily in captivity (Cromberg et al. 2007), and all are large-bodied enough that genital anatomy can be measured and studied. Further study of these two taxa might well give some insights into the factors influencing the reduction and eventual loss of the phallus in the Neognathae. Some recent work on tinamou IOs (Oliveira et al. 2003) has also provided intriguing insights into the immune function of phallus tissue,

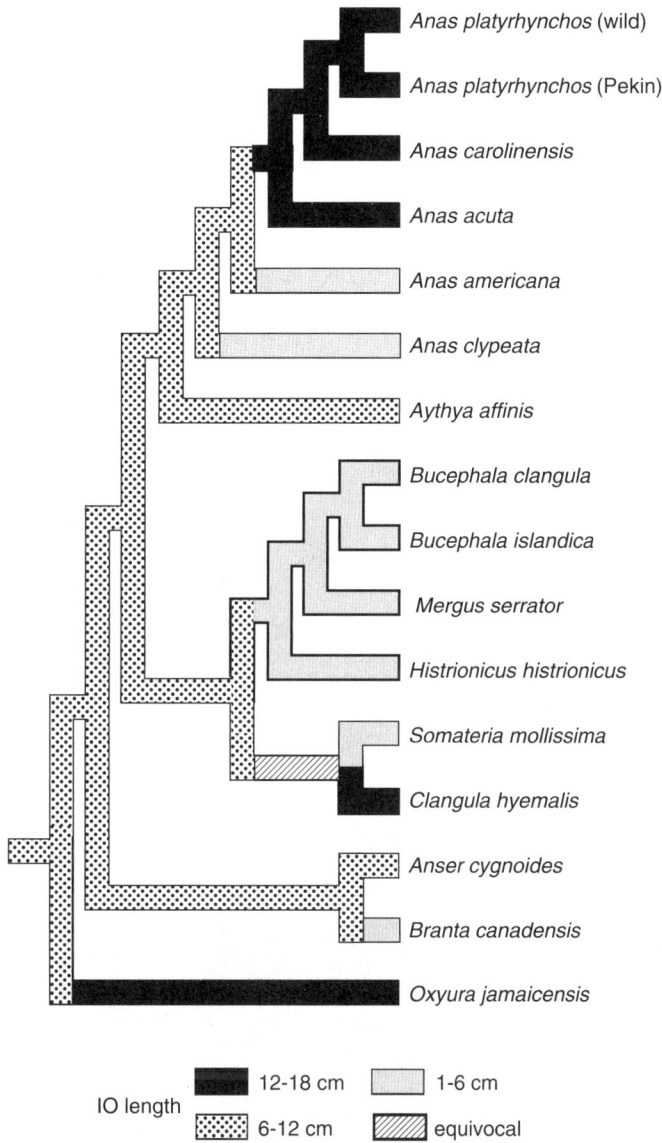

FIGURE **20.14** Phylogenetic distribution of IO size in the waterfowl. Redrawn from figure 4 in Brennan et al. (2007).

and this particularly deserves further investigation in light of the potential costs of an IO due to STDs. In that study, both spotted (*Nothura maculosa*) and red-winged tinamou (*Rhynchotus rufescens*) males were found to have plasma cells at the base of the IO and the number of these cells—known for their immunoprotective function (Slifka & Ahmed 1998)—increased dramatically at the start of the breeding season.

While information on the function of the male IO in birds has been piecemeal to date, both the pattern of IO presence/absence and the variation in its size and structure in the waterfowl are consistent with the hypothesis that sexual conflict is the most important mechanism influencing IO evolution. Since it is highly unlikely that any waterfowl species lack an IO altogether, it is quite possible that the factors affecting IO variation in that taxon have little or nothing to do with the loss of the IO in other lineages. Although assessing the fitness cost of any trait can be difficult, we are unlikely to make much more progress in understanding the loss of the IO in most birds until we understand what those costs might be. It seems to me that the study of STDs in birds and other organisms is a good place to start.

Acknowledgements I am particularly grateful to Patricia Brennan for providing me with manuscripts prior to publication; to Jim Briskie and Frank McKinney for useful discussion; to Roslyn Dakin for several excellent anatomical drawings; to Kim Cheng for providing me with some of the original data from Coker et al. (2002); and to Tim Birkhead and Doug Mock for insightful comments on a draft of this chapter. My research on avian genitalia is supported by Discovery and equipment grants from the Natural Sciences and Engineering Research Council of Canada. The preparation of this chapter was supported by grants from Queen's University and a Canada Council Killam Research Fellowship.

REFERENCES

Adler, M. 2010. Sexual conflict in waterflow: why do females resist extrapair copulations? Behavioral Ecology, 21: 182–192.

Bertelli, S. & Porzecanski, A. L. 2004. Tinamou (Tinamidae) systematics: a preliminary combined analysis of morphology and molecules. Ornitologia Neotropica, 15, 1–7.

Birkhead, T. R. 2008. The Wisdom of Birds: An Illustrated History of Ornithology. London: Bloomsbury.

Birkhead, T. R. & Hoi, H. 1994. Reproductive organs and mating strategies of the bearded tit *Panurus biarmicus*. Ibis, 136, 356–360.

Birkhead, T. R. & Møller, A. P. 1995. Extra-pair copulation and extra-pair paternity in birds. Animal Behaviour, 49, 843–848.

Birkhead, T. R., Hatchwell, B. J. & Davies, N. B. 1991. Sperm competition and the reproductive organs of the male and female Dunnock *Prunella modularis*. Ibis, 133, 306–311.

Birkhead, T. R., Stanback, M. T. & Simmons, R. E. 1993. The phalloid organ of buffalo weavers *Bubalornis*. Ibis, 135, 326–331.

Birks, S. M. & Edwards, S. V. 2002. A phylogeny of the megapodes (Aves: Megapodiidae) based on nuclear and mitochondrial DNA sequences. Molecular Phylogenetics and Evolution, 23, 408–421.

Brennan, P. L. R., Prum, R. O., McCracken, K. G., Sorenson, M. D., Wilson, A., Read, A. F. & Birkhead, T. R. 2007. Coevolution of male and female genital morphology in waterfowl. PLoS ONE, 2, e418.

Brennan, P., C. J. Clark, and R. Prum. 2010. Explosive eversion and fuctional morphology of the duck penis support sexual conflict in waterfowl genitalia. Proceedings of the Royal Society of London B 277: 1309–1314.

Brennan, P.L.R., Clark, C.J. & Prum, R.O. 2010. Explosive eversion and functional morphology of the duck penis supports sexual conflict in waterfowl genitalia. Proceedings of the Royal Society of London, B doi:10.1098/rspb.2009.2139.

Briskie, J. V. 1993. Anatomical adaptations to sperm competition in Smith's Longspurs and other polygynandrous passerines. Auk, 110, 875–888.

Briskie, J. V. & Montgomerie, R. 1993. Patterns of sperm storage in relation to sperm competition in passerine birds. Condor, 95, 442–454.

Briskie, J. V. & Montgomerie, R. 1997. Sexual selection and the intromittent organ of birds. Journal of Avian Biology, 28, 73–86.

Briskie, J. V. & Montgomerie, R. 2001. Efficient copulation and the evolutionary loss of the avian intromittent organ. Journal of Avian Biology, 32, 184–187.

Brom, T. G. & Dekker, R. W. R. J. 1992. Current studies on megapode phylogeny. Zoologische Verhandelingen (Leiden), 278, 7–17.

Burns, J. T., Cheng, K. M. & McKinney, F. 1980. Forced copulation in captive mallards I Fertilization of eggs. Auk, 97, 875–879.

Calhim, S. & Birkhead, T. R. 2007. Testes size in birds: quality versus quantity—assumptions, errors, and estimates. Behavioral Ecology, 18, 271–275.

Cary, J. A., Madill, S., Farnsworth, K., Hayna, J. T., Duoos, L. & Fahning, M. L. 2004. A comparison of electroejaculation and epididymal sperm collection techniques in stallions. Canadian Veterinary Journal, 45, 35–41.

Chiba, A. & Nakamura, M. 2003. Anatomical and histophysiological characterization of the male cloacal protuberance of the polygynandrous Alpine Accentor *Prunella collaris*. Ibis, 145, E83–E93.

Coker, C. R., McKinney, F., Hays, H., Briggs, S. V. & Cheng, K. M. 2002. Intromittent organ morphology and testis size in relation to mating system in waterfowl. Auk, 119, 403–413.

Cracraft, J., Barker, F. K., Braun, M., J, H., Dyke, G. J., Feinstein, J., Stanley, S., Cibois, A., Schikler, P., Beresford, P. et al. 2004. Phylogenetic relationships among modern birds (Neornithes): toward an avian tree of life. In: Assembling the Tree of Life (Ed. by J. Cracraft & M. J. Donoghue), pp. 468–489. New York: Oxford University Press.

Cromberg, V. U., Stein, M. S., Boleli, I. C., Tonhati, H., & Queiroz, S. A. 2007. Reproductive and behavioral aspects of red-winged tinamous (*Rhynchotus rufescens*) in groups with different sex ratios. Brazilian Journal of Poultry Science, 9, 161–166.

Darwin, C. 1871. The Descent of Man and Selection in Relation to Sex. London: John Murray.

Davies, N. B. 1983. Polyandry, cloaca-pecking and sperm competition in dunnocks. Nature, 302, 334–336.

Eberhard, W. G. 1985. Sexual Selection and Animal Genitalia. Cambridge Mass: Harvard University Press.

Eberhard, W. G. 1990. Animal genitalia and female choice. American Scientist, 78, 134–141.

Eckhard, C. 1876. Ueber die Erection der Vögel. Beitrage zur Anatomie und Physiologie, 7, 116–125.

Freckleton, R. P., Harvey, P. H. & Pagel, M. 2002. Phylogenetic analysis and comparative data: A test and review of evidence. American Naturalist, 160, 712–726.

Guzsal, E. 1974. Erection apparatus of the copulatory organ of ganders and drakes. Acta Veterinaria Academiae Scientiarum Hungaricae, 24, 361–373.

Hackett, S. J., Kimball, R. T., Reddy, S., Bowie, R. C. K., Braun, E. L., Braun, M. J., Chojnowski, J. L., Cox, W. A., Han, K.–L., Harshman, J. et al. 2008. A phylogenomic study of birds reveals their evolutionary history. Science, 320, 1763–1768.

Hosken, D. J. & Stockley, P. 2004. Sexual selection and genital evolution. Trends in Ecology and Evolution, 19, 87–93.

Höhn, E. O. 1960. Seasonal changes in the mallard's penis and their hormonal control. Proceedings of the Zoological Society of London, 134, 547–555.

Jacob, M. & Bakst, M. R. 2007. Developmental anatomy of the female reproductive tract. In: Reproductive Biology and Phylogeny of Birds. Part A Phylogeny, Morphology, Hormones, Fertilization (Ed. by B. G. M. Jamieson), pp. 149–179. Enfield, NH: Science Publishers.

Johnsgard, P. A. 1965. Handbook of Waterfowl Behavior. Ithaca, NY: Cornell University Press.

King, A. S. 1981a. Cloaca. In: Form and Function in Birds, vol. 2 (Ed. by A. S. King & J. McLelland), pp. 63–105. New York: Academic Press.

King, A. S. 1981b. Phallus. In: Form and Function in Birds, vol. 2 (Ed. by A. S. King & J. McLelland), pp. 107–147. London: Academic Press.

Lake, P. E. 1981. Male genital organs. In: Form and Function in Birds, vol. 2 (Ed. by A. S. King & J. McLelland), pp. 1–61. London: Academic Press.

Low, M., Castro, I. & Berggren, A. 2005. Cloacal erection promotes vent apposition during forced copulation in the new Zealand stitchbird (hihi): implications for copulatory efficiency in other species. Behavioral Ecology and Sociobiology, 58, 247–255.

Marchant, S. M. & Higgins, P. J. (Ed.) 1990. Handbook of Australian, New Zealand and Antarctic Birds. Volume 1, Ratites to Ducks, Part B Australian Pelican to Ducks. Melbourne: Oxford University Press.

McCracken, K. 2000. The 20-cm spiny penis of the Argentine lake duck (*Oxyura vittata*). Auk, 117, 820–825.

Montgomerie, R. & Briskie, J. 2007. Anatomy and evolution of copulatory structures. In: Reproductive Biology and Phylogeny of Birds. Part A Phylogeny, Morphology, Hormones, Fertilization (Ed. by B. G. M. Jamieson), pp. 115–148. Enfield, NH: Science Publishers.

Mulder, R. A. & Cockburn, A. 1993. Sperm competition and the reproductive anatomy of male superb fairy wrens. Auk, 110, 588–593.

Müller, J. 1836. Uber zwei verscheidene typen in dem bau der erectilen mannlichen geschleehtsorgane bei den straussartigen vogeln. Gelesen in der kgl. Akademie Wissenschaften Physikal. Abhandl., 137–177.

Müller, R. 1908. Über den Tannenberg'schen Korper. Archiv fur die gesamte Physiologie, 122, 455–483.

Nakamura, M. 1990. Cloacal protuberance and copulatory behavior of the Alpine Accentor (*Prunella collaris*). Auk, 107, 284–295.

Oliveira, C. A. & Mahecha, G. A. B. 2000. Morphology of the copulatory apparatus of the spotted tinamou *Nothura maculosa* (Aves: Tinamiformes). Annals of Anatomy, 182, 161–169.

Oliveira, C. A., Geraldo, I., Poblete, P. C. P., Macedo, G. F. & Machecha, G. A. B. 2003. Intraepithelial plasma cells in the avian copulatory organ of two tinamou species: quantitative variation during the breeding season. Anatomy and Embryology, 207, 409–416.

Owen, R. 1866. Anatomy of Vertebrates. Vol. II. Birds and Mammals. London: Longman, Green, and Co.

Owen, R. 1879. Memoirs on the Extinct Wingless Birds of New Zealand, With an Appendix on Those of England, Australia, Newfoundland, Mauritius, and Rodriguez. London: John van Voorst.

Pizzari, T. & Birkhead, T. R. 2000. Female feral fowl eject sperm of subdominant males. Nature, 405, 787–789.

Pizzari, T., Cornwallis, C. K., Lovlie, H., Jakobssen, S. & Birkhead, T. R. 2003. Sophisticated sperm allocation in male fowl. Nature, 426, 70–74.

Samour, J. H. 2002. The reproductive biology of the budgerigar (*Melopsittacus undulatus*): semen preservation techniques and artificial insemination procedures. Journal of Avian Medicine and Surgery, 16, 39–49.

Slifka, M.K. & Ahmed, R. 1998. Long-lived plasma cells: a mechanism for maintaining persistent antibody production. Current Opinion in Immunology, 10, 252–258.

Tuttle, E. M., Pruett–Jones, S. & Webster, M. 1996. Cloacal protuberances and extreme sperm production in Australian fairy-wrens. Proceedings of the Royal Society B, 263, 1359–1364.

Wesołowski, T. 1999. Reduction of phallus in birds—an avian way to safe sex? Journal of Avian Biology, 30, 483–485.

Wesołowski, T. 2001. Reduction of phallus in birds—a reply to Briskie and Montgomerie. Journal of Avian Biology, 32, 188.

Wilkinson, R. & Birkhead, T. R. 1995. Copulation behaviour in the vasa parrots *Coracopsis vasa* and *C. nigra*. Ibis, 137, 117–119.

Winterbottom, M., Burke, T. & Birkhead, T. R. 2001. The phalloid organ, orgasm and sperm competition in a polygynandrous bird: the red billed buffalo weaver (*Bubalornis niger*). Behavioral Ecology and Sociobiology, 50, 474–482.

Wolfson, A. 1954a. Notes on the cloacal protuberance, seminal vesicles, and a possible copulatory organ in male passerine birds. Bulletin of the Chicago Academy of Sciences, 10, 1–23.

Wolfson, A. 1954b. Sperm storage at lower than body temperature outside the body cavity in some passerine birds. Science, 120, 68–71.

21

Genitalic Traits of Mammals

Systematics and Variation

EDWARD H. MILLER

INTRODUCTION

Most early anatomists did not believe in evolution (Coleman 1964; Sloan 1992), but nevertheless used reproductive traits in classification. The renowned anatomist Richard Owen did not believe in Darwinian natural selection, for example, but made recommendations about classification on anatomical grounds, such as proposing that *Homo* be elevated to the rank of subclass ("Archencephala") within Mammalia. Early anatomical information about reproductive traits helped to establish higher-level mammalian classification. For example, on the basis of female reproductive anatomy, the great French systematist Henri de Blainville distinguished monotremes and marsupials from placental mammals in 1816, and then further separated monotremes from marsupials in 1834 (Huxley 1864; Simpson 1945). Resulting classifications were in the tradition of hierarchical nested classifications of the day, and superficially resembled phylogenetically based schemes that are so familiar today. However, those classifications were similarity-based and without formal reference to ancestor–descendant relationships; today, formal phylogenetic analyses permeate all aspects of comparative studies.

Adaptive interpretations of reproductive morphology also have changed. Interspecific differences in genitalic structure were long interpreted as adaptive mechanical isolating mechanisms to reduce hybridization (Mayr 1963; Dobzhansky 1970). However, interspecific diversity could not be satisfactorily explained within such paradigms, as expressed by the primate biologist and anatomist W. C. O. Hill: "It is remarkable, considering that the organs have the same rather limited functions to perform, how varied the male genitalia of primates are in their morphology" (quoted by Dixson 1998, p. 244). At present, genitalic diversity is viewed primarily as an evolutionary consequence of sexual selection[1] by mate choice (Eberhard 2004a, b, 2006; Hosken & Stockley 2004). Early literature on genitalic diversity and sexual selection was strongly biased toward males, partly because female traits simply were viewed as less interesting ("more common than elaborate, more utilitarian than bizarre"; Gowaty 1997, p. 353). This bias resulted in little attention being paid to female sexual traits

---

1. The artificiality of this concept is increasingly recognized. West-Eberhard (1983) included it within a more broadly conceived notion of social selection. Paterson (1993) pointed out that traits ascribed to sexual selection are used for multiple social purposes, and de Waal (1988, p. 232), in referring to bonobo (*Pan paniscus*) behavior, used the term sociosexual, because much of that species' so-called sexual behavior "is divorced from reproductive functions".

or to the roles of inter-sexual interactions in shaping genitalic evolution. This is no longer the case (Eberhard 1996; Arnqvist & Rowe 2005).

High interspecific variation characterizes sexually-selected systems as disparate as birdsong, primate sexual skin, and phallic morphology (Eberhard 1985; Stallmann & Froelich 2000; figure 21.1). Variation attributable to sexual selection also occurs intraspecifically (geographically, and within local populations). In this chapter I will explore the theme of variation (mainly genitalic) with reference to systematics, inter- and intraspecific patterns, and derived uses of genitalia in communication. The relative dearth of information on females unfortunately means that this chapter continues the tradition of male bias.

THE USE OF REPRODUCTIVE MORPHOLOGY IN PHYLOGENETIC INFERENCE

Like all traits, reproductive structures express both diversity and conservatism. The male reproductive system comprises penis, testes, epididymides, deferent ducts, and accessory glands, but these vary in form and function across major clades (Setchell & Breed 2006). A penis is present and delivers semen in all mammals, but penile anatomy varies greatly. The penis of marsupials and placental mammals transmits urine and sexual products, but in monotremes the urine passes to a collecting chamber for elimination via the cloaca, and the penis functions only to transmit sexual products.

Accessory reproductive glands of males also illustrate high-level variation. The main types are prostate gland, vesicular gland (= seminal vesicles), bulbourethral gland (= Cowper's gland), and ampullary gland; mucous glands (the Littre glands) and modified sebaceous glands (the preputial glands) also occur in some species (Voss 1979; Setchell & Breed 2006). The main kinds of glands are present in many species, but size, morphology, and even presence–absence vary greatly. For example, all four of the main types of glands are present in most rodents (figure 21.2A), but only prostate and bulbourethral glands occur in the blind mole rat (*Spalax ehrenbergi*; Gottreich et al. 2001). In the Carnivora, only the prostate is uniformly present, the ampullary gland is variably present (e.g., it occurs in dogs, *Canis familiaris*), and vesicular and bulbourethral glands are always absent (figure 21.2A); only the prostate is present in Cetacea (Rommel et al. 2007).

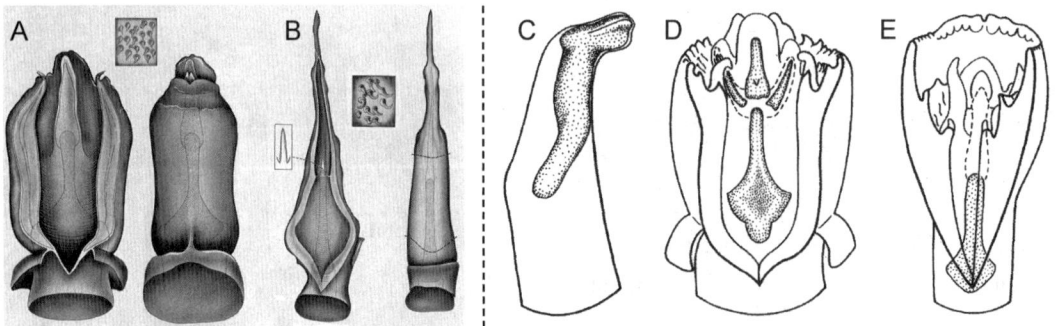

FIGURE 21.1 Sexually-selected structures typically vary greatly, even among related species. Phallic anatomy of rodents exemplifies this trend. Left of dashed line: Glans penis of white-throated woodrat (*Neotoma albigula*; A) and bushy-tailed woodrat (*N. cinerea*; B). For each species, the glans is shown (left) incised mid-ventrally to expose lumen of urethra, crater, and baculum (os penis), and (right) in ventral aspect; the insets are enlargements that show the spines which cover most of the surface of the glans. Both the bacular bone and cartilaginous apex are shown. Right of dashed line: Bacular size, morphology, and proximity to the penile surface vary across species, and influence exposure of this bone to direct selection during intromission. C, Superficial bacular position in Uinta chipmunk (*Tamias umbrinus*; left lateral view). D, Intermediate bacular position in southern red-backed vole (*Clethrionomys gapperi*; mid-ventral incised view, showing bacular shaft plus three apical processes). E, Deeply embedded bacular position in chestnut pogonomys (*Pogonomys macrourus*; incised mid-ventral view). A and B after Hooper (1960: plates I and VII); C–E after Patterson (1983: figure 1).

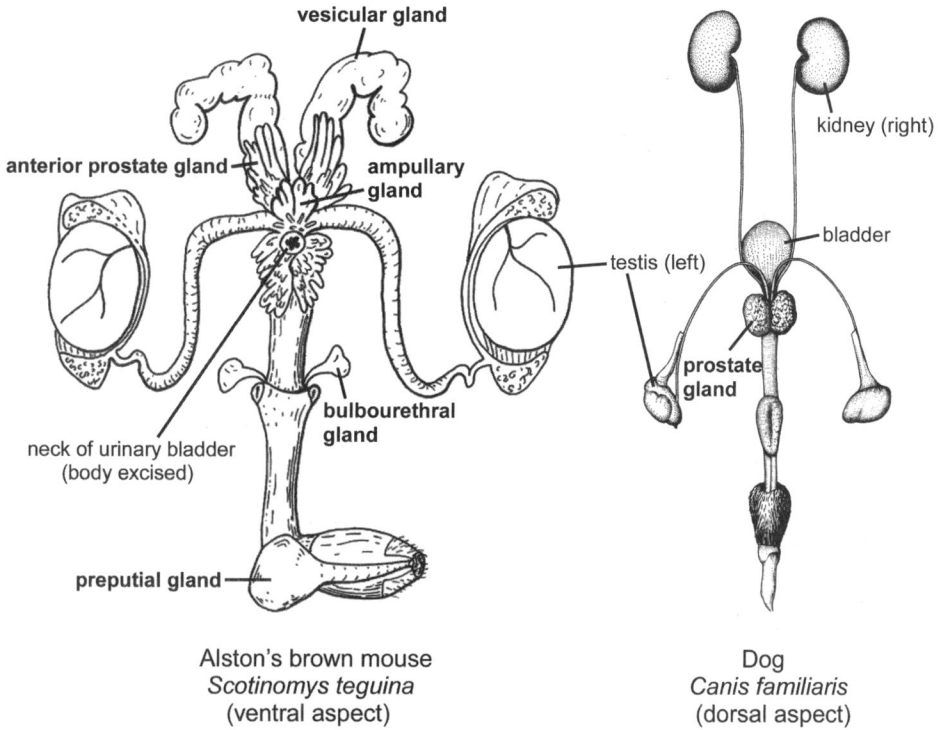

FIGURE 21.2 Diverse reproductive accessory glands occur in male mammals, but vary greatly in morphology and presence/absence in different species. All main types are present in most rodents (A, Alston's brown mouse *Scotinomys teguina*), but only one or two kinds are present in the Carnivora (prostate and ampullary glands are present in the dog *Canis familiaris*; B). (A), after Carleton et al. (1975: figure 2); (B), after Raynaud (1969: figure 441).

These joint patterns of diversity and conservatism give reproductive attributes value as high-level taxonomic traits, and these point to some clear patterns. The epididymis is present in all mammals; other structures (e.g., bulbourethral glands) presumably were present in the common ancestor to mammals, as they occur in extant monotremes, marsupials, and most placental mammals (Setchell & Breed 2006).[2] However, reliance on morphological traits also has caused considerable taxonomic instability. An example is the traditional Order Insectivora. Simpson (1945, pp. 48–53, 176; see Symonds 2005) placed varied insectivorous placental mammals in this taxon: tenrecs, elephant-shrews, tree shrews, and moles. Simpson (1945, p. 175; Symonds 2005) noted that characters of Insectivora

were "in great part primitive for all placental mammals". Candidates are sperm crypts in the oviduct, which are present in moles but absent in hedgehogs, tenrecs, and golden moles (Bedford et al. 2004); and a shallow cloaca, which is present in tenrecs and some shrews (Symonds 2005). In addition, all Insectivora except tree shrews have inguinal testes (Findley 1968). The artificial nature of the Order Insectivora has been revealed by molecular studies, which distribute its members across several clades (figure 21.3). Morphological traits now serve a subsidiary role in phylogenetic analysis, and are more valuable for elucidating patterns and rates of evolutionary change, rather than as a source of information for phylogenetic inference (Springer et al. 2007). For example, molecular data enable

---

2. Clear homology statements are needed in such discussions (Hall 1994). For example, the penis is homologous as an intromittent organ in all male mammals; but is homologous as a dual-function organ (for intromission and urination) only in marsupials and placental mammals.

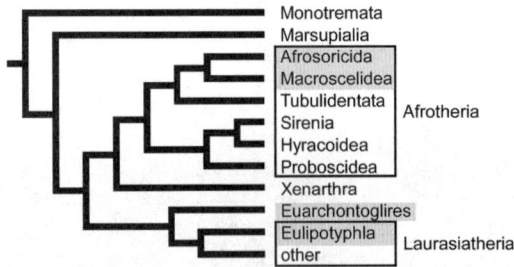

FIGURE 21.3 Terrestrial insectivores (tenrecs, golden moles, hedgehogs, shrews, etc.), and sometimes also elephant shrews (Macroscelidea) and tree shrews (Scandentia), were traditionally placed together in the Insectivora (= Lipotyphla). This is now known to be an unnatural grouping, because it included: two sister groups within the Afrotheria (Afrosoricida, Macroscelidea); several families within Laurasiatheria (solenodons, shrews, etc.); and Scandentia within the Euarchontoglires. Groups that have been included in Lipotyphla are marked by gray. After Murphy et al. (2007: figure 6) and Springer and Murphy (2007b: p. 699).

the estimation of when structures like bulbourethral glands arose: they must date to ≥ 215 Ma, when monotremes last shared a common ancestor with other mammals (Springer & Murphy 2007a, b). A synapomorphy of Afrotheria (tenrecs, golden moles, elephant shrews, sirenians, elephants, aardvarks, and hyraxes) is the trait of being primary testicond mammals, with testes remaining close to the kidney within the body cavity (Gaeth et al. 1999; Bedford et al. 2004; Setchell & Breed 2006; Seiffert 2007); this condition must be ancient, as Afrotheria and Xenarthra diverged from one another ~100 Ma (Springer & Murphy 2007a, b). Another example is the independent loss of bulbourethral glands in Cetacea and Carnivora, which can be dated minimally to the origins of those clades at ~80 and ~50 Ma, respectively (Murphy et al. 2007; Springer & Murphy 2007a, b). A final example is position of the testes in Pinnipedia. Some non-reproductive morphological data support a sister-group relationship between walruses (Odobenidae: Odobenus rosmarus) and seals (Phocidae; Wyss & Flynn 1992; Berta & Wyss 1994). Testes position is phylogenetically ambiguous: in phocids, testes are close to the ventral abdominal wall, whereas they are scrotal (pendulously scrotal in the rut) in fur seals and sea lions (Otariidae; Scheffer 1950; Stewardson et al. 1998). Testes in the walrus are

intermediate, as they "are situated outside the muscular abdominal wall, in the blubber lateral to the base of the penis, more as in the Phocidae as in the Otariidae" (Fay 1982, p. 175; figure 21.4). This situation has been clarified by molecular evidence, which shows relationships to be: (Phocidae (Odobenidae + Otariidae)) (Arnason et al. 2006; Higdon et al. 2007). Therefore the scrotal condition in Otariidae is a derived trait within the pinnipeds, and dates to at least ~25 Ma, when Odobenidae and Otariidae diverged from one another (Arnason et al. 2006).

In summary, high-level morphological and molecular phylogenies often correspond poorly. Morphological characters are most useful for revealing evolutionary rates and trends, when viewed in the context of stable molecular phylogenies (Springer et al. 2007).

Similar conclusions apply at lower levels of diversification. Many studies have tried to sort out species relationships using male genitalic traits, which often differ conspicuously between related species. Lidicker (1968) used many (66!) diverse traits of phallic soft tissue and the baculum in his phenetic study of New Guinea rodents, but reached only a few clear conclusions: monophyly of the group; the presence of two main clades; and an

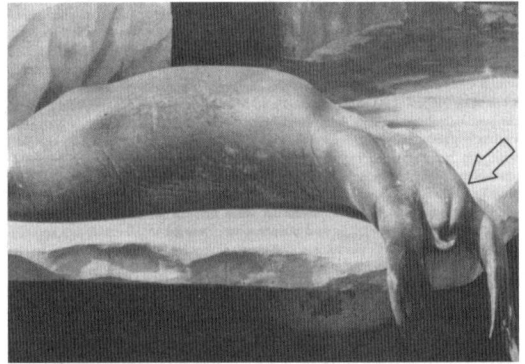

FIGURE 21.4 The walrus (Odobenus rosmarus) exhibits a tendency toward scrotal arrangement of the testes, a condition intermediate between seals (Phocidae) and fur seals and sea lions (Otariidae); the walrus is a sister group to the Otariidae. The captive male in the photograph was lying on his back, with body twisted so that the rear end is oriented obliquely toward the viewer. The photograph was taken in March, when the animal was exceptionally lean and testes enlarged. From Fay (1982: figure 108; photograph by G.C. Ray).

Anisomys-like common ancestor. Some of Lidicker's (1968) inferences have been supported by molecular analyses (e.g., *Anisomys* diverged first, within the Australia/New Guinea clade recognized by Steppan et al. [2005]). Therefore some phylogenetic signal is present in genitalic anatomy in this group of rodents; nevertheless, it clearly is too weak to establish a well resolved phylogeny. Therefore, genitalic traits at low levels of divergence are more suitable for character mapping than for phylogenetic inference, as for higher-level analyses. Two examples follow.

REPRODUCTIVE CHARACTER MAPPING AT LOW LEVELS OF DIVERGENCE: THREE EXAMPLES

The Sciuridae is a large and diverse family that includes prairie dogs (*Cynomys*), ground squirrels (*Spermophilus* and other genera), marmots and groundhogs (*Marmota*), chipmunks (*Tamias* and *Neotamias*), tree squirrels (*Sciurus* and *Tamiasciurus*), and flying squirrels (*Glaucomys*). In a detailed morphological study, Bryant (1945) identified five groups: prairie dogs plus ground squirrels; marmots; chipmunks; tree squirrels; and flying squirrels. These groups (and some higher-level groupings; e.g. "terrestrial squirrels" = prairie dogs, ground squirrels, plus marmots) are identical to those identified on molecular evidence (Herron et al. 2004; Steppan et al. 2004). Molecular analyses have resolved many other relationships: for example, *Spermophilus* ground squirrels are paraphyletic, and *Glaucomys* is the sister group to tree squirrels. Bacular morphology agrees in part with the molecular phylogeny; for example, the baculum is similar between *Cynomys*, *Spermophilus*, and related genera (Wade and Gilbert 1940; Bryant 1945; Burt 1960; figure 21.5). However, discrepancies suggest highly variable divergence rates in bacular morphology and size within some clades. For example,

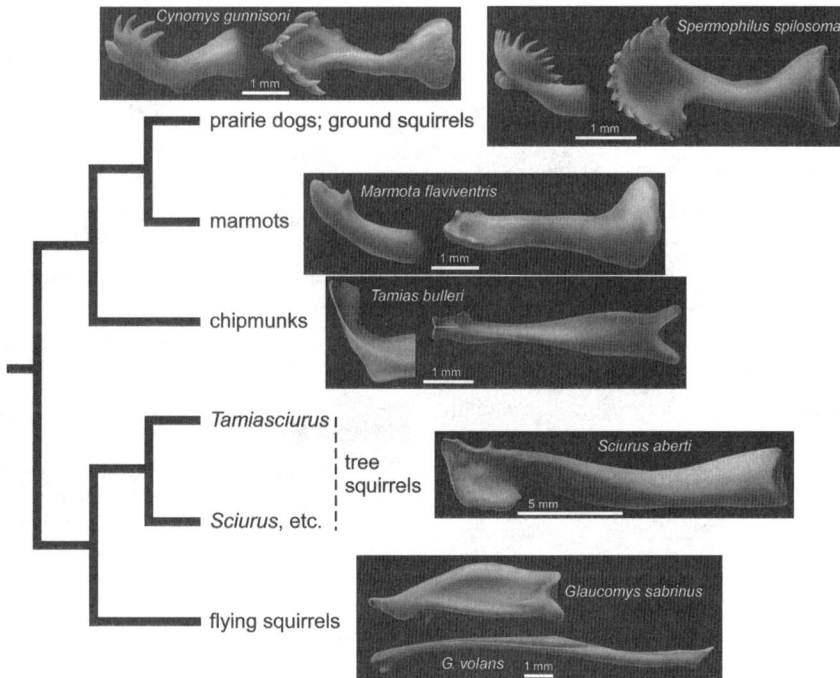

FIGURE 21.5 Character mapping of reproductive traits on a well resolved molecular phylogeny reveals both concordance and discordance. Elaborate claw-like bacula characterize the prairie dog/ground squirrel clade, and a deflected apex occurs in all chipmunks (*Tamias*). However, even bacula of fairly closely related species can differ greatly (e.g., the two species of flying squirrels, *Glaucomys*). Cladogram is based on Herron et al. (2004) and Steppan et al. (2004); illustrations of bacula are from Burt (1960).

the baculum of most tree squirrels is large and complex, but is minute and simple in *Tamiasciurus* (Layne 1952); and the baculum differs greatly between the two *Glaucomys* species (figure 21.5).

Both conservatism and variable divergence rates also characterize the spermatozoa of muroid rodents (= Muridae, Cricetidae, and Neomyidae). Breed (2004, 2005) mapped spermatozoon characters on a molecular phylogeny, and inferred that the ancestral condition was likely "a sperm head with a bilaterally flattened nucleus … acrosome-containing apical hook, and long sperm tail" (Breed 2005, p. 289), which occurs in many muroid lineages and also Heteromyidae (an outgroup; figure 21.6). The divergent sperm of *Tatera* (large naked-soled gerbils; Gerbillinae), which diverged from other

Gerbillinae 8–9 Ma, is highly derived (and in fact is unusual within the Mammalia as a whole). *Deomys* and *Lophuromys* sperm also are strongly divergent within the Muridae.

Baryshnikov et al. (2003) carried out a cladistic analysis of the baculum in the Mustelidae and relatives. They detected little phylogenetic information in bacular morphology, but through character mapping could reconstruct the ancestral state and identify some evolutionary trends. Relative size of the baculum is uniform within the group, except it is slightly shorter in the skunks and relatives (Mephitidae). The ancestral baculum was elongate and rod-shaped, with no urethral groove and with a simple apex. More complex morphology is expressed particularly in the apex, in the form of processes, openings, and spoon- or cup-shaped

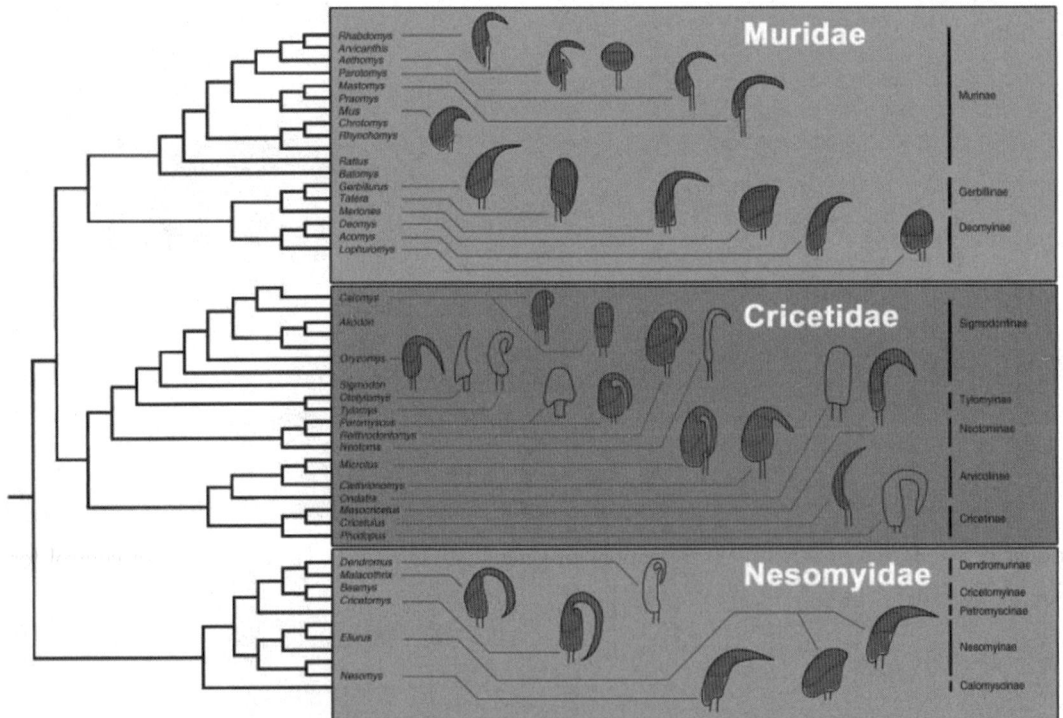

FIGURE 21.6 Character mapping of sperm traits in muroid rodents on a well resolved molecular phylogeny (from Steppan et al. 2004) reveals concordance and discordance, due to great variation in rates of evolutionary divergence. Most species have a sperm head with an apical hook. This hook is largely composed of acrosomal material in most nesomyids, cricetids, and gerbillines, but it contains a nuclear extension with a thinner two-segment acrosome in deomyines and murines. Highly divergent sperm head shapes have evolved that lack an apical hook in a few lineages of most subfamilies (e.g., *Ondatra* [Arvicolinae], two *Habromys* species [Neotominae], one *Calomys* species [Sigmodontinae], African *Tatera* [Gerbillinae], *Lophuromys* [Deomyinae]). From Breed (2005: figure 12).

processes, and the evolution of these specializations within particular lineages could be inferred.

To summarize, character mapping of reproductive traits at fairly low levels of divergence reveals both conservatism and diversification, plus highly variable rates and patterns of divergence (e.g., the baculum of *Tamiasciurus* within tree squirrels, the bacula of the two *Glaucomys* species, and the sperm of *Tatera*). Divergence patterns among related species are especially relevant to how sexual selection, mating systems, and speciation are related.

RELATIONSHIPS OF REPRODUCTIVE ANATOMY TO MATING SYSTEM

Many studies have revealed correlative trends suggesting that the size of male sexual structures is driven by sexual selection. For example, testis size is related to mating system: testes are relatively small in single-male breeding systems (e.g., monogamy), and relatively large in multi-male systems (e.g., promiscuity), presumably because of frequent copulations and high sperm competition in the latter (Kenagy & Trombulak 1986). In the Cetacea, odontocetes have notably large testes: testes of one harbor porpoise (*Phocoena phocoena*) weighed 2.7 kg, ~6% of body mass (= 45 kg), and "almost as large as … for a 50-ton fin whale" (Fontaine & Barrette 1997, p. 68). Testes of baleen whales (Mysticeti) are about the expected size for mammals of their size (Kenagy & Trombulak 1986), but within the group, mass of testes is relatively larger in large species (figure 21.7A). In the northern right whale (*Eubalaena glacialis*), combined mass of testes reaches nearly 1000 kg — the largest size both absolutely and relatively (and this is probably an

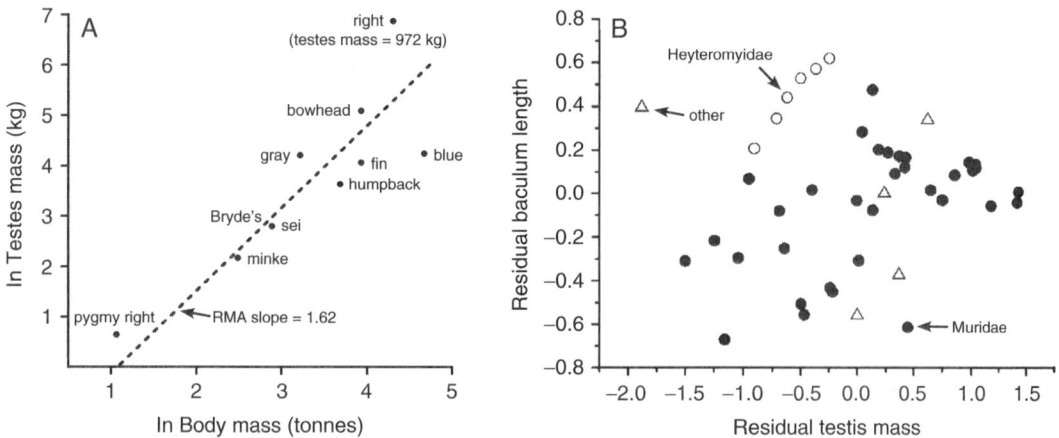

FIGURE 21.7 (A) Early allometric investigations of genital to body size in mammals did not control for phylogeny, but revealed many important patterns. In baleen whales, combined mass of testes (without epididymes) of the northern right whale (*Eubalaena glacialis*) is nearly 1000 kg, which is the largest both absolutely and relatively among baleen whales. Combined mass of testes across species is positively related to body mass, and is characterized by positive allometry (expected slope = 1, for equal proportional change [i.e., isometry] in regression). Scientific names for other species: bowhead *Balaena mysticetus*; blue, *Balaenoptera musculus*; Bryde's, *Balaenoptera edeni*; fin, *Balaenoptera physalus*; gray, *Eschrichtius robustus*; humpback, *Megaptera novaeangliae*; minke *Balaenoptera acutorostrata*; sei, *Balaenoptera borealis*; pygmy right, *Caperea marginata*. (B) Modern analyses control for phylogeny, enabling interspecific trends in relative size of reproductive structures to be evaluated without that complication. This graph is a residual plot of baculum length on testis mass (after controlling for body mass), showing that Muridae tend to have shorter bacula relative to testis size than do Heteromyidae. A few other rodent species are included for comparison. (A) After data in Brownell and Ralls (1986: table 1; those authors incorrectly reported and graphed reduced major axis regression (RMA) slope as 1.35, which is the slope in ordinary least-squares regression). (B) After Ramm (2007: figure 2).

underestimate of size in the breeding period; Brownell & Ralls 1986).[3]

Early analyses did not control for phylogenetic relationships. Allometric trends like those shown in figure 21.7A are influenced by relationship, because closely related groups tend to be morphologically and behaviorally similar (statistically speaking, the regressions are based on partially correlated data, which violates statistical assumptions and inflates the estimated degrees of freedom; O'Connor et al. 2007). Various methods to control for phylogenetic effects have been proposed (Nunn & Barton 2001; Freckleton et al. 2002). In his phylogenetically controlled analysis, Ramm (2007) found that bacular length and length of glans penis in rodents were both positively related to inferred level of sperm competition (testis mass had a significant and positive effect on both genital size measures in multiple regression analyses, when the influence of body mass was controlled for; figure 21.7B). Relationships differed across groups, suggesting important reproductive differences between Heteromyidae and Muridae.

Size of anatomical structures is related to other factors as well. In terrestrial Carnivora, relative testis size is greatest in species with brief breeding seasons, when synchrony of female estrous and sperm competition are presumed to be highest (Iossa et al. 2008). This relationship is stronger for spontaneous than for induced ovulators (Larivière & Ferguson 2003 present other perspectives). Many other reproductive structures have been investigated, for example, seminal vesicles are largest in primate species presumed to have the most intense sperm competition (with "dispersed" mating systems), and are smallest in monogamous species (Dixson 1998); and the relative size of seminal vesicles and the anterior prostate are positively related to level of sperm competition in rodents (Ramm et al. 2005).

Allometric trends are strongest at high levels of differentiation, and weaken progressively over lower levels of differentiation, such as across conspecific populations or among individuals within populations (Ramm 2007). Nevertheless, trends are apparent and require separate explanations over all scales of divergence. Intraspecific patterns can be particularly informative about relationships between sexual selection and population divergence.

INTRASPECIFIC PATTERNS: GEOGRAPHIC VARIATION

Sexual-selection theory predicts higher variation in sexually-selected traits than in non-sexually-selected traits, even across conspecific populations (Pomiankowski & Iwasa 1998). Wilkinson and McCracken (2003) investigated geographic variation in testicular size in relationship to mating system in two species of bat. In the Jamaican fruit-eating bat (*Artibeus jamaicensis*) in Panama, males attend harems in hollow trees, but female groups are labile, so females may mate with multiple males. In contrast, female group composition is stable in Mexico. In keeping with this difference, testes from Panama are more than six times the size (volume) of those from Mexico. Substantial size variation also occurs on a smaller spatial scale in bats: testes of Mexican free-tailed bats (*Tadarida brasiliensis*) from four colonies in Texas varied in average volume from ~100 to ~170 mm^3, paralleling differences in colony size (100,000 to 4–6 million), which in turn is presumed to be correlated with the intensity of sexual selection via sperm competition.

A second example of geographic variation is testicular size in the spinner dolphin (*Stenella longirostris*; figure 21.8A). A number of forms occur in this widely distributed species, including the distinctive eastern spinner dolphin (*S. l. orientalis*) of the far-eastern tropical Pacific. The so-called "whitebelly spinner dolphin" is a broadly distributed hybrid swarm morphologically intermediate between *orientalis* and spinner dolphins to the west (in Hawaii and the South Pacific). Mature male eastern and whitebelly spinner dolphins differ substantially in external appearance, and in testicular size and activity: in whitebelly spinners, testes are much larger, and more males are sexually active at any one time (as judged by sperm abundance in the epididymides). For example, for specimens with combined mass of the right testis and epididymis ~ 100–300g, only 9% of eastern spinners had copious sperm in the epididymis, versus 47% of whitebelly spinners (Perrin & Mesnick 2003; figure 21.8B). Based on these trends, Perrin & Mesnick (2003, p. 471) inferred that there is a "gradient from a more polygynous mating system in the eastern form

---

3. Accounts of Kenagy & Trombulak (1986) and Brownell & Ralls (1986) differ somewhat; I follow the latter.

FIGURE **21.8** (A) The spinner dolphin (*Stenella longirostris*) is geographically variable in body size, shape, and coloration, and in testicular size (also note the conspicuous ventral postanal hump in this large adult male eastern spinner, which is absent in the whitebelly form). (B) The whitebelly form has larger testes, and more whitebelly males have sperm than do eastern males (testis mass range 500–700 g shown as example). (A) Photo by B. Pitman/NOAA Fisheries Service, Southwest Fisheries Science Center, Protected Resources Division; (B) after Perrin and Mesnick (2003: figure 5).

to a more open, promiscuous, or polygynandrous mating system in the whitebelly spinner".

More complex patterns of geographic variation occur in traits other than size. The baculum and baubellum (os clitoridis) of *Tamias* chipmunks vary interspecifically, so might be expected to also vary geographically within species. This is not the case in two chipmunk species. The Allen's chipmunk (*T. senex*) and Siskiyou chipmunk (*T. siskiyou*) are almost identical morphometrically (in cranial features), and exhibit parallel ecogeographic variation in pelage and morphology over their largely sympatric ranges in the western United States. However, the baculum and baubellum are morphologically uniform within each species across the same range (Sutton and Patterson 2000; figure 21.9). On the surface, this finding is not in accord with conventional sexual-selection theory, but instead agrees with Paterson's (1993) theory of specific mate-recognition systems, which predicts stability in conspecific mate recognition and attraction, and in sexual behavior throughout a species' range. Evidence on this point is mixed however, as some studies point to substantial intraspecific geographic variation in sexually-selected structures of mammals and other taxa (Kelly et al. 2000; Møller 1995), in accordance with the presumed importance of sexual selection in facilitating differentiation and speciation (Arnqvist & Rowe 2005; Ritchie 2007). For example, bacula and baubella

vary between genetically distinct allopatric populations of the yellow-pine chipmunk (*T. amoenus*; Good et al. 2003). Similarly, population structure is suggested by non-reproductive traits in the European ground squirrel (*Spermophilus citellus*; Kryštufek & Hrabě 1996) and least chipmunk (*T. minimus*; Sullivan & Petersen 1988), and bacular traits vary concordantly with non-reproductive traits in both cases. More detailed analysis of population structure is needed to clarify how sexual selection contributes to the disparate patterns that have been reported in different studies.

INTRASPECIFIC PATTERNS: WITHIN-POPULATION VARIATION

Morphological variation within populations is of evolutionary interest for many reasons (Darwin 1883; Yablokov 1974; Wright 1978; Hallgrímsson and Hall 2005). In the context of sexual selection, such variation (e.g., ornament size or shape) is of special interest because, as noted above, sexually-selected traits are held to be more variable than non-sexually-selected traits (Long and Frank 1968; Long 1969; Lüpold et al. 2004). A simple example is the baculum of the harp seal (*Pagophilus groenlandicus*), which varies in size more than does the humerus (figure 21.10). Such analyses must take

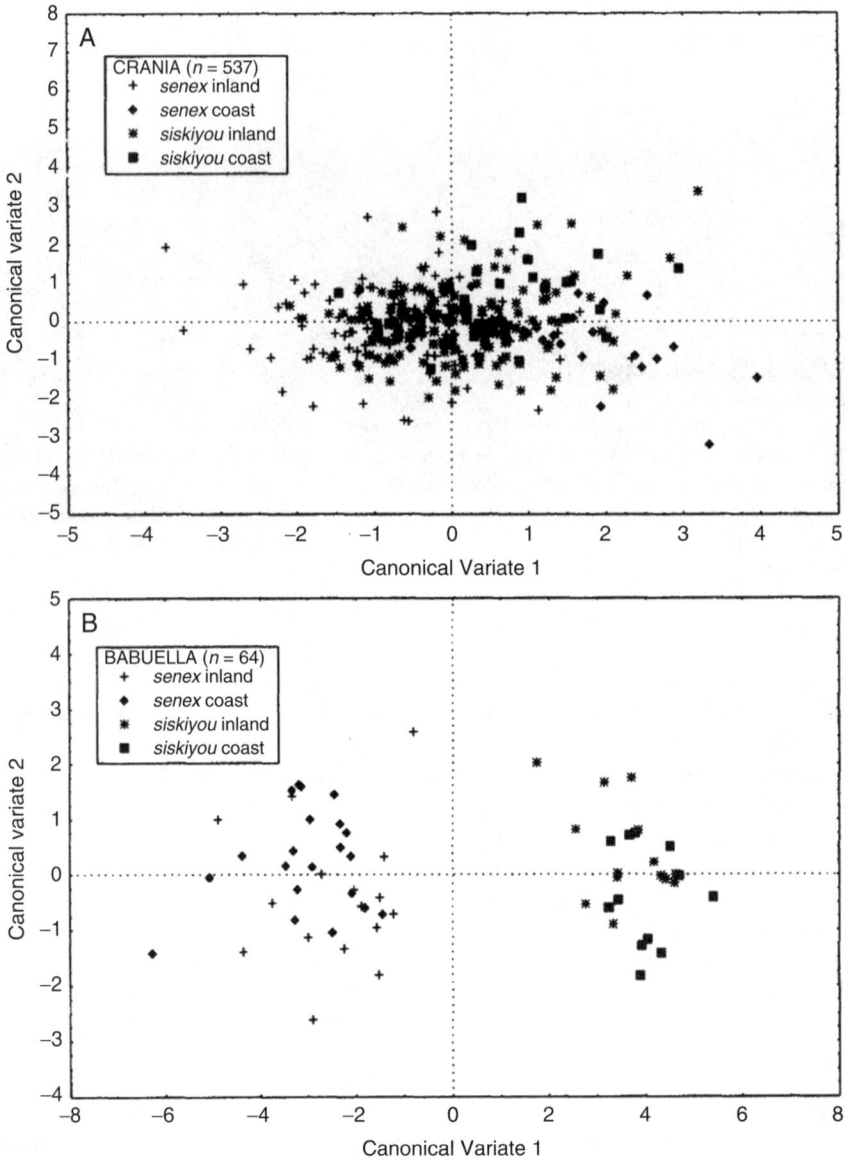

FIGURE 21.9 Two closely related chipmunks (*Tamias senex*, *T. siskiyou*) display parallel ecogeographic variation in body size and coat color across their largely sympatric ranges in the western United States, and are not distinguishable on the basis of cranial variables. However, the baculum and baubellum (os clitoridis) are uniform within and differ distinctively between the species over their distributions. Canonical plots of discriminant function scores are shown for separate analyses on cranial variables (A) and baubellar variables (B). After Sutton and Patterson (2000: figure 3).

social system into account. Two mole-rat (Bathyergidae) species have morphologically differentiated castes, and only a few males reproduce; bacular variation in this case would differ from that in a society where males are more-or-less equivalent; the same may be true of species in which males have despotic or rank-based access to females (e.g., wolf, *Canis lupus*). The more interesting question in such cases pertains to variation within reproductively active social castes or dominance ranks. Male sea

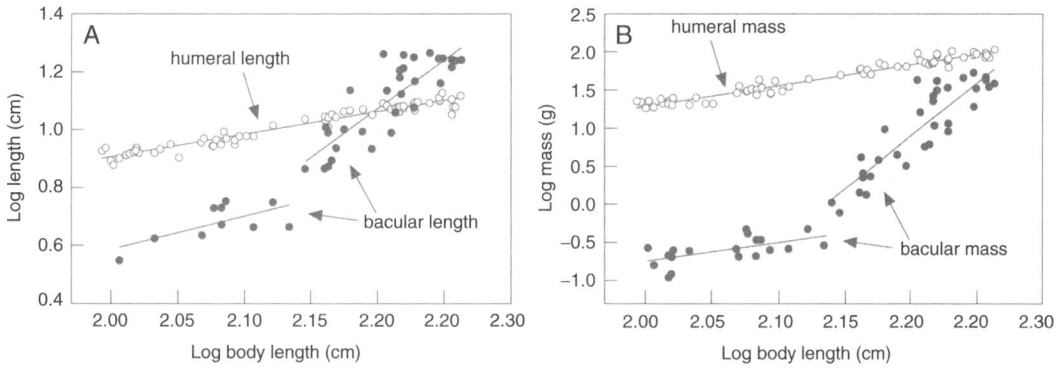

FIGURE 21.10 Sexually-selected traits often are more variable than non-sexually-selected traits within a species. In this study, variation was estimated by residuals from allometric regression (Eberhard et al. 1998; Miller 2009) for size (A, length; B, mass) of the baculum and humerus in the harp seal (*Pagophilus groenlandicus*). From Miller and Burton (2001: figure 2).

lions and fur seals exhibit deferred social maturity, entering the territorial system years after they are physiologically capable of fertilizing females (Miller 2009). Males that are large for their age may become territorial at a younger age (Miller et al. 2000; Roberts 1988; Scheffer 1950); therefore estimates of genitalic (e.g., bacular and testicular) size variation within age classes will be biased upward if territorial and non-territorial males of the same age are not represented proportionately in the samples. Wilkinson & McCracken (2003) made a similar point for bats.

Levels of variation in homologous reproductive structures may differ between males and females. The baubellum appears to be more variable than the baculum in size and morphology, although it is morphologically complex in some species (e.g., *Sciurus* tree squirrels; Layne 1954; Long and Frank 1968). As noted above, the baculum of *Tamiasciurus* tree squirrels is tiny and simple, and the baubellum is only variably present (Layne 1952). The baubellum is similarly small and variably present in the walrus, and even becomes smaller with age; yet this species has the largest baculum of any species of mammal, both absolutely and relatively (Mohr 1963; Fay 1982; Dixson 1995; figure 21.11). Such observations will remain uninterpretable until baubellar function is clarified.

Sexual recognition and mating in mammals entail all senses, and involve components of biochemistry, physiology, behavior, and morphology. It seems unlikely that sexually-selected traits would show similar patterns in variation across such a

range of systems. For example, display behavior and structures may be free to vary a fair amount, whereas size and shape of the penis and vagina must be constrained simply because of the need for morphological compatibility during intromission and copulation (Paterson 1993; Eberhard et al. 1998; Froehlich 2003; Hosken & Stockley 2004; McPeek et al. 2008). If so, one can predict a positive relationship in reproductive size traits between the sexes.

In deer mice (*Peromyscus*), bacular and vaginal lengths are positively correlated both inter- and intraspecifically (Patterson and Thaeler 1982; figure 21.12A). Kinahan et al. (2007) reported that both bacular and vaginal lengths scale positively on body size in the Cape dune mole–rat (*Bathyergus suillus*). In *Tamiasciurus*, males have a long, filiform penis, and estrous females have a long coiled vagina, unlike other tree squirrels; other examples of intersexual concordance in rodents are given by Patterson & Thaeler (1982). Some recent analyses have revealed repeated patterns of male-female coevolution. Anderson et al. (2006) investigated coevolution of sexual traits in the context of sperm competition in 48 species. They posited that length of the oviduct should increase with the intensity of sperm competition. In phylogenetically controlled analyses, they found that oviduct length was positively related to two measures known in turn to be positively correlated with intensity of sperm competition: relative testes volume and sperm midpiece volume (Anderson et al. 2005; figure 21.12B). Such examples could be multiplied and extended to other

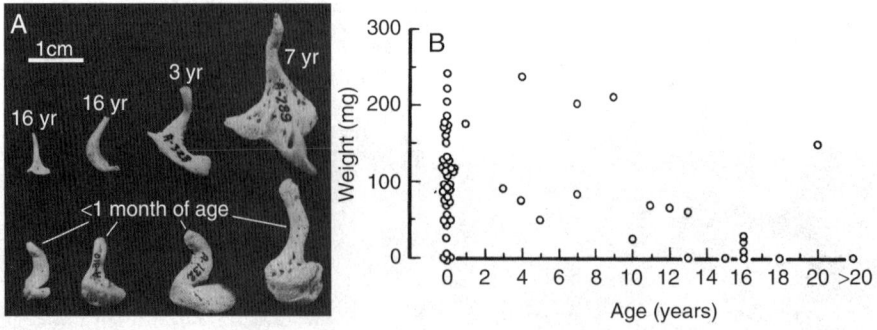

FIGURE 21.11 The baculum of the walrus (*Odobenus rosmarus*) is the largest in mammals, both in absolute size and relative to body size; as in other mammals, it increases in size with age. The species' os clitoridis is small (sometimes it is absent) and morphologically unlike the baculum, and tends to decline in size with age. (A) Clitoris bones from Pacific walruses (*Odobenus rosmarus divergens*) of various ages. (B) The clitoris bone tends to decrease in mass with age; dry mass of calcified tissue is plotted on the Y-axis. After Fay (1982: figures 26 and 27).

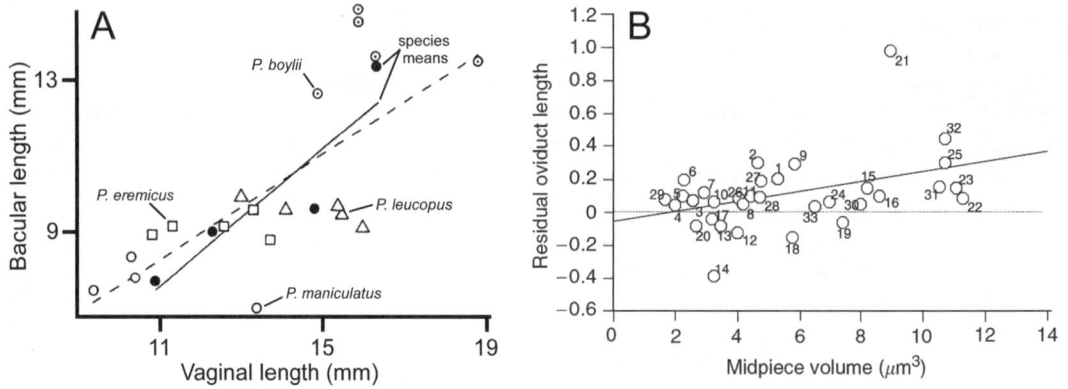

FIGURE 21.12 Concordance between male and female reproductive traits reflects coevolution between the sexes. (A) Size of male and female reproductive structures are positively correlated across populations and species of deer mice (*Peromyscus*). Multiple symbols for each species represent samples from different sites within the state sampled (New Mexico). (B) Length of the oviduct is positively related to midpiece volume of the sperm across species, suggesting coevolution through the action of sexual selection. (A) After Patterson and Thaeler (1982: figure 3); (B) After Anderson et al. (2006: figure 1b).

taxa (Eberhard 1996; Brennan et al. 2007; Rönn et al. 2007).

Parallel evolution of male and female reproductive traits can come about in various ways, but rarely through cooperative evolution (Eberhard 1996, 2004a, b, 2006; Arnqvist & Rowe 2005). Simple patterns in genitalic size, as in the examples above, sometimes must merely reflect correlated change to ecogeographic variation in body size (Kitchener et al. 1994). Others result from adaptive

changes in males to changing reproductive traits of females. In the chimpanzee (*Pan troglodytes*), the female's sexual skin swelling substantially increases the distance a male must penetrate in order to achieve fertilization, which may have led to the evolution of the male's elongated, filiform penis (Dixson & Mundy 1994). Similar explanations may apply to *Tamiasciurus* and many other species. Other evolutionary changes in males are responses to competitive conditions created by the mating system, or by females themselves (e.g., females may cluster in space or time, or have a lengthy estrus). Some such evolutionary responses are mentioned above (e.g., the large testes of the northern right whale and the whitebelly form of the spinner dolphin). Another is the enhanced mechanism of seminal coagulation and copulatory plug formation in primates with high levels of sperm competition (Dixson & Anderson 2002).

DERIVED ROLES OF GENITALIA IN COMMUNICATION

Essentially all aspects of sexual interaction between males and females entail communication, including: looking at, smelling, or touching genitals; testing urine; mounting; and physical and chemical interactions within the female during intromission (Dewsbury 1988). Each sex provides a richness of cues to the other sex. Many unspecialized morphological and behavioral cues have evolved into formalized displays and interactive behavior through ritualization (Tinbergen 1952; Immelmann & Beer 1989). For example, many endocrine-associated traits have been co-opted for signaling; the best known example is cyclical change in the sexual skin of some Old World primates. Specialized markings on or around the genitals and nipples of male and female primates also have evolved for signaling (Dixson 1998; Gerald 2003; Zinner et al. 2004). Much sexual communication takes place over short distances, and so less striking genitalic features can provide valuable information to receivers. For example, at the beginning of sexual activity in females of the greater dwarf lemur (*Cheirogaleus major*), the genital region, "including the clitoris, becomes turgescent and pink, the vaginal opening appears and the ventral side of the clitoris itself opens completely" (Petter-Rousseaux 1964, pp. 112–113; figure 21.13). Similarly subtle morphological changes take place in estrus of some pinnipeds (Miller 1991).

Some variation in morphological traits is informative about age or social rank. The penis and scrotum in the vervet monkey (*Cercopithecus aethiops*) are strikingly colored and highly variable across males, and are important in static and dynamic optical displays. In this species, males with dark scrota dominate males with paler scrota, and more antagonism occurs between males with similarly dark or pale scrota (Gerald 2001, 2003). The scrotum of the northern brown bandicoot (*Isoodon macrourus*) becomes increasingly pigmented with age, so could provide socially useful information to conspecific males or females (Gemmell 1987). The scrotum of the Geoffroy's spider monkey (*Ateles geoffroyi*) is variably pigmented, and pigmentation may increase with age in this species too (Gerald 2003). The scrotum is vivid blue in males of several marsupial species, a feature that has arisen independently at least twice (Prum & Torres 2004); presumably scrotal color is important in optical signaling, but its relationship to rank and age is unknown.

Intraspecific social mimicry of males by females is common, for example in plain-dwelling cursorial

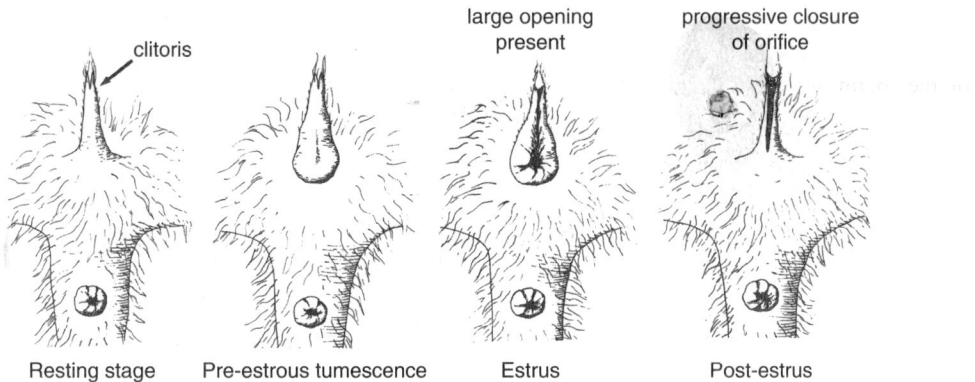

FIGURE 21.13 Physiological changes associated with reproduction often produce changes in appearance, many of which have been co-opted for purposes of optical communication. Females of the greater dwarf lemur (*Cheirogaleus major*) show conspicuous genitalic changes over the reproductive cycle that may function in optical communication. After Petter-Rousseaux (1964: figure 10).

ungulates (Wickler 1968; Geist 1998). Genitalic mimicry ("andromimicry"; Estes 1991) is part of this syndrome. The best known example is the spotted hyaena (*Crocuta crocuta*), in which andromimicry likely evolved to reduce inter-female aggression (Muller & Wrangham 2002). This has resulted in dramatic anatomical reshaping of the female, with equally dramatic functional consequences: the labia fuse during fetal life to form a pseudo-scrotum, so females lack an external vagina; and the clitoris is enlarged and approximates the size and shape of the penis, and can be erected like the penis. During intromission, the penis enters the clitoral meatus and becomes positioned in the clitoral portion of the urogenital canal (in addition, the fairly large (1.1–1.6 kg) precocial infants are born through the clitoris (Cunha et al. 2003)). Andromimicry involving the genitalia also is known for long-tailed macaques (*Macaca fascicularis*), in which females develop a pseudoscrotum (Malaivijitnon et al. 2007), and for the Bovidae (Estes 1974, 1991). Yearling female wildebeest (*Connochaetes gnu*) develop (and maintain through adulthood) "hair and adipose tissue resembling the tip of the penile sheath", and females in the Ankole breed of cattle (*Bos taurus*) "develop a conspicuous penile flap resembling that of a subadult male except for the long hairs" (Estes 1991, p. 436; figure 21.14).

Many optical, chemical, and acoustic cues are available to receivers, but are not given through specialized display behavior. Others are provided through fairly simple motor patterns such as approach, touching, or presentation — behavior which can nevertheless be highly structured. Presentation by female Old World primates is an example; it is performed even by females of species that lack sexual swellings (Wickler 1967). Similarly, tactile communication involving the genitals is

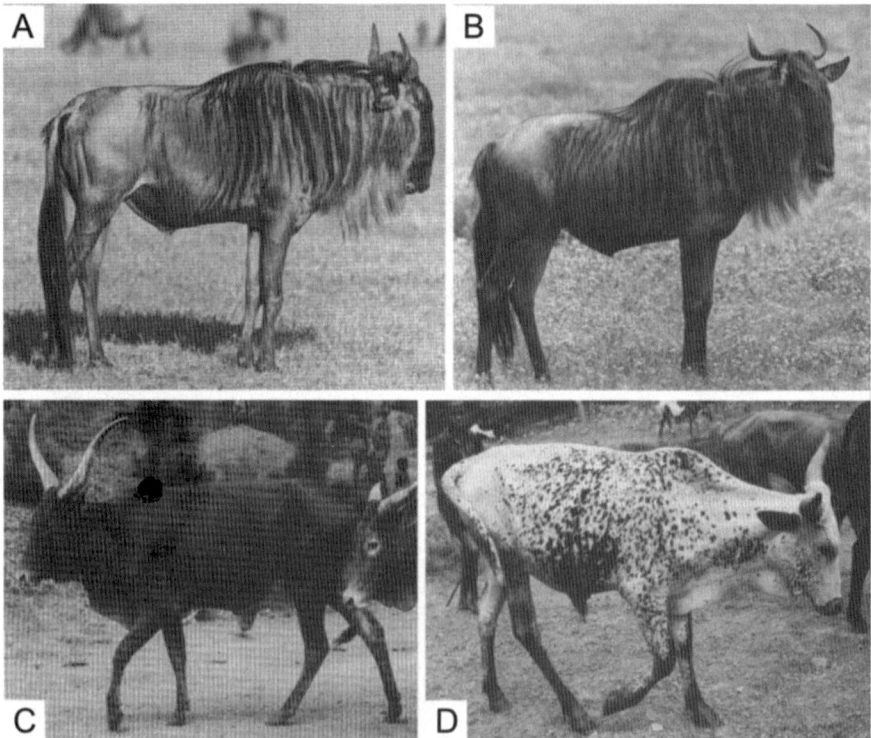

FIGURE 21.14 Mimicry of males by females ("andromimicry") occurs in many mammals, and includes examples of mimicry of the external genitalia. A and B, Adult male and female wildebeest (respectively) (*Connochaetes gnu*), showing female's well developed "penile" tuft. C and D, Subadult male and adult female (respectively) Ankole cattle (*Bos taurus*); note similarity between sexes in "penile" flap and dewlap. From Estes (1991: figures 8A, 8B, 9B, and 9A, respectively).

highly structured, and occurs in all mammals; genitals can be contacted with the manus, mouth, tongue, or nose. Extensive tactile communication occurs in the bonobo (*Pan paniscus*), including genito-genital rubbing between females (de Waal 1988; figure 21.15).

Specialized postures or movements by the sender, or complex interactions, commonly accompany morphological and other display specializations. Males of many Old World primate species display their genitals while sitting with legs spread (figure 21.16). In two species of baboon (*Papio*), the bright pink penis is conspicuous against the dark pelage, and in the vervet monkey, "the penis is usually erected … and may be repeatedly struck against the stomach with a jerky action" (Wickler 1967, p. 150), making the brightly colored genitals even more conspicuous. The genitals are used in other displays of this species as well[4]. One aggressive display between male guinea pigs (*Cavia porcellus*) is sudden extrusion of the testicles directed toward the other animal (Kunkel & Kunkel 1964). The reverse occurs in rhesus macaques: adult males may retract their testes into the inguinal canal when approached closely or frightened by a dominant individual (Altmann 1962).

To summarize, reproductive morphology of mammals is used extensively in and has become specialized in many ways for purposes of sociosexual communication. Many conspicuous optical specializations have been recognized, but far less attention has been given to less striking optical signals, to tactile signals, or to signaling and interactive behavior themselves.

CONCLUSIONS AND SUGGESTIONS FOR FURTHER RESEARCH

Mammalian genitals do not exist in isolation, and have not evolved as functionally isolated entities that serve only in reproduction. Genitals have

FIGURE 21.15 The genitals are involved in many forms of tactile communication in mammals. This photograph shows two female bonobos (*Pan paniscus*) engaged in genito-genital rubbing. From de Waal (1988: figure 11C).

multiple functions within species and divergent functions across species, and are embedded within integrated morphological–physiological–behavioral systems (Simmons & Jones 2007). Therefore to understand patterns of diversity like those touched on in this chapter, integrated research that cuts across levels of biological organization and scientific disciplines is needed.

Many significant patterns have been revealed by recent research on genitalic diversity in relation to sexual selection and mating system, such as male–female coevolution, and relationships of size and morphology of sexual structures to mating system. At the same time, the generality of many studies means that we lack insight into which mechanisms are responsible for trends and for exceptions to trends. The varied reports on intraspecific geographic variation in bacula are an example (see above). In a similar vein, why are bacular and cranial traits taxonomically concordant in identifying taxonomic groups within one subgenus (*Proechimys*) of spiny rats *Proechimys*, but discordant in another

---

4. "In *C. a. pygerythrus*, the blue scrotum is displayed to conspecific males and females during a variety of agonistic, dominance and intergroup territorial displays … The blue scrotum is featured prominently in the 'red-white-and-blue' display that combines the bright red penis, the white belly fur and skin and the blue scrotum; in the red-white-and-blue display, a dominant male walks around a submissive male with his tail raised, displaying his blue scrotum … Sometimes during the red-white-and-blue-display, a male stands upright with his erect penis bobbing up and down … frequency of performance of the red-white-and-blue display is correlated with dominance and mating success …" Prum & Torres (2004, p. 2168).

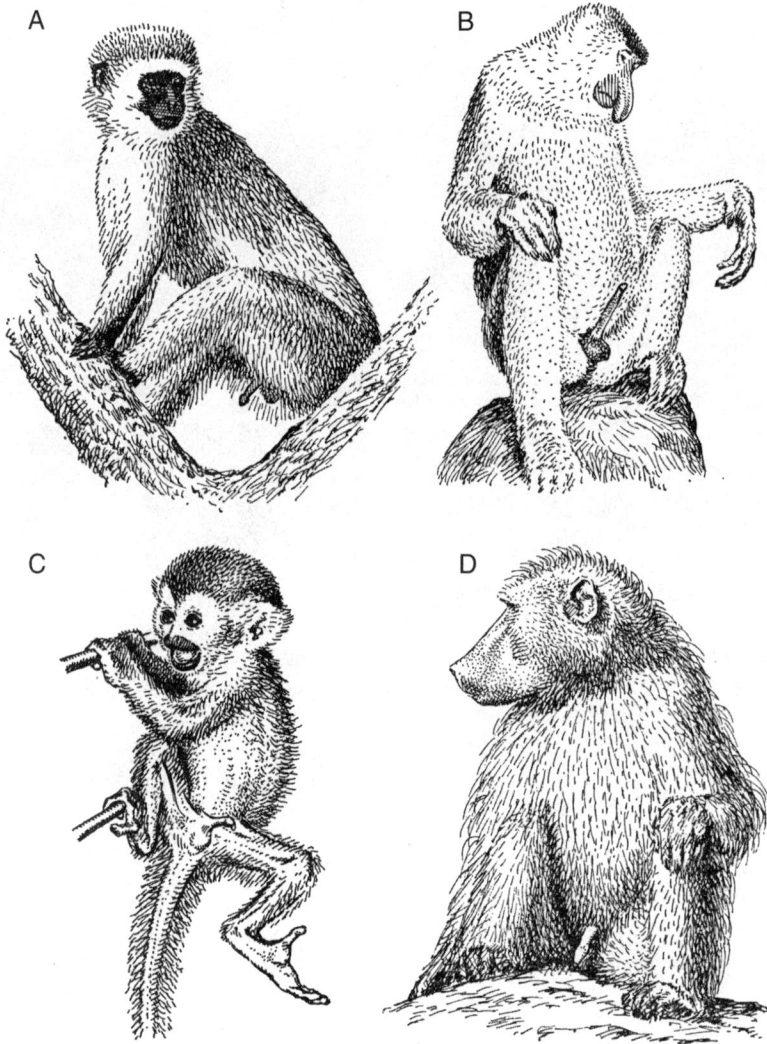

FIGURE **21.16** Many optical displays of primates involve the genitals. (A) Adult male vervet monkey (*Cercopithecus aethiops*). (B) Adult male proboscis monkey (*Nasalis larvatus*). (C) Young squirrel monkey (*Saimiri sciurus*). (D) Adult male olive baboon (*Papio anubis*). After Wickler (1967: figures 11C and 12A, B, D).

(*Trinomys*), in which only the bacula are species-distinctive (Pessôa & dos Reis 1992)? More detailed research is called for in other cases too. For example, in their study of North American voles (*Microtus*; *Clethrionomys*), Heske & Ostfeld (1990) found relatively small testes in polygynous and male-territorial species, and relatively large testes in promiscuous species, but paradoxically large testes also characterized socially monogamous species. Such findings could be illuminated through detailed functional and behavioral studies.

Recent morphological and functional analyses invite a comparative approach. The mammalian penis is inflatable and stiff (i.e., resistant to bending), design features that may be achievable in only a limited number of ways (Kelly 2002, 2004, 2008). The role of the baculum in erection has been investigated for the Norway rat (*Rattus norvegicus*;

Kelly 2000). The baculum of this species is load-bearing, and transfers forces from the distal glans to the walls of the corpus cavernosum, with the eventual effect of increasing penile stiffness. The baculum of the Norway rat lies well within penile soft tissue and is morphologically simple, unlike many species: position within the penis, morphology, and size vary greatly among rodents (Burt 1960; Patterson 1983; figure 21.1). In some species, the bacular apex is large, simple, and inflected, and lies close beneath the surface of the glans (figure 21.1C); in others, the apex is morphologically complex and multipartite (figure 21.1D), and in others it lies more deeply and a cartilaginous cap projects to beneath the glans (figure 21.1E); and so on. Such great variation must translate into appreciable functional differences interspecifically, even within single taxonomic groups. Dewsbury (1975) called for integrated studies on morphology and copulatory behavior, and this recent work offers opportunities for doing so; the role of the female in copulation needs to be included in such a program.

Comparative behavioral studies likewise are essential for revealing display functions of and phylogenetic diversification of specialized genital morphology, and in turn advancing understanding of how sexual selection operates in different populations and species. Chemical communication by products of genitalic glands or of glands that are near the genitals, or by products in urine and feces, offers many possibilities. Again, the role of the female in communication needs to be given increased attention — in addition to being involved from initial recognition through to copulation, females also play important roles in postmating events, through cryptic internal selection (Eberhard 1996) to physical removal of copulation plugs (Koprowski 1992).

Mammals have morphologically diverse, complex genitalia, whose functional diversity is just starting to be documented and interpreted in evolutionary terms. The short-beaked echidna (*Tachyglossus aculeatus*) has an anatomically remarkable penis, with a bifid glans and urethra as in monotremes and most marsupials (unifid in some derived marsupials; Smith & Madkour 1980). Each part of the urethra bifurcates further, terminating in two epidermal rosettes. All four rosettes are prominent in early erection, but those on one side retract in later erection, leaving only two to transmit semen to one of the female's oviductal ostia (Johnston et al. 2007; figure 21.17A). This pattern of erection and sperm transmission resembles the use of hemipenes by squamates. Many other mysteries about genital form, function, and evolution in mammals remain (figure 21.17B).

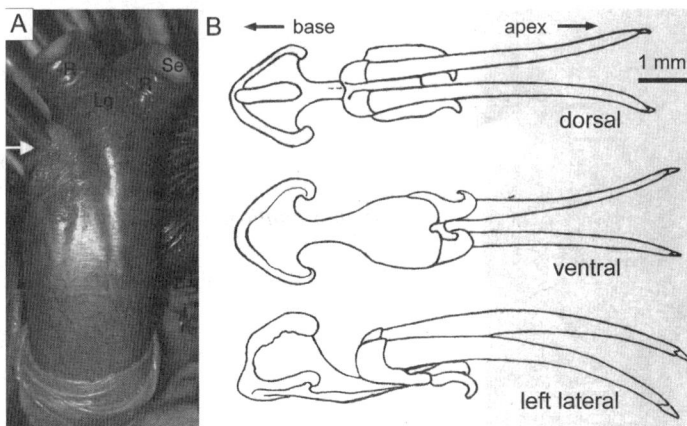

FIGURE 21.17 Mammalian genitals are complex in form and function. (A) Fully erect penis of short-beaked echidna (*Tachyglossus aculeatus*), in ventral view, showing one-sided retraction of the two terminal rosettes on the right side (arrow). The right side is erect, and semen is visible, pooled in the rosette openings (Lg, left glans; R, rosettes; Se, semen). (B) Morphologically bizarre baculum of the greater Egyptian jerboa (*Jerboa orientalis*). (A) After Johnston et al. (2007: figure 2); (B) From Didier and Petter (1960: figure 3).

Acknowledgments   I thank the many people who have helped me with this chapter, by providing literature, illustrations, criticisms, advice, support, and inspiration: W. G. Breed, W. G. Eberhard, the late F. H. Fay, M. Gonzalez-Fideli, D. A. Kelly, S. L. Mesnick, W. F. Perrin, B. D. Patterson, G. C. Ray, M. S. Springer, and S. J. Steppan. During the preparation of this chapter, my research has been supported by the Natural Sciences and Engineering Research Council of Canada.

REFERENCES

Altmann, S. A. 1962. A field study of the sociobiology of rhesus monkeys, *Macaca mulatta*. Annals of the New York Academy of Sciences, 102, 338–435

Anderson, M. J., Nyholt, J. & Dixson, A. S. 2005. Sperm competition and the evolution of sperm midpiece volume in mammals. Journal of Zoology, 267, 135–142.

Anderson, M. J., Dixson, A. S. & Dixson, A. F. 2006. Mammalian sperm and oviducts are sexually selected: evidence for co-evolution. Journal of Zoology, 270, 682–686.

Arnason, U., Gullberg, A., Janke, A., Kullberg, M., Lehman, N., Petrov, E. A. & Väinölä, R. 2006. Pinniped phylogeny and a new hypothesis for their origin and dispersal. Molecular Phylogenetics and Evolution, 41, 345–354.

Arnqvist, G., & Rowe, L. 2005. Sexual Conflict. Princeton, New Jersey: Princeton University Press.

Baryshnikov, G. F., Bininda-Emonds, O. R. P. & Abramov, A. V. 2003. Morphological variability and evolution of the baculum (os penis) in Mustelidae (Carnivora). Journal of Mammalogy, 84, 673–690.

Bedford, J. M., Mock, O. B. & Goodman, S. M. 2004. Novelties of conception in insectivorous mammals (Lipotyphla), particularly shrews. Biological Reviews, 79, 891–909.

Berta, A. & Wyss, A. R. 1994. Pinniped phylogeny. Proceedings of the San Diego Society of Natural History, 29, 33–56.

Breed, W. G. 2004. The spermatozoon of Eurasian murine rodents: its morphological diversity and evolution. Journal of Morphology, 261, 52–69.

Breed, W. G. 2005. Evolution of the spermatozoon in muroid rodents. Journal of Morphology, 265, 271–290.

Brennan, P. L. R., Prum, R. O., McCracken, K. G., Sorenson, M. D., Wilson, R. E. & Birkhead, T. R. 2007. Coevolution of male and female genital morphology in waterfowl. PLoS ONE, 2, e418.

Brownell Jr., R.L. & Ralls, K. 1986. Potential for sperm competition in baleen whales. Report of the International Whaling Commission Special Issue, 8, 97–112.

Bryant, M. D. 1945. Phylogeny of Nearctic Sciuridae. American Midland Naturalist, 33, 257–390.

Burt, W. H. 1960. Bacula of North American mammals. Miscellaneous Publications, Museum of Zoology, University of Michigan, 113, 1–76 + 25 plates.

Carleton, M. D., Hooper, E. T. & Honacki, J. 1975. Karyotypes and accessory reproductive glands in the rodent genus *Scotinomys*. Journal of Mammalogy, 56, 916–921.

Coleman, W. 1964. Georges Cuvier, Zoologist. A Study in the History of Evolution Theory. Cambridge, Massachusetts: Harvard University Press.

Cunha, G. R., Wang, Y., Place, N. J., Liu, W., Baskin, L. & Glickman, S. E. 2003. Urogenital system of the spotted hyena (*Crocuta crocuta Erxleben*): a functional histological study. Journal of Morphology, 256, 205–218.

Darwin, C. 1883. The Variation of Animals and Plants Under Domestication. Second rev. ed. New York: D. Appleton and Co.

De Waal, F. B. M. 1988. The communicative repertoire of captive bonobos (*Pan paniscus*) compared to that of chimpanzees. Behaviour, 106, 183–251.

Dewsbury, D. A. 1975. Diversity and adaptation in rodent copulatory behavior. Science, 190, 947–954.

Dewsbury, D. A. 1988. Copulatory behavior as courtship communication. Ethology, 79, 218–234.

Didier, R. & Petter, F. 1960. L'os pénien de *Jaculus blanfordi* (Murray) 1884 (sic) étude comparée de *J. blanfordi, J. jaculus et J. orientalis* (rongeurs, dipodidés). Mammalia, 24, 171–176.

Dixson, A. F. 1995. Baculum length and copulatory behaviour in carnivores and pinnipeds (Grand Order Ferae). Journal of Zoology, London, 235, 67–76.

Dixson, A. 1998. Primate Sexuality. Comparative Studies of the Prosimians, Monkeys, Apes, and Human Beings. Oxford, England: Oxford University Press.

Dixson, A. F. & Anderson, M. J. 2002. Sexual selection, seminal coagulation and copulatory plug formation in Primates. Folia Primatologica, 73, 63–69.

Dixson, A. F. & Mundy, A. I. 1994. Sexual behavior, sexual swelling, and penile evolution in chimpanzees (*Pan troglodytes*). Archives of Sexual Behavior, 23, 267–280.

Dobzhansky, T. 1970. Genetics of the Evolutionary Process. New York: Columbia University Press.

Eberhard, W. G. 1985. Sexual Selection and Animal Genitalia. Cambridge, Massachusetts: Harvard University Press.

Eberhard, W. G. 1996. Female Control: Sexual Selection by Cryptic Female Choice. Princeton, New Jersey: Princeton University Press.

Eberhard, W. G. 2004a. Male–female conflict and genitalia: failure to confirm predictions in insects and spiders. Biological Reviews, 79, 121–186.

Eberhard, W. G. 2004b. Rapid divergent evolution of sexual morphology: comparative tests of antagonistic coevolution and traditional female choice. Evolution, 58, 1947–1970.

Eberhard, W. 2006. Sexually antagonistic coevolution in insects is associated with only limited morphological diversity. Journal of Evolutionary Biology, 19, 657–681.

Eberhard, W. G., Huber, B. A., Rodriguez S., R. L., Briceño, R. D., Salas, I. & Rodriquez, V. 1998. One size fits all? Relationships between the size and degree of variation in genitalia and other body parts in twenty species of insects and spiders. Evolution, 52, 415–431.

Estes, R. D. 1974. Social organization of the African Bovidae. In: The Behaviour of Ungulates and its Relation to Management. New Series No. 24 (Ed. by V. Geist & F. Walther), pp. 166–205. Morges, Switzerland: International Union for the Conservation of Nature.

Estes, R. D. 1991. The significance of horns and other male secondary sexual characters in female bovids. Applied Animal Behaviour Science, 29, 403–451.

Fay, F. H. 1982. Ecology and biology of the Pacific walrus, *Odobenus rosmarus divergens* Illiger. North American Fauna, 74, 1–279.

Findley, J. S. 1968. Insectivores and dermopterans. In: Recent Mammals of the World. A Synopsis of Families (Ed. By Anderson, S. & J. Knox Jones Jr.), pp. 87–108. New York: Ronald Press.

Fontaine, P. M. & Barrette, C. 1997. Megatestes: anatomical evidence for sperm competition in the harbor porpoise. Mammalia, 61, 65–71.

Freckleton, R. P., Harvey, P. H. & Pagel, M. 2002. Phylogenetic analysis and comparative data: a test and review of evidence. American Naturalist, 160, 712–726.

Froehlich, J. W. 2003. Testing some theoretical expectations of sexual selection versus the recognition concept in the speciose macaques of Sulawesi, Indonesia. In: Sexual Selection and Reproductive Competition in Primates: New Perspectives and Directions. Special Topics in Primatology, vol. 3 (Ed. by C. B. Jones), pp. 539–591. Norman, Oklahoma: American Society of Primatologists.

Gaeth, A. P., Short, R. V. & Renfree, M. B. 1999. The developing renal, reproductive, and respiratory systems of the African elephant suggest an aquatic ancestry. Proceedings of the National Academy of Sciences of the USA, 96, 5555–5558.

Geist, V. 1998. Deer of the World: Their Evolution, Behavior, and Ecology. Mechanicsburg, Pennsylvania: Stackpole Books.

Gemmell, R. T. 1987. Sexual maturity in the captive male bandicoot, *Isoodon macrourus*. Australian Journal of Zoology, 35, 433–441.

Gerald, M. S. 2001. Primate colour predicts social status and aggressive outcome. Animal Behaviour, 61, 559–566.

Gerald, M. S. 2003. How color may guide the primate world: possible relationships between sexual selection and sexual dichromatism. In: Sexual Selection and Reproductive Competition in Primates: New Perspectives and Directions. Special Topics in Primatology, vol. 3 (Ed. by C. B. Jones), pp. 141–172. Norman, Oklahoma: American Society of Primatologists.

Good, J. M., Demboski, J. R., Nagorsen, D. W. & Sullivan, J. 2003. Phylogeography and introgressive hybridization: chipmunks (genus *Tamias*) in the northern Rocky Mountains. Evolution, 57, 1900–1916.

Gottreich, A., Hammel, I., Yogev, L., Bartoov, B. & Terkel, J. 2001. Structure and function of accessory sex glands in the male blind mole rat (*Spalax ehrenbergi*). Journal of Mammalogy, 82, 201–208.

Gowaty, P. A. 1997. Sexual dialectics, sexual selection, and variation in reproductive behavior. In: Feminism and Evolutionary Biology: Boundaries, Intersections, and Frontiers (Ed. by P. A. Gowaty), pp. 351–384. New York: Chapman and Hall.

Hall, B. K. (Ed.) 1994. Homology: the Hierarchical Basis of Comparative Biology. San Diego, California: Academic Press.

Hallgrímsson, B. & Hall, B. (Eds.). 2005. Variation: a Central Concept in Biology. Burlington, Massachusetts: Elsevier Academic Press.

Herron, M. D., Castoe, T. A. & Parkinson, C. L. 2004. Sciurid phylogeny and the paraphyly of Holarctic ground squirrels (*Spermophilus*). Molecular Phylogenetics and Evolution, 31, 1015–1030.

Heske, E. J., & Ostfeld, R. S. 1990. Sexual dimorphism in size, relative size of testes, and mating systems in North American microtine rodents. Journal of Mammalogy, 71, 510–519.

Higdon, J. W., Bininda-Emonds, O. R. P., Beck, R. M. D. & Ferguson, S. H. 2007. Phylogeny and divergence of the pinnipeds (Carnivora: Mammalia) assessed using a multigene dataset. BMC Evolutionary Biology, 7–216.

Hooper, E. T. 1960. The glans penis in *Neotoma* (Rodentia) and allied genera. Occasional Papers of the Museum of Zoology, University of Michigan, 618, 1–21 + 11 plates.

Hosken, D. J. & Stockley, P. 2004. Sexual selection and genital evolution. Trends in Ecology and Evolution, 19, 87–93.

Huxley, T. H. 1864. Lectures on the Elements of Comparative Anatomy. London, England: John Churchill and Sons.

Immelmann, K. & Beer, C. 1989. A Dictionary of Ethology. Cambridge, Massachusetts: Harvard University Press.

Iossa, G., Soulsbury, C. D., Baker, P. J. & Harris, S. 2008. Sperm competition and the evolution of testes size in terrestrial mammalian carnivores. Functional Ecology, 22, 655–662.

Johnston, S. D., Smith, B., Pyne, M., Stenzel, D. & Holt, W. V. 2007. One-sided ejaculation of echidna sperm bundles. American Naturalist, 170, E162–E164.

Kelly, C. D., Godin, J. J.-G. & Abdallah, G. 2000. Geographical variation in the male intromittent organ of the Trinidadian guppy (*Poecilia reticulata*). Canadian Journal of Zoology, 78, 1674–1680.

Kelly, D. A. 2000. Anatomy of the baculum-corpus cavernosum interface in the laboratory rat (*Rattus norvegicus*), and implications for force transfer during copulation. Journal of Morphology, 244, 69–77.

Kelly, D. A. 2002. The functional morphology of penile erection: tissue designs for increasing and maintaining stiffness. Integrative and Comparative Biology, 42, 216–221.

Kelly, D. A. 2004. Turtle and mammal penis designs are anatomically convergent. Proceedings of the Royal Society, Biology Letters, 271 (Suppl 5), S293–S295.

Kelly, D. A. 2008. Penises as variable-volume hydrostatic skeletons. Annals of the New York Academy of Sciences, 1101, 453–463.

Kenagy, G. J. & Trombulak, S. C. 1986. Size and function of mammalian testes in relation to body size. Journal of Mammalogy, 67, 1–22.

Kinahan, A., Bennett, N., O'Riain, M., Hart, L. & Bateman, P. 2007. Size matters: genital allometry in an African mole-rat (Family: Bathyergidae). Evolutionary Ecology, 21, 201–213.

Kitchener, D. J., Adams, M. & Boeadi. 1994. Morphological and genetic relationships among populations of *Scotorepens sanborni* (Chiroptera: Vespertilionidae) from Papua New Guinea, Australia and Indonesia. Australian Mammalogy, 17, 31–42.

Koprowski, J. L. 1992. Removal of copulatory plugs by female tree squirrels. Journal of Mammalogy, 73, 572–576.

Kryštufek, B. & & Hrabě, V. 1996. Variation in the baculum of the European souslik, *Spermophilus citellus*. Zeitschrift für Säugetierkunde, 61, 228–235.

Kunkel, P. & Kunkel, I. 1964. Beiträge zur Ethologie des Hausmeerschweinchens *Cavia aperea f. porcellus* (L.). Zeitschrift für Tierpsychologie, 21, 602–641.

Lariviére, S., & Ferguson, S. H. 2003. Evolution of induced ovulation in North American carnivores. Journal of Mammalogy, 84, 937–947.

Layne, J. N. 1952. The os genitale of the red squirrel, *Tamiasciurus*. Journal of Mammalogy, 33, 457–459.

Layne, J. N. 1954. The os clitoridis of some North American Sciuridae. Journal of Mammalogy, 35, 357–366.

Lidicker Jr., W. Z. 1968. A phylogeny of New Guinea rodent genera based on phallic morphology. Journal of Mammalogy, 49, 609–643.

Long, C. A. 1969. Gross morphology of the penis in seven species of the Mustelidae. Mammalia, 33, 145–160.

Long, C. A. & Frank, T. 1968. Morphometric variation and function in the baculum, with comments on correlation of parts. Journal of Mammalogy, 49, 32–43.

Lüpold, S., McElligot, A. G. & Hosken, D. J. 2004. Bat genitalia: allometry, variation and good genes. Biological Journal of the Linnean Society, 83, 497–507.

Malaivijitnond, S., Hamada, Y., Suryobroto, B. & Takenaka, O. 2007. Female long-tailed macaques with scrotum-like structure. American Journal of Primatology, 69, 721–735.

Mayr, E. 1963. Animal Species and Evolution. Cambridge, Massachusetts: Harvard University Press.

McPeek, M. A., Shen, L., Torrey, J. Z. & Farid, H. 2008. The tempo and mode of three-dimensional morphological evolution in male reproductive structures. American Naturalist, 171, E158–E178.

Miller, E. H. 1991. Communication in pinnipeds, with special reference to non-acoustic signalling. In: The Behaviour of Pinnipeds (Ed. by D. Renouf), pp. 128–235. London, England: Chapman and Hall.

Miller, E. H. 2009. Territorial behavior. In: Encyclopedia ofMarine Mammals. Second ed. (Ed. by W. F. Perrin, B. Würsig & H. G. M. Thewissen), pp.1156–1166. San Diego: Academic Press. (in press)

Miller, E. H. & Burton, L. E. 2001. It's all relative: allometry and variation in the baculum (os penis) of the harp seal, Pagophilus groenlandicus (Carnivora: Phocidae). Biological Journal of the Linnean Society, 72, 345–355.

Miller, E. H., Pitcher, K. W. & Loughlin, T. R. 2000. Bacular size, growth, and allometry in the largest extant otariid, the Steller sea lion (Eumetopias jubatus). Journal of Mammalogy, 81, 134–144.

Mohr, E. 1963. Os penis und Os clitoridis der Pinnipedia. Zeitschrift für Säugetierkunde, 28, 19–37.

Møller, A. P. 1995. Sexual selection in the barn swallow (Hirundo rustica). V. Geographic variation in ornament size. Journal of Evolutionary Biology, 8, 3–19.

Muller, M. M. & Wrangham, R. 2002. Sexual mimicry in hyaenas. Quarterly Review of Biology, 77, 3–16.

Murphy, W. J., Pringle, T H., Crider, T. A., Springer, M. S. & Miller, W. 2007. Using genomic data to unravel the root of the placental mammal phylogeny. Genome Research, 17, 413–421.

Nunn, C. L. & Barton, R. A. 2001. Comparative methods for studying primate adaptation and allometry. Evolutionary Anthropology, 10, 81–98.

O'Connor, M.P., Agosta, S. J., Hansen, F., Kemp, S. J., Sieg, A. E., McNair, J. N. & Dunham, A. E. 2007. Phylogeny, regression, and the allometry of physiological traits. American Naturalist, 170, 431–442.

Paterson, H. E. H. 1993. Evolution and the Recognition Concept of Species: Collected Writings (Ed. by S. F. McEvey). Baltimore, Maryland: Johns Hopkins University Press.

Patterson, B. D. 1983. Baculum-body size relationships as evidence for a selective continuum on bacular morphology. Journal of Mammalogy, 64, 496–499.

Patterson, B. D. & Thaeler, C. S. Jr. 1982. The mammalian baculum: hypotheses on the nature of bacular variability. Journal of Mammalogy, 63, 1–15.

Perrin, W. F. & and Mesnick, S. L. 2003. Sexual ecology of the spinner dolphin, Stenella longirostris: geographic variation in mating system. Marine Mammal Science, 19, 462–483.

Pessôa, L. M. & dos Reis, S. F. 1992. Bacular variation in the subgenus Trinomys, subgenus Proechimys (Rodentia, Echimyidae). Zeitschrift für Säugetierkunde, 57, 100–102.

Petter-Rousseaux, A. 1964. Reproductive physiology and behavior of the Lemuroidea. In: Evolutionary and Genetic Biology of Primates, vol. 2 (Ed. by J. Buettner-Janusch), pp. 91–132. New York: Academic Press.

Pomiankowski, A. & Iwasa, Y. 1998. Runaway ornament diversity caused by Fisherian sexual selection. Proceedings of the National Academy of Sciences of the U.S.A. 95, 5106–5111.

Prum, R. O. & Torres, R. H. 2004. Structural colouration of mammalian skin: convergent evolution of coherently scattering dermal collagen arrays. Journal of Experimental Biology, 207, 2157–2172.

Ramm, S. A. 2007. Sexual selection and genital evolution in mammals: a phylogenetic analysis of baculum length. American Naturalist, 169, 360–369.

Ramm, S. A., Parker, G. A. & Stockley, P. A. 2005. Sperm competition and the evolution of male reproductive anatomy in rodents. Proceedings of Royal Society of London B 272, 949–955.

Raynaud, A. 1969. Mammeles. In: Traité de zoology: Anatomie, systématique, biologie. Tome 16, Fascicule 6: Mammeles. Appareil génital gamétogenèse, fécondation, gestation (Ed. by P.-P. Grassé), pp. 1–853. Paris, France: Masson et Cie.

Ritchie, M. G. 2007. Sexual selection and speciation. Annual Review of Ecology and Systematics, 38, 79–102.

Roberts, W. E. 1988. Body and testis size in territorial and non-territorial northern fur seals (Callorhinus ursinus). American Zoologist, 28, 153A.

Rommel, S. A., Pabst, D. A. & McLellan, W. A. 2007. Functional anatomy of the cetacean reproductive system, with comparisons to the domestic dog. In: Reproductive Biology and Phylogeny of Cetacea: Whales, Dolphins and Porpoises (Ed. By D. L. Miller), pp. 127–169. Enfield, New Hampshire: Science Publishers.

Rönn, J., Katvala, M. & Arnqvist, G. 2007. Coevolution between harmful male genitalia and female resistance in seed beetles. Proceedings of the National Academy of Sciences of the USA, 104, 10921–10925.

Scheffer, V. B. 1950. Growth of the testes and baculum in the fur seal, Callorhinus ursinus. Journal of Mammalogy, 31, 384–394.

Seiffert, E. R. 2007. A new estimate of afrotherian phylogeny based on simultaneous analysis of genomic, morphological, and fossil evidence. BMC Evolutionary Biology, 7, 224.

Setchell, B. P. & Breed, W. G. 2006. Anatomy, vasculature, and innervation of the male reproductive tract. In: Knobil and Neill's Physiology of Reproduction, vol. 1. Third edn. (Ed. by J. D. Neill), pp. 771–825. Amsterdam: Elsevier Academic Press.

Simmons, M. N. & Jones, J. S. 2007. Male genital morphology and function: an evolutionary perspective. Journal of Urology, 177, 1625–1631.

Simpson, G. G. 1945. The principles of classification and the classification of mammals. Bulletin of the American Museum of Natural History, 85, 1–350.

Simpson, G. G. 1961. Principles of Animal Taxonomy. New York: Columbia University Press.

Sloan, P. R. 1992. Richard Owen's Hunterian Lectures at the Royal College of Physicians, May–June 1837. Chicago: University of Chicago Press/London, England: British Museum of Natural History.

Smith, J. D. & Madkour, G. 1980. Penial morphology and the question of chiropteran phylogeny. In: Proceedings of the Fifth International Bat Research Conference (Eds. D. E. Wilson & A. L. Gardner), pp. 347–365. Lubbock, Texas: Texas Tech Press.

Springer, M. S. & Murphy, W. J. 2007a. Mammalian evolution and biomedicine: new views from phylogeny. Biological Reviews 82, 375–392.

Springer, M. S. & Murphy, W. J. 2007b. Erratum. Biological Reviews, 82, 699.

Springer, M. S., Burk-Herrick, A., Meredith, R., Eizirik, E., Teeling, E., O'Brien, S. J. & Murphy, W. J. 2007. The adequacy of morphology for reconstructing the early history of placental mammals. Systematic Biology, 56, 673–684.

Stallmann, R. R. & Froehlich, J. W. 2000. Primate sexual swellings as coevolved signal systems. Primates, 41, 1–16.

Steppan, S. J., Storz, B. L. & Hoffmann, R. S. 2004. Nuclear DNA phylogeny of the squirrels (Mammalia: Rodentia) and the evolution of arboreality from c-myc and RAG1. Molecular Phylogenetics and Evolution, 30, 703–719.

Steppan, S. J., Adkins, R. M., Spinks, P. Q. and Hale, C. 2005. Multigene phylogeny of the Old World mice Murinae reveals distinct geographic lineages and the declining utility of mitochondrial genes compared to nuclear genes. Molecular Phylogenetics and Evolution, 37, 370–388.

Stewardson, C. L., Bester, M. N. & Oosthuizen, W. H. 1998. Reproduction in the male Cape fur seal Arctocephalus pusillus pusillus: age at puberty and annual cycle of the testis. Journal of Zoology, 246, 63–74.

Sullivan, R. M. & Petersen, K. E. 1988. Systematics of southwestern populations of least chipmunks (*Tamias minimus*) reexamined: a synthetic approach. Occasional Papers of the Museum of Southwestern Biology, 5, 1–27.

Sutton, D. A. & Patterson, B. D. 2000. Geographic variation of the western chipmunks *Tamias senex* and *T. siskiyou*, with two new subspecies from California. Journal of Mammalogy, 81, 299–316.

Symonds, M. R. E. 2005. Phylogeny and life histories of the 'Insectivora': controversies and consequences. Biological Reviews, 80, 93–128.

Tinbergen, N. 1952. "Derived" activities; their causation, biological significance, origin, and emancipation during evolution. Quarterly Review of Biology, 27, 1–32.

Voss, R. 1979. Male accessory glands and the evolution of copulatory plugs in rodents. Occasional Papers of the Museum of Zoology, University of Michigan, 689, 1–27.

Wade, O. & Gilbert, P. T. 1940. The baculum of some Sciuridae and its significance in determining relationships. Journal of Mammalogy, 21, 52–63.

West-Eberhard, M. J. 1983. Sexual selection, social competition, and speciation. Quarterly Review of Biology, 58, 155–183.

Wickler, W. 1967. Socio-sexual signals and their intra-specific imitation among Primates.

In: Primate Ethology (Ed. by D. Morris), pp. 69–147. London, England: Weidenfeld and Nicholson.

Wickler, W. 1968. Mimicry in Plants and Animals. London, England: Weidenfeld and Nicholson.

Wilkinson, G. S. & McCracken, G. F. 2003. Bats and balls: sexual selection and sperm competition in the Chiroptera. In: Bat Ecology (Ed. by T. H. Kunz & M. B. Fenton), pp. 128–155. Chicago, Illinois: University of Chicago Press.

Wright, S. 1978. Evolution and the Genetics of Populations, Volume 4: Variability Within and Among Natural Populations. Chicago, Illinois: University of Chicago Press.

Wyss, A. R. & Flynn, J. J. 1992. A phylogenetic analysis and definition of the Carnivora. In: Mammal Phylogeny. Placentals (Ed. by F. S. Szalay, M. J. Novacek & M. C. McKenna), pp. 32–52. New York: Springer-Verlag.

Yablokov, A. V. 1974. Variability of Mammals. New Delhi, India: Amerind Publishing Company Pvt. Ltd.

Zinner, D P., Nunn, C. L., van Schaik, C. P. & Kappeler, P. M. 2004. Sexual selection and exaggerated sexual swellings of female primates. In: Sexual Selection in Primates: New and Comparative Perspectives (Ed. by P. M. Kappeler & C. P. van Schaik), pp. 71–89. Cambridge, England: Cambridge University Press.

22

The Evolution of Primary Sexual Characters in Animals

A Summary

ALEX CÓRDOBA-AGUILAR

Primary sexual traits are possibly the most diverse traits in the Animal Kingdom only mirrored by a few other characters (e.g., ornamental traits such as coloured bird feathers, mammal horns, bird and insect songs, etc.). However, unlike those other traits, we have just started to understand the evolutionary forces that have given rise to and currently maintain form and function in primary sexual traits. This can be easily concluded from this book. In fact, in a majority of groups our knowledge is limited to a description of the main structures involved in sperm transfer and reception. Knowledge of presumable origins and current functions is limited to a handful of species (*Drosophila*, dragonflies and damselflies, a few species of beetles, butterflies, spiders, and birds). This is not at all surprising: even unraveling the current function of traits that are likely to be manipulated implies a challenge to a researcher. To a large extent, this is the reason why experiments involving manipulation of genitalic traits are extremely rare.

Given that most studies of primary sexual traits involve descriptions of genitalic characters, these traits have been used mainly to produce inferences of phylogenetic relationships (e.g., see for example Miller, Peretti this book). This approach may have its problems as genitalic traits are too variable intraspecifically as indicated by the lack of homologies in several groups (e.g., Agnarsson et al. 2007; Eberhard chapter) which leads to a lack of consensus when it comes to establish phylogenetic relationships. It is presumed, for example, that polymorphism in genital traits may be widespread (Huber 2003). There is also the question of phenotypic plasticity as demonstrated in phally polymorphism in basommatophoran snails (Jarne et al., this volume). Of course, descriptive studies are a first approach to understanding genital evolution and must be complemented with functional, developmental, ecological and genetical studies. Interestingly, even given the limited number of approaches used, most authors in this book have agreed upon sexual selection as the evolutionary force governing genital evolution, even in species whose genital evolution is less elaborate. Although the arguments used by the different authors are mainly based on verbal reasoning, this rests on well-grounded facts (e.g., the evolution of a large penis that contains nuptial gifts; Macías-Ordóñez et al. chapter). The existing results, which, by the way are limited to a handful of species, need to be corroborated in other species even in those in which other forces other than sexual selection are thought to occur. At present our knowledge of genitalic evolution is limited to a few species (mostly insects, spiders, and birds) which makes any general conclusion premature. Yet there are extremely diverse animal taxa of which we know very little in this context (e.g., Opistobranchs Valdes et al. chapter; Gastropods, Baur chapter, Jarne et al. chapter; amphibians Houck and Verrell chapter).

One preliminary generalization about the evolutionary processes governing genital evolution is that forces other than sexual selection seem unlikely to explain how genitalia have diverged among closely related taxa. The two other classical hypotheses—lock and key and pleiotropy—have not gained much support, mainly as a result of verbal arguments, since extremely few tests have actually been designed and performed. Furthermore, tests that tease apart the three hypotheses have been carried out only using a few insect species; with the usual result that the lock and key hypothesis has been discarded (e.g., water striders, Arnqvist et al. 1997; Arnqvist and Thornhill 1998; fruit flies, Andrade et al. 2009).

The following appear to be promising lines of research based on the reviews in this book:

1. Most post-copulatory sexual selection and/or studies designed to distinguish between natural and sexual selection have so far been limited to very few species. Perhaps this is because of (a) Eberhard's influence of using arthropods; (b) the fact that such studies require very detailed information and it is not feasible to cover a broad range of taxa quickly; and/or (c) the general fact that interest in sexual selection has focussed on a narrow range of taxa.

2. Female genitalia. Our knowledge of genital evolution is extremely male-biased. The fact that male genital characters are more rigid and conspicuous in many cases, may explain this bias. This also explains, for example, why researchers have found it easier to manipulate male genital traits more frequently than female genital traits. No doubt this has hampered our understanding of genital evolution in both sexes. Another factor may be the expectation, based on theory, that male–male competition and sperm competition are the most important components of sexual selection and that female choice, cryptic or otherwise is a minor issue. At best, we rely largely on evidence suggesting coevolution between the sexes in genital traits (e.g., Brennan et al. 2007; Parzer and Moczek 2008). In fact, there are several interesting cases of sexual coevolution documented in this book which

not only include taxa with separate males and females (e.g., Eady chapter) but also in hermaphrodite animals (Baur et al. chapter). However, it is not clear how and why female genitalia co-evolve (Galicia et al. 2008) in terms of the selection pressures acting on females or the female role in hermaphrodites. Of course, genetic quality of offspring is one answer to this pressure as many studies have shown that where females are allowed to choose a mate and/or they mate with multiple males, their offspring have increased fitness (Drickamer et al. 2000). However, this has been the case for pre-copulatory non-genital traits (Kokko et al. 2006). Nevertheless, whether female derive benefits by having offspring with males bearing some particular genitalic traits is not clear (Arnqvist 1997). The case of allometry is one example in which the hypoallometric values shown in male genitalia, can only be explained in terms of the female genital environment in which male genital traits have to fit.

3. Detailed studies on genital function in those groups in which form and function is known. Even accepting that sexual selection drives genital evolution, we still need to know which particular sexual selection process applies. Sexual conflict and female choice are two alternatives that although not mutually exclusive (Kokko et al. 2003), need to be tested. There are currently a number of studies that suggest that a sexual conflict scenario explains several traits (e.g., male behavior, seminal compounds in *Drosophila* males; for a recent review see Hosken et al. 2009). Choosing between these sexual selection scenarios is a necessary further step. Another issue that requires further examination is that genital traits cannot be studied as isolated traits. They act in concert with other traits such as male and female behavior. Knowing how these different traits interact with each other would permit a more complete view of the selective processes behind genital evolution.

4. Sperm allocation and prudence. A few studies mainly with fish and some insects indicate that males allocate sperm prudently as sperm production costs are not trivial (Wedell et al. 2002). There are also studies with hermaphroditic flatworms that indicate that individuals "trade sperm" matching volume transmitted (Vreys & Michiels 1998) and/or are more likely to use sperm when they have received a

spermatophore (Karlsson & Haase 2002). A few studies in *Drosophila* and butterflies (Wedell and Hosken chapter) have uncovered that such costs affect life history traits but how widespread this is, remains to be tested. Some animal taxa that would be convenient to study here are the broadcast spawners whose sexual biology grants them excellent candidates to expand this line of research.

5. Genital allometry. Allometry, the relationship of a particular trait with body size, has regained interest in relation to genital evolution (Eberhard 2009). Genital allometry in some groups (e.g., insects) usually arises as hypoallometric relationships with low phenotypic variation levels within populations (Bertin and Fairbairn 2007). Whether this is the case for many other taxa is unknown. Actually, in mammals this does not seem to be the case (Eberhard 2009). Furthermore, recently Eberhard has suggested that studying genital allometry would foster understanding of whether sexual conflict or a cryptic female choice mechanism operates to explain genitalic evolution (Eberhard 1998; Eberhard chapter; for a contrary claim see Bertin and Fairbairn 2007). Not only is studying genital allometry interesting as a way to disentangle these hypotheses, it may also be useful in understanding how sexual selection is acting on these traits compared to other sexually-selected, non-genital traits such as those that are used prior to mating. In this case, positive allometries are expected which is not the case for genital traits. Eberhard et al. (1998) has suggested that genital allometries are explained according to a one-size-fits-all pattern as the female internal fixed "environment" would not allow for deviations in genital expression. As indicated above, this pattern has its own exception, as occurs in mammals. The study of genital allometries not only would allow us to see the role played by females but patterns of genitalic evolution in general.

6. Artificial selection and quantitative genetics. Currently there a very limited number of studies that have used such approaches to answer genitalic evolution questions are available. However, these few studies have thrown light on very interesting patterns of genital evolution that are congruent with sexual selection scenarios. For example, reduced testes size when monogamy is selectively enforced in dung beetles (Simmons and García-González 2008) and sex-linked correlated responses on female genital traits when male courtship traits were artificially selected (Wilkinson et al. 2005). Some further issues that can be explored using these approaches are the costs associated with particular genital morphologies (similar to what has been found in studies in *Drosophila*), the heritability of genital traits and whether this coincides with what has been found for sexually selected traits, whether both secondary and primary sexual traits are mutually affected via selection acting on one particular set of traits, whether genital traits are pleiotropically affected via artificial selection on non-genital traits, and to what extent both sexes show correlated responses to selection.

It is clear that there is a fruitful future in genitalic evolution research. This book has identified some questions to be answered and many groups in which some model species can be found in little known taxa with different biologies. I am looking forward to reading such contributions.

REFERENCES

Agnarsson, I., Coddington, J. A., and Knoflach, B. 2007. Morphology and evolution of coweb spider male genitalia (Aranea: Theridiidae). Journal of Arachnology 35: 334–395.

Andrade, C. A. C., Vieira, R. D., Ananina, G. and Klaczko, L. B. 2009. Evolution of the male genitalia: morphological variation of the aedeagi in a natural population of *Drosophila mediopunctata*. Genetica 135: 13–23.

Arnqvist, G., Thornhill, R. & Rowe, L. 1997. Evolution of animal genitalia: morphological correlates of fitness components in a water strider. Journal of Evolutionary Biology 10: 613–640.

Arnqvist, G. & Thornhill, R. 1998. Evolution of animal genitalia: patterns of phenotypic and genotypic variation and condition dependence of genital and non-genital morphology in water strider (Heteroptera: Gerridae: Insect). Genetical Research 71: 193–212.

Bertin, A. and Fairbairn, D. J. 2007. The form of sexual selection on male genitalia cannot be inferred from within-population variance and allometry—A case study in *Aquarius remigis*. Evolution 61: 825–837.

Brennan, P. L. R., Prum, R. O., McCracken, K. G., Sorenson, M. D., Wilson, R. E. and Birkhead, T. R. 2007. Coevolution of male and female genital morphology in waterfowl. PLoS ONE 2: 1–6. http://ukpmc.ac.uk/tocrender.cgi?action=archive & journal=312

Córdoba-Aguilar, A. 2009. Seasonal variation in genital and body size, sperm displacement ability, female mating rate and male harassment in two calopterygid damselflies (Odonata: Calopterygidae). Biological Journal of the Linnean Society 96: 815-829

Drickamer, L. C., Gowaty, P. A. and Holmes, C. M. 2000. Free female mate choice in house mice affects reproductive success and offspring viability and performance. Animal Behaviour 59: 371–378.

Eberhard, W. G. 2009. Static allometry and animal genitalia. Evolution 63: 48–66.

Eberhard, W. G., Huber, B. A., Rodríguez, R. L., Briceño, R. D., Salas, I. Rodríguez, V. 1998. One size fits all? Relationships between the size and degree of variation in genitalia and other body parts in twenty species of insects and spiders. Evolution 52: 415–431.

Galicia, I., Sánchez, V. and Cordero, C. 2008. On the function of signa, a genital trait of female Lepidoptera. Annals of the Entomological Society of America 101: 786–793.

Hosken, D. J., Stockley, P., Tregenza, T., and Wedell, N. 2009. Monogamy and the battle of the sexes. Annual Review of Entomology 54: 361–378.

Huber, B. A. 2003. Rapid evolution and species-specificity or arthropod genitalia: fact or artifact? Organisms Diversity and Evolution 3: 63–71.

Karlsson, A. and Haase, M. 2002. The enigmatic mating behaviour and reproduction of a simultaneous hermaphrodite, the nudibranch *Aeolidiella glauca* (Gastropoda, Opisthobranchia). Canadian Journal of Zoology 80: 260–270.

Kokko, H., Brooks, R., Jennions, M. and Morley, J. 2003. The evolution of mate choice and mating biases. Proceedings of the Royal Society of London B 270: 653–664.

Kokko, H., Jennions, M. D. and Brooks, R. 2006. Unifying and testing models of sexual selection. Annual Reviews of Ecology, Evolution and Systematics 37: 43–66.

Parzer, H. F. and Moczek, A. P. 2008. Rapid antagonistic coevolution between primary and secondary sexual characters in horned beetles. Evolution 62: 2423–2428.

Simmons, L. W. and García-González, F. 2008. Evolutionary reduction in testes size and competitive fertilization success in response to experimental removal of sexual selection in dung beetles. Evolution 62: 2580–2591.

Wedell, N., Gage, M. J. G. and Parker, G. A. 2002. Sperm competition, male prudence and sperm-limited females. Trends in Ecology and Evolution 17: 313–320.

Vreys, C. and Michiels, N. K. 1998. Sperm trading by volume in a hermaphroditic flatworm with mutual penis intromission. Animal Behaviour 56: 777–785.

Wilkinson, G. S., Amitin, E. G. and Johns, P. M. 2005. Sex-linked correlated responses in female reproductive traits to selection on male eye span in stalk-eyed flies. Integrative and Comparative Biology 45: 500–510.

Index